International Conference on Medicine Sciences and Bioengineering (ICMSB2016)

BIO Web of Conferences Volume 8 (2017)

Guangdong, China
15 - 16 October 2016

Editors:

J.Q. Cheng
H.L. Moffitt
I. Kim
Z.T. Chi
J. Zhang

ISBN: 978-1-5108-3514-6

Printed from e-media with permission by:

Curran Associates, Inc.
57 Morehouse Lane
Red Hook, NY 12571

Some format issues inherent in the e-media version may also appear in this print version.

Printed by Curran Associates, Inc. (2017)

For additional information, please contact EDP Sciences – Web of Conferences
at the address below.

EDP Sciences – Web of Conferences
17, Avenue du Hoggar
Parc d'Activité de Courtabœuf
BP 112
F-91944 Les Ulis Cedex A
France

Phone: +33 (0) 1 69 18 75 75
Fax: +33 (0) 1 69 28 84 91

contact-edps@webofconferences.org

Additional copies of this publication are available from:

Curran Associates, Inc.
57 Morehouse Lane
Red Hook, NY 12571 USA
Phone: 845-758-0400
Fax: 845-758-2633
Email: curran@proceedings.com
Web: www.proceedings.com

TABLE OF CONTENTS

Committees

Editor

Jin Q. Cheng, Professor and Interim Chair, Department of Molecular Oncology, H. Lee Moffitt Cancer Center & Cancer Institute

IL Kim, Professor, Director, Department of Polymer Science and Engineering, Pusan National University

ZIEA Tat Chi, Chief (Chinese Medicine Department), Hospital Authority, China

Technical Program Committee

Qiqing Zhang, Doctoral Supervisor, Chinese Academy of Medical Sciences Institute of Biomedical Engineering, China

Yi Zhou, Associate Professor, Sun Yat-sen University, China

Shahina Begum, Assistant Professor, Mälardalen University, Sweden

Jiatuo Xu, Professor, Shanghai University of Trditional Chinese Medicine, China

Mingyan Jiang, Professor, Shandong University, China

Dinesh Bhatia, Associate Professor & Head, Biomedical Engineering Department,North Eastern Hill University, India

Yanqing Hu, Associate Professor Sun Yat-sen University China

Ajitkumar G. Patil, Head, S.B.M. Polytechnic, Mumbai, India (Autonomous Institute), India

Sudipta Mukhopadhyay, Associate Professor, Indian Institute of Technology Kharagpur, India

Enrico Vezzetti, Professor, Politecnico di Torino, Italy

Babita Pandey, Associate Professor, Lovely Professional University, India

V K Bairagi, Associate Professor, S P Pune University, India

Sharan Badiger, Professor, BLDE University, India

Hamed Hamid Muhammed, Researcher/ Lecturer, STH School of Technology and Health, KTH Royal Institute of Technology, Sweden

Prema T. Akkasaligar, Professor, Dr. PGH College of Engineering snd Technology, India

Guangfu Ji, Doctoral Supervisor, China Academy of Engineering Physics, China

S.Kalaivani, Associate, Professor, VIT University, India

Miguel Angel Fernandez Granero, Professor, University of Cadiz, Spain

Gautam Panda, Group Leader, [Scientist E-II {Group IV (4)}], Associate Professor in AcSIR, India

Rajesh Singla, Associate Professor, Department of ICE, India

Gylles Ricardo Ströher, Professor, Federal University of Technology-Paraná, Brazil

S. Ashok Kumar, Associate Professor, Vel Tech Dr.RR & Dr. SR Technical University, India

Gerhard A. Holzapfel, Professor & Head, Graz University of Technology, Austria

P.Venkatakrishnan, Professor & Head, K.L.N College of Information Technology, India

USHA. A Associate Professor, V.T.U, Belagaum, India

M.Ponni Bala, Assistant Professor, Anna University, India

Fahrul Huyop, Professor & Deputy Dean, Universiti Teknologi Malaysia, Malaysia

Gylles Ricardo Ströher, Professor,Federal University of Technology-Paraná, Brazil

Joseph (Yosi) Kost, Dean, Ben-Gurion University of the Negev, Israel

Ramjee Repaka, Assistant Professor, Indian Institute of Technology Ropar, India

Rakesh Kumar, Assistant Professor, BIT Mesra, India

Chi-wai Kan, Associate Professor, The Hong Kong Polytechnic University, Hongkong, China

Editor I

Jin Q. Cheng

Professor and Interim Chair, Department of Molecular Oncology, H. Lee Moffitt Cancer Center & Cancer Institute, USA

Academic Appointments and Employment

1986-1988 Attending Surgical Oncologist, Thoracic Surgical Department, Beijing Thoracic Tumor Hospital and Research Institute, Beijing, China.

1996-1997 Staff Scientist, Fox Chase Cancer Center, Philadelphia, PA

1997-2001 Assistant Professor, Department of Pathology, University of South Florida College of Medicine, H. Lee Moffitt Cancer Center and Research Institute, Tampa, FL

2001-2005 Associate Professor with tenure, Department of Pathology, University of South Florida College of Medicine, H. Lee Moffitt Cancer Center and Research Institute, Tampa, FL

2005-2007 Professor, Department of Pathology, University of South Florida College of Medicine, H. Lee Moffitt Cancer Center and Research Institute, Tampa, FL.

2007-present Senior Member/Professor, Molecular Oncology Department/Program, H. Lee Moffitt Cancer Center and Research Institute, Tampa, FL Professor, Department of Oncological Science and Pathology & Cell Biology, University of South Florida, Tampa, FL

2014.9-present Interim Chair, Molecular Oncology Department/Program, H. Lee Moffitt Cancer Center and Research Institute, Tampa, FL.

2015.9-present Distinguished Moffitt Scholar

Editor II

ZIEA Tat Chi

Chief (Chinese Medicine Department), Hospital Authority, China

Biography

Dr. Eric Ziea Tat-chi is currently Chief (Chinese Medicine Department), Hospital Authority. After his graduate from the Chung Shan Medical and Dental College (Taiwan, 1990), he became a surgeon. In 1998, Dr. Ziea came back to HK and had received his Master Degree in Medical Science (HKU, 2001), Diploma in Acupuncture and Moxibustion (HKU, 2001) and PhD in Medicine (CUHK, 2008). With his extensive exposure in both Chinese medicine and Western medicine, he was subsequently engaged in the research and education of integrative medicine. Apart from having considerable research output, he has also been actively involved in delivering lectures to CM practitioners at various universities, namely CUHK, HKBU and OUHK. Dr Ziea serves as expert of the International Classification of Diseases - Traditional Medicine - Medical Informatics Panel of the World Health Organization (WHO) since 2010, focusing on the standardization of Chinese medicine terminologies. In 2011, Dr. Ziea started to hold the post of Honorary Consultant (Traditional Medicine) and in-charge of the Electronic Health Record (Chinese Medicine) Project, which were tasked by HKSAR Government to work out standards related to Chinese medicine terminologies in medical record and medical informatics system. In the same year, he was invited to join as the ex-officio member of the Committee on Research and Development of Chinese Medicines under Innovation and Technology Commission (ITC) and Member of WP8 group (Acupuncture-Moxibustion) of Good Practice-Traditional Chinese Medicine (GP-TCM) Association. As being member of the Specialty Committee of Digestive System Disease and Internal Medicine, and Executive Council Member of the Specialty Committee of Informatics under the World Federation of Chinese Medicine Societies (WFCMS), his contribution on the safety, effectiveness and standardization of Chinese medicine and acupuncture were remarkable.

Keynote I

IL Kim

Professor, Director, Department of Polymer Science and Engineering, Pusan National University, Korea

Career

Sep. 2014–Present Director, Center for Collaborative Instruments of Engineering College, Pusan National University

Mar. 2014–Present Professor, Department of Medicine, Pusan National University Medical School

Sep. 2013–Present Chairman, Graduate School of Chemical Engineering　Polymer Science and Engineering

Sep. 2013–Present Director, Brain Korea 21 PLUS Center for Advanced Chemical Technology

Mar. 2001–Present Professor, Department of Polymer Science & Engineering, Pusan National University

Mar. 2008–Present Director, The World Class University Center for Synthetic Polymer Bioconjugate Hybrid Materials, Ministry of Education, Science and Technology

Jan. 2006 –Present Trustee member of Korea Polymer Society, Korean Institute of Chemical Engineers, and Korean Institute of Industrial Chemistry

Feb. 2005–Present Adjunct Professor, Department of Food Science, Cornell University, USA (sponsored by Yeonam Fellowship, LG Group, Korea)

Mar. 2001–Feb. 2005 Department Chairman, Professor, Department of Polymer Science & Engineering, Pusan National University

Apr. 1999–Feb. 2001 Professor, Department of Chemical Engineering, University of Ulsan

Mar. 1990–Mar. 1999 Assistant & Associate Professor, Department of Chemical Engineering, University of Ulsan

Apr. 1996–Apr. 1998 Chairman of the Department of Chemical Engineering, University of Ulsan

Jun. 1994–Jun. 1995 Post-Doc. Fellow (sponsored by Korea Science and Engineering Foundation (KOSEF) at the University of Iowa (worked with Professor Richard. F. Jordan)

Mar. 1984 – Jan. 1990 BASF Korea, Dept. of Application [technical service for styrenics (GPPS, HIPS, ABS, EPS, PS/PPO blend, etc.) applications]

Awards

1. The Professor of the Year, University of Ulsan, 1999.
2. Excellency in Technical Guidance for the Small and Medium Enterprises, Nov. 1998
3. The Professor of the Year (Research Area), University of Ulsan, March 1999
4. The Best Paper Award (with K. Bijal, C.-S. Ha, D.-E. Park), Korea Institute of Industrial and Engineering Chemistry, Dec. 18. 2006
5. Outstanding Poster Award (with V. Krishnan), Korea Polymer Society, Oct. 12. 2006
6. Yeosan Catalyst Award, Korean Institute of Chemical Engineers, April 2007
7. Outstanding paper award (with M. D. Manju, J. K. Lee, H. L.Shim, K. H. Kim, I.Kim, D.W.Park), The organizing committee of the 9th cross strats symposium on materials, energy and environmental engineering, Nov. 2007.
8. The Professor of the Year (Research), Pusan National University, Dec. 20, 2007
9. Outstanding paper award (with Y. H. Park, C. -S. Ha), Stablization of silver nanaoparticles by poly(L-tyrosine)-block-poly(ethylene glycol) in aqueous media, Pusan-Ulsan-Kyeongnam Branch, Korea Institute of Industrial and Engineering Chemistry, Dec. 17, 2007
10. Outstanding paper award (with J. M. Oh, J. H. Choi, C.-S. Ha) Synthesis of heterogeneous Ni(II)-diimine and Fe(II)pyridyl-diimine catalysts using Merrifield resin, Pusan-Ulsan-Kyeongnam Branch, Korea Institute of Industrial and Engineering Chemistry, Dec. 17, 2007.
11. Young scientist award(with K. Bijal), The 14th International Congress on Catalysis, July 13 -18, 2008
12. Young scientist award(with V. Kroshnan), The 14th International Congress on Catalysis, July 13 -18, 2008
13. Outstanding Poster Award (with V. Kroshnan), Korea Polymer Society, Oct. 10. 2008
14. Outstanding Evaluator Award, Basic Science Research, Korea Research Foundation, Dec. 17. 2008
15. Outstanding paper award (with I. Y. Ahn, J. M. Oh, C. -S. Ha), Effect of N- and O-donors of α-diimine Co(II) complexes on ethylene oligomerization/polymerization, Korea Institute of Industrial and Engineering Chemistry, Dec. 5. 2008

16. Outstanding paper award (with I. K. Lee, H. S. Suh, J. K. Lee, S. M. Huh), Ring-opening polymerization of propylene oxide by using nano-sized Zn-Co Prussian blue complexes, Korea Institute of Industrial and Engineering Chemistry, Dec. 5. 2008
17. Outstanding paper award (with K. Bijal, C. -S. Ha) "Highly active sterically and electronically modulated bimetallic late transition metal complexes for propylene oligomerizations", Korean Institute of Chemical Engineers, Dec. 16. 2008
18. Outstanding paper award (with D. Chandran, C. -S. Ha), "Bidentate pyridine alkoxide based zirconocene and titanocene complexes for ethylene polymerization", Korean Institute of Chemical Engineers, Dec. 16. 2008
19. Outstanding Poster Award (with A. Y. Hur, N. H. Shin, Y. Shchipunov, C. -S. Ha), "Bioinspired Synthesis of Polymer Polypeptide Conjugate Hybrid Materials and the Fabrication of Nanomaterials", 2009 Pusan-Kyeongnam/Kyushu-Seibu Polymer and Fiber Symposium, Oct. 26. 2009
20. The best presentation award (with H. Li), "Green synthetic strategy to carbon nanotube/SiO_2 and carbon nanotube/SiO_2/Ag nanoparticles hybrids", The 11th Cross Straits Symposium on Materials, Energy, and Environmental Sciences (CSS-11 2009), Nov. 13. 2009
21. Outstanding paper award (with K. Bijal, D. Chandran, V. Krishnan, D. W. Park), "Synthesis of spherical periodic mesoporous organosilica supported catalyst bearing late transition metal complexes by one-pot method for ethylene polymerization", Pusan-Ulsan-Kyeongnam Branch, Korea Institute of Industrial and Engineering Chemistry, Dec. 4. 2009
22. Outstanding paper award (with I. K. Lee, J. Y. Ha, H. Y. Choi), "Ring-opening polymerization of propylene oxide catalyzed by nano-sized double metal cyanide catalyst bearing different amount of co-complexing agent", Pusan-Ulsan-Kyeongnam Branch, Korea Institute of Industrial and Engineering Chemistry, Dec. 4. 2009
23. Outstanding Poster Award (with J. Y. Ha, E. H. Park, S. K. Park), "Polycarbonate polyol by alternating copolymerization of epoxides and carbon dioxide by double metal cyanide catalysts bearing alkoxy alcohols as complexing agents", International Polyurethane Forum 2010, June 1. 2010
24. Premier Professor of the Year, Pusan National University, Aug. 31. 2010
25. Outstanding paper award (with J. Wang, X. Huang, L. Zhang), "Hyperbranched polyglycidol-based multiblock copolymers: Synthesis and application for building drug delivery platforms", Pusan-Ulsan-Kyeongnam Branch, Korea Institute of Industrial and Engineering Chemistry, Dec. 20. 2010
26. Outstanding paper award (with R. P. Johnson, E. J. Choi, A. Y. Hur, X. Huang), "Synergistic hybrid materials based on synthetic polymer-*block*-polypeptide bioconjugateds", Pusan-Ulsan-Kyeongnam Branch, Korea Institute of Industrial and Engineering Chemistry, Dec. 20. 2010
27. Outstanding paper award (with X. Huang, A. Y. Hur, R. P. Johnson), "Synthesis and characterization of novel peptide-based triblock copolymers", Pusan-Ulsan-Kyeongnam Branch, Korea Institute of Industrial and Engineering Chemistry, Dec. 20. 2010
28. Outstanding paper award (with J. H. Yoon, H. Y. Choi), "Ring-opening polymerizations of propylene oxide catalyzed by double metal cyanide catalysts for the design of new functional polyol", Pusan-Ulsan-Kyeongnam Branch, Korea Institute of Industrial and Engineering Chemistry, Dec. 20. 2010
29. Outstanding paper award (with L. Zhang), "Fabrication of inerratic multichannel microtubes using pyridine-based macrocyles", Pusan-Ulsan-Kyeongnam Branch, Korea Institute of Industrial and Engineering Chemistry, Dec. 20. 2010
30. Science and Technology Award of Pusan, Pusan City, April 21. 2011
31. The Best PhD Thesis of the Year (with H. Li), "A General and Efficient rout to Fabricate Carbon Nanotube-Metal Nanoparticles and Carbon Nanotube-Inorganic Oxides Hybrids", Pusan National University, May 20. 2011
32. The Best Paper Award, "Simple and environmentally friendly preparation and stabilization of gold nano- and microcrystals using N,N',N"-tris(2-aminoethyl)phosphoric triamide", HyMaP 2011, Oct. 28, 2011
33. Outstanding Poster Award (with R. P. Johnson), "Synthesis of poly(2-hydroxyethyl methacrylate)-block-poly(L-tyrosine) by the combination of ATRP and ROP", HyMaP 2011, Oct. 28, 2011
34. Outstanding Poster Award (with D. E. Kang), "Metal-nanoparticles-encapsulated porous silica nanostructures of hyperbranched polyglycidol for catalytic applications", HyMaP 2011, Oct. 28, 2011
35. Outstanding Poster Award (with X. Huang), "Morphology-tunable architectures constructed by superamolecular assemblies of α-diimine complex", HyMaP 2011, Oct. 28, 2011
36. Outstanding Poster Award (with M. H. Kim), "Synthesis and applications of amphiphilic alkyl-terminated hyperbranched polyglycidol", HyMaP 2011, Oct. 28, 2011
37. Outstanding paper award (with E. S. Yun, J. H. Yoon), "Polycarbonate synthesis by copolymerization of carbon dioxide with epoxides over double metal cyanide catalysts bearing different complexing agents", Pusan-Ulsan-Kyeongnam Branch, Korea

Institute of Industrial and Engineering Chemistry, Dec. 19, 2011

38. Outstanding paper award (with S. I. Jang, S. H. Lee, C. -S. Ha), "Synthesis of eco friendly soybean oil polyol", Pusan-Ulsan-Kyeongnam Branch, Korea Institute of Industrial and Engineering Chemistry, Dec. 19, 2011
39. Lotte Industry-Academy Cooperation Award, the Polymer Society of Korea, Oct. 11. 2012
40. Outstanding paper award (with L. Zhang), "Synthesis of multi0amine functionalized hydrogel for reducer-free preparation of noble metal nanoparticles: Utilization as highly active and recyclable catalysts in reduction of nitroaromatics", Pusan-Kyeongnam Branch, Korea Institute of Engineering Chemistry, Dec. 10, 2012
41. Outstanding paper award (with S. Y. Song), "Controlled drug delivery of hyperbranched polyglycidol/lecithin hybrid liposome", Pusan-Kyeongnam Branch, Korea Institute of Engineering Chemistry, Dec. 10, 2012
42. Outstanding paper award (with E. S. Yun), "Biodegradable polycarbonate synthesis by copolymerization of carbon dioxide with epoxides using highly active double metal cyanide catalysts", Pusan-Kyeongnam Branch, Korea Institute of Engineering Chemistry, Dec. 10, 2012
43. Outstanding paper award (with S. J. Yu, E. S. Yun), "Preperation and characterization of super high molecular weight polyether polyols prepared by double metal cyanide catalyst", Pusan-Kyeongnam Branch, Korea Institute of Engineering Chemistry, Dec. 10, 2012
44. Outstanding paper award (with X. H. Huang), "The new photosensitizing precursors to dendritic hyperbranched polyglycidol-tagged 5-aminolevulinic acid prodrug for photodynamic cancer therapy", Pusan-Ulsan-Kyeongnam Branch, Korea Institute of Industrial and Engineering Chemistry, Dec. 17, 2012
45. Outstanding paper award (with S. I. Jang, J. Y. Shin), "Synthesis of ecofriendly antimicrobial agents used for soybean oil", Pusan-Ulsan-Kyeongnam Branch, Korea Institute of Industrial and Engineering Chemistry, Dec. 17, 2012
46. Outstanding paper award (with J. Y. Shin), "Microwave0assisted green synthesis of star-shaped poly(2-hydroxyethyl methacrylate)-b-poly-(L-Lysine) copolymers by ring opening polymerization of α-amino acid-N-carboxyanhydrides", Pusan-Ulsan-Kyeongnam Branch, Korea Institute of Industrial and Engineering Chemistry, Dec. 17, 2012
47. Outstanding paper award (with H. C. Chang, D. E. Kang), "Study of commercial polymer substrates on the formation of breath figure arrays", Pusan-Ulsan-Kyeongnam Branch, Korea Institute of Industrial and Engineering Chemistry, Dec. 17, 2012
48. Outstanding paper award (with B. K. Moon, J. V. John), "Tuning the physical properties of copoly(ether ester) thermoplastic elastomers by using the clay included in-situ polymerizaiton", Pusan-Ulsan-Kyeongnam Branch, Korea Institute of Industrial and Engineering Chemistry, Dec. 17, 2012
49. Best Presentation Award (with B. K. Moon), "Influenece of the Multifunctional Hyperbranched Polyglycidol as a Chain Extender Modifier on the Physical Properties of Copoly(ether ester) Elastomers", 2013 Kyushu-Seibu/Pusan-Gyeongnam Joint Symposium on High Polymers(16th) and Fibers(14th), Saga University, Japan, 8-10 Nov. 2013
50. Outstanding paper award (with S. J. Byeon), "Modification of Polycarbonates Synthesized by the Terpolymerization of Alicyclic Epoxides with Carbon Dioxide Using Highly Active Double Metal Cyanide Catalysts", Pusan-Ulsan-Kyeongnam Branch, Korea Institute of Industrial and Engineering Chemistry, Dec. 9, 2013
51. Outstanding paper award (with B. K. Moon, J. V. John), "Influenece of the Multifunctional Hyperbranched Polyglycidol as a Chain Extender Modifier on the Mechanical Properties of Copoly(ether ester) Elastomer", Pusan-Ulsan-Kyeongnam Branch, Korea Institute of Industrial and Engineering Chemistry, Dec. 9, 2013
52. Outstanding paper award (with M. S. Heo), "Synthesis of Hyperbranched Polyglycidols-*b*-Poly(*N*-Isopropylacrylamide) Using Nitroxide-Mediated Polymerization for Thermo-Sensitive Drug Delivery Applications", Pusan-Ulsan-Kyeongnam Branch, Korea Institute of Industrial and Engineering Chemistry, Dec. 9, 2013
53. Outstanding paper award (with J. V. John, B. K. Moon), "The influence of chain extenders on the structure and properties of the thermoplastic copoly(ether ester) elastomers and fabrication of honeycomb pattern and electrospun fiber", Pusan-Ulsan-Kyeongnam Branch, Korea Institute of Industrial and Engineering Chemistry, Dec. 9, 2013
54. Outstanding paper award (with X. H. Huang), "Palladium nanoparticles decorated mesoporous carbon spheres as catalyst for reduction of 4-nitrophenol", Pusan-Ulsan-Kyeongnam Branch, Korea Institute of Industrial and Engineering Chemistry, Dec. 9, 2013
55. Scholarship award, College of Engineering, Pusan National University, Dec. 17. 2013
56. Best Poster Presentation Award (with M. S. Heo), "Synthesis of hyperbranched polyglycidol-b-poly(N-isopropylacrylamide) using nitroxide-mediated polymerization for thermo-sensitive drug delivery system", 3rd International Symposium on Controlled/Living Polymerization from Synthesis to Application(CLP'14), (May 1-4), Turkey, 2014

Keynote II

Nor Ashidi Mat Isa

Professor, Deputy Dean, Universiti Sains Malaysia, Malaysia

Biography

Professor Dr Nor Ashidi Mat Isa received the B. Eng. Degree in Electrical and Electronic Engineering with First Class Honors from Universiti Sains Malaysia (USM) in 1999. In 2003, he went on to receive his Ph.D. degree in Electronic Engineering (majoring in Image Processing and Artificial Neural Network). He is currently a Professor and the Deputy Dean (Academic, Students and Alumni) at the School of Electrical and Electronic Engineering, USM. His research interests include intelligent systems, image processing, neural network, biomedical engineering, intelligent diagnostic systems and algorithms. As of now, he has led his Imaging and Intelligent System Research Team (ISRT) research group to publish at both national and international arena. He has published more than 120 manuscripts in ISI, SCOPUS and other indexed international journals. In addition, his contributions can be found in numerous journals, chapters in books and proceedings. Due to his outstanding achievement in research, he gained recognition, both national and internationally. Some of his achievements are appointed as visiting professor by 2 universities in China, appointed as keynote speaker in 5 international conferences, invited as invited speaker in 4 international conferences, appointed as PhD external examiners, appointed as editorial board for 4 international journals and appointed as reviewers for more than 80 ISI indexed journals. He was also appointed as mentor for 2 research grants from universities outside USM. As a research supervisor, he has successfully graduated 16 PhD and 20 Masters (by research mode) students.

Career

19 June 2003 -7 June 2007, Teaching and research, Universiti Sains Malaysia

7 June 2007 -7 July 2009, Senior Lecturer, Teaching and research,Universiti Sains Malaysia

8 July 2009 - 29 October 2015, Associate Professor , Teaching and research , Universiti Sains Malaysia

30 October 2015 -= present , Professor, Teaching and research, Universiti Sains Malaysia

Statement of Peer review

In submitting conference proceedings to BIO Web of Conferences, I certify to the Publisher that I adhere to the **Policy on Publishing Integrity** of the journal in order to safeguard good scientific practice in publishing.

1. All articles have been subjected to peer review administered by the proceedings editors.
2. Reviews have been conducted by expert referees, who have been requested to provide unbiased and constructive comments aimed, whenever possible, at improving the work.
3. Proceedings editors have taken all reasonable steps to ensure the quality of the materials they publish and their decision to accept or reject a paper for publication has been based only on the merits of the work and the relevance to the journal.

Title, date and place of the conference

2016 International Conference on Medicine Sciences and Bioengineering (ICMSB2016) have been held from October 15-16, 2016 in Guangzhou, Guangdong, China.
http://www.icmsb2016.org/

Proceedings editor(s):

Co-editors
Jin Q. Cheng, Professor and Interim Chair, Department of Molecular Oncology,
H. Lee Moffitt Cancer Center & Cancer Institute
IL Kim, Professor, Director, Department of Polymer Science and Engineering, Pusan National University
ZIEA Tat Chi, Chief (Chinese Medicine Department), Hospital Authority, China
Editor Assistant:
Jiaxing Zhang, Wuhan University, China

Date and editor's signature 2016/12/15

Preface

Held in Guangzhou, Guangdon, China from October 15-16, 2016, ICMSB2016 [The 2016 International Conference on Medicine Sciences and Bioengineering] hopes to provide an excellent international platform for all the invited speakers, authors, and participants. The conference enjoys a wide spread participation, and we sincerely wish that it would not only serve as an academic forum, but also a good opportunity to establish business cooperation. Any paper and topic around computer science and artificial intelligence would be warmly welcomed.

ICMSB2016 proceeding tends to collect the most up-to-date, comprehensive, and worldwide state-of-art knowledge on medicine sciences and bioengineering. All the accepted papers have been submitted to strict peer-review by 2-4 expert referees, and selected based on originality, significance and clarity for the purpose of the conference. The conference program is extremely rich, profound and featuring high-impact presentations of selected papers and additional late-breaking contributions. We sincerely hope that the conference would not only show the participants a broad overview of the latest research results on related fields, but also provide them with a significant platform for academic connection and exchange.

The Technical Program Committee members have been working very hard to meet the deadline of review. The final conference program consists of 94 papers divided into 3 sessions. The proceedings would be published in a volume by EDP Sciences.

We would like to express our sincere gratitude to all the TPC members and organizers for their hard work, precious time and endeavor preparing for the conference. Our deepest thanks also go to the volunteers and staffs for their long-hours work and generosity they've given to the conference. The last but not least, we would like to thank each and every of the authors, speakers and participants for their great contributions to the success of ICMSB2016.

ICMSB2016 Organizing Committee

The effects of repetitive task training combined with neuromuscular electrical stimulation on extremities for acute cerebral paralysis

Xingnan Liu[1,2] and Chunli Mei[1,a]

[1]*Department of Nursing, Beihua University, Jilin, 132013, China*
[2]*The Central Hospital of Jilin City, Jilin, 132013, China*

Abstract. In order to study the effect of repetitive task training combined with neuromuscular electrical stimulation on lower limb function for severe paralysis patients with acute cerebral apoplexy. Chose 36 cases of severe paralysis in patients with acute cerebral apoplexy and devided into treatment group and control group. 18 cases in each group, the control group was given routine rehabilitation therapy. The treatment group was given repetitive task training on the basis of control group. Including sit - stand and stand training. Evaluate the lower limb motor function by the FMA, the walking function by FAC and the activities of daily living by FIM respectively before and after 2 months of treatment.The results were showed that the scores of FMA and FIM were significantly better than before treatment after 2 months ($P < 0.05$) , and the the scores of FMA and FIM in treatment group were better than control group ($P < 0.01$). The scores changed of FMA and FIM in treatment group (△ FMA and △ FIM scores) were better than control group ($P < 0.01$).Walking function was better in treatment group than control group as the grade of FAC in treatment group was more higher. So repetitive task training combined neuromuscular electrical stimulation can more effectively improve the lower limb exercise function, walking function and daily life activity for severe paralysis patients with acute cerebral apoplexy.

1 Introduction

At present, there are few studies on the early rehabilitation therapy for severe paralysis in patients with cerebral apoplexy. In recent years, the research showed that repetitive task training can improve the walking ability of patients with cerebral apoplexy[1]. Neuromuscular electrical stimulation also could promote the recovery of motor function of paralyzed limbs[2].

The purpose of this study is to investigate the effect of repetitive task training combined with neuromuscular electrical stimulation on lower limb function for severe paralysis in patients with acute cerebral apoplexy.

[a] Corresponding author: gaoda2002@126.com
The project supported by educational department and technology department of Jilin province in China. (No. 2015153, 2013ZC029, 201339033).

2 Materials and Methods

2.1 Object of study

36 patients were selected with acute large area cerebral infarction in neurology department and stroke unit from affiliated hospital of Beihua university from June 6, 2014 to August 2015. Divided randomly into group of training task combined with electrical stimulation therapy(treatment group) and group of electrical stimulation treatment (control group) with 18 cases in each group.

Inclusion criteria: acute patients with large area cerebral infarction in middle arterial blood supply area (atherosclerotic thrombotic cerebral infarction) and there were clearly responsibility of infarction lesions on head CT or MRI. With first time occurrence and patients were under acute phase(course of disease <7d). Vital signs of patients were relatively stable, and patients were willing to cooperate with treatment as the disease condition was not getting worse. In the presence of severe limb paralysis and lower extremity muscle strength level was from 0 - 1. Patients were willing to sign informed consent.

Exclusion criteria: High mortality of patients with MCA, including the range of infarction was more than that of middle cerebral artery and the patients were under coma state. Or the patients had serious heart, lung, kidney disease who were not suitable for early rehabilitation therapy; Patients with severe pain of lower limb who could not complete treatment. Patients had cognitive or mental disorders who can not cooperate with treatment. The treatment efficacy can not be judged as patients who had poor compliance and did not comply with regulation treatment[3].

2.2 Therapeutic methods

Control group: based on conventional rehabilitation therapy including neural facilitation techniques, electroacupuncture, electric standing bed, etc. Enhanced neuromuscular electrical stimulation therapy was given to paralysis limb extensor (femoral quadriceps and tibialis anterior muscle group). The first week of treatment: 20 minutes per time, 1 time per day. The second week of treatment: 40 minutes per time, 1 time per day. The third week of treatment: 40 minutes per time, 2 times per day.Task training combined with electrical stimulation treatment group (treatment group): on the basis of the control group treatment, increase repetitive task training. First week of treatment: 10 minutes per day, 1 time per day. At the second week: 20 minutes per time, 1 time a day; At the third weeks: 20 minutes per time, 2 times a day. Given repetitive task training for relatively stable conditioned patients. We found that brain edema was reducing, consciousness getting better, blood pressure, breathing and heart rate were relatively stable. Given seat training in first week, lifting bed seat training, bed edge seat training and wheelchair seat training, repeat 3 times a day.The time were 2 weeks.sit-stand training: to begin with,the seat should be high, patient easy to stand up, which can be in a wheelchair, stool or a treatment bed, while standing, the patients' arms should place in the bed bar or treating bar.Treatment staff or family members helped patients stand up or keep standing. Place of treatment can be in the ward or treatment rooms, the treatment needed family members in, repeats 15 times a day. Observed the patient's condition, the patients tolerance range and the procedure shall be slowly. All patients selected should be treated with conventional nerve nutrition according to the condition and improve therapy of circulation, dehydration, anti-platelet aggregation[4].

2.3 Clinical evaluation

Records of general situation, including the systemic complications of patients, etc. Treatment records: the specific treatment time, completion of the situation, etc. Records of adverse effects or adverse events: including whether the patient's condition was getting worse or not, blood pressure, heart rate changing, muscle fatigue, falling down, etc. Evaluate the function of exercise by the FMA, divided

 DOI: 10.1051/bioconf/20170801001

into 17 items, a total of 34 points, the recovery of motor function of lower limb was getting better as the score was getting higher[5]. Evaluate the walking function by FAC, it was consisted of six level which from 0 to 5, the ability of walking was getting better as the level was getting higher. Evaluated the activities of daily living by FIM, using seven point scale, the independence was getting better and the dependence was getting smaller as the score was getting higher. Evaluate clinic and recovery of patients before treatment and during two months of treatment. All evaluation items shall be carried out by the trained personnel.

2.4 Statistical analysis

SPSS software was used for statistical analyses and the values are presented as the means +/- SD. ANOVA was used to compare the mean values. P values of less than 0.05 were taken to indicate statistically significant differences.

3 Results

The two groups of patients had not appear disease aggravation, recurrent stroke, cardiovascular accident, falling down and other adverse events during therapeutic process, 2 cases in the treatment group and 3 cases in the control group appeared mild dizziness and headache during the treatment, both treatment group and control group had 2 cases who appeared limb pain and other symptoms of muscle fatigue, the symptoms were alleviated after symptomatic treatment. FMA and FIM scores of these two groups had no significant difference before treatment ($P>0.05$). After 2 months treatment, the scores were significantly better than before($P<0.05$), FMA and FIM scores of treatment group were significantly better than control group($P<0.01$); FMA and FIM scores changed quantity of the treatment group ($\triangle$FMA and $\triangle$FIM scores) were significantly better than control group ($P < 0.01$) after further treatment. FAC grade of treatment group was higher, FAC grading differences between these two groups had significant meaning ($P < 0.05$) after 2 months treatment, See Table 1.

Table 1. Two groups of patients before and after treatment of FMA, FIM score

Score	FMA			FIM		
	Control Group	Treatment Group	P Value	Control Group	Treatment Group	P Value
0 month	3.78±1.48	3.83±1.38	0.06	27.30±6.30	28.60±6.30	0.07
1 month	18.11±3.66	24.06±3.90	0.04	18.11±3.66	68.67±13.09	0.03
2 months	14.33±3.40	20.22±4.35	0.01	23.67±13.5	40.11±12.22	0.01

4 Discussion

Many foreign studies had confirmed that the treatment effect was closely related to the treatment time.The quantity of patients with cerebral apoplexy, the treatment and effect of acute and subacute phase had independent dose dependent relationship, high intensity and repetitive rehabilitation could improve the motor function and the reorganization of the cerebral cortex of the patients[6].At present, many scholars at home and abroad think that the key factor for the effective rehabilitation therapy after cerebral apoplexy is to carry on the repetition, the task oriented training, and the training will be high strength and long time.

In recent years, some high quality meta analysis in foreign countries showed: within 6 months of the onset of the disease, repetitive task oriented training can improve the walking ability of cerebral apoplexy patients (including walking distance and walking speed) and ADL [7]. Foreign experts had recommend taking repetitive task oriented training, evidence based on medical class A and B. Study had proved that it was similar that neuromuscular electrical stimulation therapy which had been used wildly during the treatment of cerebral apoplexy rehabilitation can promote the recovery of motor and sensory function, and the dose of electrical stimulation was an important factor for the recovery of limb function, benefit of motor function would be greater as the duration time of the stimulus was getting longer for patients with severe paralysis[8].

In this study, control group used the treatment plan of strengthening electrical stimulation of nerve muscle for severe paralysis patients with acute cerebral apoplexy which based on clinical and research results and scientific theory in domestic and overseas. The stimulation time was 20min/d, and then increase 40 to 80 min/d.Treatment group adopt treatment methods of repetitive task orientation training combined with intense neuromuscular electrical stimulation .The results showed that the two groups of patients did not appear the stroke recurrence, disease progression and cardiovascular accidents, fell down, adverse events in the process of treatment , these situation can prove that early treatment was safe and patients can tolerate. Walking function and activities of daily living (ADL) were significantly improved of two groups of patients with lower extremity motor function after 2 months of treatment, but the benefit of the treatment group was greater than that of the control group, lower limb motor function was better, walking function classification was higher, more proportion of body achieved the walking function and the ability of daily living. The biggest task was to achieve the goal of walking function as far as possible for severely paralyzed patients with acute cerebral apoplexy, patients can not be directly carried out walking training for poor muscle strength.While sitting, standing ability and sit-stand transfer ability is the foundation to realize the walking function of patients in the future[9-10]. Treatment group used repeat seat training, sit-stand training, standing level of training and oriented task training step by step and combined with stimulation treatment of increased neuromuscular electrical stimulation to effectively improved lower limb function in advance and created a key precondition for realizing further walking function. The most important was that patients and their families who accepted the treatment learned method of self training and they was encouraged by benefit from initial treatment. Patients can keep family training with the high positivity even leaving hospital, therefore, long-term benefits can be obtained as treatment can be extended.

Combined drug therapy and early rehabilitation therapy, The state of illness will be severe for the acute cerebral apoplexy patients with large area infarction, the dehydration, nutrition, improvement of circulation, anti platelet aggregation drugs and some other drugs shall be needed in the early stage. At the same time, blood pressure, blood sugar and heart and lung function of the patient may not be stable, therefore bedside rehabilitation therapy shall be the main treatment within 1week and treatment volume shall not be large because some patients appeared small amount of hemorrhage after secondary infarction. Take active rehabilitation therapy gradually after that the vital signs of patient is stable, carry out Staged task oriented training, and strengthen the treatment of nerve muscle electrical stimulation, but the treatment plan still need to be adjusted according to the patient's condition during treatment process.

In conclusion, the preliminary clinical study confirmed that limb motor function, walking function can be improved by adopting the therapeutic schedule of repeated task oriented training combined with neuromuscular electrical stimulation. So the ability of daily life will be improved. There are still many deficiencies in this study. For example: selected samples are less, the follow-up time is shorter. Failure to comply with the requirements of ICF to conduct a comprehensive evaluation of efficacy from three levels etc. These problems will be improved in the future research.

References

1. B.French, L.Thomas, M. Leathley, et al. Does repetitive task training improve functional activity after stroke? A Cochrane systematic review and meta-analysis[J], J Rehabil Med.42(1):9-14.(2010).
2. JM.Veerbeek, M.Koolstra ,JC.Ket, et al. Effects of augmented exercise therapy on outcome of gait and gait-related activities in the first 6 months after stroke: a meta-analysis[J], Stroke.42(11):3311-3315.(2011).
3. L.Peter, B.Julie,K.Gert. Stroke rehabilitation[J], Lancet.377: 1693-1702.(2011).

4. SS.Hsu, MH.Hu, YH.Wang, et al. Dose-response relation between neuromuscular electrical stimulation and upper-extremity function in patients with stroke[J], Stroke.41(4):821-824.(2011).
5. SS.Hsu, MH.Hu, JJ.Luh, et al. Dosage of neuromuscular electrical stimulation: is it a determinant of upper limb functional improvement in stroke patients[J],J Rehabil Med.44(2):125-130.(2012).
6. AR.Fugl-Meyer, L.Jaasko, I.Leyman, et al. The poststroke hemiplegic patient. a method for evaluation of physical performance[J],Scand J Rehabil Med.7:13-31.(1975).
7. MK.Holden, KM.Gill, MR.Magliozzi , et al. Clinical gait assessment in the neurologically impaired. Reliability and meaningfulness[J],Phys Ther.64(1):35-40.(1984).
8. BE.Bates, MG.Stineman . Outcome indicators for stroke: application of an algorithm treatment across the continuum of postacute rehabilitation services[J],Arch Phys Med Rehabil.11:1468-1478.(2000).
9. JM.Veerbeek, G.Kwakkel, EE.van Wegen, et al. Early prediction of outcome of activities of daily living after stroke: a systematic review[J],Stroke.42(5):1482-1488.(2011).
10. B.French, LH.Thomas, MJ.Leathley, et al. Repetitive task training for improving functional ability after stroke[J], Cochrane Database Syst Rev.10(4).(2007).

Research progress of active polypeptides of *Cordyceps militaris*

Guangyu Xu[1], Peige Du [1], Liping An [1], Gang Lv [1], Lifeng Miao[2] and Yuan Guangxin[1,a]

[1]*College of Pharmacy, Beihua University, Jilin, Jilin, China*
[2]*First Clinical Hospital, Beihua University, Jilin, Jilin, China*

Abstract. It is recognized at home and abroad that *Cordyceps militaris* is a fungus that can be edible and used for medicinal application, and also a common cordyceps taishanensis that is widely distributed and with very high medicinal value in China. Active *Cordyceps militaris* polypeptide has shown several highlights in its absorption mechanism, such as directly absorbed without the need for digestion, and absorbed quickly and completely, and more and more international researchers are attaching great importance to it. Physiological and pharmacological activities of *Cordyceps militaris* polypeptide are mainly to activate related enzyme systems, promote intermediary metabolisms or control the DNA transcription and translation, simultaneously activate the reticuloendothelial system and macrophage, promote the transformation of lymphocytes, and as a nonspecific immune enhancer and regulator, activate the immune competent cells, especially lymphocytes, lymphokines, mononuclear macrophage system and NK cells, thereby attacking the target cells to play an antitumor role, and also exerting its anti-aging, anticoagulant, hypolipidemic to enhance the immunity, improve the liver function and delay the aging. In this paper, the immunity improvement, anticancer, antioxidation and antimicrobial activities of active polypeptides from *Cordyceps militaris* are reviewed, in order to provide a solid theoretical help for the further development and application of active polypeptides of *Cordyceps militaris*.

1 Introduction

North *Cordyceps sinensis*, also known as *Cordyceps* or *Cordyceps militaris*, and *Boschniakia rossica* in its popular name, and belongs to the medicinal fungus as does *Cordyceps sinensis*, is recognized as a type of medicinal fungi that are traditional Chinese health foods at home and abroad, can replace *Cordyceps sinensis* as a medicine commonly used in China, with some advantages such as high medicinal values, wide distribution, and easiness to be cultivated [1, 2]. The main components of *Cordyceps militaris* are *cordycepin*, polysaccharides and active polypeptides. Due to a few highlights of the active polypeptides in its absorption mechanism, it is attracting more and more researchers' attention. Active polypeptides can be directly absorbed without the need for digestion, and the

[a] Corresponding author: yuanguangxin2007@163.com
The research supported by Jilin science and technology project (201464063, 20166017) , Health planning project of Jilin Province (2014Z079), Jilin Province Science and Technology Development Program(20403110 31YY, 201603092YY) .

absorption, an initiative, forcing and preferential absorption, is faster and more complete. The process of absorption does not need consuming the energy of the body, so that it cannot increase the burden on the gastrointestinal function, in which active polypeptides can play a role of carrier for the transport. Along with the rapid development of molecular biology, biochemistry and modern protein engineering technology, the research on bioactive peptides has made an epoch-making progress and is highly valued by the researchers.

Cordyceps militaris polypeptide is a hydrolysate prepared directly from *Cordyceps militaris* by enzymolysis solution technology, and then concentrating and drying, with very strong physiological activities and functional diversity. Physiological and pharmacological activities of *Cordyceps militaris* polypeptide are mainly to activate related enzyme systems, promote intermediary metabolisms or control the DNA transcription and translation, simultaneously activate the reticuloendothelial system and macrophage, promote the transformation of lymphocytes, and as a nonspecific immune enhancer and regulator, activate the immune competent cells, especially lymphocytes, lymphokines, mononuclear macrophage system and NK cells, thereby attacking the target cells to play an antitumor role, and also exerting its anti-aging, anticoagulant, hypolipidemic to enhance the immunity, improve the liver function and delay the aging [3, 4]. In this paper, the immunity improvement, anticancer, antioxidation and antimicrobial activities of active polypeptides from *Cordyceps militaris* are reviewed, in order to provide a solid theoretical help for the further development and application of active polypeptides of *Cordyceps militaris*.

2 Brief introduction of polypeptides

As early as in the 19th century, Engels had proposed that protein is a way of life. Polypeptides are active small molecule substances between amino acids and proteins. In addition to the characteristic functions of proteins, they also have their own unique functions, such as easiness to be digested and absorbed, hypotensive effect, lowering blood lipid and cholesterol, antimicrobial activities, antioxidation, and improvement of mineral absorption. As is known to all, peptide is a biological active substance involved in the function of many kinds of cells and an indispensable participant involved in various complex biological activities in the body.

Bioactive peptides are specific protein fragments that have a positive effect on the function and state of the body and may ultimately affect the health of the body. They are the important chemical messengers between cells and organs, and can regulate the life processes, such as growth, development, reproduction, and metabolism. Bioactive polypeptides have shown several highlights, that is, they can be directly absorbed faster and more completely without the need to be digested through a process of initiative, forcing and preferential absorption, in which the process of absorption does not need consuming the energy of the body, so that it cannot increase the burden on the gastrointestinal function, and can play a role of carrier for the transport. With the rapid development of molecular biology, biochemistry and modern protein engineering technology, the research on bioactive peptides has made an epoch-making progress and is highly valued by the researchers. At present, hundreds of different kinds of biological active peptides that exhibit a variety of biological functions have been found in organisms, such as immune active peptide and hypolipidemic peptides, anti-aging peptides and so on. Studies have found that they can also be used as drugs, vaccines, diagnostic reagents and enzymes mainly for the application in medicine, health care product, food and cosmetic industry.

3 Cordyceps militaris polypeptide

Cordyceps militaris polypeptide is a composition of drug gene proteins and nucleosides, contains six compound with cyclic dipeptides, and can coordinate and balance the cardiovascular, nervous, endocrine, visceral, blood, metabolism, and immune systems. *Cordyceps militaris* polypeptide can slow the heart rate and improve the tolerance to hypoxia, thereby protecting the cardiovascular system; can inhibit a variety of pathogenic bacteria, and has a certain anti-inflammatory, antitussive and

DOI: 10.1051/bioconf/20170801002

antiasthmatic effects; has an inhibitory effect on the nervous system, playing a sedative and analgesic effect. Zhu Zhenyuan et al [5] separated and purified a acidic peptide with analgesic effect from the mycelia of *Cordyceps gunnii* by the method of physical and chemical analysis combined with biological detection method. In addition, *Cordyceps militaris* polypeptide can also improve the body's immunity and produce an anti-tumor effect. Some experts believe that *Cordyceps militaris* polypeptide will be developed into a new, efficient, broad-spectrum and safe functional peptide with anti-tumor activity. Qian Kangnan Professor et al [6] prepared a *Cordyceps* polypeptide that is a water-soluble mixture obtained by refining *Cordyceps* protein powder containing greater than 60% - 80% of high-quality *Cordyceps* protein with the specific enzymolysis engineering technology after the separation and degreasing with super critical CO_2 extraction and fractionation, using *Cordyceps militaris* and *Cordyceps sinensis* as the raw materials. The material-yield ratio of *Cordyceps* polypeptides prepared in this way was 3-5:1, which can be said to be a "miraculous polypeptide" with a price higher than gold. Experiments have demonstrated that due to the high content of active ingredients in *Cordyceps*, coupled with the unique advantages of polypeptides, such as small molecules, *Cordyceps* polypeptide can play a role in Mg levels. The results from the safety evaluation in the pharmacology and functional detection have also verified that the *Cordyceps* polypeptide has exhibited no toxicity and side effect, but a higher nutritional value and bioavaibility.

4 Applications of cordyceps polypeptides

4.1 Improvement of Cordyceps polypeptide on the immunity

Immunoreactive peptides include immune enhancing peptides and immunosuppressive peptides. Immune enhancing peptides can enhance the immune function of the body and exert a nonspecific immunoenhancement on immunodeficiencies and tumors, such as thymic peptides that are separated from thymus or serum system and can induce and stimulate immature lymphocytes, while immunosuppressive peptides are mainly used for the prevention of transplantation rejection and the treatment of autoimmune diseases, such as cyclosporine separated from the fungal culture liquid. Pharmacodynamic experiments have indicated that *Cordyceps militaris* polypeptide can significantly improve multiple immune indexes in animal experiments, promote the transformation of lymphocytes and improve the cellular immune function of the body; eliminate superoxide free radicals in the body and protect the stability and permeability of cell membranes; significantly improve the activity of NK cells. NK cells are the non-specific part in the cellular immunity, can play their roles directly as a killer cell, and have a good anti-tumor effect, indicating that they are one of the most important components in the cellular immunity.

Studies by Jin Lili at Liaoning University showed that *Cordyceps militaris* polypeptide could enhance the body immune function through multiple aspects affecting the immune system, in which a *Cordyceps militaris* polypeptide hydrolysate obtained directly from *Cordyceps militaris* by enzymolysis technology were given orally to mice successively for 30 days, and the results indicated that the *Cordyceps militaris* polypeptide hydrolysate could significantly improve the number of antibody producing cells and the level of serum hemolysin, improve the carbon clearance function and the phagocytic capacity of monocyte-macrophage, enhance the delayed allergy, and increase the activity of NK cells in mice. Thus, *Cordyceps militaris* polypeptide can enhance the immune function of mice.

4.2 Inhibition of *Cordyceps militaris* polypeptide on tumors

Malignant tumors are one of the main diseases threatening human survival, the curative effect of single-drug therapy on them is poor, and the damage caused by the radiotherapy and chemotherapy on the immune system of the host is severe. The screening of effective anti-tumor drugs from natural medicines has become an important way for developing antitumor drugs throughout the world.

Cordyceps militaris should be an ideal substitute for the traditional *Cordyceps*, and more and more scholars have paid attention n to its medicinal value. Professor Li Can [7] at Shandong Traditional Chinese Medicine University found that *Cordyceps militaris* polysaccharide could significantly inhibit esophageal cancer, breast cancer and gastric cancer cells, but not significantly inhibit lung adenocarcinoma and colon cancer cells; *Cordyceps militaris* polypeptide could significantly esophageal cancer, breast cancer, gastric cancer and lung adenocarcinoma cells, but not significantly inhibit colon cancer cells.

4.3 Antibacterial effects of *Cordyceps militaris* polypeptide

In recent years, due to the abuse of antibiotics, drug-resistant strains in clinic have been on the increase, resulting in rise in the mortality of clinical infectious diseases. The medical community has been working to find efficient and safe antimicrobial drugs. As is known to all, there are many active components with low toxicities in traditional Chinese medicines, so that to search for antibacterial resources from natural products has become a hot topic. Antibacterial peptides and antiviral polypeptide are widely distributed in nature, and can found in both prokaryotic and eukaryotic organisms, such as cyclic peptides, glycopeptides and lipid peptide obtained from bacteria and fungi. Park et al [8] separated and purified a protein CMP from fruiting bodies, Jack et al. [9] obtained a polypeptide *Cordymin* from the fermentation broth of *Cordyceps militaris* etc., and both had a certain degree of inhibitory effect on the different fungi and bacteria; Gao et al at Guizhou University [10] found that they also presented different degrees of inhibitory effects on bacteria and fungi, which were consistent with the related reports of some researchers, indicating that these substances can be well used in biological and medical fields.

4.4 Antioxidant activities of *Cordyceps militaris* polypeptide

A part of free radicals produced in the metabolism process of the body serves as the protection of the body by eliminating pathogenic microorganisms, and the other part is to maintain the balance of the body through complex changes in order to maintain the homeostasis of the internal environment. When the body's homeostasis does not remain, the free radicals generated in the body will not be cleared away in time and accumulated to a certain extent, which will pose a threat to human health, leading to the occurrence of certain diseases. Therefore, the research on the antioxidants that can eliminate the redundant free radicals appears particularly important. Shen Qiying [11] found that *Cordyceps militaris* polypeptide showed an obvious scavenging effect on oxygen free radicals and hydroxyl radicals, its scavenging hydroxyl radical was stronger than that of mannitol, a specific hydroxyl radical scavenger, and *Cordyceps militaris* extraction liquid exhibited a dose-dependent scavenging effect on both oxygen free radicals and hydroxyl free radicals.

5. Conclusions

As a species of medicinal fungi, *Cordyceps militaris* has shown an excellent medical and nourishing effect, are widely used in food, medicine, healthcare and other industries now, and has now become a topic that attracted many research scholars' attention. However, the reports on the effective components of *Cordyceps militaris* mainly focus on the research of *Cordycepic* acid, *Cordycepin* and *Cordyceps* polysaccharides, while those on the protein polypeptide are not enough, and the cost of *Cordyceps* polypeptide obtained by a specific enzymolysis biotechnology is very expensive. Therefore, this paper sets out the application of Cordyceps militaris in four aspects, such as its immunity improvement, anticancer, antioxidation, and antimicrobial effect, in order to lay the foundation and provide the necessary basis for the further research on small-molecule bioactive peptides in the fruiting body of *Cordyceps militaris*.

References

1. J.Y. Song ch., B.K. Yang, K.S. Ra, J.M Sung. J Microbiol Biotechnol. **8**, 536-9(1998).
2. T.B. Nag wh. J Pharm Pharmacol. **57**, 1509-19(2005).
3. S.Y. Won pe. Journal of Ethnopharmacology. **96**, 555-61(2005).
4. R.M. Yu yw, L.Y. Song, C.Y. Yan, Z. Zhang, Y. Zhao. Cabohydrate Polymers. **70**, 430-6(2007).
5. Z.Y. Zhu, X. Chen, J.P. Zhou, A.J. Liu, Y.J. Zhang, Y.M. Zhang. Food Science. **30**, 200-203(2009).
6. K.G. Qian. China fungal industry science and Technology Conference, (2005).
7. C. Li, B. Yan, F. Sun, L. Wang. Shandong Journal of Traditional Chinese Medicine. **32**, 744-746(2013).
8. B.T. Park, K.H. Na, E.C. Jung. Korean Journal Physiology Pharmacology. **13**, 49-54 (2009).
9. H.W. Jack, B.N. Tzi, H.X. Wang. Phytomedicine. **18**, 387-392(2011).
10. Y.Y. Gao, L.H. Zhou, Z.Y. Pan, H.F. Liao. Guangdong Agricultural Sciences. **11**, 183-185(2013)
11. Q.Y. Shen. Research Cordyceps oxygen free radical and hydroxyl free radical action. Guangxi Plant. **3**, 252(2001).

DOI: 10.1051/bioconf/20170801003

Clinical observation of early rehabilitation nursing on the prognosis of patients with acute myocardial infarction effect

Zaiying Song and Chunwen Li[a]
Beihua University, 132013, Jilin, China

Abstract. To explore the effect of early rehabilitation nursing on the quality of life of patients with acute myocardial infarction. Methods 41 cases of patients with acute myocardial infarction were selected according to the diagnostic criteria of acute myocardial infarction, stable vital signs, and no obvious complications. On the basis of routine nursing care, the system of early rehabilitation nursing care, the development of daily rehabilitation nursing plan. The cardiac function, clinical symptoms, quality of life and average length of stay were observed in 41 patients. Results During hospitalization, 41 patients had no deaths and all patients clinical symptoms significantly reduced; length of hospital stay was significantly decreased; heart function recovered well; the quality of life of patients significantly improved. Conclusion Early rehabilitation nursing care for patients with acute myocardial infarction is conducive to improving the condition, shorten the hospital stay, can greatly improve the quality of life of patients, it is worthy of clinical promotion.

1 Introduction

Cardiovascular and cerebrovascular disease is the main disease threatening the health and life of Chinese adults, including coronary heart disease and cerebral apoplexy are the cause of the Chinese people's first death. Acute myocardial infarction is a severe type of coronary heart disease (AMI), is based on coronary artery disease, coronary artery blood supply of sharp reduction or interruption of the corresponding myocardial ischemia leads to severe and persistent myocardial necrosis[1].Research shows[2],early rehabilitation nursing intervention can effectively reduce the patient's mortality rate and disability rate, improve the quality of life of patients, which is conducive to the prognosis of patients with recovery, return to society.My team of implementation of the rehabilitation nursing in 41 cases of patients with myocardial infarction during recovery period, including general nursing, activities and the rest care, diet nursing, defecation nursing, pain nursing, psychological nursing and achieved good results.

2 Methods

2.1 General information

[a] Corresponding author: Chunwen Li, chunwen9925@163.com

 DOI: 10.1051/bioconf/20170801003

41 cases of hospitalized patients from January 2014 to December in the Affiliated Hospital of Beihua University were selected and the selected of patients conforms to the standard of diagnosis of acute myocardial infarction (AMI); After the onset of 24h stable vital signs, no obvious angina pectoris without sinus tachycardia, heart failure and severe arrhythmia and cardiogenic shock, blood pressure based in the normal range. Normal body temperature and myocardial enzyme spectrum tended to turn for the better; were treated with anticoagulation, dilating coronary artery, activating blood circulation to dissipate blood stasis treatment with thrombolytic therapy for thrombolytic therapy. Male 24 cases, female 17 cases, age (52.6±13.4) years old, all the patients informed consent and signed informed consent. All patients with gender, age and educational level of baseline data index, general data comparison, no statistically significant difference ($P>0.05$).

2.2 Intervention methods

According to the patient's condition and individual condition of the formulation and implementation of rehabilitation nursing. Rehabilitation nursing should be carried out in close monitoring, Observed the rehabilitation exercise, exercise and the end of the exercise immediately after the termination (15s) heart rate, heart rate, blood pressure changes.The following conditions were required to suspend rehabilitation activities[3].a.Heart rate > resting heart rate 20 / min or heart rate ≥110 / min;b.The emergence of severe arrhythmia;c. The emergence of angina, chest tightness, shortness of breath, dyspnea, palpitations, dizziness, syncope, pale, sweating and other performance;d.ST segment depression was equaled to or larger than 0.1mv or move up is equal to or larger than 0.2mv; e. A significant increase in blood pressure>200/100 mmHg (1mmhg = 0.133kPa) or systolic blood pressure than resting levels rise was more than or equal to 30mmHg or decline was more than or equal to 10mmHg;f.The subjective fatigue (RPE was more than or equal to level 14), after exercise6~8min breathing, heart rate did not returned to pre exercise state, or cause insomnia, fatigue,edema[4];g.The heart rate was not even speeding up but slow down with the increase of load, should stop or go back to the last stage of exercise[5].

3 Patients of intravenous thrombolysis after 24h were given rehabilitation nursing, by the professional training of nurse was responsible for guiding the implementation of local departments to develop acute myocardial infarction early rehabilitation training program[6].See Table 1

Table 1. Rehabilitation content

Time (d)	Activity content and rehabilitation content
1	Bed rest, in the heart of the passive stand up, 2 times a day, each time 20 min. Activly participate in physical activity, and the introduction of cardiac rehabilitation program and the ward environment, maintain emotional stability of patients.
2-3	Semi supine or sitting position, active physical activity 2 times, each time 5-10min. Assist the bed wash, eat, use the potty bed. Introduce the advantages of early rehabilitation nursing, the attention items of rest and diet,obtain trust.
4-5	Sit and stand by the bed chair 2 /d, 10 min each time, sitting face and diet. Explain the principles of the treatment of myocardial infarction and rehabilitation procedures, allowing reading newspapers, listening to music and visiting.
6-8	Stand, walk freely walking indoor toilet, eating, Explain the importance of drug treatment and adherence to medication

 DOI: 10.1051/bioconf/20170801003

9-10	Indoor activities increased, the basic living. To explain the knowledge of emergency treatment of myocardial infarction.
11-12	Outdoor corridor slow line, each time 50-100m, 2 times /d. The relationship between exercise and rest.
13-14	Up and down the stairs movement, before the discharge guidance, regular follow-up.

2.3 Diet Care Nursing

The nursing staff assisted in the arrangement of the rehabilitation period of patients. Guidancing for patients to eat should not be too full, eat less and eat more time. The food was easy to digest with less fat and less gas production was appropriate, limiting the amount of sodium intake. If the serum lipid levels were still significantly abnormal, according to the characteristics of dyslipidemia patients, choosing appropriate lipid-lowering drugs[7].

2.4 Care of patients with disorders of urination and defecation

Constipation is common in patients with AMI, should be within 3 days of defecation in bed, do not forcibly defecate , explain to the patient in bed defecation had great significance to control the disease, not because of fear of defecate on the bed and can not eat, it would increase the risk of constipation. At the same time, from the first day of hospitalization, the daily morning honey 20ml plus 80ml temperature with boiling water to drink, abdominal massage (clockwise) 2/d, in order to promote intestinal peristalsis.

2.5 Pain nursing

Myocardial infarction patients often have severe chest pain, nursing staff should follow the doctor's advice given nitroglycerin, morphine and other drugs to prevent pain, causing shock and arrhythmia, and inquired the pain and concomitant the changes of in symptoms, to monitor the change of blood pressure in patients with attention, and attended to patients with no pulse acceleration, respiratory depression and other adverse reactions.

2.6 Mental nursing

Patients with severe chest pain should try to keep a nurse accompanied by the patient side.Because of the particularity of myocardial infarction disease, patients may produce negative emotions. The domestic and foreign studies had found that AMI was prevalent in patients with depression, anxiety, fear, pessimism, despair and other psychological. Had a bad mood would affect the treatment and rehabilitation of disease[8]. Rozanski A[9] found that there was significant correlation between the lack of emotional support and myocardial infarction mortality. The medical staff should comfort the patients and explain the bad mood would increase the heart load and myocardial oxygen consumption, which was not conducive to disease control. The nurse with the doctors rescue at the same time, should do the work to appease the patients and the families of patients[10].

3 Evaluation index

Observed in 41 patients in treadmill exercise, hospitalization duration, self-care ability etc. And observed in 41 cases of CQQC (quality of life questionnaire Chinese cardiovascular disease score). The CQQC questionnaire was written by Jiangsheng L, Chinese Rehabilitation Medical Association cardiovascular disease Specialized Committee, and cardiovascular rehabilitation medical journal. According to the definition of the quality of life of the WHO, to develop scale according to three aspects as physiological, psychological and social conditions of our country, and the appropriate reference the questionnaire[11], is divided into physical , disease , medical conditions , general life , social psychological status, working conditions and other 6 projects,24 items, for respondents according to their own situation of the corresponding answer, answer according to the score.

4 Conclusions

After 2 weeks, 41 cases of patients with the average treadmill walking distance, average hospitalization days, the recovery of heart function. See Table 2.

Table 2. Effect Analysis

Gender	Average walking distance (m/ 10 min)	Length of stay (days)	Heart function classification (n)	
			I ~II	III ~IV
male	326±140	18.9±4.4	13	4
Female	347±120	18.5±4.2	18	6

After the intervention, the CQQC score of all patients with early rehabilitation nursing was good, and the overall effect was good. Patients reflected the good recovery, quality of life was significantly improved. See Table 3.

Table 3. Patients' quality of life score

Gender	Male (24 cases)	Female (17 cases)
Physical condition	17.24±6.93	16.34±3.73
The illness	16.23±3.98	16.83±2.54
Medical condition	5.75±0.63	5.29±0.84
General life	10.02±2.89	6.53±1.72
Working condition	6.08±1.07	6.67±1.57

5 Discussion

Analysis of the effect of early rehabilitation nursing on patients with acute myocardial infarction Table 2 shows, the length of hospital stay was significantly shorter, indicating that no adverse effect of early rehabilitation nursing on patients with acute myocardial infarction, to resume working ability of patients and improve the prognosis of patients, and reduce the cost of hospitalization, reduce the burden of family and society has an important significance. It is safe and feasible. The purpose of early rehabilitation nursing is to strive for patients' life as early as possible and early discharge, and to monitor the activities of the transition to go home under the no monitoring of the security activities[12]. This stage of activity is to "not too much," will not cause a infarction or even death as the basic principles. But in the early phase of myocardial infarction, many factors affect the oxygen consumption of patients, the movement itself on patients with acute myocardial infarction (AMI) existence certain risk, the effects of various exercise on heart rate and blood pressure is inconsistent, inappropriate exercise can increase the myocardial oxygen consumption and aggravate myocardial damage, induced myocardial infarction and heart failure[12].Therefore, nursers in rehabilitation of patients with acute myocardial infarction (AMI) to evaluation of patients with personality and social psychological factors, must be fully taken into account and to minimize the adverse impact of these movements.

The research results indicate that early rehabilitation nursing patients quality of life scores higher after nursing.The study shows that the rehabilitation nursing measures is conducive to promoting the improvement of the patient's condition and the quality of life of patients[13-14].The study [15]found that the mortality and disability rate were higher in patients with acute myocardial infarction, but the early rehabilitation nursing measures could improve the quality of life of patients to a certain extent in the early rehabilitation nursing measures. Because of myocardial infarction patients with long-term bed is very easy to cause deep vein thrombosis, reduce exercise capacity, resulting in waste of the syndrome, and is not conducive to the recovery of heart function[16]. Early rehabilitation nursing by functional exercise guidance on patients, promote the patient recover as soon as possible, it can reduce caused by prolonged bed rest complications such as angina, arrhythmia and heart failure and other, eliminate fatigue of patients, improve patient self-care ability and athletic ability, and greatly ease the patient's mental burden[17]. There is conducive to the patient's condition improved[18]. And patients through early rehabilitation guidance, improve the treatment of active and active, to overcome their own bad mood, enhance the confidence to overcome the disease.

In summary, the use of early rehabilitation nursing is of very important significance in patients with acute myocardial infarction, can greatly improve the quality of life of patients, improve the patients' satisfaction degree of nursing work, is to improve the patient's condition, it is worth clinical promotion.

References

1. You LM, Sun GZ, Yuan L. People's Health Publishing House. Department of internal medicine nursing school, Third edition, Beijing.M.164-170(2002).
2. Dong ZQ. Nursing practice and research. Early rehabilitation exercise and nursing in patients with acute myocardial infarction. J. 8 (8),52 -53(2011).
3. Liu JS, Dai RZ, Cheng YZ. Journal of cardiovascular rehabilitation medicine. Percutaneous coronary artery interventional therapy after rehabilitation program.J.15 (2),125-129(2006).
4. Zhen YJ, Wang Y, Liu SY. Journal of Hebei Medical University. In the early stage of acute myocardial infarction cardiac rehabilitation.J.32 (10),1230-1232(2011).
5. Zhou SY, Zhang ZY, Tang CL, et al. Journal of modern nursing. The influence of early rehabilitation nursing on the complications after interventional therapy for acute myocardial infarction.J.15 (20),1938-1939(2009).
6. Su Q, Li J, Sun HP. China's recovery. Early rehabilitation nursing of patients with acute myocardial infarction after intravenous thrombolysis.J.1(4),283-284(2006).
7. Hu DY .Chinese Journal of Department of Internal Medicine. Chinese expert consensus on thrombolytic therapy for acute ST segment elevation myocardial infarction.J.47(2),170-174(2008).
8. L Wi, Zhou CH, Gao DS. Chinese Journal of Behavioral Medical Science. The psychological state in patients with acute myocardial infarction and complications related to correlation.J.11 (3),265-267(2002).
9. Zhang SH. China Medical Herald. Clinical nursing pathway in patients with acute myocardial infarction (AMI) application.J.8 (2): 1-17. (35),115-116(2011).
10. Xie H, Dai HQ. Mental illness and mental health.SCL-90 scale evaluation.J.6 (2),156-159(2006).
11. Dong ZQ. Nursing practice and research. Early rehabilitation exercise and nursing in patients with acute myocardial infarction.J.8(8),52-53(2011).
12. Li DM, Li JS, R WS, et al. International nursing journal. Clinical nursing pathway in acute myocardial infarction patients with pre hospital first aid in application.J.31 (10),1848-1850(2012).
13. Xiong LL. Chinese cosmetic medicine. Research progress of clinical nursing pathway in elderly patients with acute myocardial infarction.J.19 (Z3),78-79(2010).

14. Liu Q. Nursing research. Evaluation of application effect of clinical nursing pathway in patients with acute myocardial infarction.J.24 (29),2691-2692(2010).
15. Apóstolo JL, Kolcaba K. Arch Psychiatr Nurs. The effects of guided imagery on comfort depression anxiety, and stress of psychiatric inpatients with depressive disorders.J.23(6),403-411(2009).
16. Weda DT, Collin C.Int Disabil Stud. The Barthel ADL Index: a standard measure of physical disability.J.10(2),64-67(1988).
17. Guo PQ , Wang CY, Yang QY, et al. Journal of Chinese medical science. Application of early rehabilitation nursing in patients with acute myocardial infarction.J.9(35),8601-8602(2009).
18. Zhao ZH, Zhang SQ, Sun HP. China Rehabilitation. Early rehabilitation nursing care for patients with acute myocardial infarction after intravenous thrombolysis in patients with recovery and complications.J.24 (4),277-278(2009).

Teaching experience of final clinical practice by Chinese pharmacy major students based on the method combined by the learning-guiding teaching method and CAI teaching method

Shougang Wang

Beihua university affiliated hospital (First Clinic Medical College), 132011, Jilin, China

Abstract. On the gradual implementation of the new medical education reform and thoroughly applying the Educational Development Plan and the Health Care System Reform, the teaching mode of medical discipline will be changed gradually by following the law of medical education and meeting the need to boost the medical education reform. Meanwhile, the changing life-style prompts the traditional dispensing mode for Chinese traditional medicine to various modes. This changing put forward higher requirement for medicine- related professionals During the process of Chinese medicine teaching, the only method which can fulfill the new need for graduates of Chinese medicine and qualified medicine personals is to change the traditional teaching mode to the new ones which can arose the enthusiasm of working and learning by the traditional medicine students.

1 Introduction

Among the courses of the medicine discipline, Chinese pharmacy is relatively strong application and requires high practice ability. Thus, during its learning process, practice ability should be got more attentions based on strengthen the theory study[1]. Consequently, the graduates from the Chinese pharmacy can enhance their ability of linking theory to practice and working ability. In the course of learning period, final clinical practice is one of the important stage and measure to perfect the graduates' theory and practice abilities.

Learning-guiding teaching method is guided by the teacher, learn and practice by students independently. This method treats students as the learning subject, makes them gain knowledge intelligently and actively. Basing on encouraging students' autonomy penetrates the correct direction given by the teacher, which can make both teachers and students do their best to get what they deserve.

Combining CAI (Computer Aided Instruction) method with the Learning-guiding teaching method is taken during final clinical practice.

2 The implementing form of the teaching method combining learning-guiding method with CAI teaching method

The main implementing form including as follows:

Firstly, students learn independently. General speaking, in the process of final clinical practice, all internship units are relatively busy, therefore, the instructors usually cannot explain agents one by one. In order to get significant practice effect and guarantee the pharmacy to run normally, students are required to recognize agents by themselves. Thus, students will recognize the agents according to their own interests and speed, as a result of which they could learn freely and independently. To some extent, their practice interests are motivated.

Meanwhile, during the process of their final clinical practice, they will go through every step of dosage. Such experience can make them have further understanding of agents and the contradiction during dispensing medicine. Then when the students return home, they can do further study by using CAI. However, this process will lead to their unsystematic learning about the pharmaceutical knowledge after they graduate.

According to the graduates who have poor learning abilities, their learning effect will be declined a lot. Besides this, without the real-time guiding by teachers, students will be lack of the further understanding about some knowledge.

Secondly, students do focus learning. Right now, due to university's expanding their enrollment year by year, the number of graduates from Chinese pharmacy raises every year. In the process of final clinical practice, the space of the pharmacy is limited, which cannot fulfill all students practice at the same time. Hence, teachers could use the samples after collecting them. This teaching way makes students contact the real agents by themselves.

Meanwhile, teachers could apply the CAI method to teach by producing software after their data collection. Generally speaking, firstly our college will use the way of CAI to help the students to gain some understanding of the agents and secondly teach by samples, by which we can gain comparative ideal practice effect.

Thirdly, students do integrative study. In recent year, our college adopts both learning-guiding teaching method and CAI method to direct the final clinical practice. First, CAI teching method or sample teaching is used to students' focused learning, the purpose of which is to help students to understand the sharp, smell and the application site of the agents. And then, according to pharmacy's the rule of time sequence, we will choose the time for students to take part in the practice. Students who practice with problems are highly motivated and have better learning effect[2].

3 The process of combining the learning-guiding teaching method with CAI.

The process of combining the learning-guiding teaching method with CAI includes the key points as follows:

Firstly, a proper plan is drawn up. Before teaching, teachers draw a teaching plan according to the reality of the pharmacy, as well as the requirement made by Education Department and university. The teaching plan usually includes following content: first, the average level of theory and practice by undergraduates; second, to make sure the purposes and targets of the clinical practice by the reality of the pharmacy and the knowledge students have gained; third, to choose the teaching method scientifically by established goal and established target; forth, the effect of the clinical practice is evaluated by students' practice performance and their realities during and after their clinical practice

Secondly, CAI software is produced. CAI software includes many contents, of which difficult and important points are emphasized, make students get a deep understood reasonably. CAI software usually subdivides the teaching contents into general chapter and sub-chapters. General chapter concludes a general description about the history, origin and compatibility of compatibility etc. Every sub-chapter describes in detail about agents due to their different effect, such as fending off the chill, hemostasis, tonifying qi and so on. After this, agents are introduced one by one; for instance, lucid ganoderma is used for bad sleeping and purslane is applied for abdominal paining and diarrhea.

Thirdly, students are trained before practice. Relatively speaking, it is so different from practice and learning in class, for which careful training before practice is necessary. The training content includes the rules of the pharmacy, the notes during practice and the importance of pharmacy dispensing work, which can help them to practice carefully and have clear understanding for the work after graduation.

Fourthly, teachers are arranged reasonably. The choice of the teacher is in relation the effect of clinical practice. The teacher is chosen according to the following standards; first, the teacher has systematic theory knowledge and is rich in it, as well as does well in teaching; second, the teacher should do practice work seriously and explains the teaching points carefully and patiently; third, the teacher should be expressive, which can solve the problems during students' practice and learning and communicate with pharmacy well.; forth, the teacher should have advanced thoughts and accepts the new teaching pattern. They can also have a deeper understand on the syllabus and hold the practice direction exactly.

4 The combination of learning-guiding teaching method and CAI teaching method's implement conclusion

The using of the combination of learning-guiding teaching method and CAI teaching method's implement conclusion including the following:

First, in the process of practice teaching, the principle of "accordance of their aptitude" should be uphold, according to the graduates' interest, to master the theoretical knowledge level, practice ability, arrange content of the internship Reasonably. For example, for graduates who interested in pharmacological knowledge, can make them help pharmacy staff to do relevant experiments of pharmacological knowledge. In this way, we can maximum limit arouse graduate internship enthusiasm; significantly improve the effect of practice.

Second, during the internship, the teacher should base on graduate internship differences in content, targeted to guide the graduates to topic selection. During the period of topic selection, graduates can according to their own interests and hobbies, choose the study of pharmacological knowledge, or the study of the relevant components of agents, etc[3].

Third, in the process of practice, the teacher should encourage graduates to consult the relevant materials and the use of CAI courseware for learning; the related experimental materials were prepared as while. At this time, teacher should guide the graduates reasonably and teach them how to prepare the materials for the experience. When materials is prepared, teacher should guide the graduates to do the experience for helping them understand the important and difficult point, enhance their hand-working ability and encourage then operate independently as while.

5 Conclusions

Graduation practice teaching will play a practical and theoretical verification and an important role to medical students, the teaching quality of this stage will be directly related to the students majoring in pharmacy can become the target of qualified clinicians, and lay a solid foundation for students' future clinical work [4].

All in all, graduation internship is the last station of the traditional Chinese medicine learning process, and is also an important transition before the real work, is also the key to improve the ability of graduates to integrate theory with practice. In our college, through the summary of past practice and study of the new teaching mode, is committed to improve the effect of the internship, graduates for the future of the study and life lay the solid foundation.

References

1. Li XiangHong, Liu LiPing, Liang GuangRong. Research of The combination of learning-guiding teaching method and CAI teaching method during chinese medicine teaching [J]. *Guide of China Medicine*, 2012(06).
2. Wang MingJun, Cui Ying, Feng ZhiYi. Chinese materia medica graduate practice teaching experience [J]. *China's modern distance education of Chinese medicine*, 2011(11).

3. Li YueLan, Hong Yin, Song JieMin. practice and thinking in learning-guiding teaching method in teaching of Chinese materia medica [J] . *China's modern distance education of Chinese medicine*, 2010(13).
4. Shutao, Liangyuan, Wenbinzhang, Boqingmao. Medical students in clinical practice teaching the investigation of the problem[J]. *Medicine and society* 2008(01).

DOI: 10.1051/bioconf/20170801005

A comparative study on the kinematic and kinetic analyses of the different levels of axe kick among female taekwondo jins

Changtao Zhou

Shandong Sport University, Jinan, 250102, China

Abstract. This study investigated the kinematic and kinetic differences among the different levels of taekwondo axe kick among female taekwondo jins using the Vicon 3D motion analysis system and Kistler force plates. Independent t-tests were used to detect the differences. The high-level group was designated the senior group (SG), and the low-level group was designated the junior group (JG). Kinematics data showed that SG had more pelvis rotation ($p < 0.05$), earlier time of the lowest attract point ($p < 0.05$), and higher peak value of velocity curve ($p < 0.05$) than JG. No differences were observed between the two groups in terms of the lowest attack curve. Kinetics data showed that the peak ground reaction forces (GRF) in the vertical direction in SG were higher ($p < 0.05$) than those of JG; the peak force in SG was also exhibited earlier ($p < 0.05$) than that in JG. Therefore, differences were detected in the pelvis rotation, peak value in attack velocity, and GRF in the anterior–posterior direction between SG and JG. These factors also increased the attack velocity and range in SG.

1 Introduction

Taekwondo is a combat sport that uses various kicks, such as crescent, axe, spinning, and back kicks, as the basic offensive technique. Among all these kicks, the axe kick (also called axe kick) is frequently used by most well-known taekwondo jins because of its force, tactical surprise, high striking point, and long-range strike. With the implementation of new taekwondo rules in 2008, actively hitting the head has become a key factor in determining winners and losers in a tournament [1]. The axe kick is primarily used to hit the head; thus, the prevalence rate of its use is increasing[1,5,7]. In the 2008 Beijing Olympics, jins scored a high 13% rate using the axe kick, and it ranked second among all leg techniques[2]. The current author's previous study showed that the crescent kick is the most extensively used among all leg techniques, but the axe kick is the second most extensively used because of its force, tactical surprise, and high striking point[3].

To date, studies on axe kick biomechanics have considerably concentrated on kinematics. Wang Lei analyzed the axe kick technique using 3D imaging technology and determined that the change in the hip joint of the swing leg is relatively stable. In the process of knee bending, the knee joint movement is more remarkable than those of the toe and ankle joint. The change in the toe and ankle joint is significantly prominent during kicking, when the maximum value is achieved. After executing the axe kick, both toe and ankle joint displacements begin to decrease abruptly. During the entire split process, the change in the joint movement of taekwondo jins is as follows: toe > ankle joint > knee joint> hip joint. Cheng Lei performed a biomechanical analysis of female taekwondo jins during the semifinal and final rounds of the 2006 Taekwondo Championship in China using an SIMI motion analysis system. This study obtained the time, angle, velocity, gravitational changes, and other biomechanical parameters of the crescent, front, and axe kicks of female taekwondo jins. The beginning of the knee lifting, kicking, and drawing back leg stages of the swinging leg accounted for 22%, 46%, and 32%, respectively. The percentage of time of stage playing kick is higher than those of the crescent and front kicks. In summary, local studies on taekwondo sports biomechanics mainly focused on the technical movements of the kinematics description. These studies included only one research index and mainly underscored angle, speed, body weight, and other parameters. Moreover, strike speed, strike range, and other corresponding indicators were lacking. A comparison of the kinematics of the different levels of taekwondo jins has not been performed. The dynamics of these athletes at different levels is also poorly understood. Therefore, no conclusion has been formed on the foot force characteristics of the axe kick.

The Vicon (Corporation T20-s, Vicon, England) infrared 3D kinematic analysis system is proposed in the current study. By consulting the ground 3D reaction force curve (Kistler, 9287BA, Switzerland), this study synchronously analyzes the kinematics and dynamics of the axe kick and compare the different levels of data of female taekwondo jins. A summary of the technical problems of taekwondo jins and data support and scientific bases for future training is also provided.

2 Subjects and method

2.1 Subjects

This study recruited 24 female taekwondo jins as subjects. Among these athletes, 11 were national first-grade athletes and were designated as the junior group (JG) in this study. The remaining 13 players were national-level athletes and were designated as the senior group (SG).

No significant differences in height, weight, and age among members of the two groups were noted. Table 1 shows the details of the subjects.

Table 1. Details of the subjects

Groups	Number	Height(cm)	Weight(kg)	Age(y)	Training time(y)
JG	11	168.23 ± 3.12	62.14 ± 5.10	21.22 ± 2.41	3.1 ± 0.98
SG	13	169.45 ± 4.11	64.26 ± 6.61	20.10 ± 1.47	7.4 ± 3.67

2.2 Research method

2.2.1 Laboratory setting

Eight infrared cameras were placed along a laboratory platform to collect the kinematic data of the taekwondo jins. The shooting frequency of the camera was 1,000 Hz. A 3D force platform (Kistler, 9287BA, Switzerland) with a collection frequency of 1,000 Hz was installed on the platform. A Vicon system was used to synchronize the two instruments. Before the test, 34 reflectors were used to mark the joints of these athletes (full body model of the Vicon system). The diameter of the reflector is 14 mm (Figure 1).

DOI: 10.1051/bioconf/20170801005

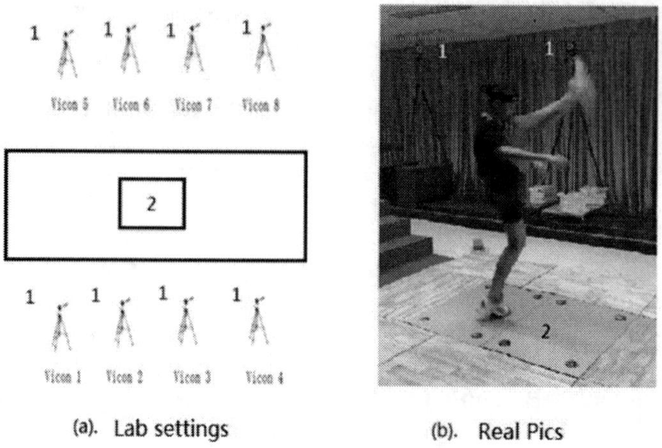

Figure 1. Laboratory setting: 1: Laboratory Setting 1: Vicon infrared cameras 2. Kistlerforce platform.

2.2.2 Experimental procedures

All subjects were informed about the details of the trial and signed the agreement. The experiment was conducted at the Laboratory of Sports Biomechanics of the Shandong Sports Institute. Upon arrival at the laboratory, the subjects were informed about the following details: 1) the objectives, significance, procedure, content, method, and notice of the experiment; 2) staff members would record and test their age, height, weight, athletic performance, and so on; 3) they would be asked to practice assessing the test before the actual experiment and to become familiar with the test environment; 4) staff members would place reflective marks on the subjects and debug the test force platform; 5) data would be collected during the experiment; 6) each subject would execute the axe kicks in their usual rhythm; and 7) each subject would repeat the action to extract five valid data; the average value will then be calculated.

2.2.3 Data processing

Changes in the angle of the ankle are considerably large. Therefore, the ankle joint of the taekwondo jin is selected as the distal end of the axe kick movement. The distance of the athlete's ankle to the body's center of gravity serves as the strike distance. Similarly, the ankle speed of the swinging leg of the taekwondo jin serves as the strike speed. The angle between the two iliac spine connection and the frontal axis in the horizontal plane is denoted as the rotation angle of the pelvis. The measured plantar reaction force data standardize the center of gravity of the athlete. Each taekwondo jin was required to provide kinematic and kinetic data five times and to be involved in the data analysis thrice.

2.3 Statistical Analysis

Data from the two groups were compared using independent t-tests. A type I error rate of below 0.05 indicates a significant difference.

DOI: 10.1051/bioconf/20170801005

3 Research results

3.1 Study of the kinematic parameters of the different grades of female taekwondo jins

3.1.1 Comparative study of the pelvis rotation

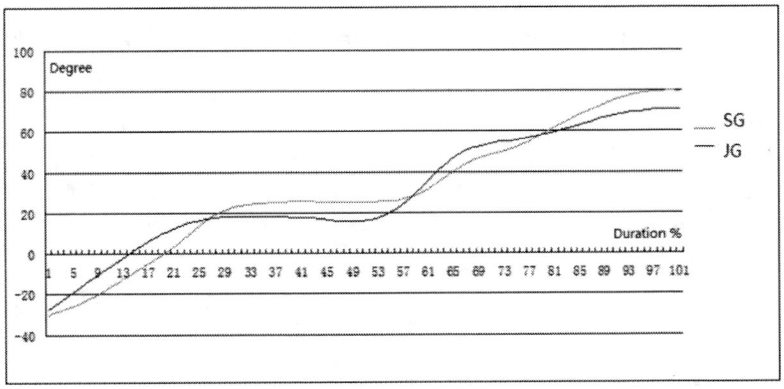

Figure 2. Pelvic rotation angle.

In general, the movement form of the pelvis during the axe kick movement in the horizontal plane is rotary–fixed–rotary.

The pelvic rotation range of SG was larger in the first rotation stage (SG = 52.65°, JG = 46.04°, $p = 0.006$) than that of JG. In the fixed phase, the pelvic rotation range of SG was relatively stable (SG = 4.12°, JG = −2.28°, $p = 0.013$). No significant difference was found in the rotation range of the two groups during the second rotation stage (SG = 53.06°, JG = 55.17°, $p = 0.365$) (Figure 2).

3.1.2 Comparative study of the swing leg range and attack speed

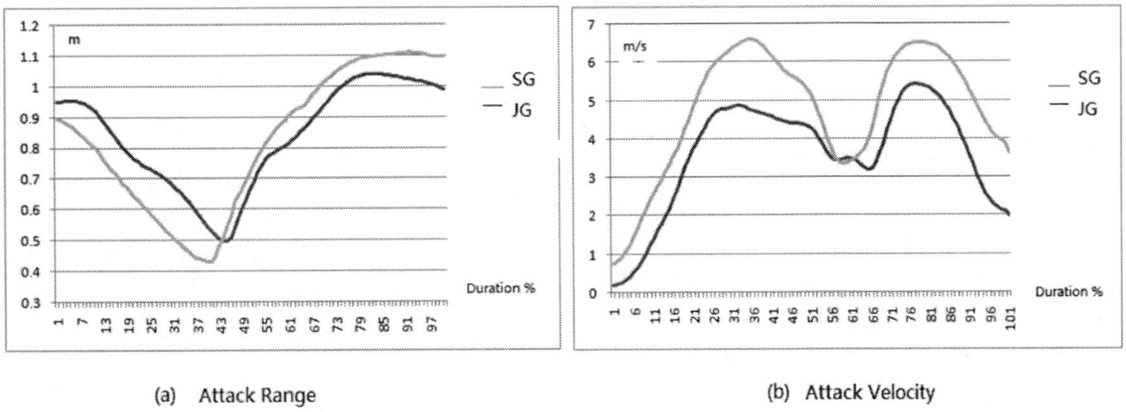

Figure 3. Kinematic parameters of the swing leg including (a) swing leg of the strike range and (b) swing leg of the strike speed.

Figure 3 (a) shows that the strike distance curve of the axe kick of the taekwondo jins is generally V-shaped. The lowest point in the SG curve was significantly earlier than that of JG (SG = 41.16%, JG = 45.24%, $p = 0.012$). The lowest point data of SG was significantly lower than that of JG (SG = 0.43 m, JG = 0.49 m, $p = 0.021$). The highest point data of SG was significantly higher than that of JG (SG = 1.11 m, JG = 1.03 m, $p = 0.000$). Thus, the swing leg range of JG was considerably small, but the

DOI: 10.1051/bioconf/20170801005

swing leg was significantly large upon reaching the lowest point.

Figure 3 (b) shows that the speed curve of the swing leg of SG has two single peaks. Both the SG first peak (SG = 6.65 m/s, JG = 4.90 m/s, p = 0.000) and the second peak (SG = 6.62 m/s, JG = 5.44 m/s, p = 0.000) were higher than those of JG. No significant difference was found at the minimum of the swing leg speed curve between the two groups (SG = 3.36 m/s, JG = 3.19 m/s, p = 0.104). SG also appeared earlier in the valley than JG (SG = 58.19%, JG = 65.34%, p = 0.009).

3.2 Study of the dynamic parameters of the different levels of female taekwondo jins

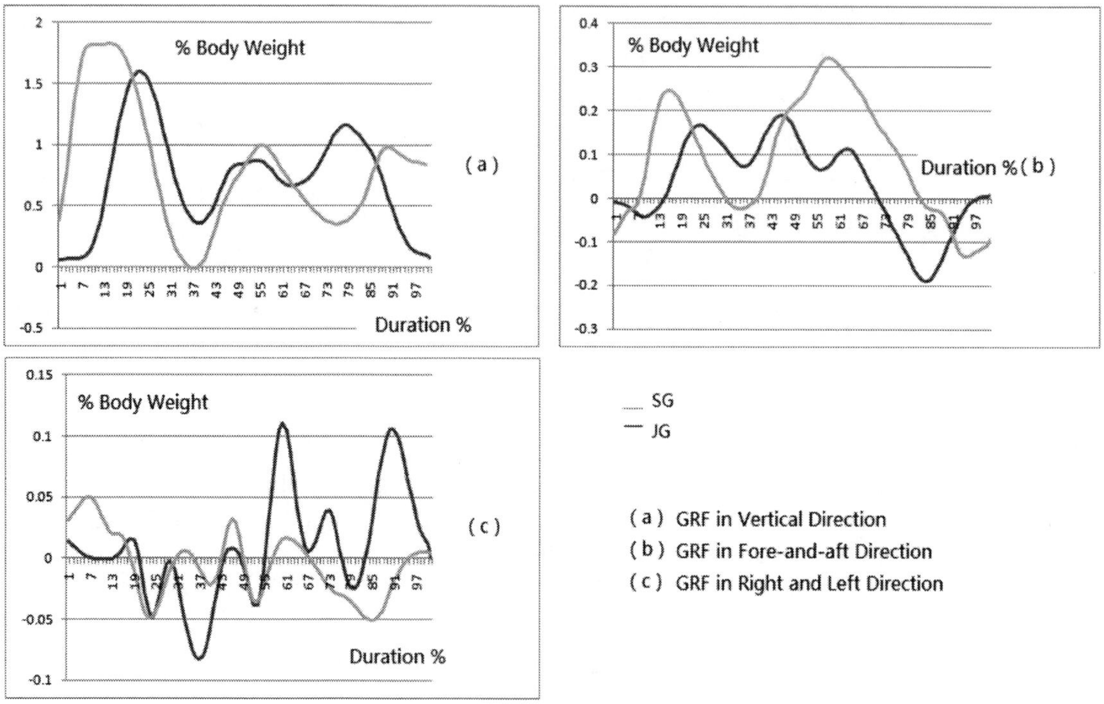

Figure 4. 3D GRF curve

Figure (4a) shows that the ground reaction force (GRF) curve is generally a three-peak curve in the vertical direction. The first peak value of SG was larger than that of JG (SG = 1.77 BW, JG = 1.60 BW, p = 0.021). The first peak of SG appeared significantly earlier than that of JG (SG = 16.54%, JG = 20.14%, p = 0.013). No significant differences were noted in the values of the second peak (SG = 1.01 BW, JG = 0.93 BW, p = 0.424) and in the appearance time of the second peak (SG = 56.94%, JG = 58.19%, p = 0.213). The third peak value of SG was significantly less than that of JG (SG = 0.99 BW, JG = 1.18 BW, p = 0.001), and the SG appearance time was significantly later than that of JG (SG = 92.74%, JG = 80.14%, p = 0.000).

Figure 4 (b) shows that the GRF direction of the fore-and-aft direction of the taekwondo jins is generally to the front of the body. The peak value of SG was higher than that of JG (SG = 0.32 BW, JG = 0.18 BW, p = 0.000). Moreover, the SG members received a considerably high impulse (force and time integral) to the front of the body.

Figure 4 (c) shows that the GRF curves of the right and left direction fluctuate up and down around the axis, but the volatility is large in this parameter. GRF to the left direction (negative value) of SG was smaller than that of JG (SG = -0.05 BW, JG = -0.08 BW, p = 0.001). By contrast, GRF to the right direction (positive value) of SG was smaller than that of JG (SG = -0.03 BW, JG = -0.11 BW, p = 0.000).

4 Discussion

Analysis of the blow range curve indicates that the value of SG in the front section of the curve is less than that of JG. However, in the back section, the value of SG is lower than that of JG, and the appearance time of the lowest point in SG is earlier than that in JG. The front section values mainly correspond to the initial ankle joint action upon reaching the top position, i.e., the preliminary swing. The values of the back end mainly correspond to the completion of the swing leg action. From the curve, the distance between the ankle joint and the center of gravity of the body in SG is smaller in the preliminary swing than that in JG. This result may be caused by high-level athletes completing preliminary swinging with a considerably large flexion angle of the lower extremity joint [4,9,11]. A large flexion angle can shorten the swing circumference of the swinging leg, thus reducing the swing movement time. At the back section of the curve, the strike range of SG is larger than that of JG. This result may be caused by high-level athletes completing a axe kick with a small flexion angle of the lower extremity joint [4,13]. Analysis of the speed curve shows that the values in the front and back of the curve of SG are less than those of JG. The appearance time of the lowest point of SG is earlier than that of JG. Speed curve shows that regardless of the preliminary swing or completion of the action stage, the SG ankle joint velocity index is greater than that of JG. The completion time of SG is lower than that of JG. SG athletes increase the flexion angle of their lower extremities and complete the swing movement rapidly. Moreover, the increased range and speed of the kick strengthens the effect.

Studies on the GRF curve in the axe kicks of taekwondo jins have limited involvement in the current study. Based on the analysis of the vertical direction of the curve, the value of the SG first peak is higher and the appearance time is earlier than those of JG. No difference exists between SG and JG in the second peak in terms of value and appearance time. The value of the SG third peak is lower and the appearance time is later than those of JG. Analysis of the peak value function indicates that the first peak is the GRF support during the swing process. The second and third peaks are supported by GRF in the splitting of the swing leg process. Figure XX shows that the SG first peak value is larger than that of JG, thus indicating that the GRF support is larger. The shadow area of the SG first peak (integral of the supporting force and supporting time) is greater than that of JG. This result indicates that the GRF impulse of SG is greater than that of JG. Therefore, the SG athletes move their center of gravity in a considerably large range during the swing movement, thus increasing the swing leg range. Increasing the gravitational potential energy of the starting position can also augment the kicking effect. The SG third peak value may be caused by the high acceleration of the swing leg. During the splitting process, other parts of the athletes' body associated with the swinging leg form a rigid system. The vertically downward acceleration of the swing leg can cause other body parts to move up and thus reduce GRF.

Analysis of the GRF curve of the fore-and-aft direction indicates that the GRF direction of the fore-and-aft direction of the taekwondo jins is generally to the front of the body. The SG peak value and impulse (force and time integral) are higher than those of JG. The considerably large GRF peak value and impulse can increase the forward movement amplitude of the SG athletes, thus effectively increasing their attack range. Analysis of the GRF curve of the left-to-right direction shows that the curve is promiscuous, as it generally fluctuates up and down around the axis. The SG GRF peak value of the left-to-right direction (negative value) is smaller than that of JG. As the opponent of a taekwondo jin is expected to be in front, the former does not need to exert the left-to-right direction force to achieve the objective of an accurate attack. Therefore, the component GRF force in the left-to-right direction plays a role only in maintaining balance in this study. The SG fluctuation range is considerably small, indicating that the taekwondo jins in this group have strong ability to balance their body. Moreover, the SG athletes no longer need the component left-to-right direction force to adjust balance. Therefore, the peak force is reduced effectively.

Analysis of the pelvic rotation range in the horizontal plane indicates that the SG rotation range is larger in the front of the curve than that of JG group. The SG pelvis rotation in the fixed stage is lower than that of JG. Moreover, no difference exists between the rotation ranges of the two groups in the back section of the curve. Values of the fore part correspond to the preliminary swing from the initial stage of the swinging leg of the top-most ankle position. Middle values correspond to the athlete completing the kicking movement after the preliminary swing. Values of the back part correspond to the stage after the kicking action, i.e., from bringing it back to letting the swinging leg touch the ground. Based on the analysis of the curve, the SG pelvic rotation range is relatively large, and thus the kicking distance of the SG taekwondo jins in the horizontal direction is considerably large. Consequently, the trunk rotation range is increased, and the attacking range of the opponent is relatively small [13,15]. As the pelvic rotation range of the SG athletes is small, the athletes can fix their pelvis better in the process of completing the kick. Consequently, the stability of the body of the SG taekwondo jins is higher than that of the JG jins.

5 Conclusion

The rotation range of the SG taekwondo jins during the preliminary swing is larger than that of the JG taekwondo jins. The pelvis is stable, and the stability of the body is improved. The kicking movement is completed rapidly by increasing the flexion angle of the swing leg, the first peak value of the vertical direction GRF and appearance time, and the peak value and impulse of the horizontal direction GRF. The axe kick increases the range and speed of attack, thus strengthening the effect of such an attack. Therefore, the training of low-level athletes should involve the enhancement of body stability, improvement of the flexion and extension angles of the swing leg, enhancement of the peak value of the vertical direction GRF and time, and improvement of the impulse and peak value of the horizontal direction GRF.

References

1. Fei Ying, Song Hua. The analysis and research of the competition rules of the taekwondo in 2008[J].Martial arts and professional boxing teaching,2009 (9):86.
2. Yang Zhijun. The study of the technical and tactical research of the current situation under the new rules of the excellent taekwondo athletes in our country[J].Sports science and technology in China,2010 ,46(6) :79-81.
3. Zhou Changtao, Yu Daifeng. The biomechanical characteristics analysis of the Roundhouse Kick of the male taekwondo athletes[J],Journal of Shandong Sports Institute,2010,26(9) : 58-63.
4. Wang Lei. The biomechanical characteristics analysis of Next Splitting Technique of the excellent female taekwondo athletes at ShanDong,2012, the master's degree thesis of Shandong Sports Institute.
5. Cheng Lei.The biomechanical characteristics analysis of the kick, front kick, chop action of the female taekwondo athletes,2005,the master's degree of Zhejiang Normal University.
6. LuWenchao. The application and the development characteristics of the taekwondo competition tips [J].Sports science and technology,1999,22(1):94-97.
7. Liu Baocheng, Zhao Qiurong, Lu Fan. The Research Report of Chinese Taekwondo Championships 1996 "Wanji Cup" [J].Journal of Xian Sports Institute,1996,14(1): 48-51.
8. Scott L. Delp, William E. Hess, David S ect. Variation of rotation moment arms with hipflexion[J]. Journal of Biomechanics. 32 (1999) 493–501.
9. Liu Baocheng, Zhao Qiurong, Lu fan etc.The technology analysis of 1997 "Kang Cup" National Taekwondo Championship [J]. Journal of Xian Sports Institute.1998,15(2):37-41.
10. Pieter W, Zemper E.D. Time-loss Injuries in Junior Olympic Taekwondo Athletes[J]. Occupational Health and Industrial Medicine Volume: 36, Issue;6,1997,pp.290.
11. Zhu Hui, Xue Xinxuan. The comparative analysis of the application of leg techniques of Chinese and foreign male's taekwondo athletes[J].Journal of Wuhan Sports Institute.2002,36(6):149-150.
12. Zhang Qingfeng, Meng Zhaofeng. Factors affecting the speed of action in Taekwondo[J]. Journal of Zhejiang Normal University(Natural Science Edition).2004,27(3):299-302.
13. Shang Hua, Wang Guanghua, Tu jun. The injury of juvenile Taekwondo, Sanda and boxing[J].Journal of Wuhan Sports Institute. 2000,34:91-92.

 DOI: 10.1051/bioconf/20170801005

14. Li Zhigan, Chen Runlin. Research on the characteristics of sports injuries of the Taekwondo Athletes in Guangdong Province[J].Journal of Guangzhou Sports Institute.2000,20(2):56-60.
15. Zhu Lihua, Liu Bai, Zhao Xin etc. Investigation and analysis of sports injuries of Chinese excellent taekwondo athletes[J].Journal of Beijing Sports Institute.2001,24(2):76-78,182.

Probe into the training mode of clinical nurse specialists in China

Yukun Zhang and Yan Song[a]
Beihua university,132013, jilin, China

Abstract. Overview of the clinical nurse specialist development history, current situation and training mode, and according to the current situation of the development of clinical nursing specialist in China, proposed the convergence of clinical nurse specialist and master of nursing specialist training, and put forward the new training mode of the "four cards and one" ,and analyzes the feasibility and difficulties.

1 Introduction

In recent years, with the rapid development of medical science and technology and the continuous improvement of people's health needs, the training of highly qualified specialists has become the trend of the development of nursing science in the new era. Clinical nursing specialist(CNS) , as an important part of advanced nurse practice(ANP), plays an important role in the development of nursing science. It can not only effectively promote the development of nursing research and teaching, but also reduce the gap between nursing research and practice, thus to improve the quality of care and reduce the cost of nursing [1]. According to the research report, China began to study the CNS in 2001 [2], and in 2010, the training of Master of nursing specialist(MNS) was started, although some progress has been made in the development of CNS and MNS in recent years, it has not formed a unified, sound training mode yet, it is still in the exploratory stage. How to explore the CNS training model which conforms to the situation of our country is the urgent problem to be solved in the development of nursing specialty in our country [3].

2 CNS development status in China

2.1 The definition of CNS

In the United States,as early as 1910, nursing experts put forward to cultivate a higher level of nursing talents in certain diseases, and officially called clinical nursing specialist (CNS) in the 1960s[4]. At present, the international concept of universal recognition of the CNS is the definition of the American Nursing Association (ANA): CNS is a registered nurse with the master or doctoral degree, with rich clinical experience and a high level of nursing skills and is an expert in a particular field of clinical knowledge [5].

[a] Corresponding author: Yan Song, 1293625334@qq.com

2.2 Development status of CNS in China

In 1996, Hong Deng first introduced the concept of clinical nursing experts in the country, and introduced the development of foreign CNS[6]. Our country began the training work in 2001[2]. Zhongshan University nursing college, University of Hongkong and the Hongkong Institute of stoma co founded the first fistula qualification certificate for school in 2001, training the first batch by the international certification of CNS in China[7], Sir Run Run Shaw Hospital Affiliated to Medical College of Zhejiang University Cultured the diabetes advanced practice nurses and wound / stoma advanced practice nurses[8]. In recent years, Beijing, Shanghai, Hangzhou and other places have begun to carry out CNS training in critical care, stoma or wound and other fields, although many areas have a certain foundation in training, the cultivation of the CNS is still in the embryonic stage. Over the past ten years, the literature on CNS has been increasing year by year, but mainly on the CNS training system, training direction, some studies on the core competence and business ability, the merger of CNS training mode and MNS system has not yet been reported in the literature. And in the development of nursing specialty, especially to the development of diabetes nurse specialist [9], nephropathy nursing experts [10], pressure sores nursing experts [11], operating room nursing experts [12], developed quickly. And cancer care, family peritoneal dialysis, coronary heart disease and other 20 professional nursing experts in the field of consultation with the patient has been in Shanghai First People's Hospital [13].

3 Preliminary exploration of the CNS training model in China

3.1 Preliminary exploration of CNS

Reading the literature found that domestic experts mainly have "specialized nurse" and "clinical nurse specialist" two kinds of translation on CNS. This is mainly due to the ambiguity caused by differences in the comprehension and translation, Only the concept of the two and the role of functional orientation,can we gradually achieve with the international standards[3]. Xiaojie Wang and etc[14] researched on the concept of specialist nurses and clinical nursing experts in our country, pointing that the specialized nurse should be referred to those who experienced some specialized training system and obtain the professional qualification certificate of nurses.Xiaojie Wang expert interviews and letters of inquiry results show: 75.6% of the experts agree the CNS translated into clinical nurse specialists, and 36 percent of nursing experts think CNS of personnel training should be nursing postgraduates [15].Feng Li and others [16] think, through the training of nursing graduate students to cultivate nursing expert talent is the inevitable development of nursing in specialist. Wenwen Zhang [17] by means of Delphi to 33 experts for consultation, and 88.9% of the experts believe that nursing graduate students will be the senior practitioner of nursing after graduation.

3.2 The assumption of "four certificates and one"

With the vigorous development of higher nursing education in recent years, Study of nursing postgraduate education scale is also expanding, by the end of 2015, there had 86 colleges and universities in our country with the master's degree in nursing science. On the one hand, the high level of nursing personnel is increasing, on the other hand, the development of specialized nursing science is also urgent for us to develop CNS as soon as possible. Under this background, The author puts forward the idea of "four certificates and one" for the training of clinical nursing specialist and master of nursing science, namely after graduation, we will acquire CNS certificate, nurse standardized training certificate, master's degree and professional master's degree certificate. Putting the CNS training and professional master's training combined closely. At the end of 2015, National Ministry of Health issued the outline of the standardized training program for nurses, the middle of which did not involve whether to continue compliance training content for professional master graduate students after graduation, this is undoubtedly not fair for the professional master's graduate students, because during the period of school they had been completed at least two years practice rotary. The implementation of the "four cards and one" can not only promote the development of nursing graduate education, but also speed up the pace of the development of CNS.

3.3 Feasibility of CNS and MNS integration

3.3.1Comparison of current situation of cultivation

The cultivation of CNS and MNS has relevance in attribute[18]. The training target localization of CNS and MNS are of high level, applied and professional nursing talents,which showed that the master of nursing and clinical nursing experts have consistency in nature. In addition, By comparison, it is found that CNS and MNS have great similarity[19] (see Table 1) in terms of enrollment standard, training mode and core competence, which provide the possibility and feasibility for the combination of the two.

Table 1. Comparison of CNS and MNS in cultivation

Project	Enrollment standard	Cultivation mode	Core ability	Cultivation direction
MNS	Bachelor degree; Or equivalent; Through nurse qualification exam;2-3 academic year.	School training; hospital training; hospital school joint training.	Educational ability; scientific research ability; clinical skill; nursing management; thinking ability; communication ability.	Emergency critical care; nursing management; surgical nursing; pediatric nursing; elderly care; psychiatric care, etc.
CNS	College degree or above; 3-5 years working experience; intermediate title; registered nurses; have certain scientific research ability.	Hospital training; hospital school joint training; joint training of hospital.	Clinical ability; scientific research ability leadership ability; cooperation ability; nursing consultation,etc.	Diabetes care; operation room nursing;ICU nursing; emergency care; wound / stoma care, etc.

3.3.2 The connect situation of professional degree and professional qualification of master degree in China

In July 2011, Shanghai Municipal Education Commission, Shanghai Municipal Health Bureau jointly promulgated the *Shanghai municipal hospital standardized training of physicians and clinical medicine master professional degree education convergence reform measures for implementation*, began a clinical hospital standardized training of physicians and clinical graduate students education merger of pilot work [20]. The professional degree graduate students after graduation can achieve

DOI: 10.1051/bioconf/20170801006

"four cards" (practitioner qualification certificate, standardized resident training certificate, Graduate Diploma, professional master degree certificate), so as to realize the organic combination of vocational qualification training (The standardized training of resident physicians) and the organic combination of master's degree in clinical medicine. In addition, the professional of the convergence of master professional degree and professional qualifications, such as Master of architecture and registered architects[21]; Master of accounting and corresponding professional qualifications of the convergence, etc. The organic connection of other major also provides reference for the combination of nursing master's degree and clinical nursing specialist. Therefore, the implementation of CNS and MNS combined with the "four cards and one" is feasible, also is the inevitable trend of the development of nursing specialist in China.

4 Analysis of difficulties in the connection between CNS and MNS

4.1 The practical ability of nursing postgraduates' training needs to be improved

Whether MNS and CNS can achieve convergence , the key lies in the quality of their own as well as social recognition. As the MNS education in our country is just at the beginning stage, the lack of a unified, specific training programs, coupled with limited staff and other reasons, China's MNS can not be very good to highlight the connotation of its "advanced practice nurse". Therefore, we need to strengthen and improve the training system of MNS “hospital school joint training” and the "dual tutorial system".

4.2 Gradually improve the CNS management system

According to the actual situation of our country, we should define the working function of CNS, establish a reasonable salary system, establish an evaluation system, and the establishment of the system of CNS and MNS convergence. The most important is the establishment of the Certification Committee, the organization of the MNS assessment, qualification, etc.

4.3 Establish pilot training institutions

Strict training is the key to ensure the quality of the CNS, so we can choose some good development of nursing science, comprehensive strength of medical schools and their affiliated hospitals to establish CNS's main train station, so as to combine the CNS with MNS organically.

4.4 CNS practice lack of legal protection

Although we continue to explore the training mode of CNS in the last ten years, it still lacks legal protection.

5 Conclusion

In summary, CNS refers to the higher level of professional nursing knowledge and skills and rich clinical experience of experts. The cultivation of clinical nursing specialist in China is still in exploration stage, CNS and MNS for organic convergence can not only promote the development of higher nursing education but also promote the improvement of the quality of clinical nursing, to achieve a win-win results, so that the domestic nursing can gradually in line with the world.

References

1. Profetto-McGrath J, Smith KB,etc. Worldviews Evid Based Nurs. Clinical nurse specialists' use of evidence in practice: A pilot study .J. 4(2),86-96(2007).
2. Wang BQ, Li J. Nursing research. Clinical nurse specialists development profile.(16),1502-1503(2007).
3. Hou XY, Zhang HM, Li R. Nursing research. Clinical nurse specialists training mode.J.27(11),961-963(2013).
4. Hunt JA. A dv Nurs. A specialist nurse: an identified professional role or a personal agenda.30(3),704-712(1999).
5. Mary MC, Maya JD. Nurse Clin of North Am. The role of the clinical nurse specialist in an adult congenital heart disease program.29(2),357-365(1994).
6. Pickergill F. Foreign medical: nursing in parts. Nurse practitioner and clinical nurse specialist is the trend of development .J.15(1),35(1996).
7. Xu CJ, Wu QL. Journal of nursing management. ICU and colostomy specialist nurses present situation investigation and analysis.J. 6(10),15-17(2006).
8. Feng JE, Hu HY, Lou QQ.Chinese Journal of nursing. A study on the role of advanced clinical nurse specialist.J.39(2),115(2004).
9. Hu XR, Liu SQ, Mo YZ. Chinese Journal of nursing. Practice and experience in the establishment of diabetes specialist nursing training.J.42(6),496-498(2007).
10. Li YJ, Zhang LY, Wang XL. Chinese Journal of nursing. Education of clinical nursing specialists of nephropathy.J.39(6),434-436(2004).
11. Xu WQ, Lou XH, Hao P. Nursing research. The effect of pressure ulcer nurse specialist on stage III severe pressure ulcer.J.20(10),2790-2790(2006).
12. Zhao LH. Family nurse. Current situation and development trend of clinical nursing specialists in China. J.6(21),1966-1967(2008).
13. Wei HI, Li P, Fang F. Journal of nursing. Effect analysis and experience of nursing specialist outpatient clinic for 8 years.J.18(9A),47-50(2011).
14. Wang XJ, Shen N. Journal of nursing management. Discussion on the concept of specialist nurses and clinical nursing specialists in our country.J.5(12),25-26(2005).
15. Wang XJ, Shen N. Journal of nursing management. Cultivation and related problems of junior college nurses in China.J.6(12):8-11(2006).
16. Li F, Han J. Nursing research. Thoughts on the cultivation of master degree of clinical nursing.J.24(3),266-267(2010).
17. Zhang WW, Wang AJ, Wang I. Nursing research. Expert investigation and Research on the training program of master of nursing master's degree.J.26(1),81-83(2012).
18. Guo J, Wang Y, Lu GH. Nursing research. Research on the cultivation of nursing master's degree from the perspective of professional embedded.J.17(6),2054-2056(2014).
19. Zhang HL, Zhao QL, Zhang J. Chinese Journal of nursing. Current situation of the cultivation of master of nursing in nursing home and abroad.J.48(6),571-573(2013).
20. Cui XY. Journal of Qilu Nursing. The feasibility and difficulties of the cultivation of nursing professional degree graduate students and training of specialized nurses.J.21(21),54-57(2015).
21. Li W.E ducation and Teaching Forum. To accelerate the promotion of professional degree education and professional qualification certification of organic convergence.J.(51),1-2(2013).

Construction and functional analysis of luciferase reporter plasmid containing connexin43 gene promoter

Hui Li, Jiapeng Li, Weilin Shi, Xiaoyu Zhang, Yuan Xiang, Xinghua Liao[a] and Tongcun Zhang[a]
Institute of Biology and Medicine, Wuhan University of Science and Technology, Hubei, China

Abstract. Connexin 43 (Cx43), the principal gap junction protein in vascular smooth muscle cells(VSMCs) resulting in modulation of gap junction. Myocardin is thought to have a key role in Vascular Smooth Muscle Cell (VSMC) development by acting on CArG-dependent genes. In this study, a human Cx43 promoter luciferase reporter construct was generated by PCR amplification of Cx43 promoter. The PCR fragment was digested and cloned into pGL3-Basic vector. The promoter sequence was verified by sequencing. The results showed that luciferase reporter plasmids of Cx43 promoter were successfully constructed. Then the effects of a key transcription factor, which plays important roles in Cx43 levels control, were investigated by luciferase reporter assays in HEK293. The results showed that myocardin can enhance transcriptional activity of Cx43. Our observations provide evidence that Myocardin targets Cx43 expression directly.

1 Introduction

Gap junctions mediate cell communication by allowing the passage of molecules from one cell to another. The major role of gap junction intercellular communication (GJIC) is considered to be the maintenance of homeostasis in organisms[1]. Connexin 43 (Cx43), the principal gap junction protein in vascular smooth muscle cells(VSMCs), regulates movement of ions and other signaling molecules through gap junction intercellular communication (GJIC) and plays important roles in maintaining normal vessel function; however, many of the signaling mechanisms controlling Cx43 in VSMCs are not clearly described, it has been identified as a potential drug target for the treatment of a broad range of diseases [2]. Cell-to-cell transfer of small molecules through gap junctions regulates diverse biologic processes[3, 4]. Moreover, the decreased connexin 43 mRNA is correlated with its protein levels[5]. As the biologic effects that occur downstream of messages communicated by gap junctions expand, it becomes imperative to be able to develop assays to define the biologically relevant second messengers that are passed between cells. Recently, a large number of studies have implicated connexin 43 (Cx43)-containing gap junctions in a complex array of biologic responses that influence bone quality in vivo[6–9].

[a] Xinghua Liao and Tongcun Zhang (zhangtongcun@wust.edu.cn) are Corresponding author.
This work was financially supported by National Natural Science Foundation of China (No. 31501149, 31401117, 31471282, 31440038, 31270837).The science and technology young training program of the Wuhan University of Science and Technology (2016)(2016xz035).

Myocardin, a coactivator of serum response factor (SRF), is highly expressed in embryonic cardiac and smooth muscle lineages plays a critical role in regulating smooth muscle cells phenotype switch [10,11]. Myocardin belongs to the SAP (SAF-A/B, Acinus, and PIAS) domain family of transcription factors and is a cardiac and smooth muscle-specific expressed transcriptional coactivator of serum response factor (SRF) which binds to the CArG box (CC[A/T]6GG) to regulate cardiac and smooth muscle-specific genes [12].

In this study, Cx43 promoter was amplificated from human genome by PCR and inserted into pGL3 (luciferase vector) basic vector successfully. Furthermore, luciferase assays were performed in HEK293. The results showed that myocardin activates Cx43 promoter transcription. Point-mutagenesis studies showed that one CArG box site was required for Cx43 activation by Myocardin. Our observations provide evidence that Myocardin targets Cx43 expression directly. Our research will help us to screen interaction between two Genes.

2 Materials and methods

2.1 Cell Culture

Human embryonic kidney (HEK)293 cells line was obtained from American Type Culture Collection. VSMCs were grown in Dulbecco's modified Eagle's medium (DMEM) (GIBCO) supplemented with 10% fetal bovine serum (FBS) at 37°C in a 5% CO_2 incubator. Cell were plated in 24-well plates, cultured to 60%-80% confluence and transfected with PEI[13] as previously described.

2.2 Plasmid Construction

The fragments of Cx43 promoter region were generated by PCR using human genomic DNA as a template and subcloned into the KpnI and NheI sites of pGL3-Basic vector. The human genome was extracted from HEK293 cells line. Cx43 promoter fragment from -643 to +103 is amplified by PCR using the primers (F: 5'- CGGGGTACCTGAAAGATGGATTGGGTAT -3'; R: 5'-TGTGCTAGCGAACTCCTTGGAGGATGA -3'). CX43-truncated promoter fragment from -486 to +103 is amplified by PCR using the primers (F: 5'-CGGGGTACCCACTCACACCTACCTCTAACTTCAC -3'; R: 5'-CGCGCTAGCTTGGAGGATGAAGTAAAATGAAAAG -3')(Figure 1). PCR condition is as follows: Pre-degeneration for 5 min at 94°C, denaturation for 30 seconds at 94°C, annealing for 30 second at 58°C, and extension for 1 min at 72°C. PCR reaction was carried out for 32 cycles and PCR products were visualized in 1% agarose gels stained with ethidium bromide under UV transillumination.

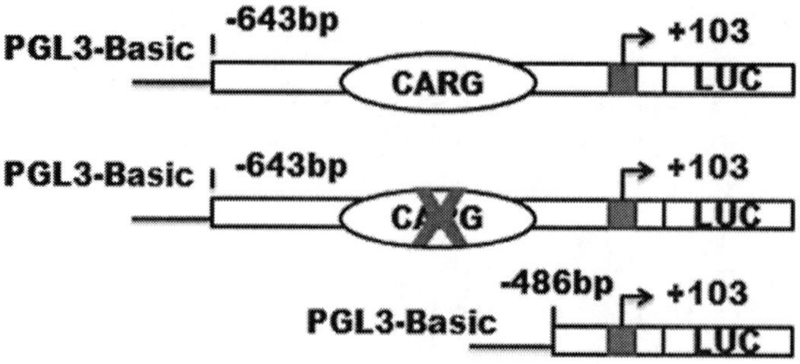

Figure 1. Sequential deletions and mutations of the human Cx43 promoter were fused to a luciferase reporter. HEK293cells in 24-well plates were co-transfected with the Myocardin expression plasmid (1μg/well) and various wild-type.

 DOI: 10.1051/bioconf/20170801007

The PCR product and pGL3-Basic vehicle plasmid were digested with restriction enzyme KpnI and NheI at 37°C for 2h. The fragment of PCR product and pGL3-Basic vehicle plasmid were mixed with 1 μL T4 ligase buffer and 1 μL T4 DNA ligase and added water to 10μL, incubated at 16°C for 2h, and then transformed into competent E.coli. Monoclonal colony was well separated by incubating in 3 mL LB which contains ampicillin at 37°C 16 hour. A monoclonal colony was picked and cultured in LB. Cx43 promoter point mutants were made with the TransStart FastPfu DNA Polymerase kit(TransGen) using Cx43 promoter(-634 to +103) as a template by the following primers: (F: 5'-TTAATTGTCCTTGCTTACAAAATTTTAAGGGAATTAGATCAT -3'; R: 5'-ATGATCTAATTCCCTTAAAAATTTTGTAAGCAAGGACAATTAA -3'). The plasmid was extracted and all constructs including mutants were verified by DNA sequencing.

2.3 Luciferase reporter assays

300ng of Cx43 promoter reporter, Myocardin (Myocardin-pCDNA3.1) expression plasmid as indicated were cotransfected. Cells were incubated for 24h before harvest. 50 μL of protein extracts (100 μL/well) were prepared for luciferase assays, which were measured by using a luciferase reporter assay system (Promega) on a Synergy™ 4 (Bioteck). Transfer 50 μL of each lysate into the wells of an opaque 96-well assay plate. Add 100 μL per well luciferase lysis buffer. Incubate at room temperature for 2 min with gentle shaking. All experiments were performed at least three times with different preparations of plasmids and primary cells, producing qualitatively similar results. Relative luciferase activity can be determined by dividing the firefly luciferase activity by the renilla luciferase activity. Next, average the replicates together and graph the data.

3 Result

3.1 Construction of Cx43 promoter luciferase reporter plasmid

To estimate the PCR amplification of Cx43 promoter, agarose gel electrophoresis was performed. As shown in Figure 2, a single band emerged at the site of 746bp, which represents the PCR products of Cx43 promoter.

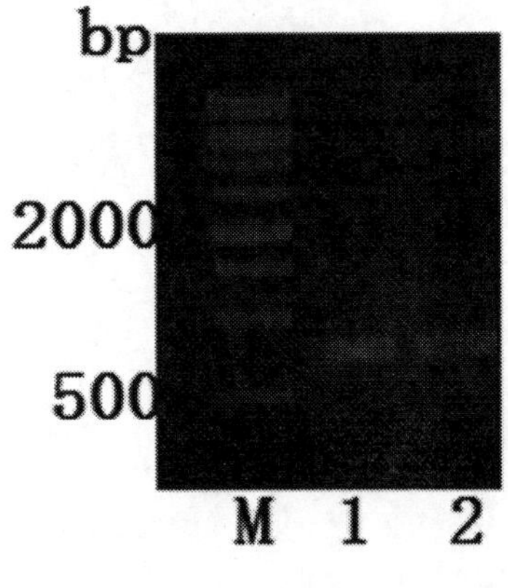

Figure 2. Agarose gel electrophoretic analysis of PCR product.
1-2: Cx43 gene promoter.

Then, the PCR product was digested by double restriction enzymes and cloned to the pGL3-Basic vector. After that, the recombinant plasmid was extracted and purified, and the agarose gel electrophoretic analysis was performed. Figure 3 showed the size of recombinant plasmid that was purified.

DOI: 10.1051/bioconf/20170801007

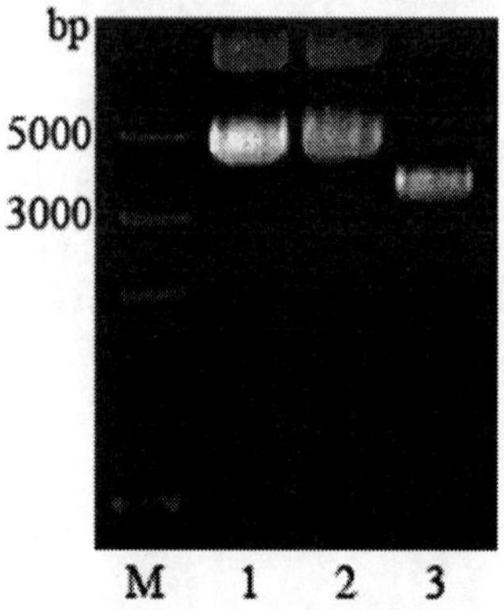

Figure 3. Agarose gel electrophoretic analysis of plasmids.
1-2: recombinant plasmid 3: pGL3-Basic vehicle plasmid.

To confirm the recombinant plasmids, we used double restriction enzymes to digest them and then electrophoresed through agarose gel. As shown in Figure 4, the recombinant plasmid was cut into two bands. One band was at about 74 bp, which represents Cx43 promoter, and another band was at 5,000 bp, which represents pGL3-Basic vehicle plasmid. The luciferase reporter plasmid of Cx43 gene promoter was further confirmed by sequencing. The result of DNA sequence alignment showed that luciferase reporter plasmid containing Cx43 promoter was constructed successfully.

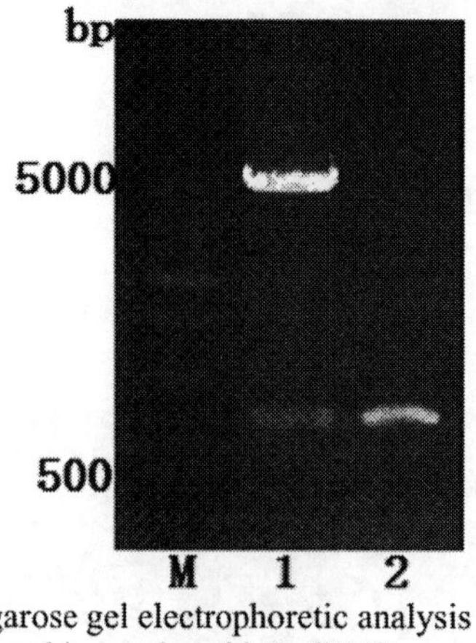

Figure 4. Agarose gel electrophoretic analysis of plasmids.
1: recombinant plasmid; 2: CX43 gene promoter.

3.2 Fluorescence of the efficiency of transduction

The transfection efficiency was demonstrated using a EGFP (enhanced green fluoresent protein) expression plasmid. As shown in Figure 5, the transfection efficiency was approximately 60%, demonstrating that DNA was transfected efficiently into HEK293.

Figure 5. The transfection efficiency of HEK293.

3.3 Myocardin can enhance the transcriptional activity of Cx43 promoter obviously

Luciferase Reporter Assays were performed after HEK293 were transfected with myocardin and Cx43 promoter. Contrasting with control group transfected with pcDNA3.1, myocardin showed a significant effect to enhance the transcription activity of Cx43 promoter (Figure 6a). Our data have demonstrated that Myocardin is able to stimulate Cx43 promoter activity, and then we asked whether CArG box site region within Cx43 promoter is responsible for Myocardin activation. Cx43 luciferase reporter constructs driven by Cx43-truncated promoter region and Cx43-mutant promoter region were generated(Figure 6b,6c). Transient transfection assays were performed. Cx43 promoter construct plasmid was cotransfected with increasing amounts of Myocardin expression plasmid. As shown in Figure 5, Cx43 promoter reporter activationby Myocardin was detected when as low as 250ng of Myocardin was transfected. Increasing amounts of Myocardin not led to higher expression of Cx43 promoter reporter. Transfection of 250ng Myocardin expression plasmid resulted in up to 170 fold activation of Cx43 promoter reporter. This indicated that Myocardin activates Cx43 gene expression transcriptionally.

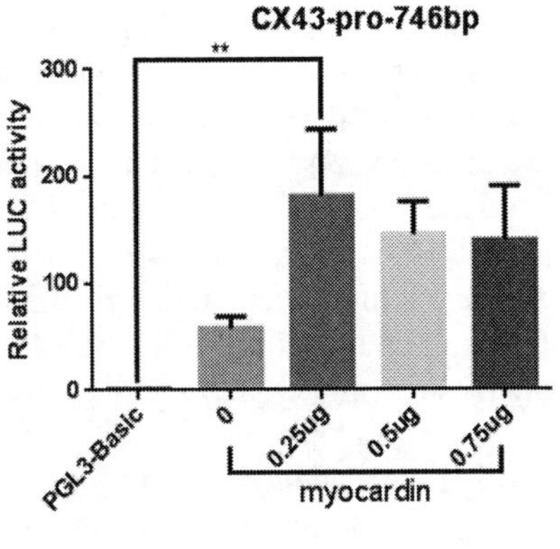

Figure 6a. Myocardin can enhance the transcriptional activity of Cx43(fold).

 DOI: 10.1051/bioconf/20170801007

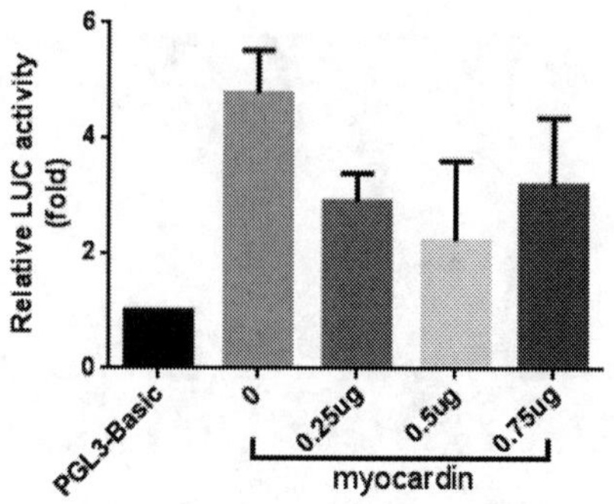

Figure 6b. Myocardin can not enhance the transcriptional activity of Cx43- truncated.

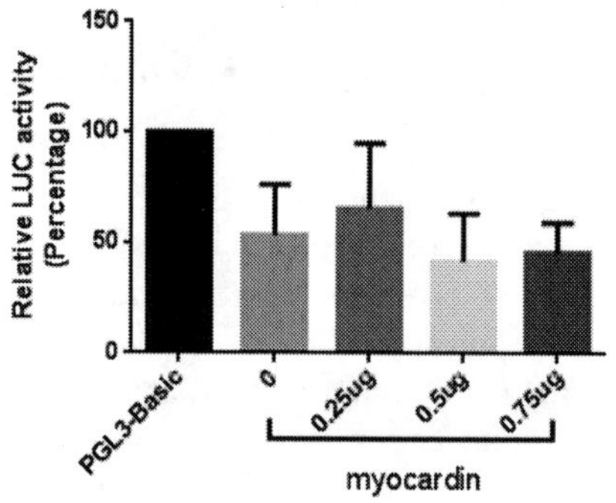

Figure 6c. Myocardin can not enhance the transcriptional activity of CX43-mutant.

4 Discussion

Gap junctions consist of channels between neighboring cells, mediating transfer of chemical molecules or electrical stimulation[14-15]. In some vascular smooth muscle diseases, Cx43 up-regulation and coupling of smooth muscle occurred, thereby inducing change in contractile property through up-regulation of cell–cell communication[16-18].

In natural development, Myocardin expression level was maintained over threshold level; thus, Myocardin would not induce significant changes in Cx43 expression. The other novel finding here is a new role for Myocardin as a repressor of Cx43 in smooth muscle. Myocardin has been known as a key regulator for VSMC and cardiac contractile genes but its regulation of non-contractile genes is less clear.Interestingly, there are conflicting data over the role of Cx43 in VSMC with one report showing an effect on Myocardin-dependent growth[19] .Myocardin was originally identified as a serum response factor (SRF) co-factor expressed only in cardiomyocytes and smooth muscle cells [20]. Myocardin increases activation of SRF-dependent genes by forming a higher order complex with SRF and facilitating its association with its CArG box DNA binding domain. Transcriptional activation is promoted by a variety of mechanisms including myocardin's own powerful transcriptional activation domain [21].

In this study, we cloned Cx43 promoter luciferase reporter plasmid successfully and found that myocardin could enhance the transcriptional activity of Cx43 promoter. Our findings support the premise from the transcriptional level that myocardin can increase the ability of Cx43 degradation in HEK293. Our observations provide evidence that Myocardin targets Cx43 expression directly. Our research will help us to screen interaction between two Genes. Research in the next decade should be focused on the development of new reagents and animal models to further understand this remarkable cofactor's functionin normal and pathological processes. In addition, there should be effort devoted towards fully elucidating transcriptomes under control of Myocardin, particularly the expanding class of long non-coding RNAs. Further, we need to define the mechanisms through which Myocardin functions independently of SRF.

References

1. A.C.Sharrow, Y.Li, A.Micsenyi, R.D.Griswold, A.Wells, S.S.Monga, H.C.Blair. Exp. Cell Res.**314**, 297–308. (2008).
2. C.N.Joshi, D.N.Martin, P.Shaver, C.Madamanchi, B.J.Muller-Borer, D.A. Tulis . Front Physiol. Jun **21**;3:220. (2012).
3. M.Vinken. Cell Mol Life Sci.Aug;**72**(15):2775-8. (2015).
4. M.S.Nielsen, L.N.Axelsen, P.L.Sorgen, V.Verma, M.Delmar, N.H.Holstein-Rathlou.Compr. Physiol **2**(3):1981–2035(2012).
5. K.Cesen-Cummings, M. J.Fernstrom, A. M.Malkinson, and R. J.Ruch. Carcinogenesis (Lond.),**19**:61– 67. (1998).
6. L.I.Plotkin. 2014 Apr 4;**5**:131. (2014).
7. S.A.Lloyd, A.E. Loiselle, Y Zhang, H.J.Donahue. J Bone Miner Res **29**:275–286(2014).
8. A.M.Buo, J.P.Stains. FEBS Lett **588**:1315–1321(2014).
9. J.P.Stains,M.P.Watkins,S.K.Grimston,C.Hebert,R.Civitelli.Calcif Tissue Int **94**:55–67(2014).
10. D.Wang ,P.S.Chang,Z.Wang,L.Sutherland,J.A.Richardson,E.Small,P.A.Krieg,E.N.Olson.Cell.Jun 29;**105**(7):851-62. (2001).
11. W.Xing, T.C.Zhang, D.Cao,Z.Wang, C.L.Antos, S.Li,Y.Wang, E.N.Olson, D.Z.Wang. CircRes. 2006 Apr 28;**98**(8):1089-97.
12. K.G.Shyu, W.P.Cheng, B.W.Wang. Mol Med. Jul 14;**21**:616-25.(2015).
13. J.M.Benns, R.I.Mahato, S.W.Kim.J. Control.Release.Feb.19;**79**(1-3):255-69. (2002).
14. C.H.Fry, G.P.Sui, N.J.Severs, C.Wu. Urology.Mar;**63**(3 Suppl 1):3-10. (2004).
15. L.Li, C.Jiang, P.Hao, W.Li, C.Song, B.Song.Am. J.Physiol.Cell.Physiol,**293**:C1627–C1635.(2007).
16. Y.Liao, C.P.Regan, I.Manabe, G.K.Owens, K.H.Day, D.N.Damon, B.R.Duling. Arterioscler Thromb Vasc Biol **27**:1037–1042. (2007).
17. M.L.Rocha, A.H.Kihara, A.P.Davel, L.R.G.Britto, L.V.Rossoni, L.M.Bendhack. Cardiovasc Res **80**:123–130. (2008).
18. M.Imamura, H.Negoro, A.Kanematsu, S.Yamamoto, Y.Kimur a, K.Nagane, T.Yamasaki, I.Kanatani, N.Ito,Y.Tabata,O.Ogawa. Am J Physiol Renal Physiol **297**:F46–F54.(2009).
19. J.Chen, H.Yin, Y.L.Jiang, S.K.Radhakrishnan, Z.P.Huang, J.Li, Z.Shi, E.P.Kilsdonk, Y.Gui, D.Z.Wang, X.L.Zheng. Arterioscler.Thromb.Vasc.Biol **31**:368–375.(2011).
20. D.Cao, C.Wang, R.Tang, H.Chen, Z.Zhang, M.Tatsuguchi, D.Z.Wang.J .Biol Chem. Nov 9;**287**(46):38495-504. (2012).
21. L.Raphel, A.Talasila,C.Cheung, S.Sinha.PLoS.One.**7**(8):e44052. (2012).

 DOI: 10.1051/bioconf/20170801008

Research on current situations of geriatric nursing education

Yujin Liu, Min Zhang [a], Wenjing Sun, Fengge Dong, Xiufeng Zhao and Bing Yu

Welfare College of Nursing of College of Humanities & Northeast Normal University, 130117, Changchun, China

Abstract. The population aging is accelerating and the aging population is growing in China. Although the geriatric nursing education has been developed for more than 20 years, geriatric nursing professionals are still insufficient and the geriatric nursing education is facing various challenges under the new situation. This paper primarily describes the developmental history and the related concepts of geriatric nursing education, and analyzes the personnel training modes and routes of geriatric nursing education, and its problems, in order to provide the basis for the reform of geriatric nursing education. The development of geriatric nursing needs a large number of outstanding nursing personnel, and the cultivation of geriatric nursing professionals depends on the development of geriatric nursing and the improvement of the teaching quality of geriatric nursing education. Front-line educators working on geriatric nursing should be committed to reforming the geriatric nursing teaching, improving the teaching quality and cultivating the high-quality nursing personnel suitable for conditions of the elderly in China.

1 Introduction

Geriatric nursing was determined as an independent specialty internationally in 1900 and geriatric nursing education has been developed for more than 20 years in China, but the geriatric nursing personnel is still insufficient, indicating that the work of geriatric nursing is facing challenges under the new situation [1] and there is still much to do in the improvement of geriatric nursing education [2]. Moreover, nurses can and should play a unique role in care of the elderly by nursing education, research and practice, which can improve not only the quality of eldercare, but also the quality of elderly life that is more important. This paper aims to understand the progress in the research on China's elderly nursing education to provide a basis for the future research on geriatric nursing education.

2 Concepts Related to Geriatric Nursing Education

Geriatric nursing is a subject to research, diagnose and treat the responses to their own present and potential health problems in elderly. Geriatric nursing was originated from theories of already existing subjects such as nursing, sociology, biology, psychology, health policy and other disciplines. Geriatric

[a]Corresponding author: Zhang Min, E-mail:1874751233 @ qq.com
Higher Education Research Key Project (2014.11.10) sponsored by Department of Education, Jilin Province.

nursing stresses to maintain and promote health, prevent and control disabilities resulting from acute and chronic diseases in elderly to develop their daily living abilities, make their body functions in the best state, and keep their living life with dignity and comfort until they leave this world [3]. Nursing education refers to activities to cultivate the professionals with a wide and deep knowledge in medicine, humanities and nursing, and able to serve for health services for the nursing discipline [4]. Geriatric nursing education is a social activity to cultivate the geriatric nursing professionals based on the social need.

3 Development of Geriatric Nursing Education

Early 20th century, the development of geriatric nursing education was slow in China. After the 1990s, the development of nursing higher education was accelerated and geriatric nursing were gradually determined as a course in many colleges and universities, but just as an elective professional course [5]. Only after 1998, was geriatric nursing established as a required course in a few higher nursing schools, but it was not popularized in our country, and geriatric nursing as an undergraduate teaching material was published in December 2000 [6]. Since then, monographs, textbooks, popular science books related to geriatric nursing were published in succession; works and papers about geriatric nursing were published in various related magazines, showing the starting of research on geriatric nursing. So far, there have been few nursing schools that are preparing to establish geriatric nursing specialty for nursing undergraduates and the research field of geriatric nursing has been also set up in the nursing graduate education. In addition, the domestic and international academic exchange in geriatric nursing has been gradually developed, and a long-term research partnership with counterparts of foreign countries in geriatric nursing has been established in some institutions.

4 Geriatric Nursing Education Status

4.1 Establishment of nursing specialty to cultivate geriatric nursing professionals

Documents showed that the "establishment of the nursing specialty to cultivate nursing professionals" was proposed in 2009, and has been proposed by many scholars since 2011. It is estimated that the elderly population will reach a maximum by 2051 in China, and after that, the elderly population will been maintained at 3 to 4 million and the total population aged 65 or older increased from 7% to 14%. The increase in aging population requires for new services in geriatric nursing institutions, such as older community service centers, rehabilitation agencies, apartments for the elderly and nursing homes, in which most work needs a large number of nurses to accomplish. However, the education for geriatric nursing is limited at all levels, even in the United States, certified geriatric nursing nurses is less than 1% of 2.5 million registered nurses. The shortage of geriatric nursing nurses and geriatric nursing specialist nurses, and the lack of funds for the long-term care for the elderly, result in the difficult to guarantee the nursing quality in home for the aged and at home. Currently the number of nursing staff is insufficient, and there are a few of nurses engaged in geriatric nursing with a low education, who have not received a systemic education in geriatric nursing, leading to the irrational structure and the lack of their professional knowledge; professional research staff with master or doctor degree is even more inadequate. Therefore, the professional in geriatric nursing is scarcer, [7] so that it is urgent to develop a geriatric nursing specialty in colleges and universities to cultivate the personnel for geriatric nursing. To strengthen the education of geriatric nursing to train more professionals for geriatric nursing is one of urgent problems currently to be considered and solved by the higher medical and nursing education institutions [8]. In 2013, Ningbo Health College has taken the lead in the establishment of geriatric nursing specialty. The establishment of geriatric nursing specialty may play a crucial role in the culture of geriatric nursing personnel, but the training of advanced geriatric nursing personnel requires for not only the establishment of geriatric nursing specialty but also the education of doctor graduates in medical colleges and universities to cultivate high-level geriatric nursing professionals with highly educated degree. In addition, the continuing education for in-service nurses should be vigorously carried out, such as teaching by correspondence, Self-learning, and seminars or online distance education, to organize clinical in-service and community nurses for further study and cultivate social practical geriatric nursing personnel to meet the needs of an aging society [7].

4.2 Emphasis on the application of teaching methods and modes to promote the teaching reform

According to the literature statistics, studies on geriatric nursing teaching methods account for 29.74%, indicating that China attaches great importance to the application of teaching methods for geriatric nursing education to improve the teaching quality of geriatric nursing. Studies have shown that the application of modular teaching [9], contract learning [10], case-based teaching and situation teaching [11], the real experience teaching [12], think aloud [13], case discussion-based salon [14], developmental case-based discussion [15] and theory and practice integration [16] can reflect the combination of theory and practice, improve the quality of geriatric nursing teaching, and make for the cultivation of new geriatric nursing professionals suitable for the needs of social development. In addition, the three dimensions of attribution theory, source of cause, stability and controllability, are integrated into the education of geriatric nursing, in which the students are trained in the adaptive attribution for improving their interest in the learning. The attribution training could guide the success in the learning of geriatric nursing to the internal attribution of their efforts and abilities, and the failure to the controllable attribution of their inadequate efforts and wrong learning way, to improve their comprehensive abilities [17]. The application of constructivist learning theory [18] may make students not passively receive knowledge but actively construct knowledge in the learning process, more interested and motivated in the learning of geriatric nursing, and learn something critically, which may bring about a revolution in the teaching or learning. Studies have suggested that the nursing education should be reformed to adapt to the needs of the future society, and more attention should be paid to the specialized training for geriatric nursing staff [19], and the six-in-one training model for nursing personnel should be emphasized to connect the quality education, expertise, teaching content, teaching team, experimental and practical training, and examination and assessment to the need of job [20], combine the theory with the practice, and improve students' presentation and communication skills, analytical and problem-solving skills, coordination and cooperation abilities, and innovative abilities [21].

4.3 Active exploration on effective ways for the nurses' in-service training and continuing education

Nurses are thinkers, designers, commanders and providers of geriatric nursing, and the improvement of the health and living standards of the elderly relies on the popularity, improvement and clinical utilization of geriatric nursing knowledge [22]. The existing clinical nurses are the main force in geriatric nursing facilities and the strengthening of clinical nurses' in-service training and continuing education is the key to improve the geriatric nursing education. The main way of continuing education in geriatric nursing is still the traditional face-to-face teaching, such as various classes for further studies, workshops, academic seminars, academic conferences, symposia, and degree courses [23]. It is recommended that the continuing education of clinical nurses should be carried by case-based teaching, sim-man education system and online knowledge base [24]. The teaching of vocational education should be carefully designed and reformed based on the characteristics of top-up students and in terms of the teaching contents, teaching methods and examination forms to achieve better teaching results [25]. The main measures to deal with it should be a diversified vocational education, including the national health education, the vocational education of nursing workers at nursing institutions for the aged, the job-transfer training of clinical nurses, and the setting-up of geriatric nursing specialty and curriculum, and it is necessary to cultivate the civil awareness of pension and provision for old age, and to bring up a pension team with strong professional skills in geriatric nursing; simultaneously to speed up the training of advanced geriatric nursing staff and managers [26].

4.4 Close attention paid to the geriatric nursing education abroad

In the process of active exploration on the geriatric nursing education, a close attention should be paid to the progress status in it in foreign countries to actively learn from their related experiences. Studies by Wang Zhangan [27] have shown that a geriatric nursing personnel should have the following capabilities: (1)core competencies to provide a high-quality care for the elderly and their families; (2)geriatric nursing capacities and suggestions on course contents and teaching strategies corresponding to " High Education Standards of Nursing Profession" (1998 revision); (3)core competencies, course contents and teaching strategies prescribed in the curriculum setting up; (4) resources/documents of the implementation of geriatric nursing course contents and teaching strategies. The continuing education in the United States provides opportunities for nurses to facilitate their improvement in geriatric nursing knowledge and skills, primarily including web-based geriatric nursing courses, distance learning and professional website [28]. The geriatric nursing education in Australia can be basically divided into four levels, secondary, vocational, undergraduate and master, nursing graduates at each level perform their own duties, and the main nursing staff is from the secondary specialized vocational education. The continuing education of geriatric nursing staff is completed mainly by their service units, and supervised and administered by Australian Board of Nurses. In Australia nursing homes, for example, its geriatric nursing continuing education system consists of three parts, orientation, in-service training and job training out of hospitals. Lin Jie [29] believed that the advanced classifying education system and the sophisticated continuing education system of geriatric nursing in Australia should be worth our learning. The main body of geriatric nursing legislation in Japan, improved geriatric nursing management network in Sweden and three-level geriatric nursing education in German (secondary, specialist training, college undergraduate) is their secondary vocational training education [30], primarily to train geriatric nursing nurses and nurse assistants [31], which is worthy of our study. Studies described above indicate that the geriatric nursing education in our country should be developed by working out a perfect law and system, establishing an improved management network and service, developing a pension mode with diversification and differentiation, and paying more attention to the diversified geriatric nursing education based on the vocational education to explore a mode suitable for the current development of geriatric nursing.

5 Currently Existing Problems in Geriatric Nursing Education

5.1 Relative lack of research on the geriatric nursing education

Geriatrics and geriatric nursing are relatively young sciences. Despite a sharp increase in demand for geriatric nursing research, but most focus on biomedicine, little attention is paid to geriatric nursing. Due to the relative lack of research on the practice of geriatric nursing education, more nurses should work on the research on geriatric nursing and apply their findings into the practice. In addition, studies have showed that the deep study on geriatric nursing education systems is still lack in colleges and universities [26]. Nurses must play an important role in the care of the elderly through the education, advanced practice and impetus of policies. Therefore, more undergraduate and graduate courses are needed to promote the education of geriatric nursing theory and its advanced practice, and at the same time the nursing research related to Chinese culture and needs of elderly should be carried out.

5.2 Lack of systematic geriatric nursing course

Most importance has been attached to the training of general nurses in the nursing education, and the specialist nursing system and specialist nursing education mode have been not developed yet in China. Geriatric nursing is not only a multi-disciplinary course, but also a practical service course. Geriatric nursing course was not opened in several school of nursing until 1998, "geriatric nursing" a planning teaching materials of Ministry of Health of People's Republic of China, was officially published in December 2000, which is a common teaching materials for both undergraduates and junior college students [32]. Currently, geriatric nursing course is taught primarily in a way of theoretical teaching and as a selective course with less contents and teaching hours compared to the other courses. The practical learning is relatively insufficient, since there is no systematic practical course of geriatric nursing in colleges and universities, and sites for the geriatric nursing internship are mainly hospitals, showing the lack of opportunities for the students to practice in different nursing-care places; all the training courses are integrated into the teaching of fundamental nursing. In this way, the students can only superficially understand the knowledge of geriatric nursing, can not fully understand the knowledge and skills necessary for geriatric nursing [24], and can not meet the actual requirements of geriatric nursing. In addition, there are other problems to be solved in geriatric nursing education, such as the students' poor willingness to working on geriatric nursing, the lack of geriatric nursing knowledge in clinical nurses, the nurses' poor awareness on the service for elderly nursing and health care [33], and the lack of geriatric nursing teachers. All the problems existing in geriatric nursing education require nurses continuously to explore effective solutions to solve them.

6 Summary

The development of geriatric nursing needs a large number of outstanding nursing personnel, and the cultivation of geriatric nursing professionals depends on the development of geriatric nursing and the improvement of the teaching quality of geriatric nursing education. Front-line educators working on geriatric nursing should be committed to reforming the geriatric nursing teaching, improving the teaching quality and cultivating the high-quality nursing personnel suitable for conditions of the elderly in China, so as to contribute to the improvement and the development of geriatric nursing.

References

1. Hu yan, Wang Yan. Practice on setting up the elderly care specialty in undergraduate nursing education. Chinese Nursing Management,**13**,2(2013).
2. Liu Ning, Guo Guifang. The status of demand conditions in elderly care in China and thinking on the cultivation of geriatric nursing personnel. Chinese Nursing Management,**11**,4(2011).
3. Hua Qianzhen. Geriatric Nursing. The 2nd Edition. Beijing: People's Medical Publishing House(2007) .
4. Zheng Xiuxia. Introduction to nursing education. Beijing: Peking University Medical Press(2011).
5. Gao Wenrong, Li Chunying, Li Caifu. The status and prospect of geriatric nursing education in China. Science and Technology Information, **36**(2011).
6. Wang Ling. Development history of geriatric nursing subject and thinking on it.Chinese Nursing Management,**7**, 4(2007).
7. Sun Shuiying, Zeng Hui, Zhang Liping. Population aging and nursing present conditions and countermeasures in China. Journal of Nursing Science, **21**, 21(2006).
8. Yu Changmei. Requirements of geriatric nursing and discussion on the setting up of geriatric nursing courses, Today Nurse, 6(2002) .
9. Xiao Xinli, Yang Xiaobin. Application of module-instruction in the teaching of geriatric nursing. Journal of Nursing Science,**23**,14 (2008).
10. Yang Zhilan, Yang Jianping, Wu Xiaohong, et al. Practice of contract learning in teaching of geriatric nursing. Journal of Nursing Science, **24**, 23(2009).
11. Sun Mengxia, Jiang Na, Liao Guixiang. Application of case teaching and situation teaching method in teaching of "geriatric nursing". Chinese Nursing Research, **25**,8B(2011).
12. Wang Ling, Xiao Shuqin. The application of experiential teaching in geriatric nursing for nursing

undergraduate students. Chinese Nursing Management,**13** , 2(2013).
13. Zou Yadong, Guo Mingxian, Cui Yan, et al. Application of think aloud discussion teaching method in teaching of "elderly nursing". Chinese Nursing Research,**27**,7B (2013).
14. Xie Xiaoyan, Li Jiayi, Zhang Chi, et al. Application of case-based discussion salon in elderly patients nursing for backbone training. Nursing Journal of People's Liberation Army,**30**, 9(2013).
15. Xiao Lily, Dong Xia, Shen Jun, Liu Hui. Application of developmental case-based discussion in continuing education of geriatric nursing in China. Chinese Journal of Gerontology, **32**, 23(2012).
16. Liu Wei, Yang Lvqing, Ding Ping. Application of theory and practice integration. Journal of Nursing,**20**,9B(2013).
17. Liu Wei. Application of attribution theory in teaching of geriatric nursing. Nursing Practice and Research,**5**,8(2008).
18. Guo Xiaoyan, Zhang Feng. The application of constructivism learning theory in geriatric nursing. China Higher Medical Education,**5**(2012).
19. Yuan Xiaoling, Kang Youle, Shi Hongni. Thinking on nursing demands of aged people and nursing education reform. Chinese Nursing Research,**19**,10A(2005).
20. Li Ruzhu, Jiang Yuehua, Zeng Xiaoying. Exploration on nursing personnel cultivation model of "six-in one" and its practice. Chinese Journal of Nursing Education,**6**,12 (2009).
21. Gao Jingquan. Evaluation on effect of carrying out community practice in geriatric nursing teavhing. Chinese Nursing Research, **23**, 5(2009).
22. Xu Xiulan. Population aging and elderly nursing development. Tianjin Journal of Nursing,**9**, 5(2001).
23. Liu Yue. Development and status of continuing medical education in China. Chinese Journal of Primary Medicine and Pharmacy,**8**,1(2001).
24. Liu Mengjie, Wu Youfeng, Zhao Jijun. Overview and propositions of carrying out geriatric nursing continuing education for clinical nurses. Chinese Nursing Research,**24**, 9(2010).
25. Zeng Lifang. Exploration on reform of geriatric nursing teaching for junior undergraduate students. Health Vocational Education,**23**,15(2005).
26. Chen Siqing. Research on diversified vocational of aged care. Journal of Nursing,**18**, 6B(2011).
27. Wang Zhangan, Zheng Honglai. Talking on the core competency standards and teaching strategies of geriatric nursing in nursing undergraduate education in the United States. Chinese Nursing Research,**25**, 11(2011).
28. Liu Xueqin, Li Li. The development of gerontological nursing in America: implications for China. Chinese Journal of Nursing,**4**, 12(2005).
29. Lin Jie. The present condition and the enlightenment of elderly nursing education in Australia. Nursing Practice and Research, **9**,21(2012).
30. Liu Xiaomin. Geriatric nursing in Geman. Chinese Journal of Nursing, 20013, 6(7): 560-561.
31. Peng Landi. Aged care experience in developed countries and prospect of our aged care in China. Chinese Nursing Research,**25**,5A (2011).
32. Wang Jing, He Guoping. Problems of nursing human resources for elderly patients in China and countermeasures. Journal of Nursing Science, **25**,23(2010).
33. Tang Fengping. Strengthening of elderly care awareness in teaching of geriatric nursing, Modern Nursing,**12**, 6(2006).

The study of nursing students' learning initiative in the course reform of aged caring

Wenjing Sun, Yujin Liu [a], Fengge Dong, Guixiu Yan and Xiufeng Zhao

Welfare College of Nursing of College of Humanities&Northeast Normal University, 130117, Changchun, China

Abstract. Purpose: Analyzing the influence of nursing students' learning initiative in the course reform of aged caring. Discuss the way of the aged care reform. Method: To reform the course of aged care in our school level 2013 88 nursing undergraduate. The specific content: learning aged care theory, learning Japanese care technology basic knowledge, adding Japanese and Taiwan's nursing concepts to the traditional aged care teaching, performing sitcoms about old people's disease and nursing way , reporting the plan of aged care by PowerPoint, organizing student volunteers to visit the nursing home and so on. The specific content lasted four months. Adopting the learning initiative (ALS) scale developed by Zang Yuli and others after course reform. Measure the students' learning initiative before and after the teaching. Result: Nursing student's self-study ability was in the middle and lower level before the course reform(59.26±7.38); After the course reform, nursing student gain higher score than before learning on the three aspects contain "Learning motivation"," Learning goals" and "Solid study". The difference has statistically significant.($P<0.05$).Conclusion: Through the aged care course reform, nursing students strengthen the study enthusiasm and initiative; enhance nursing student's self-study ability. It is conducive to improve the learning interest of aged care course for nursing students.

1 Introduction

According to the sixth national population census show that the aged 60 and above aged 60 are about 13.26% of the total population in our country. Among the aged 65 and above aged 65 are about 8.87% which is 1.19% higher than the national population census in 2000. According to the United Nations' predict that the 65 year old population proportion in china will reach 17.4% in 2030. Our country's old people population will rise to 27.8% of the total population in 2050[1]. And it will become an aging society. An aging society in our country requires us to develop aged care education fast. But the aged care education started early and development is slow. What's more, the tradition aged care education is dull. So it is hard to improve the nursing students 'enthusiasm[2]. Many a nursing student is hold informal learning mentality. They do not realize the importance of learning the aged care course in heart. Students participate in the practice of learning enthusiasm is not high. So that the teaching effect of aged care is poor and student lack the enthusiasm and initiative to learn aged care course. And they

[a] Corresponding author: Liu Yujin, E-mail:1257819211@ qq.com
This work was supported by a project grant from Education and Science Planning of Topic in Jilin Province (2014).

DOI: 10.1051/bioconf/20170801009

lack of committed to the cause of aged care initiatively and the spirit of dedication[3]. The purpose of this study is to analyze the influence of the aged care course for nursing students. Nursing students improve autonomy and motivation by the geriatric nursing course reform. Discuss a more practical way to reform the aged care course. Let nursing students realize the current situation of aged care now in China. Nursing students learn the basic skills in the aged care by the study. Offer the theoretical basis for social to develop the people who have Clinical nursing ability and aged care ability.

2 Research content

2.1 Research object

Choose our school nursing a total of 88 nursing undergraduate class two of grade 2013 as the research object.

2.2 Research methods

2.2.1 Course reform content and method

The aged care course is a required course which was arranged on the second year of the first semester. The tradition education's teaching content: aged care course has 36 class hours, theory class has 24 class hours, and experiment class has 12class hours. Adopt self-study and active learning state (ALS) to measure nursing students. Reform the aged care course for the nursing undergraduates. The course content after course reform: the aged care course hours add to 54 class hours, Adopt the PBL educate way, Take the student as the center, and develop the students' comprehensive ability and self-study spirit. Teacher used 12 class hours to teach theory, used 8 hours to report case, and students (5-6students as a group) used 10 class hours to research the information and make the nursing plan and make PowerPoint, report the nursing plan by group used 8 class hours, comprehensive and designing of nursing and Japanese nursing skills training presentation used 18 hours. Teacher concluded and evaluated the key and the nursing skills which students must learn after students' scene skills training demonstration. Organize students to visit a nursing home twice, develop the ability about talking with old people, knowledge to guide them to prevent fall down, teach them exercise about hand and brain and so on. Compared with tradition educate methods, the new methods can highlight the educate theory of soul is the classroom, students are main body, teacher is the key. Adopt Active learning state (ALS) to measure nursing students after finishing the reform of the course (Table 1).

Table 1. Comparison of teaching contents and methods after the traditional teaching and curriculum reform in the aged nursing.

Traditional teaching content	Class hours	Curriculum Reform of teaching content	Class hours
Teachers Lecture	24	Teachers Lecture	12
		Case Report	8
		Student access to information	10
		PPT Reporting care programs	8
Experiment Course	12	Melodrama presentation skills training and long-term care	16
Total	36		54

2.2.2 Research Tool

This study used the basic situation questionnaire, nursing students' autonomous learning ability scale and active learning state (ALS) scale to study the questionnaire survey .Nursing students' self-learning ability scale is compiled by the Taiwan scholar Cheng [4] (2010) Su-Fen, which is the self-rating scale, including four dimensions of learning motivation, planning and implementation, self-management, interpersonal communication, a total of 20 entries. All entries using Likert5 score, followed by 5-1 points, the scale of all entries are positive points, the higher the score of each dimension indicates that the stronger the autonomous learning ability, the scale a Cronbach's coefficient of 0.916.Active learning state(ALS)scale is writing by our country scholar Zang yuli [5] et al, this scale is a measurement tool aimed at the study of medical students and their related majors, a total of learning driving force, learning objectives, in-depth study, control learning, learning five dimensions, A total of 21 entries ,The total score is between 21 and 105 points, The scale's coefficient of Cronbach's α is 0.89.Two scales can be used in this study.

2.2.3 Arrangement and Analysis

Before and after learning to study the object of the questionnaire, the teachings of teachers in a unified guide language. The subjects completed the questionnaire, and returned the questionnaire on the spot. In this study, two questionnaires were issued before and after the study, a total of 264 questionnaires were recovered, the complete and effective questionnaires were 264, the recovery rate was 100%. Analyze the recovery of effective questionnaires into SPSS18.0 statistical software.

2.2.4 Statistical methods

Using SPSS18.0 to analyze the effective data, statistical description of the composition ratio, mean and standard deviation, the total score of the learning before and after the comparison of the paired t test, the test level of alpha =0.05.

3 Results

3.1 Basic situation investigation

In this study, a total of 88 undergraduate nursing students, to complete the study before and after the questionnaire, valid as a research object, a total of 88 people, of which 7 boys, accounting for 7.95%, girls 81 people, accounting for 92.05%; Age from 21 to 18 years of age; the number of people in the city is up to 79 people, accounting for 89.77%.

The results of this study showed that before the curriculum reform, the nursing students scored on the "learning motivation" (18.43±2.12), score on "planning and implementation" (15.74±1.85), Score in "self-management" (1.69±13.51), in the "interpersonal communication" score (11.58±1.39), Total score of autonomous learning ability of nursing undergraduates (59.26±7.38), In this study, the median value of the theory is compared with the score of 60. The mean value of the theory is in the middle and low level. (Table 2)

Table 2. The total score of nursing students' autonomous learning ability and the scores of each dimension (n=88).

The project	The highest score	The lowest score	Learning before
The learning motivation	30	6	18.43±2.12
The planning and Implementation	30	6	15.74±1.85
Self- Management	20	4	13.51±1.69
The interpersonal communication	20	4	11.58±1.39
The total score	100	20	59.26±7.38

3.2 After the reform of the course, the students' learning initiative is higher than that of the former

The results of this study showed that the scores of learning "driving force" on the learning of nursing students (19.52±2.64) were higher than the scores before the study (17.21±2.37), the difference was statistically significant (T=1.578, P<0.05);In "learning goal", the scores of nursing students (15.13±2.09) were higher than the scores before the study (14.63±2.19), and the difference was statistically significant (T=2.293, P<0.05);"Solid learning" in nursing students scores (15.69±2.33) higher than that of the former scoring (15.08±2.67), the difference was statistically significant (P < 0.05) T=2.138; in active learning total score on the learning score (76.58±8.49) higher learning before scoring (70.62±8.02), the difference has statistical significance (T=2.376, P < 0.05). There was no significant difference in the score of "deep learning" and "control learning"(Table 3).

Table 3. Comparison of learning initiative before and after nursing students (n=88)

Project	Before learning	After learning	T	P
The driving force of learning	17.21±2.37	19.52±2.64	2.461	<0.05
The goal of learning	14.63±2.19	15.13±2.09	2.293	<0.05
Depth study	11.98±2.07	12.68±2.19	1.047	>0.05
Control study	14.23±2.32	14.42±2.54	2.094	>0.05
Solid learning	15.08±2.67	15.69±2.33	2.138	<0.05
Total	70.62±8.02	76.58±8.49	2.376	<0.05

4 Discussion

With the deeply developing of nursing education revolution, the importance of self-studying and the effect of student's self-consciousness become more outstanding day by day. However, the surveys at present which about nursing student's self-studying say that their consciousness is comparability weak and their ability of self-studying is just so-so[6]. The most importance about gerontological nursing is to promote the studying positive and the ability of self-studying, if we want to promote the studying efficiency of students. Some specialist made a revolution about the practice teaching of gerontological nursing. They adjusted the tutor, the nursing students, the scheme of teaching and the contents. And it has promoted the student's positive of studying gerontological nursing. Zhou Xiaoli's investigate have said that the theory of gerontological nursing and the geriatric skill practice, and role-play and watching the movies about elderly people and visiting the gerocomium can promote the teaching effect of nursing undergraduates[7]. But the characteristic and the personality of gerontological nursing is not distinct. Both the examine theory and the ability of practice are equally importance. The teaching revolution has added the total studying time and added Japan's intervention nursing technique primary knowledge studying and the welfare concept. Taiwan's nursing notion in the contents of theory course as well. As for the practice course, the contents include the reporting of case and nursing, which by PowerPoint. The demonstration of scene play and the intervention nursing skill practice are also essential in the practice course. The theory and practice are equal in the final exam, each own the 50%.These make a benign combination of theory and practice become reality. The nursing students take part in the study with different point of views and identities. And these promote their positive and self-consciousness for and enhance the interests of learning gerontological nursing.
The investigator's point of innovation is the fusion of gerontological nursing and Japan's care skill. And it aim for paying more attention to protect and nurse elderly people's health, common disease sand frequently-occurring diseases. Borrow from Japan's care knowledge of soundmen, with the same time fusing our school's welfare concept into the contents of teaching revolution. Teach it to students by theories and practice. The "care" word can be traced back to the Japan's Old Age Welfare Law in the 1963[8]. It mean that caring about elderly people's daily life which is fundamental and helping elderly who cannot live by oneself. Some academicians have said that it is essential for our society to build up the system of caring elderly people which appropriate with national reality and dealing the complex questions about caring elderly over a long period of time. And these made by studying and quoting Japan's care skill and concept. Adding the nurse concept of the elderly people department of Taiwan Chang Geng Technology and Science into the teaching revolution contents for nurturing expert professionals who will be able to care and manage elderly's health, the spirit of team work, innovative thought and the essence of humanitarian qualities[9]. Gerontological nursing updates

student's view to aged supporting and nursing work. For example, not only do students learn to take care about elderly's mental and physiological life and daily life, but also learn to be care about elderly and relatives around themselves. And they choose to learn the gerontological nursing is for they are interested with it, but not just for the lever of salary and the position of employment. Therefore, merge Japan's care skill and concept of Taiwan's care concept into the integral procedure of teaching. These present in the practice and application of caring skill. And building up the favorable professional fundament when they working for Nursing Homes, hospital and communities after they graduation.
To furtherly promote the course revolutionary system of undergraduate's gerontological nursing. Our school will reference a leading-edge patent from foreign, which for nurturing the nursing elites. We will insist our school's instruction thought which containing the welfare notion to institute a fostering project of nursing science which presenting the characteristic of welfare according to the actual circumstance of elderly nursing expert's requirement in China, and associate with The Nurse Test and Gerontological Nursing Qualification Test[10]. In the project, we'll set up the course group of gerontological nursing which containing 10 relative courses. The course groups include the nursing of commute and the gerontological nursing, and others are lastly added. Apart from this, establishing overall practice items and adding the proportion of practice course, and adding gerontological nursing and care skill as a compulsory major course. We'll add the practice of working at Nursing Homes into student's practice and probation, at the same time students join in the health association's activity in the regional area. Its goal is to foster expert professionals who can meet the requirement of nowadays epoch and they feature with geriatric nursing.

5 Conclusion

Japan's care skill as well as Taiwan's nursing, the welfare concept are innovatively introduced into the traditional gerontological nursing in the course revolution. And at the practice has been proved that the innovative revolution which is enterprising not only promote student's enthusiasm and initiative, but enhance their ability of self-studying. It contributes to promote student's interest for studying gerontological nursing. Students also master the fundamental skill of aged supporting and nursing work. This course revolution has made it.

References

1. Statistical Bureau of the people's Republic of China .The sixth national census in 2010 the main data bulletin .(No. first)(2011).
2. Wang Ling, Wang Shifeng .Aging problem needs scientific planning ahead of time .First Financial Daily, **4**,A02(2011).
3. Mu Xiaoyun, Liu Shiying, Ding Yanli.The influence of the reform of the elderly nursing practice teaching on the learning initiative of nursing students in Higher Vocational Colleges ,**27**,11(2013).
4. Cheng S F,Kuo C L,Lin K C,et al.Development and preliminary testing of a self-rating instrument to measure self-directed learning ability of nursing students. Int J Nurs Stud,**47**,9 (2010).
5. Zang Yuli, Lou Fenglan, Chen Manhui et al. The development of medical and related professional college students' Active learning state measurement tool. Chinese Journal of Nursing Education,**3**,3(2006).
6. Lin Bangjie, Tennessee self concept scale.3 edition, Taipei: Zheng Sheng Education Science Society(1986).
7. Leftcourt, HM,Von Baeyer, CL,Ware, EE,et al. Themultidimensional-multiattributional causality scale: The development of a goal specific locus of control scale. Canadian Journal of Behavioral Science,**11**(1979) .
8. Peterson C.Semmel A, Baeyer C, etal. The Attributional Style Questionnaire. The Journal of Psychology,**97**(1982).
9. Zhou Xiaoli, Zhang Yongai. The teaching method and effect of undergraduate nursing elderly nursing. Journal of Nursing Administration,**15**,1(2015).

10. Yu Xiao, Zhao Yibo. Aged Care Services introduced in Japan under the nursing insurance.Population Journal,**3**,36(2014).

 DOI: 10.1051/bioconf/20170801010

Research on present situation of pre-hospital first-aid knowledge of the elderly in community in Jilin

Wenjing Sun, Yujin Liu[a], Guixiu Yan, Wei Meng and Fengge Dong

Welfare College of Nursing of College of Humanities&Northeast Normal University, 130117, Changchun, China

Abstract. Objective: This paper aims at understanding the present situation of pre-hospital first-aid knowledge of the elderly in community, which put forward good suggestions for developing the popularity of first-aid knowledge and training in Jilin. Method: Use convenience sampling method to select 200 elderly in community in Jilin, it makes questionnaire survey of first-aid knowledge and awareness. Result:The popularizing rate of first-aid knowledge of the elderly in community is quite low. There are obvious differences among people of different culture. Conclusion: It is urgent to develop popularity and training of first-aid knowledge to the elderly in community, so as to improve first-aid level of the elderly.

1 Introduction

Pre-hospital first-aid refers to on-the-spot rescue, transshipment and underway rescue of the injured who suffer all kinds of life-threatening emergencies, trauma and poisoning in the environment out of hospital. Although pre-hospital first-aid is temporary and meets for emergencies, effective pre-hospital first-aid can win valuable rescue time for saving the patient's life[1]. With the modernization of social life, the incidence of all traumas has been increasing year by year. As the elderly population at high risk of cardiovascular chronic diseases, first aid and mutual aid to master skills is particularly important. Pre-hospital care to grasp the precious lives of gold five minutes, can reduce the mortality of Emergency. To enable the elderly to better serve the community, but also serve their own[2]. Therefore, first-aid knowledge and awareness of peasants have directly connected with treatment and life security of the injured. In order to understand current situation of the elderly first-aid knowledge in community, the author made a questionnaire about related knowledge of the elderly in community in Jilin from Feb. to June, 2016. The report is as follows.

2 Object and method

2.1 Object

Use convenience sampling method to select Fengman district, Chuanying district, Changyi district, to

[a]Corresponding author: Liu Yujin,E-mail:1257819211@ qq.com
This work was supported by a project grant from soft science of Health planning topics of Jilin Province (2014R29).

choose 200 elder people from community as research subjects.The age is range from 60 to 94, the average age is 72. Educational level: illiteracy 24, elementary school 57, middle school 68, high school(technical secondary school) 39, junior college and above 12. Of which, 121 males and 79 females. All team members are well-informed and willing to cooperate with the investigation.

2.2 Investigation Method

According to actual situations of China's the elderly in community, the questionnaire is self-designed and recognized by expert's evaluation. It contains general information(age, educational level, gender), mastering situation of first-aid knowledge, first-aid emergency measures and demand conditions. 20 students of Nursing Specialty from our university were chosen to be investigators after training. Questionnaires were filled in and taken back on the spot by investigators. 205 questionnaires were issued, 200 validated questionnaires were taken back and the valid return rate was 97.5%.

2.3 Statistical Method

It adopts SPSS17.0 software to make a statistical treatment and applies X^2 test, the test level of alpha =0.05.

3 Result

3.1 The elderly Understanding Conditions of Pre-hospital First-aid Knowledge

It applies the popularizing rate of first-aid knowledge to evaluate the elderly understanding conditions of pre-hospital first-aid knowledge (Table 1) Popularizing rate= population of correct answers for one problem/ overall population answering problems[3].

Table 1. The elderly Understanding Conditions of First-aid Knowledge in Surveyed Areas (n=200)

Content	Population with Right Answers	Percentage (%)
1.Way to judge heartbeat and respiratory arrest	63	31.7
2.Places pressed while doing CPR	15	7.3
3.The proportion of pressing to blow while doing CPR	5	2.7
4.On-site rescue method for electric shock	63	31.3
5.How to rescue a drowning man	70	35.0
6.How to rescue the gas-poisoning injured	75	37.7
7.On-site first aid for massive hemorrhage after artery injury	18	9.0
8.How to move a patient with spinal injury	11	5.7
9.Emergence	115	57.3

As seen in Table 1, emergence call has the highest popularizing rate of first-aid knowledge(57.3%), and the second one is the on-site rescue of poisoning injured(37.7%). Places pressed of CPR(7.3%), proportion of pressing to blow(2.7%) and how to remove a patient with spinal injury(5.7%) have quite low popularizing rates.

3.2 Mastering Conditions of First-aid Knowledge of the elderly Belonging to Different Categories

It applies the mastering rate of first-aid knowledge to evaluate mastering conditions of first-aid knowledge of the elderly belonging to different categories(anyone who can answer 8 problems or more in Table 1 master the first-aid knowledge[4]), result is shown in Table 2.

Table 2. Mastering Conditions of First-aid Knowledge of the elderly Belonging to Different Categories in Investigated Area (n=200)

Item		Population	Mastering Rate(%)
Age (year)	60-69	82	53(43.1)*
	70-79	72	35(32.4)
	80-94	46	11(15.9)
Educational Level	Low(illiteracy, elementary school)	71	31(25.4)
	Middle(middle school)	68	33(32.4)
	High(high school, technical secondary school, university)	51	34 (44.7)**
Gender	Male	121	65(35.7)
	Female	79	41(34.7)

*$P<0.05$ **$P<0.05$

As seen in Table 2, the age group with highest awareness rate(43.1%) is 60-69 years old. The lowest age group(15.9%) is 80-94 years old. With higher educational level, the awareness rate is higher. Differences are statistically significant(P<0.05),gender comparison has no statistical significance(P>0.05).

3.3 The elderly Emergency Measures for Pre-hospital First-aid and Demand Conditions of First Aid

Targeted at the problem of "If sudden death happens to somebody around you, what shall you do?", people are required to answer it openly on the spot. 35.3% old people choose to ask people around them for help, 40.4% choose to call 120 emergency call, 13.4% do not know how to handle it, 3.4% choose to go away and 7.5% choose the on-the-spot first-aid. As demand conditions of first aid, investigation shows that, 92.6% think that first-aid knowledge education should be developed, among which, 21.0% consider that first-aid knowledge education should start from the elementary school; 43.2% consider that this education should regularly be developed in communities.

4 Discussion

According to the survey, most of the aged are familiar with how to call for help. It shows that emergency call has been widely publicized in community, but on-the-spot rescue ability for all kinds of incidents is still weak. For instance, the accuracy of "proportion of pressing to blowing while doing CPR" is only 2.7%, that of "how to move a patient with spinal injury" is only 5.7%.Timely and effective CPR and correct moving method have great significance for saving the patient's life and preventing further injury. Investigations reveal that, the population of high educational level have quite high awareness rate of first-aid knowledge, which maybe relate to their long time for receiving education and wide channels to receive information. However, ill-educated old people have low awareness of first-aid knowledge and poor self-rescue ability. It is possible that weak educational degree limits their mastering and understanding of first-aid knowledge[5].

During the investigation, the problem of "If sudden death happens to somebody around you, what shall you do?" was put forward to respondents. 13.4 old people did not know how to deal with it, 3.7% chose to go way who were afraid of getting into trouble and only 7.5% chose to make on-site first aid. All these show that, on one hand, cultural knowledge has already become the obstacles for pre-hospital first-aid knowledge to be popularized and developed in community; on the other hand, it shows that ideological and ethical standards of the aged need to be improved. According to this investigation, 69.6% peasants are quite willing to master first-aid knowledge. It can be seen that the aged have a quite low comprehensive first-aid ability. However, due to objective reasons of economic conditions and geographical locations, access and timeliness of professional medical emergency rescue are far away from that of cities, therefore, these groups have more urgently objective demands for first-aid knowledge and skills. It is suggested that governments and related departments should take full advantage of broadcast, TV and newspapers & periodicals to make publication and education of importance of first-aid knowledge, regularly assign organizations and personnel with training qualities to go deep into the countryside, apply more methods to regularly develop training and practicing of first-aid skills, such as simulating first-aid spot, multimedia cases and role play, so as to improve first-aid skills and awareness of the aged[6].

5 Conclusion

Through this pre-hospital First-aid Knowledge for the elderly in community surveys, found that elderly people who living alone, community health is not in place, the lack of pre-hospital emergency care knowledge etc. affect the ability of pre-hospital community of the elderly. We can update the concept, to strengthen pre-hospital care community strength, improve the quality of pre-hospital care community and other methods to improve and enhance the ability of emergency care homes for the elderly before the community, improving the quality of life of older people.

References

1. Yang Aizhen, Xu Jinggang, Song Peiqiu, Tao Xiangling. Pre-hospital emergency care for the elderly Discussion in community. Journal of Nursing Science,**11**,4(1996).
2. Yao Hongying. Comprehensive intervention for the elderly to master first aid skills Evaluation. Zhe Jiang University(2012).
3. Zhang Aiwu. 378 cases of the elderly in community pre-hospital emergency care analysis. Foreign Medical Treatment,**10**,24(2009).
4. Wang Hong, Zhu Lirong, Zhang Xuefeng, Zhu Dingfa, Zhang Lu. Residents first aid knowledge, attitude, behavior Status Survey.China Medical Herald,**9**,6(2012).
5. Huang Wenjie. 629 cases of pre-hospital emergency care in elderly community survey.Modern Medicine,**16**,14(2007).
6. Hu Suzhen. Establishment and implementation of community self-help and mutual aid system. Fu Dan University(2012).

 DOI: 10.1051/bioconf/20170801011

Effect of Tai Chi exercise for hypertension: a meta-analysis of randomized controlled trials

Yeting Zhang[1], Yan Fu[2,a], Bin Zhu[1], Chuikun Li[1] and Xueling Lin[3]

[1]*Institute of Physical Education, Chengdu University, Chengdu, China*
[2]*Institute of Physical Education, Southwest University for Nationalities, Chengdu, China*
[3]*Institute of Physical Education, Yangtze University, Jingzhou, China*

Abstract. Objectives: We designed this study to evaluate the effect of Tai Chi exercise for hypertension patients. Methods: RCTs designed to evaluate the effect of Tai Chi exercise for hypertension patients were searched from Science Direct, EBSCO, Pub Med, CNKI and Wanfang databases. Results: The meta-analysis found that (ATC) the SBP (WMD = 13.19 mmHg; 95%CI: 11.52 to 14.87; P < 0.0001) and DBP (WMD = 8.92 mmHg; 95% CI: 7.94 to 9.90; P< 0.0001) can be significantly reduced after Tai Chi exercise compared to before Tai Chi exercise (BTC). ATC significantly improved the content of NO (WMD =-7.98mmol/L; 95%CI: -10.63 to -5.33; P < 0.0001), and decreased the content of TG (WMD =0.22mmol/ml; 95%CI: 0.06 to 0.38; P =0.006) and LDL-C (WMD =0.20mmol/ml; 95%CI: 0.13 to 0.26; P < 0.0001). There was no obvious difference between ATC and BTC on HR (WMD = 1.64; 95%CI: -0.51 to 3.97; P =0.14), TC (WMD = -0.03mmol/ml; 95%CI: -0.22 to 0.17; P =0.80) and HDL-C (WMD =-0.04 mmol/ml; 95%CI: -0.09 to 0.01; P =0.13). Conclusions: As a valid treatment for hypertension patients, Tai Chi exercise can decrease SBP, DBP, TG, LDL-C and increase NO.

1 Introduction

Hypertension is observed as the most common disease in primary health care of human, severe complications can be leaded by this disease, such as atherosclerotic complications including stroke, hypertensive renal disease, and hypertensive cardiovascular disease, renal inadequacy, coronary artery disease, and even the cardiac failure[1,2]. The American Heart Association suggested that the risk of hypertension can be reduced by the effective lifestyle modifications like regular exercise[3]. Tai Chi was ancient martial arts in Oriental cultures has been used for centuries. It has also been used as an exercise form and best adapted for older individuals of Asian heritage, to enhance body awareness and body balance. Tai Chi can lowered diastolic blood pressure (DBP) and systolic blood pressure (SBP) has been proved by a recent of trials, modulated lipid metabolism, and raised the NO index of patients with hypertension, and others comprehensive beneficial effects[4-16]. Tai Chi can benefit hypertension patients which supported by increasing evidence from RCTs, but there is no strictly evaluated evidence to confirm its function in hypertension patients health according to small size individual studies samples were short of strictly designed large size and inconsistent results RCTs. Clearly, an up-to-date meta-analysis and systematic review of this topic is needed. Therefore, we designed this study to sum up the actual evidence on the curative effect of Tai Chi exercise for the

[a] Corresponding author: E-mail:fuyan1010@163.com. Tel:177-2982-8135.
This research was supported by the School Youth Foundation Project of Chengdu University (No.2080516037).

therapy of hypertension patients.

2 Methods

2.1 Strategy of search

Two reviewers carried out a scientific search in the following 5 online network databases from 2000 to 2016 to retrieve potential studies independently: Science Direct, EBSCO, Pub Med, CNKI and Wanfang databases. The following search keywords were used: ("Tai Chi Quan" or "Tai Ji Quan" or "Tai Chi") AND ("gao xue ya" or "high blood pressure" or "hypertension" or "xue ya" or "blood pressure"). On the other side, we connected colleagues and experts in this field to predicate if there are any related ongoing or unpublished researches existed, and connected the authors of these researches to get detailed data as we needed.

2.2 Eligibility criteria

Only the study which should be RCTs intended to assess the availability of Tai Chi for the therapy of hypertension patient were considered.

In this review, the participants should meet at least one of the current or past exact definitions of hypertension[1]. The research objects in the study were the same elderly people. However, hypertension patients combined arrhythmia, severe coronary heart disease(CHD), hepatic failure, severe heart failure, or kidney failures were excluded. But age, sex or ethnic origin were not restricted.

Trials were compared Before Tai Chi exercise (BTC) versus After Tai Chi exercise (ATC).Do Tai Chi exercise at least 3 times a week, not less than 50 minutes. If the studies were case reports or reporting the same results were excluded.

2.3 Data extraction

Two different reviewers got the data from the primary researches independently. The extraction data form consisted of 4 parts: (1) the study's general information, (2) the detailed information of participants,(3) the information of the method used of Tai Chi, (4) the detailed information of Blood Pressure (BP), Total triglyceride (TG), Total cholesterol (TC), Heart rate (HR),Nitric Oxide(NO), High density lipoprotein cholesterol(HDL-C), Low density lipoprotein cholesterol(LDL-C) changes.

2.4 Quality assessment

We used the revised Jadad scale to evaluate the quality of each study. The highest score was 7 points, the higher the score, the higher the quality.

2.5 Statistical analysis

Revman5.2 software was used to analyze data and publication bias. Measurable data were provided by all of the included researches, and reported the weighted mean difference (WMD) with its 95% confidence interval (CI). I squared (I^2) statistic was employed to assess heterogeneity and when $I^2 >$ 50%, the heterogeneity was significant. The fixed effects model was taken to analyze the studies. We considered $P < 0.05$ to be statistically significant.

3 Results

3.1 Document Indexing and Research characteristics

We searched 76 articles, depending on the basis of literature into and exclusion the filtered criteria, finally, we got 13 studies in this review[4-16]. Figure 1 depicts the study selection's detailed process. Tables 1 shows the essential features of the included summarized themes and trials.

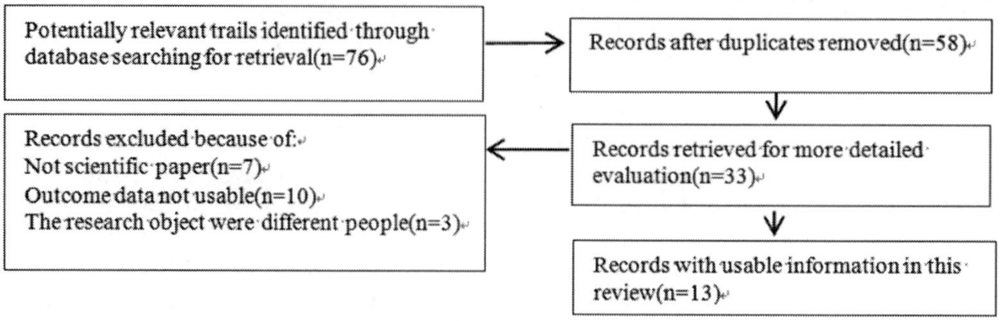

Figure 1. Study selection and identification process.

3.2 The quality of the research's methodology

The revised Jadad Scale was used to assess the quality of the studies. The score of the included subjects and trials are in Table 1.

Table 1. The included trials basic characteristics.

References	Age(yrs)	BTC (n)	ATC (n)	Male/Female	Intervention	Treatment and duration	Primary outcomes	Jadad score
JEN-CHEN et al. 2003[1]	51.6± 16.3	44	37	19/18	Yang-style Tai Chi	3times/week,50minutes /time,12weeks	(1)(2)(3)(6) (7) (8)	3
Hui-Ming Lo et al.2012[2]	58.47± 7.46	37	27	15/12	Yang-style Tai Chi	3times/week,60minutes /time,8weeks	(1)	3
Jing Sun et al. 2015[3]	45~65	150	136	117/19	Tai Chi	3h/week,12months	(1)(2)(3)(5) (7) (8)	3
WANG Xiao-jun et al.2011[4]	50~70	30	30	9/21	Tai Chi	5times/week,60minutes /time,16weeks	(1)(2)	2
JIN Zhen-yang ea al.2012[5]	61± 3.7	80	80	57/23	24-style Tai Chi	7times/week,2h/time,6 months	(2)	1
Sun Feng et al. 2014[6]	60~70	45	38	14/24	24-style Tai Chi	7times/week,2h/time,8 weeks	(1)(2)(3)	3
He You-ping et al.2012[7]	35~64	52	52	28/24	Chen-style Tai Chi	7times/week,80minutes /time,3months	(2)(4)	2
Xie Hui-juan et al.2014[8]	60~70	50	50	25/25	24-style Tai Chi	5times/week,60minutes /time,12weeks	(1)(4)	1
Mao Hong-ni et al.2008[9]	63.16± 6.43	50	50	30/20	Tai Chi	5~6times/week,60minut es/time,8weeks	(4)	1
He Jing-he et al.2011[10]	51.6± 5.3	33	33	NS	24/42/48-style Tai Chi	6times/week,40~50min utes/time,20weeks	(1)(2)(3)(6)	2
Mao Hong-ni et al.2006[11]	45~72	62	62	15/47	24-style Tai Chi	6times/week,60minutes /time,8weeks	(1)(2)(4)	2
Li Chuan-wu et al.2007[12]	58± 5	16	16	16/0	24/48-style Tai Chi	5times/week,60minutes /time,6months	(5)(6)(7)	1
Zhong Yun-jian et al.2009[13]	41~46	14	14	NS	83-style Tai Chi	6times/week,60minutes /time,16weeks	(2)(5)(6)(8)	2

Abbreviation: NS: Not given; BTC: Before Tai Chi exercise; ATC: After Tai Chi exercise;
(1)SBP; (2)DBP; (3)HR; (4)NO; (5)TC; (6)TG; (7)LDL-C; (8)HDL-C.

DOI: 10.1051/bioconf/20170801011

3.3 Meta-analysis results

3.3.1 SBP and DBP

Study or Subgroup	Before Tai Chi exercise Mean	SD	Total	After Tai Chi exercise Mean	SD	Total	Weight	Mean Difference IV, Fixed, 95% CI
He Jing-he 2011	148.4	12.3	33	132.6	14.6	33	6.6%	15.80 [9.29, 22.31]
Hui-ming Lo 2012	140.21	12.41	27	130.5	14.5	27	5.4%	9.71 [2.51, 16.91]
JEN-CHEN TSAI 2003	142.4	8.6	37	126.8	7.4	37	21.1%	15.60 [11.94, 19.26]
Jing Sun 2015	130.71	16.65	136	120.28	14.63	136	20.3%	10.43 [6.70, 14.16]
Mao Hong-ni 2006	162.74	26.36	62	140.49	23.55	62	3.6%	22.25 [13.45, 31.05]
Sun Feng 2014	142.79	12.21	38	133.8	18.41	38	5.7%	8.99 [1.97, 16.01]
WANG Xiao-jun 2011	138.59	10.86	30	124.69	5.79	30	14.5%	13.90 [9.50, 18.30]
Xie Hui-juan 2014	154.13	7.69	50	141.47	10.09	50	22.8%	12.66 [9.14, 16.18]
Total (95% CI)			413			413	100.0%	13.19 [11.52, 14.87]

Heterogeneity: Chi² = 10.93, df = 7 (P = 0.14); I² = 36%
Test for overall effect: Z = 15.42 (P < 0.00001)

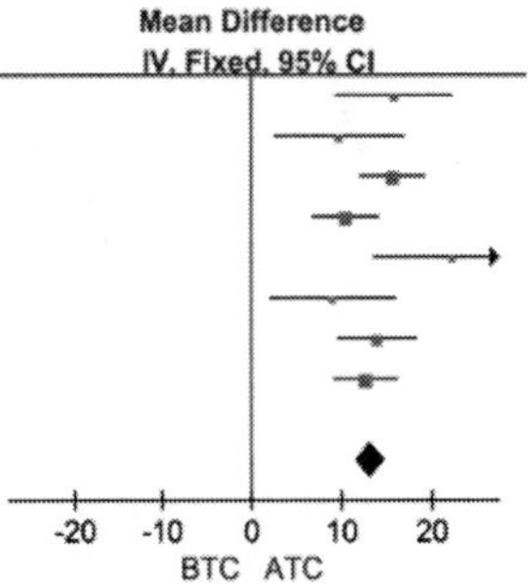

Figure 2. The forest figure of Systolic blood pressure.

Study or Subgroup	Before Tai Chi exercise Mean	SD	Total	After Tai Chi exercise Mean	SD	Total	Weight	Mean Difference IV, Fixed, 95% CI
He Jing-he 2011	93.9	7.1	33	84.2	8.3	33	7.0%	9.70 [5.97, 13.43]
He You-ping 2012	96.8	5.4	52	86.6	5.9	52	20.5%	10.20 [8.03, 12.37]
JEN-CHEN TSAI 2003	87.4	8.7	37	78.6	6	37	8.3%	8.80 [5.39, 12.21]
JIN Zhen-yang 2012	90.86	9.31	80	80.09	4.55	80	18.8%	10.77 [8.50, 13.04]
Jing Sun 2015	82.21	7.94	136	75.31	14.53	136	12.5%	6.90 [4.12, 9.68]
Mao Hong-ni 2006	93.66	13.72	62	82.74	11.97	62	4.7%	10.92 [6.39, 15.45]
Sun Feng 2014	86.42	8.56	38	80.63	10.77	38	5.1%	5.79 [1.42, 10.16]
WANG Xiao-jun 2011	85.47	12.01	60	77.34	5.92	60	8.4%	8.13 [4.74, 11.52]
Zhong Yun-jian 2009	91.8	3.6	14	84.7	3.3	14	14.8%	7.10 [4.54, 9.66]
Total (95% CI)			512			512	100.0%	8.92 [7.94, 9.90]

Heterogeneity: Chi² = 10.95, df = 8 (P = 0.20); I² = 27%
Test for overall effect: Z = 17.78 (P < 0.00001)

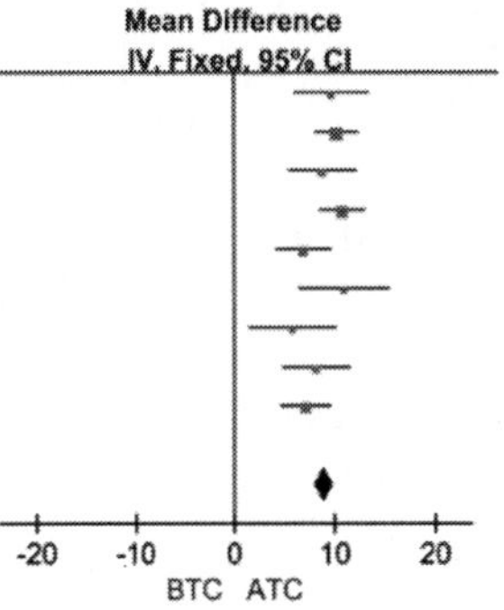

Figure 3. The forest figure of Diastolic blood pressure.

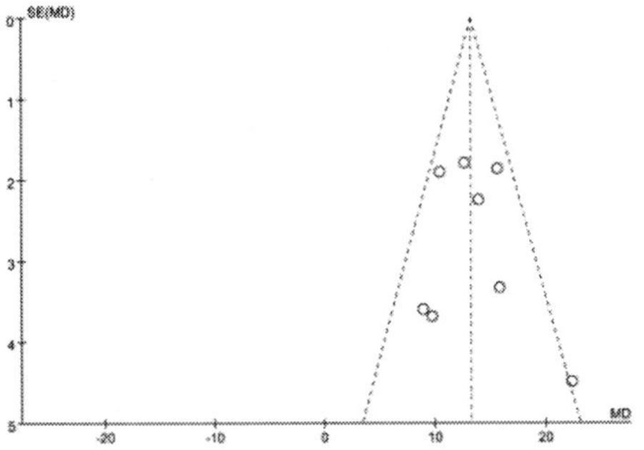

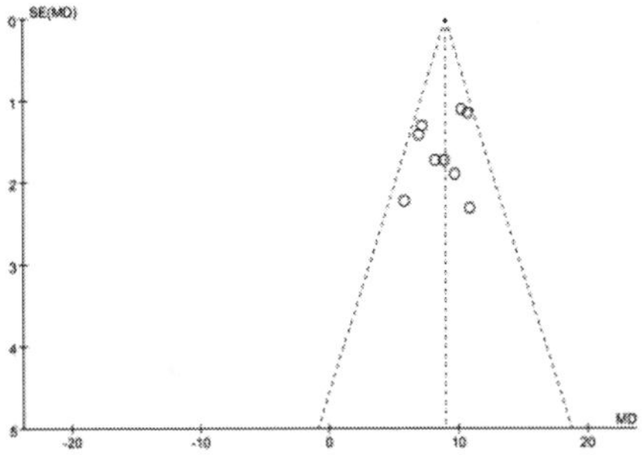

Figure 4. The funnel figure of Systolic blood pressure. **Figure 5.** The funnel figure of Diastolic blood pressure.

There is no risk of publication bias through the funnel figure (Figs.4&5).These trials demonstrated no significant heterogeneity with I^2 values ranging from 27 to 36% (Figs.2&3). The meta-analysis found that after Tai Chi exercise (ATC) can significantly reduce the SBP (WMD = 13.19 mmHg; 95%CI: 11.52 to 14.87; P<0.0001) and DBP (WMD = 8.92 mmHg; 95% CI:7.94 to 9.90; P<0.0001) compared to before Tai Chi exercise(BTC)(Figure 2& Figure 3).

3.3.2 Other outcomes

Other outcomes: HR, NO, TC, TG, LDL-C, HDL-C.(1)HR. 3 trials evaluated the effect of Tai Chi on HR. Demonstrated no significant change of HR(WMD = 1.64; 95%CI: -0.51 to 3.97; P =0.14)by ATC was identified compared BTC.(2)NO. 5 trials evaluated the effect of Tai Chi on NO.ATC significantly improved the content of NO(WMD =-7.98mmol/L; 95%CI: -10.63 to -5.33; P<0.0001). (3)TC. 3 trials evaluated the effect of Tai Chi on TC. Demonstrated no significant change of TC (WMD = -0.03mmol/ml; 95%CI: -0.22 to 0.17; P =0.80) by ATC was identified compared BTC.(4)TG. 4 trials evaluated the effect of Tai Chi on TG. ATC significantly reduced the TG content(WMD =0.22mmol/ml; 95%CI: 0.06 to 0.38; P =0.006).(5)LDL-C. 3 trials evaluated the effect of Tai Chi on LDL-C. ATC significantly decreased the content of LDL-C(WMD =0.20 mmol/ml;95%CI:0.13 to 0.26; P<0.0001).(6)HDL-C.3 trials evaluated the effect of Tai Chi on HDL-C. ATC demonstrated no significant change of HDL-C(WMD =-0.04 mmol/ml; 95%CI: -0.09 to 0.01; P =0.13).

4 Discussions

4.1 Summary of evidence

Tai Chi, as an exercise for health in a wide age range, has been used for several centuries, especially popular in the elderly[4]. Of course, we need to identify a suitable exercise regimen targeted to maintain and improve hypertension patients health[5].This study researched if Tai Chi is an safe and effective therapy methods of hypertension patients.

In conformity with the outcome measures from 13 RCTs with a total of 625 participants, this meta-analysis and systematic review study is to give an objective assessment of Tai Chi for the arrangement of hypertension patients. The meta-analysis found that after Tai Chi exercise (ATC) showed a clinically obvious improvement on the SBP (decreased by 13.19 mmHg), DBP (decreased by 8.92 mmHg), NO (increased by 7.98mmol/L), TG (decreased by 0.22mmol/ml) and LDL-C (decreased by 0.20 mmol/ml).NO is a very important biological messenger and signal molecule, also it is a formidable vasodilator factor[14].This study shows that Tai Chi exercise can increase the content of NO. Research found that hypertension and dyslipidemia are frequently existing together, and people with these conditions will have an increased cardiovascular disease risk[17-19]. This study shows that Tai Chi exercise can modulate lipid metabolism. We need to concern how long and how often Tai Chi exercise should be practiced for the therapy of hypertension patients? In this study, all of the trials duration included are ranged from 8 weeks to 1 year. So, we recommended that Tai Chi exercise should be practiced at least for 8 weeks and at least 3times/week. Hence, the combined results show that Tai Chi exercise is an effective and relatively-safe supplemental method for the therapy of hypertension patients.

4.2 Limitations

In spite of the significant active results of Tai Chi exercise for the therapy of hypertension patients, a lot of limitations also need to be noted. First, most of the trials included did not report the sample sizes calculating methods, and whether the number of participants involved met the research requirements or not is still not clear. Second, the small sample sizes and generally poor methodological quality limited the strength feasibility and generally poor methodological quality of the clinical evidence[15,16]. Third, because we only enrolled Asian hypertensive patients, there were no data about the effectiveness of Tai Chi exercise on other national's BP. So, the potential selection bias couldn't be ruled out. Fourth, the included trails used different styles of Tai Chi exercise. Finally, although the effects of Tai Chi exercise on BP outcomes were reported, but there are still a lot of recessive results that were difficult to be published. Therefore, further studies are needed.

5 Conclusions

This study shows that Tai Chi exercise is an effective treatment for hypertension patients. This can decrease SBP, DBP, TG, LDL-C and increase NO. However, more reasonable designed RCTs are still warranted before it can be used, and more methodologically rigorous researches are needed to prove Tai Chi's real clinical benefits, and needed to explore the potential mechanism of Tai Chi.

References

1. James PA, Oparil S, Carter BL, et al. 2014 evidence-based guideline for the management of high blood pressure in adults: report from the panel members appointed to the Eighth Joint National Committee (JNC 8). J Am Med Assoc311:507–20(2014).
2. Papadakis MA. Current Medical Diagnosis & Treatment 2004, 43rd ed. New York: McGraw-Hill. 401(2004).
3. Eckel RH, Jakicic JM, Ard JD, et al. 2013 AHA/ACC guideline on lifestyle management to reduce cardiovascular risk: a report of the American College of Cardiology/American Heart Association task force on practice guidelines. Circulation 129:S76–99(2014).
4. JEN-CHEN TSAI,WEI-HSIN WANG,et al. The Beneficial Effects of Tai Chi Chuan on Blood Pressure and Lipid Profile and Anxiety Status in a Randomized Controlled Trial. THE JOURNAL OF ALTERNATIVE AND COMPLEMENTARY MEDICINE 9:747-754(2003).
5. Hui-Ming Lo,et al.A Tai Chi exercise programme improved exercise behavior and reduced blood pressure in outpatients with hypertension. International Journal of Nursing Practice 18: 545–551(2012).
6. Jing Sun,et al.Community-Based MindeBody Meditative Tai Chi Program and Its Effects on Improvement of Blood Pressure, Weight, Renal Function, Serum Lipoprotein, and Quality of Life in Chinese Adults With Hypertension. Am J Cardiol 116:1076e1081(2015).
7. WANG Xiao-jun,et al. Empirical Study of Taijiquan Interventions on the Prevention and cure of Hypertension. Journal of Beijing Sport University 9:75-77(2011).
8. JIN Zhen-yang, et al. Effect of Tai Chi on the hypertension and blood lipid level among middle aged people. Modern Preventive Medicine39:4468-4471(2012).
9. Sun Feng et al. Intervention effect of taijiquan exercise on senile hypertension. Chinese journal of gerontology 34:6862-6864(2014).
10. He You-ping et al. The Effect and Mechanism of Taijiquan on Primary Hypertension. Journal of Anhui Normal University (Natural Science) 35:83-87(2012).
11. Xie Hui-juan et al. Gaseous Signal Molecular Mechanism of Taijiquan's Intervention in Aged Essential Hypertension. Journal of Wuhan Institute of Physical Education48:51-63(2014).
12. Mao Hong-ni et al. Tai chi balance cycle of NO/ET in patients with essential hypertension. Chinese journal of gerontology 28:508-509(2008).
13. He Jing-he et al. Taijiquan research on effect of intervention in patients with essential hypertension. Chinese Journal of Rehabilitation Medicine 26:968-971(2011).
14. Mao Hong-ni et al. Effect of Tai Chi exercise on blood pressure, plasma nitrogen monoxidum and endothelin in hypertensive patients. Chinese Journal of Clinical Rehabilitation 48:65-67(2006).
15. Li Chuan-wu et al. Effect of Taijiquan on endothelial diastolic function in the elder ly. Journal of Clinical Rehabilitative Tissue Engineering Research 27:5414-5416(2007).
16. Zhong Yun-jian et al. Effects of single acute Taijiquan exercise on essential hypertension. J Fourth MilMed Univ 30:3136-3139(2009).
17. Haffner SM, Miettinen H, Gaskill SP, Stern MP. Metabolic precursors of hypertension. The San Antonio heart study. Arch Intern Med 156:1994e2001(1996).
18. Oparil S,Zaman MA,Calhoun DA. Pathogenesis of hypertension. Ann Intern Med 139:761e776. (2003).
19. Stamler J, et al. Prevalence and prognostic significance of hypercholesterolemia in men with hypertension. Prospective data on the primary screeners of the Multiple Risk Factor Intervention Trial. Am J Med80:33e39(1986).

Application of sustained release microsphere in hypoparathyroidism after thyroid surgery

Ming Pan, Hong Zhang[1] and Yueshu Wang[2, a]

[1]*Department of thyroid Surgery, the Second Affiliated Hospital of Jilin University, 130041, Changchun, China*
[2]*Department of hand surgery, China-Japan Union Hospital of Jilin University, 130033, Changchun, China*

Abstract. Hypoparathyroidism (HypoPT) is one of the most common complications after thyroid surgery. The parathyroid hormone-controlled delivery system is one of the most significant therapies for hypothyroidism after thyroid and parathyroid surgery. The character of biodegradability and the slow release of PTH of the Sustained release microspheres were widely concerned. The authors present the research progress in sustained release microspheres loaded with PTH from the character, material, fabrication and application.

1 Introduction

Hypoparathyroidism (HypoPT) is diagnosed on the foundation of hypocalcemia and low plasma levels of parathyroid hormone (PTH). Recent studies from the USA and other countries have estimated that the prevalence of HypoPT is 25 per 100,000 individuals[1-3].Risk factors for permanent hypothyroidism include seropositivity for thyroid peroxidase(TPO) antibodies, high normal preoperative thyroid stimulating hormone(TSH) levels, and histologically confirmed thyroiditis, but not age, sex, family history, or weight of the resected tissue. In most instances, HypoPT is due to accidental damages to or removal of the parathyroid glands during thyroid and parathyroid surgery. Permanent HypoPT arises not only after total or subtotal thyroidectomy, but also in 11% to 28% of patients that have undergone hemithyroidectomy[4-6].Hormone replacement after thyroid and parathyroid surgery is a common and significant clinical therapy. The parathyroid Sustained release microsphere is one of the most significant therapies for hypothyroidism after thyroid and parathyroid surgery. Controlled release technology enables the delivery of a biologically active agent at specific times with defined concentrations by encapsulating the biologically active agent within a biodegradable polymer in the form of a microsphere.

2 Sustained release microsphere

2.1 Character

The material of microspheres should meet the following requirements[7]:

1)Satisfactory biocompatibility, including blood compatibility, histocompatibility and non-immunogenicit. Moreover, non-poisonous and non- teratogenic, Products after degradation have no

[a] Corresponding author:ysw_1971@163.com

toxic action, won't cause inflammation and jump reaction.

2)Wonderful biodegradability, and degradation time and drug release rate can be regulated and controled as requestment.

3)The degradation of polymer must be in a suitable time.

4)The material should possess an applicable mechanical strength, and can be shaped through conventional method.

5)The method is simple.

6)Material sourcesis is easy to obtained, and it is inexpensive and suitable for large scale application. The deactivation of protein drugs must be avoided if it is used to load protein drugs.

2.2 Material

There are many biodegradable polymer materials used for embedding proteins, peptides, nucleic acids, vaccine and other active substances or other drugs. According to the origin, it can be divided into natural polymer materials, semisynthetic polymer materials and synthetic polymer materials. The characteristics of natural polymer materials are extensive sources, low cost, stable character, non-toxic, biodegradable, prominent film-forming and ballability, such as gelatin, alginate, chitosan, etc[8-9].Semisynthetic polymer materials are mostly cellulose derivatives. They have low virulence high viscosity, increased solubility after salt [10-11].Compared with natural materials, synthetic polymer materials has rich material source, and the structure and properties can be artificially modified and regulation. Over the past 20 years these materials developed rapidly, Polylactic acid and polyglycolic acid are the most typical synthetic biodegradable polymers, with non-toxic, no immune stimulation and superior biodegradability and biocompatibility[12-13].FDA approval polylactic acid glycolic acid (PLGA) as a biodegradable materials used in pharmaceutical excipients, PLGA has become the most widely applied biodegradable polymer materials.[14-15]So far, FDA has approved eight kinds of injectable microsphere drugs. All the microsphere materials are PLGA.

2.3 Fabrication

Presently, the preparation method of polymer microspheres includes spray drying, coacervation, multi ple emulsion-solvent evaporation and microporous membrane emulsification technique. The protein d rugs need to imbed into polymer materials should be dissolved uniformly in the water, and PLGA sho uld be dissolved in organic solvent, producing W/O/W compound emulsion, then reach a desiccation b y spraying in inert air, with the solvent evaporated quickly, embedding the drugs[16-17].Condensation is also called phase separation, Drugs and materials appear two phase separation after drugs and poly mer materials mixed each other under a certain external factors stimulation, Carrier polymer materials dissolve out from solution packaging drugs. Multiple emulsion-solvent evaporation is a process that a queous solution of drug and dichloromethane solution dissolved with PLGA are mixed, following wit h fabricate primal emulsion by ultrasound and other ways. Then the primal emulsion are infused in po ur water becoming W/O/W multiple emulsion-solvent. At last, to PLGA microspheres were fabricated by a solvent evaporation technique, embedding drugs[18]. Microporous membrane emulsification tec hnique[19] is a new technology developed from emulsion preparation. Disperse phase were pressed in fenestra under nitrogen pressure. When the forming emulsion drops reach a certain size at the fenestra exit, they were divorced from the surface of emulsion under the action of forces, entering into continu ous phase.

2.4 Defect

The size was not uniform, which is not only poor repeatability in preparation that make the process more complex, but also lead to inconsistent drug release behavior, causing unsteady blood drug concentration of patients. [20]The energy consumption of spray drying process is high that is easy to cause the inactivation of protein drugs, reducing the effect. Condensation need more organic solvents, which is difficult to wipe off the residual organic reagents. It is easier to become agglomerate by phase separation and more difficult to conduct mass production.

3 Application of Sustained release microsphere loaded with PTH

3.1 Hormone replacement therapy

Microspheres can be used to provide a bioactive calcium sensitivity control PTH, and provides a potential way in hormone replacement therapy. Biodegradable poly(lactide-co-glycolide) (PLGA) microspheres loaded with PTH were made by Christopher Breuer[21] using a modification of the double emulsion (water/oil/water) solvent evaporation technique. Controlled release of PTH was demonstrated in vitro over a 3-week period and release of physiological significant concentrations of PTH was achieved using this methodology. In 2007,Christopher K.Breuer[22] used the spontaneous emulsion technique and fabricated PTH microspheres by encapsulating PTH(1–34) in PLGA. Another kind of microspheres were generated by mixing calmodulin with PTH(1–34) in the PLGA microspheres. It caused a significant increase in serum calcium levels when PTH released from the PTH/calmodulin microspheres.

3.2 The growth and repairment of bone

The developed PLGA microspheres with sustained release and long-term effect show great potential as a promising delivering and releasing carriers for inducing a biologically active response in bone, treating early osteoarthritis and osteoporosis. Rajalakshmanan Eswaramoorthy[23] developed an effective controlled-release system for prolonging the treatment duration of an intra-articular injection for osteoarthritis treatment in rats. The effect of PTH/PLGA microspheres on suppressing osteoarthritis progression was similar to that of a once-every-three-day injection of PTH(1–34), indicating that both the sustained and intermittent action of PTH(1–34) effectively suppress osteoarthritis progression. In the same year, Peter X. Ma[24] prepared poly(lactic-co-glycolic acid) (PLGA) microspheres containing bovine serum albumin (BSA) or human PTH1–34 by using a double emulsion method with high encapsulation efficiency and controlled particle sizes. And both in vitro and in vivo assays demonstrated that the bioactivity of PTH was maintained largely during the fabrication of PLGA microspheres and upon release. Recently, in Wei-yan Meng[25] prepared PLGA microspheres by a water-in-oil-in-water double emulsion solvent evaporation and PTH1-34 was encapsulated to the microspheres by a technique with good sphericity and high efficiency. The polyester microspheres had a good spherical shape.PTH1-34 loaded microspheres exhibited a slow release behavior for about 25 days. Cell proliferation experiments demonstrated that PTH1-34 loaded PLGA microspheres can enhance the proliferation and differentiation of MC3T3 cell effectively. The cells incubated with PTH1-34 loaded PLGA microspheres exhibited higher ALP level.

4 Conclusion and expectation

Sustained release microspheres are still in the laboratory research stage, with a lot of problems need to be solved. However, the character of biodegradability and the slow release of PTH of the Sustained release microspheres were widely concerned. It is a hopeful role in physiologically controlled drug-delivery way of PTH in clinical practice, especially, in the therapy of hypoparathyroidism after thyroid surgery.

References

1. L. Underbjerg, T. Sikjaer, L. Mosekilde, L. Rejnmark. J Bone Miner Res.28,2277-2285(2013).
2. J. Powers, K. Joy, A. Ruscio, H. Lagast. J Bone Miner Res,28,2570-2576(2013).

DOI: 10.1051/bioconf/20170801012

3. L. Underbjerg, T. Sikjaer, L. Mosekilde, L. Rejnmark. J Bone Miner Res,30,1738-44(2015).
4. R. Gärtner, M. Reincke. Der Internist.49,538–544(2008).
5. A. Schäffler, Diabetologie und Stoffwechsel. 36(2009).
6. W.M. Wiersinga. Horm Res, 56(Suppl 1), 74–81(2001).
7. A. Kumari, S.K. Yadav, S.C. Yadav. Colloid Surfaces B, 75(1),1-18(2010).
8. M.S. Soliman , T. Behzad. Pharm Dev Tech, 2(3), 33-35(2014).
9. X Hua,S Tan,HM Bandara,Y Fu,S Liu.Plos One, 9(12),e114271(2014).
10. S. Butu,F.G. Ince,H. Erdugan, N. Sahiner. Carbohydrate Polymers, 86(2), 636-643(2011).
11. J. Shaji,V. Patole. Indian J Pharm Sci, 70(3), 269-277(2008).
12. V. Lassall,M.L. Ferreira. Macromol Biosci, 7(6),767-783(2007).
13. J.A. Floyd, A. Galperin, D.R. Buddy. Adv Drug Deliver Rev, 91,23-37(2015).
14. K.M. Hirenkuma, J.S. Steven. Polymers, 3(3), 1377-1397(2011).
15. G. Orive, A.R. Gascón, R.M. Hernández, A. Domínguez-Gil,J.L. Pedraz. Trends Pharmacol Sci, 25(7), 382-387(2004).
16. T. Petreska-Ivanovska,L. Petrushevska-Tozi,A. Grozdanov,R. Petkovska,J. Hadjieva. Chem Ind Chem Eng Q, 20 (4),549-564(2014).
17. F. Sansone, T. Mencherini, P. Picerno, T. Esposito, P.D. Gaudio. J Pharm pharmacol, 2 (4),100-109(2014).
18. J. Naik, R.D. Kulkarni, A.B. Lokhande, S. Mishra. Int J Pharm Bio Sci, 3(4),573 -590(2012).
19. C. Charcosset, I. Limayem, H. Fessi. J Chem Technol Biot, 79(3), 209-218(2004).
20. L.Y. Wang,G.H. Ma,Z.G. Su. J Controlled Release, 106(1-2), 62-75(2005).
21. T. Anthony,P. Fong,A. Goyal,W.M._Saltzman,R.L. Moss. Journal of Pediatric Surgery, 40(1),81-85(2005).
22. P. Fong, A. Goyal, M. Brennan, J. Park, L. Moss. Journal of Surgical Research, 143(2),195-199(2008).
23. R. Eswaramoorthy, C.C. Chang, S.C. Wu, G.J. Wang, J.K. Chang. Acta Biomaterialia, 8(6),2254–2262(2012).
24. G. Wei, G.J. Pettway, L.K. Mccauley, P.X. Ma. Biomaterials, 25(2):345-352(2004).
25. S.B. Zhang, Z. Zhang, M.N. Yu, T.B. Liu, J.J. Wang. Acta Polymerica Sinica, 014(2),270-275(2014).

Effects and mechanism of juglone in combination with 5-FU on colon cancer CT-26 cells in vivo

Tianyu Liu, Yan Li[a] and Na Li

School of Life Science, 100081 Beijing Institute of Technology, China

Abstract. The effects and mechanism of juglone alone or combined with 5-fluorouracil on colon cancer CT-26 cells were examined in vivo. Starting at 5 weeks of age, male Balb/c mice were transplanted the CT-26 cells and treated with juglone alone or combined with 5-fluorouracil. Observed the morphologic characteristics of tumor tissue by HE staining after treatment of 9 days. We detected the content of IL-2, VEGF and IFN-γ in serum and the protein expression of ki-67, E-cad and Bcl-2 in tumor tissue. The inhibition rate of animals given low-dose or high-dose juglone alone and combined with 5-fluorouracil is 31.73%, 40.38%, 56.73% and 75.01% respectively. Animals given juglone alone or combined with 5-fluorouracil could increase the content of IL-2 and IFN-γ in serum and the expression of E-cad and decrease the content of VEGF in serum and the expression of ki-67 and Bcl-2 significantly compared with those injected 2% ethanol NS ($P<0.05$ in each). The present data suggest that the juglone could reduce the risk of CT-26 cells metastasis and proliferation and enhance the function of the immune system. The possible anti-tumor mechanism of juglone is down-regulated expression of Bcl-2 to induce the apoptosis of tumor cells.

1 Introduction

Colon cancer as one of the malignant digestive tumor is a major global health problem[1]. Using extensively surgical remove, combined with chemotherapy and radiotherapy after operation is effective in treating colon cancer currently. 5 – fluorouracil, a metabolic type of antitumor drugs, is recognized in clinical to treat colon cancer and other digestive tract cancer [2]. Because of the significant effect on the treatment of colon cancer, 5 – fluorouracil has been applied widely as the first-line drug of colorectal carcinoma chemotherapy. Recent evidence has shown that patients have developed a resistance which potentially has decreased the proportion of people who could be cured [3]. The other main side effects of 5 – fluorouracil are myelosuppression and hemorrhagic enteritis and its low lipophilicity, bad selectivity and evident side effects also limit its application. Juglone (5-hydroxy-1,4-naphthoquinone), a hydroxy naphthoquinone compound, is derived from the fresh root bark, branch bark or green peel of Juglansmandshurica [4]. Many data derived from laboratory studies have suggested that juglone has a good antitumor activity both in vitro or vivo. It has obvious inhibitory effect on the proliferation of Hela cells and esophageal cancer cells in vitro and breast cancer cells, stomach cancer cells and S180 solid cells in mice [5-8]. However, there is no report

[a]Corresponding author: 2120151319@bit.edu.cn

about the effect of juglone on colon cancer in vivo. In this study, the effects and mechanism of juglone alone or combined with 5-fluorouracil on colon cancer CT-26 cells were examined in vivo. In addition, the effects of juglone and juglone combined with 5-fluorouracil on immune function in vivo is discussed.

2 Materials and methods

2.1. Animals, cells, test drugs and chemical reagents

60 male Balb/c mice, which were purchased from Peking University Health Science Center (Beijing, China), were used. CT-26 cells were purchased from Peking Union Medical College (Beijing, China). Juglone was purchased from Sigma. 5-fluorouracil was purchased from Beijing Coal General Hospital (Beijing, China).

All animals were housed in plastic cages (ten mouse per cage). They had free access to water and diets under controlled environmental conditions of humidity (50 ± 10%), lighting (12 h light/dark cycle) and temperature (23 ± 2°C). CT-26 cells of a suitable concentration were inoculated in culture flasks containing RPMI-1640 solution with 10% of fetal bovine serum. The cultures were then incubated at 37°C, with 5% CO_2 and saturated humidity; culture transfer was performed once every 2-3 days. The test drugs and chemical reagents were stored in the freezer (4°C).

2.2. Experimental procedure

The study was performed after the mice were allowed to acclimate for 5 days. CT-26 colon cancer mouse models were made on the sixth day by transplanting 0.2mL CT-26 cell suspension (1×10^7/mL) in the mouse left anterior subcutaneous part. Juglone and 5-fluorouracil were injection via abdominal according to the following group once a day. A total of 60 mice of 5 weeks of age were divided into six groups as follows: group 1, 10 mice for 2% ethanol NS as model group; group 2, 10 mice for 5-fluorouracil (20mg/kg 5-FU) alone as positive drug group; group 3, 10 mice for high-dose juglone (5mg/kg) alone; group 4,10 mice for low-dose juglone (2.5mg/kg) alone; group 5, 10 mice for juglone (5mg/kg) combined with 5-fluorouracil (20mg/kg 5-FU); group 6, 10 mice for juglone (5mg/kg) combined with 5-fluorouracil (10mg/kg 5-FU). After 9 days, complete autopsies were performed on all animals after weighed and took blood by removalling eyeball. At autopsy, the size (length, width and height) of tumors were measured and the maximum size was selected for analysis. Thymus and spleen were examined cautiously. Tumor tissues were fixed in 10% buffered formalin, embedded in paraffin blocks and processed by routine histological methods with hematoxylin and eosin stains, then Calculated the expression rate of nuclear factor Ki-67, E- cadherin (E-cad), B-cell lymphoma 2 gene (Bcl-2). Calculated tumor inhibition rate, thymus and spleen indexes and determined the content of interleukin -2 (IL-2), interferon-γ (IFN-γ), vascular endothelial growth factor (VEGF) by Elisa.

2.3. Statistical analysis

Statistical evaluation was done using the SPSS 17.0 and ANOVA. Results presented as mean ± standard deviation ($\bar{x}\pm SD$).

3 Results

3.1 Effect of juglone and juglone combined with 5-fluorouracil on proliferation of CT-26 in mice

Table 1 indicates that The inhibition rate of animals given low-dose or high-dose juglone alone and combined with low-dose or high-dose 5-fluorouracil is 31.73%, 40.38%, 56.73% and 75.01% respectively. These data showed that juglone combined with 5-fluorouracil was more cytotoxic than juglone alone and the effect of juglone combined with 5-fluorouracil (20mg/kg 5-FU) was better than 5-fluorouracil (20mg/kg 5-FU), indicating that juglone may have a good anti-cancer effect on CT-26 cells. The expression rate of nuclear factor ki-67 is closely related to the cell proliferation. Compared with the model group, the expression rate of ki-67 were significantly decreased when animals were treated with juglone or 5-fluorouracil (p<0.05). Juglone combined with 5-fluorouracil (20mg/kg 5-FU) could better inhibit the expression of ki-67 than 5-fluorouracil (20mg/kg 5-FU) (p<0.01). By observing the morphological characteristics of CT-26 cells (Figure 1), we found both juglone and 5-fluorouracil could make tumor necrose obviously, such as some nucleus becoming smaller or cells becoming shrinkage even cytolysis. These results are consistent with the experiment of anti-tumor rate.

Table 1 Effect of juglone and 5-fluorouracil on the tumor proliferation in mice ($\bar{x}$±SD, n=10)

Group	Dose of juglone (mg/kg)	Dose of 5-FU(mg/kg)	Tumor weight(g)	anti-tumor rate(%)	expression rate of Ki-67(%)
Model	-	-	1.04±0.21	-	83.28±10.40
5-FU	-	20	$0.32±0.19^{b}$	69.23^{b}	$42.85±11.07^{b}$
Juglone	2.5	-	$0.71±0.19^{a,d}$	$31.73^{a,d}$	$66.23±12.39^{a}$
	5	-	$0.62±0.17^{b,c}$	$40.38^{b,c}$	$59.67±6.16^{b}$
Juglone combined with 5-FU	5	10	$0.45±0.18^{b}$	56.73^{b}	$43.20±7.76^{b}$
	5	20	$0.26±0.20^{b}$	75.01^{b}	$20.84±9.72^{b,d}$

$^{a,\,b}$ Significantly different from the model group (^{a}P<0.05, ^{b}P<0.01)
$^{c,\,d}$ Significantly different from the 5-FU group (^{c}P<0.05, ^{d}P<0.01)

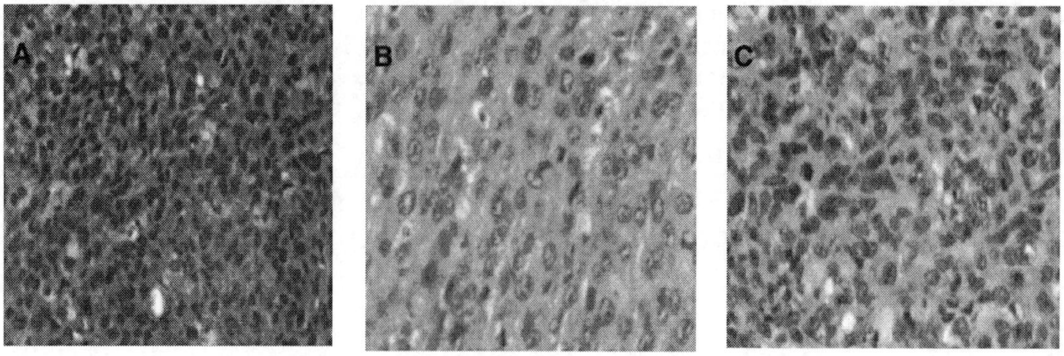

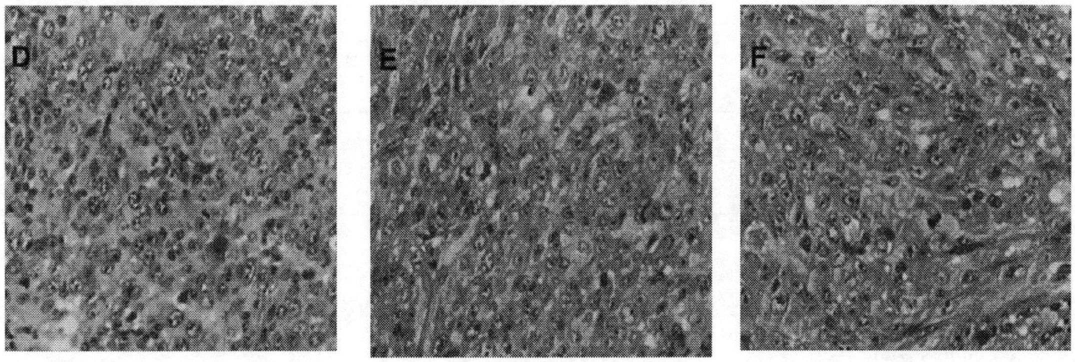

Figure 1. Colon tumour tissues stained with hematoxylin and eosin. A: model ; B: 20 mg/kg 5-FU; C: 2.5 mg/kg juglone; D: 5 mg/kg juglone; E: 5 mg/kg juglone with 10 mg/kg 5-FU. F: 5 mg/kg juglone with 20 mg/kg 5-FU.

3.2 Effect of juglone and juglone combined with 5-fluorouracil on metastasis of CT-26 in mice

VEGF as the main regulator induces tumor angiogenesis and the protein expression of E-cadherin may suggest a cell adhesion. Compared with the model group, the mouse serum VEGF levels of the groups treated with juglone or juglone combined with 5-fluorouracil is significantly decreased ($P<0.05$) and similar with the control group which treated with 5-fluorouracil alone. The protein expression of E-cadherin of the groups treated with juglone or juglone combined with 5-fluorouracil is significantly increased ($P<0.01$) (Table 2). These results suggest that juglone could inhibit proliferation, invasion and metastasis of the tumor in mice.

Table 2 Effect of juglone and 5-fluorouracil on the tumor metastasis in mice ($\bar{x}\pm SD$, n=10)

Group	Dose of juglone (mg/kg)	Dose of 5-FU(mg/kg)	VEGF (pg/mL)	E-cad integrated optical density(IOD)
Model	-	-	745.42±18.07	37868±5234
5-FU	-	20	693.77±25.94[b]	93436±10646[b]
Juglone	2.5	-	704.67±15.47[a]	90419±21256[b]
	5	-	697.80±15.94[b]	88934±8396[b]
Juglone combined with 5-FU	5	10	704.57±17.88[b]	92139±8025[b]
	5	20	681.80±17.95[b]	92777±15937[b]

[a, b] Significantly different from the model group ([a] $P<0.05$, [b] $P<0.01$)

3.3 Effect of juglone and juglone combined with 5-fluorouracil on apoptosis of CT-26 in mice

Table 3 shows that expression rate of Bcl-2 of the groups treated with juglone or juglone combined with 5-fluorouracil is significantly decreased ($P<0.01$). It indicates that juglone could down-regulate the expression of Bcl-2 in CT-26 cells to promote apoptosis.

Table 3 Effect of juglone and 5-fluorouracil on the tumor apoptosis in mice (x̄±SD, n=10)

Group	Dose of juglone (mg/kg)	Dose of 5-FU(mg/kg)	expression rate of Bcl-2(%)
Model	-	-	73.02±4.87
5-FU	-	20	40.29±6.18[a]
Juglone	2.5	-	55.71±5.31[a]
	5	-	51.29±4.72[a]
Juglone combined with 5-FU	5	10	44.29±4.43[a]
	5	20	38.04±4.79[a]

[a, b] Significantly different from the model group ([a] $P<0.01$)

3.4 Effect of juglone and juglone combined with 5-fluorouracil on immunity of mice

Table 4 shows that thymus index of the groups treated with juglone are significantly increased comparing with the model group and control group ($P<0.01$). Comparing with the control group, thymus index of the groups treated with juglone combined with 5-fluorouracil are also increased ($P<0.01$). Juglone and juglone combined with 5-fluorouracil could also increase the spleen index. The mice serum IL-2 and IFN-γ level could be raised after using juglone or juglone combined with 5-fluorouracil.

Table 4 Effect of juglone and 5-fluorouracil on immunity of mice (x̄±SD, n=10)

Group	Dose of juglone (mg/kg)	Dose of 5-FU(mg/kg)	Thymus index(mg/g)	Spleen index(mg/g)	IL-2 (pg/mL)	IFN-γ (pg/mL)
Model	-	-	2.42±0.71	4.15±1.04	112.92±7.32	285.83±22.95
5-FU	-	20	0.64±0.15[b]	3.72±0.61	131.34±5.77[b]	328.57±17.85[a]
Juglone	2.5	-	3.47±0.58[b,d]	4.29±0.37	132.29±6.27[b]	331.43±16.38[b]
	5	-	3.64±0.38[b,d]	6.81±1.13[b,d]	131.25±5.78[b]	332.92±17.64[b]
Juglone combined with 5-FU	5	10	2.10±0.76[d]	6.19±1.34[b,d]	138.23±7.43[b]	342.14±29.63[b]
	5	20	2.01±0.43[d]	6.31±1.54[b,d]	134.69±6.32[b]	362.92±20.76[b,c]

[a, b] Significantly different from the model group ([a] $P<0.05$, [b] $P<0.01$)
[c, d] Significantly different from the 5-FU group ([c] $P<0.05$, [d] $P<0.01$)

4 Discussion

In this study, juglone could inhibit the tumor proliferation in mice obviously. Juglone could reduce the risk of proliferation and metastasis of tumor cells by decreasing the level of VEGF in serum and increasing the protein expression of E-cadherin. Furthermore, juglone may improve immune function by increasing some immune factors such as IL-2, INF-γ to improve antitumor activities. The effect of juglone combined with 5-fluorouracil on antitumor is enhanced apparently. And this measure could reduce the side effects of 5-fluorouracil like impaired immune function. The antitumor mechanism of juglone may be involved in blocking Bcl-2 protein expression to induce tumor cells apoptosis. In conclusion, juglone may be a new promising chemotherapeutics for colon cancer.

References

1. M. R. Buunen, W. Hop, et al. "Survival after laparoscopic surgery versus open surgery for colon cancer: long-term outcome of a randomised clinical trial." *Lancet Oncology* **10**.1(2009):44-52.
2. J. Li, et al. "Inhibition of autophagy augments 5-fluorouracil chemotherapy in human colon cancer in vitro and in vivo model." *European Journal of Cancer* **46**.10(2010):1900-1909.
3. R. L. Zhai, X. M. Zhang, et al. "Relationship between O~6 -methylguanine-DNA methytransferase expression in colon cancer stem cells and acquired 5-Fu chemotherapy drug resistance in colon cancer cells." *Chinese Journal of Experimental Surgery* **31** (2014):248-250.
4. A. Ohta, P.M. Sivalingam, et al. "Isolation of naphthazarin from walnut 'Onigurumi', and its inhibitory action on oxidative phosphorylation in mitochondria." *Toxicon* **11**.3(1973):235-241.
5. W. Zhang, X. Y. Zhao, et al. "Effects of different concentrations of juglone on the proliferation and apoptosis of HeLa cells." *Medical Journal of Chinese Peoples Liberation Army* **38**.2(2013).
6. L. Li, P. Chen, et al. "Juglone inhibits the proliferation of human esophageal carcinoma EC1 cells." *World Chinese Journal of Digestology* **18**.11(2010):1147-1151.
7. M. T. Paulsen, M. Ljungman. "The natural toxin juglone causes degradation of p53 and induces rapid H2AX phosphorylation and cell death in human fibroblasts."*Toxicology & Applied Pharmacology* **209**.1(2005):1-9.
8. Y. B. Ji, et al. "Studies on Inhibition of Juglone in Exocarpium Juglandis Immaturum on S(180) Sarcoma in Mice." *Journal of Chinese Pharmaceutical Sciences* **44**.3(2009):195-199.

A study on regional comprehensive performance evaluation indicator system of rational use of drugs

Tongda Sun[1,a] , Wenjie Zhou[2] , Yu Zhu[3], Jinguo Wu[4] and Luning Fan[1]

[1]*Department of Health Information Management, School of Health Service and Health Management, Ningbo College of Health Sciences, Ningbo City, Zhejiang Province, 315100, China*
[2]*Department of Foreign Languages, School of Humanity and Social Science, Ningbo College of Health Sciences, Ningbo City, Zhejiang Province, 315100, China*
[3]*Health and Family Planning Bureau of Yinzhou District, Ningbo City, Zhejiang Province, 315100, China*
[4]*Jintang Software Company Limited, Ningbo City, Zhejiang Province, 315041, China*

Abstract. The current research presents the design of a 4-degree-3-level performance evaluation indicator system of rational use of drugs for health care institutions accord with the Balanced Score Card (BSC) method. Financial index, patient index, professional process index, and development and blazing new trials index are adopted in the light of scientific, guiding, operable and generalizable principles. The index weight is based on the analytic hierarchy process, and comprehensive performance evaluation indicators are calculated by a linear integrated weighting method. Its practical application in 21 state-run health care institutions in Ningbo, from 2008 to 2012, has arrived at the finding that the comprehensive performance evaluation indicator system offers a scientific, practical and effective performance management quantification and is thus worth popularizing.

1 Introduction

How to promote the level of medical staffs' rational use of drugs is the key link in the application of essential medicine system. Most of the previous researchers at home and abroad focus more on rational use of drugs or performance and quality evaluation on a certain level [1-4], while it is obviously insufficient in the respect of representation and completion with the abolishment of drug-dependent medical policies. In order to standardize the supervision mechanism of rational use of drugs for health care institutions, it is meaningful to research so as to construct an integrative performance evaluation indicator model which can generalize and reflect the numerous elements and influential factors of rational use of drugs under the background of new medical reform. The current research presents the design of a regional comprehensive performance evaluation indicator system of rational use of drugs accord with the Balanced Score Card (BSC) method.

---

[a] Corresponding author: suntongda@foxmail.com
This study was supported by a grant from the Science and Technology Planning Project of Social Development in the year of 2013 of Ningbo city, Zhejiang province, China (Project number: 2013C50028).

2 Subjects and methords

2.1 Subjects

21 state-run health care institutions in a certain city, China, have been chosen as subjects. Institutions' original data from 2008 to 2012 has been collected with a method of field investigation and used to calculate the actual value of evaluation indicators after proper statistical treatments. The average value of the actual data from 2008 to 2012 has been calculated to be the evaluation indicator for each health care institution.

2.2 Choice of evaluation indicators

Based on the Balanced Score Card (BSC) method, financial index, patient index, professional process index, and development and blazing new trials index have been adopted to the performance evaluation indicator system of rational use of drugs in the light of people-oriented, sustainable, policy-type, guiding, safe, operable and generalizable principles.

2.3 Calculation and standardization of evaluation indicators

(1) Calculate the index value of valuation indicator.

(2) Identify the properties of the indicators: the indicators were divided into three types: Positive Index (the higher was the better), Negative Index (the lower was the better), Neutral Index (target the value around or on a moderate level).

(3) Calculate the standardized value: relatively process the indicator values according to the linear interpolation method [5] so as to get the standardized value.

2.4 Define the indicator weight

The Analytic Hierarchy Process (AHP) [6] was adopted to define the weight of the screening indicators on each level.

2.5 Build a mathematical model of performance evaluation indicator system of rational use of drugs

The linear comprehensive weighted index method [6] was adopted to build a mathematical model of performance evaluation indicator system of rational use of drugs. The computation formulas are as follows:

$$A_i = \sum_{j=1}^{m} (W_j \times Z_{ij}) \quad \text{(j=1,2,...,m)} \tag{1}$$

$$Z(A_i) = \sum_{i=1}^{n} A_i \quad \text{(i=1,2,...,n)} \tag{2}$$

$$U(A_i) = \frac{Z(A_i)}{E(A_i)} \times 100\% \tag{3}$$

In the above formula, $Z(A_i)$ stands for the comprehensive index value of the performance evaluation indicator system of rational use of drugs for each health care institution. $E(A_i)$ stands for

 DOI: 10.1051/bioconf/20170801014

the benchmark index of the performance evaluation indicator system for each health care institution. *U(Ai)* stands for the realization degree of the comprehensive index value of the performance evaluation indicator system for each health care institution. A_i refers to the comprehensive evaluation indicator value of different levels. i refers to the number of health care institutions to be evaluated. j refers to the indicators of performance evaluation on each level. Z_{ij} equals the target value (standardized value) from Ai to Z_j. W_j is the weight value of Z_j on different levels, thus, $\sum_{j=1}^{m} W_j = 1$.

The regional comprehensive performance evaluation on rational use of drugs in a certain department or health care institution can thus be made as well as ranked according to the value of Z. A higher Z means a better performance management on rational use of drugs.

2.6 Data processing

The research is statistically processed with the software SPSS 16.0 for Windows.

3 Results

3.1 Evaluation indicator system

The research group has put forward 103 preselected group indexes, after being screened, 36 of which have been ascertained as indexes on the infrastructure layer. Financial index, patient index, professional process index, and development and blazing new trials index have been adopted as the four dimensions, meanwhile, 4 first level index, 12 second level index and 36 third level index have been adopted as the three levels of the performance evaluation indicator system of rational use of drugs in health care institutions (Table 1). There are 8 financial indexes, 12 patient indexes, 11 professional process indexes, and 5 development and blazing new trials indexes respectively. 17 positive indexes, 4 neutral indexes and 15 negative indexes have been set up according to the index attribute as well.

3.2 Results of index weighting calculation (Table 1).

Table 1. Comprehensive performance evaluation indicator system of rational use of drugs and its weight based on BSC

No.	First Level Index (weight, %)	Second Level Index (weight, %)	Third Level Index (weight, %)	Unit	Index Attribute	Synthetic Weight
1	Financial module (20)	1.Earning power(35)	Ratio of regular fiscal aid revenue to operating revenue (60)	%	positive	0.0420
2			Operating income and expenditure balance ratios (40)	%	neutral	0.0280
3		2.Spending power(35)	Outpatient's average medicine cost per doctor visit (25)	RMB	negative	0.0175
4			Average daily medicine cost of hospitalization (20)	RMB	negative	0.0140
5			Average medicine cost of each discharged patient (25)	RBM	negative	0.0175

6			Average ratio of pharmaceutical affairs cost price to each hundred-yuan drug sales revenue (30)	RMB	negative	0.0210
7		3.Support capability(30)	Coverage of health care institutions' management system of rational use of drugs (60)	%	positive	0.0360
8			Rate of qualified prescriptions (40)	%	positive	0.0240
9	Patient module (35)	1.Medicine attributes (35)	Average medicine variety on each prescription (25)	type	neutral	0.0306
10			Average variety of essential drugs in each prescription (25)	type	Positive	0.0306
11			Average variety of antibiotics in each prescription (25)	type	negative	0.0306
12			Percentage of antibiotic prophylaxis applied on operated patients (15)	%	negative	0.0184
13			Percentage of antibiotics applied on inpatients (10)	%	negative	0.0123
14		2.Image and prestige (25)	Patients' awareness rate of rational use of medicine (35)	%	Positive	0.0306
15			Utilization of generic names of drugs (25)	%	Positive	0.0219
16			Utilization of essential drugs or formularies (40)	%	Positive	0.0350
17		3.Patients' needs	Average cost of each prescription (35)	RMB	negative	0.0490
18			Percentage of essential drugs in each prescription (25)	%	positive	0.0350
19			Cost of antibiotics in each prescription (30)	%	negative	0.0420
20			Average expenses on antibiotics from patients who have accepted antibiotic treatment (10)	RMB	negative	0.0140
21	Project process module (35)	1.Quality of medicine use (45)	Utilization of antibiotics (20)	%	negative	0.0315
22			Utilization of injections (15)	%	negative	0.0236
23			Utilization of steroids (10)	%	negative	0.0158
24			Proportion of prescriptions with over two kinds of antibiotics (30)	%	Negative	0.0473
25			Reports rate on adverse drug reactions (25)	%	Positive	0.0394
26		2.Efficiency	Average consultation time (40)	Min.	Positive	0.0420

DOI: 10.1051/bioconf/20170801014

27		of medicine use (30)	Average time of compounding (40)	Sec.	Positive	0.0420
28			Average days of antibiotic prophylaxis for operating patients	Day	Negative	0.0210
29		3.Medicine follow-up (25)	Percentage of medicine labeling integrity (30)	%	Positive	0.0263
30			Percentage of antibiotic treated patients being sent to pathogenic examinations and drug sensitive tests (10)	%	Positive	0.0088
31			Patients' degree of satisfaction in rational use of drugs (60)	%	Positive	0.0525
32	development and creative module (10)	1.Diathesis and capabilities (40)	Percentage of health technicians with middle and senior titles (60)	%	Neutral	0.0240
33			Thesis published by health technicians per capita (40)	article	Positive	0.0160
34		2.Information capacity (35)	Information process degree of rational use of drugs (100)	%	Positive	0.0350
35		3.Ability to inspire (25)	Coverage of on-the-job vocational training on rational use of drugs (70)	%	Positive	0.0175
36			Annual earnings per capita of the staff members on the payroll (30)	RMB	neutral	0.0075

3.3 Actual application of the comprehensive performance indicator system of rational use of drugs in health care institutions from 2008 to 2012 in a Chinese city

The average composite performance index of rational use of drugs among 21 health care institutions from 2008 to 2012 in a Chinese city was 61.15. The average index of the U Institution was the highest 72.11 in 2012, while in the year 2009, that of the K Institution was merely 44.26 as the lowest. Judging from the average index from 2008 to 2012, the T Institution, scoring 66.84, achieved the highest composite performance index of rational use of drugs, while the B Institution only accounted for 50.23. The top three were the T Institution, the Q Institution and the E Institution in proper sequence with scores 66.84, 66.48 and 65.38 respectively. The last three were the B Institution, the K Institution and L Institution with scores 50.23, 52.15 and 55.93 in file (Table 2).

Table 2. Results of the performance evaluation of rational use of drugs in health care institutions from 2008 to 2012 in a Chinese city

INSTs	Financial module indicator		Patient module indicator		Project process module indicator		Develop and creative module indicator		Comprehensive performance evaluation	
	Performance index	ranks	Performance index	ranks	Performance index	ranks	Performance index	ranks	Total Performance index	ranks
A	11.73	17	16.92	19	21.70	6	6.75	9	63.49	6
B	9.24	21	14.81	21	21.17	7	9.32	1	63.29	7
C	11.96	16	16.89	20	22.47	4	7.05	8	65.19	4
D	11.57	18	19.09	18	23.31	2	6.70	10	66.48	2
E	14.14	4	24.79	1	18.76	17	8.02	4	58.37	17
F	12.77	10	24.43	5	19.10	15	4.04	19	59.53	15
G	13.39	9	23.51	7	18.59	18	2.62	21	56.36	18
H	13.39	8	20.66	15	17.40	20	4.10	18	52.15	20
I	14.20	2	23.26	10	18.07	19	4.49	16	55.93	19
J	12.22	15	23.34	9	19.94	11	5.82	13	60.83	11

DOI: 10.1051/bioconf/20170801014

K	12.43	13	22.22	13	13.40	21	5.01	14	50.23	21
L	11.11	19	20.64	16	19.68	12	7.11	7	60.72	12
M	14.17	3	23.03	11	20.94	9	3.33	20	62.29	9
N	12.51	12	22.42	12	22.72	3	7.69	5	65.38	3
O	12.57	11	19.94	17	21.10	8	6.59	11	62.88	8
P	14.07	5	24.77	3	20.11	10	4.37	17	62.02	10
Q	13.85	7	23.92	6	22.01	5	8.27	3	63.80	5
R	14.25	1	23.45	8	19.05	16	7.33	6	58.65	16
S	13.88	6	21.99	14	19.14	14	8.38	2	59.84	14
T	12.39	14	24.78	2	23.48	1	6.20	12	66.84	1
U	9.96	20	24.54	4	19.23	13	4.85	15	60.36	13

4 Discussion

In China, National Essential Drug System (NEDS) is an important and difficult part of the new health care reform. To set up and manage an effective NEDS, a scientific and effective performance evaluation system of rational use of drugs is much needed. The research integrates experts' suggestions with the current practice of rational use of drugs in health care institutions, so as to set up a representative, guiding, practical, independent and sensitive indicator system. What's more, its practical application in 21 state-run health care institutions from 2008 to 2012 in a Chinese city, has arrived at findings accord with the operational state, which has further revealed the correctness, effects, reliability and scientific nature of the evaluation findings. Thus, being repeatable and feasible, the performance indicator system has been proven a practical and effective performance management quantization tool which is worth of populating.

However, under the background of new national health care reform, it is a huge systematic project to set up a performance evaluation system of rational use of drugs for health care institutions. The various levels of health care institutions and their unique developmental characteristics have directly led to the complexity and difficulty in setting up a performance evaluation system of rational use of drugs. Despite the fact that the current research have managed to choose fewer but better indicators in an objective and overall way, we believe there are significant improvements still to be made. Abundant, repeated investigations and applications wait to be planned and then carried out step by step. Aimed at an ambition of digitalization of performance evaluation indicator system of rational use of drugs, the construction of residents' electronic health files (HER) should be progressively refined, so as to make the evaluation system a valid method in measuring the management effects of health care institutions' performance evaluation of rational use of drugs.

References

1. K. Holloway, L. v. Dijk. Rational use of drugs, Geneva, WHO/EMP/MIE, *The world medicines situation* (2011). http://www.who.int/medicines/areas/policy/world_medicines_situation/WMS_ch14_wRational.pdf.
2. National Monitoring Office for rational use of drugs. *The profile of rational use of drugs in foreign countries*, China Licensed Pharmacist, **8**, 14-16 (2011).
3. Ministry of Health of P.R.China, *Progress in essential drug system in China* (2011). http://www.moh.gov.cn/publicfiles/business/htmlfiles/mohywzc/s7652/201110/53137.htm
4. T. D. Sun, Zh. Q. Gu, L.Wang. etc. *A study on the impact of essential medicines system on medical staff's behaviour*, Chinese Hospitals, **16**, 23-25 (2012).
5. L. M. Zhang, L. J. Huang, J. L. Xia. etc. *Study on target standardization of comprehensive evaluation*, Chinese Journal of Health Statistics, 11, 1-4 (1994).
6. Y. R. Ren, Q. O. Liu, (eds.), *Fundamental technology of health management --- pragmatic operational research and systems engineering*, Joint publishing house of BMUC&PUMC, 177-190(1993).

Study of the influence on emotional work of contract nurses on Job Burnout

Changping Song, Weiwei Fan, Chenxu Zhao and Ling Gao[a]
Beihua University, Jilin City, Jilin Province, 132021 China

Abstract. Discuss the influences of contract nurse' emotion work on job burnout and provide theoretical basis for reducing the job burnout of perception level which caused by the emotion work of contract nurse. **Methods:** The data which comes from the questionnaire survey about the Chinese edition of Maslach job burnout scale for 298 contract nurse is used for statistical analysis by SPSS11.5 software. **Results:** The current contract nurse is more inclined to "surface acting" on working and the nurses have a lower level on "depersonalization" ($P < 0.05$, the difference is statistically significant) and the more "deep acting" are used, the lower level of job burnout. There is a various factors' correlation between contract nurse's emotion work and job burnout. **Conclusions:** The "surface acting" could increase job burnout, "deep acting" and emotion expression requirements could increase personal job satisfaction.

1 Introduction

Emotional work refers to the individual who produces a kind of facial or body expressions which can be observed by public that made by the emotional management, including the surface acting, deep acting and emotional expression requirements. Job burnout is individual who cannot successfully deal with the working pressure after the extremely reaction and the individual is under pressure for a long time experience with the emotion, attitude and behavior of state failure, including emotional exhaustion, depersonalization and diminished personal accomplishment[1]. The development of market economy and the increase of social competition make our country is entering the peak of job burnout, more happened in the service staff, among which the medical staff in the second place. Contract system nurses, namely, the employment of nurses referring to the hospital are not a member of the official establishment, but the hospital signed a labor contract with the labor contract nurses. There is a gap between the contract nurses and formal preparation nurses in wages, insurance, promotion and so on. resulting in Contract nurses' job satisfaction levels drop, turnover rate increase, Seriously affect the stability of the nursing team, also bring adverse effect to the quality of care. This project investigates and analysis about the effect of contract nurses' emotional work on job burnout, in order to provide theoretical basis for reducing the perceived level of occupational burnout caused by the emotional work of contract nurses.

[a] Corresponding author : 364760038@qq.com
This subject is the social development project of Jilin science and Technology Bureau, Jilin province.

DOI: 10.1051/bioconf/20170801015

2 Research objects

In June 2014, a questionnaire survey was used to investigate the contract system of 298 nurses from different departments in the Affiliated Hospital of Beihua University. were female, age 22-37 years old average 29 years old, department of internal medicine nurses 158, accounting for 53.2%, surgical nurses 140, accounting for 46.8%.

3 Method of investigation

First of all, trained the project team, to grasp the Emotional Workload Scale, the assessment method of nurses' Job Burnout Scale Questionnaire and the main content. The nurses' Emotional Workload Scale and nurses' Job Burnout Questionnaire were distributed to the 298 contract nurses who were investigated. The nurses' Emotional Workload Scale and nurses' Job Burnout Questionnaire were distributed by researchers to the 298 contract nurses who were investigated, and explained the purpose, details and filling method of the survey with consent of the person being investigated. The survey subjects completed the questionnaire in an anonymous manner, were collected questionnaire right away to ensure accuracy. Last Statistical analysis the obtained data by SPSS11.5 software package

4 Investigation tool

Questionnaire is the Chinese version of Maslach Burnout Inventing[2]. Referring to Ye Zhihong[3] and other domestic norm of the diagnosis standard of occupational burnout emotional exhaustion is greater than or equal to 27,depersonalization is greater than or equal to 8,personal accomplishment is less than or equal to 24 as the critical value. Li YongXin[4] divides burnout degree into four types:(1) zero burnout: in the assessment of occupational burnout, three factor scores were lower than the critical value; (2) mild burnout: in the assessment of burnout, a factor score higher than the critical value; (3) moderate burnout: in the assessment of burnout, two factor scores were higher than the critical value; (4) high burnout: in the assessment of occupational burnout, three factor scores were higher than the critical value.

5 Results and discussion

5.1 Nurses emotional work condition

Table 1. Emotional work of contract nurses (n= 298)

Emotional labor scores	Maximum	Minimum	Average+/-Standard deviation
Surface acting	2.53	4.14	4.58±0.36
Deep acting	1.47	3.48	3.09±0.44
Emotional expression requirements	1.67	4.78	3.76±0.34

Table 1 shows that current contract nurses are more inclined to work with surface acting, especially when nurses were suffer from severe emotional events, or at a high degree of job burnout or excessive emotional state. They don't want to do the positive cognitive adjustment on the subjective, but adopt appropriate emotion by surface acting. There are also some contract nurses tend to a pleasant mood on surface because of the needs of nursing work, but no really treat patients from deep inside.

5.2 Comparison nurses' occupational burnout scores with Maslach(n=309)

Table 2. Comparison Contract nurses' occupational burnout scores with Maslach

	The number of	emotional exhaustion	depersonalization	Loss of personal accomplishment
Maslach	309	21.19±9.53	7.12±2.21	36.54±7.34
experiencegroup	149	29.53±11.02	9.37±2.14	39.02±7.98
control group	149	28.98±11.84	9.65±2.23	39.12±8.06
P		0.041	0. 018	0. 023

Table 2 shows that contract nurses performance moderate emotional exhaustion, personal work without a sense of accomplishment and low depersonalization. Compared with the Maslach, the depersonalization of the nurse group is seriously low. ($P<0.05$) Main performances are emotional fatigue, lack of sympathy for patients, be incompetent for the job, and lack of personal accomplishment, etc..

5.3 The relationship between emotional work and job burnout of contract nurses

Table 3. The relationship between emotional work and job burnout of contract nurses

Project	emotional exhaustion	depersonalization	Loss of personal accomplishment
Surface acting	0.165	0.121	0.144
Deep acting	0.015	0.021	0.147
Emotional expression requirements	0.161	0.272	0.212

Table 3 shows that there is a correlation between emotional work and job burnout of contract nurses. The more contract nurses use the deep acting, the lower gain the level of job burnout; The more contract nurses use the surface acting, the higher gain the level of job burnout.

6 Conclusion

The surface acting of emotion work has a significant predict effect on contract nurse's emotional exhaustion of job burnout and the expression of depersonalization. The more use of surface acting means the nurse in the work often camouflage the emotions and show excessive professional nursing behaviors. This expression behavior is inconsistent with cognitive evaluation and emotional experience will decrease self-sense of reality. The individual needs to pay more psychological resources which make it easier to fall into mental fatigue and to some extent is easier to burnout, aggravate emotional exhaustion and depersonalization. However deep acting pays more attention to inner experience, more adjustment from the perspective of cognitive evaluation to achieve the request of organizational emotional expression to individuals, which reflects that contact nurses coordinate individual inner feelings with their external emotions, the psychological processing will make people have more kinds of moods, it is helpful for nurse to manage emotions in the work better, alleviate burnout, and improve the nurse's personal achievement. Therefore, the "surface acting" could increase job burnout, "deep acting" and emotion expression requirements could increase personal job satisfaction.

References

1. Elloy DF, Terpening W, KohlsJ. Aeausal model of burnout Among selfmanaged-work teammembers[J].*The Journal of Psychology*,2001,135(3):321-334.
2. Feng Ying, Luo Hong, Ye Zhihong,ect. Job Burnout of Nurses[J].*Chinese Mental Health Journal*,

2004,18(7):477-479.
3. Ye Zhihong, Luo Hong, Jiang Anli. Diagnostic standard and norms of Maslach Burnout Inventory for nurses In Hongzhou[J]. *Chinese Journal of Nursing*, 2008,43(3):207-208.
4. Li Yongxin, A comparative study of job burnout among two groups: based on the perspective of integration [D].Doctoral Dissertation of East China Normal University,2005:74-76.

Survey on demand of the aged people for college volunteers in home nursing care service

Meng Tian, Wenjuan Zhong[a] and Jia Guo

Health Science and Nursing College of Wuhan Polytechnic University, Hubei 430023 China

Abstract: Objects: To study the demand of the aged community residents for college volunteers in home nursing care service and its influence factors. **Methods:** Questionnaires on the aged residents' expectation and requirements for college volunteers in home nursing care was prepared on basis of studies of reference books, study materials and pre-investigation, and then the questionnaires was completed by means of convenient sampling of 200 aged residents. **Results** The demand and expectation of the elderly for college volunteers in health care and spiritual comfort was strong. The old people who have more children, living alone and often feel lonely are more likely to need college volunteers. **Conclusions:** Volunteers should be arranged reasonably according to the needs of the aged people, and priority should be given to those who have more children, living alone and often feel lonely.

1 Introduction

Home nursing care service is the key to solve the pension problem in China and also the direction for future development of China's pension. Home nursing care service refers to the government and social communities arrange care workers with basic nursing skills to provide home nursing care services for the aged, including daily life care, housekeeping service, rehabilitation nursing and spiritual comfort services. More researches have been made on home care services, but not much in respect of home care service provider. Some scholars believe that home care solely provided by the government is far beyond adequacy, we should give full play to the market, private non-enterprise units and social workers [1-3]. Several scholars point out that, based on qualitative analysis on home care, it is imperative to introduce volunteers[4-6]. Voluntary activity is a kind of selfless activity contributed by individuals or a group of people in absence of financial or material benefits with the purpose to improve the quality of life of human beings [7]. Home care for the aged people provided by the volunteers for free, thus the community can save money for the more difficult groups. As a result, more old people can access and enjoy home care service[8]. This research is conducted to investigate the needs of the elderly for college volunteers to participate in home care services. The details are as follows.

[a] Corresponding author: wjzhong8616@126.com
The content of this essay is one of the achievements in relation to philosophy and social science from major research project *Interactive Research on Community Development and College Education* by Hubei Provincial Department of Education, the item number is 2015z33

 DOI: 10.1051/bioconf/20170801016

2 Object and Method

2.1 Object of research

The object of research was chosen from four residential communities of Changqing Garden of Wuhan. Standard: (a) above 60 years old; (b) aware and easy to communicate with; (c) volunteer to be the object of the research. Exclusion: (a) below 60 years old; (b) unconsciousness or difficult to communicate; (c) unwilling to be the object of the research.

2.2 Method of study

The purpose and significance of the investigation were explained by trained investigators so as to obtain their consent and cooperation. The questionnaires were filled by the object of the research independently, and the researchers filled the questionnaires for those who have difficulty in writing according to their answer. Totally 200 questionnaires were handed out and 195 valid questionnaires were collected back, 97.5% valid.

2.3 Means of research

The items in the questionnaire were drawn up in reference to a large amount of research papers and work as well as pre-investigation among the object group of people, which include: (a) general information of the aged people, such as gender, age, marital status, educational level, occupation, living independence, economic conditions, relationship with family members, caregivers, feeling of loneliness and so on; (b) the needs for college volunteers: including life care, medical care, spiritual comfort and legal consulting; (c) other: degree of expectation of volunteers for home care and positive aspects of the volunteers, etc..

2.4 Statistical Methods

A database was created by using SPSS22.0 software to make statistical description over the data collected and the independence of the qualitative data was proved by means of chi square test. The factors that affect the choice will of the aged people were analyzed by means of two element regression analysis, and the statistical data was useful in case $P < 0.05$ for variance.

3 Results

3.1 General information of the aged people

The age of people subject to the research ranged from 60 to 96 years old, or 8.5 years above or below 76.5. And the general information of the aged group was shown as follows. **Gender:** 74 males; 121 females. **Marital status:** 110 live with spouses; 2 divorced; 83 bereft of their spouses. **Education level:** 59 of or below primary school; 99 of high school or secondary specialized school; 32 of college or above. **Occupation before retirement:** 20 farmers; 110 enterprise employees; 46 public institutional staff; 19 others. **Living independence:** 153 live independently; 42 partially-independent. **Status of economics:** 60 live with surplus money; 100 live with roughly enough money; 34 live with slight economic difficulty; 1 live with economic difficulty. **Payment of medical treatment:** 186 covered by medical insurance; 7 pay by themselves or their spouses; 2 pay by their children. **Children:** 1 without child; 34 with one child each; 70 with two children each; over 80 with three or more children. **Way of living:** 48 live alone; 76 living with spouse; 63 live with children; 8 live in pension institutions; **Whether have caregivers:** 54 unattended, 81 persons by their spouses, 56 persons by children, 4 persons by other ways; **Feelings of loneliness:** often 30 people, occasionally 69 people, never 97.

3.2 The needs of the aged people for college volunteer in respect of home care

Among expected services by college volunteers, medical care or nursing was up to 36.1% in proportion, while spiritual comfort 33.33%, daily living care 20.8% and legal consulting 9.8%.

3.2.1 The need of the aged people for college volunteers in respect of medical care

The need for health lecture was of highest proportion of 55.9%, followed by rehabilitation training 51.3%, regular physical examination 37.9%, emergency relief 34.9%, and accompany to see a doctor 30.7%.

3.2.2 The need of the aged people for college volunteers in respect of spiritual comforts

Art Gala (60%) accounted for the highest in need, followed by tea Forum (54.4%), calligraphy and painting contest (33.3%), chat to relieve boredom (31.8%), psychological counseling (18%) and reading (7.2%).

3.2.3 The aged people's demand for college volunteers in life care

Housekeeping (40%) accounted for the highest in proportion of demand, followed by food delivery (30.8%), wash clothes (18.5%), shopping (15.4%), pick out (13.3%) and bath (3.6%).

3.2.4 The need of the aged people for college volunteers in respect of law service

Legal assistance was of 19.5%, legal advice 17.9%, legal lectures 13.4%.

3.3 Influential factors of the aged people's demand for college volunteers to participate in home nursing care service

3.3.1 Single factor analysis on the aged people's demand for college volunteers to participate in home nursing care service

The research indicated that factors affecting the aged people's demand for college volunteers to participate in home nursing care include: ways of living, number of children, loneliness, self-care ability in daily life, with their discrepancies having statistical significance. Details are shown in Table 1 as following.

Table 1. Single Factor Analysis on the Demand of the Aged people's for College Volunteers to Participate in Home Nursing Care Service(n=195)

Item		wish(%)	Not wish(%)	χ^2	P
number of children	1	21(61.8)	13(38.2)		
	2	60(85.7)	10(14.2)	13.841	0.008
	3	46(90.2)	5(9.8)		
	Above 3	34(87.2)	5(12.8)		
Ways of living	Alone	47(97.9)	1(2.1)		
	With spouse	58(76.3)	18(23.7)	16.585	0.001

	With children	53(84.1)	10(15.9)		
	Endowment inst.	4(50)	4(50)		
Loneliness	Often	29(96.7)	1(3.3)	11.471	0.003
	Seldom	61(89.7)	7(10.3)		
	Never	72(74.2)	25(25.8)		
Self-care ability	entirely	121(79.1)	32(20.9)	8.052	0.002
	partially	41(96.7)	1(2.4)		

3.3.2 Binary logistic analysis on the aged people's demand for college volunteers to participate in home nursing care service

Taking whether the aged people's wish college volunteers to participate in the home nursing care as dependent variable and ages, gender, marital status, number of children, ways of living, loneliness, self-care ability in daily life, educational background as independent variable (assignment of number of children: 1 child=0, 2 children=1, 3 children=2, above 3 children=3; assignment of ways of living: living alone=0, living with spouse=1, living with children=2, live in endowment institutions=3; assignment of loneliness: often=0, seldom=1, never=2), binary logistic analysis was carried out with 3 variables in this binary regression equation. The results showed that the elderly who have more children, live alone and often feel lonely are more likely to need college volunteers. For details, please refer to the following Table 2.

Table 2. Binary Logistic Analysis on the Aged People's Demand for College Volunteers to Participate in Home Nursing Care Service(n=195)

Item	B	S.E.	Wald	df	Significance	Exp(B)
Number of children	0.519	0.214	5.863	1	0.015	1.681
Ways of living	-0.644	0.231	7.755	1	0.005	0.525
Loneliness	-1.039	0.374	7.722	1	0.005	0.354
Constant value	4.041	1.345	9.031	1	0.003	56.886

4 Discussion

The aged people are in great demand for volunteers in health care and spiritual comfort, which is in line with the research results concluded by Guo Guanjun[9], Gao Jing[10], etc. It can be seen that the aged people became more aware of health care along with the improved living standard of people. The research indicated that the aged people had more demand for college volunteers to provide health knowledge explanation and rehabilitation exercises in respect of health care, almost the same with the research results made by Zhao Rong[11]. However, the demand for periodic physical examination, first aid service, seeing doctor accompanied by someone was not so high, which is probably due to the fact that most of the aged people are covered by medical care insurance and have access to periodic physical examination every year. Meantime, affected by traditional ideas, many aged people prefer looking doctor accompanied by their family members and obtaining help by acquaintance surroundings in case of emergency. Furthermore, as the decline of body function, partial loss of social function and narrow social relationship circle, the aged people yearn for college volunteers to conduct some recreational activities to enrich their life. Regarding spiritual comfort, the aged people have higher demand for arts festival and tea party, while have lower demand for chatting and psychological consultation, which are possibly due to the fact that 78.5% of the aged people participating in this questionnaire can entirely look after themselves, can communicate with people surrounding them and thus have little interest in chatting with college volunteers who have generation gap with the aged people.

The aged people who have more children, live alone and often feel lonely are more likely to need college volunteers. The self-care ability was not in the regression equation, but the single factor analysis indicated that it had connection with the aged people's demand for college volunteers. Study conducted by Xing Fengmei[12] also suggested that self-care ability is one of the reasons why the aged people need home nursing care service. From our point of view, 78.5% of the elderly participating in this questionnaire can entirely look after themselves, only a few of the elderly are people who have partial self-care ability and have no self-care ability at all. Therefore, we should also consider the self-care ability. The aged people who have more children are more eager to receive home nursing care by college volunteers, which is probably due to the fact that it is widely believed the more children the aged people have, the more care the aged people received from their children thus they have little need for college volunteers. However, the survey conducted by Chen Zhike[13] indicates that the hypothesis of "the fewer children the aged people have, the more demand they have for home nursing care" failed in statistics verification($P=0.884$). We view that the more children the aged people have, the easier the custody dispute brought, the more home nursing care services are wanted: children themselves would shuffle off the custody responsibility for their parents and the elderly could not be taken good care. Given this, further study is needed.

Compared with other aged people, those who live alone and often feel lonely are more eager for college volunteers for home nursing care, which is in conformity with the research result conducted by McCallion[14]. Besides, the survey also indicated that 79.5% of the elderly wish college volunteers to participate in the home nursing care service. As you see, college volunteers play a non-negligible role in home nursing care service, especially for the aged people who have more children, live alone and often feel lonely. Among the home care services by college volunteers, the top three the aged people valued most are service attitude, specialized knowledge, mainly means medical science knowledge and duty on time. Therefore, training relating to those three aspects should be enhanced to college volunteers.

The aged people's need for daily life care is lower than that of medical care and spiritual comfort. The survey conducted by Wu Jialin[15] shows that daily life care by their family member is more satisfactory than that of volunteers. We viewed this as one of the reasons why the aged people have little demand for volunteers and 78.5% of the aged people in this questionnaire are those can look after themselves entirely.

It was found through the survey that the aged people have little demand for legal advice, which is probably because the aged people seldom encounter legal disputes in their daily life.

5 Conclusions

To make a long story short, the demand by the aged people for college volunteers in home nursing care is great, especially for medical health care and spiritual comfort. Number of children, ways of living and loneliness are the main factors affecting demand of the aged people for college volunteers to participate in home nursing care service. In view above, the community can arrange college volunteers nearby to provide regular service for the aged people and give priority to those who have more children, live alone and often feel lonely. To work out guideline and policy for college volunteers in home nursing care, a great deal of data involving big samples and more regional areas are required for further study.

References

1. Cui Yan. Government roles in the pattern of home endowment[J]. *Business Culture*, 2008(8):101.
2. Ru Mengdan, Ji Huaxiong. City of a people-run non-enterprise unit endowment service function

DOI: 10.1051/bioconf/20170801016

analysis[J] .Journal of Hulunbeier University,2015,**2**(23): 16-19.
3. Cheng Wei, Liu Haiying, Zhang Yuqi. Towards a Programme about Social Workers' Involvement in Home-Based Care Services for the Elderly in Community[J].*Theory and Modernization*,2013(1):58-61.
4. Zheng Shixiang. Thinking based on the aging society under the background of the arts students to participate in the home endowment services[J].*Learning week*,2015(5): 238.
5. Shi Wei, Xie Yu. The Urban Elderly' Choice of Home Care Providers and the Influencing Factors: From the Perspective of Welfare Pluralism[J]. *Northwest Population Journal*,2015,**1**(36): 48-54.
6. Xu Fengyun. Introduction to the cultivation of community elderly home care services volunteers[J].*Seek Medical and Ask the Medicine*,2012,**10**(7):614.
7. Maranta A, Sladowski PS (2010) The Changing Culture of Volunteering: A Skills-based Approach to Volunteer Involvement. A discussion paper by Volunteer Benevolves, Canada.
8. Yuan Xiaoxiao\, Keyonghong, Wu Minlun, et al. Probe into feasibility of application of university resources in community home care service[J]. *Chinese Nursing Research*,2013,**27**(6):1788.
9. Guo Guanjun, Wang Huimin, Wang Fuguo, et al. The old people life quality of yanji city and the present situation of old-age care demand that occupy the home[J].*Journal of Qilu Nursing*, 2015,**21**(6):56.
10. Gao Jing, Bai Dingxi, Wu Chenxi,et al. Investigation and study of the status and demand on home endowment service in four cities of sichuan province [J].*Chinese Journal of Convalescent Medicine*,2015,**24**(1): 92-95.
11. Zhao Rong, Tang Fengping, Li Ya.Survey of home aged care needs of urban elderly in Zhuzhou city and its countermeasure research[J]. *Chinese Nursing Research*,2015,**29**(4):136.
12. Xing Fengmei, Wang sudong, Wang Liwei,et al. Influencing factors of home care service demand in housebound elderly people[J].*Chinese Journal of Public Health*,2014,**30**(5):641-645.
13. Chen Zhike, Ma Shaozhen. The research of factors about the elderly demanding home care service-Based on the social investigation in Hunan Province[J].*Journal of Central South University*,2012,**18**(3):26-30.
14. McCallion, G.M.& Stout, R.W.(1992).A comparison of community care of elderly people in Wales and Northern Ireland.Aging Clin. Exp. Res,4,348.
15. Wu Jialin, Wang Junqiao, Lu Meiling, et al. Care needs of community-dwelling adults aged 80 years and older: meeting of the needs[J].*Journal of Nursing Science*,2013,**28**(12):89-91.

Fluorescence spectral studies on interaction of SEGS with BSA

Fuchun Si[1], Wenbin Wang[1,2,a], Xuejie Song[1], Xin Zhao[1] and Lulu Hao[1]

[1]*Laboratory of molecular biology for TCM, Henan University of Chinese Medicine, Zhengzhou 450046, China;*
[2]*National Key Laboratory of Cryospheric Sciences/Tien Shan Glaciological Station, Cold and Arid Regions Environmental and Engineering Research Institute, Chinese Academy of Sciences, Lanzhou 730000, China*

Abstract. The seeds of Gleditsia sinensis is one of the most effective drugs for the treatment of esophageal carcinoma. To definite the active ingredients of the seeds of gleditsia sinensis, SEGS was extracted from the seeds of Gleditsia sinensis. Investigations the interaction between the active ingredient of traditional Chinese medicine and BSA is helpful for explaining the medicine mechanism. Under the simulative human physiological conditions, the interaction between SEGS and bovine serum albumin (BSA) was investigated by fluorescence spectroscopy. The quenching mechanism of SEGS with BSA is dynamic quenching, SEGS bound with BSA by van der Waals forces or hydrogen bond. The binding constants and the number of binding sites between SEGS and BSA at different temperatures (298 and 308 K) were obtained.

1 Introduction

Bovine Serum Albumin (BSA) is the most abundant soluble protein in the blood plasma, and can help to keep colloidal osmotic blood pressure and act as the transportation and disposition of many drugs, and is capable of binding, delivering an extraordinarily diverse range of endogenous and exogenous compounds like fatty acids, nutrients, steroids, certain metal ions, hormones, enzymes, surfactants and a variety of therapeutic drugs, through the bloodstream to their target organs[1].Therefore, BSA plays an important role in the transportation, distribution, free concentration, excretion, metabolism and interaction with the target tissues of these ligands[2]. Investigations the interaction between the active ingredient of traditional Chinese medicine and BSA is helpful for explaining the medicine mechanism. It is especially useful for select anti-medicine vitro, and can giving theory guide for anti-medicine design[3].

Gleditsia sinensis is a widely used traditional Chinese herb that belongs to the Caesalpinioideae family[4]. For the past 10 years, our group have carried out purposeful studies of anti-esophageal

[a] Corresponding author: contributepaper@126.com
This research was supported by the special grade of the financial support from the China Postdoctoral Science Foundation (Grant No.2016T90669), the National Science Foundation for Post-doctoral Scientists of China (Grant No.2015M582189), the National Science Foundation (Grant No.81550014), the Henan Postdoctoral Sustentation Fund, China (Grant No. 00104256), Jinshui District Science and Technology Bureau Social Program (Grant No. 20132602).

carcinoma medicine from over 2000 kinds Chinese traditional herbs. The seeds of Gleditsia sinensis (Chinese honey locust) is one of the most effective drugs for the treatment of esophageal carcinoma[5]. The seeds of Gleditsia sinensis were performed extraction by 95% ethanol. The crude extract was dried and suspended in water and successively treated with chloroform, ethyl acetate and n-butanol. The n-butanol fraction showed good cytotoxic activity to the esophageal carcinoma cells, Eca109, Ec9706 and TE1, by MTT assay. And then the n-butanol fraction subjected to Dianion HP-20 column and successively eluted with H_2O, 30%, 50%, 70%, 90%(v/v)EtOH-H_2O, respectively. The 70% EtOH fraction was chromatographed on silica gel. Sub-fraction was subjected to repeating column chromatography on silica geland further purified with recrystallization, obtained colourless power(SEGS).

2 Experimental

2.1 Apparatus and Reagents

F-7000 fluorescence spectrophotometer (Hitachi, Japan), pHS-3C digital pH-meter (Shanghai REX Instrument Corp., China) Bovine serum albumin, BSA (Beijing Olympic nemesis Biological Technology Co. Ltd.China, relative molecular mass 65000), was dissolved daily in 0.1 mol/L Tris–HCl buffer solution (pH =7.3) to prepare a solution (1×10^{-5}mol/L), SEGS to prepare a solution (80mg/L), All chemicals were analytical-reagent grade. All solutions were prepared with double-distilled water.

2.2 Methods

Fluorescence spectra were carried out on a Hitachi F-7000 fluorescence spectrophotometer using 1.0 cm quartz cells. The excitation wavelength of BSA studied in this work was 280 nm.

3 Results and discussion

3.1 Fluorescence quenching of BSA

Fluorescence technique is wide spread application in investigating of interactions between the active ingredient of traditional Chinese medicine and protein molecules[6]. It is well known that BSA has intrinsic fluorescence due to the presence of amino acids, mainly tryptophan (Trp) and tyrosine (Tyr) residues. The interaction between SEGS and BSA has been mornitored upon exciting BSA at 280 nm where both Trp and Tyr residues get excited. The variations of emission spectra was shown in Figure 1. The fluorescence quenching spectra of solutions containing a BSA fixed concentration and different concentrations of SEGS at 298 K and 308 K, It can be observed that the fluorescence intensity of BSA decreases regularly with the increase addition of SEGS, under the experimental condition, the SEGS does not produce any emission in the specified range of study, and there is no significant emission wavelength shift. These results suggest that SEGS interact with BSA can quench BSA intrinsic fluorescence.

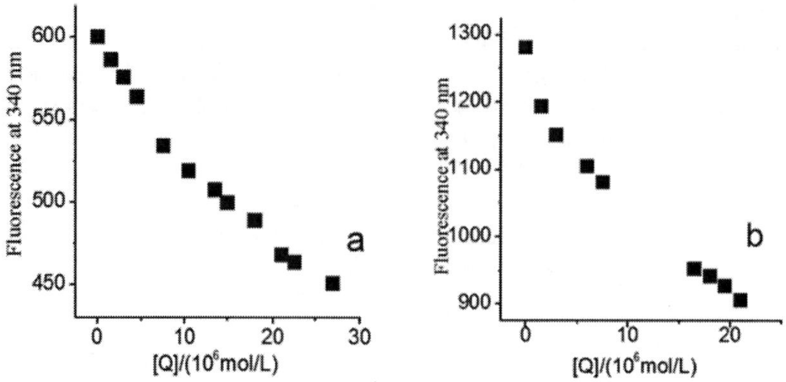

Figure 1. Changes in the emission spectra (λex = 280 nm) of the BSA (1.0×10^{-5} mol/L) upon increasing concentrations of SEGS (a) 298 K and (b) 308K

3.2 Determination of quenching mechanism

There are two quenching types in characterizing the mechanism of the interaction between quenchers and macromolecules: static and dynamic quenching. Static quenching refers to the formation of a non-fluorescence fluorophore-quencher complex. With temperature increasing, the stability of fluorophore-quencher complex reduced, the quenching constant decreased. However, opposite feature exists in dynamic quenching, Dynamic quenching refers to the quencher diffusing towards the fluorophore during the lifetime of the excited state and, upon contact, the fluorophore returns to the ground state without emission of a photon. Dynamic quenching mechanism was probed using the Stern-Volmer equation[7]:

$$F_0/F=1+K_q\tau_0[Q]=1+K_{sv}[Q] \quad (1)$$

Where F_0 and F represent the fluorescence intensities of BSA in the absence and in the presence of the external quencher, Kq is the bimolecular quenching rate constant, K_{SV} is the Stern–Volmer quenching constant, τ_0 is the average lifetime of biopolymers without quencher, the fluorescence lifetime of the biopolymers is 10^{-8} s [8], [Q] is the concentration of the quencher.

This work studied the fluorescence spectrum of the action between BSA and SEGS, obtained the plot of F_0/F against [Q] at different temperature shown in Figure 2.

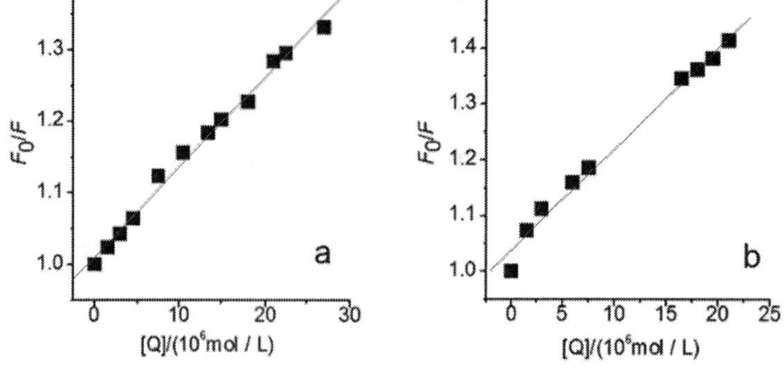

Figure 2. Stern-Volmer plot for the quenching of BSA by SEGS at different temperatures (a) 298 K and (b) 308K

Equation (1) was applied to determine K_{SV} by linear regression from a plot of F_0/F against [Q], hence, Kq can be caculate by $Kq=K_{SV}/\tau_0$. As shown in Table 1.

Table 1. Constant of the systems of interaction between SEGS and BSA at different temperatures

T/K	Ksv/(L/mol)	Kq/(L/mol•s)
298	1.25×10^{-2}	1.25×10^{6}
308	1.81×10^{-2}	1.81×10^{6}

For the SEGS–BSA systems, the values of K_{SV} and $Kq=K_{SV}/\tau_0$ obtained from the plots at 298 K and 308 K are shown in Table 1. According to the literatures[9-10], which suggest that the fluorescence quenching mechanism is mainly dynamic quenching.

3.3 The binding constant Kb and the number of binding sites n

The relationship between fluorescence quenching intensity and the concentration of quenchers can be used to obtain the binding constant and the number of binding sites[11].

$$\mathrm{Lg}(F_0\text{-}F)/F=\mathrm{Lg}K_b+n\mathrm{Lg}[Q] \quad (2)$$

The values of K_b and n at 298K and 308K were obtained from the double logarithm regression curve (Figure 3).
The values of K_b and n at 298K and 308K are listed in Table 2.

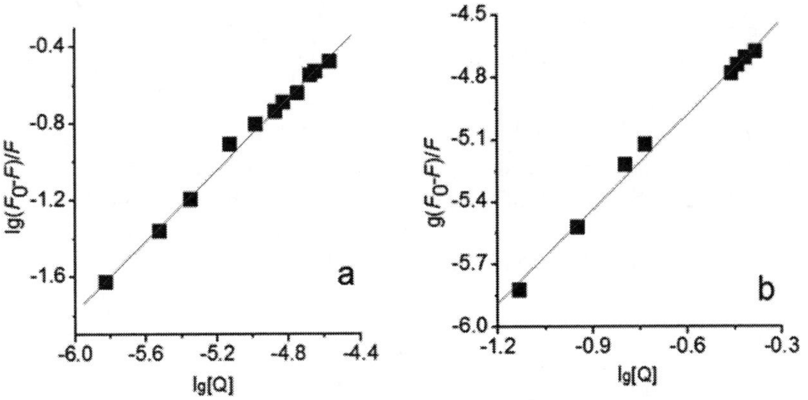

Figure 3. The plots of $\lg(F_0\text{-}F)/F$ vs. lg[Q] at different temperatures (a) 298 K and (b) 308K

The values of K_b and n at 298K and 308K are listed in Table 2.

Table 2. The binding constant K_b, and the number of binding sites n

T/K	double logarithm regression curve	r	k_b	n
298	$\lg(F_0\text{-}F)/F=3.7811+0.9273\lg[Q]$	0.99746	6.041×10^{2}	0.9273
308	$\lg(F_0\text{-}F)/F=2.6680+0.6565\lg[Q]$	0.99648	4.656×10^{2}	0.6565

The n values are nearly 1, and thus indicate the existence of a single binding site in BSA for SEGS. The interactions of SEGS with BSA decreased when rising the temperature, which suggest that interactions of SEGS with BSA were exothermic[12].

3.4 Binding modes

The acting forces between drug and biomolecule are composed of weak interactions of molecules such as hydrogen bond formation, van der Waals forces, electrostatic forces, and hydrophobic interaction. The thermodynamic parameters could be calculated by equation (3), (4) and (5).

$$\mathrm{Ln}(K_2/K_1)=\Delta \mathrm{rHm}(1/\mathrm{T1}-1/\mathrm{T2})/\mathrm{R} \quad (3)$$

$$\Delta \mathrm{rGm}=-\mathrm{RTLnK} \quad (4)$$

$$\Delta \mathrm{rSm}=(\Delta \mathrm{rHm}-\Delta \mathrm{rGm})/\mathrm{T} \quad (5)$$

The thermodynamic parameters are presented in Table 3.

Table 3. Thermodynamic parameters for the association of SEGS with BSA

T/K	ΔrHm/(KJ/mol)	ΔrSm/(J/K)	ΔrGm/(KJ/mol)
298	-19.78	-13.42	-15.87
308		-13.44	-15.73

The negative signs for ΔrHm, ΔrSm and ΔrGm reveal that the binding processes are influenced by temperature, low temperature are spontaneous, high temperature are non-spontaneous. According to the literature[13,14,15], The negative ΔH and ΔS values suggest the van der Waals forces or hydrogen bond formation exist in SEGS and BSA.

4 Conclusion

The interaction between SEGS and BSA has been investigated under simulated physiological conditions (pH 7.3, ionic strength 0.1 mol/L) using fluorescence methodology. The constant Kq of the SEGS and BSA is less than 2×10^{10} $\mathrm{L{\cdot}mol^{-1}{\cdot}s^{-1}}$, which suggest that the fluorescence quenching mechanism of BSA by SEGS was consistent with dynamic quenching. The binding reaction was influenced by temperature for Δ rHm, Δ rSm and Δ rGm is negative. ΔH and ΔS is negative suggest that Van der Waals forces or hydrogen bond play a major role in the SEGS–BSA interaction.

References

1. A. Jahanban-Esfahlan, V. Panahi-Azar, Food chemistry, 202 (2016).
2. J. Zhang, H.H. Sun, Y.Z. Zhang, J. Solution Chem, 41 (2012).
3. Y. Li, Y. Gu, H. He, Y. Zhou, Acta Chimica Sinica. 70, 143-150 (2012).
4. D. N. Oleynikov, A. V. Rohin, Applied Biochemistry and microbiology. 46, 103-107(2010).
5. F. C. Si, World Chinese Journal of Digestology. 28, 3139-3145 (2008).
6. R. Suparna, D.Somnath, P. Chiranjit. Spectrochimica acta. Part A, Molecular and biomolecular spectroscopy, 166, 84-94(2016) .
7. X. Zheng, M.P. Ye, Y. Li, Journal of China University of Metrology, 1 (2015).
8. S. Y. Bi, Y. T. Sun, C. Y. Qiao, H. Q. Zhang, C. M. Liu, J. Luminescence.129, 541–547 (2009).
9. H. Cao, D. Wu, H.Wang, M. Xu, Spectrochim. Acta A 73, 972–975 (2009).
10. F. Cui, Y. Yan, Q. Zhang, X. Yao, G. Qu, Y. Lu, Spectrochim. Acta A 74, 964–971 (2009).
11. L. N. Hordge, K. L. McDaniel, D. D. Jones, et al, Talanta, 152 (2016).
12. F. Y. Zhang, Y. N. Ni, Acta Chimica Sinica. 70, 1379-1384 (2012).
13. F.L. Cui, J.L. Wang, Y.R. Cui, J.P. Li, Anal. Chim. Acta, 571 (2006).
14. N. Nanjundan, P. Selvakumar, R. Narayanasamy, et al. J. Photoch. Photobiolo. B, 141 (2014).
15. L.Wang, D. W. Zhao, L. J. Zi, et al. J. Henan Normal University. 40, 87-90 (2012).

 DOI: 10.1051/bioconf/20170801018

Study on the Correlation between job adaptation obstacle and perceived social support of community nurses in Changchun

Wei Meng, Yujin Liu[a], Wenjing Sun, Xiufeng Zhao, Fengge Dong, Guixiu Yan, Bing Yu and Zhuang Cong

Welfare College of Nursing of College of Humanities&Northeast Normal University, 130117,Changchun, China

Abstract. Objective: To investigate the present situation of job adaptation and perceived social support of community nurses in Changchun, and to explore the relevance between them, for the purpose of providing the basis for community nursing managers to implement effective human resource management. Methods: A general demographic information questionnaire, job adaptation obstacle scale and perceived social support scale were used to investigate 290 community nurses in Changchun. Results: The score of job adaptation obstacle was 20.85±5.18; the score of perceived social support was 64.25±10.32, the score of support in the family was 20.01±3.58, and the score of the support out of family was 42.57±6.86; the job adaptation obstacle was negatively correlated with the perceived social support, the support in the family, and the support out of family. Conclusion: The job adaptation situation of the nurses in the survey communities was generally poor and the perceived social support was at a moderate level. Therefore, community nursing managers should actively understand the situation of nurse job adaptation, and then take effective measures to improve the community nurses social support, improve the current situation of the poor job adaptation of the community nurses, and prevent loss of nursing talents, for the improvement of the quality of nursing service.

1 Introduction

In recent years, with the progress and development of community medical service in China, the community nurse attracts more and more people's attention. Foreign studies on community nurses are the issues of the job satisfaction and work motivation, job stress and job burnout, work environment and safety, et al[1-3]. Domestic studies on community nurses mainly involve the occupational values, job involvement, job stress and job satisfaction and so on[4-7].The above studies have revealed that the nurse job adaptation affects the quality of nursing service, and a poor nursing job adaptation is inclined to causing a series of problems, such as the lowered work enthusiasm, the tense relationship between nurses and patients, and declined quality of nursing service[8].The concept of perceived social support was put forward by Ausubel first as early as 1958, which has caught the attention of scholars in recent years and has been widely used in various fields[9].At present, the research about

[a] Corresponding author: Liu Yujin, E-mail:1257819211@ qq.com
This work was supported by a Social Sciences planning project of Department of Education of Jilin Province (2014).

the situation of community nurse job adaptation and perceived social support is fewer in China, and moreover, there is fewer research on the correlation between job adaptation obstacles and perceived social support in nurses. Therefore, this study is aimed at understanding the present situation of community nurses job adaptation and perceived social support through the survey of community nurses, and exploring the correlation between them, which may provide references for the management of the community nursing human resources, thereby stabilizing the community nursing teams, improving the quality of community nursing service, and promoting the progress and development of community nursing career.

2 Research content

2.1 Research object

By using the method of cluster sampling, a questionnaire survey was carried among nurses of the health service centres (stations) of five districts (Chaoyang District, Kuancheng District, Luyuan District, Erdao District, and Nanguan District) in Changchun. Inclusion criteria: (1) community registered nurses; (2) Nurses with Chinese nurse qualification certificate; (3) those who volunteered to fill in the questionnaires and participate in the survey. Exclusion criteria: (1) Community nursing interns;(2) Those who were not on duty due to going out for study, pregnancy and puerperium during the survey.

2.2 Research methods

2.2.1 Survey method

A questionnaire survey was used and there were three questionnaires. (1) General demographic information questionnaire, including age, cultural level, professional rank, etc.; (2) Job adaptation obstacle scale: The job adaptation obstacle scale Wa compiled by Tyollaska[10] was adopted, the scale could reflect a series of individuals' views and interests in reality and their own motivation. The job adaptation obstacle scale in Chinese version should have a good reliability and validity. In this research, the job adaptation obstacle scale' Cronbach's alpha was 0.78. The scale included 37 items, in which answering "yes" was scored 1 point for 29 items and answering "no" was scored 1 point for eight items, its total scores were 37 points, and the higher the score, the more the lack of work motivation and the maladaptation. (3) Perceived social support scale (PSSS): The scale compiled by Zimet,et al in 1987 was used[11].The scale was translated and amended by Chinese scholars Jiang Qianjin[12] in 2001, and the revised scale with a good reliability and validity has been widely used in China. Self-perception is emphasized in PSSS, and what is assessed with the scale is the support degree that an individual can perceive in the different social relations. There were 12 items in the scale, which was divided into two dimensions, namely the support in the family (parents, relatives, and spouse, etc.) and the support out of family (colleagues, leaders, friends, etc.). The score of support in the family included the cumulative scores of item 3, 4, 8, and 11, and the score of support out of family was made up of cumulative score of the other items. Each item was evaluated using a seven-level scoring method (Likert scale), which meant that the higher the score, the more the perceived social support. 300 Survey questionnaires were given out and 290 valid questionnaires were recovered, with a valid response rate of 96.7%.

2.2.2 Statistical method

Statistical Package for the Social Sciences (SPSS) 21.0 for Windows was used to perform the statistical analyses. Percentage was used to describe count data, $X \pm S$ was used to describe measurement data, Pearson correlation analysis was used to measure the correlation between job adaptation obstacle and perceived social support, and a significance level of 0.05 or lower was considered statistically significant.

3 Results

3.1 General demographic information

Nurses' age was more concentrated in 25 to 29 years old, the majority had 2-5 years of working years, nurses with the professional title of nurse accounted for 70.7%, the number of them in a contract system of employment form accounted for 56.9%, and most of them had the college degree and were unmarried (see Table 1).

Table 1. General demographic information (%)

Variables	Groups	n	(%)
Age(year)	≤24	70	24.1
	25-29	160	55.2
	30-34	31	10.7
	≥35	29	10.0
Working year(year)	≤1	70	24.1
	2-5	130	44.8
	6-10	43	14.8
	≥11	47	16.2
Marital status	unmarried	150	51.7
	married	138	47.6
	divorce	2	0.7
Educational level	technical education	44	15.2
	associate degree	198	68.3
	bachelor's degree	48	16.5
Technical title	junior RN	205	70.7
	senior RN	50	17.2
	nurse in charge	30	10.3
	associate professor nurse	5	1.7
Employment form	formal-employed nurses	105	36.2
	contract-employed nurses	165	56.9
	others	20	6.9

3.2 Job adaptation situation

The results showed that the nurses' job adaptation obstacle score was 20.85±5.18, significantly higher than the Chinese female norm 17.28±5.14 reported by Ji Shumao[13] with a statistically significant difference ($p<0.05$).In addition, the results also showed that score differences in the job adaptation obstacle among the nurses with different educational levels, technical title, and employment form were statistically significant. The results of pairwise comparison showed that the job adaptation situation of nurses with bachelor's degree was worse than that of nurses with technical education and associate degree, that of nurse-in-charge and associate professor nurse was better than that of junior RN and senior RN, and that of contract-employed nurses was worse than that of formal-employed nurses, as shown in Table 2.

Table 2. Job adaptation situation (n=290)

Variable	Groups	job adaptation obstacle	F	P
Age (year)	≤24	20.56±5.59	1.386	0.709
	25-29	20.82±6.22		
	30-34	20.44±6.07		

DOI: 10.1051/bioconf/20170801018

	≥35	19.11±5.56		
Working year (year)	≤1	20.13±5.80	5.641	0.131
	2-5	20.01±6.35		
	6-10	20.65±4.68		
	≥11	20.78±6.31		
Marital status	unmarried	20.75±6.30	2.703	0.259
	married	21.02±5.57		
	divorce	19.50±10.61		
Educational level	technical education	19.00±4.08	16.875	0.000
	associate degree	19.73±4.33		
	bachelor's degree	22.22±6.54		
Technical title	junior RN	22.96±5.09	10.410	0.015
	senior RN	22.63±7.07		
	nurse in charge	16.33±5.52		
	associate professor nurse	15.50±3.62		
Employment form	formal-employed nurses	18.50±3.21	7.577	0.023
	contract-employed nurses	23.03±4.08		
	others	22.29±3.27		

3.3 Perceived social support

The total score of community nurses' perceived social support was 64.25±10.32, the total score of family support was 20.01±3.58, and the total score of support out of family was 42.57±6.86.

3.4 Correlation between job adaptation obstacle and perceived social support

The results showed that there was a negative correlation between job adaptation obstacle and perceived social support ($p<0.01$). The detailed results are shown in Table 3.

Table 3. Correlation between job adaptation obstacle and perceived social support (r-values)

	Job adaptation obstacle
Total perceived social support	-0.325
Support in family	-0.357
Support out of family	-0.261

4 Discussion

4.1 Analysis on the present situation of job adaptation obstacle of community nurses in Changchun

The results showed that the score of job adaptation obstacle of community nurses in Changchun was 20.85±5.18, significantly higher than the Chinese female norm 17.28±5.14 reported by Ji Shumao, it indicated that the community nurses might lack the work motivation and be poor in the job adaptation. The reasons may be related to the community nurses' single working mode, fuzzy self-positioning, and low professional identity. Studies have revealed that community nurses' wages are lower, social benefits are relatively poor, social status is not high, but their working time is longer, workload is heavier , and they often engage in trivial work[14]. All these likely lead to the lack of work motivation, reduced work enthusiasm, lower job satisfaction, and even job adaptation obstacle gradually in community nurses[15]. Studies have also showed that the low level of wages is one of the important reasons for the turnover of community nurses, which can hinder the development of community nursing[14]. Therefore, the relevant departments should improve the community nurses' payment and stabilize the community nursing team.

In addition, this research revealed that the job adaptation situation of nurses with bachelor's degree was worse than that with technical education and associate degree, which may be due to the reason that the nurses with bachelor's degree had a high expectation of future career in school, but after working in the community, they found that the gap between the actual work situation and the expectation was bigger, which lead to their apparent psychological imbalance and the generation of their contradicted emotion about the profession of community nurses, so it is easy for them to adapt their job poorly[16].Studies have showed that the expectation of nursing profession of nurses with technical education and associate degree was lower, and when they join in work after graduation, they find that the gap between the actual work situation and expectation is not so significant, so that their psychological gap feeling is not so strong, and they are easy to produce the love of nursing work and have the work motivation, which may be the reason why the nurses with technical education and associate degree are not easy to appear job adaptation obstacle[17]. The job adaptation situation of nurse-in-charge and associate professor nurse was better than that of junior RN and senior RN, which is different from the conclusion of clinical nurses' job adaptation obstacle made by Shao Xiaozhen[18]. The key reasons may be related to the following two aspects: (1) The investigation groups are different, and there are some differences in the group characteristics; (2) There is a certain gap in their working environment, working intensity, working pressure, etc. between clinical nurses and community nurses. The conclusion made in this study might be related to the shorter working time of junior RN' and senior RN', the lack of work experience and the ability to adapt themselves to changing circumstances, and the incapacity to realize their self-value when they engage in the community nursing profession, etc. The job adaptation situation of contract-employed nurses was worse than that of the formal-employed nurses, which might be because community contract-employed nurses' social benefits, chances of promotion, and further education outside were worse than formal-employed nurses'. All of them easily result in the lack of work motivation and the job adaptation obstacle in the community contract-employed nurses. Due to these reasons, community managers should pay attention to the mind and body change of community nurses, especially nurses with bachelor's degree, those working for shorter time, junior RN and senior RN, and contract-employed nurses, and take effective measures positively to promote community nurses to adapt the community health service work.

4.2 Analysis on the present situation of community nurses' perceived social support in Changchun

Perceived social support focuses on the emotional experience and (or) satisfaction of individual subjective feeling about understood, respected and supported[19].The results showed that the community nurses' perceived social support was at a moderate level, and the score of support out of family was higher, indicating that when the nurses had difficulties and needed help, the community colleagues, leaders, and friends gave them much support. On the one hand, it might be linked to most of nurses' time distributed on work, more contact with the colleagues and leaders than their families, in which the colleagues and leaders tended to be the first people to offer them help when problems occurred; on the other hand, it might be linked to the nurses likeness to pour out their thoughts with their friends, no intention to bring negative emotions to their families. Therefore, more care and help given to the nurses by community managers will help to improve the community nurses' perceived social support.

4.3 Correlation between job adaptation obstacle and perceived social support of community nurses in Changchun

This research results revealed that the job adaptation obstacle was negatively correlated with the perceived social support, support in the family, and support out of family, namely the lower perceived social support, the worse community nurses' job adaptation obstacle. For this reason, community nurses' families should properly give nurses understanding and support. Community nursing managers should often focus on the nurses' psychological change and work adaptation situation, and when they have emotional problems, managers should listen to nurses patiently and provide emotional support for them[20]; when nurses have work difficulties, nursing managers should give them guidance and help positively to enhance their self-confidence; the relevant departments should properly improve the payment of community nurses, give them more opportunities of further education and advanced studies [21], arouse the nurses' initiative to study and work, arouse their working enthusiasm, improve their sense of professional worth, and prevent their job burnout, all of which may contribute to improve nurses' perceived social support degree[22], thereby promoting community nurses to adapt to community nursing work, preventing the loss of community nurses, improving the quality of community nursing service, and promoting the progress and development of community nursing career.

References

1. Sacks E, Alva S, Magalona S,et al. Examining domains of community health nurse satisfaction and motivation: results from a mixed-methods baseline evaluation in rural Ghana. Human resources for health,13(2015).
2. McTiernan K, McDonald N. Occupational stressors, burnout and coping strategies between hospital and community psychiatric nurses in a Dublin region. *Journal of psychiatric and mental health nursing*,22,3(2015).
3. Terry D,Le, Q,Nguyen, U et al. Workplace health and safety issues among community nurses: a study regarding the impact on providing care to rural consumers. BMJ open,5,8(2015).
4. Xu Ping, Zhu Lei, Lv Zhuohong et al. Community nurses' professional values and its influencing factors. Nursing research,30,5B(2016).
5. Wang Ruifang. Study on present situation and relationship between working environment and job involvement of community nurses. Disease surveillance and control,10,3(2016).
6. Guo Yan, Jiang Wenhui, Li Xiaomei, et al. Research about community nurses' job stressors influence on job satisfaction in Xi'an. Nursing management journal,16,1(2016).
7. Zheng Dan. Research progress of community nurses' job satisfaction. Nursing research,29, 1B(2015).
8. Xie Lixia, Quan Suqin, Jia Hongli, et al. Present situation research about nurses job adaptation obstacle in one Three-grade hospital in Hengyang. Chinese nursing research,9,3(2015).
9. Xiao Shuiyuan. The influence of social support to health of body and mind. Chinese journal of mental health,1,4 (1987).
10. Zhang Haiwei. Status of job adaptation obstacle and its influencing factors of male nurses. Chinese nursing research,25,9(2011).
11. Blumenthal JA, Burg MM, Barefoot J, et al. Social support, type A behavior, and coronaryartery disease. Psychosomatic Medicine,49(1987).
12. Huang Li, Jiang Qianjin, Ren Weihong. Correlation research of coping styles, social support, and the psychosomatic symptoms in cancer patients. Chinese journal of mental health,10,4(1996).
13. Ji Maoshu. Minnesota multiphase personality inventory, the latest research and multi-class scale explanation. Beijing:science press,1(2004).
14. Yang Rongfang, Ma Jinghua, Wang Hui, et al. Current situation survey about nursing human resources of community health service center in Shijiazhuang. Journal of community medical journal,14,5(2016).
15. Hu Haixia, Zhang Minmin, Gao Yuxia. Correlation between the professional values and job burnout status of community nurse in part of community in Changchun. *Journal of liberation army nursing*,29,2B(2012).
16. Lu Qinrui, Wang Yanli, Zhao Xiaohong, et al. Current investigation about clinical nurses' professional attitudes and influencing factors. Chinese nursing research,3,30(2016).
17. Lu Xuya, Zhu Xiuwen. Nursing undergraduate professional attitudes influence factors analysis. Education of traditional Chinese medicine,24,1(2005).

18. Shao Xiaozhen, Wang Cong. Correlation research between nurse' job adaptation obstacle and personality characteristics. Human resources management, 9 (2013).
19. Li Yingying. Relationship research about Junior high school students' gratitude and perceived social support, family functioning. Changchun: the Northeast Normal University(2013).
20. Zhou Haiping, Ma bo. Correlation research between alexithymia and perceived social support of ICU nurses. The nurse education magazine, 31,4 (2016).
21. Xin Li. Current analysis of community nurses. Traditional Chinese medicine in Inner Mongolia, 18(2012).
22. Zhou Zhili, Zhang Hui. Relationship between job burnout and perceived social support of nurses. Journal of occupational and health,27,13 (2011).

 DOI: 10.1051/bioconf/20170801019

Assessment of DNA interstrand crosslinks in NIH/3T3 cells induced by Chloroethylnitrosoureas

Xiaotong Zheng, Xuechai Chen[a], Linna Zhao, Minjun Guo and Rugang Zhong

College of Life Sciences and Bioengineering, Beijing University of Technology, Beijing Key Laboratory of Environment and Viral Oncology, Beijing, China

Abstract: Aim: To investigated the mechanism of Chloroethylnitrosoureas (CENUs) leading to secondary tumors after therapy, three kinds of CENUs, namely, Nimustine (ACNU), Carmustine (BCNU) and Semustine (MeCCNU) were used to investigated drug-induced DNA interstrand crosslinks(ICLs). **Method:** The alkaline comet assay was adopted to compare DNA damages induced by CENUs under different concentrations. **Results:** With the increase of drug's concentration, DNA migration exhibited a positive concentration-dependent relationship in all three drugs. ACNU was shown to cause significant crosslinking. The other two drugs also induced crosslinking, but this could not be analyzed at high concentrations due to the high cytotoxicity of the drugs which caused cells death.

1 Introduction

Recently, cancer has become a major health issue in the world because of its high mortality and incidence. Although many pathogenic mechanism of cancer have been proposed, its real mechanism remains unclear. Epidemiological studies have shown that almost 80%-90% of human cancers are caused by the environment[1], being chemical pollutants the most important cancer-causing factors. Nitrosoureas are a class of DNA alkylating-agent anticancer drugs that include Chloroethylnitrosoureas (CENUs), such as 1-((4-amino-2-methyl -5-pyrimidinyl) methyl)-3-(2-chloroethyl)-3-nitrosourea (Nimustine, ACNU), 1,3-bis(2-chloroethyl)-1-nitrosourea (Carmustine, BCNU), 1-(2-chloroethyl)-3- cyclohexyl-L-nitrosourea (Lomustine, CCNU), 1-(2-chloroethyl)-3- (4-methlycyclohexyl) -1-nitrosourea (Semustine, MeCCNU) and 1-[N-(2-chloroethyl)-N-nitrosourea] ethylphosphonic acid diethyl ester (Fotemustine, FUNU)[2]. Previous studies have indicated that the cytotoxicity of CENUs is related to the formation of DNA interstrand crosslinks (ICLs)[3]. ICLs are extremely toxic to cells since they prevent the separation of the two strands of a DNA double helix for

[a] Corresponding author: Chen xuechai. Email: chenxuechai@bjut.edu.cn
The authors thank the support from China Postdoctoral Science Foundation (No. 2012M510302), and partial support from the Beijing Key Laboratory of Environment and Viral Oncology of China (No. PXM2012_014204_00_000181) and Youth Fund of the National Natural Science Foundation of China (NO. 81400935). The authors also acknowledge the support provided by the professors in the College of Life Sciences and Bioengineering, Beijing University of Technology and all colleagues in the laboratory.

cellular processes, such as replication and transcription, by coordinating chemical reactions with bases on opposing strands[4]. The formation of ICLs between complementary base pair is supposed to be the critical step for the anticancer activity of CENUs[5]. Zhao's group found that the ICLs between the N^1 atom of guanine and the N^3 atom of the complementary cytosine through an ethylene is theoretically determined to be the main crosslinking damage in the reactions of DNA and CENUs[6]. According to the clinical observation, CENUs can also lead to the generation of secondary tumors after the treatment. However, the carcinogenic mechanisms of CENUs are still poorly understood[7]. Reports showed that the pathogenetic mechanism of secondary tumors during therapy due to formation of ICLs[8]. Therefore, studies of drug-induced ICLs to reveal their anticancer mechanism and carcinogenic role are of great significance. To assess genotoxicity of drugs, there are different genetic markers and tests such as chromosomal aberrations, micronucleus test and alkaline comet assay[9]. Alkaline comet assay, also known as single cell gel electrophoresis (SCGE) assay, has been increasingly used in the areas of genotoxicity studies on account of its rapid, simple, sensitive and versatile superiority[10].

In this study, normal NIH/3T3 cells were used for a preliminary study of ICLs induced by three kinds of CENUs: ACNU, BCNU and Me-CCNU. The alkaline comet assay was performed to detect DNA breaks and interstrand cross-links, in order to further study the relationship between CENUs and ICLs.

2 Materials and Methods

2.1 Materials and cell culture

ACNU, BCNU and Me-CCNU, low melting agarose (LMA), and dimethyl sulfoxide (DMSO) was purchased from Sigma-Aldrich. *Tert*-butyl hydroperoxide (tBHP) was obtained from Sinopharm Chemical Reagent Co. Ltd. Propidium iodide was supplied by MERCK and normal melting agarose (NMA) was purchased from Shanghai yito bio-instrument company limited, China. Dulbecco's modified eagle's medium and fetal bovine serum were purchased from Hyclone.

The NIH/3T3 mouse fibroblast cell line was obtained from Cell Resource Center (CRC). The cells were maintained as monolayer in Dulbecco's modified Eagle's medium (DMEM) supplemented with 10% fetal bovine serum, 100 units/ml of penicillin and 100 μg/ml of streptomycin. Cells were grown at 37°C in an incubator with a humidified atmosphere of 5% CO_2.

2.3 Cytotoxicity assay

Cytotoxicity of NIH/3T3 cells against CENUs was performed according to *K. Ishitsuka et al* [11]. NIH/3T3 cells were harvested with 0.25% trypsin and 0.02% EDTA, then were seeded (1×10^4 cells per well) on a 96-well plate. After incubated for 24 h, cells were treated with various concentrations of CENUs in serum-free medium for 24 h. To evaluate the cytotoxicity, 10 μl of Cell Counting Assay Kit-8 (Dojindo molecular Technologies, Japan) solution was added to each well, afterwards the microplates were incubated at 37°C for 1-2 h. Absorbance was measured at 450 nm using a microplate reader.

2.4 The alkaline comet assay

The alkaline comet assay described by Günter Speit[12] was used to assess CENUs -induced DNA damage and DNA interstrand cross-links in NIH/3T3 cells. Cells were cultured at 5×10^4 cells /well into a 12-well plate. The next day, cells were exposed to concentrations of 15, 30, 45, 60 or 75 μg/ml CENUs at 37°C for 2 h. After treatment, approximately 2500 NIH/3T3 cells in 50 μl DMEM were mixed with 5 μl 1% low melting point agarose. The mixture was added to microscope slides precoated with 1% nomal melting point agarose. The slides were immersed overnight in lysis solution (2.5 M

NaCl, 100 mM EDTA, 10mM Tris, pH10, 1% Triton-X and 10% DMSO were added fresh) at 4°C, and then placed in electrophoresis buffer (300 mM NaOH and 1 mM EDTA, pH >13) at 4°C for 30 min to unwind DNA. Electrophoresis was performed at 25 V and 300 mA for 30 min. The slides were then washed in neutralizing solution (0.4 M Tris-HCl, pH7.5) three times, stained with Gel-Red (10,000×, Biotium) and dried with 75% ethanol. The results were analyzed using a fluorescence microscope (Leica DM3000) and image analysis software I.A.S (Delta Sistemi), and the comet parameter (mean percentage of DNA in the tail) was used to demonstrate the damage of DNA.

3 Results

3.1 Concentration-dependent death of NIH/3T3 cell caused by CENUs

To investigate the effect of CENUs on normal cell death, the NIH/3T3 were exposed to increasing concentrations of CENUs for 24 h, and cell viability was measured using the CCK-8 assay. As shown in Figure 1, all of the three kinds of CENUs induced a concentration-dependent decrease in cell viability, and half maximal inhibitory concentration (IC50) of ACNU, BCNU, Me-CCNU were 750, 60 and 65 μg/ml respectively. These results indicated that the cytotoxicity for normal cells of both BCNU and Me-CCNU is more than 10 fold higher than ACNU. Based on this, several concentrations nearby the IC50 of BCNU or Me-CCNU were further analyzed, and cells were treated with concentrations of 15, 30, 45, 60 and 75 μg/ml for all three drugs.

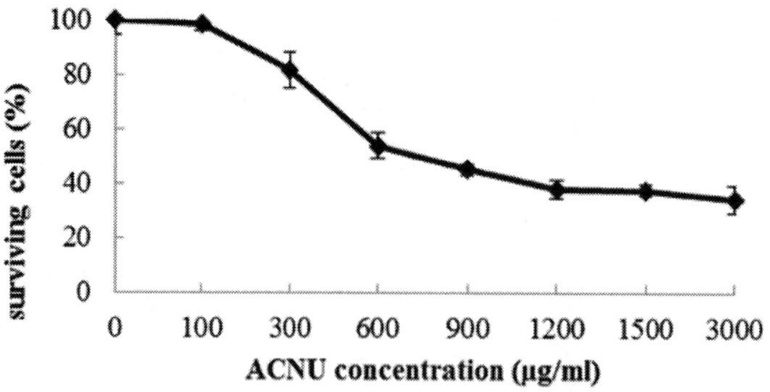

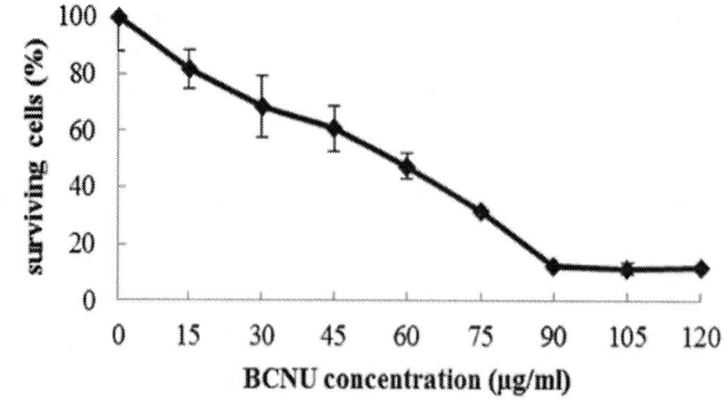

 DOI: 10.1051/bioconf/20170801019

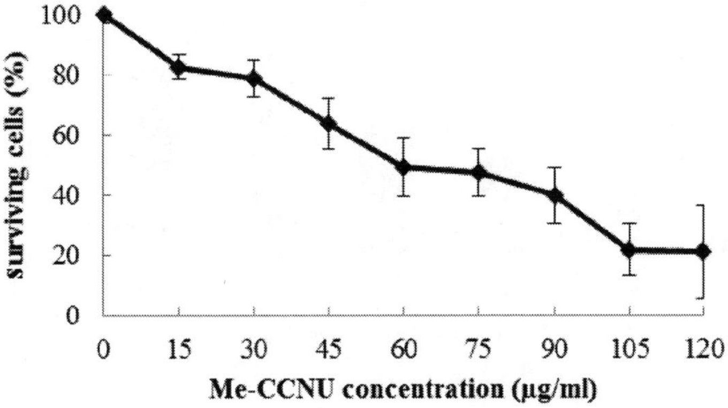

Figure 1. Cytotoxicity of CENUs in different concentrations in NIH/3T3 cells.

3.2 CENUs -induced DNA breaks

The concentration-dependent DNA damage in NIH/3T3 cells treated with CENUs (15, 30, 45, 60, 75 μg/ml) was measured using a modification of the comet assay. As shown in Figure 2, all of the three drugs showed a positive concentration-dependent relationship, DNA migration is directly proportional to the concentration of drugs. In the given concentrations, DNA breaks caused by ACNU were in a lesser degree than the other two drugs, the rate of tail to total being lower than 50%. The result indicates that at lower concentration of BCNU, DNA breaks are in lower level, however, at higher concentration, BCNU induced DNA breaks are predominant.

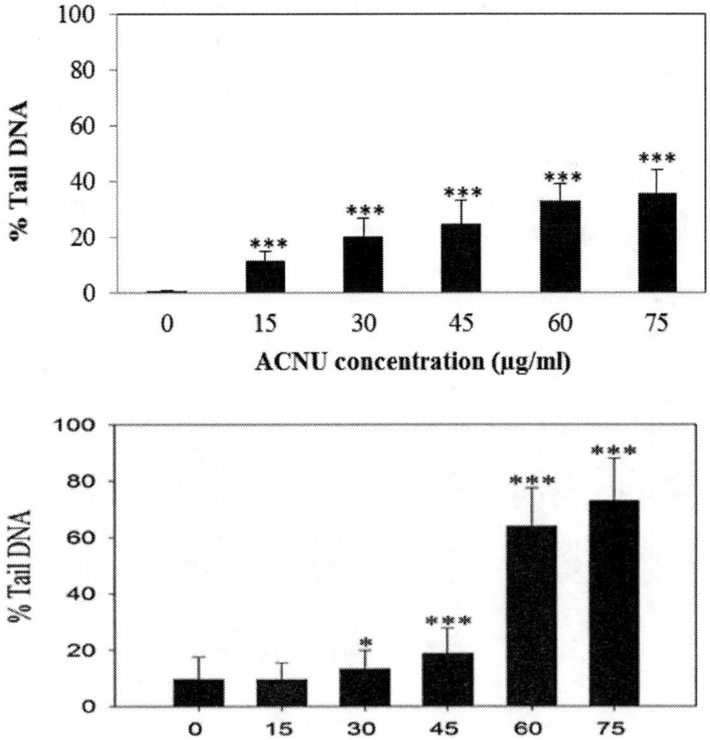

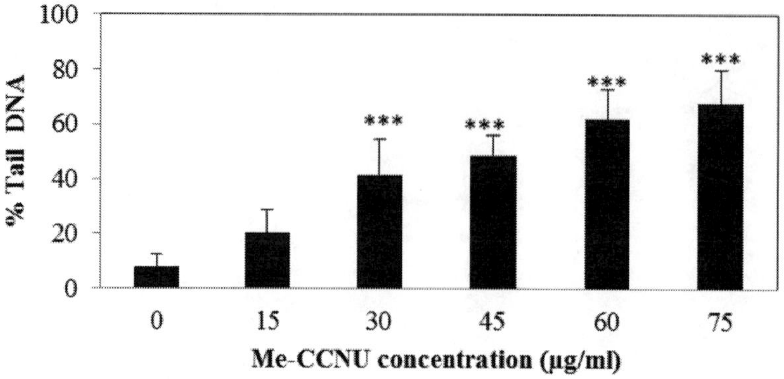

Figure 2. DNA migration caused by different concentrations of CENUs as determined by alkaline comet assay. Results were shown as mean±S.D. of three independent samples(n=3). (*p<0.05; **p<0.01; ***p<0.001)

3.3 CENUs -induced DNA interstrand crosslinks

To observe the DNA migration-retarding effect of cross-links, a co-treatment experiment using a second genotoxic agent was performed. In the present research, *tert*-butyl hydroperoxide (tBHP) was used as the second genotoxic agent[13]. This genotoxic agent can cause DNA breaks, and if a substance induces cross-linking, one will expect a decrease in DNA migration caused by the second genotoxic agent. In this experiment, NIH/3T3 cells were exposed to 400 μM tBHP for 1h, and then different concentrations of CENUs were added. As shown in Figure 3, with the increase concentration of ACNU, DNA migration decreased, which indicated that ACNU could cause the DNA ICLs. Because of the low cytotoxicity of ACNU, higher concentrations around IC50 were also selected for further analysis. The result showed that a significant rate of ICLs were observed especially at 150 and 300 μg/ml. However, the DNA breaks increased at 450 and 600 μg/ml, which indicated that at 150 and 300 μg/ml of ACNU the rate of interstrand cross-link is higher than the rate of DNA breaks. For BCNU, a similar result was observed at concentrations of 15 and 30μg/ml. BCNU led to statistically significant decreases in DNA migration. However, the decreases were not statistically significant at 45 μg/ml, and the DNA migration showed a small increase at 60 and 75 μg/ml. This result indicates that the effect is very strong, confirming the presence of cross-links at concentrations of 15 and 30 μg/ml. Nevertheless, ICLs weren't observed under the given concentration for Me-CCNU. Its high cytotoxicity causes cell death at higher concentrations, making it difficult to conclude whether Me-CCNU could cause ICLs.

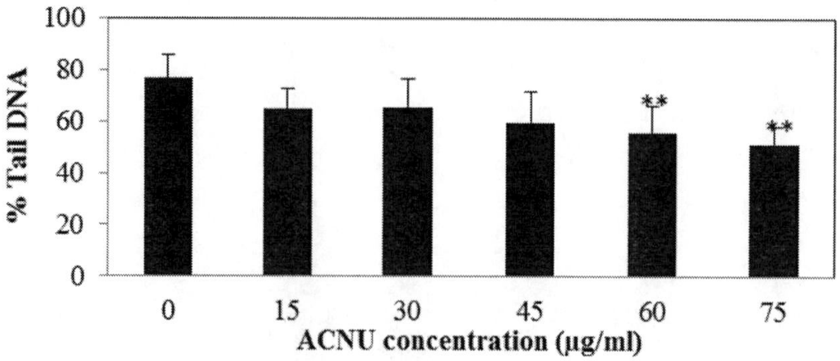

 DOI: 10.1051/bioconf/20170801019

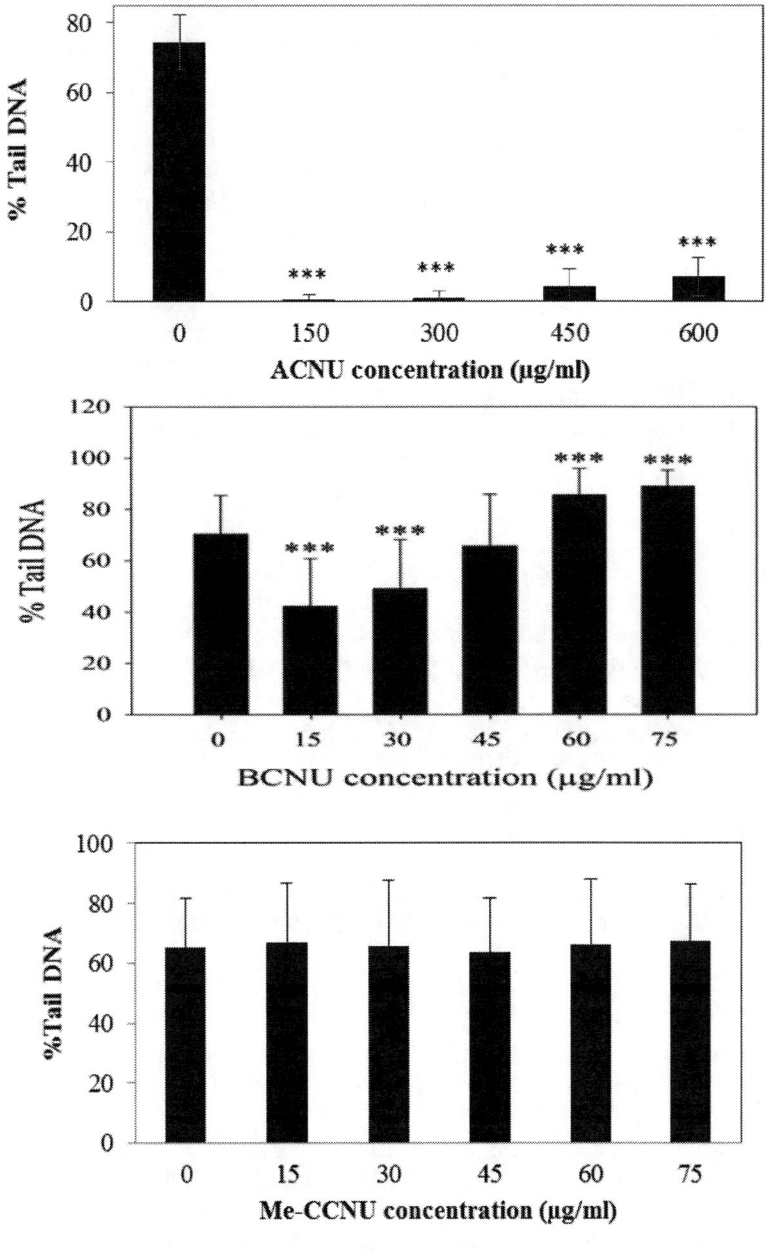

Figure 3. Confirmation of the presence of cross-links. NIH/3T3 cells were first treated with 400 μM tBHP at 37°C for 1h and then with different concentrations of CENUs at 37°C for 2 h. Results were shown as mean ±S.D. of three independent samples(n=3) (*p<0.05; **p<0.01; ***p<0.001)

4 Discussion

CENUs are widely used in the treatment of different cancers. They damage DNA by alkylation of bases, single/double strand breakages and interstrand crosslinks[14-15]. DNA damage caused by different nitrosoureas has been studied via different methods of detection, such as agarose gel electrophoresis assay[16], fluorescence assay[17] and lipid chromatography/mass spectrometry (LC/MS)[18]. Being highly sensitive and specific, the comet assay proved to be a useful technique to evaluate the degree of DNA damage[19].

Our study compared the interstrand crosslinking induced by different kinds of CENUs via single cell gel electrophoresis (SCGE) assay firstly, and it showed that SCGE assay is a convenient method to detect the DNA breaks and interstrand crosslinks. Clinical observations showed that CENUs may also cause secondary tumors during the treatment of cancer. For this reason, it is important to establish a DNA cross-links cell model for normal cells, therefore NIH/3T3 cells from the primary mouse embryonic fibroblast were used in our research.

From the results of the cytotoxicity test, IC50 for both BCNU and Me-CCNU are similar, and ACNU displayed the lowest toxicity with 10 fold lower than the other two. Furthermore, ACNU is water soluble with Me-CCNU slightly soluble, and BCNU has a higher solubility in organic solvents. In the same series concentrations, all of the three drugs showed a positive concentration-dependent relationship, DNA breakage increased in direct proportion concentration of drugs. The results of interstrand crosslinks indicated that the treatment with BCNU at concentrations lower than 100 μg/ml, in particular at 15 and 30 μg/ml, present the most obvious ICLs. At concentration higher than 100 μg/ml, ACNU could induce a big rate of ICLs, but cells treated with the other two drugs died because of their high cytotoxicity. Based on these results, ACNU is the most suitable CENUs for further study of DNA interstrand crosslinks since its water-solubility, low cytotoxicity and high DNA crosslinking.

References

1. C. Espina, K. Straif, S. Friis, M. Kogevinas, R. Saracci, H. Vainio, J. Schuz, CANCER EPIDEMIOL. J. **39**, S84-S92(2015).
2. T. Nikolova, F. Hennekes, A. Bhatti, B. Kaina, Cell Cycle. J. **11**, 2606-19(2012).
3. C.J. Wilds, F. Xu, A.M. Noronha, CHEM RES TOXICOL. J. **21**, 686-695(2008).
4. Y. Kee, A.D. D'Andrea, Genes Dev J. **24**, 1680-94(2010).
5. T. Yamaguchi, H. Kanemitsu, S. Yamamoto, M. Komatsu, H. Uemura, K. Tamura, T. Shirai, J TOXICOL PATHOL . J. **23**, 25-30(2010).
6. L. Zhao, R. Zhong, Y. Zhen. J THEOR COMPUT CHEM. J. **6**, 631-639(2011).
7. A. Stankiewicz-Kranc, W. Miltyk, E. Skrzydlewska, DRUG CHEM TOXICOL. J. **33**, 55-63(2010).
8. W.P. Tong, M.C. Kirk, D.B. Ludlum, Cancer Res. J. **8**. 3102-3105(1982).
9. M. Chandrasekhar, P.V. Rekhadevi, N. Sailaja, M.F. Rahman, J.P. Reddy, M. Mahboob, P. Grover, Mutagenesis. J. **21**. 249-54(2006).
10. Z.X. Gao, X.L. Song, S.S. Li, X.R. Lai, Y.L. Yang, G. Yang, Z.J. Li, Y.H. Cui, H.W. Pan, Invest Ophthalmol Vis Sci. J. **57**. 3093-3102(2016).
11. K. Ishitsuka, T. Hideshima, M. Hamasaki, N. Raje, S. Kumar, H. Hideshima, N. Shiraishi, H. Yasui, A.M. Roccaro, P. Richardson, K. Podar, S. Le Gouill, D. Chauhan, K. Tamura, J. Arbiser, K.C. Anderson, Blood. J. **106**. 1794-800(2005).
12. G. Speit, A. Rothfuss, Methods Mol Biol. J. **920**. 79-90(2012).
13. H. Heck, M. Casanova, REGUL TOXICOL PHARM. J. **40**. 92-106(2004).
14. A. Naghipur, M.G. Ikonomou, P. Kebarle, J.W. Lown, J Am Chem Soc. J. **112**.389-95(2007)
15. W.J. Bodell, MUTAT RES-FUND MOL M. J. **522**.85-92(2003).
16. M.T. Hayes, J. Bartley, P.G. Parsons, G.K. Eaglesham, A.S. Prakash, Biochemistry. J. **36**. 10646-54(1997).
17. H. Ueda-Kawamitsu, T.A. Lawson, P.R. Gwilt, BIOCHEM PHARMACOL. J. **63**.1209-1218(2002).
18. L. Li, L. Zhao, R. Zhong, RAPID COMMUN MASS SP. J. **28**.439–447(2014).
19. R.A. Furtado, F.R. de Araujo, F.A. Resende, W.R. Cunha, D.C. Tavares, J APPL TOXICOL. J. 30. 254-9(2010).

Aerobic exercise ameliorates learning and memory deficits of aging rats induced by D-galactose via promoting SYP and BNDF expression in hippocampus

Yan Fu[1,2], Yeting Zhang[3] and Qiongjia Yuan[2,a]

[1] *Department of Physical Education, Southwest University for Nationalities, Chengdu, China*
[2] *Department of Sports Medicine, Chengdu Sport University, Chengdu, China*
[3] *Department of Physical Education, Chengdu University, Chengdu, China*

Abstract. The aim of this study was to explore the effects of aerobic exercise on learning and memory of aging brain and the underlying potential mechanism. We adopted the method of intraperitoneal administration for 6 weeks to induce aging rat model and gave rats swimming exercise intervention in the process of aging modeling. Then the Morris Water Maze test, immunofluorescence, western blotting and real time PCR technology were adopted to estimate rats' learning and memory abilities and the expression levels of BDNF and SYP in hippocampus respectively. Compared with the aging model rats induced by D-gal administration, the rats subjected to swimming exercise in the process of aging modeling showed not only faster acquisition and better retention of the maze but also higher expression levels of BDNF and SYP in hippocampus. These findings provide evidence that aerobic exercise could ameliorate learning and memory deficits of aging rats induced by D-gal, which is related to aerobic exercise' promoting hippocampal BNDF and SYP expression.

1 Introduction

The decline in cognitive functions, especially learning and memory deficits, is prevalent among older adults, which results in degraded quality of life for them[1,2]. Therefore, how to slow down the impairments of learning and memory in the aging process to promote the development of cognitive function is one of the goals of public health. It has been demonstrated by both epidemiological and experimental studies that physical exercises contribute to achieving the goal[3,4], however, the underlying mechanism remains incompletely unclear.

It is well known that the learning and memory formation are based on the neuron synaptic plasticity[5]. Synaptophysin(SYP), one of the most widely used protein markers of synaptic plasticity in the brain, existing in almost all pre-synaptic membranes and vesicles, is thought essential for stimulating synapse formation and synaptic transmission process[6,7]. Previous studies have confirmed that the decrease of SYP in hippocampus, an important brain area responsible for learning and memory, is relevant to age-dependent impairments in learning and memory[8,9]. Whether aerobic exercise has influence on SYP in aging hippocampus still remains unknown.

Brain-derived neurotrophic factor (BDNF) plays a critical role in the survival, growth, differentiation and synaptic plasticity of mammalian neurons[10,11]. It is well documented that low

[a] Corresponding author:yqj1225@163.com
This work was supported by grants from the National Natural Science Foundation of China (No.31371202), the Scientific research project of Sichuan Education Department (No. 15ZA0294),

BDNF expression in brain involved in the learning and memory deficits in the process of aging [12]. Available data in the literature indicate that regular aerobic exercise may promote BDNF release in budding and adult hippocampus[13-15], but few experiments focus on the effects of exercises on BDNF expression in aging hippocampus.

The free radical theory of aging indicates that increase of free radicals and oxidative stress and decline of anti-oxidation function are the main factors leading to aging and even to shortening lifespan[16]. D-galactose (D-gal), a reducing sugar, is a naturally occurring metabolic substance in body. At high levels, it can result in the excessive production of oxygen free radicals that damage cellular lipids, proteins, and DNA, which eventually impairs normal function of macromolecules and cells[17]. It has been shown that prolonged systemic exposure of rodents to D-gal makes oxidative stress increase and anti-oxidation function decline, which causes brain cell biological membrane peroxide reaction, and finally results in degenerative changes of brain[17,18]. These changes are considerably similar to the symptoms in natural aging brain. Thus, the aging rodent model induced by D-gal is recognized as an ideal model for studying the mechanism underlying brain aging and the approach for anti-aging.

In the study, we have used the aging rat model induced by d-gal to verify directly the hypothesis that regular aerobic exercise intervention would ameliorate learning and memory deficits of aging rats via promoting SYP and BNDF expression in hippocampus.

2 Materials and methods

2.1 Animals and groups

Clean grade healthy adult male Sprague-Dawley rats, each weighing 280–320g, were purchased from the experimental animal center of Chengdu Dasuo Biological Technology Co. LTD(Chengdu, China; Animal production license: CSXK, 2012-24). All rats were kept in conditions of temperature (22-25°C) and humidity (45-65%) and 12h/12h light-dark cycle, with food and water available *ad libitum*. After acclimatization to the laboratory for 1 week, the rats were randomly divided into 3 groups: control(C) group, D-gal administration (D) group, D-gal administration with exercise (DE) group, 12 rats in each group. All procedures were performed in accordance with Animal Care and Use Guidelines in China.

2.2 D-gal administration

According to literature published[19], the study adopted the method of systemic injection of D-gal (sigma, USA) to establish aging rat model, with rats in group D and DE intraperitoneally injected withD-gal at dose of 100mg//kg. d(dissolved in 0.9% saline to 5% concentration, injection volume was 2 ml/kg. d) daily for 6 weeks to induce aging, whereas those in group C were intraperitoneally injected with 0.9% saline in the same volume with D-gal.

2.3 Exercise protocol

During the 6 weeks of D-gal administration, the rats in group DE were subjected to exercise intervention. The endurance swimming exercise training in a transparent glass tank (150cm in length, 60cm in width, 50cm in depth, with the water temperature kept at 33 ± 1°C) was used in our study[19]. On the first day of training, the rats swam for 20min, and then the training duration was lengthened daily by 10min until each rat could swim continuously for 60min. Subsequently, rats swam 1h per day for 6 times a week (9:00–10:00 PM from Monday to Saturday).

2.4 Morris Water Maze (MWM) test

The MWM test, consisting of 6 consecutive days of oriented navigation trial and a 60s of probe trial, was performed to evaluate spatial learning and memory as described previously by Morris[20] with some modifications. The water maze was a black circular pool (150cm in diameter, 50cm in height) filled with opaque water at 23 ±1°C and surrounded by various visual cues. Moreover, the tank was divided into 4 equal quadrants (marked respectively as E, S, W and N) with two imaginary perpendicular lines. During the oriented navigation trial, each rat participated in 4 trials per day to learn to locate the platform (9cm in diameter) that fixed in the quadrant E and submerged 1.5cm underneath the surface of water. The animal was gently placed into the pool at the midpoint of the edge of quadrant, changed for each trial, and was allowed a maximum of 120s to locate the platform. Rats failing to find the platform within 120s were guided towards it and remained there for 15s, the same period as was allowed for the successful animals. The time that it took the rat to arrive at the submerged platform (referred to as "escape latency") and the swimming speed were recorded with a digital camera to evaluate spatial learning ability. 24hrs after the last navigation training trial, the platform was removed and each rat was subjected to the probe trial, during which the rats were placed into the tank at the midpoint of the edge of quadrant W and allowed to swim in the tank for 60s. The time spent in the target quadrant (quadrant E) and the time of crossing the location where the platform had been fixed were recorded as a measure of spatial memory.

2.5 Tissue sampling

Tissue samples were collected in the 24hrs after the MWM test, with all rats sacrificed by decapitation and the brains quickly removed from the skull. 6 whole brain samples per group were immersed in 4% paraformaldehyde for immunofluorescence assay, while the bilateral hippocampus of the other 6 brains were quickly dissected on ice and stored at –80°C for western blot and real-time PCR analysis.

2.6 Brain section and immunofluorescence assay for SYP

The brain tissue samples fixed in 4% paraformaldehyde were embedded in paraffin and coronally sectioned at 50μm thickness using a microtome (Leica, Wetzlar, Germany), and five tissue sections were selected at 30μm apart between +1.7 and +3.0mm posterior to the bregma referring to a mouse atlas[21]. SYP immunofluorescence staining was performed in accordance with the standard protocols. DAPI was used for nuclear staining. In the end, 20μl of glycerole was applied to each slide and a cover slip was sealed in place. In order to obtain accurate data, immunofluorescence staining was carried out carefully under the same conditions and fluorescence images were obtained by the OLYMPUS microscope and analyzed by using Image-Pro Plus 6.0 software. SYP immunoreactivity was evaluated by measuring the integral optical density (IOD) of SYP immuno positive contents.

2.7 Protein extraction and western blot analysis for SYP and BDNF

The right hippocampus samples stored at –80°C were used for BDNF and SYP protein quantification using western blot. Total protein in the hippocampus was extracted as previously described[19]· Protein content was measured by BCA assay (Thermo Scientific) according to the standard protocols. Sample containing 50μg proteins was electrophoresed on 5–20% gradient SDS-PAGE gels to separate and then blotted onto low-fluorescent polyvinylidene fluoride (PVDF) membranes. After having been blocked with 5% non-fat milk for 1h, the PVDF membrane was incubated overnight at 4°C with the anti-BDNF or anti-SYP antibody(both are polyclonal rabbit anti-rat antibody, diluted at 1:500, Santa Cruz, USA) and then incubated for another 2hrs with appropriate horseradish peroxidase-conjugated secondary antibody(polyclonal goat anti-rabbit IgG, diluted at 1:2000, Santa Cruz, USA). The membranes were visualized by using the enhanced chemiluminescence detection system (Pierce,USA). β-actin (polyclonal, diluted at 1:5000; Abcam, Cambridge, MA) was used for internal control. Integral optical density (IOD) of immunoblots was quantified by using Quantity One Gel Imaging System (Bio Rad, USA) and was normalized to β– actin.

2.8 RNA Extraction and Real-Time PCR Analysis for BDNF mRNA

Remaining hippocampus samples stored at –80°C were used for BDNF mRNA using real-time PCR analysis. Total RNA in hippocampus was extracted using TRIZOL Reagent (TaKaRA, Japan). After removing genomic DNA residual using DNase I, RNA samples were reversely transcribed into complementary DNA (cDNA) using the iScript™ cDNA Synthesis Kit(BIO-RAD,USA)in accordance with the manufacturer's protocols. Then the real-time PCR was carried out using GoTaq® qPCR Master Mix (Promega, USA) and CFX96 Touch™ Real-Time PCR Detection System (Bio-Rad, USA). After an initial incubating at 94°C for 10min, DNA samples were amplified for 40 cycles of 95°C denaturation for 15s, annealing at 60°C for 60s, 72°C extension for 1min. For each sample, the levels of BDNF mRNA were normalized against β-actin mRNA levels using the $2^{-\Delta\Delta Ct}$ method[22]. The sequences of primers for RT-PCR amplification were as follows: BDNF(forward primer: 5′-AGC GTG TGT GAC AGT ATT AGC GAG-3′ and reverse primer: 5′-ATG GGA TTA CAC TTG GTC TCG TAG A-3′), β-actin(forward primer: 5′-CGT AAA GAC CTC TAT GCC AAC A-3′ and reverse primer: 5′-TAG GAG CCA GGG CAG TAA TC-3′).

2.9 Data analysis

All values were reported as the means ±standard deviation (M±SD). In terms of the data from the probe trials, western blot analysis and real-time PCR were made using a one-way analysis of variance (ANOVA) by SPSS (SPSS Inc, Chicago, IL, USA) 19.0 for windows, while the data from oriented navigation trial was analysed with a two-way ANOVA. A value of $P < 0.05$ was considered statistically significant.

3 Results

3.1 Effects of aerobic exercise on spatial learning and memory ability

In the MWM navigation task, all animals' escape latencies gradually shortened (Figure 1a), indicating that each rat was able to learn to locate the submerged platform. The rats in the group C and DE showed maximal performance on day 3 and did not significantly improve thereafter, whereas those in group D improved at a lower pace, and reached maximal performance on day 5(Figure 1a). Analysis of the average escape latencies demonstrates that the group D were significantly longer than the group C and DE on days 2, 3, and 4 of training ($p < 0.05$), while there was no statistical difference on days 1, 5 and 6 among these groups as regards ($p > 0.05$). There was no significant difference in swimming speed among 3 groups during the 6-day training period (Figure 1b), suggesting that the escape latencies were not influenced by swimming velocity. In the probe trial, the rats in group D spent less time in the target quadrant compared with both group C and DE ($p < 0.05$), however, there was no remarkable difference in that between group DE and C (Figure 1c). In addition, rats in group C and DE crossed where the platform was fixed more often than group D ($p < 0.05$) (Figure 1b). Taken together, these data suggested that D-gal administration of 6 consecutive weeks impaired spatial learning and memory, while the aerobic exercise ameliorated the learning and memory impairment caused by D-gal administration.

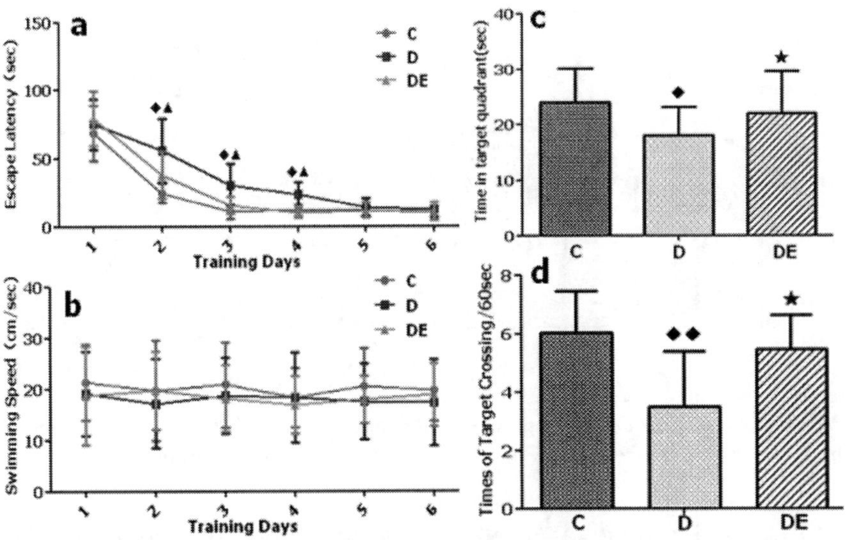

Figure 1. Changes in spatial learning and memory using the MWM test.
(a) Escape latency on each training day. (b) Mean swimming speed on each training day.(c) Number of crossings platform in the target quadrant during 60 seconds probe trail. (d) The time spent in the target quadrant during probe trial. ◆$P< 0.05$, ◆◆$P< 0.01$,*vs*. the group C; ▲ $P< 0.05$, *vs*. the group DE; ★$P< 0.05$, *vs*. the group D.

3.2 Effects of Aerobic Exercise on SYP Protein Expression in Hippocampus

Under the laser confocal microscope, the immune positive products of SYP were observed to scattered distribution in hippocampus, and compared with group D, the fluorescence intensities of group C and group DE were more intense (Figure 2). Further analysis indicated that the IOD value of SYP immunofluorescence positive products in group D was lower than group C ($P < 0.01$), while that of group DE was significantly higher than group D ($P < 0.01$) (Figure 3a).

Consistent with the results of immunofluorescence assay, western blot analysis revealed that SYP protein in group D was significantly lower than both group C ($P < 0.01$) and DE ($p > 0.05$). There was no statistical difference between the group C and DE. The present results show that chronic administration D-gal caused the SYP expression levels in hippocampus decline, while exercise may upregulate SYP, a marker for synaptic plasticity in hippocampus thus reversed the downward to some extent caused by the D-gal treated (Figure 3b).

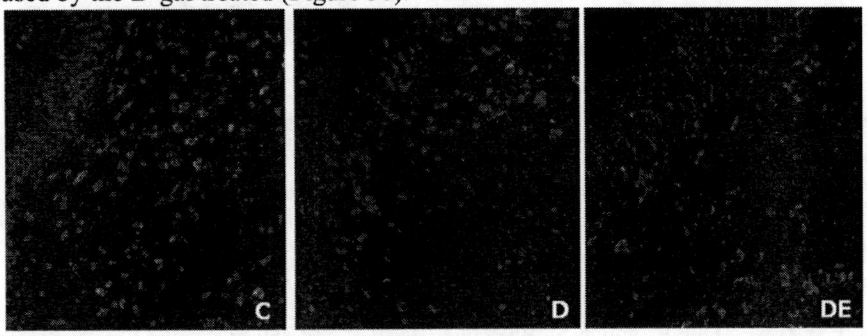

Figure 2. Immunofluorescence micrographs of SYP in hippocampus.
Note: Blue-fluorescence: nucleus tagged by DAPI; Green fluorescence: SYP tagged by FITC.

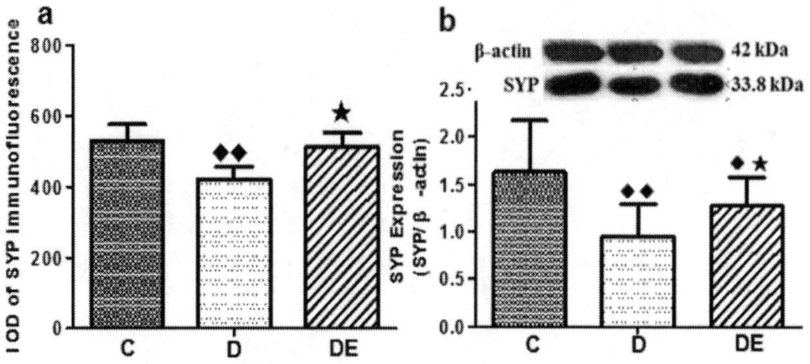

Figure 3 Changes in hippocampal SYP protein.
(a) IOD of SYP immunofluorescence positive products in hippocampus. (b) SYP protein levels in hippocampus using western blot. ◆*P*< 0.05, ◆◆*P*< 0.01,*vs*. the group C; ★*P*< 0.05, *vs*. the group D.

3.3 Effects of Aerobic Exercise on BDNF protein and mRNA Levels in Hippocampus

According to Western blot analysis, the rats in group D had obviously less BDNF level in hippocampus than group C and DE (P<0.01, P<0.05; Figure 4a), and that the BDNF levels slightly declined in group DE compared with group C, however, these values did not reach statistical significance (p>0.05; Figure 4a). Consistent with those of the west blot analysis, the results from real-time PCR showed that the levels of hippocampal BDNF mRNA were significantly declined in the group D compared with both group C and group DE (p>0.05; Figure 4b). These findings demonstrated that D-gal administration restrains BDNF genetic transcription as well as protein synthesis, however, exercise intervention attenuated the inhibitions induced by D-gal administration.

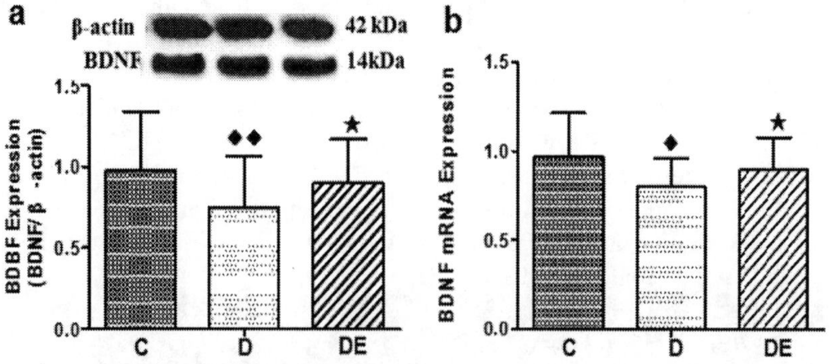

Figure 4 Changes in hippocampal BDNF protein and mRNA.
(a) BDNF protein levels in hippocampus. (b) BDNF mRNA levels in hippocampus. ◆*P*< 0.05, ◆◆*P*< 0.01,*vs*. the group C; ★*P*< 0.05, *vs*. the group D.

4 Discussion

Nowadays, systemic D-gal administration is one of the most extensively used brain aging model building methods[17, 18]. Our previous studies have confirmed that rats injected with D-gal for 6 weeks resulted in deterioration of learning and memory capacity and increase of free radicals in the brain[19], which is similar to the performance in natural aging. In order to test the influence of aerobic exercise on the hippocampal SYP and BDNF of aging rats' and to explore the effects of aerobic exercise on learning and memory of aging brain and the potential mechanism underlying, we built the aging model in this way again and gave rats swimming exercise intervention in the process of aging modeling in the current study.

It is generally acknowledged that synapse is the key structure of neuron networking and information transmission and that neuronal synaptic plasticity is the important part of neurobiology for learning and memory ability[10,11]. Many studies have implicated a contribution of synaptic plasticity reduction in learning and memory dysfunction in the aging process[23, 24]. SYP, a membrane protein with 38kD molecular weight, distributes nearly in all pre-synaptic terminals and is involved in synapse formation, neurotransmitter release and synaptic vesicle cycle[6,7], etc. At present, SYP is recognized as a reliable marker protein for synaptic plasticity in the brain and often used to qualitatively or quantitatively analyze synaptic density and distribution[6]. It has been shown that SYP levels in hippocampus decline with aging and present an age-dependent trend[25]. In the aging model induced by D-gal in our study, the hippocampal SYP expression is also down-regulated, suggesting that synapse degenerative changes in hippocampus, which may be an important cause of learning and memory impairment induced by D-gal treated.

Despite the fact that there is no way to stop the physiological brain aging process, yet numerous epidemiological surveys and experimental studies indicate that appropriate sports could delay the process of aging brain and improve the learning and memory ability[3, 4]. Consistent with what has been previously reported[19], rats' performance in MWM demonstrates that regular swimming exercise could prevent learning and memory impairment against brain senescence in the present study, and it is possible that this prevention effect for learning and memory impairment may be a consequence of improved synaptic plasticity and neurotransmission through the increase of SYP expression. This point is based on the observation that both immunofluorescence and western blot assay of hippocampal SYP in the present study showed significantly higher expression level in rats that did exercise in the process of D-gal injection than rats that didn't exercise.

BDNF, one of the most abundant neurotrophin factors, widely distributing in the central nervous system, has been confirmed by animal studies that it could regulate hippocampal synaptic plasticity and cognitive function through multiple mechanisms[26,27], and its' expression level in hippocampus is closely related to the ability of learning and memory[10,11]. BDNF knockout mice exhibit decreases in the number of docked vesicles and in the levels of SYP and impairments in neurotransmitter release[28]. A lot of experimental results show that BDNF amount in brain, especially in hippocampus, is reduced in the aged[12,29,30]. Consistent with these results, we found that both BDNF mRNA and BDNF protein levels in hippocampus of aging model rats were significantly lower than the adult rats in the current study. Exercise has been found to increase hippocampal BDNF both in the growth period and adulthood. In the present study, we investigated the effect of aerobic exercise intervention on BDNF expression in hippocampus during the aging and the results demonstrated that aerobic exercise intervention in the aging process could prevent age-related hippocampal BDNF gene transcription and translation levels decrease in a certain extent. Further statistics show that the hippocampus BDNF levels were highly positively correlated to SYP. These results indicate that exercise induced-BDNF express involve in preventing aging-related hippocampal synaptic plasticity impairment and learning and memory decline.

In summary, the findings provide evidence that aerobic exercise ameliorates learning and memory deficits of aging rats induced by D-gal, which may be related to promoting hippocampal BNDF and SYP expression by aerobic exercise. However, further study is also needed to probe which BDNF signaling pathway is involved in the strengthening of synaptic plasticity and the improvement of learning and memory.

References

1. Subramaniam M, Chong SA, Vaingankar JA, et al. Prevalence of dementia in people aged 60 years and above: results from the WiSE study [J]. *J Alzheimers Dis.***45**(4):1127-1138(2015).
2. Ritchie SJ, Dickie DA, Cox SR, et al. Brain volumetric changes and cognitive ageing during the eighth decade of life [J].Hum Brain Mapp,36(12):4910-4925(2015).
3. Łojko D, Pałys W, Czajkowska A, et al. Association of cognitive performance with the physical activity and body mass index in middle-aged and older rural inhabitants [J].*Eur Rev Med Pharmacol Sci*, **18**(23):3645-3652(2014).

4. Chang M, Jonsson PV, Snaedal J, et al. The effect of midlife physical activity on cognitive function among older adults:AGES–Reykjavik Study[J].J *Gerontol A Biol Sci Med Sci*,65:1369-1374 (2010).
5. Ales S. Dynamic learning and memory, synaptic plasticity and neurogenesis:an update[J].*Front Behav Neurosci* 8: 106(2014).
6. Yang L, Zhanhua L, Jing L, et al. Downregulation of caveolin-1 contributes to the synaptic plasticity deficit in the hippocampus of aged rats [J]. *Neural Regen Res*,**8**(29):2725-2733(2013).
7. Reddy PH, Mani G, Park BS, et al. Differential loss of synaptic proteins in Alzheimer's disease: implications for synaptic dysfunction[J].J *Alzheimers Dis*,**7**(2):103-117(2005).
8. Davies HA, Kelly A, Dhanrajan TM, et al. Synaptophysin immunogold labelling of synapses decreases in dentate gyrus of the hippocampus of aged rats [J].*Brain Res*, **986**(1-2):191-195(2003).
9. Smith TD, Adams MM, Gallagher M, et al. Circuit-specific alterations in hippocampal synaptophysin immunoreactivity predict spatial learning impairment in aged rats [J].J Neurosci, 20(17):6587-6593(2000).
10. Cowansage KK, LeDoux JE, Monfils MH. Brain-derived neurotrophic factor:a dynamic gate keeper of neural plasticity [J].*Curr Mol Pharmacol*, 3:12–29(2010).
11. Edelmann E, Lessmann V, Brigadski T. Pre and postsynaptic twists in BDNF secretion and action in synaptic plasticity. Neuropharm,76 (Pt.C),610–627(2014).
12. Rothman SM, Mattson MP. Activity-dependent,tress-responsive BDNF signaling and the quest for optimal brain health and resilience throughout the lifespan [J].*Neurosci*,**239**(7):228-240 (2013).
13. Erickson KI, Gildengers AG, Butters MA. Physical activity and brain plasticity in late adulthood [J].*Dialogues Clin Neurosci*,**15**(1):99-108(2013).
14. Alomari MA, Khabour OF, Alzoubi KH, et al. Forced and voluntary exercises equally improve spatial learning and memory and hippocampal BDNF levels[J].*Behav Brain Res*,247:34-39 (2013).
15. Bechara R, Kelly M.Exercise improves object recognition memory and induces BDNF expression and cell proliferation in cognitively enriched rats [J].*Behav Brain Res*, 245:96-100(2013).
16. Liochev SI. Reactive oxygen species and the free radical theory of aging [J].*Free Radic Biol*, 60:1-4(2013).
17. Haider S, Liaquat L, Shahzad S, et al. A high dose of short term exogenous D-galactose administration in young male rats produces symptoms simulating the natural aging process [J]. *Life Sci*,**3**(124):110-119(2015).
18. Yu YH, Ren GP, Liu YN. Effect of FGF-21 on learning and memory ability and antioxidant capacity in brain tissue of D-galactose-induced aging mice[J].*Yao Xue Xue Bao*,**49**(7):1000-1006(2014).
19. Qiongjia Y, Jinmei Z, Wenqian D, et al. The effect of aerobic exercise on spatial learning and memory ability and expression of NCAM in hippocampus during the aging process of rats [J]. *Chine Sport Science*. **32**(8):85-90(2014).
20. D'Hooge R, De Deyn PP. Applications of the Morris water maze in the study of learning and memory[J].*Brain Res Rev*,**36**(1):60–90(2001).
21. Franklin KBJ, Paxinos G. The mouse brain in stereotaxic coordinates. San Diego: Academic Press, (1997).
22. Kenneth JL, Thomas DS. Analysis of Relative Gene Expression Data Using Real-Time Quantitative PCR and the $2^{-\Delta\Delta}$CT Method. Methods, **25**(4):383-481(1997).
23. Hatanpää K, Isaacs KR ,Shirao T, et al. Loss of proteins regulating synaptic plasticity in normal aging of the human brain and in Alzheimer disease. J *Neuropathol Exp Neurol*.**58**(6):637–643 (1999).
24. Patterson SL. Immune dysregulation and cognitive vulnerability in the aging brain: Interactions of microglia,IL-1β,BDNF and synaptic plasticity[J].*Neuropharm*.**96**(Pt A):11-18(2015).
25. Smith TD, Adams MM, et al. Circuit-specific alterations in hippocampal synaptophysin immunoreactivity predict spatial learning impairment in aged rats[J].J *Neurosci*.**20** (17):6587–6593 (2000).
26. Shin MK, Jung WR, Kim HG, et al. The ganglioside GQ1b regulates BDNF expression via the NMDA receptor signaling pathway[J].*Neuropharm*,77:414-421(2014).
27. Hussam J, Mohamed K. Acute BDNF treatment upregulates GluR1-SAP97 and GluR2-GRIP1 interactions: implications for sustained AMPA receptor expression[J].*PLOS ONE*,**8**(2):1-13 (2013).
28. Heldt SA, Stanek L, Chhatwal, JP, et al. Hippocampus specific deletion of BDNF in adult mice impairs spatial memory and extinction of aversive memories[J].*Mol.Psy*,12: 656–670 (2007).
29. Petzold A, Psotta L, Brigadski T, et al. Chronic BDNF deficiency leads to an age-dependent impairment in spatial learning. Neurobiol Learn Mem.120:52-60(2015).
30. Zeng Y, Tan M, Kohyama J, et al. Epigenetic enhancement of BDNF signaling rescues synaptic plasticity in aging[J].J *Neurosci*,31:17800–17810(2011).

Study on interaction between salicylaldehyde l-serine schiff base and human serum albumin by fluorescence spectroscopy

Yanqiu Yang, Yanju Liu, Jundi Zhang and Huaixia Yang[a]

School of Pharmacy, Henan University of Chinese Medicine, Zhengzhou 450046, China

Abstract. The interaction of salicylaldehyde L-serine Schiff base (L) with human serum albumin (HSA) was examined by fluorescence emission spectra at the excitation wavelength 290 nm. Through fluorescence quenching experiments, it was confirmed that the combination of L with HSA was static quenching process. Thermodynamic parameters, such as ΔG, ΔH and ΔS, were calculated at different temperatures, showing that van der Waals force or hydrogen bond interaction were mostly responsible for the binding of L to HSA. The experiments results showed that the microenvironment and the conformation of HSA changed during the binding reaction.

1 Introduction

Amino acid Schiff base attracted attention of researchers, for their interesting structure and unique biological action. Serum albumin can be effectively combined with many type endogenous substances in the human body and exogenous drugs, which play an important role in storage and transportation [1].Therefore, the study of the interaction between proteins and drug molecules are a meaningful topic in the field of life and chemical sciences, which help to study the strength of the combination and the bonding mechanism between proteins and small molecules.

In the present work, the interaction between L and HSA was investigated at two temperatures under physiological conditions, utilizing the fluorescence method. We probed into the fluorescence quenching mechanism of HSA by L and calculated the binding constant K_b and the number of binding sites n for L-HSA complex.

[a] Corresponding author: yanghuaixia886@163.com

This study was supported by HenanProvince basic and frontier technology researchprojects (152300410214), Program for the key scientific research project of colleges and universitiesin Henan Province (16A350002), and Science & Technology Innovation Talents in Henan College of Chinese Medicine (2014XCXRC04).

DOI: 10.1051/bioconf/20170801021

2 Experimental

2.1 Apparatus and reagent

2.1.1 Apparatus

970 CRT fluorometry (Shanghai Scientific Instrument Co., Ltd. China), MP521 Lab pH/Conductivity Meter (Shanghai San-Xin Instrumentation, Inc. China).

2.1.2 Reagents

Salicylic aldehyde (Sinopharm Group Co., Ltd. China), L-serine (Biochemical reagents, Sinopharm Group Co., Ltd. China), HSA (Beijing Olympic nemesis Biological Technology Co. Ltd. China, relative molecular mass 66000), was dissolved daily in 0.1 $mol \cdot L^{-1}$ Tris-HCl buffer solution (PH = 7.4) to prepare a solution (1×10^{-5} $mol \cdot L^{-1}$), salicylaldehyde L-serine Schiff base to prepare a solution (1.0×10^{-3} $mol \cdot L^{-1}$), All chemicals were analytical-reagent grade. All solutions were prepared with double-distilled water.

2.2 Methods

L is synthesized on the basis of reference [2]. The synthesis route shown as Figure 1.

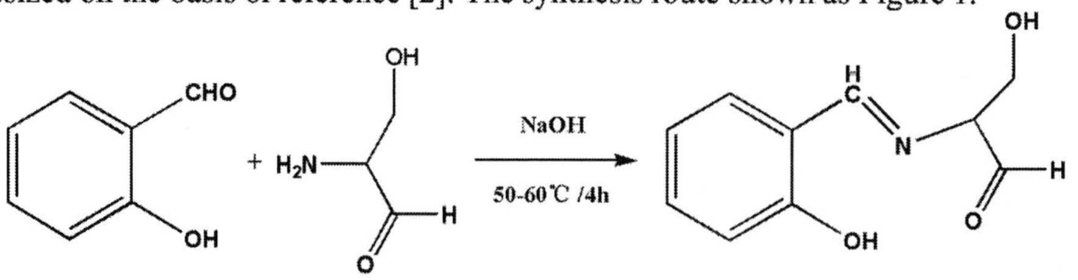

Figure 1. The synthesis route of L.

An appropriate concentration of HSA was titrated by successive additions of a 1.0×10^{-3} $mol \cdot L^{-1}$ doubly distilled water stock solution of L. Titrations were done manually by using trace syringes. The fluorescence and Synchronous fluorescence spectra was measured on 970 CRT fluorometry, using 1.0 cm quartz cells. The excitation wavelength was 290 nm. the $\Delta\lambda$ is set at 60 nm or 15 nm. Experiments were measured under two temperatures (298K, 308K).

3 Results and discussion

3.1 Fluorescence quenching of HSA

The fluorescence of HSA originates from the amino acid residues of tryptophan (Trp) and tryosin (Tyr), which are sensitive to the environment around them [3]. It emits a strong fluorescence peaked at 352 nm with excitation wavelength of 290 nm. Figure 2 shows the fluorescence quenching spectra of solutions containing HSA fixed concentration and different concentrations of L at 298K and 308K, it can be observed that the fluorescence intensity of HSA decreases regularly with increasing of L, but there is no significant emission wavelength shift. These results suggest that L can interact with HSA and quench its intrinsic fluorescence [4].

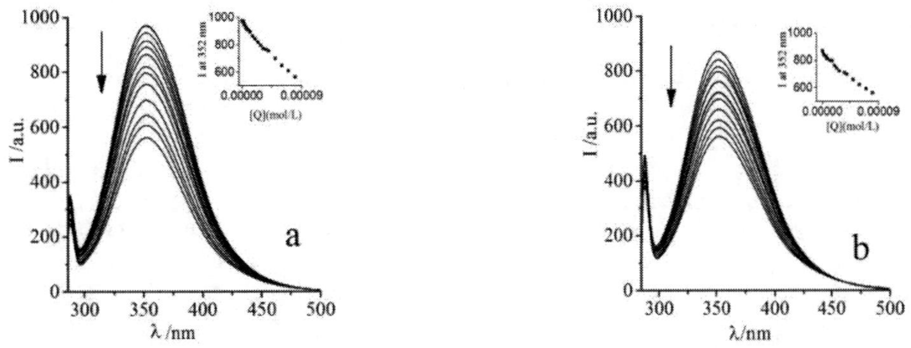

Figure 2. Changes in the emission spectra (λex = 290 nm) of the HSA (1×10^{-5} mol/L) upon increasing concentrations of L (a) 298 K and (b) 308K.

3.2 Determination of quenching mechanism

There are two quenching types in characterizing the mechanism of the interaction between quenchers and macromolecules: static and dynamic quenching. Static quenching refers to the formation of a non-fluorescence fluorophore-quencher complex. Dynamic (collisional) quenching refers to the quencher diffusing towards the fluorophore during the lifetime of the excited state and, upon contact, the fluorophore returns to the ground state without emission of a photon. Dynamic quenching mechanism was probed using the Stern-Volmer equation [5].

$$F_0/F = 1 + K_q\tau_0[Q] = 1+K_{sv}[Q] \quad (1)$$

where F_0 and F represent the fluorescence intensities of HSA in the absence and presence of the external quencher, K_q is the bimolecular quenching rate constant, K_{sv} is the Stern-Volmer quenching constant, τ_0 is the average lifetime of biopolymers without quencher, the fluorescence lifetime of the biopolymers is 10^{-8} s [6], [Q] is the concentration of the quencher.

This work studied the fluorescence spectrum of the action between HSA and L, obtained the plot of F_0/F against [Q] at different temperature. As it was shown in Figure 3. Equation (1) was applied to determine K_{sv} by linear regression from a plot of F_0/F against [Q], hence, K_q can be caculate by $K_q = K_{sv}/\tau_0$. As shown in Table 1.

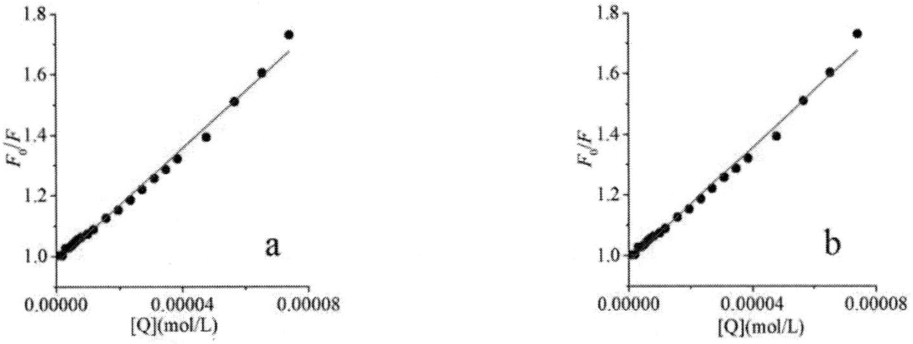

Figure 3. Stern-Volmer plot for the quenching of HSA by L at different temperatures (a)298K and (b)308K.

Table 1. Constant of the systems of interaction between L and HSA at different temperatures.

T/(K)	K_{sv}/(L•mol^{-1})	K_q/(L•mol^{-1}•s^{-1})
298	9.375×10^3	9.375×10^{11}
308	9.397×10^3	9.397×10^{11}

For the L-HSA systems, the values of K_{sv} and $K_q = K_{sv}/\tau_0$ obtained from the plots at 298 K and 308 K are shown in Table 1. According to the literature [7] for dynamic quenching, the maximum scatter collision quenching constant of various quenchers with the biopolymer is 2.0×10^{10} L·mol^{-1}·s^{-1}. The values of K_q are larger than 2.0×10^{10} L·mol^{-1}·s^{-1}, which suggest that the fluorescence quenching is caused by a specific interaction between HSA and L, and the quenching mechanism mainly arise from complexes formation, while dynamic collision could be negligible in the concentration range studied [8].

3.3 The binding constant K_b and the number of binding sites n

For static quenching, the relationship between fluorescence quenching intensity and the concentration of quenchers can be used to obtain the binding constant and the number of binding sites [9].

$$\mathrm{Lg}[(F_0\text{-}F)/F] = \mathrm{Lg}K_b + \mathrm{nLg}[Q] \quad (2)$$

The values of K_b and n at 298K and 308K were obtained from the double logarithm regression curve (Figure 4). The values of K_b and n at 298K and 308K are listed in Table 2.

The n values are nearly 1, and thus indicate the existence of a single binding site in HSA for L. The interactions of L with HSA increased when rising the temperature, which suggest that interactions of L with HSA were endothermic.

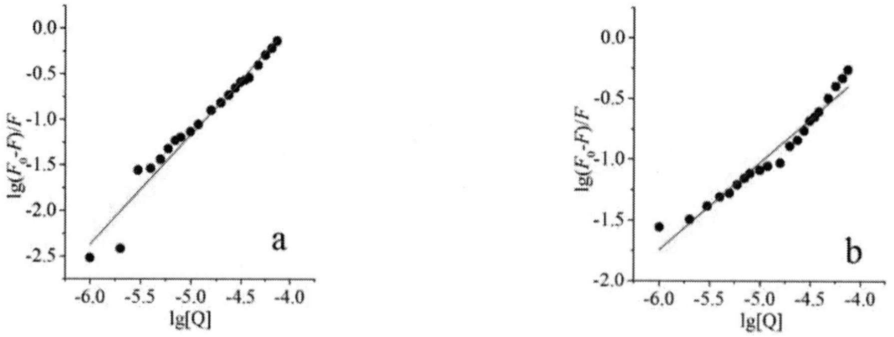

Figure 4. The plots of lg[(F_0-F)/F] vs. lg[Q] at different temperatures (a)298K and (b)308K.

Table 2. The binding constant, K_b, and the number of binding sites, n

T/(K)	double logarithm e quation	R	K_b/(L/mol)	n
298	Lg[(F_0/F)/F=4.8474+1.202Lg[Q]	0.3211	7.0372×10^4	1.202
308	Lg[(F_0/F)/F=2.5554+0.716 Lg[Q]	0.1269	0.0359×10^4	0.716

 DOI: 10.1051/bioconf/20170801021

3.4 Binding modes

The acting forces between a drug and a biomolecule are composed of weak interactions of molecules such as hydrogen bond formation, van der Waals forces, electrostatic forces, and hydrophobic interaction [10]. The thermodynamic parameters could be calculated by eqation (3), (4) and (5),

$$\mathrm{Ln}\ (K_2/K_1) = \Delta rHm\ (1/T_1 - 1/T_2)\ /R \quad (3)$$

$$\Delta rGm = -RTLnK \quad (4)$$

$$\Delta rSm = (\Delta rHm - \Delta rGm)/T \quad (5)$$

The thermodynamic parameters are presented in Table 3.

Table 3. Thermodynamic parameters for the association of L with HAS.

T/K	ΔrHm/(kJ.mol^{-1})	ΔrSm/(J.K^{-1})	ΔrGm/(kJ.mol^{-1})
298	-402.80	-1258.86	-27.66
308			-15.07

The negative signs for ΔG reveal that the binding processes are spontaneous, the formation of drug-HSA coordination compounds are spontaneous. The thermodynamic parameters, enthalpy change (ΔH) and entropy change (ΔS) are the main evidence to confirm binding modes. From the thermodynamic standpoint, $\Delta H > 0$ and $\Delta S > 0$ imply a hydrophobic interaction; $\Delta H < 0$ and $\Delta S < 0$ reflect the van der Waals forces or hydrogen bond formation; and $\Delta H \approx 0$ and $\Delta S > 0$ suggest an electrostatic force [11]. The both negative ΔH and ΔS values indicate that the L-HSA binding is an exothermic and entropy decreasing process. And van der Waals force or hydrogen bond interaction plays a major role in this process. The negative signs for ΔG reveal that the formation of L-HSA coordination compounds are spontaneous.

3.5 Conformation investigation

The synchronous fluorescence spectra are frequently used to characterize the interaction between the molecule probe and proteins since it can provide information about the molecular microenvironment in the vicinity of the chromophore molecules. HSA has intrinsic fluorescence due to there are three intrinsic fluorophores, namely, Trp residues, Tyr residues, and phenylalanine (Phe) residues. Trp and Tyr have the similar excitation spectrum, use the synchronous fluorescence can distinguish them. The synchronous fluorescence spectra reflect the different nature of chromophores: when the Δλ between excitation wavelength and emission wavelength is set at 60 nm, the synchronous fluorescence spectrum gives the characteristic information of tryptophan residues and when Δλ value is 15 nm, the synchronous fluorescence spectrum is characteristic of tyrosine residues. Synchronous fluorescence spectra of HSA at various L concentrations when $\Delta\lambda = 60$ nm and $\Delta\lambda = 15$ nm are shown in Fig. 5. As it is known, nearly all hydrophobic amino acids are embedded in the cylinders forming hydrophobic cavities. When $\Delta\lambda = 15$ nm, The blue-shift effect suggested that the polarity around the tryptophan residues decreased and the hydrophobicity increased. When $\Delta\lambda = 60$ nm, the emission peaks show essentially no shift over the investigated concentration range. which indicate that L have little effect on the microenvironment of tryptophan residues in HSA.

DOI: 10.1051/bioconf/20170801021

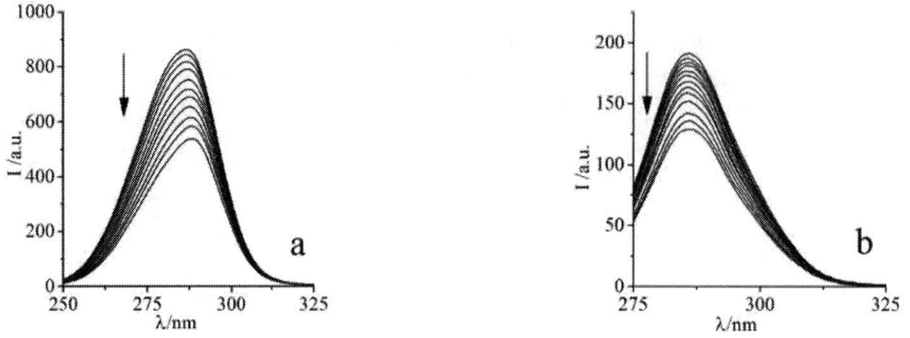

Figure 5. Changes in the synchronous fluorescence spectra of HSA upon increasing concentrations of L (a) Δλ = 60 nm and (b) Δλ = 15 nm.

4 Conclusions

The interaction between L and HSA has been investigated under simulated physiological conditions (pH 7.4, ionic strength 0.1 mol/L) using fluorescence methodology. The quenching rate constant, the binding constants, the number of binding sites, the thermodynamic parameters and the binding forces were obtained. The experimental results indicate that the fluorescence quenching mechanism of HSA by L was consistent with the static model, and the binding reaction was spontaneous. Van der Waals force or hydrogen bond interaction played a major role in the L-HSA interaction. The results of synchronous fluorescence spectra initiative indicate that the secondary structure of HSA molecules was affected significantly in the presence of L.

References

1. J. Zhang, H.H. Sun, Y.Z. Zhang, J. Solution Chem, **41** (2012).
2. S.G. Dong, H.M. Cao, L.J. Zhao, Journal of Jilin Medical College, **5** (2014).
3. Y.J. Liu, X.F. Li, H. Fan, L. N. Lu, China Measurement & Test, **9** (2015).
4. A. Jahanban-Esfahlan, V. Panahi-Azar, Food chemistry, **202** (2016).
5. X. Zheng, M.P. Ye, Y. Li, Journal of China University of Metrology, **1** (2015).
6. W.Y. He, G.Y. Chen, J. Du, Acta Chim. Sinica, **66** (2008).
7. F.L. Cui, J.L. Wang, Y.R. Cui, J.P. Li, Anal. Chim. Acta, **571** (2006) .
8. L. N. Hordge, K. L. McDaniel, D. D. Jones, et al, Talanta, **152** (2016).
9. N. Nanjundan, P. Selvakumar, R. Narayanasamy, et al. J. Photoch. Photobiolo. B, **141** (2014).
10. J. J. Xue, Q. Y. Chen. Spectrochim. Acta A, **120** (2014).
11. F. Samari, M. Shamsipur, B. Hemmateenejad, T. Khayamian, Eur. J. Med. Chem., **54** (2012).

Clinical application research on the use of the Photon Therapeutic Apparatus combined with Hops extract compound ointment in treatment for breast cancer patients

Yunzhu Chen, Liying Qun and Guang Sun[a]

Breast Surgery Department of China-Japan Union Hospital of Jilin University, No.169,Xiantai Street Changchun Jilin Province,130021, China

Abstract. Objective: To research Carnation-87C photon theraapeutic apparatus combine with compound hops extract ointment in the clinical effect of breast cancer patients after operation with unhealed incision of skin flap necrosis. Method: The experimental group is treated by the conventional therapy and Carnation-87C photon therapeutic apparatus combine with compound hops extract ointment, while the control group is treated by the conventional therapy only, such as regular dressing change. The other treatments for two groups of patients are the same. Through 4 weeks' treatment, the growth of fresh granulation tissue and healing time were observed between two groups. Results: There were statistically significant differences ($P<0.05$) in the growth of fresh granulation tissue and healing time between two groups of patients with non-healing skin flap necrosis after the operation. Conclusion: The treatment with the Carnation-87C photon therapeutic apparatus combine with compound hops extract ointment for breast cancer patients after operation with unhealed incision of skin flap necrosis is easy to operate with good curative effect, and it is worth popularizing in clinical application.

1 Introduction

Since May 2011, Carnation-87C photon therapeutic apparatus (Shenzhen Pumen science and Technology Co., Ltd. production) combine with compound hops extract ointment (Jilin Haifa pharmaceutical Limited by Share Ltd Jilin Aodong group) were used in 31 cases of breast cancer postsurgical patients with skin flap necrosis in our department and their wound healing finally acquired definitive curative effect.

2 Method

2.1 Inclusion criteria

[a] Corresponding author: guangsun2013@163.com
The study was supported by a grant from the key scientific and technological project for Jilin Province of China (No. 20150204081SF).

DOI: 10.1051/bioconf/20170801022

Patients from China-Japan Union Hospital of Jilin University during May, 2011 and May, 2016, female, accepted the breast cancer modified radical mastectomy or radical mastectomy, diagnosed with breast cancer by pathology and immunohistochemistry examination, had long-term non-healing (>2 weeks) skin flap necrosis after radical mastectomy or modified radical mastectomy.

2.2 Exclusion criteria

Patients had serious infection, autoimmune disease or basic diseases of other organs, such as liver, kidney, heart, cerebrovascular, etc. Patients had other serious complications, such as severe pneumonia, pulmonary embolism, myocardial infarction, etc. Patients were allergic to compound hops extract ointment.

2.3 General information

We selected the patients from China-Japan Union Hospital of Jilin University during May, 2011 and May, 2016. Those patients with skin flap necrosis after breast cancer surgery required a long recovery cycle (more than 2 weeks). Following the strict inclusion and exclusion criteria, all the patients we chose were female who had been taken the breast cancer modified radical mastectomy or radical mastectomy. Using the method of random number table, the 61 cases of breast cancer patients in the group of number were divided into two groups: 31 cases in experimental group while 30 cases in control group.

The experimental group, aged between 36-73 (51.48±9.06) years old, median age is 51 years old, is treated by the conventional therapy and Carnation-87C photon therapeutic apparatus combine with compound hops extract ointment. Their necrotic skin flap healing period is between 15 and 45(23.10±7.92) days. The average wound area is 20.48±19.45 cm 2 and 13 tumours are on the left breast while 18 tumours are on the right breast.

The control group, aged between 31-69 (52.87±7.71) years old, median age is 54 years old, is treated by the conventional therapy only, such as regular dressing change. Their necrotic skin flap healing period is between 28and 60(39.73±8.20) days. The average wound area is 19.97±18.34cm2 and 19 tumours are on the left breast while 11 tumours are on the right breast.

The two groups of cases showed no statistical difference in age and the average wound area ($P > 0.05$), but showed significance difference in treatment days ($P < 0.05$). General clinical information is shown in Figure 1 below treatment days.

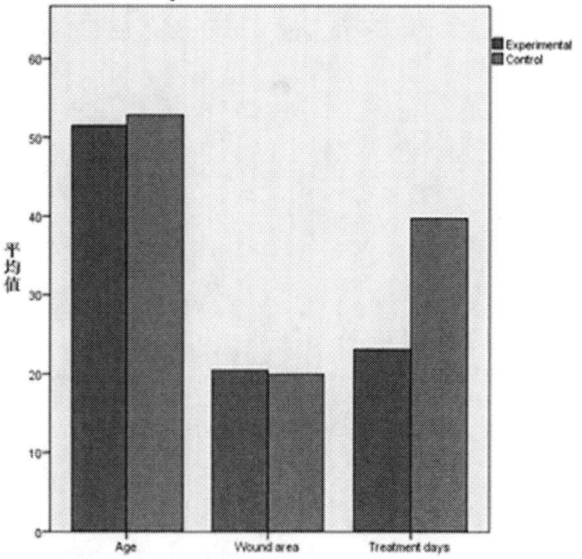

Figure 1. The General Information of the two groups which show no statistical different in age, the average wound area (($P > 0.05$)), but show significance difference in treatment days ($P < 0.05$).

3 Treatment

The experimental group is treated by the conventional therapy and Carnation-87C photon therapeutic apparatus combine with compound hops extract ointment. After using the iodine tampons to disinfect surrounding skin, washing the wound by physiological saline repeatedly and removing the necrotic tissue completely, we use the photon therapeutic apparatus to irradiate the wound for 15 minutes each time with 10- 12cm of distance. Then we apply compound hops extract ointment to the wound duly; keep appropriate negative pressure drainage unobstructed and use external pressure dressing to make the compound hops extract ointment integrated with the wound closely. The dressing change should be done once a day. The fresh granulation tissue growth and healing status of the wound can be observed stage by stage.

The control group is treated by the conventional therapy only, such as regular dressing change. After using the iodine tampons disinfect surrounding skin, washing the wound by physiological saline, use scissors to cut off the wound necrotic tissue. Then use physiological saline cotton ball to wipe the wound purulent secretion off. The last step is to wash the wound by physiological saline repeatedly, keep negative pressure drainage unobstructed and use external pressure dressing to make sterile gauze integrated with the wound closely. The dressing change should be done once a day. Observe the fresh granulation tissue growth and healing status of the wound. The other treatments for two groups of patients are the same.

3.1 Standard of curative effect and evaluation method

Healing: The wound should be covered by the newborn epithelial tissue. The healing area is more than 50% with no exudation and the local tissue should be repaired successfully.

Effective: Part of the wound should be covered by the newborn epithelial tissue. The healing area is more than 50% with no exudation and the new granulation tissue can be seen.

Invalid: There is no newborn epithelial tissue covering the wound. The healing area is less than 50% with exudation or bleeding and erosion appears at local wound.

If the wound is still unhealed after 4 weeks' treatment, we should continue the same treatment and statistics the time when the wound heals completely.

3.2 Statistical processing

SPSS23.0 statistical software is applied; t test analysis is used for measurement data while count data is used for chi-square test. $P<0.05$ for differences is statistical significance.

4 Results

Through our comparative analysis of two groups' clinical curative effect after treatment. The results showed that in experimental group including 31 patients, 27 cases showed healed (87%), only 1 case showed invalid (3%), healing period was between 15 and 45(23.10±7.92) days; and in control group including 30 patients, 20 cases showed healed (67%), 3 cases showed invalid (10%), healing period was between 28and 60(39.73±8.20) days. Two groups of patients were statistical significance ($p<0.05$).

Table 1. Comparison the two groups of patients' clinical curative effect after treatment.

Group	Case number	Healed	Effective	Invalid	Recovery Rate	Healing time
Experimental	31	27	3	1	87	23.10±7.92
Control	30	20	7	3	67	39.73±8.20

5 Discussion

Not only in China but also in the other countries all over the world, the breast cancer is one of the most malignant tumours which are harm to women's health. Taking surgery is still the key part in breast cancer treatment, and that makes postoperative subcutaneous effusion, infection and skin flap necrosis become common postoperative complications in most breast cancer patients, and that common postoperative complications can directly affect the development of subsequent chemotherapy and radiotherapy. The postoperative complications' occurrence rate is between 10% and 61% according to domestic and foreign comprehensive reports[1].

There are a lot of factors including the incision's location selection, the thickness of the flap, the reasonable use of electric knife; complete wound hemostasis and effective drainage can affect the healing process of the incision. High tense surgical incision is not conducive to healing of the wound. The skin flap necrosis often caused by the absolute shortage of tissue volume and the blood supply obstacle. Most surgeons are used to radio-knife for skin flap decoherence in the operation of breast cancer and because a lot of patients' fat layer is too thick, postoperative patients often suffer from skin flap necrosis or small area of the wound become ulcer and scabby. Uncompleted wound hemostasis is easy to cause subcutaneous effusion or accumulate blood which is not conducive to the wound healing. Inoperative drainage is easily to cause dead space below skin flap, which increase the leakage of the wound and the risk of infection. So the wound healing won't be good enough even the wound will be disunion. Therefore, surgeons should pay more attention to the five key points below to help postoperative patients' wound healing. Firstly, surgeons should select moderate skin tension of incision. Secondly, surgeons should keep a small amount of subcutaneous fat layer to ensure the blood supply. Third, surgeons should use low-power radio-knife as far as possible. Fourth, patients' wound hemostasis should be complete. Fifth, clinical medical stuff should keep drainage smooth [2].

As shown in Figure 2, the Carnation - 87 - c red light therapeutic apparatus can output the monochrome photon beam whose wavelength is between 600 and 700 nm. After irradiating high intensity photon beam irradiation on the wound, photon energy is absorbed by mitochondria in cells, stimulate a variety of the enzyme activities such as cytochrome oxidase in mitochondrial, promote the synthesis of adenosine triphosphate (ATP), improve the function of cell respiration and circulation, increase the amount of vascular endothelial cell factors and basic fibroblast growth factors in blood to facilitate the proliferation of fibroblasts and endothelial cells. Red light therapy can enhance leukocyte phagocytosis, improve the cellular immunity and humoral immunity functions, strengthen the ability to resist infection, get inflammation of resolution [3]. Red light treatment can obviously improve local blood circulation, reduce blood vessel permeability, reduce wound fluid seepage, improve the absorption of the drainage [4], so as to promote wound healing and shorten the length of hospital stay. In short, Carnation - 87 - c red light therapy can cure local tissue inflammation, get detumescence and analgesic effects and can obviously shorten the healing time, improve hospital's quality of medical therapy. In summary, red light irradiation has effect on local anti-inflammatory, detumescence, analgesia, which can significantly shorten the healing time to improve the quality of life of patients with breast cancer after operation and curative effect, and it has good expand prospect improve hospital quality of care.

Compound hops extract ointment contains, hops extract, gamma - evening primrose oil, vitamin B12, etc. Its main functions are antipruritic, analgesic and repairing vascular endothelial cells. It also has almost the same topical analgesic effect compared with cocaine and procaine [5]. It is reported that hops extract is antibacterial, antiviral, antitumor, antioxidant and sedative which can obviously inhibit the growth of the gram-positive bacteria such as staphylococcus aureus [6-8]. Hops extract compound ointment solves the problems such as drying easily, no persistent action time which are caused by external traditional drugs. When early injured skin micro vascular expansion happens, hops extract compound ointment can be fully absorbed by skin and directly be involved in RNA synthesis in skin cells, so blood vessel endothelial injury can be constantly repaired and vascular endothelial cells can be protected. According to the functional characteristics of hops extract compound ointment, the wound healing and good results were obtained successfully. Apply hops extract compound ointment to the wound duly; keep appropriate negative pressure drainage unobstructed and use external pressure dressing to make the hops extract compound ointment integrated with the wound closely. This method can improve the anti-inflammatory and analgesic therapeutic effect of hops extract compound ointment on saprophytic muscle and accelerate the growth of epithelial cells. Because the medicine source is rich, the production technology is mature, the method of application is convenient and the curative effect is good without side effects, it is worth popularizing and applying in clinic.

31 patients have no obvious side effects in the process of using Carnation-87C photon therapeutic apparatus combine with compound hops extract ointment and it has obvious advantages in the treatment of skin flap necrosis after breast cancer operation.

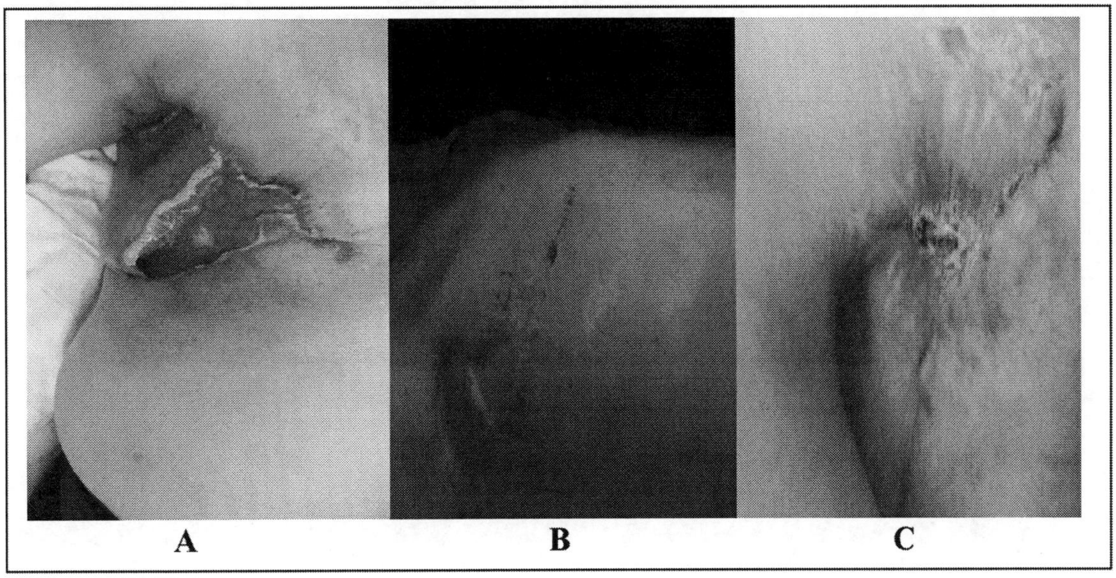

Figure 2. Headings 15 days after breast surgery. There is a 12.58×5.32cm2 skin flap necrosis appear on this patient right chest. A: The incision before treatment. B: The photon therapeutic apparatus was working. C: The incision got healed after 4 weeks by using he photon the rapeutic apparatus combined with compound vitamin B12 ointment II.

References

1. Komorowski A, Zanini V, Regolo L,et al.Necrotic complications after nipple and areola sparing mastectomy J[J].*World J Surg*,2006,**30**(7):1410.
2. Olsen MA, Nickel KB, Margenthaler JA, et al. Increased risk of surgical site infection among breast-conserving surgery reexcisions [J]. *Ann Surg Oncol*, 2015,**22**(6):2003-2009.
3. Yinghong Xia, Chuanlong Miao, Shao Wang, et al. The inhibition of pain caused by Vitamin B12 ointment on inflammation. [J] *Journal of Bethune medical university*, 1996, 22 :597.
4. Yumei Liu, Chemical Composition and Pharmacological Effect of Hops: A Review. [D]. Xinjiang: Xinjiang University College of Chemistry and Chemical Engineering,2009. 23-30.
5. YU Xue-ming, LI Juan, Clinical observation of the photon therapeutic apparatus in preventing its complications after flap surgery.Chinese Journal of Aesthetic Medicine,2015,20:1008-6455.
6. Danbin Jia, Yu Zhu, Shan Liu, et al. Effect of red light irradiation on wound healing [J] *J Fourth Mil Med Univ*,2008,**29**(13):1195-1197.
7. Aifeng Zhu, Hongbin Yang, Xuefeng Wan， et al. Curative effect observation of red light in the treatment of postoperative wound of condyloma [J]. *Practical clinical medicine*, 2010,**11**(1):67-69.

DOI: 10.1051/bioconf/20170801022

8. Bin Zhang, Bin Gao. Analysis of high energy and narrow spectrum photon therapy in the treatment of third degree pressure ulcers. [J] *Chinese medical equipment*,2013,**10**(2):15-17.

 DOI: 10.1051/bioconf/20170801023

Clinical observation on the effect of Hops extract compound ointment in the treatment of breast cancer patients

Miao Miao, Liying Qun and Guang Sun[a]

Breast Surgery Department of China-Japan Union Hospital of Jilin University, 130021, China

Abstract. The purpose of this study is to investigate the therapeutic effect of hops extract compound ointment in treating skin flap necrosis after breast cancer surgery. 48 cases of after various types operation of breast cancer patients with postoperative skin flap necrosis were divided into two groups according to random number table, 24 cases of treatment group, 24 cases of the control group. In treatment group, there are 24 patients, among them, 20 cases show healed (83.33%), only 1 case shows invalid (4.17%). At the control group, there are 24 patients, among them, 12 cases show healed (50%), 5 case shows invalid (20.83 %). The treatment group compared with control group, with statistical significance($p<0.05$). Experiment results show that in the treatment group, the total efficiency was obviously higher than that of the Control group. And the time of the wound healing is significantly shorter than the control group. Visible, treating skin flap necrosis after breast cancer surgery with hops extract compound ointment could anti-inflammatory, promote skin lesions healing, analgesia, and reduce the exudation, which is worth in clinical application

1 Introduction

Breast cancer is the most common malignancy in women worldwide [1-3]. In recent years, studies have indicated that breast cancer is a systemic disease that requires surgery combined with chemotherapy, radiotherapy and endocrine therapy. The preferred method of treatment is surgical treatment, in the current clinical applications, the five most common surgical methods: breast cancer radical resection, extended radical mastectomy of breast cancer, modified radical mastectomy of breast cancer, breast cancer mastectomy, breast conserving surgery. Each operation method is still not completely avoiding the occurrence of postoperative complications, especially modified radical mastectomy wound area is relatively large, even if the bleeding wound thoroughly, but because of bleeding wounds or poor lymph circulation, skin flap necrosis of the breast one of the most common complications after surgery. Each action method still cannot completely avoid the occurrence of postoperative complications, especially breast modified radical operation wound area is relatively large, even if the wound bleeding completely, but due to wound ooze blood or lymph circulation is not smooth, skin flap necrosis is one of the most common complications after breast cancer. It not only

[a] Corresponding author: guangsun2013@163.com
The study was supported by a grant from the key scientific and technological project for Jilin Province of China (No. 20150204081SF).

DOI: 10.1051/bioconf/20170801023

delayed breast cancer patients with incision healing, it will also increase with the time of hospitalization, cost of hospitalization, it will also increase the patient's hospital stay, hospital costs, allowing patients to bear a greater economic and mental stress, more serious is delayed radiotherapy and chemotherapy time, thus affecting radical mastectomy the comprehensive treatment. However, the hops extract compound ointment has anti-inflammatory ointment to promote healing of skin lesions [4-6]. In recent years, the treatment of skin flap necrosis after breast cancer operation by using this new ointment has a good clinical effect, which greatly reduces the incidence of subcutaneous effusion and skin flap necrosis after breast cancer surgery. We will do the following summary on it.

2 Clinical data

2.1 General information

A total of 48 cases of skin flap necrosis after breast cancer surgery were collected. the patients we selected from October, 2010 to April 2016 in China-Japan Union Hospital of Jilin university. All patients were female. According to the principle of random divided into two groups,24 cases of treatment group, 24 cases of the control group. In the treatment group, the average age 42-70(56. 19 ±10. 21) years old, duration14-25 (19.95 ±4. 29) days the average wound area(20. 23 ± 19. 42) cm2. 18 patients are stage II, 6 patients with stage III; modified radical mastectomy of breast cancer is 21 cases, breast cancer radical resection is 3 cases; On the left side of the 17 cases of breast cancer, 5 cases on the right side; In the control group,the average age46-68(55.12±9. 15) years old, duration16-27 (22. 1 ±3. 17) days the average wound area(19. 81 ±20. 05) cm2. 19 patients is stage II,5 patients with stage III; modified radical mastectomy of breast cancer is 22 cases, breast cancer radical resection is 2 cases; On the left side of the16cases of breast cancer, 8 cases on the right side. After t test, the age, course of disease, wound area, disease, surgery and other aspects of the comparison, the difference was not statistically significant ($P > 0.05$), the two groups were comparable.

2.2 Inclusion criteria

1 The patient diagnosed with breast cancer by pathology examination.
2 The breast cancer patients have a postoperative wound skin flap necrosis not healing for a long time after operative treatment.

2.3 Exclusion criteria

1 Serious infection or autoimmune disease.
2 Allergic to hops extract compound ointment
3 Such basic disorders as severe liver, kidney, heart, cerebrovascular, etc.

3 Treatment

The treatment group required in strict accordance with the principles of surgical dressing, 2% povidone-iodine disinfection wound, cut off necrotic tissue, and the wound was rinsed with hydrogen peroxide, saline, sterile cotton ball dipped in dry wounds. coated 1g hops extract compound ointment, and covered with sterile gauze, and then provided the appropriate pressure bandage, which could facilitate the absorption of wound exudates and promote drug absorption. Dressing change once a day. It was observed the new born skin appearing and the healing time and wound healing time.

Patients in the control group were treated with the principles of surgical dressing, 2% povidone-iodine disinfection wound, cut off necrotic tissue, and the wound was rinsed with hydrogen peroxide, saline, sterile cotton ball dipped in dry wounds. Conventional alcohol gauze on the incision, and covered with sterile gauze, and then provided the appropriate pressure bandage, which could facilitate the absorption of wound exudates and promote drug absorption. Dressing change once a day. It was observed the new born skin appearing and the wound healing time.

The patients need to dress timely if the dressing exudate. Then according to the future growth of the wound dressing exudation every other day or every 2d dressing.

3.1 Efficacy determination

Healed: The symptoms were completely relieved. The wound healed completely, and no scar or no scar was 1/2.

Improved: The symptoms were relieved. The wound was reduced by more than 25%, or a large number of scar formation

Invalid: The symptoms were not relieved, narrowing of less than 25% of the wound.

3.2 Statistical processing

Application SPSS17.0 statistical software, measuring data to X±S said, using t test analysis; Count data used chi-square test, $P < 0.05$ for the difference was statistically significant.

2.5 Observe the new born skin appearing and the healing time and wound healing time. Two groups would be observed curative effect after 4weeks.

4 Results

4.1 Comparison of clinical efficacy between the two groups

After we compare the two groups of patients' clinical curative effect after symptomatic treatment. In Table 1, the results show that: In treatment group, there are 24 patients, among them, 20 cases shows healed (83.33%), 3 case shows improved (12.50%), only 1 case shows invalid (4. 17%).At the control group, there are 24 patients, among them, 12 cases show healed (50.00%),7case shows improved (29.17%),5 case shows invalid (20.83%). The treatment group compared with control group, with statistical significance($p<0.05$).

Table 1. Comparison the two groups of patients' clinical curative effect after symptomatic treatment.

Group	Case number(n)	Healed		Improved		Invalid	
		n	Rate (%)	n	Rate (%)	n	Rate (%)
Treatment	24	20	83.33	3	12.50	1	4.17
Control	24	12	50.00	7	29.17	5	20.83

After we compare the two groups of patients' clinical curative effect after symptomatic treatment, we found that the treatment group compared with control group, with statistical significance ($p<0.05$).

4.2 Comparison of the new born skin appearing and wound healing time between the two groups

In Table 2, the new born skin averagely appeared after (5.03 ±1.05) days, and the healing of wound took (23. 92 ±4.01) d days. And at the control group, the new born skin averagely appeared after (8.61±2. 01) days, and the healing of wound took (31.90±7. 33) days. There were significant differences between the treatment group and the control group ($P <0. 05$).

Table 2. Comparison of the new born skin appearing and wound healing time between the two groups.

Group	Case number	the new born skin appearing time	the wound healing time
Treatment	24	5. 03 ±1.05	23. 92 ±4. 01
Control	24	8. 61 ± 2. 01	31. 90 ±7. 33

5 Discussion

5.1 Analysing the cause of skin flap necrosis after breast cancer surgery

Breast cancer incidence rate has jumped to the first place in female malignant tumor.4 Surgical operations in the treatment of breast cancer is particularly important, breast cancer surgical trauma is relatively large, easy to produce a variety of complications, especially of skin flap necrosis, long healing complications and sequelae of axillary scar caused, a serious impact on patients with work, study and life. After breast cancer surgery flap necros is over a long period of healing the pathogenesis is unclear, according to our clinical observation that has the following several reasons: unreasonable choice of surgical incision, tension, too thin flap trimming, electrosurgical coagulation excessive tissue ablation, drainage barrier free form hematoma, pressure tight bandage or uneven pressure, its low immunity and systemic malnutrition. Typical isolated flap surgery range of separation, lateral to the latissimus dorsa, inside to the midline of the sternum, to the clavicle, issued to the costal margin, even with reduced radical operation, scope of flap separation did not reduce, because the breast tissue often extends far edge [7]. Due to the extensive dissection especially axillary stripping are more likely to have potential of die cavity caused by liquid deposition, and chest wall movements, and breathing, shoulder movement generated shear and delay the healing of the skin flap.

5.2 Introduction of hops extract compound ointment

The main components of the ointment are extract, auxiliary ingredients have vitamin B12, gentamicin sulcate and the like. This ointment can inhibit microbial growth due to destruction of cellular transmembrane pH gradient [8]. Topical administrations have accelerated ulcer healing, and promote skin RNA synthesis, having itching and analgesic effect. Moreover, the ointment has anti-bacterial and anti-inflammatory, inhibit the growth of bacteria, improve microcirculation, promote capillary regeneration, wound supply epidermal cells proliferation, differentiation required nutrients, immunity and significantly improve local tissue regeneration, enhance local microcirculation and promote growth clumping factor, accelerate healing.

Hops extract compound ointment is an type of ointment formulations, can moist wound, so that the wounds will be in a moist environment, which would keep the wound lasting effect, prompt the drug would be easy to penetrate the skin, reduce the wound tissue, inflammatory exudate, effective drainage through the intact liquefaction timely rot discharge exudate, removal of metabolites is to promote the growth of granulation tissue that is the basis of myogenic, long leather flat muscle that is new epithelial crawling.

5.3 Hops extract compound ointment applied to breast cancer

As shown in Figure 1, flap necrosis is one of the most common complications after the breast cancer surgery, which will increase economic and mental pressure of patients. More serious is to delay radiotherapy and chemotherapy time, will affect the comprehensive treatment of breast cancer. Hops extract compound ointment has accelerated ulcer healing, analgesic antipruritic, antibacterial effect. The results of this study has be shown that the total effective, time of occurrence and the new epithelial healing time of the treatment group was better than those of the control group, indicating that the ointment can be applied to the treatment of breast cancer after skin flap necrosis. However,

 DOI: 10.1051/bioconf/20170801023

this study focused primarily on the analysis of clinical cases, the mechanism still needs further research to explore.

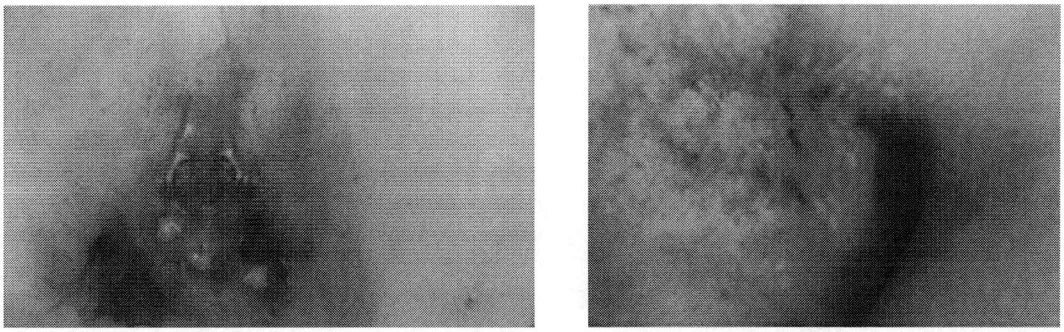

A B

Figure 1. Photo A. 15 days after breast surgery. There is a 9.87×5.45cm2 skin flap necrosis appear on this patient left chest. Photo B. Photo C. 4 weeks after use drug treatment, the incision got healed.

References

1. Badar F, Faruqui ZS, Ashraf A and Uddin N: Third world issues in breast cancer detection. J Pak Med Assoc 57: 137-140, 2007.
2. Jamal S, Mamoon N, Moghal S, Mushtaq S and Luqman M: Carcinoma breast: A histopathological audit. J Coll Physicians Surg Pak 16: 117-119, 2006.
3. Jemal A, Bray F, Center MM, Ferlay J, Ward E and Forman D:Global cancer statistics. CA Cancer J Clin 61: 69-90, 2011.
4. Patel KM, Hill LM, Gatti ME, et al. Management of massive mastectomy skin flap necrosis following autologous breast reconstruction[J]. *Annals of Plastic Surgery*, 2012, **69**(2):139-144.
5. Pinell White, Ximena A, Kolegraff Keli, et al. Skin-Sparing Mastectomy and Immediate Tissue Expander Breast Reconstruction in Patients With Macromastia Using the Passot Breast Reduction Pattern[J]. *Annals of Plastic Surgery*, 2014, 72:158-164.
6. CJ Baldwin. The variation in breast density and its relationship to delayed wound healing: A prospective study of 40 reduction mammoplasties[J]. *Journal of plastic, reconstructive and aesthetic surgery*, 2010, **63**(4): 663-665.
7. Jarman JH, Etchells TA, Martin JD, et al. An integrated framework for risk profiling of reast cancer patients following surgery [J].*Artificial intelligence in Medicine*, 2008, **42**(3): 165-188.
8. Leite IR, Faria JR, Marquez L D S, et al. Evaluation of hopextract as a natural antibacterial agent in contaminated fuel ethanol fermentations [J]. *Fuel Process Technol*, 2013, **106**(2): 611-618.

BIO Web of Conferences **8**, 01024 (2017) DOI: 10.1051/bioconf/20170801024
ICMSB2016

Comparison of clinical efficacy between decitabine combined with half the amount of CAG regimen with CAG regimen alone in patients with inermediate to high-risk myelodysplastic syndrome

Zhenwei Fan[1,a], Xuan Wang[2], Chunshan Zhao[1], Wei Wang[2] and Tingting Yu[2]

[1]*College Nursing of Beihua University, 132000, jilin city, jilin province, China*
[2]*Hematology of Affiliated Hospital of Beihua University, 132000, jilin city, jilin province, China*

Abstract. Objective: To investigate the efficacy and safety of decitabine combined with half the amount of CAG and CAG regimen alone. Methods: Comparison the effectiveness, overall survival and incidence of adverse reactions of 42 cases of MDS used decitabine combined with half the amount of CAG regimen (decitabine 20mg/m^2, once a day, d1-3, Accra neomycin 7mg/m^2,intravenous injection, once a day, d4-7, cytarabine 10mg/m^2,every 12 hours,d4-10, granulocyte colony stimulating factor 300μg, once a day, d4-10, white blood cell count>20×10^9/L when deactivated for four courses) and 48 patients in MDS were treated with chemotherapy alone CAG (Accra neomycin 14mg/m^2,once a day, d1-3, cytarabine 10mg/m^2, every 12 hours, d1-14, granulocyte colony stimulating factor 300μg, once a day, d1-14, white blood cell count>20×10^9/L when deactivated). Results: Compared with pure CAG, clinical efficacy of decitabine combined with half the amount of CAG regimen is better, the treatment does not increase the risk.

1 Introduction

Myelodysplastic syndrome (MDS) are a group disease, begins at hematopoietic stem cells, In blood cell dysplasia of the main features, there may be abnormal of refractory blood cytoplasm and the amount, prone to transformation of acute myeloid leukemia. Since the majority of patients with the disease are elderly, poor physical function, and the ways of low disease malignant clonal proliferation, bone marrow dysplasia, lead to the treatment of drug sensitivity and the poor resistance, so the prognosis is poor. With the progress of research on epigenetics, DNA methyltransferase inhibitorto,in decitabine represented, make MDS patients get better control and mitigation in clinical application manipulation, especially in patients with RAEB or have been converted into leukemia. In this study, in order to find a more rational and effective treatment, survival, the remission rate and incidence of adverse reactions were compared between two groups patients, who suffer high-risk MDS, the application of decitabine joint half the amount of CAG and simple application of CAG regimen.

[a] Corresponding author: Zhenwei Fan: 22289671@qq.com

2 Materials and method

2.1 Object

Observation group patients have been 31patients with MDS treated in the Hematology,from January 2010 to December 2014, in Affiliated Hospital of North China University. Inclusion criteria: All patients underwent bone marrow puncture, press MICM classification inspection, are diagnostic for MDS in RAEB-1 and RAEB-2 in line with 2008 WHO revised classification criteria. Of which 14 males and 17 females; median age is 61 (41-70) years; RAEB-1 13 cases, RAEB-2 18 cases; treatment median absolute neutrophil before 0.73 (0.21-1.81) × 10^9 / L; median hemoglobin value of 72 (51-88) g / L; median platelet count is 28 (3-64) × 10^9 / L; median proportion of bone marrow blasts is 13 (6-18)% ; karyotype groups: good 10 groups, the middle 13 groups, poor 8 groups; IPSS risk groups: intermediate risk I 9 cases, the intermediate risk II 15 cases, high-risk 7 cases, agreed after communicating with the family application decitabine combined with half the amount of CAG regimen.

Patients in the control group have been 37 patients with MDS treated in the Hematology,from January 2008 to December 2014, in the Affiliated Hospital of North China University hospitalized. Inclusion criteria with the former, of which 18 males and 19 females; median age 57 (36-77) years; RAEB-1 16 Li, RAEB-2 21 cases; treatment median absolute neutrophil before 0.62 (0.18-1.97) × 10^9 / L; median hemoglobin value of 73 (48-92) g / L; median platelet count of 22 (4-73) × 10^9 / L; median proportion of bone marrow blasts 11 (6-17) %; karyotype groups: good 12 groups, 16 groups of intermediate, poor 9 groups; IPSS risk groups: patients with intermediate risk I 12, intermediate risk II 17 cases, 8 cases of high risk, all the families of patients have agreed with CAG regimen.

Two groups of data,for statistical analysis, patient's age, sex, the median white cell count, the median hemoglobin values, the median platelet values karyotype groups, RAEB-1 and RAEB-2 are two $P > 0.05$, the difference was not statistically significant.

2.2 Treatment

Decitabine joint half the amount of CAG: decitabine 20mg/m^2, once a day,d1-3, Accra neomycin 7mg/m^2, intravenous injection, once a day,d4-7,cytarabine 10mg/m^2, every 12 hours, d4-10, granulocyte colony stimulating factor 300μg, once a day, d4-10, white blood cell count>20×10^9/L when deactivated for four courses.

CAG simple scheme: Accra neomycin 14mg/m^2,once a day, d1-3, cytarabine 10mg/m^2, every 12 hours, d1-14, granulocyte colony stimulating factor 300μg, once a day, d1-14, white blood cell count>20×10^9/L when deactivated.

2.3 Treatment of adverse events occurred during and principles

According to WHO, acute and subacute adverse drug reactions grading criteria during chemotherapy adverse reactions were graded evaluation. Before treatment, review of blood, liver function, renal function, periodic testing during treatment, suppression of bone marrow during chemotherapy for taking blood component transfusion, granulocyte colony stimulating factor, thrombopoietin methods such as supportive care, patients infected antibiotic therapy.

2.4 Evaluation criteria

Evaluation Criteria Evaluation Criteria reference IWG MDS International Working Group evaluated. At the end of each course of chemotherapy review of blood, blood rise when underwent bone marrow biopsy to assess the efficacy.

2.5 Follow-up

All patients were followed until December 1, 2015. Mainly through telephone communication of patient follow-up. Follow-up data from the medical records. Cases of death during follow-up based on medical records or family members of patients with confirmed phone. Overall survival (OS) time was diagnosed with MDS to the time of death or clinical diagnosis of MDS to December 1, 2015.

2.6 Statistical

Description of the measurement data is represented by the median and range; statistical analysis with SPSS18.0 software. Effectiveness between the two groups were compared using $\chi 2$ test, agranulocytosis time and supportive treatment were compared using t test, $\chi 2$ test was used to compare the incidence of infection and bleeding incidence, Kaplan-Meier method to estimate time OS, OS rates between the two groups Log-rank test was used to compare.

3 Results

3.1 Efficacy

Observation of 31 cases of patients after one course of treatment clinically effective number of cases to 27 cases (87.1%) achieved complete remission (CR) 22 cases (71.0%), partial remission (PR) 5 cases (16.1%), completely ineffective (hematology and bone marrow no improvement) 4 cases (12.9%), with a median 1.7 courses of treatment, an average of 1.9 courses, the median OS time of 18.4 (6-56) months, an average of 28.6 months.

The control group of 37 patients with treatment after a course of treatment clinically effective number of cases to 26 cases (70.3%) achieved complete remission (CR) 19 cases (51.4%), partial remission (PR) 7 patients (18.9%), completely ineffective (hematology and bone marrow no improvement) in 11 patients (29.7%), with a median 1.8 courses of treatment, an average of 2.3 months of treatment, the median time OS 10.6 (4-30) months, an average of 18.4 months (Table 1).

Table 1. The efficacy and survival of the two groups of patients with myelodysplastic syndrome.

groups	Number of cases	Efficacy (Example)			ORR	OS Time
		CR	PR	R		
DAC+ half the Amount of CAG	31	22	5	4	89.4	18.4(6-56)
CAG	37	19	7	11	80.1	10.6(4-30)

Observation group and control group response rate was higher, $P<0.05$, the difference was statistically significant, overall survival was observed in the control group leader, the difference was statistically significant.

3.2 Adverse reactions observed

Adverse reactions occur after suppression of the bone marrow caused by infection, bleeding, selected 68 cases during treatment were inhibited bone marrow, the hematologic adverse reactions of grade 3-4 neutropenia average time of 13.3 days, on average the amount of red blood cells and blood cell transfusion need supportive treatment is 8 and 80U, enrolled patients with varying degrees of infection, including 42 cases of infection 3-4, 1-2 grade infection in 12 cases, mainly for lung infection (56.3%), followed by perianal infection (29.3%), there are some patients with skin and soft tissue infections, urinary tract infection, 22 patients with varying degrees of gastrointestinal bleeding in some patients skin, mucous membranes, bleeding gums and other parts. All patients after anti-infection, bleeding,

 DOI: 10.1051/bioconf/20170801024

blood component transfusion and other symptomatic and supportive treatment, were safely through bone marrow suppression, see Table 2.

Table 2. Two adverse reactions in patients with myelodysplastic syndrome conditions and supportive care compared to the situation.

groups	Number of cases	Adverse reactions			Supportive care	
		The average time of agranulocytosis(d±s)	The incidence of infection(%)	The incidence of bleeding(%)	RBC transfusions (U±s)	Platelet transfusions (U±s)
DAC+ half the Amount of CAG	31	14.4±1.8	78	31	7.5±1.9	78±14
CAG	37	13.2±2.4	81	33	7.6±1.4	72±18

Lack of observed group and control group the mean granulocyte time difference was not statistically significant (P> 0.05), two groups of patients the incidence of infection and the incidence of bleeding was no significant difference (P> 0.05), two groups of red blood cells transfusions of platelets and the difference was not statistically significant (P> 0.05).

Remission rate was significantly higher in patients in the observation group, adverse reactions and therapeutic support there was no difference.

4 Discussion

In this study were selected two groups of patients with MDS IPSS risk groups for the intermediate- or high-risk patients, patients with a high risk of this type of transformation to leukemia [1-2]. Because such patients, mainly in the elderly, although it still can not give a positive chemotherapy get a higher response rate and longer lifetime. In particular, the conventional CAG program, due to cardiac toxicity of anthracyclines and cytotoxic chemotherapy drugs, often resulting in older patients can not tolerate. With the gradual deepening of epigenetic research, we found that in the process of hematopoiesis high-risk MDS patients, some start factor methylation occurred, resulting in some important gene silencing, inactivated, thereby causing disease gradually progress to leukemic transformation. Bacher U, Mihara k [3-4] and put forward a higher proportion of P15 gene MDS patients in cell cycle regulation (INK4b) abnormal methylation. On the basis of these theoretical studies, in 2006, decitabine as a specific DNA methylation inhibitor approved by the US FDA listed, Zagonel V [5], etc. The study shows that decitabine effective treatment of MDS and thus widely used in clinical, but dose decitabine did not reach a consensus. Thereafter through a lot of studies have found that high concentrations of decitabine main play its cytotoxicity, while low concentrations allows DNA methyltransferase enzyme inactivation does not lead to cell death [6]. Many scholars through a lot of tests to explore the application of decitabine dose, while exploring single land decitabine MDS effect is more ideal. Finally, Chowdhury S [7] found that application of low-dose decitabine with gemtuzumab combined treatment of relapsed or refractory acute myeloid leukemia achieved better results. Since then many researchers have studies with decitabine in combination can improve the efficacy of the program MDS. Qin [8] study found that decitabine in inhibiting DNA demethylation, while also improving the cytotoxicity of cytarabine, G-CSF or GM-CSF can increase the sensitivity of cell cycle dependent drug so in theory decitabine for MDS patients with joint CAG program should be a good effect. Qian [9] reported that patients with high-risk MDS applications to decitabine joint CAG regimen, complete remission rate reached 82.6%. High Su [10] pointed out that decitabine monotherapy and combination regimens half and full CAG myelodysplastic syndrome and acute myeloid leukemia have a good effect; Zhang Pingyun [11] found that a small dose of diazepam Combined CAG gemcitabine chemotherapy in high-risk patients with myelodysplastic syndrome more effective than chemotherapy alone in patients with CAG. This study showed that decitabine Joint

CAG half the amount in high-risk myelodysplastic syndrome exception has a good effect, this result is consistent with the HL group [12] and other findings alone with no difference in efficacy of CAG regimen; in survival time aspect, decitabine joint leader in half the amount of CAG CAG regimen alone group, but the difference was not statistically significant, given the small sample size related. But we should pay attention to, such as the corresponding incidence of infection, bleeding and other complications are also higher incidence of adverse reactions and supportive care aspects of the research groups studied, decitabine joint half dose group and CAG CAG regimen alone no difference in the treatment group; Steensma [13] and so noted in the ADOPT trial, patients with pulmonary infection probability 11% probability of pulmonary infection in this study is much higher than this result, consider the patient to recognize laminar flow ward inadequate, poor economic conditions, more and more people in our hospital general wards, increasing the chance of cross-infection related. Recommended in the early chemotherapy, proper use of drugs to prevent fungal and anti-bacterial treatment to reduce the risk of infection may be to a certain extent, this view is consistent with the Jain [14] of.

References

1. Nimer SD. Myelodysplastic syndromes.Blood,2008;111(10):4841-4845.
2. Nolte F, Hofnann WK. Molecular mechanisms involved in the progression of myelodysplastic syndromes.Future Oncol,2010;**6**(3):445-455.
3. Bacher U, Haferlach T, Kern W,et al. A comparative study of molecular mutations in 381 patients with myelodysplastic syndrome and in 4130 patients with acute myeloid leukemia. Haematologica, 2007;**92**(6):744-752.
4. Mihara K, Takihapa Y, Kimura A. Genetic and epigenetic alterations in myelodysplastic syndrome. Cytogenet Cenome Res,2007;**118**(2-4):297-303.
5. Zagonel V, Lo Re G, Marotta G,et al. 5-Aza-2′ -deoxycytidine(Decitabine) induces trilineage reponse in unfavourable myelodysplastic syndromes. Lukemia,1993;7(suppl 1):30-35.
6. OKI Y, AOKI E, ISSA J P, et al. Decitabine-Bedside to bench[J].*Crit Rev Oncol Hematol*,2007, 61(2):140-152.
7. Chowdhury S, Seropian S, Marks PW. Decitabine combined with fractionated gemtuzumab ozogamicin therapy in patients with relapsed or refractory acute myeloid leukemia[J].*Am J Hematol*, 2009,**84**(9):599-600.
8. Qin T, Youssef EM, Jelinek J,et al. Effect of cytarabine and decitabine in combination in human leukemic cell lines[J].*Clin Cancer Res*,2007,**13**(14):4225-4232.
9. Qian S, Duan L. Effect of low-dose cytarabine and aclarubicin in combination with decitabine on the outcome of patients with high-risk myelodysplastic syndromes[G].12th International Symposium on Myelodyspl- astic Syndromes.Berlin, Germany: Leukemia Research,2013:S152-S153.
10. The high Su, Chou Huiying, Jin Zhengming etc. decitabine monotherapy and combination programs CAG half and full observation [J] Efficacy in treating abnormal myelodysplastic syndrome, leukemia and acute myeloid. *Journal of Hematology* 2014,35 (11):961-964 (Chin J Hematol, November2014, Vol.35, No.11)
11. Tibet, Wu Wenzhong, Cui Guoxing, decitabine and other programs and joint CAG CAG regimen alone, and compare the efficacy of high-risk myelodysplastic syndrome [J] Chinese Journal of Experimental Hematology, (Journal of Experimental Hematology) 2014;.. 22 (5) :. 1341-1344
12. HL group, Dengyin Fen, Zhang Lei, et decitabine sequential half the amount of CAG regimen in high-risk myelodysplastic syndrome and leukemia transformation of four cases [J], leukemia. Lymphoma 2012; 21 (7);. 433-435 (Journal of Leukemia & Lymphoma, July2012, Vol, 21, No.7)
13. Steensma DP, Baer MR, Slack JL,et al. Multicenter study of decitabine administered daily for 5 days every 4 weeks to adults with myelodysplastic syndromes: the alternative dosing for outpatient treatment(ADOPT)trial [J].J *Clin Oncol*,2009,**27**(23):3842-3848.

14. Jain N, Mattiuzzi GN, Cortes J, et al. Benefit of anti-infectious prophylaxis in patients with acute myeloid leukemia or high-risk myelodysplastic syndrome receiving frontline"targeted therapy"[G].49th American Society of Hematology Annual Meeting. Atlanta, Georgia: Blood,2007:2858.

DOI: 10.1051/bioconf/20170801025

A combination of gangliosides and nerve growth factor alleviates lipopolysaccharide-induced neuronal cells damage and its mechanism

Ying Song[1,a], Yun Bei[1], Dongsheng Mao[2], Wei Ding[1] and Zhiyun Lin[1]

[1]*College of Pharmaceutical Science, Zhejiang University of Technology, 310014 Hangzhou, China*
[2]*Community Health Services Center of Mishixiang, Gongshu District, 310005 Hangzhou, China*

Abstract. Objective: To evaluate the effect of gangliosides (GM1) in combination with nerve growth factor (NGF) against neuronal cells damage evoked by lipopolysaccharide (LPS), and tries to uncover its probable mechanism. **Methods:** (1) Cell viability was measured using Methyl thiazolyl tetrazolium (MTT) method, which was also determined the optimum concentration of LPS for the damage models; meanwhile, cell morphology was observed by microscope. (2) The expression level of NF-κB was detected by RT-PCR. (3) Finally, NF-κB inhibitor pyrollidine dithiocarbamate (PDTC) was treated for the research of NF-κB pathway. **Results:** (1) MTT results shown that the LPS injury was dose-dependent, and 100nmol/L was selected as the optimum damage concentration. (2) Through the morphological observation, MTT and RT-PCR analysis, we found that GM1 and NGF both can protect cells against LPS injury; interestingly, combination of GM1 and NGF had a slighter LPS injury than GM1 administration alone. Moreover, the expression of NF-κB in combination group was lower than that in GM1 group, indicated that blockage of NF-κB pathway was better for cells living. **Conclusion:** Combination of GM1 and NGF has a better protective act on LPS injury than GM1 alone. The mechanism may have some connections with NF-κB pathways.

1 Introduction

Ganglioside (monosialoteterahexosyl gangliosides, GM1) is an important nerve protective agent in the brain [1]. In previous studies, several *in vivo* and *in vitro* models have been demonstrated that GM1 play an important role in cell-cell recognition and signal transduction pathways. GM1 has varieties of functions, such as stability maintaining the Na^{+}-K^{+}-ATP enzyme activity and cellular membrane, promoting the maturation of neuronal cells, regulating the proliferation of neural cells by polypeptide growth factor, protecting nerve cells from damage and promoting repairing after injury. Furthermore, GM1 can inhibit the toxicity of excitatory amino acid with enhanced role of neurotropic factor by blocking the translocation of protein kinase of PKC-α [2-4]. Studies have also proved that GM1 can relief the apoptosis of nerve cells after injury by the regulation of PKC signal pathway, but its protective effect on the LPS induced cell inflammatory injury need to be further researched [5]. In

[a] Corresponding author: songying@zjut.edu.cn
This work was supported by Grants-in-aid from the natural science foundation of Zhejiang province (No. LY17H090018) and the fund of national clinic pharmacology study (No. L2011079).

recent years, with the increasing numbers of bacterial infections, especially the infection of Gram-negative bacterium, more and more researchers have focused on this issue. Some researches illustrated that GM1 have connections with other molecules, receptors or signal pathways, such as phospholipase Cγ (PLCγ)[6], tropomyosin receptor kinase (Trk) [7] and so on. However, little is known about the relationship between GM1 with NF-κB, particularly the joint action of GM1 and NGF with NF-κB.

Nuclear factor (NF-κB), as a multifarious transcription regulatory proteins, participate in regulating the expression of various inflammatory cytokines and cell apoptosis [5, 8], is attracting rising concerns and the research of its signal system is gradually becoming the hotspot in the field of neural biological science. Besides, some studies suggested that the activation of NF-κB signaling pathway is sometimes dualistic, and the mechanism of how to induce apoptosis is not fully understand [9].

Both primary neurons and rat pheochromocytoma cell line (PC12 cells) were studied in this research, the protective effect of GM1 and NGF on LPS damage was observed by the alterative content of NF-κB. In the current study, we found that LPS-induced cell damage had some contact with the activation of NF-κB in both PC12 cells and primary neurons; we also observed the effect of combined treatment of NGF and GM1, and tried to understand the mechanism of the action.

2 Materials and methods

2.1 Materials

2.1.1 Chemicals

GM1, NGF, LPS, PI (propidium iodide), PDTC (Pyrrole aldehyde disulfide carbamic acid) and Hoechst33342 were purchased from Sigma-Aldrich (Steinheim, Germany). 3-(4,5-dimethyl-2-thiazolyl)-2,5-diphenyl-2-H-tetrazolium bromide (MTT) was purchased from Shanghai Biological Technology (Shanghai, China). DMEM and horse serum were purchased from Thermo Fisher Scientific (MA, USA). Fetal bovine serum was purchased from Sijiqing (Hangzhou, China).

2.1.2 Primary neurons and PC12 cells

Primary neurons were made according the reference from the Academy of Military Medical Sciences. Neonatal Sprague-Dawley (SD) rats were brought from the Laboratory Animal Centre of Zhejiang Academy of Medical Sciences. All the procedures were carried out under aseptic conditions [10]. In brief, the neonatal rat was sterilized by 70% ethyl alcohol and then conducted sacrifice; the brain was collected on ice plate to prepare tissue. The brain was put into ice-bathed dissecting fluid; cortex was separated by ophthalmic forceps, removed meningeal and vessels and put into the plate which is covered with dissecting fluid. The cortex was cut into the size of $1mm^2$ and then washed by dissecting fluid twice. Put the tissue into pancreatic enzyme solution, digested in the temperature of 37°C. 20 minutes later, discarded pancreatic enzyme, douched sediment with 3ml culture solution for two times. Added 2ml culture solution, beat upon for 10~15 times, made the cell suspension, standing the cell suspension for 1~2 min, Absorbed supernatant into another tube, added 2ml cultivation liquid, repeated percussion, and then combined the supernatant. The supernatant was filtered through 200-mesh sieve. Take 25μl filtered liquid, added 475μl cultivation liquid, diluted for 20 times, then used for counting. Cells concentrate was diluted to 1×10^6/ml by cultivation liquid and inoculate into the plate (96-well plates, 100μl per well). On the first day after inoculation, Cell cultivation liquid was medium changed for every 3 days or 2 times a week. After cultivated for five days, nerve cells were deal with 3~5μg/ml cytarabine (ARA-C), which is used for inhibited the excessively growth of glial cell. 48h later, medium changed, and cultivated for another 10 days then can be used for the study.

PC 12 cells were obtained from Shanghai Institute of Cell (Shanghai, China).

 DOI: 10.1051/bioconf/20170801025

2.2 The effect of GM1 on cells

Both primary neurons and PC12 cells were seeded into 96 well plates and incubated with 100μmol/L, 10μmol/L, 1μmol/L, 0.1μmol/L, 0.01μmol/L or 0μmol/L GM1 for 24h. Cell viability was assessed using MTT assay. Opacity density (OD) was measured at 570 nm using a spectrophotometer.

$$\textit{cell viability}(\%)=(\textit{OD of cells with different treatment})/(\textit{OD of cells with no treatment})\times 100 \quad (1)$$

2.3 Establishment of LPS damage model

Primary neurons and PC12 cells were plated into 96 well plates, treated with different concentration of LPS (0nmol/L 0.1nmol/L, 1nmol/L, 10nmol/L, 100nmol/L or 1000nmol/L) for 24h. Cell viability was determined by MTT assay as **2.2**, and the optimum concentration of damage model were figured out by the MTT assay results.

2.4 The protective effect of GM1, GM1 combined with NGF on LPS damaged cells

Primary neurons and PC12 cells were separately divided into four groups: a control group (no specific treatment), a damage group (treated with 100nmol/L of LPS, damage model group), a GM1 treatment group (treated with GM1 in damage model) and a combination treatment group [treated with GM1 and NGF (50ng/ml) in damage model group]. The OD of each group was determined by MTT assay after treated for 24 h.

2.5 Morphological observation

Inverted light optical microscopy (Nikon TS100, Tokyo, Japan) was performed to assess cell morphologies. Characteristic morphological features of in vitro PC12 cells and primary neurons were observed such as their characteristic shape and dendrite formation.

2.6 The relative expression of NF-κB was detected by RT-PCR

PC12 cells were seeded into 6 well plates, and four groups were separated as **2.4**. The RNA of each group were extracted respectively and cultured for 24 h, NF-κB of each group was detected by RT-PCR (F: CTTCTGGGCCATATGTGGAGAT; R: TCGCACTTGTAACGGAAACG), Reference is GAPDH (F: GGTGGACCTCATGGCCTACAT; R: GCCTCTCTCTTGCTCTCAGTATCCT). Primers sequences were designed by Beacon designer Express Primer 2.0, synthesized by Shanghai Biological Engineering Co., Ltd.

2.7 LPS damage is related to NF-κB pathway or not

Pyrollidine dithiocarbamate (PDTC) is a kind of inhibitor of NF-κB. In order to determine the LPS damage on cells is related to NF-κB pathway or not, cells were divided into three groups: a control group, a LPS damage group and a PDTC+LPS group. MTT assay was used to measure the role of NF-κB in PC12 cells or primary neurons.

2.8 Statistical analysis

Values were expressed as mean ± SD. Statistical analysis was performed with One-way ANOVA using Graph Pad Prism software. P values less than 0.05 were considered statistically significant (*$P<$ 0.05; **$P<$ 0.01).

3 Results

3.1 MTT assay on PC12 cells

Viability of PC12 cells were assessment by MTT assay. The viability of control group was seen as 100%. Although the viability of PC12 cells which treated with 0.001~10μmol/L of GM1 were higher than control group, there was no significant difference (Figure1-a). That is to say GM1 alone cannot influence the living of PC12 cells. In Figure 1-b, cells incubated with a concentration gradient (0.1~1000μmol/L) of LPS have revealed dramatically reduced survival rate. The mortality rate of cells were almost 50% while the dosage of LPS were between 10~1000μmol/L. According to the previous studies and our results, 100μmol/L was chosen for put forward the damage model. The survival rate of LPS damaged PC12 cells was obviously increased with the administration of GM1, suggesting that GM1 can mitigate the injury of LPS (Figure 1-c), and 1μmol/L of GM1 can relieve the LPS injury into minimum. In order to make the results more clear, we have concluded the above results together. Five groups have been divided: a LPS group (100μmol/L LPS, damage group), a GM1 group (1μmol/L GM1 + 100μmol/L LPS) and a NGF group (100μmol/L LPS+50ng/ml NGF), a GM1+NGF group (1μmol/L GM1 + 100μmol/L LPS +50ng/ml NGF) and a control group. Compared among these groups, we found that cells cultured with combination of NGF and GM1 have higher cell vitality, as is shown in figure1-d.

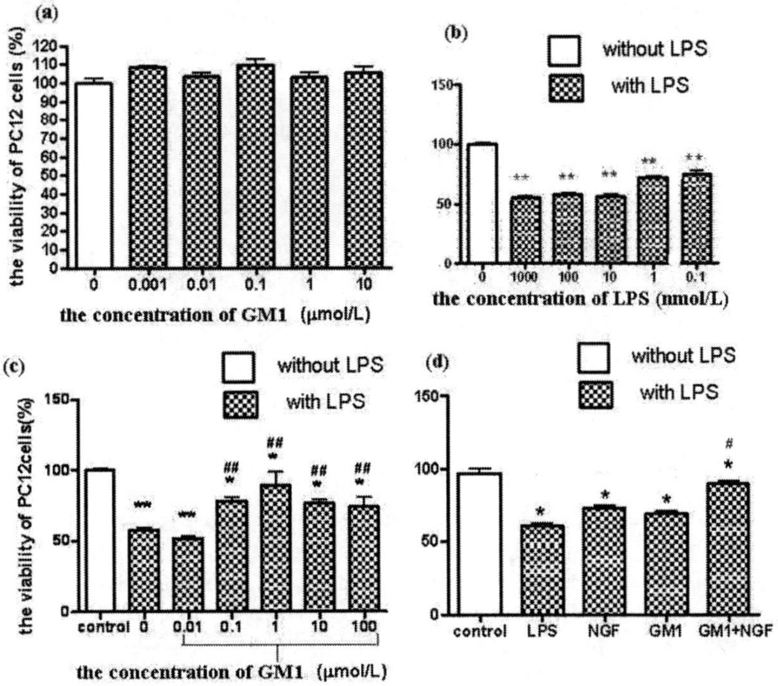

Figure 1. Effects of GM1/NGF on viability of PC12 cells. (a) The effect of GM1 on PC12 cells. GM1 (0-10μmol/L) alone have no significant influence on the viability of PC12 cells. (b) The effect of LPS on PC12 cells. With the incremental concentration of LPS, the sever injury has been caused on PC12 cells and 100nmol/L has been chose for the damage model. (c) The protective effect of GM1 on LPS injured PC12 cells. When the concentration of GM1 is higher than 0.1μmol/L; there shows a better protective effect. (d) The effect of GM1 and NGF on PC12 cells. The combination administration of GM1 and NGF can be sharply reduced the LPS damage than NGF or GM1 alone. Values were expressed as mean± SD (n=6). **P* <0.05 and ***P*< 0.01, compared to control group; #*p*<0.05 and ##*p*<0.05 compared to GM1 group; Student's t-test.

3.2 MTT assay on primary neurons

In primary neurons, survival rate was also assessed through MTT assay and all the method or procedure were the same as the study of PC12 cells. Similar results were obtained from the primary neurons: a. GM1 alone cannot influence the living of primary neurons (Figure 2-a); b. A concentration gradient (0.1~1000μmol/L) of LPS can obviously reduce the cell survival rate in primary neurons, the optimum concentration for the damage model in primary neurons was also 100μmol/L (Figure 2-b); c. 0.1~100μmol/L of GM1 can notably protect the cells against LPS damage (Figure 2-c) and d. combination treatment of GM1 and NGF can remarkably enhanced the viability in LPS damaged primary neurons (Figure 2-d).

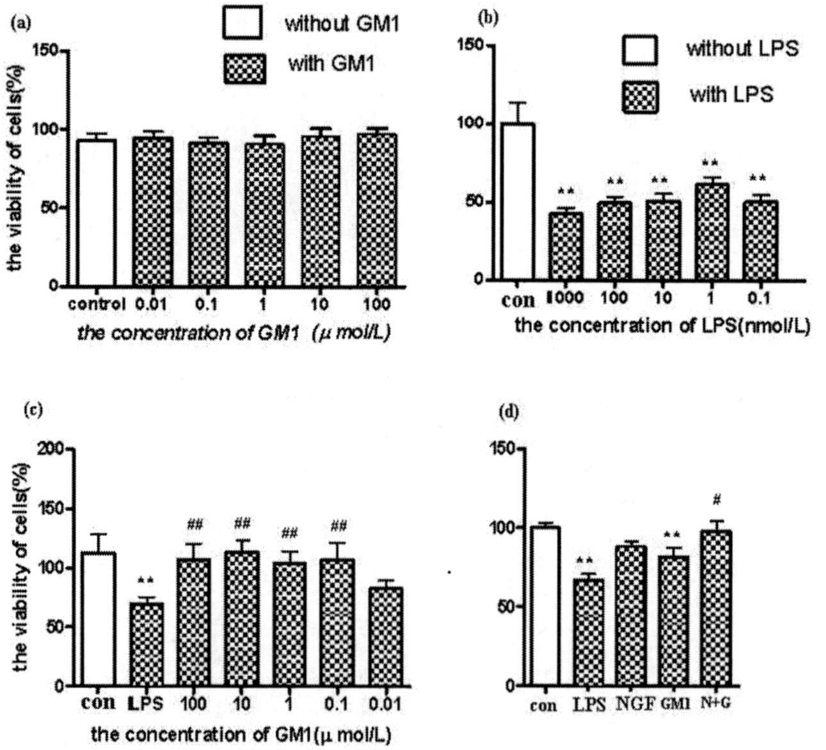

Figure 2. Effects of GM1/NGF on primary neurons viability. (a) The effect of GM1 on primary neurons. GM1 (0-10μmol/L) alone also have no significant influence on the viability of primary neurons. (b) The effect of LPS on primary neurons. With the incremental concentration of LPS, the sever injury has been caused and we also chose 100μmol/L for the establishment of damage model. (c) The protective effect of GM1 on LPS induced injury in primary neurons. When the concentration of GM1 was higher than 0.1μmol/L; there was an obviously increasing in the survival rate. (d) The cooperatively effect of GM1 and NGF on primary neurons. The results on both PC12 cells and primary neurons have been corporately shown that combination administration of GM1 and NGF has a stronger effective on the protection of LPS damage. Values were expressed as mean± SD (n=6). *P <0.05 and **P< 0.01, compared to control group; #p<0.05 and ##p<0.05 compared to GM1 group; Student's t-test.

3.3 Morphological observation of PC12 cells and primary neurons

The morphological of control group, LPS damage group, GM1 treated LPS damage group and combination medication of GM1 and NGF in PC12 cells and primary neurons have been observed by an inverted microscope. The results were showed as the two pictures below: Figure 3 and Figure 4.

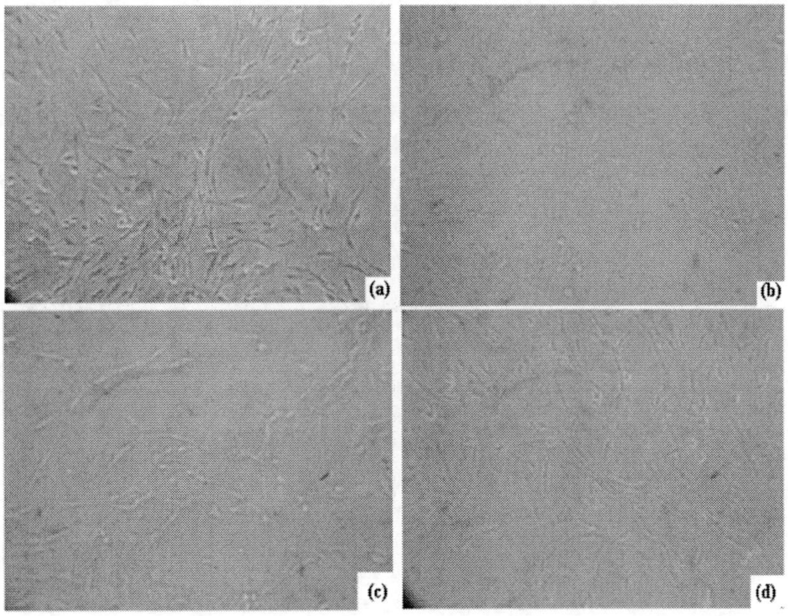

Figure 3. The morphology of PC12 cells. Normal differentiation of PC12 cells were clear in outline, and their shapes were round, spindle and ellipse, cells were also showed distributed like cluster, and dendrites were clearly to be seen (Figure 3-a). While the cells were damaged by LPS for 24 hours, the disappearances of cells were shrinking, flat and protuberances were disappeared (Figure 3-b); the damages of cells were reduced if we added GM1 into the LPS (Figure 3-c); in Figure 3-d, the protective effect of GM1 and NGF were much more significantly than GM1 alone, and the cell morphology in this group were similar to the control group.

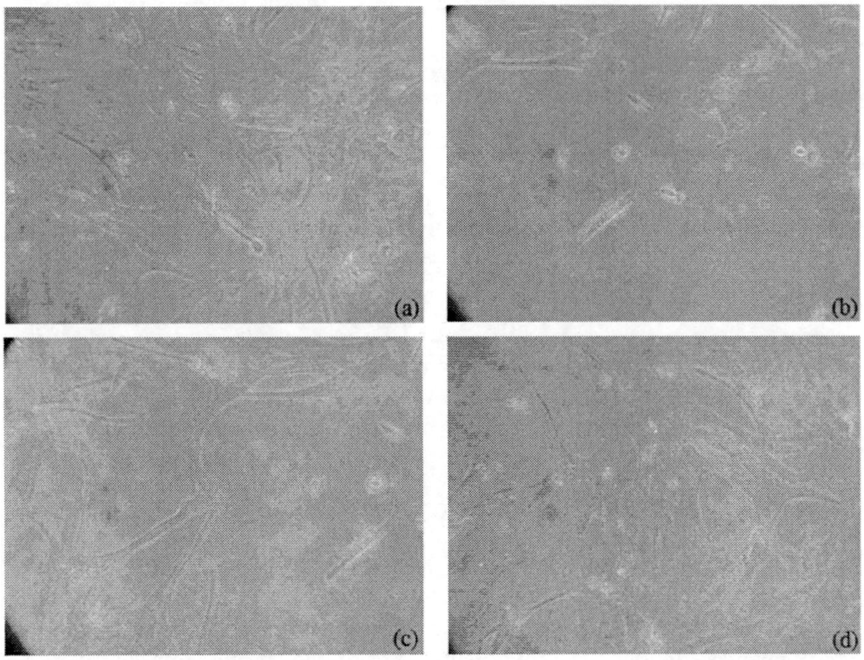

Figure 4. The morphology of primary neurons. Figure 4-a have described the normal living status of primary neurons. LPS damaged primary neurons were represented in Figure 4-b. The cells were less to see in the picture. As is shown in Figure4-c, GM1 have a certain protective effect on LPS induced damage; however, the combinations of GM1 and NGF have a stronger effect on LPS induced damage (Figure 4-d).

DOI: 10.1051/bioconf/20170801025

3.4 The relative expression of NF-κB was detected with RT-PCR

The RT-PCR results showed that the relative expression of NF-κB in LPS injured cells was significantly higher than that in control group. It is indicated that LPS damage can increased the expression of NF-κB in PC12 cells. The relative expression of NF-κB in the drug treatment group was significantly lower than that of the injured cells, drug treatment decreased the expression of NF-κB, treatment with NGF and GM1 can recovered the injury to some extent, it is illustrated that NGF had a protective effect of damage by LPS, and the protective effect may related to the inhibiting activation of NF-κB signaling pathway. The relative expression levels of NF-κB in PC12 cells were detected by RT-PCR. The results were shown as Figure 5.

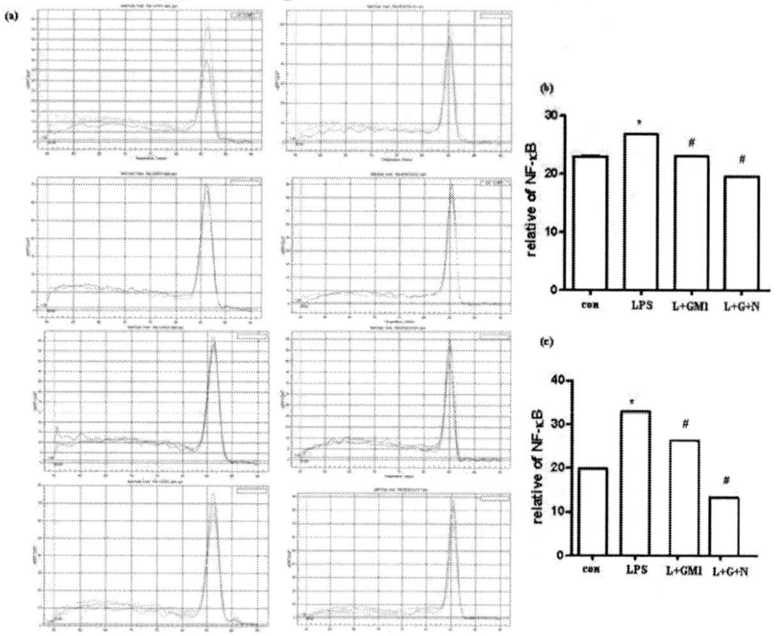

Figure 5. Relative expression levels of NF-κB in each group. (a) RT-PCR melting curve in different groups. The melting curve analysis revealed a mean melting temperature (Tm) of 86.5±0.6°C, and the samples showed only a small variation in the parameters. (b) Relatively expression of NF-κB in PC12 cells. NF-κB has been highly expressed in LPS group. The expression of NF-κB in GM1 and LPS group were the same as the normal group. The expression of NF-κB has been highly decreased in the combination of GM1 and NGF in LPS damaged PC12 cells. (c) Relatively expression of NF-κB in primary neurons. Compared with (b) and (c), it is demonstrated that the protective effect of GM1 and NGF were both effective in PC12 cells and primary neurons. Values were expressed as mean ± SD (n=4); **p*<0.05 and ***p*<0.01, compared to control group, #*p*<0.05 and ##*p*<0.01, compared to GM1 group; Student's t-test.

3.5 The relationship between LPS damage and NF-κB pathway

The results of RT-PCR have showed that the LPS damage and the protective of GM1 and NGF may relate to the expression of NF-κB. So we tried to figure out the deeper connection between NF-κB and the viability of LPS damaged PC12 cells and primary neurons by the using of NF-κB inhibiter PDTC. The results were shown as Figure 6. The treatment of PDTC can significantly improve the viability of LPS injured PC12 cells and primary neurons.

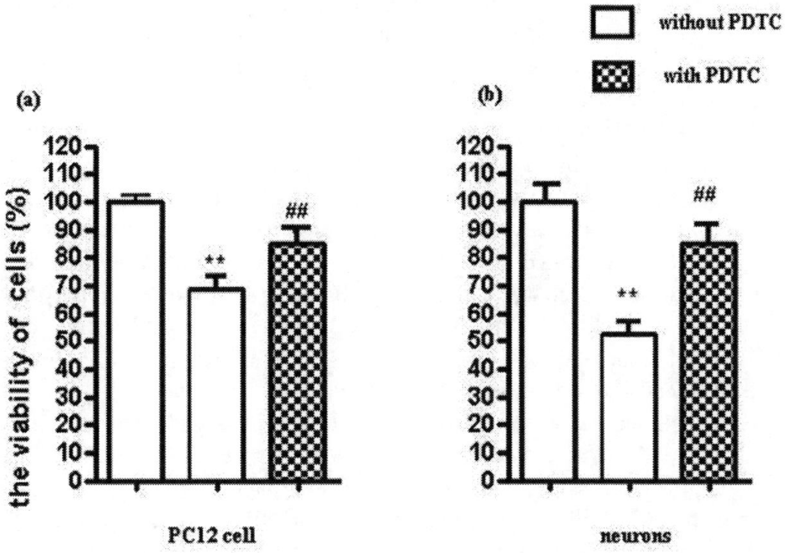

Figure 6. The effect of NF-κB inhibitor PDTC on LPS damaged PC12 cells and primary neurons. (a) The damaged cells which added PDTC had a higher viability than no-treatment damaged cells. That is to say the damage of LPS may have up-regulated the expression of NF-κB. (b) In the study of primary neurons, result also showed that there may have a connection between LPS damage and NF-κB expression. Values were expressed as mean ± SD (n=6); **p*<0.05 and ***p*<0.01, compared to control group, #*p*<0.05 and ##*p*<0.01, compared to GM1 group; Student's t-test.

4 Discussions

In this experiment, we had known that PC12 cells can be damaged by LPS; and this injury was dose-dependent, that is to say, the injury will be increased with the increasing concentration of LPS in a certain case. When the LPS concentration is as high as 100nmol/L, the cell survival rate was only 55%; according to the previous study, we select 100nmol/L as the concentration to build the injury model. The morphological observation results proved that LPS induced PC12 cell damage is not only manifested in the cell death, but also effect cells structures and functions. At the early stage of cell death, even if there is no exist of necrosis in cells, the synapses of the cells are disappeared; shrinkage is appeared in round injured cells. When the concentration of LPS is as high as 1mol/L, the number of death cells increased significantly.

GM1, a group of sphingolipids which included sialic acid, are nutrients for the survival, growth, differentiation and function of neurons. Animal and clinical experiments showed that GM1 has certain protective and repairing effects for the injury of both central nervous system (CNS) and peripheral nervous system, including cerebrovascular diseases, diabetes, alcoholic and uremic neuropathy. GM1 also have some certain functions on biochemical index, symptoms improvement, delay of diseases progression, and protecting the neurons for CNS degenerative disease which including Parkinson's disease, senile dementia and so on. Our experiments have shown that GM1 has no toxic effects on neuronal cells, and has a protective effect on LPS damage. Its effective concentration is 0.1, 1, 10, 100μmol/L, and 1μmol/L is the most effective concentration. Morphological observation also revealed that GM1 can obviously improve the survival cells induced by LPS, which is consistent with the previous studies. In our research, we have studied on both PC12 cells and primary neurons, which made our research much more precise and convictive.

NF-κB is a central substance that mediates a number of immune and inflammatory responses. It is also an important transcriptional regulator of apoptosis signaling pathway, but its effect on neuronal cells is in the dispute [11-12]. Our experiments have used RT-PCR to detect the expression levels of NF-κB, the results showed that the expression levels of NF-κB in LPS injured PC12 cells and primary neurons were significantly higher than those in control group, which indicated that the expression of NF-ϰB is associated with LPS induced injury, and the more serious damage it takes, the higher expression level of NF-ϰB. This is accordance with the previous study which the AD patients with severe neuronal apoptosis in brain tissue have a higher activation level of NF-ϰB [13]. In addition, the expression level of NF-ϰB in the treatment group was lower than that in the injury group, which was similar to the normal cells, indicated that decreased injury can reduce the expression of NF-ϰB, and which is also an indicator for the protective effect of LPS damage. But there are some studies showed that apoptosis will be occurred when PC12 cell were in the condition of the decreased content of NF-ϰB, this indicate that the content of NF-ϰB in cells needs to be maintained at a certain amount, the increased or decreased amount will lead the cells to apoptosis [14]. And the experimental results shown that treated with GM1 combined with NGF have a better protective effect in cells. It is illustrated that in PC12 cells, the expression level of NF-ϰB in normal cells or combination administrated GM1 and NGF cells are approximately the same.

In Conclusion, LPS can induce the damages of PC12 cells, which will lead cells to death. The change of expression of NF-ϰB was detected in damage models. We can speculate that there is a certain relationship between the expression of NF-ϰB and LPS damage. Instead, GM1 has a protective effect on LPS injury, and the protections will much more obviously if we combined administration with NGF. It should be mentioned that previously studies had reveal that GM1 may play an important role in the activating the functions of NGF [15]. Our results showed that the phenomenon of death cells was decreased, so, GM1 may play a role to slow down or block the process of cell death. This phenomenon illustrated that the LPS damage and the protective effect of GM1 on LPS injury may relate to the activation of NF-ϰB signaling pathway. Finally, MTT assay confirmed the conclusion that there is a relationship between LPS damage and NF-ϰB signaling pathway by the using of NF-ϰB inhibitor PDTC.

References

1. Sandhoff K., Harzer K., J. Neurosci. *Gangliosides and gangliosidoses: principles of molecular and metabolic pathogenesis.* **25**, 10195-10208 (2013).
2. Giordano G., Sanchez-Perez A.M., Burgal M., et al. J. Neurochem. *Chronic exposure to ammonia induces isoform reselective alterations in the intracellular distribution and NMDA receptor–mediated translocation of protein kinas C in cerebellar neurons in culture.* **1**, 143-157 (2005).
3. Lallemend F., Hadjab S., Hans G., et al. J. cell. Sci. *Activation of protein kinase Cbetal constitutes a new neurotrophic pathway for deafferented spiral ganglionneurons.* **19**, 4511-4525 (2005).
4. Yates A. J., Saqr H.E., Van Brocklyn J. J. Neurooncol. *Ganglioside modulation of the PDGF receptor. A model for ganglioside functions.* **1**, 65-73 (1995).
5. Beni S.M., Tsenter J., Alexandrovich A.G., et al. J. Cereb Blood Flow Metab. *CuZn SOD deficiency, rather than overexpression, is associated with enhanced recovery and attenuated activation of NF-kappa B after brain trauma in mice.* **4**, 478-490 (2006).
6. Duchemin A. M., Ren Q., Neff N.H., Hadjiconstantinou M. J. Neurosci. *GM1-induced activation of phosphatidylinositol 3-kinase: involvement of Trk receptors.* **104**, 1466-1477 (2008).
7. Duchemin A. M., Ren Q., Mo L., et al. J. Neurosci. *GM1 ganglioside induces phosphorylation and activation of Trk and Erk in brain.* **81**, 696-707 (2002).
8. Israel A. Trends Cell Biol. *The IKK complex: an integrator of all signals that activate NF-ϰB.* **10**, 129-133 (2000).
9. A. Campos, V. Vasconcelos. Int. J. Mol. Sci. *Molecular mechanisms of microcystin toxicity in animal cells.* **11**, 268-287 (2010).
10. Wiemerslage L., Ismael S., Lee D. Mitochondrion. *Early alterations of mitochondrial morphology in dopaminergic neurons from Parkinson's disease-like pathology and time-dependent neuroprotection with D2 receptor activation.* (2016).
11. Zhen Y., Wang B., Zhou W.Q. *Chinese Journal of Clinical Rehabilitation. The apoptosis effect of salicylic acid on P12 cells.* **6**, 70-72 (2006).

12. Ayumi T., Katsumi H., et al. J. Neurochem. *Lysosomal accumulation of Trk protein in brain of GM1-gangliosidosis mouse and its restoration by chemical chaperone*. **3**, 399-406 (2011).
13. Ji H., Zhang X., Du Y., Liu H., Li S., et al. Brain. Res. Bull. *Polydatin modulates inflammation by decreasing NF-κB activation and oxidative stress by increasing Gli1, Ptch1, SOD1 expression and ameliorates blood-brain barrier permeability for its neuroprotective effect in pMCAO rat brain*. **1**, 50-59 (2012).
14. Mattson M.P. J. Neurochem. *NF-kappa B in the survival and plasticity of neurons*. **6**, 883-893 (2005).
15. Dragicevic N., Delic V., Cao C., et al. Neuropharmacology. *Caffeine increases mitochondrial function and blocks melatonin signaling to mitochondria in Alzheimer's mice and cells*. **8**, 1368-1379 (2012).

DOI: 10.1051/bioconf/20170801026

Objective evaluation of acupuncture treatment in patients with cervical spondylosis by pulse diagnosis device

Haifang Zhang, Zhifeng Yu[a], Linjie Fu and Xiaozuo Lu

School of Chinese Medicine and Engineering, Tianjin University of Traditional Chinese Medicine, Tianjin, 300193, China

Abstract. To study changes in pulse diagram parameters (PDP) in patients with cervical spondylosis (CS) before and after acupuncture treatment, explore the characteristics of PDP and the relationship between PDP changes and therapeutic effectiveness, and provide evidence for outcome prediction and objective evaluation of CS treatment before and after acupuncture treatment.Patients with CS were treated with acupuncture and measured with a pulse acquisition device based on image (PADBI) before the first and after the tenth acupuncture sessions. PDP changes from before until after the acupuncture sessions and patient impressions were analyzed to judge the effect of acupuncture treatment.The PDP values in effective patients were closer to normal values. This indicated that Qi stagnation and blood stasis of the patients were improved. The PDP changes from before to after the first acupuncture treatment were more obvious than those from before to after the tenth acupuncture treatment. This result indicates that the speed of symptom improvement decreased significantly after several acupuncture courses. Analysis of correlation between efficacy and PDP showed that the changes in PDP in five patients was abnormal, which mainly manifested as values of h1, u, p, Pp, and t1, and no significant changes or differences were increased with standard values. This indicated that the symptoms of CS were not improved in these patients.PADBI can provide evidence for outcome prediction of acupuncture treatment in patients with CS. PADBI can provide evidence for objective evaluation of acupuncture treatment of CS.

1 Introduction

Pulse feeling is one of the important diagnostic methods in Traditional Chinese Medicine (TCM) for obtaining physiological and pathological information. It is difficult to identify pulse patterns accurately because pulse conditions contain a lot of information. The traditional method of pulse feeling with the fingers is subjective and the results can vary from person to person.

Cervical spondylosis (CS) is caused by degeneration and prolapse of the cervical intervertebral discs and cervical joints, which directly compress cervical spinal nerves, vertebral arteries and the spinal cord. The major clinical manifestations are pain along the neck and shoulder, upper extremity numbness, and radiating pain[1-3]. CS is similar to that of Qi stagnation and blood stasis in terms of TCM theory. The etiology of CS is multifactorial and involves poor posture, anxiety, depression, neck strain, and sports or occupational activities[4]. Conventional medical treatments for CS neck are

[a] Corresponding author: fengzhiu86@hotmail.com

DOI: 10.1051/bioconf/20170801026

limited by their modest effectiveness. TCM interventions have been widely used for the management of CS neck pain[5-8] and acupuncture is one of the most popular managements. Many clinical studies have been conducted to evaluate the efficacy of acupuncture[9-11]. While, modern medicine has many diagnostic methods for CS, TCM diagnoses mainly rely on auscultation-olfaction, interrogation, and palpation, especially the pulse feeling. Because of differences in understanding and clinical experience in traditional pulse feeling, the objectification of pulse information is more important.

To study the significance of pulse acquisition device based on image (PADBI) in the objective evaluation of acupuncture treatment for CS, we used TD-III-Type PADBI to measure PDP changes before and after acupuncture treatment in 50 patients with CS.

2 Materials and methods

2.1 Study samples

This study included 50 patients suffering from CS being treated with acupuncture at the Chinese Medicine outpatient clinic, Affiliated Baokang Hospital of Tianjin University of TCM between October 2013 and March 2014. The group was composed of 38 women and 12 men, ranging from 18 to 78 years old, with an average age of 44.08. Among the 50 patients, 22 patients were under 40, 19 patients were 41-60 years old, and 9 patients were aged 61 years or older. All patients understood and agreed with the study items. The study design was approved by Medical Ethics Committee of Tianjin University of Traditional Chinese Medicine.

2.2 Inclusion criteria

Patients were included if they were men or women aged between 18 and 80, fit the diagnostic criterion established in the 2nd Symposium Meeting on CS[12] and agreed to sign the informed consent and complete treatments on time.

2.3 Exclusion criteria

Patients with allergic constitution and needle syncope were excluded. Patients were excluded if they had a serious primary disease such as severe hypertension, cardiovascular and kidney diseases, hematopoietic system diseases, and other life-threatening diseases, or if they were suffering from mental disorders. Patients unwilling to participate in the study or were participating other medical trials were excluded.

2.4 Diagnostic criteria for CS

According to the 2nd Symposium Meeting on Cervical Spondylosis, the general diagnostic principle for CS was diagnosed by clinical manifestations and X-rays. If patients presented with typical CS clinical manifestations, but X-ray did not reveal abnormalities and other diseases were excluded, then patients could still be diagnosed as having CS. Patients without clinical chief complaints and signs, but with abnormal X-rays should not be diagnosed with CS.

2.5 Treatment

Acupoints selected were: Jiaji (EX-B2), Dazhui (BU 14), Fengchi (GB 20), Xinshe (EX-HN 21), Baihui (BU 20), Sishencong (EX-HN 1), Quchi (LI 11), Shousanli (LI 10), and Hegu (LI 4). Sterile single-use acupuncture needles (40 mm in length and 0.30 mm in diameter; Hwato brand, Suzhou Medical Products Factory Co., Suzhou, China) were inserted into Jiaji (EX-B2) points to a depth of 20 to 30 mm at an angle of 75 degrees. The direction of needle insertion in Fengchi (GB 20) was

DOI: 10.1051/bioconf/20170801026

orientated towards the contralateral canthus (to a depth of 25 to 30 mm). Points in the head were inserted into the skin to the subcutaneous tissue at an angle of 15 degrees to the point surface. Other points were perpendicularly inserted. The inserted needles were manually manipulated until the patient felt numbness or the acupuncture sensation known as "De Qi." Needles were retained in the points for 30 min. Sessions were administered once every other day with 10 sessions constituting one course.

2.6 Trial design

Fifty subjects with CS neck pain from outpatient clinics were recruited and treated with acupuncture. Then, their PDP was measured with PADBI before and after the first and the tenth acupuncture treatments. The assessment scale for cervical spondylosis used in the study was based on the score diagram of treatment on diseases of lumbar vertebrae[13] by the Japanese Orthopaedic Association in 1984 which can objectively reflect improvements in cervical function, and quantitatively evaluate the efficacy of patients with CS.

One course of treatment with acupuncture was given to each patient. The doctors performing all therapies had extensive traditional Chinese clinical experience. PDP was recorded before and after the first acupuncture treatment and before and after the tenth acupuncture treatment. The parameters recorded and analysis were extracted independently by other researchers.

Essential information including age, case history, and treatment history were collected. All of the findings in the history and physical examination were recorded in a standard TCM assessment form. Patients were asked to remain relaxed and were pulse-diagnosed by a TCM doctor. Their pulse type of the left hand was recorded before acupuncture treatment. PADBI was placed on the left radial artery and the detection software was started. Pulse signals were recorded for 30 s before acupuncture treatment. Then, the acupuncture treatment was carried out for 30 min. Patients rested for 1-2 min after withdrawing the needle, and pulse signals were recorded for another 30 s.

2.7 Statistical analyses

SPSS 18.0 statistical software (IBM Corp., Armonk, NY, USA) was used for statistical analyses. Pulse measurements are expressed as mean ± standard deviation. Comparisons for pulse were analyzed by time domain analyses. The inner-group comparison was analyzed by t-test. A P value < 0.05 was considered statistically significant.

3 Results

We used TD-III-Type PADBI to measure changes in the PDP (p, u, t1, h1, Pp) before and after acupuncture treatment in 50 patients with CS. The P angle is the angle of the main wave, and can reflect changes in vascular elasticity and condition of blood flow. Decreased P angle indicates ameliorated vascular elasticity, normal circulation of Qi and blood, unobstructed blood flow, and little myocardial ischemia. The U angle is the angle of ascent, and can reflect changes in vascular elasticity and blood viscosity. Increased U angle indicates better vascular elasticity and healthier Qi and blood circulation. t1 corresponds to the rapid ejection period of left ventricle. Decreases in t1 after treatment indicate that Qi and blood are sufficient and indicate normal function. h1 is the main wave height, and can reflect the ejection function of the left ventricle and arterial compliance. Increases in h1 indicate that the left ventricular ejection function is enhanced, arterial compliance is improved, Qi and blood are sufficient, and Qi and blood circulation are healthy. The Pp value is the numerical difference between two peaks in the pulse wave. The numerical difference in normal human is commonly less than 0.12 s. Decreases in Pp after treatment indicate stable pulse wave rhythm and stable circulation of Qi and blood.

The pulse biology before and after the first treatment, before and after the tenth treatment, and before the first treatment and after the tenth treatment were analyzed (Tables 1-3). The PDP values of effective patients were closer to normal. This indicated that the Qi stagnation and blood stasis of these patients were improved. The data showed that the changes in PDP values from before to after the first acupuncture treatment were more obvious, than those from before to after the tenth acupuncture treatment. Furthermore, patients entered a consolidating adjustment period after one course of treatment, which was also effective for acupuncture treatment, although the speed of symptom improvement decreased significantly. Therefore, PADBI has important significance on the objective evaluation of acupuncture treatment in CS.

We found that the PDP (h1, u, p, Pp, t1) values were closer to healthy after the tenth treatment in patients whose PDP tended to be healthy after the first treatment. Although the changes in PDP from before to after the tenth acupuncture treatment were less pronounced, the values after the tenth treatment were closer to normal than the values after the first treatment. This is because of increasing differences value in the assessment scale.

The analysis of correlation between efficacy and PDP showed changes of PDP in five patients that indicated ineffective treatment. This manifested as the values of h1, u, p, Pp, t1 remaining constant or increasing (Table 4). This indicated that the circulation of Qi and blood were not adjusted. The symptoms of Qi stagnation and blood stasis and arthralgia were not improved.

Table 1. Comparison of pulse diagram parameters before and after the first session acupuncture treatment (x±s).

Index	Before treatment	After treatment	t value	P value	Standard value
Pp (s)	0.22±0.24	0.12±0.16a	2.748	0.008	<0.12
t1 (s)	0.13±0.06	0.10±0.03a	3.019	0.004	0.07-0.11
h1 (mm)	13.52±9.10	17.10±7.78b	-2.429	0.019	9-22
u (deg.)	65.11±2.55	86.51±1.15a	3.693	0.001	80-87
p (deg.)	69.82±43.23	31.87±21.73a	5.734	0.000	19-42

Notes: Pp value is the numerical difference between adjacent peaks of the pulse wave. The numerical difference of normal human is commonly less than 0.12 s. t1 corresponds to the rapid ejection period of the left ventricle. h1 is the main wave height, it can reflect the eject function of left ventricle and arterial compliance. U angle is the angle of ascent, and can reflect changes in vascular elasticity and viscosity of blood. P angle is the angle of the main wave, and can reflect changes in vascular elasticity and condition of blood flow. a $P < 0.01$, vs before the first session treatment, b $P < 0.05$, vs before the first session treatment.

Table 2. Comparison of pulse diagram parameters before and after the tenth session acupuncture treatment (x±s).

Index	Before treatment	After treatment	t value	P value	Standard value
h1 (mm)	17±7	21±10a	-2.651	0.011	9-22
p (deg.)	34±26	29±21a	1.409	0.165	80-87

Notes: h1 is the main wave height, it can reflect the eject function of left ventricle and arterial compliance. P angle is the angle of the main wave, and can reflect changes in vascular elasticity and condition of blood flow. a $P < 0.05$, vs before the tenth session treatment.

Table 3. comparison of pulse diagram parameters before the first session and after the tenth session acupuncture treatment (x±s).

Index	Before treatment	After treatment	t value	P value	Standard value
Pp (s)	0.22±0.24	0.08±0.10a	3.743	0.000	<0.12
t1 (s)	0.13±0.06	0.11±0.03a	2.849	0.006	0.07-0.11
h1 (mm)	13.52±9.10	20.56±9.67a	-3.906	0.000	9-22
u (deg.)	65.11±2.55	86.53±0.96a	4.016	0.000	80-87
p (deg.)	69.82±43.23	28.68±21.07a	6.143	0.000	19-42

Notes: Pp value is the numerical difference between adjacent peaks of the pulse wave. The numerical difference of normal human is commonly less than 0.12 s. t1 corresponds to the rapid ejection period of the left ventricle. h1 is the main wave height, it can reflect the eject function of left ventricle and arterial compliance. U angle is the angle of ascent, and can reflect changes in vascular elasticity and viscosity of blood. P angle is the angle of the main wave, and can reflect changes in vascular elasticity and condition of blood flow. a $P < 0.01$, vs before the first treatment.

Table 4. analysis of correlation between efficacy and pulse diagram parameters (x±s).

Item	Pp (s)	t1 (s)	h1 (mm)	u (deg.)	p (deg.)
Marked effect	0.06±0.10	0.11±0.01	22.49±3.51	86.72±0.68	22.96±5.83
Effect	0.07±0.06	0.11±0.03	19.84±5.16	86.46±1.36	32.62±3.38
No effect	0.23±0.24	0.13±0.06	13.53±9.20	65.09±2.58	70.61±43.31
Standard value	<0.12	0.07-0.11	9-22	80-87	19-42

Notes: a total of 50 subjects with CS neck pain were treated with acupuncture. The needles were retained in the points for 60 min once every other day with 10 sessions constituting one course. The pulse diagram parameter was measured with pulse acquisition device based on image before and after the first acupuncture treatment and compared. Five patients (the difference value was 5 to 0 in the assessment scale) did not see an effect, 27 patients (the difference value was 13 to 6 in the assessment scale) had significant changes, and eighteen patients (the difference value was 23 to 14 in the assessment scale) had very significant changes.

4 Discussion

For CS, TCM considers that Qi and blood stasis and arthralgia are the predominant disease changes. Therefore, the clinical treatment for CS is focused on these two aspects. In China, the use of acupuncture in the treatment of CS neck pain has been suggested[14-17]. TCM pulse diagnosis is one of the major clinical diagnostic methods in TCM, but the descriptions of pulses are subject to the interpretation of individual doctors, and this lacks quantitative standardization. King et al [18] tried to quantify TCM pulse diagnosis but were unable to reflect pulse qualities adequately. We aimed to study objective PDP parameters before and after acupuncture treatment that reflect the blood-Qi changes of the body.

In our study, we found that the pulse wave in patients was disorganized and irregular before the first treatment, especially in the anterior half, which indicated the influence of the abnormal of Qi and blood caused by CS. However, the pulse wave became regular after treatment. Based on the physiological significance represented by the PDP changes, we found that the tone of the great vessels was lower, the arterial elasticity was improved, and the pulse normalized. This indicated that acupuncture can regulate vascular function, effectively relieve the compression of spinal nerves and vasculature. These changes can improve the dysfunction of Qi and blood caused by CS. Therefore, PADBI has important significance in the objective evaluation of acupuncture treatment for CS.

In addition, we found that the PDP was normalized after the tenth treatment in patients whose PDP was closer to normal after the first treatment. The PDP after the tenth treatment did not improve compared with that before the first treatment when the PDP had no obvious change from before to after the first acupuncture treatment. Therefore, acupuncture treatment was effective for patients if it was effective after the first treatment. The effect of acupuncture in the treatment of CS was positive if patients continued treatment. The acupuncture treatment was ineffective for patients if it was ineffective after the first treatment and it may not develop favorable therapeutic effects if patients continued treatment. After the tenth treatment, changes in PDP did not improve in five patients, which indicated that their CS symptoms were not improved.

Changes in pulse information can reflect changes in Yin-Yang, Qi-blood, the function of viscera, and it can reflect hemodynamic changes in the cardiovascular and circulatory system. The transition of traditional sphygmology from qualitative to quantitative was an important aspect of the objective and standardization of sphygmology. Standardization has important significance to the development of pulse diagnosis in TCM. According to analysis of PDP by time domain analysis, we found that the PDP significantly changed from before to after acupuncture treatment. The main manifestations of the time domain signals were disorganized in patients, but they became regular after treatment. This is because CS had relatively obvious influences on the central nervous system. Time domain analysis mainly analyzes the relationship between amplitude of pulse signals and phase of pulsation. Therefore, disorganized signals before the first treatment became regular after treatment. This change indicated that the compressive symptoms of the neck were improved and the central nervous system symptoms were relieved after acupuncture treatment[19].

In conclusion, changes in PDP parameters and symptoms in patients with CS after treatment were consistent, so PDP could be used to study standard treatment CS. Changes in PDP from before until after the first acupuncture treatment and the efficacy of late treatment were consistent. Therefore, PDP before and after an initial acupuncture treatment could be a predictive index of efficacy.

References

1. Rao RD, Currier BL, Albert TJ, et al. Degenerative cervical spondylosis: clinical syndromes, pathogenesis, and management. J Bone Joint Surg Am. 89(6) , 1360- 1378 (2007).
2. Butler JS, Oner FC, Poynton AR, et al. Degenerative cervical spondylosis: natural history, pathogenesis, and current management strategies. Adv Orthop.2012 , 916987(2012).
3. Ferrara LA. The biomechanics of cervical spondylosis. Adv Orthop .2012, 493605(2012).
4. Binder A. Cervical spondylosis and neck pain. BMJ. 334(7592) , 527-531(2007).
5. Trinh K, Cui X, Wang YJ. Chinese herbal medicine for chronic neck pain due to cervical degenerative disc disease. Spine. 35(24) , 2121-2127(2010).
6. Avery RM. Massage therapy for cervical degenerative disc disease: alleviating a pain in the neck? Int J Ther Massage Bodywork . 5(3) ,41-46(2012).
7. Lauche R, Langhorst J, Dobos GJ, et al. Clinically meaningful differences in pain, disability and quality of life for chronic nonspecific neck pain - a reanalysis of 4 randomized controlled trials of cupping therapy. Complement Ther Med . 21(4) ,342-347(2013).
8. Bertozzi L, Gardenghi I, Turoni F, et al. Effect of therapeutic exercise on pain and disability in the management of chronic nonspecific neck pain: systematic review and meta-analysis of randomized trials. Phys Ther. 93(8) , 1026-1036(2013).
9. Fu WB, Liang ZH, Zhu XP, et al. Analysis on the effect of acupuncture in treating cervical spondylosis with different syndrome types. Chin J Integr Med. 15(6) , 426-430(2009).
10. Liang ZH, Di Z, Jiang S, et al. The optimized acupuncture treatment for neck pain caused by cervical spondylosis: a study protocol of a multicentre randomized controlled trial. Trials.13, 107(2012).
11. Willich SN, Reinhold T, Selim D, et al. Cost effectiveness of acupuncture treatment in patients with chronic neck pain. Pain.125(1-2) , 107-113(2006).
12. Sun Y, Chen Q. The 2nd symposium meeting on cervical spondylosis. China J Surg. 31(8) , 472-476(1993).
13. Yu W, Huo S. The paper introduces a Evaluation Method of lumbar function. Zhong Guo Kang Fu Yi Xue Za Zhi.13(5) , 214-215(1998).
14. Guo YQ, Chen LY, Fu WB, et al. Clinically randomized controlled study on abdominal acupuncture for treatment of cervical spondylosis. Zhong Guo Zhen Jiu.27(9) , 652-656(2007).
15. Liang ZH, Yang YH, Yu P, et al. Logistic regression analysis on therapeutic effect of acupuncture on neck pain caused by cervical spondylosis and factors influencing therapeutic effect. Zhong Guo Zhen Jiu.29(3) , 173-176(2009).
16. Dong WK, Lin XH. Clinical observation on cervical spondylosis of neck type treated with acupuncture at original and terminal points of trapeziust. Zhong Guo Zhen Jiu.32(3),211-214(2012).
17. Xu SJ, Liang ZH, Fu WB. Chronic neck pain of cervical spondylosis treated with acupuncture and moxibustion in terms of the heart and kidney theory: a randomized controlled trial. Zhong Guo Zhen Jiu .32(9) ,769-775(2012).
18. King EI, Cobbin D, Ryan D. The reliable measurement of radial pulse: gender differences in pulse profiles. Acupunct Med. 20(4) , 160-167(2002).
19. Liu ZR, Li XX. On analysis of sphygmogram. Sci Sin B. 25(12) , 1286-1295(1982).

Obstructive sleep apnea-hypopnea syndrome and cognitive impairments in the elderly

Shuling Song[1], Jin Tan[1], Yuyang Miao[2] and Qiang Zhang[1,a]

[1]*Department of Geriatrics, Tianjin Medical University General Hospital; Tianjin Geriatrics Institute, Tianjin, China*
[2]*Tianjin Medical University, Tianjin, China*

Abstract. Obstructive sleep apnea-hypopnea syndrome (OSAHS) is a common sleep-related breathing disorder that is associated with significant morbidity and mortality. It has received increasing attention that neurocognitive deficits occur with a high frequency in OSAHS. However, it is rarely known that OSAHS impacts on cognition in the elderly in whom an increased prevalence of OSAHS is present. In this review we consider recent studies in the association between OSAHS and cognitive impairments, with specific interest in the older population. Firstly, we elucidate the characteristics of OSAHS and OSAHS-related cognitive impairments in the older patients. Many studies have showed that the prevalence of OSAHS increases with age and it is higher in the elderly than other population. Moreover, OSAHS is associated with higher incidence of comorbidities and increased risk of clinical deterioration in the elderly, especially the neurocognitive impairments which even can develop dementia. Subsequently, we discuss the possible reasons of cognitive impairments that caused or aggravated by OSAHS in the elderly. The intermittent hypoxia (IH)-related disturbances of homeostasis such as oxidative stress, inflammation, and age-related changes such as the changes of sleep architecture, the declined expression level of anti-aging gene, medical comorbidities and polypharmacy, may be both contribute to the increased risk of cognitive impairments in the older patients with OSAHS.

1 Introduction

OSAHS is a sleep-related breathing disorder characterized by repetitive episodes of airflow cessation resulting in IH and sleep fragmentation which disrupts the sleep architecture and causes an overall poor quality of sleep. Subsequently, daytime functioning is significantly impaired, including excessive sleepiness, fatigue and mood problems [1]. Moreover, OSAHS is always accompanied by multisystem diseases, such as hypertension, arrhythmia, ischemic stroke, diabetes mellitus and so on [2]. Therefore, OSAHS and OSAHS-related multisystem injuries have attracted more and more attention, especially

[a] Corresponding author: Zhang Qiang, zhangqiangyulv@163.com
This work was supported by National Natural Science Foundation of China (Grant NO.81370183), Tianjin Natural Science Foundation (Grant No.14JCYBJC27800), National Clinical Key Subject Construction Project of NHFPC Fund and National Training Program of Innovation and Entrepreneurship for Undergraduates (Grant No.201510062005).

the neurocognitive impairments. A wide range of neurocognitive impairments have been identified in OSAHS patients, from attention to memory and executive functions, someone even develop dementia [3].

Recently, a study researched the relevance of sleep disordered breathing (SDB), hypoxia, and risk of mild cognitive impairment (MCI) and dementia in older women [4]. It is a prospective study performed in 298 older women without dementia, founding that 44.8% of women with SDB developed MCI or dementia after a mean of 4.7 years of follow-up, compared to 31.1% of women without SDB. After multivariate adjustments, the presence of SDB was still associated with increased risk of developing subsequent MCI or dementia (odd ratio [OR]: 1.85; 95% confidence interval [CI]: 1.11-3.08). These results suggest that SDB is associated with increased risk of developing cognitive impairments in the older women. It is noteworthy that what on earth the characteristics of OSAHS and OSAHS-related cognitive impairments in the elderly are. In this review, we summarize the current knowledge about characteristics of OSAHS and OSAHS-related neurocognitive impairments in the elderly, and particularly discuss the possible reasons that increase the risk of developing neurocognitive impairments in the older patients with OSAHS.

2 Characteristics of OSAHS in the elderly

2.1 Higher prevalence

OSAHS, the most common sleep apnea-hypopnea syndrome, can be suffered by population of all ages including children, youth, middle-aged and old people. Although estimates of OSAHS prevalence vary by age, sex, and diagnostic criteria, OSAHS tends to be more common in older adults. A higher prevalence of OSAHS of 38% up to 63.7% was reported in the elderly [5-7], compared with the middle-age adults whose prevalence of OSAHS ranged from 2% to 4% [8]. The prevalence of OSAHS increases most proportionally with age. In a study performed in 1042 volunteers aged 20 to 80 years old, the prevalence of OSAHS of 20-29y, 30-39y, 40-49y, 50-59y, 60-69y, 70-80y was 7.4%, 24.2%, 37.7%, 49.2%, 60.2% and 86.9%, respectively. And the risk of OSAHS also increased with age, reaching OR = 34.5 (95%CI=18.5-64.2, $P < 0.001$) for 60-80 years old compared with the 20-29 years old [9].

The higher prevalence of OSAHS in the elderly may be explained by changes in the anatomy or function of the upper airway. The pharyngeal collapsibility is increased with age [10], and the muscular endurance of upper airway is decreased in the elderly [11]. In addition, ventilatory control instability [12] and disordered sleep structure [13], predisposed by sleep fragmentation and resultant frequent shifts from sleep to arousal and vice versa, might contribute to the occurrence of OSAHS in elderly.

2.2 Longer duration of apnea-hypopnea

It has been demonstrated that the apnea-hypopnea duration of OSAHS increases with age, and it is longer in the elderly with OSAHS. One study reported that the duration of apnea-hypopnea events was longer in the elderly compared to young and middle-aged OSAHS patients, and the age-related changes were speculated to affect the long duration of apnea-hypopnea events in the older patients, such as increased propensity for pharyngeal collapse, increased deposition of parapharyngeal fat and decreased partial arterial pressure of oxygen [14]. The other study also found that the duration of apnea-hypopnea events was longer in the older patients and it increased significantly with age [15]. The authors explained these findings by breakdown in sensing apnea-hypopnea events and reduction in the stimulus triggering arousals in the older patients with OSAHS. These changes related to age could be associated with the longer duration of apnea-hypopnea in the older patients with OSAHS, but they require large studies.

2.3 Severe multisystem injuries

OSAHS is associated with cardiovascular, cerebrovascular, metabolic, respiratory, digestive and urinary diseases, resulting in multisystem injuries, especially in the elderly whose impairments are more common and more severe.

OSAHS has been related to several cardiovascular comorbidities, such as ischemic heart disease, congestive heart failure, cardiomyopathy and arrhythmias. The older patients with OSAHS have higher risks of occurrence cardiovascular diseases, severe OSAHS even lead to cardiovascular death. One study followed up 372 normotensive older subjects (mean age, 68.2y) 3 years, showed that an AHI ≥ 30 per hour was independently related to incident arterial hypertension (P=0.02; OR=1.8; 95% CI=1.1-2.8), which suggested that severe OSAHS was associated with new-onset arterial hypertension in normotensive elderly [16]. Another study about the relevance between SDB and arrhythmias, showed that complex ventricular ectopy was associated most strongly with OSAHS [17]. In a prospective study performed in 939 elderly (≥ 65yr), the fully adjusted hazard ratios for cardiovascular mortality were 2.25 (CI=1.41-3.61) for the untreated severe OSAHS group, and 1.38 (CI=0.73-2.64) for the untreated mild to moderate OSAHS group when compared with the control group [18]. These results suggest that severe OSAHS not treated is associated with rised risk of cardiovascular death in the elderly.

In a retrospective cohort study performed in 250,907 patients hospitalized for pneumonia, 15,569 (6.2%) of patients were diagnosed with OSAHS, who had a higher prevalence of chronic pulmonary disease when compared with who not with OSAHS (68% vs 47%). After adjustments for demographics, comorbidities, admitting physician specialty, hospital type and the effects of clustering, OSAHS was associated with a rised risk of transfer to intensive care (OR=1.54; 95% CI=1.42-1.68) and intubation (OR=1.68; 95% CI=1.55-1.81) on or after the third hospital day [19]. Collectively, these findings provide evidence to support the view that OSAHS increases the susceptibility to respiratory disease, and hypoxia becomes more severe when OSAHS coexists with these diseases, which increases the risk of clinical deterioration.

Evidences from population, clinic-based, and laboratory studies had indicated that OSAHS can increase the risk of developing of insulin resistance, glucose intolerance and type 2 diabetes [20]. Moreover, OSAHS can contribute independently to the development of nocturia, and the frequency of nocturia increases as the severity of OSAHS increase [21]. It is reported that OSAHS patients also suffer from nocturnal gastroesophageal reflux and laryngopharyngeal reflux, which can lead to injury of esophageal mucosal and ulcer [22]. In conclusion, multisystem diseases associated with OSAHS can aggravate the damage that is related to ageing.

3 Characteristics of cognitive impairments in the elderly with OSAHS

3.1 High risk of occurrence

IH and sleep fragmentation, the major physiopathologic characteristic of OSAHS, can cause ischemia and hypoxia of brain resulting in increased risk of cerebrovascular accidents, such as ischemic stroke [23], brain infarction [24], mild cognitive impairment and dementia [4].

Multiple studies have been demonstrated that SDB is significantly associated with cognitive impairments. In a study performed in 1389 community-dwelling individuals 60 to 70 years old [25], 10.8% of subjects of this older sample reported apnea during sleep, and 49.5% reported snoring. After adjustments for age, gender, educational level, tobacco status, alcohol consumption, depressive symptomatology and number of medications, logistic regression models showed that both snoring and apnea were associated with low scores in the Digit Symbol Substitution Test (OR= 1.80, 95% CI=1.09-2.99; OR=1.58, 95% CI=0.87-2.89, respectively), and in the Trail Making Test (OR=2.14, 95% CI=1.24-3.69; OR=1.88, 95% CI=1.04-3.39, respectively). This study indicated that SDB was a risk factor for poor cognitive performance in tests requiring visual attention skills. Yaffe et al. [4] also found that the adjusted OR for developing cognitive impairments was 1.85 (95% CI=1.11-3.08) in

DOI: 10.1051/bioconf/20170801027

older women with SDB. All these findings suggest that SDB is associated with a rised risk of developing cognitive impairments in elderly.

3.2 Severe cognitive impairments

OSAHS-related cognitive impairments in the elderly are more severe, which are reflected by poor performance in cognitive tests and changes of brain structure. A study comparing the cognitive performance in attention-alertness tests among subjects of different ages, showed that the older patients had poor performance compared with their age-matched controls, while young patients' cognitive performance was similar to their age-matched controls [26]. These findings suggest that cognitive dysfunctions of OSAHS patients are age dependent. Ayalon et al. [27] also showed that older patients had poor performance for immediate word recall and slower reaction time during sustained attention in comparison with older controls, young patients and young controls, and detected a decreased activation of the frontal gyrus, hippocampus and parietal lobe during the tasks by functional magnetic resonance imaging. These results are supported by studies that used structural and functional neuroimaging to examine the brain regions affected in patients with OSAHS with cognitive impairments.

3.3 More underlying diseases

It is common that the elderly have more underlying diseases, such as chronic obstructive pulmonary disease (COPD), coronary heart disease and heart failure, which can worsen ischemia/hypoxia of OSAHS patients, and increase the risk of cognitive deficits in the older patients. For example, the occurrence of both COPD and OSAHS in an individual patient has been described as "overlap syndrome", which has a high risk for comorbidities, such as arterial hypertension and diabetes [28].

4 Possible reasons of cognitive impairments that caused or aggravated by OSAHS in the elderly

4.1 Oxidative stress

IH is now being recognized as a main factor of contributing to occurrence of OSAHS-related comorbidities. IH promotes oxidative stress by increasing production of reactive oxygen species (ROS) which is associated with cell damage and apoptosis. It has led to the hypothesis that oxidative stress induced by IH is associated with neurocognitive impairments in patients with OSAHS. Xu et al. [29] developed chronic intermittent hypoxia (CIH) mouse model to study the mechanisms of CIH-mediated brain cortical neuronal cell damage. They detected significant increase of ROS level and oxidative stress response markers (c-Fos, c-Jun and NF-κB) in brain cortex of mice exposured to CIH. Moreover, CIH induced caspase-3 activation in cortical neuronal cell. However, transgenic mice, overexpressing Cu,Zn-superoxide dismutase exposed to CIH, had a low level of ROS and reduced neuronal apoptosis in brain cortex. These results demonstrated that increased level of ROS was associated with neuronal cell damage. The other study also demonstrated that antioxidants intervention can prevent CIH-induced decline in spatial learning during the Morris water maze as well as CIH-induced decrease in superoxide dismutase levels [30]. Taken together, all of these findings indicate that CIH-mediated increased ROS production and subsequently oxidative stress propagation contribute, at least partially, to neuronal apoptosis and neurocognitive impairments.

4.2 Inflammation

Recently, lots of studies have showed that OSAHS patients have increased plasma markers of systemic inflammation, and IH is also associated with increased levels of inflammatory cytokines, adhesion molecules, and activation of circulating neutrophils in IH rodent model. One study explored the association between OSAHS and inflammation through measured the activity of NF-κB by electrophoreticmobility shift assay, and the concentration of soluble E (sE)-selectin and soluble vascular cell adhesion molecule-1 (Svcam-1) by enzyme-linked immunosorbent assay in OSAHS patients and control subjects [31]. They found that the activity of NF-κB and the concentration of sE-selectin and Svcam-1 were higher in OSAHS patients than control subjects. Another study also identified a strong association between OSAHS and increased levels of TNF-α and IL-8 [32]. And these inflammatory markers were reduced after continuous positive airway pressure (CPAP) therapy. What's more, it was demonstrated that IH caused low-grade neuroinflammation in the hippocampus of mice, including early but transient elevation of cytokine (IL-1β and RANTES/CCL5), and delayed but long-term changes of microglia which are major inflammatory cells in the central nervous system [33]. Peripheral cytokines, which can penetrate the blood-brain barrier directly via active transport mechanisms or indirectly via vagal nerve stimulation, combine with central cytokines cause endothelial dysfunction and aggravate brain damage [34]. These results suggest that IH-mediated inflammation is associated with neurocognitive deficits.

4.3 Changes of sleep architecture

Sleep architecture underwent significant alterations throughout the aging process in the general population. In other words, sleep architecture, especially the sleep-wake cycle, changes significantly in the elderly, which include the decrease of total sleep time, sleep efficiency, percentage of rapid eye movement (REM) sleep and slow wave sleep, and the increase of WASO (night-time spent awake after sleep onset), arousal index, sleep latency, REM sleep latency and the percentage of stages 1 and 2 [35]. It is noteworthy that OSAHS causes REM sleep fragmentation, while REM sleep may play a major role in selective emotional memory processing and recalibrating the brain's responses to emotional events [36]. REM sleep disturbance caused by OSAHS may affect the cognitive functions through the emotional disturbance. Moreover, slow wave activity (SWA) during the non-REM has been postulated to play a role in the abstraction of rules and the integration of knowledge into existing schemas [37], while OSAHS decreases or fragments causes slow wave sleep. Thus, OSAHS can cause the cognitive impairments through such effects on SWA during the non-REM stages. From the above results, we can speculate that sleep stage dysfunction and fragmentation may play an important role in the development of cognitive impairments in the older patients with OSAHS.

4.4 Anti-aging gene

Some of diseases are more common in the elderly, such as coronary heart disease, heart failure, stroke and so on, whose common pathophysiological characteristic is ischemia/hypoxia. This phenomenon indicates that the older people have unique vulnerability to ischemia/hypoxia. Moreover, the increased susceptibility to IH in aging rats was also demonstrated through developing IH model [38]. Anti-aging genes, such as sirtuin 1 (Sirt1), Klotho and so on, decrease with age, and they are declined in ischemic/hypoxic diseases. Collectively, all theses findings suggest that vulnerability of the older people to ischemia/hypoxia may be associated with declined expression of anti-aging gene.

Sirt1, the mammalian NAD+-dependent histone deacetylase, is involved in the regulation of different cellular processes, ranging from cell cycle, metabolism, and apoptosis to the cellular energy and redox status. It has been demonstrated that the expression level of Sirt1 decreases with age [39], and low level of Sirt1 is associated with increased susceptibility to ischemic/hypoxic diseases [40]. Shinji et al. detected the expression of Sirt1 mRNA and serum cystatin C, and prevalence of mutations in mitochondrial DNA in young mice and aged mice, respectively [41]. The result showed that the expression of Sirt1 was reduced in aged mice, which was linked to age-associated kidney dysfunction and mitochondrial oxidative damage. However, after long-term calorie restriction, the expression level of Sirt1 was increased, and kidney dysfunction and mitochondrial oxidative damage were attenuated. Collectively, these findings indicate that the susceptibility of the elderly to hypoxia is associated with decreased level of Sirt1.

As a recently discovered anti-aging gene, Klotho was originally identified in a mutant mouse that exhibited multiple disorders resembling human premature-aging syndrome, including shortened life span, arteriosclerosis, skin atrophy, osteoporosis and emphysema [42]. Yamamoto et al. detected the urinary 8-hydroxydeoxyguanosine (8-OHdG, a biological marker of in vivo oxidative DNA damage) of Klotho-overexpressing transgenic mice and wild-type mice, and found that the average urinary 8-OHdG excretion in transgenic mice was one-half of that in wild-type mice [43]. Furthermore, the authors found that the transgenic mice survived a challenge with a lethal dose of paraquat significantly longer than wild-type mice. All these results indicate that Klotho can increase resistance to oxidative stress.

The declined expression level of anti-aging gene is not only associated with increased susceptibility to ischemic/hypoxic diseases, but also related to cognitive impairments. Michán et al. detected that Sirt1 was expressed in hippocampal neurons by using a mouse specific anti-Sirt1 antibody [44]. Then they found that Sirt1 knockout mice exhibited poor cognitive performance, including immediate memory, classical conditioning and spatial learning. These findings suggest that Sirt1 plays an important role in maintaining normal learning, memory and synaptic plasticity. Recently, it also has been demonstrated that Klotho can increase the expression of erythropoietin receptor (EpoR) in neuronal cell and enhance Epo-mediated neuroprotection [45].

Based on above findings, we speculate that the more frequent and more severe OSAHS-related cognitive deficits in the elderly may be associated with the declined expression level of anti-aging gene.

4.5 Others

Some diseases that are common in the elderly, such as obesity, hypertension, diabetes, are both connected with increased risk of OSAHS and cognitive impairments. The relationship between obesity and OSAHS is bidirectional, that is, obesity and OSAHS reinforce each other [46]. In a study aimed to evaluate the impact of obesity on cognition of patients with OSAHS, the Total intelligence quotient and the Performance intelligence quotient scores were significantly lower in obese OSAHS patients with poor performance on cognitive tests, compared to OSAHS patients and normal controls [47]. These results suggest that obesity has a synergic and additive action with that exerted by OSAHS, aggravating the cognitive impairments. The effects of hypertension on cognition and OSAHS are similar to the impacts of obesity, which interact with OSAHS and both are related to cognitive impairments. It also has been found that both OSAHS and hypertension are associated with worse auditory verbal memory and executive function in the elderly [48]. The coexistence of many diseases can increase risk of underlying OSAHS and aggravate cognitive impairments, which provide further support to the hypothesis that medical comorbidities may play an important role in the occurrence of cognitive impairments in older patients with OSAHS.

Since the multisystem diseases are common in the elderly, so polypharmacy has become common in the elderly who are more likely to develop cognitive impairments associated with medications. Psychoactive drugs, antidepressants and anticonvulsants can lead to dementia and delirium. Moreover, non-psychoactive drugs, such as antibiotics, cardiac medications and histamine H2 receptor antagonists may lead to acute or chronic cognitive impairments [49]. It is noteworthy that certain medications can worsen underlying OSAHS. For example, the patients who taken atypical antipsychotics (AAs) had significantly elevated OR compared with the patients who didn't take AAs (4.5 vs 1.9) [50]. The impact of OSAHS on cognition is greater in the elderly than in other OSAHS populations may be associated with polypharmacy, but it requires large studies to delineate.

5 Conclusion

The presence of OSAHS is higher in the elderly. And the damage of nervous system, especially cognitive impairments, is more frequent and more severe in the older patients. However, the effectiveness of treatments of OSAHS and OSAHS-related cognitive impairments is little known in the older patients. So the study of pathogenesis of OSAHS and OSAHS-related cognitive impairments is of great significance for clinical treatment in the older patients. IH-related disturbance of homeostasis and age-related changes may provide more space for the exploration of new treatments in the elderly who have OSAHS. But these mechanisms described above of cognitive impairments induced by OSAHS in the elderly are unclear. Further studies are needed to determine the correlation between OSAHS and cognitive impairments, which might accelerate the development of therapeutic interventions in OSAHS and OSAHS-related cognitive impairments in the elderly.

References

1. T.J. Vaessen, S. Overeem, M.M. Sitskoorn, Sleep. Med. Rev **19**, 51 (2015).
2. N.A Dewan, F.J. Nieto, V.K. Somers, Chest **147**, 266 (2015).
3. R.S. Bucks, M. Olaithe, P. Eastwood, Respirology **18**, 61 (2013).
4. K. Yaffe, A.M. Laffan, S.L. Harrison, S. Redline, A.P. Spira, K.E. Ensrud, S. Ancoli-Israel, K.L. Stone, JAMA **306**, 613 (2011).
5. P. Johansson, U. Alehagen, E. Svanborg, U. Dahlström, A. Broström, Age. Ageing **41**,468 (2012)
6. E. Sforza, F. Chouchou, P. Collet, V. Pichot, J.C. Barthélémy, F. Roche, Eur. Respir. J **37**, 1137 (2011).
7. S.S. Ng, T.O. Chan, K.W. To, K.K Chan, J. Ngai, A. Tung, F. W. Ko, D.S. Hui, PLoS One **10** (2015).
8. T. Young, M. Palta, J. Dempsey, J. Skatrud, S. Weber, S. Badr, N. Engl. J. Med **328**, 1230 (1993)
9. S. Tufik, R. Santos-Silva, J. A. Taddei, L. R. Bittencourt, Sleep. Med **11**, 441 (2010) .
10. M. Eikermann, A. S. Jordan, N. L. Chamberlin, S. Gautam, A. Wellman, Y. L. Lo, D. P. White, A. Malhotra, Chest **131**, 1702 (2007).
11. C. Worsnop, A. Kay, Y. Kim, J. Trinder, R. Pierce, J. Appl. Physiol **88**, 1831 (2000).
12. S. Celle, R. Peyron, I. Faillenot, V. Pichot V, M. Alabdullah, J. M. Gaspoz, B. Laurent, J.C. Barthélémy, F. Roche F, Hum. Brain. Mapp **30**, 2090 (2009).
13. W. Moraes, R. Piovezan, D. Poyares, L. R. Bittencourt, R. Santos-Silva, S. Tufik, Sleep. Med **15**, 401 (2014).
14. E. George, V. Katerina, S. Maria, B. Lambros, N. Konstantina, G. Dimitrios, Sleep. Disord, **2012** (2012) .
15. J. C. Ware, R. H. McBrayer, J. A. Scott, Sleep **23**, 165 (2000) .
16. M. Guillot, E. Sforza, E. Achour-Crawford, D. Maudoux, M. Saint-Martin, J. C. Barthélémy, F. Roche, Sleep. Med **14**, 838 (2013).
17. R. Mehra, K. L. Stone, P. D. Varosy, A. R. Hoffman, G. M. Marcus, T. Blackwell, O. A. Ibrahim R. Salem, S. Redline, Arch. Intern. Med **169**, 1147 (2009).
18. M. A. Martínez-García, F. Campos-Rodríguez, P. Catalán-Serra, J. J Soler-Cataluña, C. Almeida-Gonzalez, I. De la Cruz Morón, J. Durán-Cantolla J, J. M. Montserrat, Am. J. Respir. Crit. Care. Med **186**, 909 (2012).
19. P. K. Lindenauer, M. S. Stefan, K. G. Johnson, A. Priya, P. S. Pekow, M. B. Rothberg, Chest **145**, 1032 (2014).
20. R. N. Aurora, N. M. Punjabi, Lancet. Respir. Med **1**, 329 (2013).
21. S. Ayik, K. Bal, G. Akhan, Turk. J. Med. Sci **44**, 249 (2014).
22. S. K. Wise, J. C. Wise, J. M. DelGaudio, Otolaryngol. Head. Neck. Surg, **135**, 253 (2006).
23. S. Redline, G. Yenokyan, D. J. Gottlieb, E. Shahar, G. T. O'Connor, H. E. Resnick, M. Diener-West, M. H. Sanders, P. A. Wolf, E. M. Geraghty, T. Ali, M. Lebowitz, N. M. Punjabi, Am. J. Respir. Crit. Care. Med **182**, 269 (2010).

24. K. Minoguchi, T. Yokoe, T. Tazaki, H. Minoguchi, N. Oda, A. Tanaka, M. Yamamoto, S. Ohta, C. P. O'Donnell, M. Adachi, Am. J. Respir. Crit. Care. Med **175**, 612 (2007).
25. M. J. Dealberto, N. Pajot, D. Courbon, A. Alpérovitch, J. Am. Geriatr. Soc **44**, 1287 (1996).
26. M. Alchanatis, N. Zias, N. Deligiorgis, I. Liappas, A. Chroneou, C. Soldatos, C. Roussos, Sleep. Breath **12**, 17 (2008).
27. L. Ayalon, S. Ancoli-Israel, S. P. Drummond, Am. J. Respir. Crit. Care. Med **182**, 413 (2010)
28. E. H. Steveling, C. F. Clarenbach, D. Miedinger, C. Enz, S. Dürr, S. Maier, N. Sievi, S. Zogg, J. D. Leuppi, M. Kohler, Respiration **88**, 451 (2014).
29. W. Xu, L. Chi, B. W. Row, R. Xu, Y. Ke, B. Xu, C. Luo, L. Kheirandish, D. Gozal, R. Liu, Neuroscience **126**, 313 (2004) .
30. L. Hui-guo, L. Kui, Z. Yan-ning, X. Yong-jian, Sleep. Med **11**, 205 (2010).
31. A. K. Htoo, H. Greenberg, S. Tongia, G. Chen, T. Henderson, D. Wilson, S. F. Liu, Sleep Breath **10**, 43 (2006).
32. S. Ryan, C. T. Taylor, W. T. McNicholas, Am. J. Respir. Crit. Care. Med **174**, 824 (2006).
33. E. Sapin, C. Peyron, F. Roche, N. Gay, C. Carcenac, M. Savasta, P. Levy, M. Dematteis, Sleep **38**, 1537 (2015).
34. C. J. Wilson, C. E. Finch, H. J. Cohen, J. Am. Geriatr. Soc **50**, 2041 (2002).
35. W. Moraes, R. Piovezan, D. Poyares, L. R. Bittencourt, R. Santos-Silva, S. Tufik, Sleep. Med **15**, 401 (2014).
36. A. N. Goldstein, M. P. Walker, Annu. Rev. Clin. Psychol **10**, 679 (2014).
37. N. Landmann, M. Kuhn, H. Piosczyk, B. Feige, C. Baglioni, K. Spiegelhalder, L. Frase, D. Riemann, A. Sterr, C. Nissen, Sleep Med. Rev **18**, 531 (2014).
38. Gozal D, Row BW, Kheirandish L, R. Liu, S. Z. Guo, F. Qiang, K. R. Brittian, J. Neurochem **86**, 1545 (2003).
39. H. Gong, J. Pang, Y. Han, Y. Dai, D. Dai, J. Cai, T. M. Zhang, Mol. Med. Rep **10**, 3296 (2014)
40. C. Tong, A. Morrison, S. Mattison, S. Qian, M. Bryniarski, B. Rankin, J. Wang, D. P. Thomas, J. Li, FASEB. J **27**, 4332 (2013).
41. S. Kume, T. Uzu, K. Horiike, M. Chin-Kanasaki, K. Isshiki, S. Araki, T. Sugimoto, M. Haneda, A. Kashiwagi, D. Koya, J. Clin. Invest **120**, 1043 (2010) .
42. M. Kuro-o, Y. Matsumura, H. Aizawa, H. Kawaguchi, T. Suga, T. Utsugi, Y. Ohyama, M. Kurabayashi, T. Kaname, E. Kume, H. Iwasaki, A. Iida, T. Shiraki-Iida, S. Nishikawa, R. Nagai, Y. I. Nabeshima, Nature **390**, 45 (1997).
43. M. Yamamoto, J. D. Clark, J. V. Pastor, P. Gurnani, A. Nandi, H. Kurosu, M. Miyoshi, Y. Ogawa, D. H. Castrillon, K. P. Rosenblatt, M. Kuro-o, J. Biol. Chem **280**, 38029 (2005).
44. S. Michán, Y. Li, M. M. Chou, E. Parrella, H. Ge, J. M. Long, J. S. Allard, K. Lewis, M. Miller, W. Xu, R. F. Mervis, J. Chen, K. I. Guerin, L. E. Smith, M. W. McBurney, D. A. Sinclair, M. Baudry, R. de Cabo, V. D. Longo, J. Neurosci **30**, 9695 (2010).
45. M. F. Cheng, L. J. Chen, H. S. Niu, T. T. Yang, K. C. Lin, J. T. Cheng, Acta. Neurobiol. Exp (Wars) **75**, 60 (2015).
46. S. Ryan, S. J. Crinion, W. T. McNicholas, QJM **107**, 949 (2014).
47. O. Vitelli, A. Tabarrini, S. Miano, J. Rabasco, N. Pietropaoli, M. Forlani, P. Parisi, M. P. Villa, Sleep Med **16**, 625 (2015) .
48. L. M. Kinoshita, J. A. Yesavage, A. Noda, B. Jo, B. Hernandez, J. Taylor, J. M. Zeitzer, L. Friedman, J. K. Fairchild, J. Cheng, W. Kuschner, R. O'Hara, J. E. Holty, B. K. Scanlon, Sleep Breath **16**, 1201 (2012).
49. M. Shinohara, M. Yamada, Brain Nerve **64**, 1405 (2012).
50. A. Shirani, S. Paradiso, M. E. Dyken, Sleep Med **12**, 591 (2011).

Antioxidant ofsmallmolecular weightchitosan oligosaccharidein vitro

Shang Kong[#], Peiqiu Cao[#], Jiao Guo[a] and Zhengquan Su[a]

Key Research Center of Liver Regulation for Hyperlipidemia SATCM/Class III Laboratory of Metabolism SATCM, Guangdong TCM Key Laboratory for Metabolic Diseases, Guangdong Pharmaceutical University, Guangzhou, 510006, China

Abstract. In recent years, there has been increasing interest in finding natural antioxidants, since they can protect the human body from free radicals and retard the progress of many chronic diseases. This study is to evaluate antioxidant of a certain molecular chitosan oligosaccharide (Mw 1000 Da, COST) in vitro. The antioxidant activities of COST were investigated in vitro by several antioxidant assays, including DPPH radical scavenging, hydroxyl radical scavenging, superoxide radical scavenging, reducing power and metal chelating activities.In vitro studies found that COST had noticeable scavenging activities on 2,2-diphenyl-1-picryl-hydrazyl (DPPH) radical, hydroxyl radical, and superoxide anion, and possessed reducing power and metal chelating activities, and its capability of anti-oxidation enhance with the increase of concentration.COST presents anexcellent biological rolein scavenging free radical, which could be explored as a potential antioxidant or dietary supplement to retard diseases caused by peroxide.

1 Introduction

The increase of mean life in the industrialized world will contribute to 5% of the population older than 85 yr of agein 2050[1].With the research about the molecular mechanism of aging, the 'free radical theory of ageing' is showing promise in helping to understand the process of ageing and in finding effective anti-ageing agents[2]. The theory postulates that aging is caused by excessive reaction of free radicals in living things, which leads to cell death and tissue injury by causing cellular damage[3]. Excessive accumulation of ROS increases the chances of the onset of diseases, indeed, it is linked to biological wear and tear leading to certain types of age-related diseases including cardiovascular disease[4], Type II diabetes[5], the decline of female reproductive function[6] and so on. So people have serious implications with respect to interventional strategies scavenging free radical to delay senescence. Chitosan oligosaccharides (COS), as degradation product of chitosan (CTS), have been

[a] Corresponding author. E-Mails: suzhq@scnu.edu.cn (Z.-Q.S.) and wshxalb@163.com (J.G.).
[#]Shang Kong and Peiqiu Cao make equal contributions to this work.
This project was financially supported by the National Science Foundation of China (no. 81173107), the Science and Technology Planning Project of Guangdong, China(2013B021100018).

extensively applied in pharmacy, food and other fields becauseof good biocompatibility, degradation, non-toxicity and easy-absorbability[7].Several studies have demonstrated that COS and related products have beneficial effects on weight loss[8] and anti-hyperlipidemia[9]. In recent years, some studies have shown that COS had obviously inhibitory effects on oxidation and glycation, its activities were linked with degrees of deacetylation (DD)[10]. Furthermore different molecular weight of COS was likely to contribute significantly towards the antioxidant effect, there was inversely proportional relationship between them[11,12]. But there are currently no found one research that evaluates antioxidant activities a certain molecular chitosan oligosaccharide (Mw1000 Da, DD≥91%, COST) in vitro.In vitro the antioxidant activities of COST were investigated in vitro by several antioxidant assays, including hydroxyl radical scavenging, superoxide radical scavenging, reducing power and metal chelating activities.

2. Materials and methods

2.1 Materials

COST (Mw1000 Da; DD 95.6%) was purchased from Shandong Aokang Biotech Ltd., Shandong, China. 2,2-diphenyl-1-picryl-hydrazyl (DPPH) was obtained from Shanghai Mai Kelin Biochemical Technology Ltd., Shanghai, China. Nitro blue tetrazolium (NBT), phenazinemethosulfate (PMS), nicotinamide adenine dinucleotide-reduced (NADH) were purchased from Shanghai Aladdin Reagent Ltd., Shanghai, China. All other reagents and solvents were of analytical grade.

2.2 In vitro antioxidant activities of COST

2.2.1 DPPH assay

Free radical scavenging ability was determined by testingthe change of the absorption by 517 nm before and after the reaction of DPPH and antioxidant substances.COST samples was dissolved in 1 mL of distilled water at the different concentrations taken in test tubesof Sample group and was added 1.0 ml of 0.1mMethanol solution of DPPH, mixed well and allowed to incubate at room temperature for 30 min in the dark. At the same time, one test tubes of Control group was mixed with 1.0mL DPPH and 1.0mL distilled water, and the sample solution of Blank group was mixed with 1.0 mL COST and 1.0 mL anhydrous alcohol. The absorbance value of the sample was measured at 517 nm. The percentage of DPPH radical scavenging activity was theoretically calculated by the following formula:

$$\text{DPPH radical scavenging activity (\%)} = \left(1 - \frac{A_{sample} - A_{blank}}{A_{control}}\right) \times 100\% \quad (1)$$

2.2.2 Hydroxyl radical scavenging assay

The hydroxyl radical scavenging ability of COST was measured according to a method described previously with some modification[13]. Briefly, the different concentration of COST was mixedwith 1.0mL $FeSO_4$(6.0 mM), 1.0mL H_2O_2(2.5μM), respectively. The mixture was then incubated at room temperature for10 min. The reaction was initiated upon addition of 1.0 mL salicylic acid solution (6.0mM). The absorbance was determined at 500 nm after 30 mins reaction time at 37°C. The ability of hydroxyl radical scavenging (%) was calculated as follows:

$$\text{Scavenging rate (\%)} = \frac{A_0 - (A_1 - A_2)}{A_0} \times 100\% \quad (2)$$

Where $A_1$was the absorbance of the sample and A_0 was the absorbance of the solvent control, whereas

 DOI: 10.1051/bioconf/20170801028

A_2 was the absorbance of the reagent solution without sodium salicylate.

2.2.3 Superoxide anion radical scavenging assay

The scavenging assay of superoxide anion radical was carried out according to the method with a little modification. COST samples with different concentrations were prepared with distilled water. The reaction mixture containing of 1.0 mL of test samples, 1.0 mL NADH (338μM), 1.0mL NBT (72μM) and 1.0mL PMS (300μM) was reacted at room temperature for 5 min. Absorbance of the resulting solution was measured at 560 nm on a spectrophotometer against a blank. The scavenging ability of superoxide anion radical was calculated by the following equation:

$$\text{Scavenging activity } (\%) = 1 - \frac{A}{A_0} \times 100\ \% \tag{3}$$

Where Asample was the absorbance of sample group and A_0 was the absorbance of blank control group (Tris-HCl buffer instead of NADH).

2.2.4 Measurement of reducing power

The experiment was performed by the modified method. Briefly, 1ml COST with different concentrations were added to 1.0 ml of potassium ferricyanide (1% w/v) and 0.2 mL phosphate buffer (0.2 M pH6.6), respectively. The mixed solution was shaken well and incubated at 50°C for 20 min. Then 1.0 ml of trichloroacetic acid solution (10% w/v) was added to the mixture with the purpose of terminating the reaction. After 6000 r/min centrifuge for 10 min,0.35 ml of 3% ferric chloride was added to 1ml supernatant of mixture. Absorbance of final reaction product was read at 700 nm on a spectrophotometer against a blank after 15 min at room temperature.

2.2.5 Metal chelating activity(11)

The different concentrations of COST samples were added to a mixture of 0.05ml FeCl2 (2 mM) and 1ml deionized water, then stewing 2 min. The reaction was initiated by the addition of 5 mMferrozine (1 mL). After the mixture had reached equilibrium, the absorbance of the solution was then measured spectrophotometrically at 562 nm. The ability of COST to chelate ferrous ion was calculated using the following equation:

$$\text{Chelating ability}(\%) = \frac{A_2 - (A_0 - A_1)}{A_2} \times 100\% \tag{4}$$

where A0 and A2 were the optical density at 562 nm without and with samples, respectively. A1 was the absorbance of the group without $FeSO_4$

3. Results

3.1 DPPH scavenging activity

DPPH is a free radical which has unpaired electron in its middle nitrogen atom. DPPH can pairs with a free radical scavenger to yield a colored change gradually[13]. Therefore, antioxidative ability of COST can be evaluated by the assay of scavenging DPPH radical. The scavenging effects of COST on DPPH radical were shown in (Figure 1) exhibited COST has obvious scavenging effect to DPPH, and the scavenging activity is growing with increasing the concentration. The active amino and hydroxyl groups in COST molecules canprovide the hydrogen ions to bind with DHHP, which can achieve the purpose of removing the DPPH free radical.

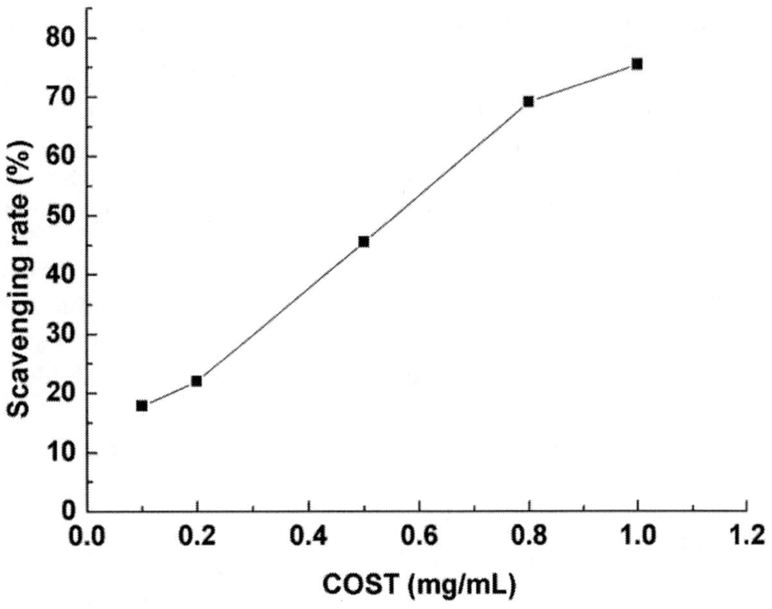

Figure 1. Scavenging effect COST with different concentrations towards DPPH radicals.

3.2 Hydroxyl free radical (·OH) scavenging activity

Hydroxyl radical is one of the most reactive and dangerous free radicals among the reactive oxygen species, which can cause serious damage to the adjacent biological molecules by reaction of Fe (II) complex with H_2O_2 in the presence of salicylic acid[14]. In this study, we evaluated the scavenging activity of COST on HO• generated through the Fenton reaction. As shown in Figure 2, the scavenging capacities of COST were elevated quickly with theincrease of concentration. Fenton reaction is one of the important mechanisms in · OH. The removal rate of ·OH is an important indicator of antioxidant capacity[15]. The results of this experiment can be used as evidence of the antioxidant capacity of COST, and provide clues for mechanism of its antioxidant in the further study.

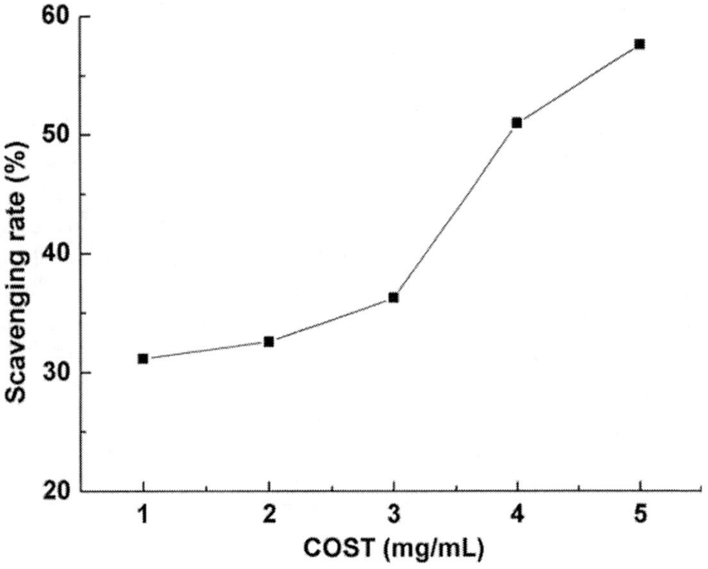

Figure 2. Scavenging effect of COST with different concentrations towards hydroxyl radicals

DOI: 10.1051/bioconf/20170801028

3.3 Superoxide anion radical scavenging activity

Compare to other radicals, superoxide anionradical is a relatively weak oxidant in most organisms, but it has much longer lifetime and could form secondary radicals such as hydrogen peroxide and hydroxylradicalthat causetissue damages and various diseases[16]. Therefore, the scavenging effect on superoxide radical is an important way to illustrate the mechanism of antioxidant activity. In Figure 3, we found that when the concentration of COST was less than 0.5mg/ml, the scavenging activity is weak. Once the concentration of COST was more than 0.5mg/ml, the scavenging effects of COST at various concentrations on superoxide radical were dose-dependent. Superoxide radical is the first oxygen free radicals formed by ground state oxygen received an electron, and may be other radicals formed by a series of reactions. COST can react with oxygen free radicals, so that oxygen radicals will be completely removed. The result shows that COST has rather highability of scavenging superoxide radical.

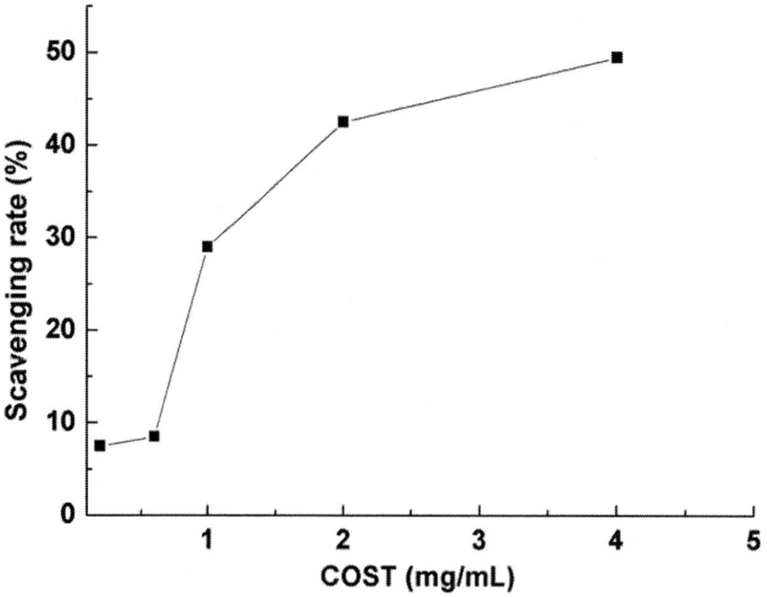

Figure 3. Scavenging effect of COST with different concentrations towards superoxide radicals.

3.4 Reducing power

Usually reductant provides electron to reduce free radicals produced by oxidative stress. Antioxidant reduces Fe^{3+} to Fe^{2+} under a certain condition and absorbance ofreaction mixture is read at 700 nm[17]. Higher absorbance of antioxidant indicates it has greater reducing power. The reducing powers of COST were shown in Figure 4, reducing power increased with increasing concentration of COST. The reducing properties are generally associated with the presence of reductones, which have been shown to exert antioxidant action by breaking the free radical chain by donating a hydrogen atom[11].Reductones are also reported to react with certain precursors of peroxide, thus preventing peroxide formation. Our result indicates it was likely to contribute significantly towards the observed potential antioxidant activity.

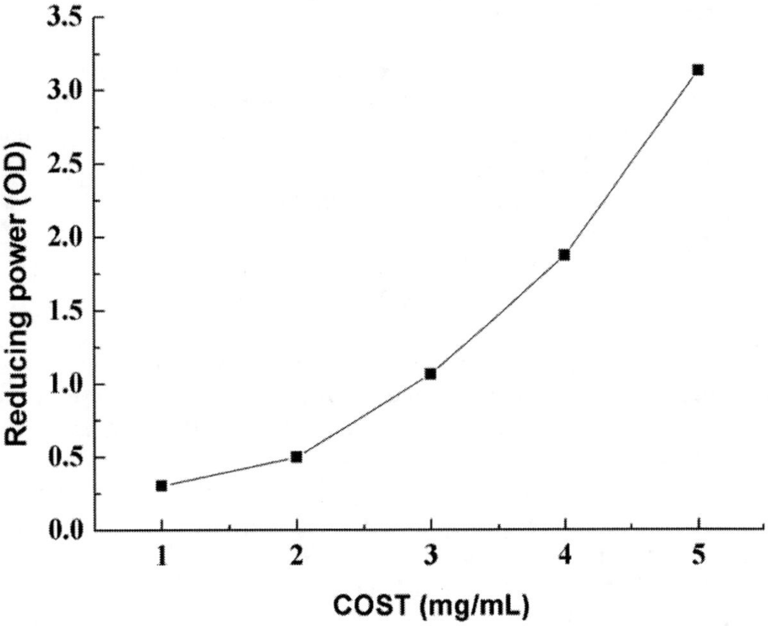

Figure 4. Reducing power of COST with different concentrations

3.5 Chelating ability

Ferrous ion chelating ability is another important indicator to evaluate an antioxidant's capacity in vitro. The chelating ability of COST against $Fe^{2+}$is shown in Figure 5 that the chelating ability were elevated quickly with the increase of concentration of COST. Fe^{2+} was thought to be one encouragement of causing oxygen free radicals and promoting oil peroxide, it can prevents oxidative damage by reducing the concentration of Fe^{2+} generated in the Fenton reaction[18]. Our dates demonstrate that the form of COST and Fe^{2+} complex can greatly improve the stability of Fe^{2+}, which may contribute to the prevention of cardiovascular disease.

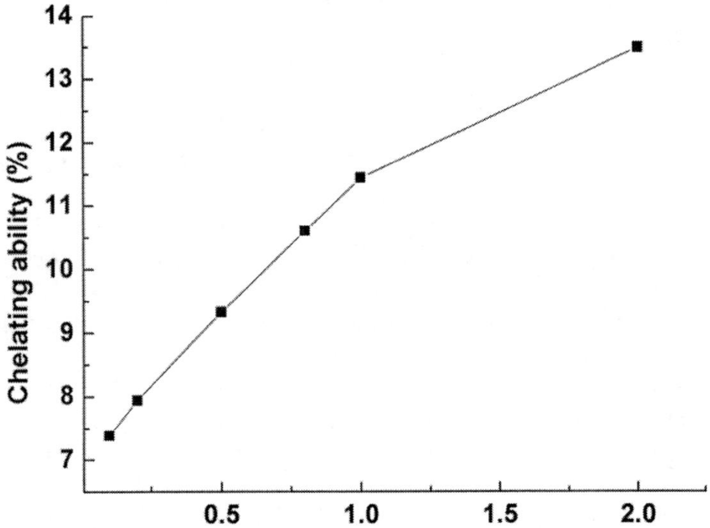

Figure 5. Chelating effect of COST with different concentrations on ferrous ions

4 Discussion and Conclusion

In normal condition, Aging was usually known asa process that when the biological development is mature, with the increase of age, its own physiological functionwill decrease accompanying with the decline in the stability of internal environment, and the structure, group will gradually degenerated in change, leading to die, which is anirreversible phenomenon[19]. Once excessive accumulation of free radicals such as superoxide anion radical, hydroxyl radical and so on in the body, that results a series of free radical reaction, which accelerate the aging process resulting in high incidence of hyperlipidemia, hypertension, diabetes, Alzheimer's disease, Parkinson's disease and cancer [20-22].It is thus rapidly becoming an urgent necessity to develop safe and effective types of antioxidants that may scavenge free radical todelay senescence.

Chitosan (CTS) is a polysaccharide material extracted from shell; it is generally considered a good excipient because of its biocompatibility and show a variety of activities such as anti-hyperlipidemia[23], lose weight[24], antioxidant[25]. As the small molecular derivatives of CTS, COS have shown the better water-solubility and thus can be digested and absorbed more easily by the gastrointestinal tract in animals and humans, which may make them better biological activity[26-28]. COST, one type of COS, is known byits certain molecular weight 1000Da, which may possess more significant biological activity of antioxidant.

In this study, we firstly explore its antioxidant function in intro.Reactive oxygen species (ROS), including superoxide anion radicals, hydroxyl radicals, and hydrogen peroxide etc., are often generated by oxidation product of biological reactions or exogenous factors. The study of its antioxidant activity in vitro, has shown that COST can scavenge DPPH radical, hydroxyl radical, and superoxide anion, and possess reducing power and metal chelating activities, which expressed high antioxidant activity and the activity depended on its molecular weight and concentration.

Excessive generation of ROS, induced by various stimulating factors such as certain pollutants, tobacco smoke, and which exceed the antioxidant capacity of the organism will lead to aging, cancer, and other many diseases. The results of experiment in vitro was given to us that COST may play an important role in the elimination of ROS and protect the cells against toxic effects of ROS.

References

1. Tuljapurkar, S., Li, N., and Boe, C., A universal pattern of mortality decline in the G7 countries, *Nature,* **405**, 789 (2000).
2. Harman, D., Free-radical theory of aging. Increasing the functional life span, *Ann N Y AcadSci,* **717**, 1 (1994).
3. Ho, Y. S., Magnenat, J. L., Gargano, M., and Cao, J., The nature of antioxidant defense mechanisms: a lesson from transgenic studies, *Environ Health Perspect, 106 Suppl***5**, 1219 (1998).
4. Cao, P., Pan, H., Xiao, T., Zhou, T., Guo, J., and Su, Z., Advances in the Study of the Antiatherogenic Function and Novel Therapies for HDL, *Int J MolSci,* **16**, 17245 (2015).
5. Masoro, E. J., Physiology of aging, *Int J Sport NutrExercMetab,* **11***Suppl*, S218 (2001).
6. Tatone, C., Di Emidio, G., Vitti, M., Di Carlo, M., Santini, S. J., D'Alessandro, A. M., Falone, S., and Amicarelli, F., Sirtuin Functions in Female Fertility: Possible Role in Oxidative Stress and Aging, *Oxid Med Cell Longev,* **2015**, 659687 (2015).
7. Hossain, S., Rahman, A., Kabir, Y., Shams, A. A., Afros, F., and Hashimoto, M., Effects of shrimp (Macrobraciumrosenbergii)-derived chitosan on plasma lipid profile and liver lipid peroxide levels in normo- and hypercholesterolaemic rats, *ClinExpPharmacolPhysiol,* **34**, 170 (2007).
8. Cao, P., Huang, G., Yang, Q., Guo, J., and Su, Z., The effect of chitooligosaccharides on oleic acid-induced lipid accumulation in HepG2 cells, *Saudi Pharmaceutical Journal* (2016).
9. Pan, H., Yang, Q., Huang, G., Ding, C., Cao, P., Huang, L., Xiao, T., Guo, J., and Su, Z., Hypolipidemic effects of chitosan and its derivatives in hyperlipidemic rats induced by a high-fat diet, *Food Nutr Res,* **60**, 31137 (2016).

10. Zhang, C. M., Yu, S. H., Zhang, L. S., Zhao, Z. Y., and Dong, L. L., Effects of several acetylated chitooligosaccharides on antioxidation, antiglycation and NO generation in erythrocyte, *Bioorg Med ChemLett,* **24**, 4053 (2014).
11. Xing, R., Liu, S., Guo, Z., Yu, H., Wang, P., Li, C., Li, Z., and Li, P., Relevance of molecular weight of chitosan and its derivatives and their antioxidant activities in vitro, *Bioorg Med Chem,* **13**, 1573 (2005).
12. Fang, I. M., Yang, C. H., Yang, C. M., and Chen, M. S., Chitosan oligosaccharides attenuates oxidative-stress related retinal degeneration in rats, *PLoS One,* **8**, e77323 (2013).
13. Liu, X., Sun, Z., Zhang, M., Meng, X., Xia, X., Yuan, W., Xue, F., and Liu, C., Antioxidant and antihyperlipidemic activities of polysaccharides from sea cucumber Apostichopusjaponicus, *CarbohydrPolym,* **90**, 1664 (2012).
14. Sheng, J., and Sun, Y., Antioxidant properties of different molecular weight polysaccharides from Athyriummultidentatum (Doll.) Ching, *CarbohydrPolym,* **108**, 41 (2014).
15. Herraiz, T., and Galisteo, J., Hydroxyl radical reactions and the radical scavenging activity of beta-carboline alkaloids, *Food Chem,* **172**, 640 (2015).
16. Sheng, J., and Sun, Y., Antioxidant properties of different molecular weight polysaccharides from Athyriummultidentatum (Doll.) Ching, *CarbohydrPolym,* **108**, 41 (2014).
17. Liu, X., Sun, Z., Zhang, M., Meng, X., Xia, X., Yuan, W., Xue, F., and Liu, C., Antioxidant and antihyperlipidemic activities of polysaccharides from sea cucumber Apostichopusjaponicus, *CarbohydrPolym,* **90**, 1664 (2012).
18. Liu, W., Wang, J., Zhang, Z., Xu, J., Xie, Z., Slavin, M., and Gao, X., In vitro and in vivo antioxidant activity of a fructan from the roots of Arctiumlappa L, *Int J BiolMacromol,* **65**, 446 (2014).
19. Hill, S. M., Hao, X., Liu, B., and Nystrom, T., Life-span extension by a metacaspase in the yeast Saccharomyces cerevisiae, *Science,* **344**, 1389 (2014).
20. Bulterijs, S., Hull, R. S., Bjork, V. C., and Roy, A. G., It is time to classify biological aging as a disease, *Front Genet,* **6**, 205 (2015).
21. Greer, D. B., Hermanns, M., and Cooper, C. M., Making Lemonade Out of Life's Lemons: A View Into the World of Aging With Parkinson's Disease, *J PsychosocNursMent Health Serv,* **53**, 20 (2015).
22. Miedema, M. D., Lopez, F. L., Blaha, M. J., Virani, S. S., Coresh, J., Ballantyne, C. M., and Folsom, A. R., Implications of the Eighth Joint National Committee Guidelines for the Management of High Blood Pressure for Aging Adults: Atherosclerosis Risk in Communities Study, *Hypertension,* **66**, 474 (2015).
23. Tao, Y., Zhang, H. L., Hu, Y. M., Wan, S., and Su, Z. Q., Preparation of Chitosan and Water-Soluble Chitosan Microspheres via Spray-Drying Method to Lower Blood Lipids in Rats Fed with High-Fat Diets, *Int J MolSci,* **14**, 4174 (2013).
24. Tan, S., Gao, B., Tao, Y., Guo, J., and Su, Z. Q., Antiobese effects of capsaicin-chitosan microsphere (CCMS) in obese rats induced by high fat diet, *J Agric Food Chem,* **62**, 1866 (2014).
25. Feng, T., Du, Y., Li, J., Wei, Y., and Yao, P., Antioxidant activity of half N-acetylated water-soluble chitosan in vitro, *European Food Research and Technology,* **225**, 133 (2007).
26. Huang, L., Chen, J., Cao, P., Pan, H., Ding, C., Xiao, T., Zhang, P., Guo, J., and Su, Z., Anti-obese effect of glucosamine and chitosan oligosaccharide in high-fat diet-induced obese rats, *Mar Drugs,* **13**, 2732 (2015).
27. Qiao, Y., Bai, X., and Du, Y., Chitosan oligosaccharides protect mice from LPS challenge by attenuation of inflammation and oxidative stress, *International Immunopharmacology,* **11**, 121 (2011).
28. Fang, I., Yang, C., and Yang, C., Chitosan oligosaccharides prevented retinal ischemia and reperfusion injury via reduced oxidative stress and inflammation in rats, *Experimental Eye Research,* **130**, 38 (2015).

BIO Web of Conferences **8**, 01029 (2017) DOI: 10.1051/bioconf/20170801029
ICMSB2016

Investigation of therapeutic effect of Longchai Formula Granule on chronic hepatitis B patients: A double-blind, controlled and randomized clinical trial

Zhanpeng Yan[1, 2], Tingting Xu[1, 2], Kelei Su[1, 2], Chi Wang[3], Tong Yan[4], Zhong Hua[5], Erli Gu[6] and Fangshi Zhu[1, 2,a]

[1]*Affiliated Hospital of Traditional Chinese and Western Medicine, Nanjing University of Chinese Medicine, Nanjing 210028, China*
[2]*Jiangsu Province Institute of Traditional Chinese Medicine, Nanjing 210028, China*
[3]*Taizhou Integrated Hospital of Traditional Chinese and Western Medicine, Taizhou 225300, China*
[4]*Changzhou Third People's Hospital, Changzhou 213001, China*
[5]*Wuxi Infectious Diseases Hospital, Wuxi 214005, China*
[6]*Nantong Third People's Hospital, Nantong 226000, China*

Abstract. Objective: To objectively evaluate the clinical efficacy of Longchai Formula Granule (LCFG) on chronic hepatitis B (CHB). Methods: According to GCP, 100 CHB patients who meet enter criterion were encoded and divided randomly into the test group (50 cases) and the control group (50 cases). The patients in test group were treated with LCFG and matrine capsule simulation agent, the patients in control group was treated with matrine capsule and LCFG simulation agent. Both groups have 12-week treatment period. The changes of Chinese medicine symptoms, ALT, AST, graduation of liver function, HBeAg, HBsAg, HBV-DNA in CHB patients were observed and compared. Results: After unblinding, statistical data showed that there were 82 effective actual CHB cases composed of 40 CHB patients in test group (10 cases dropout) and 42 CHB patients in control group (8 cases dropout). The total efficient of Chinese symptoms improvement and ALT in test group is better than the control group ($P<0.05$). The contents of AST, the graduation of liver function, HBeAg, HBsAg, and HBV-DNA in two groups were totally improved than before treatment, but there was no significant differences between groups ($P>0.05$). As a whole, it shows a tendency that the change of these values in test group induced by LCFG is better than these values in control group induced by matrine capsule. Conclusions: In summary, LCFG is effective for CHB patients and has no significant difference with matrine cpasules. However, LCFG is better than matrine capsule in the improvement of symptomatology and liver function.

1 Introduction

Hepatitis B virus (HBV) infection witch infected is a serious global health problem, and about 2 billion people were infected worldwide. Each year more than 300 thousand HBV carriers die of acute

[a] Correspondence to: Fangshi Zhu, E-mail: zhufangshi@126.com, Tel: +86-25-52632761, Address: No.100, Shizi Street, Hongshan Road, Nanjing City, Jiangsu Province, China.
This work was supported by Scientific and Technological Innovation and Achievements Transformation Special Projects of Jiangsu Province (BM2009903)
Trial registration: This study is registered at ClinicalTrials.gov, ChiCTR-IOR-14005585.

hepatitis, liver fibrosis, liver cirrhosis and hepatocellular carcinoma (HCC) induced by chronic hepatitis B (CHB) [1-4]. Antiviral and liver-protecting therapy is a standardized treatment method that can reduce morbidity and mortality HBV infected patients[5-6]. Since a significant proportion of CHB patients fail to get satisfactory treatment from conventional medicine, so many CHB patients seek help from complementary and alternative medicine (CAM), especially traditional Chinese medicine (TCM)[7]. The Chinese pharmacopoeias contain numerous specific medicines and formulas, and numerous studies show that several herbs have antiviral activity and can prevent viral infection. Recently some clinical reports have suggested that the appropriate usage of TCM in combination with modern western medicine can prevent HBV infection and ameliorate symptoms of CHB patients, TCM have shown great potential prospects for treating CHB[8-10].

Longchai Formula (LCF) is an experienced and effective Chinese medicine formula developed by Professor JIN Shi, which is PhD tutor of Nanjing University of Chinese Medicine and the instructor inheriting the academic experience of Chinese famous physicians. LCF has affirmative treatment efficacy on CHB patients in clinical applications for many years, and LCF is good at improving symptoms, recovering liver function, decreasing HBV surface markers in CHB[11-12]. In our previous study, we have elaborated and discussed the theoretical mechanism of LCF on treating CHB[13]. LCF and LCF-containing serum could inhibit the growth of HepG2.2.15 cells and reduced the secretion level of HBsAg and HBeAg in vitro[14-15].Recent experiments show that LCF could increase drug sensitivity of nucleoside analogues (NA) treat CHB[16]. In order to objectively further evaluate the clinical efficacy of LCF on the treatment for CHB, a randomized, double-blinded, double-dummy and parallel-controlled trial was designed and implemented. The results are as follows.

2 Methods

2.1 Patients

One hundred CHB patients were selected from 5 hospitals, including Affiliated Hospital of Traditional Chinese and Western Medicine (Nanjing University of Chinese Medicine), Taizhou Integrated Hospital of Traditional Chinese and Western Medicine, Changzhou Third People's Hospital, Wuxi Infectious Diseases Hospital and Nantong Third People's Hospital, date from 30, December 2010 to 31, December 2012. 100 patients were divvied randomly into the test group and test group by statistics software SAS9.1. After unblinding of this clinical trail, there were 18 cases of dropouts and withdrawals, containing 10 cases in test group and 8 cases in control group (the expulsion rate <20%). The trial group was comprised 17 women and 23 men, age range 20-63 years (mean age is 38.45±14.28 years), course of disease range 14-78 months (mean course is 39.55±16.57 months). The control group was comprised 16 women and 26 men, age range 25-61 years (mean age is 38.45±14.28 years), course of disease range 12-70 months (mean course is 34.78±15.79 months). The χ^2 and t test showed no significant difference in age, sex and course of disease. (P >0.05).

2.2 Inclusion criteria

The inclusion criteria were as follows: (a) patients conforming to the diagnosis standard of chronic hepatitis B that defined in the Guidance of Prevention and Treatment of Chronic Hepatitis B updated in 2010;[17] (b) patients conforming to the TCM standard of "zheng" differentiation for chronic hepatitis B according to Guiding principle of clinical research on new drugs of Chinese medicine;[18] (c) patients with the age of 18-65 years and gen; (d) patients do not taking related medicine of treating CHB within half of month.

2.3 Exclusion criteria

The exclusion criteria were as follows: (a) patients aged below 18 or above 65 years old; (b) patients infected other type hepatitis virus at the same time; (c) patients had autoimmune hepatitis or hepatitis caused by drug intoxication and ethylism; (d) patients had been diagnosed with severe hepatitis, hepatic failure, cirrhosis, hepatocellular carcinoma or serious primary disease in cardiovascular, cerebrovascular, kidney, lung, endocrine system, hematopoietic system and mental disease; (e) patients were pregnant or lactating women; (f). patients were allergic physique or allergic to multiple drugs.

2.4 Dropouts and withdrawals

Participant in this clinical trail who did not complete the treatment programs had been excluded during the observation period.

2.5 Treatment

The CHB patients in test group were treated with LCFG (consisting of six medicinal herbs: Solanum nigrum L., Bupleurum chinense DC., Hedyotis diffusa Willd, Sedum sarmentosum Bunge, Gardenia jasminoides Ellis and Scutellaria baicalensis Georgi, and produced by Jiangyin Tianjiang Pharmaceutical Co., Ltd.), 2 times a day, 1/2 dose a time, infused with water. Meanwhile patients in test group were administrated with mimetic matrine capsules used for placebo-controlled (Jiangsu Province Engineering Research Center), 3 times a day, 0.2g a time. The CHB patients in control group were treated with matrine capsules (containing 5%medicine, starch, food coloring, bittering agent, and produced by Chia Tai Tianqing Pharmaceutical Co., Ltd.), 3 times a day, 0.2g a time. Meanwhile patients in control group were administrated mimetic LCFG used for placebo-controlled (Jiangyin Tianjiang Pharmaceutical Co., Ltd.), 2 times a day, 1/2 dose a time, infused with water. Both groups have therapeutic course for 12 weeks, according to the rule that minimum observation time is no less than 3 months from the Guiding principle of clinical research on new drugs of Chinese medicine[18].

2.6 Evaluation of efficacy

At the time point of 0, 6 and 12 weeks, the efficacy of drugs on CHB patients was evaluated comprehensively. Blood routine examination, urine routine examination, defecate routine examination, kidney function examination, electrocardiogram and chest examination were detected in order to drug safety. The contents of HBV-DNA, HBsAg, HbeAg, ALT and AST in serum of CHB patients were detected.

Symptom score was evaluated and composed of fatigue, liver pain, jaundice, hepatosplenomegaly, tongue, pulse and so on. According to the degree of symptoms CHB patients were divided into none, mild, moderate, and severe patients, and the symptom score were respectively recorded as 0, 1, 2 and 3. In the end of trail the total scores of each CHB patient was calculated.

The assess standard of the damage degrees of liver function as follows: (a) mild damage, there are no or mild clinical symptoms, and only 1 or 2 kind liver function index appears mildly abnormal; (b) moderate damage, symptoms, signs and index from observe and laboratory tests is between mild damage and severe damage; (c) severe damage, there are significant or persistent hepatitis symptoms, such as fatigue, anorexia, abdominal distension, loose stools, hepatic face, liver palms, spider, splenomegaly, etc. The contents of serum ALT and/or AST in CHB patients were repeatedly or persistently elevated, content of serum albumin is decreased or the ratio of albumin/globulin (A/G) is reversed, content of gamma globulin is significantly increased. In addition to the above abnormal changes, albumin is less than 32g/L, total bilirubin is 5 times higher than the normal, prothrombin activity is more than 40% and less than 60%, cholinesterase is less than 2500U/L, any one in the above four indexes happened could be diagnosed as chronic severe hepatitis. These principles were referenced to the Guiding principle of clinical research on new drugs of Chinese medicine.[18]

The assessment standard of efficacy on CHB as follows: (a) marked effect, the main symptoms negative; (b) normal effect, the main symptoms alleviated or disappeared, liver function is normal or has 50% reduction than before, one index among HBV-DNA, HBeAg and HBsAg was negative; (c) no effect, related symptoms and indexes have not ameliorated or even worsen. These principles were referenced to the Guiding principle of clinical research on new drugs of Chinese medicine[18].

The assessment standard of syndrome curative effect of TCM as follows: (a) clinical fully recovered, the main symptoms and signs disappeared or almost disappeared, and reduction of symptom score is above or equal 95%; (b) marked effect, the main symptoms and signs improved significantly, and reduction of symptom score is between 70% and 95%; (c) normal effect, the main symptoms and signs improved mildly, and reduction of symptom score is between 30% and 70%; (d) no effect, main symptoms and signs no improved or worsen, reduction of symptom score is below 30%. These principles were referenced to the Guiding principle of clinical research on new drugs of Chinese medicine[18].

2.7 Statistical analysis

Statistical software SAS 9.1 (SAS Institute Inc., USA) was used for data processing. Quantitative data are expressed as mean ± standard deviation. For comparison of measured data between the two groups, the t test of independent samples was used for date of homogeneity of variance. The Wilcoxon test was used for classified data. Statistical significance was assumed at $P<0.05$.

3 Results

3.1 Comparison of TCM symptom improvement between two groups

Table 1. Comparison of TCM symptom improvement between two groups.

Group	n	Clinical recovery	Marked effect	Normal effect	No effect	Total effective rate
Test	40	11	16	10	3	92.50%
Control	42	9	6	18	9	78.57%

Notes: After statistical analysis there was significant difference between the two groups of total effective rate. ($P=0.038<0.05$)

3.2 Comparison of ALT and AST in the serum of patients

Table 2. Comparison of the serum ALT and AST between the two groups ($\bar{x}\pm s$ U/L).

Liver function	Test group			Control group		
	n	Pre-treatment	Post-treatment	n	Pre-treatment	Post-treatment
ALT	40	103.18±40.04	46.91±29.17	42	97.14±54.89	52.57±31.75
AST	40	82.01±40.79	43.25±13.52	42	67.79.±36.81	50.67±22.89

Notes: After 12 weeks, the level of ALT and AST in the two groups were significantly improved than before ($P<0.05$), and the reduction level of AST in the test group was better than the control group ($P<0.05$). There was no significant difference between the two groups in ALT ($P>0.05$).

3.3 Comparison of degrees of liver function between two groups

Table 3. Comparison of degrees of the liver function between the two groups.

Group	Pre-treatment			Total	Post-treatment			Total	*P*
	I	II	III		I	II	III		
Test	32	8	0	40	37	3	0	40	
Control	32	10	0	42	36	6	0	42	0.85
Total	64	18	0	82	73	9	0	82	

Notes: I represents mild damage, II represents moderate damage, III represents severe damage.
After 12 weeks in test group, the number of patients with mild liver function damage increased from 32 to 37, and the number of patients with moderate damage reduced from 8 to 3. While in the control group, the number of patients with mild liver function damage increased from 32 to 36, and the number of patients with moderate damage reduced from 10 to 6. There was no significant difference between the two groups in the liver function improvement rate ($P>0.05$). In general, it seems to show a tendency that the change of these values in test group is better than that in control group.

3.4 Comparison of quantitative reduction of HBeAg and HBsAg in serum of CHB patients

Table 4. Comparison of amount of HBeAg ($\bar{x}\pm s$, PEI U/ml) and HBsAg($\bar{x}\pm s$, ng/ml) between two groups.

Items	Test group			Control group		
	n	Pre-treatment	Post-treatment	n	Pre-treatment	Post-treatment
HBeAg	40	96.67±73.15	53.50±42.96	42	93.57±99.26	67.88±53.13
HBsAg	40	307.73±114.58	257.47±109.85	42	324.44±125.27	264.33±157.45

Notes: After 12 weeks, HBeAg and HBsAg in serum of CHB patients were improved partly than before treatment, but there was no significant difference between groups ($P>0.05$).

3.5 Comparison of quantitative reduction of HBV-DNA

Table 5. Comparison of quantity of HBV-DNA between the two groups(log10(IU/ml)).

Items	Test group (n=40)		Control group (n=42)	
	Pre-treatment	Post-treatment	Pre-treatment	Post-treatment
HBV-DNA	$(2.17\pm3.62)\times10^7$	$(0.92\pm1.75)\times10^7$	$(3.23\pm9.67)\times10^7$	$(1.23\pm2.45)\times10^7$

Notes: After 12 weeks, HBV-DNA was all inhibited partly by drugs than before treatment in both groups, but there was no significant difference between groups ($P>0.05$).

3.6 Comparison of total efficacy of drugs on CHB

Table 6. Comparison of total efficacy between the two groups.

Group(n)	Marked effect	Normal effect	No effect	Total effective rate	*P*
Test(40)	10	18	12	70.00%	0.83
Control(42)	9	18	15	64.29%	

Notes: The relieve rate of liver function, HBeAg, HBsAg, and HBV-DNA of the two groups were all improved than before, but there was no significant difference between the two groups ($P>0.05$). Overall, it seems to show a tendency that the change of these values in test group is better than that in control group.

4 Discussion

Oxymatrine is an active ingredient extracted from herb medicine Sophora. According to reports oxymatrine plays an important role in protecting liver cells and stimulating the body cells to produce interferon[19]. Oxymatrine also has the effects of anti-inflammatory, anti-hepatic injury, inhibition of HBV replication and immunomodulatory[20-21]. Since 2005, matrine was recommended to treat CHB according to Guideline on chronic hepatitis B prevention and treatment, which was made by Chinese Medical Association Hepatology and Chinese Medical Association infectious diseases[17]. So far, many clinical studies have shown that matrine possesses potential advantages in treat CHB[22-25]. In this trail we have choosed matrine as control drug and expected to further confirm and evaluate the efficacy of LCFG for CHB patients and HBV. As shown in Table 1, TCM symtom improvement in test group is better than in control group ($P<0.05$). The results of this trail showed that there was no significant difference between the LCFG and matrine capsule in trems of reduction of serum HBeAg, HBsAg, HBV-DNA and ALT in CHB patients($P>0.05$)(Table2, 3, 4, 5). However, the reduction level of AST in the test group was better than the control group ($P<0.05$)(Table2). At the last, the comparison of total efficacy (the summary of comparision of AST, ALT, HBV-DNA, HbeAg and HBsAg) between the two groups is not significantly (Table6). In a word, it seems to show a tendency that the change of these values in LCFG group is better than that in matrine group.

LCF consists of seven commonly used herbs including Solanum nigrum L., Radix Bupleuri, Hedyotis diffusa Wild, Sedum Sarmentosum, Scutellaria baicalensis, Fructus Gardeniae, Radix Glycyrrhizae. According to the prescription composition of TCM theory, Solanum nigrum L. and Hedyotis diffusa Wild are considered as sovereign medicine, their medicinal property of bitter-cold exerts a certain function of heat-clearing and detoxifying. Sedum Sarmentosum and Fructus Gardeniae are considered as minister medicine, their medicinal property of sweet-cool and bitter-cold can assist minister medicine to exert function removing dampness, detoxifying and clearing heat. Radix Bupleuri and Scutellaria baicalensis are considered as assistant medicine, the combination of two herbs can unobstructe liver-Qi, harmonize stomach-Qi, dry dampness and clear heat, this function is similar to Xiaochaihu Tang recorded in Shanghanlun. Radix Glycyrrhizae was used to be guide medicine in this formula and could reconcile the property of other drugs.

The whole formula possesses the function of heat-clearing, detoxifying, removing dampness and soothing liver-Qi. Recently, many TCM studies suggest that the appearance of CHB is most attributed to infection of dampness-heat and epidemic toxin, coupled with deficiency of healthy Qi. Multiple pathogenic factor could cause dampness-heat mixed with epidemic toxin, and gradually lead to imbalance of viscera, Qi and blood, and dysfunction of body's defenses. Some theories on pathogenesis of CHB in TCM are got more attention[26], including toxin theory, theory of deficiency of vital Qi, theory of stagnancy of damp-heat and theory of multiple reason (toxin, vital Qi deficiency, Qi stagnation, blood stasis tangled with each other). In our opinion[27], dampness-heat and epidemic toxin are the leading factors that cause CHB, liver-stagnation and spleen-deficiency is an important step to CHB. LCF has the effect of heat-clearing and detoxifying, harmonizing spleen and stomach precisely basing on the CHB pathogenesis of toxin invasion, vital Qi deficiency and Qi stagnation in TCM.

So far, it was widely reported that single medicine herb from LCF has the hepatoprotective and antiviral effect at home and abroad. For example, Solanum nigrum extract can protect liver against the CCl_4-induced oxidative damage in rats, and has the antiviral potential against HCV, HBV and HIV[28-30].Ursolic acid extracted from Hedyotis diffusa Wild possesses hepatoprotective ability and can inhibit HBV replication by blocking the HBx[31].Saikosaponins, the main active constituents of Radix Bupleuri, have been shown to possess immunomodulatory, hepatoprotective, anti-tumor and anti-viral activities, and Saikosaponin c exhibits anti-HBV activity and Saikosaponin d possesses potent cytotoxicity against human hepatocellular carcinoma cells[32]. The extracts of Sedum sarmentosum showed strong hepatoprotective effect on D-galactosamine-induced cytotoxicity in primary cultured mouse hepatocytes[33]. Baicalin extracted from Scutellaria radix showed hepatoprotective effect on acetaminophen-induced liver injury in mice[34]. A kind of glycoprotein isolated from Gardenia jasminoides Ellis can ameliorate liver function[35].

The above papers have reported that the hepatoprotective effect and antiviral activity of single medicine herb from LCF in terms of modern pharmacological mechanism. The microscopic mechanisms and therapeutic targets of LCF may be closely relation to the active ingredients included in LCF, and pharmacological mechanism of LCF treating CHB remains will be further studied in the future.

References

1. Lavanchy D. J Viral Hepat. **11**(2): 97-107(2004).
2. Perz JF, Armstrong GL, Farrington LA, Hutin YJ, Bell BP. J Hepatol. **45**(4): 529-538(2006).
3. Lai CL, Ratziu V, Yuen MF, Poynard T. The Lancet. **362**(9401): 2089-2094(2003).
4. Wu GY, Chen HS. World J Gastroenterol. **13**(6): 830-836(2007).
5. Alvarado-Esquivel C, de la Ascensión Carrera-Gracia M, Conde-González CJ, Juárez-Figueroa L, Ruiz-Maya L, Aguilar-Benavides S, Torres-Valenzuela A, Sablon E. J Antimicrob Chemother. **57**(2):221-223(2006).
6. Lu JJ, Chen EQ, Yang JH, Zhou TY, Liu L, Tang H. Virol J vol. **9**: 10-15(2012).
7. Zeng XX, Bian ZX, Wu TX, Fu SF, Ziea E, Woon WT. Am J Chin Med. **39**(6): 1061-1074(2011).
8. Ceylan-Isik AF, Fliethman RM, Wold LE, Ren J. Curr Diabetes Rev. **4**(4): 320-328(2008).
9. Wang G, Zhang L, Bonkovsky HL. Chin J Integr Med. **18**(4): 253-255(2012).
10. Cui X, Wang Y, Kokudo N, Fang D, Tang W. Biosci Trends. **4**(2): 39-47(2010).
11. Lu SH. Clinical Research on Treating chronic hepatitis B of Liver depression and spleen deficiency, heat resistance permits with Longchai decoction [D]. Nanjing University of Chinese Medicine(Chin). (2009).
12. Liu QZ. Journal of Nanjing University of TCM(Chin). **22**(1):55-57(2006).
13. Wu XY, Zhu FS. Chinese Journal of Basic Medicine in TCM(Chin). **15**(6): 469-470(2009).
14. Xue BY, Wu XY, Yao L, Lu SH, Zhu FS. Journal of Liaoning University of TCM(Chin). **12**(12): 10-11(2010).
15. Su KL, Zhu FS. 2012 International Symposium on Information Technology in Medicine and Education [A]. 845-850(2012).
16. Zhu FS. Modem Digestion&Intervention(Chin). **17**(6): 334-337(2012).
17. Chinese Society of Liver Disease and Infectious Diseases, Chinese Medical Association. Guidance for prevention and treatment of chronic hepatitis B. Chin J Hepatol(Chin). **13**: 881-891(2005).
18. Zheng XY. Beijing: China Medico-pharmaceutical Sciences and Technology Publishing House(Chin) [A]. 143-151(2002).
19. Li P, Si WK, Wang Y, Yao J. Chinese Journal of Biochemical Pharmaceutics(Chin). **25**(5): 261-264(2004).
20. Liu M, Liu XY, Cheng JF. China Journal of Chinese Materia Medica(Chin). **28**(9): 801-804(2003).
21. Liu N, Kan QC, Zhang XJ, Xv YM, Zhang S, Zhang GX, Zhu L. Exp Mol Pathol. **97**(3):470-476(2014).
22. Huang XL. China medicine and pharmacy(Chin). **2**(4): 96-97(2012).
23. Gu XB,Yang XJ,Wang D. Pharmacology and Clinics of Chinese Materia Medica(Chin). **26**(2): 75-77(2010).
24. Wang ZY, Li GS, Huang HX. Chinese Journal of Integrated Traditional and Western Medicine (Chin). **26**(8): 681-684(2006).
25. Zhang XW, Hu GQ, Li CY. World Chinese Journal of Digestology(Chin). **13**(3): 317-320(2005).
26. Liang LR, Liang JX. Haerbin Medical Journal(Chin). **24**(4): 42(2004).
27. Wang P, Zhu FS. Jiangsu Journal of Traditional Chinese Medicine. **43**(11): 76-77(2011).
28. Lin HM, Tseng HC, Wang CJ, Lin JJ, Lo CW, Chou FP. Chemico-Biological Interactions.

171(3): 283-293(2008).
29. Javed T, Ashfaq UA, Riaz S, Rehman S, Riazuddin S. Virology Journal. **8**: 26-32(2011).
30. Halim VA, Muck A, Hartl M, Ibáñez AJ, Giri A, Erfurth F, Baldwin IT, Svatos A. Proteomics. **9**(1): 171-181(2009).
31. Wu HY, Chang CI, Lin BW, Yu FL, Lin PY, Hsu JL, Yen CH, Liao MH, Shih WL. Journal of Agricultural and Food Chemistry. **59**(5): 1713-1722(2011).
32. Chiang LC, Ng LT, Liu LT, Shieh DE, Lin CC. Planta Medica. **69**(8): 705-709(2003).
33. Ninomiya K, Morikawa T, Zhang Y, Nakamura S, Matsuda H, Muraoka O, Yoshikawa M. Chem Pharm Bull(Tokyo). **55**(8): 1185-1191(2007).
34. Jang S, Kim HJ, Hwang KM, Jekal SJ, Pae HO, Choi BM, Yun YG, Kwon TO, Chung HT, Kim YC. Immunopharmacol Immunotoxicol. **25**(4): 585-594(2003).
35. Lee SJ, Oh PS, Lim KT. Clinical and Experimental Pharmacology and Physiology. **33**(10): 925-933(2006).

The clinical value of HbA1c in combination with FPG in the early screening of the elderly with type 2 diabetes

Lihua Liu[1,#], Wenging Chen[2], Minghua Dong[3,#], Lixia Jiang[4], Wei Qiu[3], Jian Li[5], Xiaoting Luo[1], Zhengchun Huang[5], Qin Wu[4], Qinfeng Wu[3], Shuiqin Chen[1], Lu Ou-Yang[3] and Shumei Li[3,a]

[1]*Department of Biochemistry and Molecular Biology, Gannan Medical College, Ganzhou, Jiangxi, China*
[2]*Department of Anesthesiology GanZhou Municipal Hospital, Ganzhou Jiangxi, China*
[3]*Department of Epidemiology in Preventive Medicine, Gannan Medical College, Ganzhou, Jiangxi, China*
[4]*Department of Clinical laboratory of 1'th affiliate hospital, Gannan Medical College, Ganzhou, Jiangxi, China*
[5]*Department of Anatomy Gannan Medical College, Ganzhou, Jiangxi, China*

Abstract ***Objective:*** To investigate the FPG associated with HbA1c diagnosis of T2DM, to clear the best point of tangency HbA1c diagnosis of T2DM. ***Methods:*** Using simple random sampling method, from this group of T2DM field epidemiological investigation data in the database we randomly selected 300 cases of T2DM patients, 300 cases of healthy people. The sensitivity and specificity of FPG, HbA1c and HbA1c/FPG combination were calculated by diagnostic screening test; Using ROC curve evaluation FPG, HbA1c and HbA1c in combination with FPG diagnosis value of T2DM. ***Results:*** The cutoff point of FPG for diagnosing T2DM was 6.19 mmol/L, the sensitivity and specificity were respectively: 83.00%, 93.67%, area under the curve (AUC) was 0.928; the cutoff point of HbA1c for diagnosing T2DM was 6.40%, the sensitivity and specificity were respectively: 87.00%, 90.33%, area under the curve (AUC) is 0.935;Combined use of HbA1c and FPG in the diagnosis of T2DM, simultaneous detection of HbA1c and FPG, when HbA1c≥ 6.40% or FPG≥6.19 mmol/L diagnosed with T2DM, the sensitivity and specificity were respectively: 97.79%, 84.61%, area under the curve (AUC) is 0.960. ***Conclusions:*** The cutoff point of HbA1c for diagnosing T2DM was 6.40% , Combined HbA1c and FPG in the T2DM diagnosis, which can greatly improve the sensitivity, the combined application is more conducive to the early screening of T2DM.

1 Introduction

With the development of society by leaps and bounds, the prevalence of diabetes is in a rising trend. According to the International Diabetes Federation (IDF) latest statistics, by 2040, nearly 642 million people are expected to suffer from Diabetes, 90% to 95% of whom are in Type 2 Diabetes Mellitus(T2DM). China has now the most T2DM patients in the world [1].

Fasting Plasma Glucose (FPG) or 2 hour Postprandial Glucose(2hPG) is the most commonly used

[a] Corresponding author: Shumei Li, gnyxylsm@163.com. Phone: 008615083787928.
[#] These authors contributed equally to this study and share first authorship.
This study was supported by scientific fund from National Natural Science Fund in China (No.81360445), Science and technology support program of Jiangxi Province (NO.20132BBG70086).

method in clinical diagnosis of T2DM, But it only represents the instant moment of blood Glucose levels, and chronic hyperglycemia is the main characteristic of T2DM, So only FPG or 2hPG on the diagnosis of T2DM has a certain missed diagnosis. Glycated hemoglobin A1c(HbAlc) can reflect the 2-3 months of blood Glucose levels, it can be used as indicators of T2DM patients with long-term control blood Glucose. In 2010, the American Diabetes Association(ADA) took HbA1c as a new diagnostic method for T2DM[2]. However, domestic research on HbA1c as the diagnostic standard of T2DM is still in its initial stage. Because of different sample population, people's life style and the environment are different, the value of HbA1c is different. Therefore, The elderly T2DM early screening in our region and the optimal cut-off point are studied, and results will now be reported as follows.

2 The research object and methods

2.1 The object of the research

Using simple random sampling method, From this group of T2DM field epidemiological investigation database we randomly selected 300 cases of T2DM patients, 300 cases of healthy people.

2.2 The research methods

(1) questionnaire: Main content includes gender, age, history of smoking and drinking, and other general information; (2) laboratory tests: subjects in the morning on an empty stomach phlebotomize, subjects were unified for breakfast after blood, continue to blood after 2 hours, Biochemical parameters such as FPG, 2hPG and HbA1c were determined by Hitachi 7600 automatic biochemical analyzer.

2.3 Statistical analysis

Using Stata12.0 statistical software for data analysis, The sensitivity and specificity of FPG, 2hPG and HbA1c in the diagnosis of T2DM was calculated, using ROC curve evaluation FPG, 2hPG, HbA1c diagnosis value of T2DM.

3 Results

3.1 The basic situation of the research object

From the T2DM field epidemiological survey data of this research group, 300 cases of T2DM patients were randomly selected, and 300 healthy people were studied, average age was 60.020 ± 9.684 years, The maximum is 86 years old, the minimum is 38 years old, The age distribution of the study subjects was normal distribution (Table 1); The object of study the age distribution of normally distributed (Figure 1), the distribution of FPG, 2hPG is skewed distribution (Figure 2, Figure 3), the distribution of HbA1c is basic normally distributed (Figure 4).

Table 1. The basic data of subjects.

variable	n	$(\bar{x} \pm s)$	minimum	maximum
age	600	60.020±9.684	38	86
FPG	600	6.813±2.359	4.280	20.130
2hPG	600	11.310±5.428	2.700	32.120
HbA_1c	600	6.723±1.496	3.400	15.100

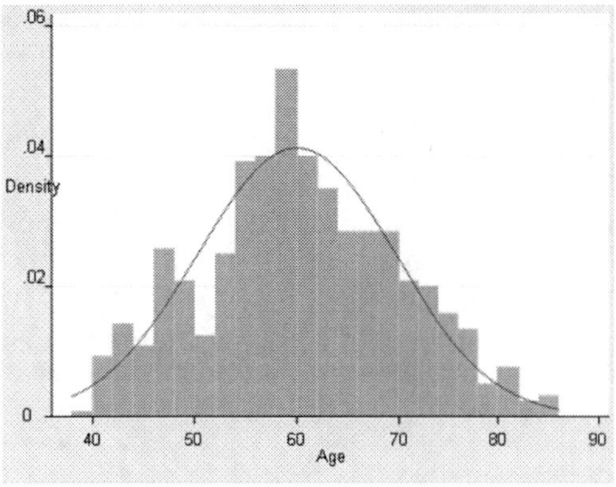

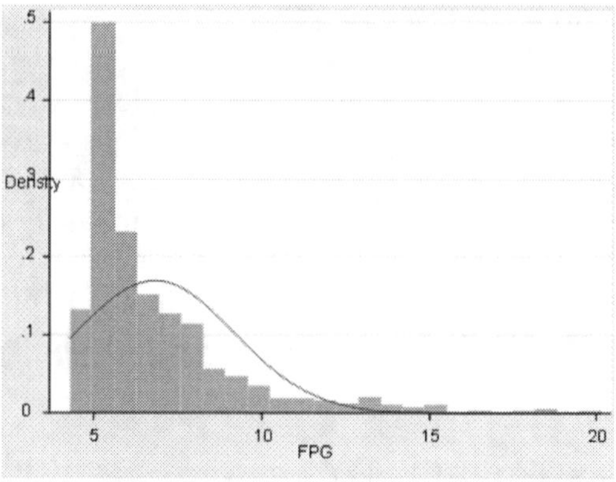

Figure 1. The age distribution of subjects.

Figure 2. Fasting plasma glucose distribution of subjects.

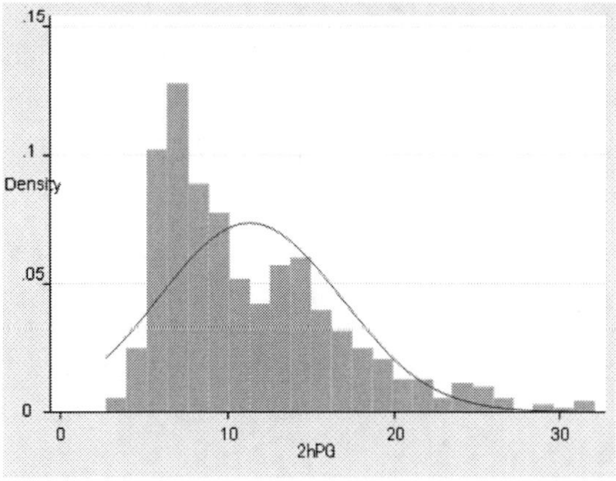

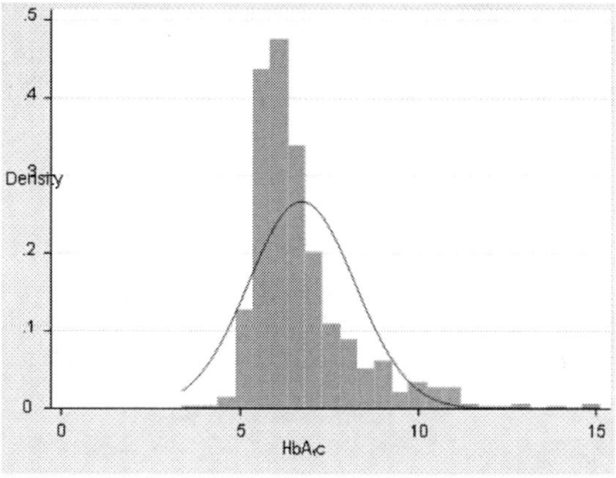

Figure 3. 2h Postprandial glucose distribution of subjects.

Figure 4. HbA1c distribution of subjects.

3.2 To evaluate the diagnostic value of FPG, 2hPG, HbA1c, HbA1c and combined with FPG in T2DM

3.2.1 The sensitivity and specificity of FPG in the diagnosis of T2DM

Application of FPG alone in the diagnosis of T2DM, when FPG≥6.19mmol/L, the sensitivity and specificity were respectively: 83%, 93.67% (Table 2).

Table 2. Sensitivity and specificity of diagnosis for T2DM using FPG

Cutpoint	Sensitivity	Specificity	Correctly Classified	LR+	LR-
(>= 4.28)	100.000%	0.000%	50.000%	1.000	
(>= 4.39)	100.000%	0.330%	50.170%	1.003	0.000
…					
(>= 6.17)	83.000%	92.670%	87.830%	11.318	0.184
(>= 6.18)	83.000%	93.000%	88.000%	11.857	0.183
(>= 6.19)	83.000%	93.670%	88.330%	13.105	0.182
(>= 6.20)	82.670%	93.670%	88.170%	13.053	0.185
(>= 6.22)	82.330%	93.670%	88.000%	13.000	0.187
…					
(>= 20.13)	0.330%	100.000%	50.170%	0.997	
(> 20.13)	0.000%	100.000%	50.000%	1.000	

3.2.2 The sensitivity and specificity of 2hPG in the diagnosis of T2DM

Application of 2hPG alone in the diagnosis of T2DM, when 2hPG≥11.18mmol/L, the sensitivity and specificity were respectively: 84.33%, 100.00% (Table 3).

Table 3. Sensitivity and specificity of diagnosis for T2DM using 2hPG.

Cutpoint	Sensitivity	Specificity	Correctly Classified	LR+	LR-
(>= 2.70)	100.000%	0.000%	50.000%	1.000	
(>= 2.92)	100.000%	0.330%	50.170%	1.003	0.000
...					
(>= 10.96)	84.330%	99.330%	91.830%	126.500	0.158
(>= 11.00)	84.330%	99.670%	92.000%	252.998	0.157
(>= 11.18)	84.330%	100.000%	92.170%	0.157	
(>= 11.22)	84.000%	100.000%	92.000%	0.160	
(>= 11.25)	83.670%	100.000%	91.830%	0.163	
...					
(>= 32.12)	0.330%	100.000%	50.170%	0.997	
(> 32.12)	0.000%	100.000%	50.000%	1.000	

3.2.3 The sensitivity and specificity of HbA1c in the diagnosis of T2DM

Application of HbA1c alone in the diagnosis of T2DM, when HbA1c≥6.40%, the sensitivity and specificity were respectively: 87.00%, 90.33% (Table 4).

Table 4. Sensitivity and specificity of diagnosis for T2DM using HbA1c.

Cutpoint	Sensitivity	Specificity	Correctly Classified	LR+	LR-
(>= 3.40)	100.000%	0.000%	50.000%	1.000	
(>= 4.10)	100.000%	0.330%	50.170%	1.003	0.000
...					
(>= 6.20)	90.330%	78.670%	84.500%	4.234	0.123
(>= 6.30)	88.330%	85.670%	87.000%	6.163	0.136
(>= 6.40)	87.000%	90.330%	88.670%	9.000	0.144
(>= 6.50)	83.000%	93.000%	88.000%	11.857	0.183
(>= 6.60)	76.330%	94.670%	85.500%	14.313	0.250
...					
(>= 15.1)	0.33%	100.00%	50.17%	0.9967	
(> 15.1)	0.00%	100.00%	50.00%	1.0000	

3.2.4 The sensitivity and specificity of HbA1c combined with FPG in the diagnosis of T2DM

When HbA1c≥ 6.40% , HbA1c screening test is positive, or FPG ≥ 6.19 mmol/L for screening test is positive, combined use of HbA1c and FPG, namely HbA1c≥ 6.40% or FPG≥6.19 mmol/L in the two indicators as long as there is a positive, and then combined screening test is positive. from the table(2-4), combined use of HbA1c and FPG diagnosis of T2DM, when HbA1c≥6.40% or FPG≥6.19 mmol/L, which can calculate the sensitivity and specificity were respectively: 97.79%, 84.61%.

3.3 FPG, 2hPG, HbA1c, combination HbA1c and FPG ROC curve

From application of FPG, 2hPG, HbA1c diagnostic screening test, you can see that separate application of FPG, 2hPG, HbA1c, the area under the curve AUC were respectively : 0.928, 0.959, 0.935, 95% ci respectively: (0.905, 0.950), (0.943, 0.975), (0.915, 0.955); HbA1c combined with FPG area under the curve AUC was 0.960, 95% CI (0.944, 0.976); Single application of FPG, 2hPG, HbA1c, and HbA1c combined with FPG diagnostic screening test area under the curve orderly was HbA1c+FPG > 2hPG > HbA1c > FPG (0.960 > 0.959 > 0.935 > 0.928), They have statistical significance (Table 5, Figure 5).

Table 5. ROC area in FPG, 2hPG, HbA1c and combination HbA1c and FPG in diagnosis of T2DM.

variable name	Obs	ROC Area($\bar{x} \pm s$)	-Asymptotic Normal— [95% Conf. Interval]	
FPG	600	0.928±0.012	0.905	0.950
2hPG	600	0.959±0.008	0.943	0.975
HbA1c	600	0.935±0.010	0.915	0.955
HbA1c+FPG	600	0.960±0.008	0.944	0.976

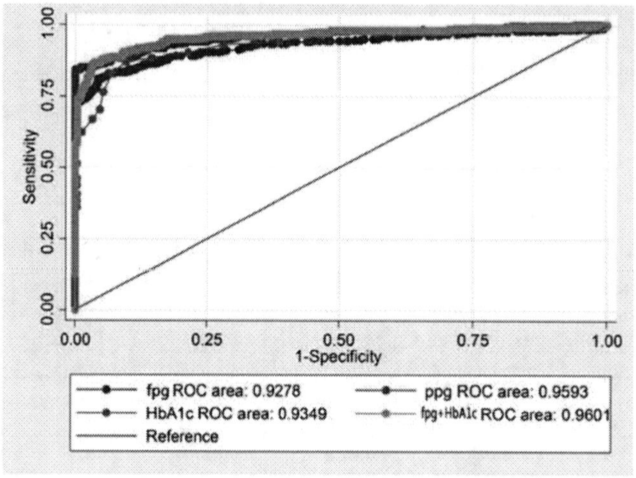

Figure 5. ROC curve of FPG, 2hPG, HbA1c and combination HbA1c and FPG in diagnosis of T2DM.

4 Discussion

4.1 HbA1c the clinical diagnostic value and the best point of contact

Though T2DM in China has increased dramatically in recent years, yet a lot of T2DM patients were not diagnosed early. HbA1c as T2DM in Europe and the United States has been one of the standards of diagnosis. Selvin [3] research shows that the coefficient of variation of 2hPG, FPG and HbA1C shows a decreasing trend (16.7% > 5.7% > 1.7%), The advantages of HbA1c coefficient of variation are small, so HbA1c is more reproducible than FPG as a tool for the diagnosis of T2DM. HbA1c can be done at any time, and do not require patients to fast or do any preparation. Furthermore FPG, 2hPG lifestyle changes would be a short time, and HbA1c status is not affected by diet, movement and whether the effects of taking certain medications, so it can be a reliable basis for monitoring of blood glucose control is good or not. So ADA in 2010 and in diabetes guidelines issued the HbA1c≥6.5% as one of the diagnostic criteria for T2DM.

Although HbA1c in the diagnosis of T2DM has more advantages than the FPG, 2hPG, there are ethnic and regional differences, which followed by a lot of T2DM researchers according to different ethnic regions, HbA1c application study on a large number of experiments in early diagnosis of T2DM, HbA1c and ADA in the diagnosis of the T2DM patients the best cut-off points were compared.

In the multi-ethnic population of Canada, the best cut-off value of HbA1c was 5.9%, the sensitivity and specificity were 75% and 79.1%[4]. In South Korea in 2012 a study also showed that people in the area of diagnosed T2DM HbA1c and the best cut-off value was 5.95%. The sensitivity of 77.0% and a specificity of 89.4% [5]. In some areas of our country, HbA1c has been used in the early diagnosis of T2DM in community population. BaoY et al[6] Study on the community population in Shanghai showed that when the HbA1c was 6.3%, the sensitivity and specificity were 62.8% and 96.1% respectively. Zhou XH[7] on the Qingdao community population study found that when HbA1c was 6%, the sensitivity and specificity of screening T2DM reached the best balance point. Ye Xiangjun et al [8] research shows that using HbAlc in patients with T2DM diagnosed the best cut-off value was 6.35%. The sensitivity and specificity were 80% and 89%.

In this study, 300 male and female subjects were studied , the number of subjects who are 40 to 64 years accounted for 68.45%, the number of subjects who are 65 years old and above accounted for 31.55%. the cutoff point of FPG for diagnosing T2DM was 6.19 mmol/L, with sensitivity and specific degrees respectively: 83.00% and 93.67%; the cutoff point of 2hPG for diagnosing T2DM was 11.18 mmol/L, with sensitivity and specific degrees respectively: 84.33% and 100.00%; the cutoff point of HbA1c for diagnosing T2DM was 6.40%, with sensitivity and specific degrees respectively: 87.00% and 90.33%; The corresponding rate of missed diagnosis was 17%, 15.67% and 13%, which further showed that the rate of missed diagnosis of HbA1c was lower than that of FPG and 2hPG in the diagnosis of T2DM. The Relationship of ROC area for $AUC_{2hPG} > AUC_{HbA1c} > AUC_{FPG}$ (0.935> 0.959> 0.959), Tips for early diagnosis of T2DM, HbA1c than application FPG has higher diagnosis rate, Although application 2hPG to diagnosis sensitivity and specificity of T2DM more high, 2hPG has operation problem, poor repeatability and long time, detection result is easy to change, considering application HbA1c as T2DM index of early diagnosis is more effective.

In this study, when HbA1c is 6.4%, its sensitivity and specificity were the best, which was close to the research results of BaoY[6] and so on, which may be due to we are in the old revolutionary base areas, compared to other developed cities, economic development is relatively backward here, the majority of samples are older workers and farmers, and they like salty, spicy as well as pickled products and their health consciousness is not strong. Some may early suffer from T2DM, because they lack regular physical examination, they did not find T2DM in time to prevent intervention as soon as possible.

4.2 Value of combined application of HbA1c and FPG in the diagnosis of T2DM

2hPG is currently the "gold" standard in the diagnosis of T2DM, but it consumes long time and more energy, and it is tedious in operation and poor in repeatability, so it does not apply to large-scale epidemiological survey. Separately using HbA1c to early diagnosis of T2DM may also cause misdiagnosis, but if combine HbA1C with FPG for early diagnosis of T2DM will greatly improve the sensitivity of diagnosis and reduce the rate of misdiagnosis. At present, many abroad studies Has also suggested that a combination of HbA1c and FPG for early diagnosis of T2DM. Such as Yoshihik [9] according to a study in the samples of 109 patients diagnosed with T2DM combined HbA1c and FPG test, finding that 18 patients with T2DM FPG is normal, and HbA1c is higher than normal, so a separate use FPG misdiagnosis in early diagnosis of T2DM has certain, Combination use of HbA1c and FPG in early diagnosis of T2DM, which can reduce the rate of missed diagnosis of T2DM. besides, the results of Kim KS et al[10] show that the sensitivity can reach 95.5% when combine HbA1c ≥6.1% and FPG≥6.1mmol/L for early diagnosis of T2DM, which is much higher than any single application of HbA1c ≥6.1%or separate application of FPG≥6.1mmol/L, thus the rate of misdiagnosis will be reduced misdiagnosis. China survey found that heart disease of coronary heart disease an important with disease or diabetes, mostly T2DM, but for many patients with coronary heart disease FPG is normal or near normal, but the rise in blood glucose after load, that is to say after the OGTT experiment 2hPG blood glucose is very high, and the patients with coronary heart disease only by FPG to diagnose T2DM or pre diabetes would result 75 percent of the rate of missed diagnosis, for the people of the disease have to be tested by combination with HbA1c and FPG [11].

Hu Y [12] to high-risk groups such as screening for T2DM research shows that 2298 from ruijin hospital check-up crowd as the research object, when the HbA1c was 6.1% and a separate FPG was 6.1 mmol/L, the sensitivity and speciality was also 81%; But if HbA1c and FPG combined application in the early diagnosis of T2DM, which greatly improved the sensitivity and specificity. When HbA1c was 6.1% and FPG is 6.1 mmol/L, the sensitivity and specificity of early diagnosis of T2DM were increased to 96.5%, 96.3% respectively, Their combination can also be used to screen for impaired glucose, with a sensitivity of 82.4%. There are a lot of research is also supported by the combination of HbA1c and FPG to diagnose T2DM [13, 14].

Our study separately analyzes the sensitivity of FPG and HbA1c when diagnosing T2DM , respectively83.00%, 87.00%, It is suggested that both FPG and HbA1c may cause missed diagnosis and the missed diagnosis rate is 17% and 13% respectively; Combined use of HbA1c and FPG diagnosis of T2DM, simultaneous detection of HbA1c and FPG, when HbA1c ≥6.40% or FPG≥6.19 mmol/L diagnosed with T2DM, the sensitivity was 97.79%, the missed diagnosis of just 2.21%. So united HbA1c and FPG can improve the sensitivity of T2DM in early diagnosis, reduce missed diagnosis. In this study area under the ROC curve is greater than 0.9, and the area of the largest is the combined use of HbA1c and FPG, so the combined use of HbA1c and FPG diagnosed T2DM diagnosis rate is highest.

The combined application of HbA1c and FPG has better sensitivity and specificity in the diagnosis of T2DM, On early detection, early diagnosis and early treatment of T2DM combined use of HbA1c and FPG has good effect, and can once pump blood and get two a different of results, in other words, it can early diagnosis whether with T2DM, which reduces patients' and medical personnels' trouble of two times pumping blood, which also avoids separately application FPG in the diagnosis of T2DM which needed more trouble of repeatly pumping blood. Combined application can reflect the situation of glucose metabolism from two different levels of the instant instant blood glucose level and the relative long-term blood glucose level, and it is a convenient and effective method for early diagnosis of T2DM in the elder.

References

1. Yang WY, Lu JM, Weng JP, Jia WP, Ji LN, Xiao JZ, N Englo J Med, *Prevalence of diabetes among men and women in China*, **362**,1090-1101(2010).
2. Saudek CD, Herman WH, Sacks DB, Bergenstal RM, Edelman D, Clin Endoerinol Metab, *A new look at screening and diagnosing diabetes mellitus*, **93**,2447-2453(2008).
3. Selvin E, Crainiceanu CM, Brancati FL, Coresh J, Arch Intern Med , *Short-term variability in measures of glycemia and implications for the classification of diabetes*, **167**,1545-1551(2007).
4. Herman WH, Ma Y, Uwaifo G, Haffner S, Kahn SE, Horton ES, Diabetes Care , *Differences in A1C by race and ethnicity among patients with impaired glucose tolerance in the Diabetes Prevention Program*, **30**,2753-2757(2007).
5. Kim JH, Kim GW, Lee MY, Shin YG, Koh SB, Diabetes Metab J , *Role of HbA$_{1}$c in the screening of Diabetes Mellitus in a Korean Rural Community*, **36**,37-42(2012).
6. BaoY, Ma X, Li H, Zhou M, Hu C, Wu HY, BMJ, *Glycated hemoglobin A$_{1}$c for diagnosing diabetes in Chinese population: cross sectional epidemiological survey*, **340**,c2249(2010).
7. Zhou X, Pang Z, Gao W, Diabetes Care , *Performance of an A1C and fasting capillary blood glucose test for screening newly diagnosed diabetes and pre-diabetes defined by an oral glucose tolerance test in Qingdao, China*, **33**,545-550(2010).
8. Ye XQ, Lan XC, Xu GL, Prevention and Treatment of Cardio-Cerebral-vascular Disease, *The value of HbA1c in the diagnosis of T2DM*, **12**,306-308(2012).
9. Yoshihik U, Steffes MW, Clin Chem, *A primary predictor of the development or reversal of complications of diabetes mellitus*, **47**,1157(2001).
10. Kim KS, Kim SK, Lee YK, Diabetic Medicine , *Diagnostic value of glycated haemoglobin HbA$_{1}$c for the early detection of diabetes in high--risk subjects*, **25**,997-1000(2008).
11. Diabetes branch of Chinese Medical Association, Chinese Journal of Diabetes Mellitus ,*Guidelines*

for prevention and treatment of type 2 diabetes in China (2013 Edition), **6**,447-498(2014).

12. Hu Y, Liu W, Chen Y, Zhang M, Wang LH, Zhou H, Wu PH, Acta Diabetol , *Combined use of fasting plasma glucose and glycated hemoglobin Alc in the screening of diabetes and impaired glucose tolerance*, **47**,231-236(2010).
13. Herman WH, Fajans SS, Pol Arch Med Wewn , *Hemoglobin A1c for the diagnosis of diabetes: practical considerations*, **120**,37-40(2010).
14. Inoue K, Matsumoto M, Kobayashi Y, Diabetes Res Clin Pract , *The combination of fasting plasma glucose and glycosylated hemoglobin predicts type 2 diabetes in Japanese workers*, **77**, 451-458(2007).

Reduced gray matter volume of left superior temporal gyrus in schizophrenia with auditory verbal hallucinations: a voxel-based morphometry study

Huawang Wu[1, 2], Fengchun Wu[2], Xiaoyin Ke[2], Ripeng Li[2], Xiaobing Lu[2] and Yuping Ning[2,a]

[1]*Key Laboratory for NeuroInformation of Ministry of Education, School of Life Science and Technology, University of Electronic Science and Technology of China, Chengdu, 625014, China*
[2]*The Affiliated Brain Hospital of Guangzhou Medical University (Guangzhou Huiai Hospital), Guangzhou, 510370, China*

Abstract: Auditory verbal hallucinations (AVHs) which are the core psychopathology symptoms in schizophrenia patients cause amounts of distress, functional disability and behavioral dys-control. However, the neural mechanism with the onset of AVHs in schizophrenia remains unclear. In the present study, we recruited 30 schizophrenia patients with AVHs, 20 schizophrenia patients with non-AVHs (nAVHs), and 50 healthy controls. Structure magnetic resonance imaging was obtained, and voxel-based morphometry (VBM) was used to analyse the gray matter volume (GMV) in the three groups. We found that the GMV of anterior cingulate cortex (ACC), left superior temporal gyrus (STG), and left middle temporal gyrus (MTG) significantly decreased in schizophrenia patients compared to healthy controls. In addition, the left STG with significantly decreased volume was found in nAVHs compared to AVHs, but no significant changes of GMV in ACC and left MTG were found between AVHs and nAVHs. Our current study identified the abnormal GMV of the left STG in schizophrenia patients with AVHs. Our findings further supported the view that, whilst mild reduced GMV of the left STG might still enable the emergence of AVHs, more severe alterations may prevent the occurrence of AVHs experience.

1 Introduction

Auditory verbal hallucinations (AVHs) were typically defined as hearing a voice or other sound in the absence of an external stimulus. The life-time prevalence was about 7.25% in general population, and the age of the highest prevalence rate was less than 30 years and 1.1% of individual sought professional help for AVHs[1]. AVHs which was the core psychopathology symptom affected about 60-80% schizophrenia patients and caused amounts of distress, functional disability and behavioral dys-control[2,3].

Structural magnetic resonance imaging has begun to provide insights into the neuroanatomy of AVHs in schizophrenia. A meta-analysis revealed that the severity of AVHs was associated with gray

[a] Correspondence Author: ningjeny@126.com
This work was supported by the National Basic Research Program of China (973 program, 2011CB707801).

matter volume (GMV) reductions in left superior temporal gyrus (STG) but without significance changed of GMV between AVHs and nAVHs group[4]. Another study showed that the volume of the putamen was lower in AVHs compared to nAVHs and no other brain regions with significantly changed GMV were found between the two groups [5]. A recent study of AVHs found that decreased bilateral thalamic GMV was associated with the severity of AVHs [6]. These findings indicated that the neuroanatomical mechanism of AVHs remains controversial.

In the current study, we recruited 30 schizophrenia patients with AVHs and 20 schizophrenia patients without history of AVHs and 50 aged, gender matched healthy controls. The two patients groups were well matched in clinical variables and without combined any other hallucinations except for AVHs. We employed the voxel-based morphology (VBM) to evaluate the relationships between the GMV and AVHs symptom.

2 Methods

2.1 Participants

In this study, 50 schizophrenia subjects and 50 healthy controls were recruited at the Affiliated Brain Hospital of Guangzhou Medical University and all were right-handed and Han Chinese ancestry with the age between 18 and 45 ages. Schizophrenia patients were assessed by two experienced psychiatrist and diagnoses with schizophrenia using the structured clinical interviewer of DSM-IV (SCID) Patient Edition. The Positive and Negative Syndrome Scale (PANSS) was used to evaluate clinical symptoms of psychosis. A score of 4 points or more must be at least three items among the positive symptoms of PANSS and the total scores of PANSS were at least of 60 points. According to the experience of AVHs, schizophrenia patients were subdivided into two groups: AVHs group including patients who experienced AVHs in the past 24 hours with or without history of AVH, and nAVHs group including patients who had never experienced AVH during the course of illness before the MRI scanning. And all of schizophrenia patients were never experienced any other hallucinations including visual hallucination, olfaction hallucination, tactile hallucination, gustatory hallucination, visceral hallucination and motor hallucination. The Auditory Hallucination Rating Scale was used to assess AVHs on seven characteristics: frequency, reality, loudness, number of voices, length, attention dedicated to hallucinations, and hallucination-induced arousal. All healthy controls were screened using the SCID Non-Patient Edition to confirm the lifetime absence of Axis I illness and selected control subjects had no known history of psychiatric illness in the any two lines of first–to third degree biological relatives. All schizophrenia patients and healthy controls were without a lifetime history of seizures, head trauma, serious medical or surgical illness, substance abuse or dependence, or contraindications for MRI. Potential participants were excluded if discovered gross abnormalities signals on the cerebral. The written informed consent was obtained from all subjects and approved by the local ethics committees of the Affiliated Brain Hospital of Guangzhou Medical University.

2.2 Data acquisition

MRI data was acquired on a 3.0-Tesla MR imaging system (Achieva X-series, Philips Medical Systems, Best, the Netherlands) with an eight-channel SENSE head coil, in the Department of Radiology, the Affiliated Brain Hospital of Guangzhou Medical University, Guangzhou, China. Whole brain structural images were acquired with a three dimensional T1-weighted turbo field-echo (TFE) sequence. The following parameters were used during the scanning: repetition time (TR)= 8.2ms; echo time (TE)=3.7ms; flip angle (FA)= 7^0; FOV= 256×256 mm^2; 256×256 acquisition matrix; 1 mm slice thickness without inter-slice gap; voxels size= 1×1×1 mm^3 , and 188 continuous sagittal slices.

2.3 Data Analysis

All the 3D T1 weighted images were performed by using VBM8 toolbox (http://dbm.neuro.uni-jena.de/vbm8) for SPM8 package (http://www.fil.ion.ucl.ac.uk/spm). The default procedures in the VBM8 toolbox were used for preprocessing. All T1 Images were corrected for bias-field inhomogeneity; segmented into GM, white matter and cerebral spinal fluid; and normalized to standard space using high dimensional DARTEL normalization. The normalized GM images were then modulated account for volume changes resulting from the normalization process. The normalized and modulated images were smoothed with 8mm full-width at half-maximum (FWHM) smoothing kernel for final statistical analysis. After this preprocessing, we obtained smoothed, modulated, normalized data that were used for the statistical analysis. To achieve maximal sensitivity, to optimize voxel residual smoothness estimation and to exclude false positives in non-GM tissue, voxel-wise comparisons were masked using a comparison specific explicit optimal threshold gray matter mask created using the Masking toolbox (www.cs.ucl.ac.uk/staff/g.ridgway/masking).

2.4 Statistical Analysis

Statistical group difference maps were constructed using a general linear model (GLM) by performing ANOVA to test for the main effect of group with sex and age as covariates. The threshold for voxel-wise comparisons of GMV was set at $p < 0.05$, family-wise error (FWE)–corrected at the cluster level (corrected for non-stationarity of smoothness for VBM data) with an initial voxel threshold of $p < 0.001$. We then extracted the mean GMV of clusters showing significant changes in main effect of group and then exported these to SPSS for post-hoc analysis with Bonferroni corrected ($p < 0.05$). A correlation analysis between the mean GMV of clusters with clinical variables in schizophrenia patients with AVHs also performed.

3 Results

3.1 Demographic and Clinical Characteristics

Demographic and clinical data for the subjects are shown in Table 1. The three groups were well-matched in gender, age and years of education. There were no significant differences in durations of illness, PANSS positive score, PANSS negative score, PANSS general score and PANSS total score between the two schizophrenia patients groups.

Table 1. Demographic information for schizophrenia patients and healthy controls.

Characteristics	AVHs(n=30)	nAVHs(n=20)	HC(n=50)	*P*-value
Age(years)	27.30±5.27	30.5 ±6.77	28.56±6.35	0.191
Sex(male/female)	18/12	14/6	34/16	0.700
Education(years)	11.70±3.29	10.65±3.08	11.94±2.71	0.260
Duration of illness(months)	45.97±56.26	54.25±45.80	-	0.586
PANSS				
Positive subscale	25.10±5.23	23.95±3.41	-	0.390
Negative subscale	22.67±8.20	22.25±6.10	-	0.847
General psychopathology	39.80±9.67	42.55±7.90	-	0.296
Total score	87.57±19.27	88.75±10.52	-	0.803
AHRS total scores	28.23±5.22	-	-	-

 DOI: 10.1051/bioconf/20170801031

3.2 Group differences in GMV

A main effect of group was observed in the anterior cingulate cortex (ACC), the left superior temporal gyrus (STG) and the left middle temporal gyrus (MTG). Compared to healthy controls, the schizophrenia patients with or without AVHs showed lower GMV of the ACC, the left STG and the left MTG, and significant lower GMV in left STG in nAVHs group than AVHs groups. However, no significant differences of GMV in ACC and left MTG was found between AVHs and nAVHs groups.

3.3 Correlation between alterations in GMV and AVH symptom severity

We don't find any correction between alterations in GMV and AHRS total scores in AVHs group.

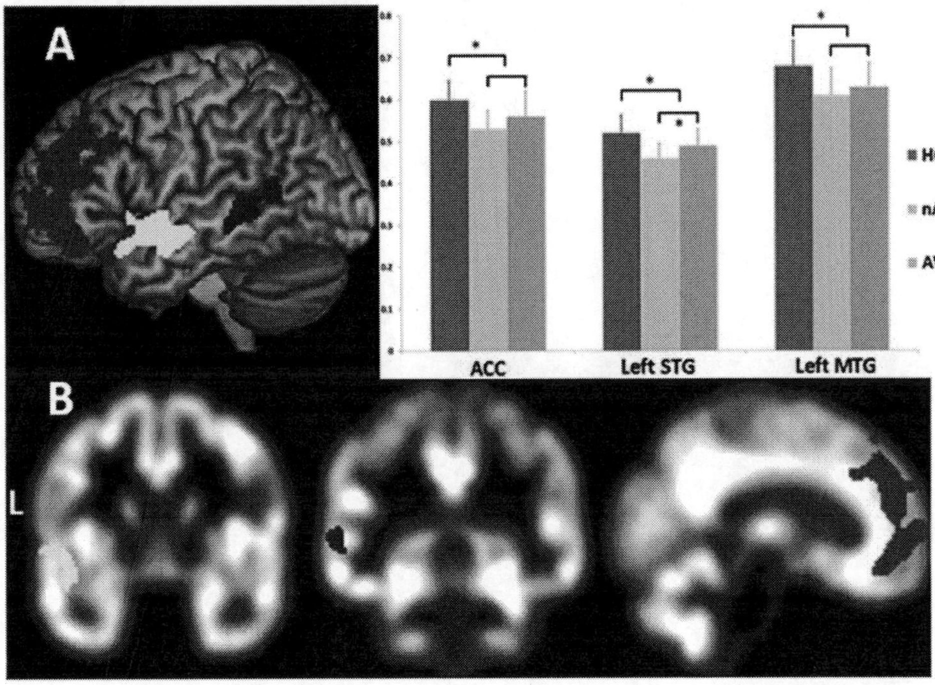

Figure 1. Changed GMV in schizophrenia patients with AVHs. Main effects of mapping showed significant GMV differences in ACC, left STG and left MTG (The threshold was set at $p < 0.05$, FWE corrected at the cluster level) (see A and B). The bar chart displayed different GMV of these clusters between three groups. * Significant difference between groups using post-hoc analysis with Bonferroni corrected, $P<0.05$.

4 Conclusions

In this study, we used VBM to evaluate the changes of GMV of schizophrenia with or without AVHs. We find that the GMV of the ACC, the left STG and the left MTG significantly decreased in schizophrenia patients compared to healthy controls. The AVHs group showed intermediate GMV in the STG compared to the nAVHs group and healthy controls, but no significantly changed of GMV in the ACC and left MTG between the AVHs and nAVHs groups. Our findings indicate that the left STG plays an important role in the neural underpinnings of AVHs in schizophrenia, and further support to the hypothesis that a certain degree of preserved the GMV of the left STG may contribute to the onset of AVHs in schizophrenia.

References

1. B Krakvik, F Laroi, AM Kalhovde, K Hugdahl, K Kompus, O Salvesen, TC Stiles, E Vedul-Kjelsas Scand J Psychol **16**:12236 (2015).
2. TH Nayani, AS David Psychol Med **26**:177-189 (1996).
3. SS Shergill, RM Murray, PK McGuire Schizophr Res **32**:137-150 (1998).
4. G Modinos, SG Costafreda, MJ van Tol, PK McGuire, A Aleman, P Allen Cortex **49**:1046-1055 (2013).
5. MJ van Tol, L van der Meer, R Bruggeman, G Modinos, H Knegtering, A Aleman Neuroimage Clin **4**:249-257 (2013).
6. P Huang, Y Xi, ZL Lu, Y Chen, X Li, W Li, X Zhu, LB Cui, Q Tan, W Liu, C Li, D Miao, H Yin Scientific report(2015).

The Application of 4C Training Model in the Operating Room Nurse Training

Feng Zhao, Cui Wang and Ruofeng Yin[a]

China-Japan Union Hospital of Ji Lin University, Changchun, Jilin, China

Abstract. The operating room is an important place for the treatment of patients undergoing surgery, the quality level of clinical nursing has a direct influence in the therapeutic effect of patients. In the present study, four components instructional design mode (4C training mode) was applied in the training of professional nurses in the operating room, it involved the following aspects: regular skill knowledge analysis, goals setting, as well as the construction and implementation of the project. Targeted training methods and training content were formulated in accordance with the learning ability and knowledge reserve of different professional nurses, and different skills required to master, to fully mobilize the enthusiasm and initiative of nurses, thereby to improve the operation and comprehensive level of nurses in the operation room to achieve good results.

1 Introduction

Operating room is the most important place for operative treatment of patients and the quality of nursing work within the operating room is also one of the most critical standards measuring the overall medical level and service quality of a hospital. Work in the operating room is not only complicated, but involved with multiple disciplines. Hence, it is required to be provided with a well-qualified nursing team. How to improve the overall quality of nurses in the operating room, rationally utilize limited human resources, sufficiently arouse the enthusiasm of nurses and improve work efficiency and nursing quality of nurses in the operating room is the rigorous problem confronted with by operating room nursing managers.

With the rapid development of contemporary surgery, surgical operations have grown into diversification from the original simplification. In addition to continuous expansion of surgical fields and scopes, new surgical methods, new instruments and new apparatus also keep emerging, which asks operating room nurses (ORNs) to be transformed into those intensive, highly specialized and refining constantly from conventional simple single and crude coordination. In other words, requirements for level of expertise and overall quality of ORNs become increasingly higher. Only well-qualified ORNs are able to coordinate with specialized surgeries that are developing towards superior difficulty and refinement. Considering this, effective training for nurse specialists is particularly important[1]. In this study, the 4C training pattern is applied into operating room nurse specialist training that has achieved preferable effects according to the relevant observations. As for this pattern, it is not only effective in improving operating room nurse specialist training effects, but also beneficial to cultivate modern ORNs highly specialized and versatile. As a result, operating room nursing work complies with the development of surgery and work of ORNs can also be comprehensively and systematically improved. Concerning the associated details, they are described below.

[a] Corresponding author : Ruofeng Yin, Email:50703463@qq.com

2 Information and methods

2.1 General information

160 nurses working in the operating room of our hospital from March 2014 to March 2015 are selected and randomly divided into a control group and a study group by means of random number table. For each group, there are 80 nurses. As for the study group, the 4C training pattern is adopted to carry out specialized trainings for nurses here; in comparison, the regular operating room specialist training is employed by the control group. Differences in general data such as age, sex, education background and working hours, etc. of nurses from such two groups have no statistical significances during comparisons ($P>0.05$), in addition to being comparable.

2.2 Methods

2.2.1 Conventional specialist training

Dependent on job demands of departments, operating rooms are categorized into 7 specialist groups including general surgery, cardiothoracic surgery, orthopedics, neurosurgery, gynaecology and obstetrics, and urinary surgery, etc.; in which, personnel coordinating with specialized operations take their turns regularly and the corresponding nurse specialists of diverse specialist groups are trained targeted at basic knowledge, fundamental theories and skills that should be mastered by them to cooperate with various specialized operations. In addition, the relevant training contents and modes are uniformly arranged by the head nurse in charge of teaching. After the training, these nurses are also assessed[2].

2.2.2 4C Training pattern

In line with conventional specialist training methods, 7 specialized groups are allocated with nurse specialists according to practical surgical operation quantities and workload, and each group comprises 1 leader and 9-11 members. To be specific, members consist of fixed team members (2-3) and team members taking turns (6-8). As for the former, they assist the team leader to formulate and implement working plans. Members of every group are allocated proportionally to form a talent pyramid so as to guarantee stable operation and future development of these specialist groups. Regarding skilled team leaders with strong work abilities and senses of responsibilities, they should be nurse-in-charge or more qualified personnel with a certain management capability. The development direction of each specialist group should be organized by the head nurse in charge of teaching and also discussed in detail among fixed members of it. At the same time, the corresponding training plans, goals and elaborate head nurses should be established together with the relevant implementation schemes. Then, according to these schemes, management objectives can be decomposed and become practical in terms of each related post. Nurses of specialist groups are trained so that they have the capability to fulfill the established training goals based on the existing plans and schemes.

2.2.2.1 Prospect analysis

Such a stage can be classified into three aspects. First is the demand analysis on specialized operation cooperation. Nurse specialists are cultivated with an aim to cooperate with demands of various specialized operations in a better manner, and make ORNs able to rapidly and effectively master working skills corresponding to these specialized operations and qualified for specialized operation cooperation in rapid development. Considering this, demands of diverse specialized operations should be analyzed elaborately to precisely and comprehensively decompose key points of specialized operation cooperation. Second is feature analysis on training objects. As the subject of learning, training objects are crowd that should be changed. In order to achieve good training effects, education backgrounds and working experiences of them should be mastered in the first place. Then, individualized specialist trainings are formulated corresponding to their different learning and working experiences, etc. In this way, training demands can conform to characteristics and styles of diverse trainees. Thirdly, it is about the analysis on department development demands. Due to the rapid growth of surgery, new techniques and new devices keep emerging for the operating room and the present operating room nursing is also at a stage of rapid development. As for training of ORNs that comply to cooperation needs of rapid surgical operation development and satisfy rational construction demands of department personnel echelon, perfect feasible training plans should be formulated with taking physical truths of these departments, time, costs and staff composition, etc. into consideration.

2.2.2.2 Objective setting

Objective setting is a greatly important link, because objectives are theoretical basis for training plan preparation and can provide directions for training. According to the 4C training pattern, objectives are divided into two categories. One is the objective that can be repeatedly applied, that is, the content that should be mastered for a long time during practical work. Combining particulars of work in the operating room, it includes the common conventional operation cooperation, as well as use, maintenance and simple troubleshooting of apparatus and equipment in common use. The other is the objective that cannot be applied over and over again. It refers to skills that should be mastered during temporary jobs. In combination with specifics of work in the operating room once more, it consists of operating room management. In detail, it is involved with management methods that should be changed flexibly according to different circumstances at different periods. Besides, it also involves treatment of emergencies and problems arising in the operating room, etc.

2.2.2.3 Construction of scheme

After the fulfillment of objective setting, we duly set about establishing training schemes. In these schemes, prominence should be given to the key points and emphasis is also laid on cultivating specialized abilities of ORN specialists as well as their comprehensive qualities. Thus, the overall nursing level and synthetic abilities in an operating room can be improved. From the perspective of scheme contents, the status quo of training objects should be grasped through self-assessment by trainees and preliminary assessment by specialist groups, so as to prepare scheme contents appropriate to these objects in accordance with their present situations and objective design. During scheme establishment, learning resources and environment of the department are reasonably utilized to carry out multimedia teaching, professional lectures, operation cooperation practice demonstration and interrelated resource learning in a variety of forms and contents. In detail, with regard to the multimedia teaching, knowledge points and cutting-edge technologies of the department are taught by means of multi-media network teaching. Regarding the professional lectures, diseases and the corresponding therapeutic methods, surgical procedures and nursing cooperation, etc. of the department are systematically explained by surgeons and senior nurse specialists. About the operation cooperation practice demonstration, it is not elaborated here. Finally, for interrelated resource learning, self-knowledge is supplemented through books, database resources and network teaching videos.

2.2.2.4 Implementation of scheme

This section covers three aspects described below. Firstly, a perfect hierarchical management system under the control of head nurses of departments should be set up. Various specialized departments are planned as a whole by the head nurse in charge of teaching; while, the responsibility to implement specialist trainings in concrete should be borne by the specialist group leaders. Such a vertical hierarchical management system is more beneficial to define objectives clearly and improve the abilities of specialist group leaders to play a dominant role, exert all capabilities fully, and actively participate into formulating and supervising trainings for the corresponding nurses in all specialist groups. Secondly, each specialist group leader takes the responsibility to settle and summarize operation techniques and theories that should be mastered by the corresponding department, so as to form an integral knowledge and skill learning system. Furthermore, all kinds of skills and knowledge are made into courseware for seminars and lectures as theoretical trainings on a regular basis. Professional technical operations are performed in a sequence of specialist operation-specialist emergency operation-new technique. Thirdly, technical files should also be established to improve the assessment system of a department. In addition to completing tests arranged and required by the nursing department, contents of assessment are additionally added by the specialist department.

2.2.3 Evaluation criterion

The evaluation criterion is constituted by assessment of theoretical knowledge, assessment of operating room nursing manipulation and assessment of professional ability. The manipulation assessment refers to the uniform standard and manipulations are selected randomly to be rated. Both theoretical and manipulation assessments are 100 points, and it is deemed to be qualified if 60 and above points are obtained. With regard to the professional ability assessment, it is carried out by the Comprehensive Quality Evaluation Table of ORNs prepared by the department, covering clinical competency, learning time management ability, solidarity cooperation ability, communication skills, specialty knowledge level and operation coordination ability. Each of them is endowed with 10 points. That graded to be 8 points and above is qualified, while below 8 points is unqualified.

3 Statistical analysis

Statistical software SPSS 20.0 is employed to perform statistical analysis on all data. Enumeration data are described by rate and tested by χ^2. By contrast, measurement data are accounted by Mean ± Standard Deviation and verified by t. Besides, those with $P<0.05$ is considered to be different and have statistical significances.

4 Results

Regarding nurses from the study group qualification rate, qualification rates obtained from assessment of theoretical knowledge, assessment of operating room nursing manipulation and assessment of professional ability are respectively 97.5%, 92.5% and 95%, which is prominently higher than 82.5%, 75.0% and 77.5% of the control group. Differences comparisons of such two groups are statistically significant ($P<0.05$), as shown in Table 1.

Table 1. The Compare two groups of nurse training results.

Group	qualification rate of basic knowledge	Operation qualification rate	Business capability assessment
Control group	82.5%	75.0%	77.5%
Observation group	97.5%	92.5%	95.0%
χ^2	5.00	4.50	5.16
p	P < 0.05	P < 0.05	P < 0.05

5 Discussion

The operating room is an important place where patients are treated. With the continuous development of science and technology and constant improvement of medical level, higher requirements have been put forward by the entire society for operating room nursing. Integrating multiple related disciplines, operating room nursing has higher requirements of nurses. Therefore, trainings for them are of great importance. On one hand, by virtue of such trainings, skill level of them can be improved; on the other hand, guarantees are provided for the improvement of nursing quality in the hospital. How to exploit potentials of nurses in a better manner and make the trainings yield twice the result with half the effort is a subject that all managers should face [3-5].

The 4C training pattern includes regularly decomposing skills that should be mastered by surgery nurses, analyzing skill elements and relevant knowledge, selecting teaching strategy, and assembling training strategy. The team responsible for the project should decompose skills that should be grasped by nurses in the operating room in the first place, mainly covering clinical care, nursing supervision, nursing scientific research and teaching. Then, contents at all levels are decomposed further. To be specific, the clinical care is involved with: basic knowledge of operating room nursing, placement of operative position, anesthesia preparation and cooperation, cooperation with mini-middle-operations in diverse departments[6]; mastering particularities associated with surgical methods, instrument and dressing preparations; comprehending new surgical techniques and key points for cooperation with salvage surgery; fostering emergency rescue capabilities; and, precautions to operating room nursing errors; etc.. In summary, clinical care training is about the repeated application and strengthening of skills during trainings for nurses in diverse departments. In terms of operating room management, nursing scientific research and teaching trainings, they should be conducted combining personal interest and competency on the basis of senior nurses with certain experiences. 4C training pattern applied into nurse specialist trainings should be adopted to prepare targeted training methods and contents according to learning abilities, knowledge reserves and skills required mastering of nurses from diverse departments, so that learning enthusiasm and initiative of them can be sufficiently mobilized and the goal to improve manipulation and comprehensive levels of ORNs is fulfilled. In addition, thanks to the employment of 4C training pattern, systematicness of operating room nursing is reinforced, which is conducive to improvement of operating room nursing quality and efficiency in the hospital. This study indicates that, in the study group taking advantage of such a 4C training pattern, qualification rates relate to all assessments of nurses are obviously higher than those of nurses from the control group receiving conventional specialist training.

In conclusion, operating room nurse specialist training that adopts the 4C training pattern can significantly enhance various competencies of nurses, and improve qualities of nurse specialists. Considering such a preferable application effect, it is worth popularization and application.

References

1. Ying Qiu, Jing Zhang. Study on setting up of training courses of operation room nurse specialists[J]. *Chinese Nursing Research* 2014, **28**(8B) :2859-2862.

2. Yunmiao Ma, Anling Li, Huimin Zhao. Research status quo of core competence of specialist nurses in operating rooms at home and in abroad[J]. *Chinese Nursing Research* 2013, **29**(3): 1029-1031.
3. Yingyu Zeng, Xiaodan Wu, Guiying Xu, et al. Correlation study between core competence and fixed specialty in operating room nurses[J]. *Journal of Nursing Administration* 2014, **14**(11):778-788.
4. Lianying Li. Specialist in the operating room nurses training mode[J]. *Nursing Practice and Research* 2013, **10**(11):142-143.
5. Feihong Lan. Specialist in the operating room nurses the cultivation of the status quo[J]. *Today Nurse* 2015, 2:16-18.
6. Yan Ding, Weijing Liu. Investigation on core competency and curriculum design for operating-room nurse speclalists[J]. *Journal of Nursing Seience* 2013, **28**(22):4-5.

DOI: 10.1051/bioconf/20170801033

Application of precise training to surgical endoscopic nurses training

Feng Zhao, Yu Jin and Xueying Zhou[a]
China-Japan Union Hospital of Ji Lin University, Changchun, Jilin, China

Abstract. Minimally invasive surgery is an inevitable trend of the development of surgery, the scope of endoscopy surgery application has been covering the whole departments of surgery, which also put forward higher requirements on the nursing coordination during surgery, training high-quality endoscopy nurses is a focus of the researchers. In the paper, the precise training mode was applied to the professional training process of endoscopy nurses, the systematic training of endoscopy nurse was conducted from the following three aspects, including the precision of teaching, the precision of operative cooperation, and the precision of assessment, which had gained good effects. The research indicated that the precise training mode can contribute to upgrading various capacities of endoscopy nurses in the operating room, and improving the teaching quality and effect of specialized training, suggesting a good application effect.

1 Introduction

With the development of endoscopy, the applicability of endoscopic surgery has gradually broadened and now covers almost all disciplines of surgery, with the inevitable trend of minimally invasive surgery in the development of surgical operations. The application and update of advanced equipment and devices have also accelerated the development of the quality of endoscopic surgery. Rapid development of endoscopic surgery has also put forward higher requirements for the cooperation of operating room (OR) nursing [1]. OR nurses should not only have a strong sense of responsibility, but also should have a wealth of knowledge and specialist skills. Also, the demand of high-quality specialized endoscopy nurses in operating room continues to grow. How to efficiently train the specialized OR nurses, has always been the issue of concern for managers. In our hospital, the specialized OR management was implemented in May 2013, and the specialized endoscopic group was established at the same time. In August 2014, the precision training was applied in the training of specialized endoscopy nurses and achieved good results. And this project of precision training will be introduced and reported as follows in this article.

2 Information and methods

2.1 General information

Our hospital is now a comprehensive Grade III Level A Teaching hospital with a total of 36 operating rooms and more than 3,000 cases of endoscopic surgeries annually. These endoscopic surgeries include: laparoscopy in general surgery, urological endoscopy, thoracoscopy, gynecological

[a] Corresponding Author: Xueying Zhou, Email:50703463@qq.com.

endoscopy, micro-endoscopic discectomy (MED), arthroscopy and laparoscopic thyroidectomy, etc. Currently, there are 1 group leader and 8 group members in the specialized endoscopic group, including 3 nurses-in-charge and 6 nurse practitioners. 30 subjects were selected from the nurses in the rotating training in different specialized departments of operating rooms and the nursing students of the training base for specialized operating rooms during May 2013 - July 2014 as the control group, and received conventional specialized training for operating room nursing, instead of precision training. And 34 subjects were selected from the nurses in the rotating training in different specialized departments of operating rooms and the nursing students of the training base for specialized operating rooms during August 2014-October 2015 as the experimental group, and received precision training. The differences in age, gender, education level, working hours and other general information between the two groups of nurses were not statistically significant ($P>0.05$), while the data were comparable.

2.2 Methods

2.2.1 Conventional specialized training

The training period was 7 months, including 1.5 months of theoretical classes with centralised teaching of the knowledge of the types of endoscopic equipment and the subspecialties of endoscopic surgeries, and 5.5 months of practices and exercises in the form of conventional teaching and rotating in different departments to practice the operations of all endoscopic equipment and the cooperation in the subspecialties of endoscopic surgeries.

2.2.2 The framework of precision training

2.2.2.1 The precision in teaching and learning

(1)The precision of the training content and scheduling. The teaching and learning contents were made explicit for each training period: 1.5 months of theoretical classes totally broke away from the classroom type of teaching. During Week 1 and 2, the supervisor of the department of medical devices and the technicians of major brands of endoscopic equipment gave detailed introductions of the type, model, size and purpose, the methods of use, installation and disassembly, cleaning and disinfection, as well as the methods of maintenance of each equipment. During Week 3 and 4, the students were grouped to learn the properties, functions, intraoperative applications and simple trouble shooting solutions of all equipment, and to observe the whole processes of all the subspecialties of endoscopic surgeries. In Week 5 and 6, the clinicians and anesthesiologists taught the students the knowledge of anatomy, medicine, surgery, therapy, anesthesia and other related areas, and introduced the latest treatments and technologies, as well as the training for scientific innovations and writing. During the remaining 5.5 months, the student took rotations in various subspecialties of endoscopic surgeries. Group members of the specialized endoscopic group taught the perioperative cooperation, and the students should become familiar and skilled in the methods of operative cooperation and procedures of all subspecialties of endoscopic surgeries. The schedule of the rotations in the subspecialties of endoscopic surgeries was: 8 weeks for laparoscopy in general surgery, 3 weeks respectively for urological endoscopy, thoracoscopy and gynecological endoscopy, 2 weeks respectively for micro-endoscopic discectomy (MED) and arthroscopy, as well as 1 week for laparoscopic thyroidectomy. (2) The precision of the mentors: Before the start of training, the group members of the specialized endoscopic group were assessed by group leader and the head nurses of all departments, for the evaluation of the specialized knowledge and skills for the subspecialties, and the mentors were determined according to the results of the assessment in the areas that they were good at. Meanwhile, the whole-process mentor responsibility system was implemented with 1 mentor responsible for 2-3 students, to help the students develop learning plans, to keep up with the students' learning status and results, to timely solve the problems of the students encountered in training [2], and to participate in

DOI: 10.1051/bioconf/20170801033

the evaluation of students during the whole process. (3) The precision of the utilization of learning resources With the department resources, various forms of learning and multimedia teaching were carried out: multimedia network teaching was applied to introduce the specialized knowledge and cutting-edge technologies. Lectures and Seminars: the surgeons and specialized senior nurses gave lectures on the diseases of the specialties, the treatments, surgical procedures and nursing knowledge, with the database resources, network teaching videos as the supplement to specialized knowledge. Also, the illustrated brochures of the equipment and devices were made [3, 4], together with the graphics of the operation flow charts stored in the operating rooms, easy for students to learn and check at any time.

2.2.2.2 The precision of operative cooperation

All the procedures and processes of operative cooperation were integrated into precision operation: (1)Preoperative period: the preparation of the operation with standardized procedures of material preparation. All the things needed for the subspecialties of endoscopic surgeries were arranged and displayed according to the unified standards. The specifications of the standards of the preparation procedures were made accordingly for the preparation of the endoscopic equipment, devices, dressings and disposable supplies. The practices of all subspecialties of endoscopic surgeries were all standardized accordingly, so as to quickly and accurately complete the preparation and to improve efficiency. The standardized procedures of patient positioning: the surgical positions of the subspecialties of endoscopic surgeries were standardized with the specifications on the placement of the position mats and the restraints; the surgical positioning methods were completed into details and made into illustrated brochures for the students to check. The standardized procedures of patient positioning could improve the standardization and qualification level of surgical positioning. (2) Medium-term: the intraoperative phase. The standardized procedures of operative cooperation were established, and the devices were placed and passed in accordance with intraoperative processes accordingly. Meanwhile, the emergency response procedures were also set up to cope with the sudden changes of operative methods during the endoscopic surgeries such as the conversion from laparotomy to laparotomy. The sudden changes in the situation, the surgical instruments and the surgical positioning require accurate and quick responses of the nurses for smooth coordination. Besides, the failures of the endoscopic equipment and devices need timely and accurate solutions by the nurses. Therefore, the emergency response measures should be established for timely and accurate responses in case of emergencies, to improve the response capacity of endoscopy nurses, so as to ensure smooth operations. (3) Later stage: postoperative equipment and device processing. The standardized procedures of the cleaning and disinfection methods of endoscopic equipment and devices were established to standardize the disassembly, cleaning and disinfection, as well as the maintenance of the endoscopic equipment and devices [5], in order that the endoscopy nurses could quickly master the effective methods of cleaning and disinfection as well as routine maintenance measures.

2.2.2.3 The precision of assessment and evaluation

The precision assessment system was established by formulating the evaluation sheets by the group leader and group members of the specialized endoscopic group, including the evaluation of professionalism, expertise, preoperative preparation, equipment & device usage, operative cooperation, surgical positioning, and emergency responses, and so on. (1) The assessment by the mentor: during the theoretical learning stage, the mentors should conduct regular student assessment within the responsibilities, mainly in the form of theoretical questions, and should help the students with the weak parts. During the period of practice and exercise, after the completion of one subspecialty, the responsible mentor should conduct the subspecialty assessment of the students, with the records given objectively according to the professional knowledge and the operative cooperation abilities. (2) Self-assessment by the students: after the completion of the studies in one subspecialty, the students should do self-assessment according to the evaluation sheets, for the review and summary of the studies during this stage. The students should also make presentations with PowerPoint documents to share the experience of the studies in this subspecialty.

2.2.3 Assessment methods

After the training of the two groups of trainees, the evaluations of theoretical knowledge, practical skills and operational abilities should be conducted, together with the satisfaction questionnaire of the students on the teaching. (1) The evaluations of theoretical knowledge and practical skills included 100 points respectively, and the evaluation of practical skills was conducted according to the SOP of Endoscopy Nursing with the exam items selected randomly. (2) The evaluation of operational abilities was conducted by the evaluation sheets filled by the surgeons after the completion of the training of each student, including the aspects such as: nursing professionalism, preoperative preparation, intraoperative cooperation, intraoperative communication skills, asepsis and no-touch isolation techniques, and emergency responses. Evaluation sheets were designed in unified form, with a total of 30 points, and 5 points for each item: 1=bad; 2=acceptable; 3= good; 4= very good and 5=excellent. (3) After the completion of the training, the students should fill out the teaching satisfaction questionnaire with the following contents: training contents, training methods, training effects, teaching quality and mentor responsibility, and so on. The questionnaire was designed with the full marks of 25 points, and 5 points for each item: 1= dissatisfied; 2=acceptable; 3= satisfied; 4= very satisfied and 5=excellent.

3 Statistical methods

The data were collected and analysed by statistical software SPSS 19.0. Mean ± standard deviation was used to describe the measurement data with the t-test. The counting data were described by rates and was tested by the χ^2 test, with $P<0.05$ indicating that the difference was significant.

4 Results

The scores of the evaluations of theoretical knowledge, practical skills and operational abilities were significantly higher in the experimental group than those in the control group ($P<0.05$), and also the results of teaching satisfaction were also significantly higher than those in the control group ($P<0.05$), as shown in Table 1.

Table 1. The Compare two groups of nurse training results.

Group	n	Theoretical knowledge Assessment	Operation skills assessment	Business capability assessment	Teaching satisfaction score
Control group	30	85.5±6.0	89.5±4.0	21.0±4.0	18.0±3.0
Observation group	34	92.8±5.0	94..6±5.0	25.0±3.0	22.0±3.0
t		5.31	4.47	4.56	5.32
p		P < 0.05	P < 0.05	P < 0.05	P < 0.05

5 Discussion

5.1 The significance of specialixed training of endoscopy nurses

The specialized operative cooperation is the inevitable trend of development of OR nursing [6], also is in line with the needs of the development of modern surgical techniques. The specialized nursing cooperation for the specialized surgery, can improve the initiative, accuracy and understanding of the OR nurses, so as to improve the quality of operative cooperation and OR nursing. Compared with

DOI: 10.1051/bioconf/20170801033

other surgeries, endoscopic surgeries have the features of various surgical instruments and devices and fast updates of the techniques, hence putting forward higher requirements of the specialized endoscopy nurses. Therefore, high-level endoscopy nurses with high quality are needed to be trained for the OR nursing. Efficient methods of training have always been an issue of concern of the managers.

5.2 The precise aspects of precision training

Precision training means the precise standards and specifications of all parts of training to for the effective, satisfactory results of the training. (1) In the conventional specialized training, the theoretical training is prior to the practice training, and limited in the classroom, with the knowledge explained and memorized without complete understanding or master. Also, there is no specific system of the theoretical teaching, nor a systematic integration of knowledge, and is just a superposition of single points of knowledge, falling short of good training effects. In the precision training, the training content and schedule are precisely arranged, with contents, purposes and estimated effects of each parts specified clearly. The theoretical training is totally out of the conventional classroom mode, with the introduction of the endoscopic equipment and devices with the real products. The endoscopic equipment and devices are various with complex operations. The conventional training only started the usage, installation, disassembly, cleaning, disinfection and maintenance of the devices and equipment in the period of practical training, while there is only presentation with PowerPoint documents in the theoretical training. In this way, the students cannot fully master the knowledge early in the training, and cannot skilfully use the endoscopic equipment and device in the training of the subspecialties of endoscopic surgeries. As a result, the students pay more attention to the points of the endoscopic equipment and devices, ignoring the operation cooperation and emergency responses or other key skills. Precision training can avoid such problems, by teaching the students the usage of the equipment and devices for full mastery in the beginning of the training, resulting in the corresponding skills in the follow-up stages. In the design of the training contents of precision training, the students first observe the surgical processes, and then listen to the introductions of the systematic knowledge of the subspecialties. Compared with the conventional training, this arrangement can make the students have a preliminary understanding of the surgeries before the theoretical knowledge, knowing the key points for the theoretical knowledge, so as to achieve high efficiency of learning. (2) In the precision training, the precision of the teaching arrangement with the mentors of all subspecialties selected by the specialized group can put the mentors in the subspecialties that they are actually good at, so as for a better teaching effect and the maximum optimization of the teaching resources. Meanwhile, the whole-process mentor responsibility system can help the students develop learning plans, keep up with the students' learning status and results, timely solve the problems of the students encountered in training, and participate in the evaluation of students during the whole process. In the conventional training, the theoretical training is conducted together by the theory teachers, and the practical training is the rotating in the subspecialties with one mentor for one student, and the learning effects are totally controlled by the student oneself, with no one else to be responsible for the whole-process control of the learning effects except the final evaluation. With the whole-process mentor responsibility system, the mentor can always keep up with the students' learning progress, participate in the evaluations with sufficient understanding of the weaknesses of the students and supplemental learning in a timely manner This mode can also create a supervising effect of the learning process, so that students feel the pressure of learning and motivation, so as to mobilize the enthusiasm of the students [1]. (3) The standardization of surgical processes is also an important part of precision training. The precision of operative cooperation can make the specialized OR nursing easier in the operative cooperation, so as to achieve standardization and accuracy. By the establishment of the standard processes of material preparation, patient positioning, operative cooperation, equipment cleaning and disinfection, and emergency responses, the operative cooperation for endoscopic surgeries can become more accurate, detailed, standardized and efficient. (4) Meanwhile, in the precision training, the establishment of the precision of evaluation system, can combine the mentor assessment and student self-assessment for the studies of each period and each subspecialty. Compared with the final exam in the conventional training, this system can timely find out the weaknesses of the students and make up for them, so as to reach the corresponding level of teaching and learning.

5.3 The effects of precision training

The level of satisfaction is an effective standard to evaluate the teaching quality and effects. In this study, the scores of the teaching satisfaction and physician satisfaction were significantly higher in the experimental group than those in the control group, while the scores of the evaluations were also significantly higher than those in the control group. Causal Analysis: (1) Precision training can make the training structure more scientific, with a clear setting of the learning purposes of the students, so as to improve the initiative and enthusiasm for better learning effects. (2) The precision of the setting of the training mentors can put the mentors in the areas that they are good at, to improve the effect of teaching, the responsibility of the mentors, the interaction between mentors and students and the teaching quality. (3) In the studies in the subspecialties, the students can receive standardized evaluations, timely solutions to the learning problems, comparatively fixed physician-nurse setting and a better degree of teamwork. In conclusion, precision training can greatly improve various abilities of the specialized endoscopy nurses, enhance the teaching quality of specialized training, and have better training effects.

References

1. Hongmei Tan, Hongxia Dai, Houchan Chang, et al. Effects of specialized training program for nurses of laparoscopy team in operating room[J]. *Journal of Nursing Seience* 2011, **26**(12) :18-19.
2. Lixue Chen, Donghua Liu. Tutorial system application of specialist nurse training of cavity mirrors in the operating room[J]. *Journal of Qilu Nursing* 2014, **20**(24): 120-121.
3. Nian Qin, Aiying Zeng, Hao Huang. Map recognition system in the practice of quality control in the special surgical instruments packaging[J]. *Journal of Nurses Training* 2011, **26**(8):693-694.
4. Liping Guo, Shihua Liu, Juan Liu, et al. Application of instruments map teaching for nursing intems in sterilization and supply center [J]. *Chinese Nursing Research* 2013, **27**(10C):3439-3440.
5. Yuanyuan Li, Huifen Yang, Shiyan Li, et al. Process management of endoscopic instruments in laminar-flow operating room[J]. *Journal of Nursing Seience* 2011, **26**(18) :6-7.
6. Yan Ding, Weijing Liu. Investigation on core competency and curriculum design for operating-room nurse speclalists[J]. *Journal of Nursing Seience* 2013, **28**(22):4-5.

DOI: 10.1051/bioconf/20170801034

Down-regulation of cystathionine-γ-lyase/H_2S system inhibits cell growth in human breast cancer MDA-MB-231 cells

Jing You, Mingming Ma, Juan Ye, Xiaoyan Shi and Tianxiao Wang[a]

College of Pharmacy, Henan University, Kaifeng 475004, Henan Province, China

Abstract. Hydrogen sulfide (H_2S), as the third gasotransmitter, plays important roles in cancer biological processes. Endogenous H_2S can possess pro-cancer functions. Cystathionine-γ-lyase (CSE), one of main metabolic enzymes synthesizing H_2S, possesses important roles in the development and progression of cancer. In the present study, the roles of CSE/H_2S system in human breast cancer cells were investigated. It was observed that CSE was expressed in human breast cancer MDA-MB-231 cells and inhibition of endogenous CSE/H_2S significantly reduced the cell viability and inhibited cell growth and proliferation in MDA-MB-231 cells. Meanwhile, CSE knockdown induced apoptosis of MDA-MB-231 cells and inhibited migration of MDA-MB-231. In conclusion, CSE/H_2S system possesses important roles in human breast cancer cells.

1 Introduction

Cystathionine-γ-lyase (CSE) is a key H_2S-producing enzyme in many tissues except for the central nervous system [1]. CSE catalyzes L-cystine to yield H_2S depending on pyridoxal-5'- phosphate as cofactors. Endogenous H_2S, as the third gasotransmitter signaling molecule alongside nitric oxide (NO) and carbon monoxide (CO), is found in various tissues and organs and plays important roles in many physiological processes, such as vasorelaxation, angiogenesis, cellular energy production and cytoprotection, via multiple mechanisms including activation of signal pathways and stimulation of potassium channels [2-7].

Recently the research of H_2S is also focused on the cancer field and it has been demonstrated that endogenous H_2S produced by CSE promotes proliferation of human cancer cells. Studies indicate that administer of exogenous NaHS or production of endogenous H_2S induce the proliferation of human colon cancer HCT116 cells and human hepatocellular carcinoma cells [8-10].

Currently, few of works on the roles of CSE/H_2S in breast cancer have been reported. Breast cancer is the most frequent malignancies and the sixth leading cause of cancer death for females in China. The identification of new biomarkers and new genes involved in breast cancer progress may provide novel approaches for diagnostic and prognostic evaluation. So in this study, we investigated the effect of down-regulation of CSE/H_2S system on MDA-MB-231 cell growth to explore the roles of CSE/H_2S system in human breast cancers. Our loss-of-function studies demonstrated that the

[a] Corresponding author: wtx1975@126.com
The authors would like to acknowledge the financial assistance provided by a grant from the Key Science and Technology Fund of Henan Province (No. 142300410128) in China.

down-regulation of CSE/H_2S system induced apoptosis accompanied by inhibiting proliferation of breast cancer cell lines in vitro.

2 Materials and methods

2.1 Reagents and antibodies

The CSE inhibitor DL -propargylglycine (PAG) was purchased from Sigma Aldrich (Saint Louis, MO, USA). Anti-CSE antibody was purchased from Abcam (Cambridge, MA, USA). Rabbit polyclonal anti-β-actin (1:2,000; cat. no. 20536-1-AP) primary antibodies and peroxidase- conjugated Affinipure goat anti- rabbit/mouse secondary antibodies (1:10,000; cat. no. SA00001 -2 and SA00001-1) were obtained from Proteintech Group, Inc. (Chicago, IL, USA).

2.2 Cell culture

Human breast cancer MDA-MB-231cells and human normal breast MCF-10A cells were cultivated in Dulbecco's modified Eagle's medium (DMEM) supplemented with 10% fetal bovine serum in a 37 °C incubator with 5% CO_2.

2.3 Western blot

Total cell lysates were prepared in RIPA buffer (50 mM Tris-HCl, pH 8.0, 150 mM sodium chloride, 1.0% NP-40, 0.5% sodium deoxycholate and 0.1% SDS) with 10μg/ml of the protease inhibitor PMSF and their lysates were obtained by centrifugation at 12,000 × g for 10 min. Protein concentration of cells lysates was determined and equal amounts of protein were separated by SDS-PAGE followed by electrophoretically transferred to a PVDF membrane (Millipore, Corp.) at 70 mA for 2 h. Then the membrane was blocked in 5 % nonfat dry milk in TBST. Afterwards, the membrane was probed with specific primary antibodies at 4°C overnight. After washing in TBST, the membranes were incubated in anti-mouse IgG or anti-rabbit IgG secondary antibody for 2 h at room temperature. After the membranes were washed with TBST, the proteins were visualized using ECL detection system (Alpha, USA).

2.4 RNAi treatment

Cells were seeded in 6-well culture dish and transfected at approximately 40% confluent. Transfection was performed with LipofectamineTM^{2000} transfection reagent, according to the manufacturer's instructions (Invitrogen, CA, USA). Results were checked by western blot or cell viability at 48 h after transfection. si RNA was synthesized by Invitrogen (Shanghai, China) and the sequences were the following, CSE-specific si RNA (sense: 5'-GGUUUAGCAGCCACUGUAAdTdT-3'; antisense: 5'-UUACAGUGGCUGCUAAACC dTdT-3').

2.5 H_2S detection

The production of H_2S from cancer cells was measured spectrophotometrically as described below. Briefly, MDA-MB-231 cells were transfected with CSE siRNA or exposed to DL -propargylglycine (PAG, Sigma Aldrich, Saint Louis, MO, USA) and treated with 2 mM L-cysteine and 0.5 mM pyridoxal phosphate. Meanwhile 1 % (w/v) zinc acetate (500 μl) was added on the filter papers stick on the cover of plate to absorb H_2S. After 48 h, the filter papers were put in the tubes containing 0.2 % (w/v) N, N-dimethyl-p-phenylenediamine-dihydrochloride dye (500 μl), 10 % (w/v) ammonium ferric sulfate (50 μl) and 3 ml H_2O and incubated for 20 min at room temperature and absorbance was subsequently monitored at 670 nm. The production of H_2S was determined using a standard curve of NaHS (0-1 mM ; R 2 = 0.9995) and presented as nmol·min $^{-1}$ per 1×10^{6} cells.

2.6. Cell viability, proliferation, migration and apoptosis assays

Cells were classified into CSE knockdown group and control group. In the CSE knockdown group, cells were pretreated with CSE si RNA or 2 mM PAG for 48 h. Each sample was tested at least three replications. Cell viability was performed via MTS assay (Sigma-Aldrich, St. Louis, MO, USA) according to the manufacturer's instructions. Cell proliferation was detected with EdU assay (RUIBOBIO, Guangzhou, China). The EdU assay was performed by plating cells into 96-well dish and staining the cells according to the protocol of the EdU assay kit. The scratch wound assay was used to determine the cell migration. Cells were seeded into 6-well plate and scraped with 10 µl pipette tip at approximately 90% confluency to generate scratch wound and rinsed twice with PBS. Then cells were cultivated in the medium with 5% FBS and the distance was measured at the beginning and after 12 h, 24 h and 48 h. Meanwhile, cell cycle and cell apoptosis were investigated with flow cytometry.

2.7 Statistical analysis

Statistical analyses were performed with the SPSS 17.0 software (SPSS, Inc., Chicago, IL, USA). The results are expressed as the mean±s.d. Differences between two groups were analyzed using the Paired Samples T-test. $P<0.05$ was considered to indicate a statistically significant difference.

3 Results

3.1 Knockdown of CSE decreases cell viability and inhibits proliferation in MDA-MB-231 cells

To explore the relationship between the CSE/H_2S system and the cell viability and proliferation of breast cancer cells, firstly, we checked the intrinsic level of CSE protein in breast cancer cell line MDA-MB-231 as well as the normal breast cell line MCF-10A. Figure 1A showed that CSE protein was slightly over-expressed in both MDA-MB-231 cells compared with MCF-10A cells. Then we knocked down CSE using siRNA transfection or inhibited the activity of CSE protein by PAG and investigated the changes of the cell proliferation using cell viability and EdU assay. Interestingly, inhibition of intrinsic CSE/H_2S with CSE siRNA or PAG significantly reduced the growth (Figure 1B-D) and proliferation (Figure 2) of MDA-MB-231 cells. The results suggested that the CSE/H_2S system could regulate the survival and proliferation of human breast cancer MDA-MB-231 cells.

DOI: 10.1051/bioconf/20170801034

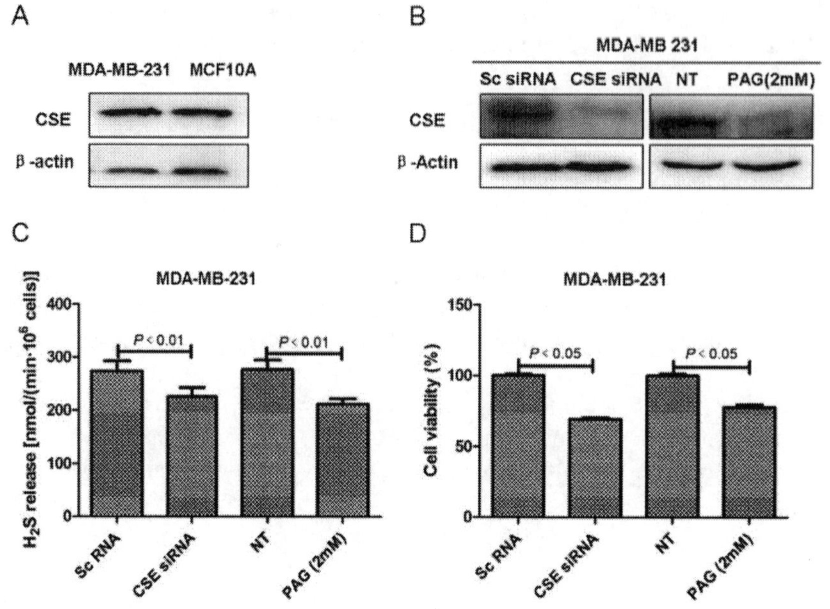

Figure 1. Expression of CSE and analyses of cell growth associated with the inhibition of CSE/H_2S system in human breast cancer cells. (A) Expression level of CSE in human breast cancer and normal breast cells were detected by WB. (B) CSE siRNA transfection and CSE inhibitor PAG treatment in MDA-MB-231 cells were determined by WB. (C) The effects of CSE siRNA and CSE inhibitor PAG on the production of endogenous H_2S in MDA-MB-231 cells. 1×10^6 cells were seeded in 6-well plate and then treated with CSE siRNA and CSE inhibitor PAG for 48 h. The production of endogenous H_2S was assessed spectrophotometrically using N, N-dimethyl-p-phenylene diamine-dihydrochloride. (D) MTS assays were used to examine the effect of CSE siRNA and CSE inhibitor PAG on cell growth in MDA-MB-231 cells. 1×10^5 cells were seeded in 6-well plate and then treated with CSE siRNA and CSE inhibitor PAG for 48 h. Cell viability was evaluated in triplicate by Microplate Spectrofluorometer.

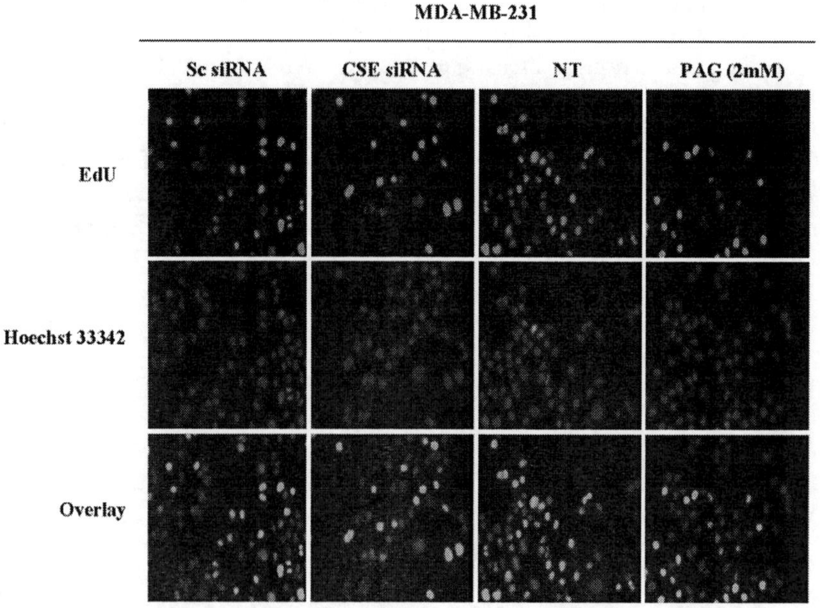

Figure 2. Analyses of cell proliferation associated with the inhibition of CSE/H_2S system in human breast cancer cells. EdU assays were used to test the effect of CSE siRNA and CSE inhibitor PAP on cell proliferation in MDA-MB-231 cells (20×). EdU, 5-ethynyl-2'-deoxyuridine. NT, no treatment.

 DOI: 10.1051/bioconf/20170801034

3.2 Knockdown of CSE in MDA-MB-231 cells arrests cell cycle in S phase

To further investigate the effect of knockdown of CSE on cell proliferation, we detected the changes of cell cycle in MDA-MB-231 cells treated with CSE siRNA or PAG. Flow cytometry results showed that MDA-MB-231 cells exposed to CSE siRNA or PAG were arrested in the S phase (Fig. 3). The results further proved that knockdown of CSE inhibited cell proliferation.

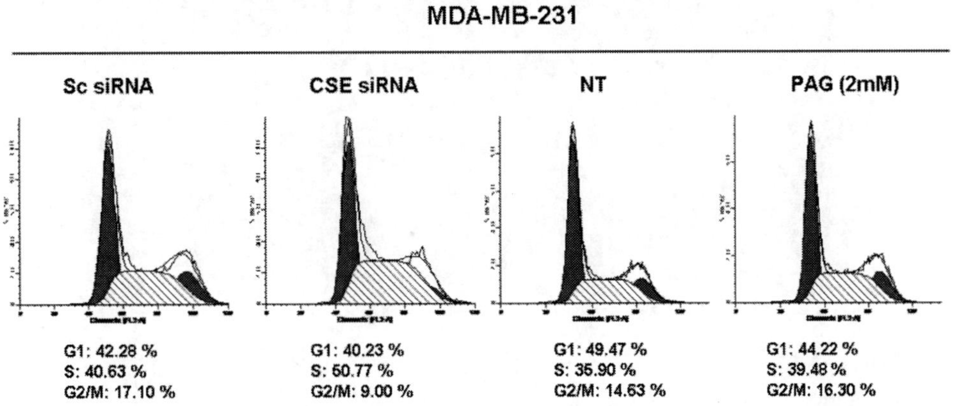

Figure 3. Analyses of cell cycle associated with the inhibition of CSE/H_2S system in human breast cancer cells. Flow cytometry was used to test the effect of CSE siRNA and CSE inhibitor PAP on cell cycle in MDA-MB-231 cells. NT, no treatment.

3.3 Knockdown of CSE in MDA-MB-231 cells induces cell apoptosis

Apoptosis is one of the major mechanisms of inhibiting cell proliferation or inducing cell death. To explore how does the CSE/H_2S system regulate the proliferation of hepatoma cells, we investigated the effect of CSE knockdown on cell apoptosis. Flow cytometry results showed that CSE siRNA transfection or CSE inhibitor PAG treatment caused the increased apoptosis rate in MDA-MB-231 cells (Figure 4).

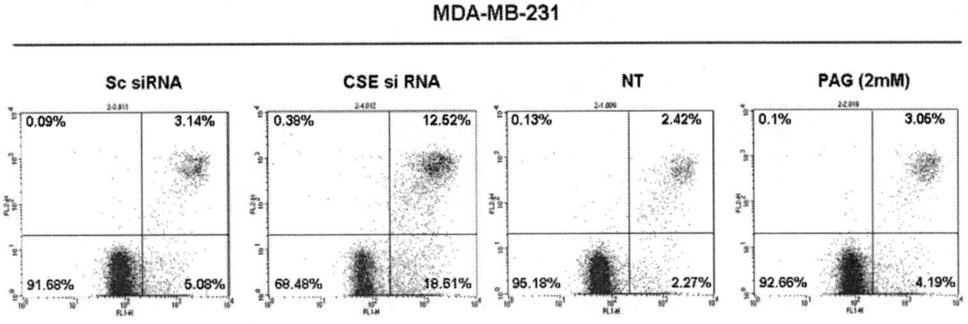

Figure 4. Analyses of cell apoptosis associated with CSE downregulation in MDA-MB-231 cells. Flow cytometry assay with Annexin V/FITC double staining. NT, no treatment.

3.4 Knockdown of CSE in MDA-MB-231 cells inhibits cell migration

Meanwhile, we performed the scratch wound assay and observed that the application of CSE siRNA or PAG inhibited the migration in MDA-MB-231 cells (Figure 5).

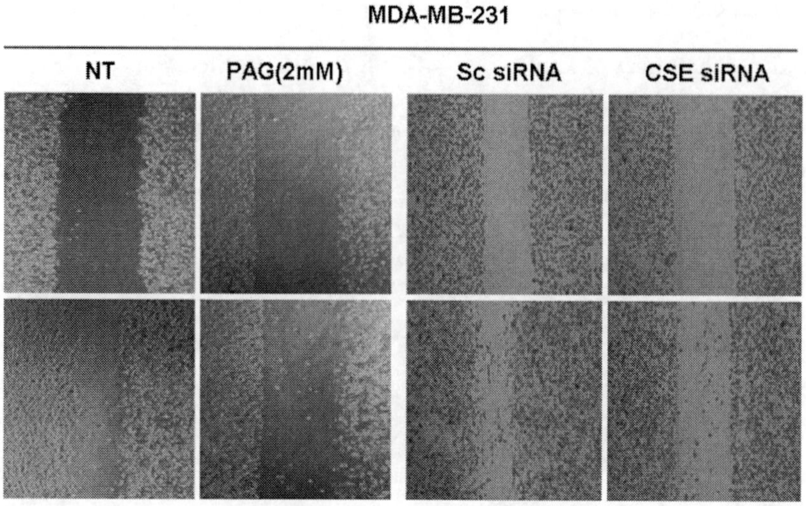

Figure 5. Analyses of cell migration associated with CSE downregulation in MDA-MB-231 cells. The scratch wound assay was performed to detect cell migration. NT, no treatment.

4 Discussion

H_2S, as the third gasotransmitter, has academically got increasing concern. Studies on H_2S are involved in many scientific fields, including many physiological and pathological processes Current studies on H2S also mainly focus on its functions in physiological and pathophysiological processes [11]. We do not have a good understanding on the functions of H_2S in cancer field and previous studies possess a little controversy, which is exogenous H_2S could induce HCT116 cell proliferation, but inhibit WiDr cell growth [12-13]. So the functions of H_2S in cancer field need to be further investigated.

Few of works on the functions of H_2S in tumor were performed, especially breast cancer. Therefore, in this study we investigated the functions of CSE/H_2S system in breast cancer to further explore the functions of H_2S in cancer cells. We found that the knockdown of CSE/H_2S system significantly inhibited cell growth and proliferation, induced cell apoptosis and inhibited cell migration. The results showed that the CSE/H_2S system possessed vital roles in breast cancer cells.

The following things will be the researches about the mechanism of CSE/H_2S system regulating cell proliferation and apoptosis and the effect of CSE/H_2S system on tumor xenograft growth in vivo.

References

1. R.Wang, Antioxid Redox Signal, **12**,1061-1064(2010).
2. C.Coletta, A.Papapetropoulos, K.Erdelyi, G.Olah, K.Módis, P.Panopoulos, A.Asimakopoulou, D.Gerö, I.Sharina, E.Martin, C.Szabo, Proc Natl Acad Sci USA, **109**, 9161-9166 (2012).
3. Y.Kimura, Y.Goto, H.Kimura, Antioxid Redox Signal, **12**,1-13(2010).
4. J.Sheng, W.Shim, H.Wei, SY.Lim, R.Liew, TS.Lim, BH.Ong, YL.Chua, P.Wong, J Cell Mol Med, **17**,1345-1354(2013).
5. D.Popov, Arch Physiol Biochem, **119**,189-194(2013).
6. P.Yin, C.Zhao, Z. Li, C.Mei, W.Yao, Y.Liu, N.Li, J.Qi, L.Wang, Y.Shi, S.Qiu, J.Fan, X.Zha, Cell Signal, **24**,1229-1240 (2012).
7. G.Yang, Y.Pei, Q.Cao, R.Wang, J Cell Physiol, **227**, 3192-3200 (2012).
8. C.Szabo, C.Coletta, C.Chao, K.Módis, B.Szczesny, A.Papapetropoulos, M.R.Hellmich, Proc Natl Acad Sci USA, **110**, 12474-12479(2013).
9. W.J. Cai, M.J. Wang, L.H. Ju, C. Wang, Y.C. Zhu, Cell Biol Int, **34** ,565-572 (2010).
10. K.Fan, N.Li, J.J Qi, P.Yin, C. Zhao, L.Y. Wang, Z.X. Li, X.L. Zha, Cellular Signalling, **26**, 2801-2808 (2014).

11. G.Kumar, A.Chhabra, S.Mishra, H.Kalam, D.Kumar, R.Meena, Y.Ahmad, K.Bhargava, D.N.Prasad, M.Sharma, EBioMedicine, **6**, 171-189 (2016).
12. K.Módis, C.Coletta, A.Asimakopoulou, B.Szczesny, C.Chao, A.Papapetropoulos, M.R.Hellmich, C.Szabo, Nitric Oxide, **41**,146-156 (2014).
13. Q.Cao, L. Zhang, G.Yang, C.Xu, R.Wang, Antioxid Redox Signal, **12**, 1101-1109 (2010).

Research and Design No Light Spot in Front Located Hysteroscopy

Kun Dai[1,a], Bo Xiang[2] and Zhaomin Jia[2]

[1]*Electrical Department, Tangshan Polytechnic College , Tangshan, 063020,china*
[2] *Intelligence and Information school of Tangshan University, Tangshan 063020, Heibei, P.R.China*

Abstract. This paper first introduces the recent development of the popular hysteroscopy, the international minimally invasive diagnosis instrument, and then analyzes the lighting and video of hysteroscopy. According to high energy consumption of common hysteroscopy, front located hysteroscopy system is put forward. The no light spot principles, the lens structure, the manufacture process of lens prototype, the trial manufacturing process of power box are all discussed in this paper. It is the first time that prototype has been design and tested in domestic. In the normal conditions, the energy consumption decreased by 298/300. And the volume, the weight, and the price of the designed front located hysteroscopy are greatly reduced. Moreover, tests show that it can apply to special conditions such as the disaster relief field and field operation.

1 Introduction

Hysteroscopy is the advanced minimally invasive medical instrument [1], used for examination and treatment of gynecology uterine cavity. In October, 2010, the 64th session of China international medical equipment exposition, numerous hysteroscopy produced by Chinese and foreign manufacturers are exhibited. In spite of the different in manufacturer and brands, the working principle and composition are the same. The working principle and composition of hysteroscopy are shown in figure 1, the front of the lens body is inserted to the uterine and enlarge the to-be-observed parts. Because of its intuitive and preciseness, it becomes the first choice in examining the gynecological hemorrhagic diseases and intrauterine lesions. It is a kind of fiber illuminant endoscope, mainly including chamber lens, energy system, illuminant systems and imaging system.

A 300W xenon lamp is used as the power source. Light is transmitted by optical fiber and the lighting is realized by the light of the conduction body. Image can be gained through imaging system by light transmitting to cavity via video conversion on the monitor. Energy systems provide energy for the light source system and the imaging system. The combination device has a characteristic of high power consumption, big volume, high price and rigid using environment. Researches show: light shall finally focused on local lighting of the irradiated content, light loss in optical cable and optical transmission conduction; the long transmission distance by light transmission to cavity lens imaging also brings certain loss. Therefore, the front located hysteroscopy is first developed and tested, and the effect is satisfactory.

[a]Corresponding author: daikun943@126.com

DOI: 10.1051/bioconf/20170801035

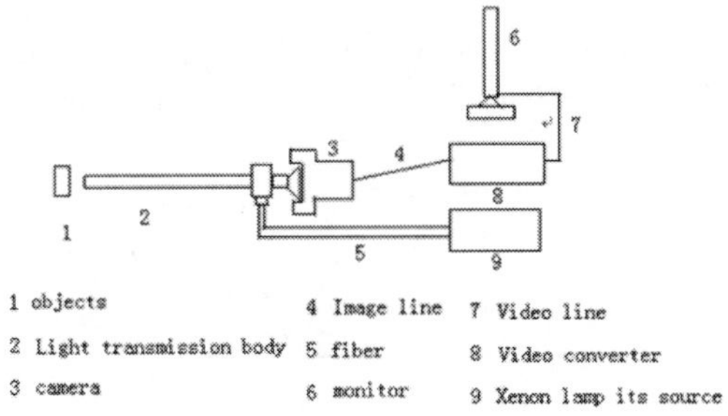

Figure 1. Hysteroscopy principle diagram

2 Structure of the no light spot front located hysteroscopy system

A. The Principle Of No Light Spot Front Located Hysteroscopy System

The principle of No Light Spot front located hysteroscopy system is shown in figure 2.

B. By installing the camera and lighting together, directly lighting and videoing on the subject of local, decoding through the link of guide bar, connection box at the tail and the 4 core connecting plug, linking to the power box with the wire, realizing the power source lighting and video signal transmission, video signals are transmitted by the power box through video line and being displayed as the image on the monitor. The tail connecting box is functioned as a handle to change the angle and distance between the camera and the subject to be lighted in order to get the best perturbation state. Under appropriate conditions, the connecting box can be used to fix or support something.

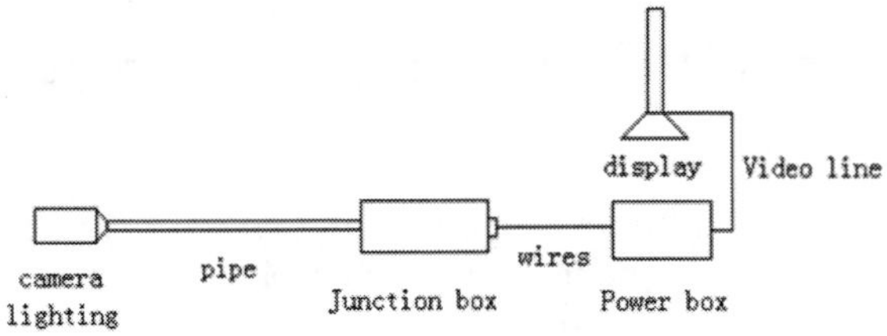

Figure 2. New hysteroscopy system

C. The Constitute of the Front Located Hysteroscopy Lens

The constitute of the front located hysteroscopy lens is shown in figure 3.On the left, it is the face Schematic. And on the right of Figure 3, it is the side schematic diagram. The light source systems adopt the six HLEDs to realize the lighting [2] [3] [4]. The total power of luminous is less than 0.5W.The cavity lens adopt a φ2mm high-definition camera and a φ3.5mm decoding video circuit. A 0.2 mm crystal abrasive disc is used to protect the video circuitry. All of the devices are installed in a space with φ5mm and 15mm-long, The silver line which can be resistant to high temperature is used to draw the control video out with 2 piece of wires. one is for videoing, and the other one is for lighting.

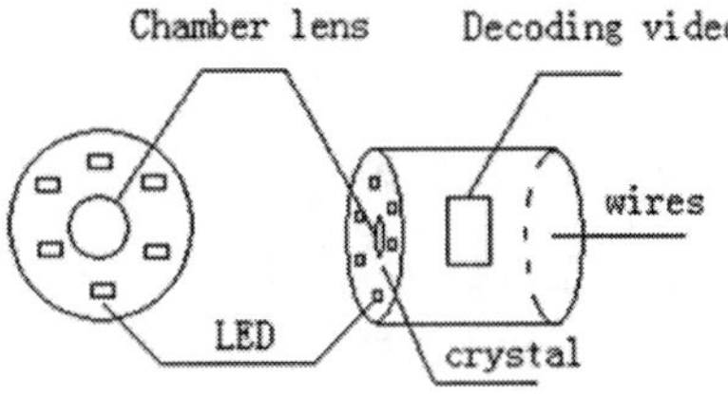

Figure 3. Front hysteroscopy lens

 DOI: 10.1051/bioconf/20170801035

D. The Constitute of no light spot the Front Located Hysteroscopy Lens

The constitute of the no light spot front located hysteroscopy lens is shown in figure 4.

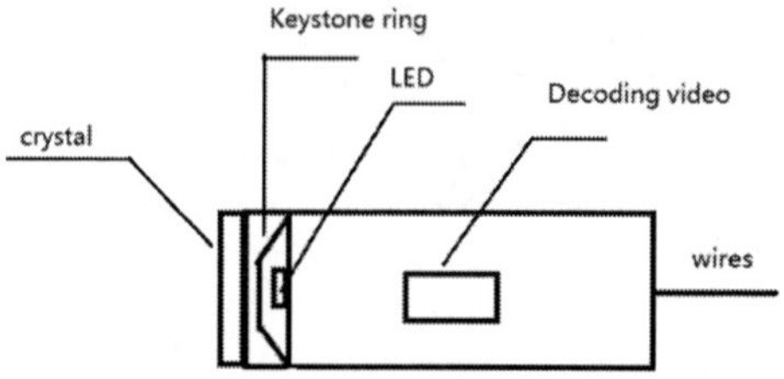

Figure 4. No light spot the Front Located Hysteroscopy Lens

After the test, Keystone ring Angle is 5 degrees.

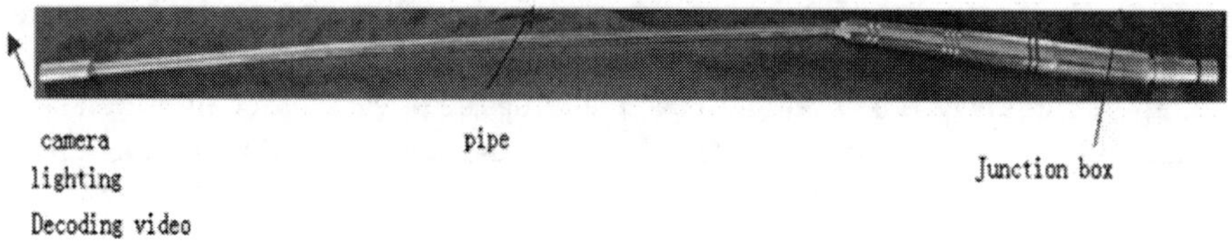

Figure 5. Camera shell

3 No light spot front hysteroscopy system trial-production process

A. Lens Prototype Trial Producing Process

1. Test the parameter of the voltage current of LED [2],[4], and measure the working properties, determine the control parameters, control scheme and connection schemes based on the special requirements of surgical instruments.
2. parameters, control scheme and connection schemes based on the special requirements of surgical instruments. Determine cameras and decoding working circuit voltage range, also combined with surgical instruments special requirements to determine control parameters, control scheme.
3. Manufacture the camera shell in machining method, welding method as shown in Figure 5
4. Weld clear cameras and decoding circuit with the power line, video line. Weld six miniature HLEDs and draw the lighting line, the length of the wire should be exceeded the connection plug tail 10mm.
5. Make the camera and LED lead on the left and thread through the steel body on the right. The wire should be through the right and until to the tail part.
6. Detect the working situation of the cameras and the LED. If it is normal, go to the next step.
7. Fix the cameras, the LEDs and the decoder circuit.
8. Test circuit.
9. Fix crystal chips.
10. Install and fix the connecting plugs

B . Trial Production of Power Box

The constitute of the power box as shown in Figure6.

1. Design and install the camera circuit and the LED control circuit using the multi-function PCB with 30mm–wide and 50mm-long [5] [6]. By measuring parameters through the instrument, their parameters should be adjusted to correspondent with the need of lens prototype.

 DOI: 10.1051/bioconf/20170801035

2. Punch two holes at the front and back of the power box, respectively, and install the power switch, the plug implement, the LED which can display the electric power, the video interface, and the charging interface.
3. Connect the LED with the circuit which can display the low voltage.
4. Fix the 12V high-power Li-ion batteries and multifunctional PCB.
5. Connect the plug sensor and the multi-function circuit. Connect the circuit between the battery and the low voltage circuit. Connect the circuit between the video joint and the multi-function PCB. Connect the circuit between the charging interface and the battery. Connecting the circuit between the battery and multi-functional PCB.
6. Coverer and charge.

4 Realization of no light spot front located hysteroscopy camera and its effect

Disinfect the lens prototype and attachment with formalin, connect lens prototype and power box with wire, and connect power box and monitor with video cable. Adapt the monitor with 12v DC power source, provide electricity for the monitor, preheat and transfer to the video.

By connecting the power supply, and lighting the six LEDs luminescence of the prototype lens box, the image can be displayed on the monitor. The displayed images can be seen on Figure 6. It is noted that the image shown in Figure 6 is the palm print image under other external lighting conditions.

Via testing the parameters of lens prototype in the air of 20°C, 40°C, and in the water of 20°C, 40°C, respectively, the parameters of lens prototype of 40°C air and 40°C water are obtained and list in Table I and Table II, respectively. Here the temperature in cavity of the body is nearly 40°C.

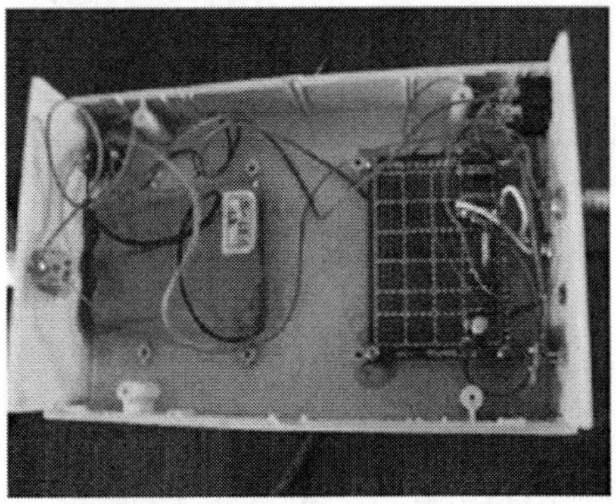

Figure 6. power box

Table 1. PARAMETERS IN 40 °C AIR

Time	Power box of Voltage(V)	Camera Voltage(V)	LED voltage (V)	LED Current (mA)	20mm Illuminance (Lux)
1min	12.59	4.98	2.56	45.2	82.4
5min	12.58	4.98	2.53	45.3	82.5
1h	12.36	4.96	2.57	45.6	81.3
5h	12.31	4.97	2.54	44.6	80.2
12h	12.13	4.96	2.49	46.1	80.3
24h	11.09	4.98	2.48	45.1	80.6
48h	11.05	4.92	2.48	46.2	80.8
72h	11.03	4.96	2.47	46.1	80.1

Table 2. PARAMETERS IN 40 °C WATER

Time	Power box of Voltage(V)	Camera Voltage(V)	LED voltage (V)	LED Current (mA)
1min	12.52	4.98	2.54	45.6
5min	12.53	4.98	2.56	45.1
1h	12.23	4.96	2.52	45.2
5h	12.25	4.97	2.52	44.8
12h	12.13	4.92	2.47	46.2
24h	11.01	4.93	2.47	43.2
48h	11.06	4.90	2.46	45.5
72h	11.07	4.92	2.45	43.7

The working stability can be seen through working parameters on the two tables. Even the intensity of illumination can not be measured in water, it can be deduced from the voltage current condition of led. From these Tables, the distinguish determination can be decided to distinguish 0.2 mm objects.

Experimenting with 1000mA charger, which has been in charging for 8 hours, it can guarantee the continuous 80—hour working time. No Light Spot Front Located Hysteroscopy Camera Found no spot.

5 Summary

By installing the light and the video at the front, the designed hysteroscopy system can reduce its energy consuming by eliminating intermediate transmission without decreasing the image quality. The power box with a 12V DC input has small dimension and light weight. The entire system is easy to carry and can be operated mostly anywhere. A 12V car battery with a simple connector is sufficient to turn the system to operational mode in extreme case such as disaster relief.

References

1. The 64th session of China international medical equipment exposition exhibition [M/CD].Beijing. National Medicine Lilan Group. 2010,10
2. Q. S. Zhang. Selected LED Applied Circuit [M].Beijing. mechanical industry press, 2010,2-21
3. GB50034-2004, building lighting design standard [S]
4. Instruction for YB4180 series transistor characteristics graphics instrument [Z]. Yang: Jiangsu Yangzhong Lvyang electrical equipment Co. LTD, 2004
5. W.P Na. The Selection and Application of General Integrated Circuit [M]. Beijing. people's wiley&sons press, 2004,129-151
6. China metrology press. New Edition for Collection of Electronic Circuit [M] .Beijing. China Met

The Application of Enhanced Recovery after Surgery on Pancreaticoduodenectomy

Xiaoxiao Yao and Dan Zhang

Department of Hepatobiliary&PancreasSurgery,The No.2Hospital, Jilin University,Changchun

Abstract. enhanced recovery after surgery was firstly proposed by Danish surgeon Kehlet in 2001. It can reduce stress and trauma as well as postoperative complications and accelerate rehabilitation of patients with preoperative, intraoperative and postoperative application of a series of effective measures based on evidence-based medicine. Its philosophy conforms to modern biology-psychology-social medical model and medical ethics standards while relieving patients' pain, reducing costs and shortening hospitalization time. Its contents can be summarized as: (1) preoperative propaganda and education; (2) effective anesthesia and analgesia; (3) accurate surgical operation; (4) postoperative rehabilitation. With theproposal and gradual recognition of ERAS concept, it has been well applied in the surgeries of gastrointestinal tract, thyroid, breast, urinary system, etc. Corresponding diagnosis standards have been summarized. Especially in the colorectal surgery, the application effect is the most significant. But the concept of ERAS still has great controversy in surgical fields because of complicated operation, great trauma, various complications such as liver surgery and pancreatic operation, so it has not been widely recognized.

1 Instruction

The PD involves pancreas, biliary tract, stomach, duodenum, jejunum and many other organs. Not only the tumor needs excision, but pancreatic anastomosis, bilioenteric anastomosis and gastrointestinal anastomosis are required for the reconstruction of the digestive tract, so PD is the most difficult surgical method with great trauma in the field of general surgery. Literature reports that the incidence of postoperative complications of PD reaches 48% [1], such as pancreatic fistula, biliary fistula, bleeding, delayed gastric emptying, incision infection, abdominal infection, stress ulcer, MODS, etc. Abroad treatment centers of general surgery, which applied ERAS concept to pancreatic operation earlier, reported its curative effects. The results of the retrospective study on 41 pancreatic-operation patients from 2005 to 2011of the University of Melbourne showed: the LOS after surgery and the total LOS of the 20 cases in the ERAS group was significantly lower than the 21 patients in the traditional treatment group ($P < 0.05$); but 3 patients in the ERAS group returned to the hospital due to stomachache while no patients in the traditional treatment group returned [2]. The results of the retrospective study of ERAS application on 82 PD patients of Deoliveira et al. [3] showed: the LOS after surgery and the total LOS were 9d and 11d respectively and the rate of readmission was 14.6%. In this group, the overall incidence of postoperative complication was 47.6%; as for the patients placed with abdominal drainage tube, the incidence of pancreatic fistula was 8.5%. However, the majority of the ERAS clinical researches related with pancreatic surgery are monocentric retrospective researches, so there lacks a unified evaluation system. Meanwhile, the ERAS scheme they apply in the pancreatic surgery usually refers to the pancreatic surgeon. There is no unified scheme aiming at pancreatic disease and pancreatic surgery accepted by most of the pancreatic surgeon. In 2011, ESPEN and ASMEN conducted a systematic review on high-quality random clinical study, cohort study of large samples and META analysis of ERAS after PD. They classified relevant treatment measures and summarized different recommended levels [4]. According to the statistical results and combining the clinical working experience of our department, the application of ERAS on PD is discussed here.

2 Preoperative Treatment Scheme

With the application of ERAS on gastrointestinal surgery, clinical practice, evidence-based medicine, preoperative propaganda and education and smoking cessation is confirmed by clinical practice and evidence-based medicine; routine mechanical bowel preparation and long fasting before operation is unnecessary; liquid diet containing glucose 2h before anesthesia is one of the reliable measures. It can reduce trauma and stress stimulation, having positive influence on rapid recovery of patient, so it is widely accepted in practice. Meanwhile, it is also applicable to pancreatic operation and other surgical operations. Preoperative jaundice-reducing treatment before PD surgery is controversial. The results of META analysis of PD perioperative treatment shows: preoperative biliary drainage will not reduce postoperative mortality; on the contrary, it increases the incidence of complications related to drainage [5]. According to our previous clinical experience, we believe preoperative biliary drainage should not serve as the preoperative preparation but should be applied to specific patients. Regarding the patients whose total bilirubin is greater than 171 umol/L with severe malnutrition or cholangitis, we can consider to conduct preoperative biliary drainage.

3 Intranperative Treatment Scheme

Compared with relatively normative and mature intraoperative treatment scheme of gastrointestinal surgery, the complexity of PD leads to its slow progress in the aspect of minimally invasive, so it is still necessary to shorten surgery time and reduce operative stress so as to achieve greater progress.

(1) Prophylactic reference of antibiotics: single dose of prophylactic antibiotics is used 30 minutes before operation. If the operation time exceeds 3 hours or the bleeding volume is greater than 1500 ml, we can add a dose (according to the half-life of the drug).

(2) Intraoperative insulation: avoiding hypothermy during the operation can reduce the incidence of postoperative incision and the complications of cardiopulmonary system [6]. The measures adopted include rising the indoor temperature of the operating room, warming infusion liquid in advance, washing abdominal cavity with warm saline, covering insulation blanket during and after operation, avoiding unnecessary body exposure, etc.

(3) Controlling stress and reducing trauma: the guideline mentions epidural analgesia during operation: compared with opioid drugs of intravenous application, epidural analgesia in the middle part of the chest has good analgesia with less respiratory depression [7]. But it requires the recognition and cooperation of the anesthetist. As for the laparoscopic surgery, in the gall bladder, stomach, colon and rectum surgeries, endoscopic surgery can significantly relieve the trauma and stress of patients and enhance recovery comparing with laparotomy. But PD not only requires tumor separation and excision around important blood vessels after peritoneum, but also requires complicated digestive tract reconstruction. It has higher requirements of the experience, operation level and endurance of doctors as well as the endoscope instrument. An operation usually takes seven to eight hours. Long-time anesthesia itself increases the stress and trauma of patients, which goes against the concept of ERAS. In clinical work, we regard "minimally invasive surgery" as a concept rather than an operation, so we don't insist on performing PD through endoscope, but reduce the trauma to patients by shortening the operation time, reducing intraoperative bleeding with gentle and careful operation.

4 Postoperative Treatment Scheme

In the postoperative ERAS treatment scheme of PD, the focus of the debate is the indwelling drainage tube, the indwelling time of the drainage tube and early-stage feeding of patients.

(1) Blood glucose control: the rising of blood glucose and insulin resistance after PD will lead to anastomotic fistula and other serious complications and rise perioperative mortality.

(2) Postoperative analgesia: establishing standardized painless wards with individualized, multi-mode and preventive analgesia, such as PCIA pump, epidural analgesia, NSAIDs analgesics so as to reduce or avoid systemic application of opioid drugs.

(3) Shortening service time of antibiotic after surgery. For postoperative patients of PD, we combine blood routine examination, C-reactive protein examination, procalcitonin examination as well as body temperature changes to monitor postoperative infection in our clinical work. Antibiotic is usually stopped when procalcitonin falls to 0.5 ng/ml.

(4) Restrictive fluid therapy: postoperative transfusion should follow the principle of transfusion according to demands in order to avoid tissue edema, microcirculation disturbance, hydrothorax, seroperitoneum, abnormity of cardiopulmonary function which will aggravate trauma and stress.

(5) Peritoneal drainage: pancreatic fistula is the most dangerous postoperative complications of PD. Part of the foreign research results show that the incidence of complications caused by postoperative indwelling abdominal drainage or not has no statistically significant difference [8]. However, as there is no completely unified diagnosis critical, the incidence of postoperative pancreatic fistula reported by different centers has great difference. The drainage tube is of great importance to postoperative observation and treatment of complications such as pancreatic fistula and severe bleeding. Our experience is to indwell a drainage tube at the sides of both pancreas-intestine anastomotic stoma and bilioenteric anastomosis stoma. The key point is that the drainage tube should be placed at a distance of 1-1.5 cm from the anastomotic stoma to avoid the injury of the anastomotic stoma in the process of tube drawing; in the meanwhile, if there are no complications such as anastomotic fistula or hemorrhage, the tubulization time should not exceed 5 days.

(6) Early-stage enteral nutrition: early-stage feeding can protect the barrier function of gastrointestinal mucosa and promote postoperative recovery of gastrointestinal functions. Our approach is to place the intestinal feeding tube 10-15 cm away from the gastrointestinal anastomotic stoma and provide small dose of enteral nutrition since the first or second day after operation. Through the retrospective analysis of 127 patients of PD in our department from 2006 to 2013, we draw a conclusion that early-stage enteral nutrition after PD can actively accelerate the recovery of digestive function after operation. The method is secure and effective for it not only promotes the recovery of patients' gastroenteric function but also avoids the adverse impact of food on anastomotic healing.

The study of ERAS, especially in the aspect of PD, started quite late. There are much controversy and few literature reports at home and abroad, but all the research results show its feasibility, security and effectiveness to some degree. European counterparts brought forward the application guide of ERAS in pancreatic operation, but in view of different national conditions, cultural background, ways of life and medical models, western standards cannot be adapted by Chinese patients completely, so blind imitation may have negative effect on patients' recovery and even lead to medical disputes on the contrary. But as practitioners of pancreas surgery, we should catch sight of the significance of ERAS in the development of modern surgery, make efforts to break old ideas and accept new things. In the pursuit of best prognosis and enhanced recovery, we should find an appropriate balance. Customizing ERAS scheme for different patients can truly push forward the cause and bring good news to patients.

References

1. LimongelliP,KhorsandiS.E., Pai M,etal.Management of delayed postoperative hemorrhage after pancreaticoduode-nectomy: a meta-analysis.Arch Surg 2008; 143: 1001-1007; discussion 1007
2. Nikfarjam M. Weinberg L,Low N,et al. A fast track recovery program significantly reduces hospital length of stay following uncomplicated pancreaticoduodenectomy. JOP,2013,14 (1):63—70
3. DeOliveira ML,Winter JM,Schafer M,et al. Assessment of complications after pancreatic surgery:A novel grading system applied to 633 patients undergoing pancreaticoduodenectomy. AnnSurg,2006,244 (6):931—939
4. Guyatt GH,Oxman AD,Vist GE,et al. GRADE:an emerging consensus on rating quality of evidence and strength of recommendations. BMJ,2008,336 (7650):924—926
5. Sewnath ME,Karsten TM,Prins MH,et al. A meta-analysis on the efficacy of preoperative biliary drainage for tumors causing obstructivejaundice. AnnSurg,2002,236 (1):17-27
6. Kurz A,Sessler DI,Lenhardt R. Perioperative normothermia to reduce the incidence of surgical--wound infection and shorten hospitalizationStudy of Wound Infection and Temperature Group. N Engl J Med,1996,334 (19):1209—1215
7. Werawatganon T,Charuluxanun S. Patient controlled intravenous opioid analgesia versus continuous epidural analgesia for pain afterintraabdominal surgery. Cochrane Database Syst Rev,2005, (1):CD004088
8. Conlon KC,Labow D,Leung D,et al. Prospective randomized clinical trial of the value of intraperitoneal drainage after pancreaticresection. AnnSurg,2001,234 (4):487—493

BIO Web of Conferences **8**, 01037 (2017) DOI: 10.1051/bioconf/20170801037
ICMSB2016

SVM classification model in depression recognition based on mutation PSO parameter optimization

Ming Zhang[1,2,3,4], Shengfu Lu[1,2,3,4], Mi Li[1,2,3,4,a], Qian Zhai[5,6], Jia Zhou[5], Xiaofeng Lu[1,2,3,4], Jiying Xu[1, 2, 3, 4], Jia Xue[1, 2, 3, 4], and Ning Zhong[1, 2,3,4,7]

[1]*Beijing Advanced Innovation Center for Future Internet Technology, Beijing University of Technology, Beijing, China*
[2]*Laboratory of Intelligent Science & Technology, International WIC Institute, Beijing University of Technology, Beijing, China*
[3]*Beijing International Collaboration Base on Brain Informatics and Wisdom Services, Beijing, China*
[4]*Beijing Key Laboratory of MRI and Brain Informatics, Beijing, China*
[5]*Beijing Anding Hospital, Capital Medical University, Beijing, China*
[6]*Beijing Key Laboratory for Mental Disorders, Beijing, China*
[7]*Department of Life Science and Informatics, Maebashi Institute of Technology, Maebashi, Japan*

Abstract. At present, the clinical diagnosis of depression is mainly through structured interviews by psychiatrists, which is lack of objective diagnostic methods, so it causes the higher rate of misdiagnosis. In this paper, a method of depression recognition based on SVM and particle swarm optimization algorithm mutation is proposed. To address on the problem that particle swarm optimization (PSO) algorithm easily trap in local optima, we propose a feedback mutation PSO algorithm (FBPSO) to balance the local search and global exploration ability, so that the parameters of the classification model is optimal. We compared different PSO mutation algorithms about classification accuracy for depression, and found the classification accuracy of support vector machine (SVM) classifier based on feedback mutation PSO algorithm is the highest. Our study promotes important reference value for establishing auxiliary diagnostic used in depression recognition of clinical diagnosis.

1 Introduction

With the rapid development of society, people's living standard is constantly raising, the pressure on living for people is bigger and bigger. The mental diseases are more common, such as depression. Depression is mainly characterized by psychological disorder, and the prevalence of depression is increased year by year. Depression has become a serious harm to people's physical and mental health. According to the report of the WHO in 2001, the harm caused by depression ranked first among all kinds of diseases [1]. Data shows that 70% of people are under a condition of unhealthy or subhealthy, and psychological stress related diseases accounted for about 5% -10% of the population, physical and mental disorders has become a common disease gradually. At present, the identification and treatment of depression are mainly based on the way of face-to-face interview with the doctor according to the diagnostic manual (DSM- IV)[2]. When patients are in severe depression, the misdiagnosis rate of the doctors will be low. However, the rate misdiagnosis of the doctors is likely higher when patients are in

[a] Corresponding author: limi@bjut.edu.cn

mild depression. What's more, even some healthy people may be misdiagnosed and they get wrong medicine. The reason is that the experience of doctors has great impact on the identification of depression. Therefore, the auxiliary diagnosis of depression is very important.

At present, many medical devices have been used in auxiliary diagnosis depression, such as EEG (electroencephalo-graph) and fMRI (Functional Magnetic Resonance Imaging) [3]. The EEG can acquire the signal data of the cerebral cortex. Some sensors will be placed on the patient head while collecting data. After receiving the signal from the sensors, the signal will be amplified and converted into the waveform by amplifying system. Under normal circumstances, the waveform of the EEG follows certain regularity. The regularity will be destroyed and the waveform will be changed, when the cortical lesions. Accordingly, through analysing the waveform of the EEG, it can be used in auxiliary diagnosis depression. In addition, the fMRI is also a common auxiliary diagnosis method. The fMRI can observe the brain activity in during the cognitive task process, when combing with the blood oxygen level dependence and the echo planar imaging technology, which is different from the EEG. What's more, the fMRI is a non-invasive and no radiation exposing auxiliary diagnostic method. However, the fMRI can't achieve the purpose of diagnosis depression.

The eyes are the source of emotion. The reason why eyes can express emotion is because of the expansion and narrowing of the pupil, the rotation of the eyeball and the gaze of the eyes. People's thoughts and emotions are related to the changes of pupils closely, and an unpleasant stimulus can invoke pupillary constriction [4]. Contrarily, the pleasant stimulus will make the pupil expand. Therefore, the changing of pupil is a sign of the activity of central nervous system. Data shows that the patients of depression with the features of gazing the positive emotional information (such as happy expression) less and gazing negative emotional information (such as sad expression) frequently [5]. Accordingly, we can know people's emotional state by tracking the attention of a person's bias attention, the rate of pupil diameter, the saccadic slope and other visual information.

We will use an infrared eye tracking device Tobii T120 to obtain the visual information and the pupil size changes related to cognition, and then we use support vector machine (SVM) and the particle swarm algorithm to extract and classify the collected features to achieve the purpose of diagnosis of depression. The parameters in the SVM have great influence on the accuracy of data recognition, and setting the reasonable parameters of the SVM will improve the recognition accuracy effectively.

Based on above description and discussion, the main purpose of this paper is to optimize the parameters of the SVM by using an improved particle swarm optimization algorithm, so as to improve the recognition accuracy of the SVM. First, a feedback mutation particle swarm optimization algorithm is proposed in this paper. Then, we use the proposed particle swarm optimization algorithm to optimize the parameters based on radial basis function function (RBF) of the SVM. Last, we use the optimized SVM to classify the data of depression patients, and the result proves our proposed method is effective.

2 PSO algorithms

Particle swarm optimization (PSO) algorithm is a swarm intelligence algorithm which is proposed by Eberhart and Kennedy [6]. It is a kind of optimization algorithm based on iteration. At the beginning of the algorithm, the speed and position of the particles are required to be randomly initialized. Then the particles try to find the optimal solution by iterating in the solution space. In PSO, each particle is flying at a certain speed in D dimension of the search space. The position of the ith particle is expressed as $x_i=(x_{i1},x_{i2},\ldots,x_{iD})$, where $x_{id}\in[popmin,popmax]$, d is the dimension of the search space and $d\in[1,2,\ldots,D]$. pop_{min} and pop_{max} are the minimum and maximum value of the search space. The velocity of the ith particle is expressed as: $v_i=(v_{i1},v_{i2},\ldots,v_{id})$, where $v_{id}\in[v_{min},v_{max}]$, $d\in[1,2,\ldots,D]$, and v_{min} and v_{max} are the minimum and maximum value of the velocity of the particle.

The position x_i and velocity v_i of the particles are updated by the following equations:

$$v_{ij}(t+1)=v_{ij}(t)+c_1r_1[pbest_{ij}(t)-x_{ij}(t)]+c_2r_2[gbest_j(t)-x_{ij}(t)] \quad (1)$$

$$x_{ij}(t+1) = x_{ij}(t) + v_{ij}(t+1) \tag{2}$$

where i denotes the ith particle, j denotes the jth dimension of particles, t denotes iteration, v_{ij} denotes jth dimension velocity of ith, x_{ij} denotes jth dimension position of ith, $pbest_{ij}$ is the jth dimension of the best previous position of the particle itself and $gbest_j$ denotes the jth dimension of the best previous position of all particles in the swarm. c_1 and c_2 are the acceleration parameters, both r_1 and r_2 are randomly numbers and they change in [0,1].

2.1 Mutation PSO algorithm

The PSO algorithm need set only few parameters, and it convergences fast, but it falls into local optimum easily. In order to avoid the shortcomings, many new methods are proposed. Such as Natsuki and Higashi promoted a Gaussian mutation PSO algorithm [7]. The Gaussian mutation of the basic idea is that the position of the particle will be changed when the particle i fall into the local optimum. The change method for the position of particle is adding a distribution of Gaussian values to original position. Through the Gaussian mutation, particles jump in the vicinity of the original position, to a certain extent, improving the particles' ability to jump out of the local optimal position. In addition, there are many variants of Gaussian mutation, such as Zhan D et al. introduced Gaussian cloud learning strategy [8].

The probability density function image of Gaussian distribution is characteristic with both ends are narrow, high in the middle so that the mutation particles will not have large change when the Gaussian mutation is used. For this reason, Wang H et al. proposed a Cauchy mutation algorithm [9]. The Cauchy mutation also disturbs the position of particles, but the difference is the method of disturbance which adds a value of Cauchy distribution to the original position.

The probability density function of the Cauchy distribution is a relatively smooth strip, which has the characteristics of both ends of large and middle small. This leads to return a larger probability of Cauchy distribution, which makes the position of particles have a larger change on the original basis. Therefore, the Cauchy mutation has a strong ability of disturbance, which is more suitable for long-distance jump. In order to reduce the variation of long range hopping and to better control the jump distance of the particles, Zhang L et al. added a scaling factor on the basis of Cauchy PSO mutation [10].

Dirk Brockmann found Levy flight pattern [11]. From the distribution point of view, Levy flight is most of the value changes in a small range, but it also produces a long distance occasionally. The small probability time may bring great influence. Based on this feature, Hakl H and Uğuz H applied Levy flight pattern to particle swarm optimization [12]. Because most cases of the Levy flight pattern moves in a small range, and produce a long-distance movement occasionally. On the one hand, it is advantageous for the fine search in a small range; on the other hand, it can generate long range hopping for the particles which trapped in the local optimum. However, such a large range of jump phenomenon is accidental, and the stability of the algorithm needs to be improved.

The Gaussian mutation changes the range of particle position smaller, and the Cauchy mutation have great change to the particles position, Levy mutation can also bring a large range to change the location of the particle but with chance. Wang H et al. used an adaptive method combines the three mutation algorithm to balance the influence of this three mutation algorithm on the movement of the particle position [13]. Although this method combined with three mutation algorithm, in the process of, it also changed the particle population optimal location. As we all known, the particles move toward the optimal position of the population in the process of iteration, the convergence speed of the algorithm is decreased because of the change of the population optimal position. Similarly, Nishio T et al. used the individual best position on the mutation strategy [14].

Andrews, P. S. adapted a random mutation algorithm [15]. The mutation idea was that when the population of the global optimal particle continuous N generation is not improved, then use the particle mutation operation to generate a new "social part" learning sample. However, the position of the particles in the mutation operation is not in accordance with a certain distribution, but randomly selected location in the search area. Through the mutation operation, particles distribution is more extensive and location diversity has been promoted. To some extent, avoiding the particles fall into local optimal solution. However, it's so random on the particles position mutation, the speed of particles convergence slow down.

Beyond all that, there are also many other improvement methods. Such as Dong w et al. [16] used adaptive mutation strategy, Ngo T T et al. [17] proposed extraordinary motion for particles in the PSO all got better results.

2.2 Feedback mutation PSO algorithm (FBPSO)

All the above mutation algorithms, to a certain extent, improve the ability of particles jump out of the local optimum solution [7, 9, 12, 13, 15]. However, these algorithms don't consider the information of the particles in the population when the particles are mutating. Therefore, this paper presents a feedback mutation particle swarm optimization algorithm. We use the fitness's feedback of particles to balance the global exploration and local search.

The fitness value particles can reflect the current state of particles, and fitness is also a kind of judgment whether particle swarm convergence property. Therefore, this paper uses particles' average fitness value as the condition to decide when particles need to mutate. The mutation thought is that if the current particle's fitness less than the average fitness, indicating where the particle position is better, but the better position may be a local optimization. So select a portion of the particles from these particles to mutate.

The formula of the particle's average fitness is defined as follows:

$$fitness(\mathrm{i}) \le fitness\,\mathrm{avg} \tag{3}$$

$$fitnessavg = \frac{\sum_{i=1}^{s} fitness(\mathrm{i})}{s} \tag{4}$$

where s denotes population size, $fitness(i)$ denotes fitness of ith particle, $fitnessavg$ denotes average fitness of all particles in swarm.

Then, we select some of particles mutate, and the selection method is as follows:

$$rand() \le P_u \tag{5}$$

$$P_u = 1 - t/T \tag{6}$$

where t denotes current iteration, T denotes total number of iterations, $rand()$ is a random value uniformly distributed in [0, 1].

The condition (3) is used to judge the convergence of the population, The condition (5) is used to choice which particle to make mutate. If the two conditions are met,we will mutate the particle. The mutation formula is defined as follows:

$$x_i^{'} = x_i + \beta \cdot x_i \tag{7}$$

$$\beta \sim N(0, \sigma^2) \tag{8}$$

$$\sigma = \sqrt{\left|\frac{fitness(\mathrm{i}) - fitnesszbest}{fitnessavg - fitnesszbest}\right| + 0.1} \tag{9}$$

where $x_i^{'}$ denotes the after mutation position of ith particle. x_i is position of current particle. β obeys Gaussian distribution the expectations of 0, the standard deviation is σ. $fitness(i)$ denotes fitness of ith particle. $fitnesszbest$ denotes the best fitness in the whole swarm. If the $fitness(i)$ become larger, we can find σ is larger in formula (9), it means the larger standard deviation of Gauss distribution. At this time, the probability density image of Gauss distribution is relatively flat, the distribution is more dispersed, and the possibility of obtaining greater value is increased. The position of the particle will change in a large range, which can complete the long distance movement of the particle position to improve the ability to jump out of the local optimal solution. On the contrary, if the $fitness(i)$ become smaller, we can find σ is smaller in formula (9), at this time, Gauss distribution probability density image is taller, distribution is relatively concentrated, the possibility of getting a larger value is declined. The position of the particle will change in a small range, and the particles can complete fine search near the optimal solution.

The mutation method that we propose controls the global exploration and local search ability by using the fitness feedback of current particle.

3 Depression Recognition Methods Based on PSO and SVM

3.1 Methods

In this study, 48 subjects, including 24 patients with depression and 24 healthy subjects whose gender, age and education level are similar, participate in the study. The experiment materials are obtained from the international standard expressions library, the NimStim set of facial expressions [18]. We selected 36 people's (18 males, 18 females) three types (happy, sad and neutral) expressions. During the experiment, we use infrared vision eye tracking systems (Tobii T120) to record eye related information. We identified a total of five features, including bias positive attention,bias negative attention, the rate of positive pupil diameter, the rate of negative pupil diameter and the positive saccadic slope. The total number of subjects is 1728, including 864 depressed and 864 healthy subjects.

We used SVM classifiers based on the RBF to classify depressed and healthy subjects, with the SVM penalty coefficient C and parameter g. For any two models, different (C, g) values will produce a different recognition promotion. During parameter selection and optimization, we chose the highest recognition promotion of (C, g) as the optimal parameter values.

3.2 Feature Selection

There are many methods for selecting features, including entropy, probability distribution, and reparability criterion based on statistical tests. In this paper, we used statistical methods to judge the classification indicators. In the features of the normal group and the patient group, if the feature is significant difference, it can be used as the feature of classification. Before selecting feature, we also cleared exception data, for example, we cleared beyond three times the standard deviation range data. What's more, in order to unify the dimension of data, we normalized the data.

Generally, training sets should account for 70%-85% of the total sample, and we use 75%. We divided the data into training set and test set. The training set is 1296, including 648 depressed and healthy subjects. The test set is 432, including 216 depressed and healthy subjects.

The experiment platform is lenovo G470 PC, 64-bit Windows 7 operating system, the CPU is Intel core i32350M, and it's frequency is 2.30 GHz, memory is 4GB. In experiment, we use LIBSVM toolkit to test the ability of the six kinds of PSO algorithm in optimizing the parameters of the SVM base on MATLAB R2010b.

3.3 Parameter Optimization based on Mutation Method

In order to analyze the effect of FPSO, in this paper, we use different mutation algorithm to compare with FBPSO. In the experiment , we compare PSO (particle swarm optimization)[7], GPSO(Gaussian mutation particle swarm optimization)[9], HPSO(Cauchy mutation particle swarm optimization)[12], LFPSO(Levy flight particle swarm optimization)[13], RPSO(Rand mutation particle swarm optimization)[15] and FBPSO(Feedback mutation particle swarm optimization) total six algorithms to optimize the parameters of SVM. The six mutation algorithms are shown in Table 1:

Table 1 different mutation algorithms

Algorithm	Mutation
PSO	—
GPSO	$x_i^{'} = x_i(1 + Gaussian(\sigma))$
HPSO	$x_i^{'} = x_i + Cauchy(\sigma)$
LFPSO	$x_i^{'} = x_i + S$ $S = \alpha + Levy(\beta)$
RPSO	$x_i^{'} = x_i + (\mathrm{pop}_{max} - \mathrm{pop}_{min}) \cdot r$
FBPSO	$x_i^{'} = x_i + Gaussian(\sigma) \cdot x_i$ $\sigma = \sqrt{\left\lvert \frac{fitness(\mathrm{i}) - fitnesszbest}{fitnessavg - fitnesszbest} \right\rvert + 0.1}$

In this paper, the FBPSO algorithm process is as follows:
Step 1: Initialize the position and velocity of the particles, set the size of the population and the number of iterations.
Step 2: Train the SVM model based on the parameters C and g, compute the classification accuracy.
Step 3: Compute the fitness values of all the particles.
Step 4: Update the personal best and global best values.
Step 5: Update the particle's speed and position.
Step 6: Calculate $fitness(i)$ and $fitnessavg$, if $fitness(i) \le fitnessavg$ then continue, else goes to Step9.
Step 7: If $rand() \le P_u$ then continue, else goes to Step9.
Step 8: Calculate σ according to the formula (9) and make the mutation according to the formula (7).
Step 9: If the maximum number of iterations is reached then continue, else jump to step 2.
Step 10: Output the parameters C and g.

4 Comparisons of the Results from Different Mutation Algorithms

The experiment results of different mutation algorithms are shown in Table 2.

4.1 Comparison of the classification accuracy

From the Table 1, we can see that the classification accuracy of the original PSO classifier without mutation is lowest, it's 85.65%. After adding the mutation strategy, the recognition accuracy of all the classifiers is improved, which shows that the mutation strategy improve the ability of jumping the local optimal solution. In all the mutation algorithms, the classification accuracy of RPSO is lowest. The main reason is that the RPSO adapt random mutation strategy, which leads to the position of particles disperse after the mutation, and affects the final classification accuracy. In this paper, the classification accuracy of FBPSO algorithm is the highest, and it is proved that the improved algorithm is beneficial to balance the global exploration and local search.

4.2 Comparison of the programs running time

By comparing the running time of PSO and other mutation algorithms, the without mutation PSO algorithm and other mutation algorithms except RPSO of the runing time is almost the same. The running time of RPSO algorithm is the longest, which shows that the convergence time of the algorithm is affected by the random mutation strategy. The running time of GPSO algorithm is the shortest, which shows that the *Gaussian* mutation, to a certain extent, accelerate the convergence of algorithm. In this paper, the running time of improved FBPSO is the second. It shows that using fitness feedback of to adjust the position of the particles can improve the convergence of the algorithm.

Table 2 Results of different algorithms

Algorithm	C	g	Accuracy (%)	Running time (s)
PSO	2.2141	3.6930	85.65	559.66
GPSO	2.3327	3.2975	86.34	541.52
HPSO	2.3526	3.2948	86.13	585.77
LFPSO	2.6811	3.2562	86.36	578.53
RPSO	2.3448	3.3352	86.11	892.71
FBPSO	**1.9884**	**3.5110**	**88.56**	**552.07**

5 Discuss and Conclusion

In this paper, the feedback mutation particle swarm optimization (FBPSO) improves the classification accuracy after it's applied to the SVM classification model. Furthermore, the running time of the algorithm needs less time than other algorithms except the *Gaussian* mutation (including *Cauchy* mutation, *Levy* mutation and *Rand* mutation algorithms). On the one hand, this paper improves the PSO algorithm through the mutation of the position of particles. In addition to the mutation PSO algorithm, the inertia weight, the accelerating factor and other parameters can also influence the updating formula of particles directly. If these parameters are selected properly, they will also have a great influence on convergence accuracy and convergence time of particles [19]. Therefore, the particle velocity updating formula can also be improved [20]. On the other hand, this paper classifies the data of depression recognition based on the SVM classification model, if we use the other classification models that are popular in the last few years, such as deep learning, may improve the classification accuracy further. For the above problems still need further study.

Acknowledgment

This work is supported by the National Basic Research Programme of China (2014CB744600), the National Natural Science Foundation of China (61602017), the Beijing Natural Science Foundation (4164080), the Beijing Outstanding Talent Training Foundation (2014000020124G039), the National Natural Science Foundation of China (61420106005), the International Science & Technology Cooperation Program of China (2013DFA32180), the Special fund of Beijing Municipal Science and Technology Commission (Z151100003915117), Beijing Municipal Administration of Hospitals Clinical Medicine Development of Special Funding Support (ZYLX201607 and ZYLX201403), and Beijing Municipal Administration of Hospitals' Ascent Plan (DFL20151801).

References

1. Guilbert J J. The world health report 2002 - reducing risks, promoting healthy life.[J]. Education for Health, 2003, 16(16):230-230.
2. Segal, Daniel L. Diagnostic and Statistical Manual of Mental Disorders (DSM-IV-TR)[M]// Diagnostic and statistical manual of mental disorders :. American Psychiatric Association, 2013:4189-4189.
3. Christmann C, Koeppe C, Braus D F, et al. A simultaneous EEG–fMRI study of painful electric stimulation[J]. Neuroimage, 2007, 34(4):1428-37.
4. Hess E H. ATTITUDE AND PUPIL SIZE.[J]. Scientific American, 1965, 212(4):46-54.

DOI: 10.1051/bioconf/20170801037

5. Duque A, Vázquez C. Double attention bias for positive and negative emotional faces in clinical depression: Evidence from an eye-tracking study[J]. Journal of Behavior Therapy & Experimental Psychiatry, 2015, 46:107-14.
6. Y.H. Shi, R.C. Eberhart, Parameter selection in particle swarm optimization, in: Annual Conference on Evolutionary Programming, San Diego, 1998
7. Higashi N, Iba H. Particle swarm optimization with mutation[C]//Swarm Intelligence Symposium, 2003. SIS'03. Proceedings of the 2003 IEEE. IEEE, 2003: 72-79.
8. Zhan D, Lu H, Hao W, et al. Improving particle swarm optimization: Using neighbor heuristic and Gaussian cloud learning[J]. Intelligent Data Analysis, 2016, 20(1): 167-182.
9. Wang, H., Li, C., Liu, Y., & Zeng, S. (2007). A Hybrid Particle Swarm Algorithm with Cauchy Mutation. Swarm Intelligence Symposium, 2007. Sis (pp.356-360).
10. Zhang L, Tang Y, Hua C, et al. A new particle swarm optimization algorithm with adaptive inertia weight based on Bayesian techniques[J]. Applied Soft Computing, 2015, 28: 138-149.
11. Brockmann, D., & Sokolov, I. M. (2002). Lévy flights in external force fields: from models to equations. Chemical Physics, 284(1), 409-421.
12. Hakl H, Uğuz H. A novel particle swarm optimization algorithm with Levy flight[J]. Applied Soft Computing, 2014, 23(5):333-345.
13. Wang H, Wang W, Wu Z. Particle swarm optimization with adaptive mutation for multimodal optimization[J]. Applied Mathematics & Computation, 2013, 221(9):296-305.
14. Nishio T, Kushida J, Hara A, et al. Adaptive particle swarm optimization with multi-dimensional mutation[C]//2015 IEEE 8th International Workshop on Computational Intelligence and Applications (IWCIA). IEEE, 2015: 131-136.
15. Andrews, P. S. (2006, July). An investigation into mutation operators for particle swarm optimization. In 2006 IEEE International Conference on Evolutionary Computation (pp. 1044-1051). IEEE.
16. Dong W, Kang L, Zhang W. Opposition-based particle swarm optimization with adaptive mutation strategy[J]. Soft Computing, 2016:1-10.
17. Ngo T T, Sadollah A, Kim J H. A cooperative particle swarm optimizer with stochastic movements for computationally expensive numerical optimization problems[J]. Journal of Computational Science, 2016, 13:68-82.
18. Nim Tottenham, et al., The NimStim set of facial expressions: Judgments from untrained research participants. Psychiatry Research 168 (2009) 242-249.
19. Lim W H, Isa N A M. Two-layer particle swarm optimization with intelligent division of labor[J]. Engineering Applications of Artificial Intelligence, 2013, 26(10): 2327-2348.
20. Mandal S, Ghoshal S P, Kar R, et al. Design of optimal linear phase FIR high pass filter using craziness based particle swarm optimization technique[J]. Journal of King Saud University-Computer and Information Sciences, 2012, 24(1): 83-92.

BIO Web of Conferences **8**, 01038 (2017) DOI: 10.1051/bioconf/20170801038
ICMSB2016

Clinical research of effects of retaining the uterine blood supply hysterectomy on ovarian function

Yufei Cai [1] and Hongxia Sun [2,a]

[1]*Obstetrics and Gynecology Department,Affiliated Hospital of Beihua University*
[2] *Pharmacological Department, Pharmacy, Beihua University jilin jilin 132011, China*

Abstract. Objective To evaluate the effect of hysterectomy for reserving the uterine blood supply on ovarian endocrine function and on symptoms of menopausal transition. Methods Uterine benign lesions should be line the uterus times total resection in 100 patients were randomly divided into hysterectomy group of retaining uterus vascular supply group(research group,n=50) and traditional total hysterectomy group (the control group, n=50), comparing two groups in operation time, intraoperative bleeding ,postoperative fever and residual polyp, blood tests were taken to check the serum sex hormone levels change and clinical observation for perimenopausal symptoms before and postoperative three months, six months, one year and two years at the same time respectively. Results There was no significant difference between two groups in operation time, intraoperative blood loss, postoperative fever and residual polyp ($P>0.05$). There was no significant difference among research group before and after operation in serum sex hormones($P>0.05$),the symptoms of the menopausal transition hardly appear; postoperative FSH, LH increased significantly in control group ($P<0.05$),E2 decrease ($P<0.05$), perimenopausal symptoms appeared more often. Conclusion The effect of uterus hysterectomy for retaining vascular supply on ovarian endocrine function is less than the traditional total hysterectomy, this operation has a certain importance to preserve ovarian function and delay the occurrence of premature ovarian aging.

1 Instruction

The endocrine function of ovary can be affected by hysterectomy. In traditional hysterectomy, cutting off the ovarian branch of uterine blood vessels could reduce the ovarian blood supply by fifty to seventy percent, leading to ovarian follicle degeneration to reduce hormone production and ovarian function declined four to five years after hysterectomy ,which increased the occurrence of complications such as peri-menopausal symptoms, osteoporosis, high blood lipids and coronary heart disease and so on. Subtotal hysterectomy for maintaining uterine vessels (STHMUV) is a kind of surgery that does not affect ovarian hormone secretion, which can ensure the blood supply to the ovaries after removal of uterus, STHMUV can delay the natural aging of the ovary. So that the ovary is still able to secrete estradiol and other hormones, which delayed the emergence of ovarian physiological decline [1].

[a] Corresponding authorsunhongxiajl@126.com

2 Materials and methods

2.1 General information

One hundreds cases of patients which had sign of subtotal resection of uterus (SROU), aging between thirty to forty-five years old, was selected from the Department of gynecology in the affiliated hospital of Beihua University during the period from December 2014 to October 2010. and were randomly divided into STHMUV group (research group,n=50,indicating R group) and traditional SROU group (control group, n=50,indicating C group). There was no significant difference in age, types of diseases in the two groups (P>0.05).Cervix was examined and serum lipid levels were detected before operation. All women were admitted to hospital within three to five days after the last menstrual period. There was no significant difference in serum TC, HDI, TG and LDL in the two groups (P>0.05).

2.2 Choice of operation mode

The control group with reference to Li Guangyi, editor in chief of "Practical Gynecology laparoscopic surgery" for the traditional subtotal hysterectomy [10]. The study group adopted the reservation of the uterine vessels subtotal hysterectomy. The study group and the control group were the same, both of which were combined with the spinal epidural anesthesia. The oxytocin and saltwater were mixed to be injected into the uterus, its vasoconstrictive effects lead to the whole uterine tissue became pale due to ischemia, the remaining part of the liquid was dumped into the contralateral body layer during the surgery; cross-cutting and episiotomy surgery could be taken in lower abdomen, each layer was cut to be checked in order by open view according to its anatomical position, the uterus was maintained in a raised position during the operation , both sides of the uterine appendages and uterine blood vessels was clamped with tissue forceps. The position for uterine resection was about 1 cm interior to uterine horn, toward the direction of uterine isthmus, resection of uterine body by the shape of triangle. The lower position was close to the inside of uterus cervix, with about 2 cm thickness of both side of uterine body.The muscle and endometrial layers can be clearly seen at this point, if there was,such as the obvious hysteromyoma, which can be removed, with hemostatic forceps or electric hemostasis. The uterine wound was sutured by U-shaped, and check whether there is staxis or not meanwhile.

2.3 Observation index

Operation time, bleeding during operation, postoperative fever and residual polyps were compared between the two groups. Serum lipid levels were taken to detected three months, six months, one year and two years after operation and compare the serum lipid levels, hemodynamic of uterine artery and survey of sexual satisfaction in different time periods.

2.4 Statistical methods

SPSS 17.0 statistical software was used for statistical analysis, all data were expressed as mean±the standard deviation ($\bar{x}$ ±s), single factor analysis of variance was used for comparison among groups, LSD-t test was used for comparison between two groups, the difference was statistically significant when P<0.05.

 DOI: 10.1051/bioconf/20170801038

3 Results

3.1. Operation effects

Operation time, intraoperative blood loss, postoperative body temperature and postoperative residual segment polyps were compared between two groups of, the results showed that there was no significant differences (P>0.05), see Table 1.

Table 1. Comparison of two groups of patients with operation($\bar{x} \pm s$)

Group	*n*	Operation T/min	Blood loss/ml	BT/°C	Residual polyps /%
Research(R)	50	53.2±7.6	90±12.6	37±0.5	0/0
Control(C)	50	55.1±8.3	95±16.7	37.5±0.2	28/56

3.2. Sexual hormones level comparision in serum

Sexual hormone in serum before and after operation is shown in Tab 2

Table 2. The comparison results of serum sexual hormone levels before and after two groups operation($\bar{x} \pm s$)

Group	n	FSH/U·L^{-1}				
		before	3m after operation	6m after operation	1y after operation	2y after operation
Research	50	17.65±1.02	18.10±1.31	18.67±1.74*	19.45±1.68*	20.17±1.25*
Control	50	16.32±1.41	16.05±1.06	26.35±1.89$^{\triangle}$	53.36±1.74$^{\triangle}$	89.01±1.96$^{\triangle}$
Group	n	LH/U·L^{-1}				
		before	3m after operation	6m after operation	1y after operation	2y after operation
Research	50	17.31±1.57	17.36±1.91	18.45±1.73$^{\triangle}$	20.15±1.43$^{\triangle}$*	23.18±1.73$^{\triangle}$*
Control	50	18.47±1.94	18.32±1.62	22.03±1.87$^{\triangle}$	23.19±1.78$^{\triangle}$	27.15±1.69$^{\triangle}$
Group	n	E2/(pg·mL^{-1})				
		before	3m after operation	6mafter operation	1y after operation	2y after operation
Research	50	346.12±16.42	337.21±13.45*	321.04±10.36*	312.15±12.47*$^{\triangle}$	298.34±12.08*$^{\triangle}$
Control	50	343.24±17.64	320.07±12.37$^{\triangle}$	297.34±15.49$^{\triangle}$	256.48±11.67$^{\triangle}$	203.17±12.09$^{\triangle}$

□P<0.05 vs group compared in the previous observation time; □P<0.05 vs Control Compared in the same observation time.

3.3. Comparison of incidence of peri-menopausal symptoms in the two groups

Follow-up team followed up 50 cases, control group 50 cases; Two groups of postoperative incidence of perimenopausal symptoms is shown in table 3.

Table 3. The comparison between incidence of peri-menopausal symptoms in the two groups

Group	n	Hot and sweating(%)	palpitation (%)	Insomia and weak(%)	Loss of libido(%)	Genital atrophy(%)
Research	50	2(4)	2(4)	5(10)	4(8)	1(2)
Control	50	15(30)	12(24)	26(52)	23(46)	6(12)
P		<0.05	<0.05	<0.01	<0.01	>0.05

4 Discussion

The uterus is not only the target organ of ovarian hormone, but also has a close connection with the ovary in anatomical and endocrine function, both accept double blood supply. Uterine blood supply comes from the uterus branch of the uterine and ovarian arteries; Ovarian blood supply comes from the ovarian branch of the ovary and uterus artery. The literature reported that blood supply of the ovarian branch of the uterine artery accounted for 50%~70% of ovarian blood supply[2]. Hysterectomy cut off the bilateral ovarian blood supply for preserved uterine arteries and veins, resulting in follicle degeneration, reduced hormone production,and the ovarian function declined after 4~5 years of operation[3].The sub-total hysterectomy to maintain uterus blood vessels retained its normal blood supply of residual ovary, so ovarian secretion is not affected by the operation which can delay premature ovarian functiona aging, at the same time, the remaining uterus can secrete hormones, cytokines and maintain the endocrine balance between uterine and ovarian[4], maintain the normal function of the ovaries and play a role.

Siddle N[5] found that ovarian failure occured four years earlier than that of natural menopause after hysterectomy, ovarian failure rate is 34% with two years, increased to 54% within four years, the incidence of severe menopausal symptoms were significantly higher than the normal population. The experiment results showed that the serum follicle stimulating hormone (FSH) levels increased ,but estradiol (E2) levels droped in 50 patients for total hysterectomy compared with the control group during preoperative and postoperative periods, and perimenopausal symptoms were obvious; The amplitude change in serum sex hormone levels was less than the control group for 50 cases of subtotal hysterectomy maintaining uterine blood vessels during postoperative and preoperative times, and the incidence of perimenopausal symptoms were significantly lower than the control group.Research results showed that the ovarian branch of retaining uterine artery and vein affect ovarian blood supply less, reduced the effects of operation on ovarian hormone secretion; At the same time, there were no significant differences($P>0.05$)in operation time, intraoperative blood loss, postoperative fever and postoperative residual polyp compared with the control group,which illustrated that uterus Hysterectomy of preservation uterine vessels did not increase the operation time and blood loss, also did not increase the postoperative infection and postoperative complications.

A number of studies have confirmed that estrogen has a protective effect on cardiovascular system, the low level of serum sex hormone levels and atherosclerosis is associated with high cardiovascular events [6]. Research has shown that decreased estrogen levels cause atherosclerosis, elevated blood lipids can enhance incidence of coronary heart disease by tow-folds, which constitued one of the main risk factors of the coronary heart disease(CHD). E2 decrease of normal postmenopausal women cause blood lipid disorders, dyslipidemia is the important cause of the occurrence of coronary heart disease in postmenopausal women; Similarly, E2 levels drop can also lead to blood lipid disorders in premenopausal women after hysterectomy[6].This study result showed that there was no obvious changes in hormone levels in serum for retension of uterus blood vessels of sub-total hysterectomy group, FSH increased significantly in traditional sub-total hysterectomy.Study results suggested the traditional total hysterectomy retained the ovaries, but reduced ovarian blood supply caused by cutting off uterine arteriovenous branches, affected ovarian endocrine function, droped E2 levels, which may affect blood lipid level, its effect on blood lipid specific remained to be confirmed by next step research); Reserved with uterine arteriovenous ovarian surgery can reduce the influence on ovarian blood supply, reduced the influence on ovarian hormone secretion, there were no significant decline in serum E2 level in Subtotal hysterectomy for maintaining uterine vessels (STHMUV) compared with the preoperative group, meanwhile reduced incidence of perimenopausal symptoms. people have higher requirement of life quality with the progress of the society. we should choose hysterectomy that has smaller influence on ovarian function for the non-menopause women with gynecological benign diseases , retain ovarian branch of uterine arteriovenous supply as far as possible,which can reduce the effect on ovarian hormone secretion, slowed the emergence of ovarian premature aging, and at the same time also can reduce the incidence of postoperative coronary heart disease; the operation is simple,safe,easy to master,less postoperative complications, all these provide a good surgical option [7-10].

References

1. Xu Xiaofeng, Zhou Huandi, Li Miao. J.Nanchang Univer, **50**,18 (2010)
2. Wu Fengying, Deng Zhaohong, Tian song. Xinjiang med, **39**,86 (2009)
3. Cicinelli E,Einer-Jensen N, Galantino P.. Ann NY Acad Sci,**1034**,19 (2004)
4. Chen Lu, Li Guangyi, Han Yubin. Chinese J of mini-trauma surgery, **9**, 609 (2009)
5. Siddle N,Sarrel P,Whitehead H..Fertil Steril,,**47**,94 (1987)
6. Mohamad M J,Karayyem M,Mohammad M A,et al.,Am J Phys Anthropol,,**132**,151(2007)
7. Li Kaijun, Wu CongCong ,Xiang Yijia,et al. prevention and treatment of cardiovascular and cerebrovascular disease, **14** ,222 (2014)
8. Wang Jingjing, He Ruhua, Wu Ling. Ningxia med J,**36**,13 (2014)
9. Sheng Li, Ye Ping. Chin J of prac Int Med, **34**,15 (2014)
10. Gimbel H..Acta Obstet Gynecol Scand,**86**, 133 (2007)

BIO Web of Conferences **8**, 01039 (2017) DOI: 10.1051/bioconf/20170801039
ICMSB2016

Research of ddi based on multi-label conditional random field

Yangzhi Yu [1], Hongtao Deng [1,a] and Xun Zhu [1]

[1]*Mathematics and Computer Science Department of JiangHan University, China*

Abstract. The detection of drug name and drug-drug interaction(DDI) is considered as a sequence labeling task in this paper. We present the multi-label CRF method to complete it. Compared to the traditional method, our method can not only identify drug names, but also can identify drug-drug interaction. According to the characteristics of medical texts, this paper extracts the good features of the description of DDI. The proposed method has good performance in DDIExtraction 2013 evaluation corpus.

1 Introduction

The detection of DDI is an important research task in ensuring patient safety since these DDI can be so dangerous and increase health care costs. The phenomenon of drug combination is becoming more and more common. It can enhance the efficacy and improve the therapeutic effect. However, it also may result in increased frequency of adverse drug reactions and increased severity of adverse reactions. The DDI has become a vital factor affecting the drugs rational use. More and more researchers pay their attention to the DDI extraction research.

To address the extraction of DDI from biomedical texts, two main steps were proposed: the detection and classification of pharmacological substances and the detection and classification of drug–drug interactions.

There is a large number of entity relationship extraction researches, such as protein extraction, RNA molecules relationship extraction, DDI extraction and the drug adverse effects event extraction and so on. Hasegawa T[1] used the method based on co-occurrence to extract entity relationships. They argue that all entities that co-occur in a text fragment should have a relationship. Deng B[2] used the method based on semantic pattern to identify entity relationships in Chinese. They extract the rules by hand, and then match with the actual situation to determine whether the two entities are related. Now the mainstream method is based on pattern matching[3], which requires less manual intervention.

2 Task description

DrugBank[4] is a constantly updated database of saving chemical informatics, bioinformatics resources and so on, which holds a large number of drugs and drug targets.

[a] Hongtao Deng: hongtaodeng@qq.com

2.1 Drug entity recognition

Name entity recognition estimates whether a character sequence represents a name entity, and determines its types, namely to find name entities and tagging name entity. Drug entity recognition refers to the identification of name entities in the text in the description of pharmacological substances. The task includes the determination of the boundary of the entity name and the determination of the entity type. The types of drug entities are classified into four types: Drug, Brand, Group, and No-Human. The following is a brief introduction of the four types of entities:

1) Drug type: This type refers to all have been approved for treatment, prevention and diagnosis of chemicals or drugs.

2) Brand Type: This type refers to original drug firstly developed by a pharmaceutical company.

3) Group Type: This type represents a class of drug having a certain function.

4) No-Human Type: This type represents the drug not approved for using as a clinical medicine, and has an effect on the body's organs.

2.2 Drug-drug interaction detection

The ultimate goal of the task is to extract DDI, which refers to the relationship between the two drug entities within a sentence. The types of DDI are divided into four types: Advice, Effect, Mechanism, and Int.

1) Advise: When two drug entities are suggested to use together in the text, the relationship between the two drugs is defined as Advise type.

2) Effect: Mechanisms of drug action is generally divided into pharmacodynamic mechanism and pharmacokinetic mechanism. Pharmacodynamics mechanism refers to the effect of a drug affected by another drug. When the article mentioned the existence of such effects of the two drugs, it is defined as Effect type.

3) Mechanism: Pharmacokinetics mechanism refers to that the absorption, diffusion, metabolism and excretion of drugs are affected. When there is pharmacokinetics mechanism effect between the two drugs, this relationship is called the Mechanism.

4) Int: When the text only mentioned the existence of a relationship between the two drugs, but did not make a specific description of the relationship, unable to determine the type of relationship, it is defined as Int.

3 CRF method

3.1 Introduction of CRF algorithm

CRF is a kind of algorithm based on probabilistic graph model, which is used to label and segment sequence data. Lafferty, J., McCallum and Pereira[5] proposed CRF model in 2001. CRF is a discriminative undirected probabilistic graphical model. It is often used for labeling or parsing sequential data, such as natural language text, biological sequences[6] and computer vision[7]. CRF acquires good performances in shallow parsing[8], named entity recognition[9] and gene finding, among other tasks, being an alternative to the related hidden Markov models. We use CRF to solve drug NER as sequence annotation problem.

Lafferty[5] define a CRF on observations X and random variables Y as follows: Let G=(V,E) be a graph such that $Y=(Y_V)$, $_V \in V$, so that Y is indexed by the vertex of G. Then (X, Y) is a conditional random field when the random variables Y_V, conditioned on X, obey the Markov property with respect to the graph:

$p(Y_V | X, Y_{W}, _{W} \neq {}_V) = p(Y_V | X, Y_W, _{W \sim V})$.

Where W~ V means that w and v are neighbours in G.

3.2 CRF drug entity recognition

We need to apply CRF to recognize drug entity first, then apply the multi-label CRF to extract DDI. CRF is often used in named entity recognition. And it has been successfully applied in the field of biomedical RNA, DNA, protein and other entities identifications. There are four types of drug entities that need to be identified in this task, which not only identify the boundary of the entity, but also identify the type of the entity.

3.2.1 Encoding mode

The BEISO label is used to encode the sentences in the paper. Label B represents the beginning word of drug entity, label E stands for the ending word of drug entity, label I represents the middle word of drug entity, S represents a single-word entity, label O stands for the word of non-entity. B, E, I, S, each label has four types respectively (for example B_drug, B_brand, B_group, B_drug_n), corresponding to the four types of drug entity (drug, brand, group, drug_n) .

3.2.2 Feature design

(1) POS feature: getting POS information by Stanford tool, which uses unigram, bigram, trigram feature.

(2) Ignore cases feature： using unigram, bigram, trigram feature.

(3) Word feature: 'A' replacement for uppercase letters, 'a' replacement lowercase letters, '0' to replacement for numbers;

(4) Words Brown cluster prefix： using unigram, bigram, trigram feature.

3.3 DDI based on multi-label CRF

3.3.1 Multi-label CRF algorithm

The traditional method is only applied for CRF sequence labeling， which can be successfully used in the NER task. However its drawback is that it can only identify the continuous entity. In order to avoid this shortcoming, the multi-label CRF is proposed, which adopt 6 classes labels to mark the sentence. They are "B_val", "I_val", "O", "E_val", "S_val" and "number" labels. Since the multi-label method has the ability to mark one symbol with more labels, which can view non-adjacent tokens as an entity. It has achieved good results in the recognition of discrete entities mission[10]. And then we made some changes to the algorithm so as to identify the entity relationship. One is to package multiple consecutive drug entities. Another is to improve the label coding. In this paper, we removed the original I label, which represents the internal word of the entity. Because of identifying various types, label B and E is divided into four types, corresponding to the four types of relationship.

Label coding algorithm. We adopt the multi-label to mark two entities which have relations with label B, E. If an entity has more than one label, it is required to add the label to the overlay algorithm. According to the characteristics of the task, we removed the I type labels. Multi-labels of entity relationship recognition contain five types of labels: "B_val", "E_val" and "O", "S_val", "number". The B type label represents the previous entity of DDI, and the E type label represents the latter entity of DDI. O type label represents a type of non drug substance. S type label represents drug entity of no DDI. Binary values are represented by the number and val.

Label decoding algorithm. The output of CRF method is just a label sequence, which we need to decode the relationship we need. Decoding process is divided into two steps. First the coordinate drug package is separated to several drugs, and then the drugs with interacted relation should be combined as a pairs.

3.3.2 Packing algorithm

The phenomenon is very common that different drug entity have interaction with the same entity in a sentence. For example,"Benzthiazide may interact with alcohol, bloodthinners, decongestantdrugs (allergy, cold, and sinus medicines), diabetic drugs, lithium, norepinephrine, NSAIDs like Aleve or Ibuprofen, and high blood pressure medications." In the sentence, Benzthiazide has interaction with other entities. If we match the two drugs with relation directly, the label value will be too large. The sparsity of large value label will lead to the reduction of identification effect. The multi-label CRF algorithm limits that the number of interaction entity pairs is less than six in one sentence, so we should pack parallel drug as a drug entity by some rules we specified. The parallel entities are generally connected by a high frequency words, such as 'comma', 'and', 'or', 'such as', and 'like' so on. We first pick up 30 words which had higher frequency before and after the drug from the training corpus. If the frequency word is between two drug entities, they will be packaged together and represented by the first drug. If an entity's type is determined in decoding process, all coordinate entities with it should have the same type of relationship.

3.3.3 Feature design

According to the characteristics of biomedical text, we selected a number of targeted features. There are POS feature, word feature ignored case, dependency feature and two entities' distance feature, as shown in Table1.

Table 1. Multi-label CRF training data example.

word	POS	Entity label	Dependcy word	Dependcy word location	Entity location	Output label
concurrent	JJ	N	therapy	2	0	O
therapy	NN	N	recommended	9	0	O
with	IN	N	ORENCIA	4	0	O
orencia	NN	brand	therapy	2	1	B_{advise}_1
and	CC	N	ORENCIA	4	0	O
antagonists	NNS	group	ORENCIA	4	2	E_{advise}_1
is	VBZ	N	recommended	9	0	O
not	RB	N	recommended	9	0	O
recommended	VBN	N	null	0	0	O
.	.	N	recommended	9	0	O

4 Experiment

DrugBank corpus is divided into training data and testing data. The training data contains 572 articles, a total of 5675 sentences. The testing data contains 158 articles, a total of 973 sentences. Statistical results of drug entity and DDI are shown in Table 2.

Table 2. Quantity of drug substance and drug relation in the corpus.

	Training	Testing
documents	572	158
sentences	5675	973
Drug	8197	1518
Group	3206	626
Brand	1423	347
Drug_n	103	21
Mechanism	1260	279
Effect	1548	301
Advice	819	215
Int	178	94

We analyze the corpus.Table3 scribes the quantity of each type drug. As shown, the majority are Drug and Group, and there are 82 entities of Drug_n type. The length of the entity token of Drug and Group type are less than 3. Almost all type entity's name are made up of single token.

Table 3. The number of each type of entity name.

Drug	Group	Brand	Drug_n
1583	1403	488	82

We mainly utility the Tokenizer, Part-Of-Speech Tagger[11], Dependency Parser[12], Constituency Parser[13] in the feature designing. The steps of multi-label CRF to classifying the drug and DDI are file parsing, pre-process, entity coding, entity packing, relational coding, decoding, and analyzing results.

Table4 is the recognition result. It is considered that the entity is correctly identified, only when the entity boundary and the entity type are identified correctly.

Table 4. The result of recognition.

	Accuracy	Recall	F value
Drug	0.857	0.888	0.872
Brand	0.836	0.811	0.823
Group	0.829	0.774	0.801
Drug_n	0.764	0.567	0.651
Total	0.840	0.823	0.831
Advise	0.483	0.421	0.449
Effect	0.468	0.444	0.456
Mechanism	0.394	0.386	0.390
Int	0.351	0.364	0.357
Total	0.461	0.411	0.439

5 Conclusion and future work

It is significant of NER and RE task from huge medical text. In this paper, we mainly research the extraction of the interaction between two drugs utilizing multi-label CRF model in medical text. The multi-label CRF model realized the extraction of non-adjacent named entity. We also extracted a variety of features to improve the model efficiency from the medical text. Our task may provide a reference for future work.

Acknowledgment

We thank all reviewers for the insightful comments. This work is supported by the Graduate Innovation Foundation, key disciplines "Management Science and Engineering" of Jianghan University, Hubei Province, China.

References

1. Hasegawa T, Sekine S, Grishman R. Discovering relations among named entities from large corpora[C]//Proceedings of the 42nd Annual Meeting on Association for Computational Linguistics. Association for Computational Linguistics, 2004: 415.
2. Deng B, Fan X, Yang L. Entity relation extraction method using semantic pattern[J]. Jisuanji Gongcheng/ Computer Engineering, 2007, 33(10): 212-214.
3. Tikk D, Thomas P, Palaga P, et al. A comprehensive benchmark of kernel methods to extract protein–protein interactions from literature[J]. PLoS Comput Biol, 2010, 6(7): e1000837.

4. Wishart D S, Knox C, Guo A C, et al. DrugBank: a comprehensive resource for in silico drug discovery and exploration[J]. Nucleic acids research, 2006, 34(suppl 1): D668-D672.
5. Lafferty, J., McCallum, A., Pereira, F. (2001). "Conditional random fields: Probabilistic models for segmenting and labeling sequence data". Proc. 18th International Conf. on Machine Learning. Morgan Kaufmann. pp. 282–289.
6. He, X.; Zemel, R.S.; Carreira-Perpiññán, M.A. (2004). "Multiscale conditional random fields for image labeling". IEEE Computer Society. CiteSeerX:10.1.1.3.7826
7. Sha, F., Pereira, F. (2003). "shallow parsing with conditional random fields"
8. Settles, B. (2004). "Biomedical named entity recognition using conditional random fields and rich feature sets". Proceedings of the International Joint Workshop on Natural Language Processing in Biomedicine and its Applications. pp. 104–107.
9. Tsuruoka Y, Tsujii J. Boosting precision and recall of dictionary-based protein name recognition[C]//Proceedings of the ACL 2003 workshop on Natural language processing in biomedicine-Volume 13. Association for Computational Linguistics, 2003: 41-48.
10. Lin Wutao. Research of medical information extraction based on multi label 2015[D]. CRF Wuhan: Wuhan University, 2015
11. Kristina Toutanova, Dan Klein, Christopher Manning, and Yoram Si nger. 2003. Feature-Rich Part-of-Speech Tagging with a Cyclic Dependency Network. In Proceedings of HLT-NAACL 2003, pp. 252-259.
12. Chen D, Manning C D. A Fast and Accurate Dependency Parser using Neural Networks[C]//EMNLP. 2014: 740-750.
13. Zhu M, Zhang Y, Chen W, et al. Fast and Accurate Shift-Reduce Constituent Parsing[C]//ACL (1). 2013: 434-443.

Analysis of clinical characteristics of non-allergic rhinitis of children

Feng Liang[1], Fanglei Han[1], Xiaodan Jiang[2] and Lin Li[3,a]

[1]*Department of Anesthesiology, China- Japan Union Hospital, Jilin University, 130033, China*
[2]*Otorhindary,Affiliated Hostipal og Qingdao University,Qingdao,266555,china*
[3]*Dept. Otolaryngolgy and head & Neck, China- Japan Union Hospital, Jilin University, 130033, China*

Abstract. Purpose: This study aims to discuss clinical manifestations of non-allergic rhinitis (non-allergic rhinitis, NAR) children patients aged 15 years old or younger and their inducement, comorbid diseases, family medical history and quality of life and to analyze clinical characteristics of children of different age groups so as to provide clinical evidence for the improvement of NAR children patients' comprehensive diagnosis and treatment. Method: Questionnaires were issued to 74 children patients diagnosed with NAR between June, 2014 and June, 2015 and all of them had the specialized examination of nasal cavity. The standardized 100-point visual analogue scale (VAS) was used as the symptoms survey scale to evaluate the severity of symptoms. Those patients surveyed were divided into two groups: Group A of preschool children under 6 years old and Group B of school-age children under 15 years old. Lateral comparison was done for severity of symptoms of single sample so as to reduce the subjective factors. After that, symptom characteristics of both groups were compared and SPSS 13.0 was also used to finish the statistical analysis. Results: The occurrence rate of sneezing, nasal mucus, nasal obstruction and nasal blockage of NAR children patients was 86.49%, 84.60%, 91.20% and 73.65% respectively. The most serious symptom of children of Group A was nasal obstruction and its occurrence rate was higher than that of Group B. The difference had statistical significance (x2=19.194, P<0.05). These two groups had no significant difference in terms of sneezing and nasal blockage (x2=0.474, x2=0.048, P>0.05). The most serious symptom of children of Group B was nasal mucus and its occurrence rate was higher than that of Group A. The difference had statistical significance (x2=16.92, P<0.05). Conclusion: Characteristics of NAR clinical symptoms of children under 6 years old (including 6 years old) and children above 6 years old were different. The acquisition of concomitant symptoms, inducements and other clinical data of children NAR can provide a reference for improving the comprehensive diagnosis and treatment level of children NAR.

1 Instruction

The prevalence rate of non-allergic rhinitis (NAR) of children gets higher year by year in the world. According to the clinical diagnostic standard, its allergen test is negative. During the attack, the nasal cavity will be in a high reaction condition, showing some clinical symptoms such as nasal blockage, sneezing, watery nasal mucus and nasal obstruction, which can strongly affect physical and psychological health of the children [2]. However, NAR children patients usually do not show atypical clinical manifestations and they rarely tell that to their parents. Besides, parents or children often overlook that, which will lead to the delayed treatment. This study adopts the way of questionnaire survey and the standardized visual analogue scale (VAS) to evaluate clinical features of NAR children

[a] Corresponding author: lilin01@jlu.edu.cn

DOI: 10.1051/bioconf/20170801040

patients and to analyze clinical characteristics of children at different ages. This will provide clinical basis for more accurate diagnosis.

2 Materials and methods

2.1 Clinical data

74 cases of NAR children patients were included in our study from June 2014 to June 2015. 50 male cases and 44 female cases were included. The study was approved by the institutional review board of China-Japan Union Hospital. All the participants provided written informed consent before the study.

Inclusion criteria: (1) Main symptoms include nasal blockage, sneezing, much watery nasal discharge and nasal obstruction. They have suffered the disease for four weeks. (2) The specialized examination of their nasal cavities show that they have pale nasal mucosa, oedematous and watery nasal discharge. (3) Their skin prick test is negative. (4) They are aged under 15 years old (including 15 years old) and the youngest child is 3 years old.

Exclusion criteria: (1) They have suffered from acute upper respiratory infection, infantile pneumonia and other infectious diseases for the nearest two weeks. (2) Some patients are suffering from congenital heart disease, chronic bronchitis, amygdalitis, autoimmunity disease and other diseases.

Groups divided according to the age sections: Children patients are divided into two groups: one is for preschool children under 6 years old and the other is for school-age children under 15 years old.

2.2 Survey methods

Questionnaire survey. Refer to researches of international children asthma and allergic diseases (ISAAC) and concrete conditions. Children patients or their parents shall fill in basic situations, symptoms and signs, concomitant diseases, inducement, family disease history and living quality evaluation. Besides, standardized VAS (0—100 points) method is used to evaluate the severity of symptoms.

2.3 Statistical method

The highest score of all symptoms in standardized VAS of single case is evaluated as positive. In order to reduce influence of subjective factors, SPSS 13.0 is used for analysis and the experimental results are shown as percentage. x^2 test is adopted to compare results of the two groups. $P<0.05$ means that the result is meaningful.

3 Results

3.1 Clinical symptoms

The occurrence rate of sneezing, nasal mucus, nasal obstruction and nasal blockage of NAR children patients is 86.49%, 84.60%, 91.20% and 73.65%respectively.

II. Clinical symptoms of preschool children and school-age children

The most serious symptom of children patients in preschool group is nasal obstruction and its occurrence rate is higher than that of Group B. The difference shows statistical significance (x2=19.194, $P<0.05$). Both groups have no significant difference in terms of nasal blockage and sneezing (x2=0.474, x2=0.048, $P>0.05$). The most serious symptom of children patients in school-age group is nasal mucus and its occurrence rate is higher than that of Group A. the difference has statistical significance ($x^2=16.92$, $P<0.05$).

Table 1. The relationship between AR clinical symptoms of children of ~ 6 years old and those of children of ~ 15 years old

Group Total	cases	Nasal obstruction	Nasal mucus	Sneezing	blockage
~6 years old	26	22	19	18	21
~15 years old	48	42	25	28	40
x2		19.194	16.92	0.474	0.048
P		<0.05	<0.05	>0.05	>0.05

Discussion

NAR can greatly damage living quality and daily activities of children [1]. Moreover, NAR of children is closely related to their asthma. Researches have proved that the occurrence rate of asthma in hyper-responsiveness rhinitis children patients is 2 to 7 times higher than that of common people [2]. However, great variation exists in NAR clinical symptoms of children and those symptoms are often ignored by their parents and children patients, which will bring certain difficulty to NAR diagnosis and treatment. In 2004, some scholars conducted related investigations to NAR of children aged 9~10 years old in Nanjing and the results showed that the occurrence rate of sneezing, nasal discharge, nasal obstruction and nasal blockage was 37.7%, 44.4%, 29.0%, 15.6% respectively [3]. This study investigated children under 15 years old (including 15 years old) and made statistics about their nasal symptoms, eye symptoms, coughing and chest symptoms. Statistical results showed that the occurrence rate of main clinical symptoms was high and the occurrence rate of sneezing, nasal mucus, nasal obstruction and nasal blockage was 89.19%, 93.24%, 91.89% and 73.65% respectively. In addition, researches of Kemp and other experts also showed that the common symptom of preschool children was nasal obstruction [4]. This survey also suggested that clinical symptoms of NAR children patients at preschool stage and school-age stage were also different. Except for sneezing and nasal blockage, the main symptom of preschool children patients is nasal obstruction while that of the school-age children patients is nasal mucus. Children with NAR shall be diagnosed and treated as early as possible so that their NAR development can be controlled and the occurrence of complications can be avoided or decreased.

Acknowledgement

All authors declared no conflict of interest in the research. The work presented in this report is the subject of a pending patent filed by the China-Japan Union Hospital of Jilin University (201510382781.X). This study was supported by the National Natural Science Foundation of China (811007020), the Health and Family Planning Foundation of Jilin Province (20152046), the Science and Technology Development Plan Foundation of Jilin Province (20160101070JC), and the Norman Bethune Program of Jilin University (2015305).

References

1. Demoly P, Bousquet PJ, Mesbah K, Bousquet J, Devillier P.Clin Exp Allergy. 43 (8) 881-8 (2013)
2. Burgess JA, Walters EH, Byrnes GB, Matheson MC, Jenkins MA, Wharton CL, Johns DP, Abramson MJ, Hopper JL, Dharmage SC. J Allergy Clin Immunol. 120(4) 863-869(2007)
3. Qian H, Zheng X, Zhang M, Weschler L, Sundell J. PLoS One. 18;11(5): e0155742(2016)
4. KempA, Bryan L. Med J 141 640-643 (1984)

BIO Web of Conferences **8**, 01041 (2017) DOI: 10.1051/bioconf/20170801041
ICMSB2016

Repair of Nasal Soft Tissue Defects with Paramedian Forehead Skin Flap with Base at the Supraorbital Artery Region

Feng Liang[1], Fanglei Han[1], Xiaodan Jiang[2] and Lin Li[3,a]

[1]*Department of Anesthesiology, China- Japan Union Hospital, Jilin University, 130033, China*
[2]*Otorhinolary,Affiliated Hostipal og Qingdao University,Qingdao,266555,china*
[3]*Dept. Otolaryngolgy and head & Neck, China- Japan Union Hospital, Jilin University, 130033, China*

Abstract. Purpose: This study is used to explore the method for repairing nasal soft tissue defects with paramedian forehead skin flap with base at the supraorbital artery region and its curative effect. Method: Retrospective case study. Clinical data of one case of a patient with defects of most nasal soft tissue in China–Japan Union Hospital of Jilin University were used. The forehead flap with base at the supraobital artery will be transferred to nasal soft tissue defects so as to repair the nose of the patient. Result: All forehead flaps with the blood supplied by supraobital artery of one patient case all survived. The blood supply of the flap is stable because quality, color and luster are all similar with those of the nose. No bleeding, flap hypertrophy or overstaffed pedicle appeared after the operation. They all healed up in Phase I. Conclusion: Paramedian forehead skin flap with base at the supraorbital artery region is an ideal flap supply for repairing most nasal soft tissue defects.

1 Instruction

Nose, which is in the center of our face, plays many important functions, such as breathing, smelling and beauty. The damage of nasal defects not only includes the loss of nasal functions but also some immense psychological trauma. Thus, the functional recovery and reconstructive beauty appearance are main aims of nasal repairing reconstructive surgery. When repairing nasal soft tissue defects, doctors shall consider both the color and thickness of the flaps. The typical patient with serious nasal defects in our department is the one experiencing the resection of external nasal malignant tumors. We used paramedian forehead skin flaps with base at the supraorbital artery region for the repairing and reconstruction of his nose and we obtained relatively satisfactory results.

2 Materials and Methods

2.1 Clinical data of typical cases

Clinical data of one typical case with serious nasal defects was accepted in Otolaryngology & Head and Neck Surgery Department of China–Japan Union Hospital of Jilin University. The study was approved by the institutional review board of China-Japan Union Hospital. The participant provided written informed consent before the study.

[a] Corresponding author: lilin01@jlu.edu.cn

2.2 A typical case

Male, 34 years old, with a nasal neoplasm found for 40 days. The neoplasm with the size of a soybean is on the septum of the left nasal vestibule. The patient can feel a bit of bitterness. After the neoplasm is pierced, sanguinopurulent secreta flow out. The neoplasm grows progressively and it stands out on the left side of the external nasal tip. After the neoplasm, which has the size of 5X3.5X3.5cm, is cut, most parts of the nasal tip and the nasal septum lose. Then, mark the forehead according to the size and distance of the nasal soft tissue defects. Besides, the pedicle shall be appropriately widened when they are separated and 2-4mm of the subcutaneous fat shall be left. Operations near the flap root shall be done with care so as not to damage the blood supplying vessels. If the skin donor sites are defective, surrounding tissues can be appropriately loosed before they are sutured directly. After the flap design, the nose will be repaired from the paramedian forehead skin flap with base at the supraorbital artery region. The flap survives at Phase I after the surgery.

2.3 Postoperative pathologic findings

Well-differentiated and moderately differentiated squamous cell carcinoma with keratinization.

3 Discussion

Nose is in the center of our face and the nasal soft tissue defects will affect facial contours and features, which will bring both mental and physical damages to patients. Nasal tissues have bad elasticity, small looseness and thin subcutaneous soft tissues. Generally speaking, they cannot directly heal the wound after the appearance of large tissue defects. Nasal aesthetic subunits divide the nose into the dorsal nasal area, lobular areas on both sides of the nose and triangles and nasal tip areas on both sides of the soft tissue. The nasal defect repair shall abide by the nasal subunit aesthetic principle so as to reshape the nasal aesthetic structure[1]. In general, if the defects do not exceed 50% of the subunit, remaining normal mucosa shall be retained as much as possible. If the defect areas exceed 50%, the defect scope will be extended to the whole subunit [2]. Forehead flaps, which are adjacent to the nose and have similar skin color and luster with the external nose, are the first choice for the supply of flap for nasal defects. The traditional median muscle flap of the forehead uses the trochlea artery as the feeding artery. However, it has insufficient length and needs to be designed to median flap. After the operation, some side effects, such as narrow forehead, scars, bad structural form of external nasal subunits (alar base and nasal tip) or overstaffed structure, often appear, thus, patients are not quite satisfied with it after the surgery.

3.1 Applied anatomy of supraobital arteries, veins and nerves:

Supraobital artery is an ophthalmic artery from the internal carotid and it comes with the supraorbital nerve. It runs forward between the superior levator muscle and the superior orbital wall in the orbital cavity. The supraobital artery goes out of the supraorbital hole (incisura) when it is 25.0±0.6mm(24.4mm-25.6mm) away from the anterior median line. The distance between that place and the place where the supratrochlear artery goes out of the orbital cavity is 1.0±0.4(0.6-1. 4) cm. The external diameter of the supraobital artery at the place where it goes out of the orbital cavity is 1.4±0.1 (1.3-1.5) mm[3]. After that, the artery will pass the supraorbital margin and run to the fronto-partial. It has anastomosis branches with the superficial temporal artery at the forehead [4]. After the supraobital artery goes out of the supraorbital hole or the incisura, it will be divided into the deep and superficial branches: (1) The external diameter of the starting point of the superficial branch is 0.9±0.1 (0.8-1. 0) mm and this branch mainly anastomose into meshes. Then, it will send branches inwards to anastomose with the superal trochlea artery and will send branches outwards to directly anastomose with frontal branch of superficial temporal artery [5]. They will be anastomosed in the frontalis muscle in outer half-side of the forehead. In most cases, the supraobital artery will be directly connected with the end of the frontal branch of superficial temporal artery. Sometimes, they may also survive by forming a communicating branch between them. Then, they will send several branches upwards to anastomose a mesh with the forehead artery and will send some branches downward which will spread over the upper eyelid. (2)The deep branch of the supraobital artery will run upward and backward tightly along the periosteum after it separates from the supraobital artery. It will not be connected with the frontalis muscle[6-7]. Suprorbital vein goes together with the supraobital artery to collect the venous blood in the arterial distribution area.

3.2. Advantages of forehead flaps with the base at supraobital artery

(1) Supraobital artery is the main longitudinal blood supply artery of the central forehead flap [8]. It is crossed with the angular artery into a net in the inner canthus. It has reliable blood supply and there is no need to dissect blood supply arteries. Besides, it has enough venous returns. Stage I shows that the survival rate of flaps is high. Thus, the pedicle skin flap with the blood supplied by the supraobital artery can be narrower than that of the median forehead flap [9]. (2) Supraobital artery comes from the side and its pedicle is both wide and thick. It can transfer some sensory nerves to the repair region simultaneously and can partially recover some sensory functions. (3) The color, quality, texture and thickness of forehead flaps are similar to those of the nasal skin. Besides, it becomes thin with the running of the supraobital artery and thus, the thickness of flaps can be designed according to the need. It shows high beauty values. (5) Aesthetic subunits can be totally covered by the flaps in the design and the scars can be hidden in the junction of the subunits (which is usually the turn-back site). (6) There can be narrower pedicles and thus the flaps can rotate and stretch freely. (7) It can reduce the occurrence of eyebrow malformation after the flap transposition. (8) The vertical projection of the supraobital artery is 2cm away from the central line, which means that it is located in the central zone of the eyebrow. The pedicle can be designed to be 1.5cm because this will not affect the twist operations of the flap.

3.3. Operative Indications

1. Repair and reconstruct defects of soft tissues at face, cheek and chin, including the immediate repair of defects after the tumor operation; 2. Reconstruct organs of some maxillofacial regions and the defects of jaw, nose, eyelids and auricular. 3. Seal or cover deeper tissues (such as muscle tendon, muscles, nerves, large blood vessels and bones, etc.) or exposed wound surfaces. 4. Repair and reconstruct penetrating defects of cheek and nose.

3.4 Surgical Considerations

During the lateral stripping of flaps, the subdermal vascular network shall be intact so as to prevent necrosis of wound margins. Protection of vascular pedicle and skin branches: Supraorbital transverse veins run within one centimeter of the supraorbital rim while nasal dorsum venous arch runs in the deep epidermis within 1cm above the inner canthus connection line of the osseous of both sides. Below the eyebrow, the median flap pedicles near the forehead had better be separated to the superficial layer. Besides, do not hurt upraorbital transverse veins so as to increase venous returns of the flap and avoid postoperative hematoma. Nasal dorsum venous arch runs in the deep epidermis within 1cm above the inner canthus connection line of the bone so as to protect the vein, increase venous returns of the flap and avoid postoperative hematoma.

References

1. Ward PD, Collar R.Facial Plast Surg. 29(5):340-5 (2013)

2. Jiang H, Li QF, Gu B, Liu K, Shen GX, Zheng DN. Zhonghua Zheng Xing Wai Ke Za Zhi. 23(1):40-2 (2007)
3. Rodrigues ML, Köhler HF, Faria JC, Ikeda MK, Vartanian JG, Kowalski LP.Int J Oral Maxillofac Surg. 38(8) 850-4 (2009)
4. Hou D, Fang L, Zhao Z, Zhou C, Yang M.Exp Ther Med. 5(3) 751-756(2013)
5. Kim YJ, Kim HR, Jun YJ, Seo BC.J Craniofac Surg. 22(5):1968-71(2011)
6. Abul- Hassan HS,Ascher GD,Acland RD. Plast Reconstr Surg. 77 17(1986)
7. Skaria AM.Dermatol Surg. 41(7) 859-62 (2015)
8. Baker SR, Aflord ED. Otolaryngol Head Neck Surg. 424-30 (1993)
9. Menick FJ. The paramedian forehead flap. Clin Plast Surg. 17 607-22 (1990)

Therapy efficacy of hyperreative rhinitis by the combination of intranasal steroid and oral H1-antihistamine

Feng Liang[1], Fanglei Han[1], Xiaodan Jiang[2] and Lin Li[3,a]

[1]*Department of Anesthesiology, China- Japan Union Hospital, Jilin University, 130033, China*
[2]*Otorhinolary,Affiliated Hostipal og Qingdao University,Qingdao,266555,China*
[3]*Dept. Otolaryngolgy and head & Neck, China- Japan Union Hospital, Jilin University, 130033, China*

Abstract. Background Hyperreative rhinitis (HRR) is identified as rhinitis which has at least two symptoms such as sneeze, itches, watery secretion and nasal blockage. Little is known about the pathology of HRR, as well as its clinical features and treatment. Objective To investigate efficacy HRR by combined therapy with intranasal steroids and antihistamines Methods The age, gender, predisposing factors and clinical symptoms of 200 HRR patients were analyzed. Both intranasal steroids and antihistamines were applied for 8 weeks. Symptoms were assessed with VAS score before and after the treatment. Result: Before the therapy, 87% of HRR patients showed persistent symptoms; 69% complained about severe and persistent effects on the quality on the life; sneezing (96%) and rhinorrhea (78%) were the top 2 symptoms reported; All patients completed the 8-wk combined therapy with intranasal steroids and antihistamines; 96% patients were satisfied with the therapy, with all symptoms except wheezing have been clear relief ($p < 0.05$); the remaining 4% reported the unsatisfied result of the therapy. Conclusion The combined therapy with intranasal steroids and antihistamines controlled most of clinical symptoms of HRR.

1 Instruction

Rhinitis is an extremely common rhinopathy. Rhinitis has been divided into four main categories: infectious rhinitis, allergic rhinitis (AR), structural rhinitis, and non-allergic rhinitis (NAR). Differential diagnosis of AR from NAR is often difficult because of their similar clinical presentation. NAR is now a diagnosis of exclusion characterized by similar symptoms to AR, but with no allergic sensitization-related symptoms and signs of infection [1]. Hyperreactive rhinitis (HRR) is identified as the combination of AR and NAR because of similar symptoms.

HRR affects a significant number of patients in clinical practice. Studies have shown an increase in the prevalence of asthma in patients with HRR[2] it has been hypothesized that there was a link between upper and lower airways beyond allergy associated inflammation. Our current study focused on the efficacy of one combined therapy with intranasal steroid and oral antihistamine on HRR.

2 Materials and methods

[a] Corresponding author: lilin01@jlu.edu.cn

2.1 Clinical data

200 consecutive untreated patients (110 females and 90 males), who were diagnosed of HRR, participated the study. The cases were collected from the files recorede between January, 2015 and December, 2015, from the Department of Otolaryngology, Head and Neck Surgery of the University of China-Japan Union Hospital attached to Jilin University, China. The study was approved by the institutional review board of China-Japan Union Hospital. All the participants provided written informed consent before the study.

2.2 Clinical questionnaire and medical interview

Data of the initial evaluation of NAR patients were extracted from the medical history obtained during their first visit to our center. A clinical questionnaire, a complete file of physical examination and symptoms were included. Visual Analogue Scale (VAS) was used to evaluate the severity of nasal symptoms before and after the therapy. The range of the score was 0-3: higher scores indicated more severe symptoms, and score 0 indicated no symptoms. One year follow-up evaluation was also conducted after the treatment. NAR was defined by the presence of two or more nasal symptoms (sneezing, itching, rhinorrhea, or nasal obstruction) plus a negative SPT response and serum IgE level specific to aeroallergens.

2.3 Therapy procedures

Patients were treated with once-daily intranasal budesonide for 8 weeks. 200 mg Budesonide were given to patients aged <11 years at study entry and 400 mg to patients aged >11 years at study entry. Simultaneously, Loratadine was administered orally to patients (>12 years old) with 10 mg per day before sleeping for 8 weeks

2.4 Statistical analysis

Descriptive statistics (frequency, mean, median, SD, and range) were used to analyzed the clinical features of HRR. Paired t-test was used to analyze the therapy results. The data were analyzed with SPSS for Windows 17.0 (SPSS, Inc, Chicago, Ill). P-values < 0.05 were considered statistical significant.

3 Result

3.1 Age and sex distribution

A total of 200 patients were survey in our study. They aged from 4 to 76, with an average of 38 ± 14) years old. 87% of patients showed persistent symptoms and 69% complained about severe and persistent effects on the quality on the life; there was no significant difference in the age between male and female patients (p=0.983)

3.2 The efficacy of an 8-wk combined therapy of intranasal steroid and oral antihistamine

All 200 patients were contacted by telephone for follow-up evaluation after 8-wk therapy. All the symptoms were significantly decreased after the 8-wk therapy (p<0.05) 96% patients were satisfied with the therapy, with all symptoms have been clear relief ($p < 0.05$); the remaining 4% reported the unsatisfied result of the therapy but still has decreased VAS after the combined treatment. (Table 1).

Table 1. All the HRR symptoms were significantly decreased after the 8-wk combined therapy

	before the treatment				after the treatment				z	p
VAS	0	1	2	3	0	1	2	3		
sneezing(case)	8	24	64	42	80	52	20	0	-5.12	0
secretion(case)	16	20	36	80	104	44	4	0	-5.18	0
blockage(case)	20	32	42	44	128	20	4	0	-5.07	0
ithcy(case)	44	36	24	48	124	28	0	0	-4.46	0

4 Discussion

Both NAR and AR are associated with the nasal mucosa hyper-responsiveness and resulted in similar symptoms and physical signs, so we identified the hyperreactive rhinitis (HRR) according to symptoms. Currently, the diagnosis of HRR relies on the clinical symptoms.

Standard treatment reference of HRR has not been established yet. Similar treatment designed for AR has been used to treat to HRR due to the shared symptoms and morbidity between this two types of rhinitis. Studies showed that intranasal steroids such as Budesonide, Fluticasone and Mometasone, could relieve the edema condition of nasal cavity and decrease the symptoms of NAR. Rinne[3] reported that intranasal steroids could successfully control the recurrence of the NAR. Patients with sneezing as a predominant symptom responded well to oral antihistamines. The combination of Flunisolide and Loratadine showed better clinical efficacy than intranasal steroid alone [4]. In our study, 96% patients showed alleviated symptoms of nose, eye and lower airway after a combined therapy of intranasal steroids and oral anti-histamines for 8 weeks, indicating the therapy was effective in HRR treatment. The histamine receptor 3 was the decisive factors in nasal neural system, whereas the target of Loratadine is histamine receptor 1 that mainly responsible to the allergic reaction. Studies have showed that Azelastine decreased all the symptoms of NAR [5] and AR. Azelastine is the intranasal antihistamine commercially available. Studies showed that Azelstine interfered with vasoactive neuroperptide substance P and inhibited the production of leukotrienes B4 and C, which might possibly decrease the expression of nuclear factor-kB, which was a transcription factor for multiple proinflammatory substances[6]. The efficacy of intranasal Azelastine might be further explored to replace Loratadine in the combination therapy to treat NAR and AR.

In conclusion, HRR had it own clinical feature, such as high risk population is adults (30-39), slightly more in female, The combined therapy of intranasal steroid and oral histamines was effective in controlling HRR symptoms. In addition, the individualized therapy procedure needs further consideration for the inter-individual variability of the therapy efficacy observed in our study.

Acknowledgement

All authors declared no conflict of interest in the research. The work presented in this report is the subject of a pending patent filed by the China-Japan Union Hospital of Jilin University (201510382781.X). This study was supported by the National Natural Science Foundation of China (811007020), the Health and Family Planning Foundation of Jilin Province (20152046), the Science and Technology Development Plan Foundation of Jilin Province (20160101070JC), and the Norman Bethune Program of Jilin University (2015305).

References

1. Bousquet J, Fokkens W, Burney P, et al.. Allergy. **63** 842-853, (2008)
2. Chawes BL, Bønnelykke K, Kreiner-Møller E, et al. J Allergy Clin Immunol. 2010; **126 (3)** 567-573.e1-8(2010)
3. Rinne J, Simola M, Malmberg H, et al. J Allergy Clin Immunol. **109(3)** 26-432 (2002)
4. Purello-D'Ambrosio F, Isola S, Ricciardi L, et al. Clin Exp Allergy.**29(8)** 143-1147 (1999)
5. Banov CH, Lieberman P. Vasomotor Rhinitis Study Groups. Ann Allergy Asthma Immunol.**86** 28-35 (2001)
6. Greiner AN, Meltzer EO. J Allergy Clin Immunol. **118(5)**985-996 (2006)

The observation of the sedation effects of intranasal methods of dexmedetomidine for magnetic resonance imaging in children

Shuang Qi [1], Suming Wu [2] and Feng Liang [1,a]

[1]*China–Japan Union Hospital of Jilin University*
[2]*The 4th Hospital of Jilin University(The General Hospital of FAW)*

Abstract. purpose: To compare the sedation effects of two intranasal methods of dexmedetomidine for magnetic resonance imaging (MRI) in children. Method: 40 children for MRI were randomly divided into two groups, namely 20 cases for each group: nasal drip group(group A)and nasal spray group(group B). Group A and group B received intranasal dexmedetomidine of 1.5 ug/kg by two methods. Monitoring the HR, SpO_2, RR and inspection time of the two groups so as to observe the sedation effects and Ramsay sedation score while recording the adverse reactions. Results: There was no significant difference in the HR, SpO_2, RR between the two groups at the same time point($P>0.05$). Compared with group A,the time-to-fall-asleep in group B was shorter and the adverse reactions significantly decreased($P<0.05$).Conclusion: Compared with nasal drip,nasal spray dexmedetomidine by MAD can shorten the time-to-fall-asleep and reduce adverse reactions for MRI in children.

1 Introduction

Magnetic resonance imaging (MRI) is an effective diagnostic method to identify diseases, but its testing time is long with large instrument noise, so most of the children patients show poor obedience. Light activity of children will lead to artifacts or false appearances which will affect the accuracy of diagnosis. Besides, the cost of MRI is relatively high and repeated inspections will increase economic burden to parents, so rational use of sedative drugs is an important part to guarantee a smooth MRI for children. Mason et al. [1] studied and confirmed the sedative advantages of dexmedetomidine for MRI in children. The research of domestic scholars [2] confirms that dexmedetomidine nasal drip of 2-5 years old children can achieve good sedative effect. But the optimal dose and methods of nasal administration still need further discussion. Therefore, this study aims at comparing the sedation effects of two intranasal methods of dexmedetomidine for MRI in children.

2 Materials and Methods

2. 1 General Data

From January to December of 2015, 40 children accepted MRI examinations in China-Japan Union Hospital of Jilin University. Those children were aged at 2-5 years old and their sex was unlimited.

[a] Corresponding author: 295253720@qq.com

Their BMI is 17.0 to 24.0 kg/m^2, classified as grade I to II by ASA. There was no abnormality with their nasal cavity structure and no damage to their nasal mucosa. The children had no sedation taboos, allergic history of dexmedetomidine or respiratory and circulatory system diseases. They were randomly divided into the group of dexmedetomidine administration by nasal drip (group A) and group of dexmedetomidine administration by nasal spray (group B).

2. 2 The Administration Methods

The two groups of children patients got fasted routinely before medication. They were placed at the supine position. Extracting 1 ml dexmedetomidine (the batch number: 15051332; Jiangsu Hengrui Pharmaceutical Co., Ltd; the concentration is 100 ug /ml) and administrating the dose of 1.5ug/kg by their parents to the nasal cavity of children at the presence of anesthesiologists. Group A used 1ml syringes without needles for nasal inhalation; group B used MAD (model: MADl00, Wolfe Tory Pharmaceutical Co., Ltd.) for nasal spray. Parents administrated to both left and right nostrils alternately with equal drug. Observing the children patients' reactions until they fell asleep and then making positioning scanning. Keeping monitoring their HR, SpO_2 and RR after medication.

2. 3 Observation Index

Observing the cooperative degree of the two groups of children during medication. Using Ramsay grading method to measure the sedation effects.1point: restlessness; 2 points: cooperating quietly; 3 points: having responses to commands; 4 points: drowsiness with sensitive reactions to loud acoustic stimulation; 5 points: drowsiness with slow reactions to loud acoustic stimulation; 6 points: drowsiness without any response. 2-5 points is moderate sedation. Recording the fall-asleep time of both groups of children (administrating drugs to the time graded by 2-5 points according to Ramsay). Comparing the HR, SpO_2 and RR before medication(t_0) and 20 min after medication (t_1). Recording the incidence of crying, bucking, respiratory depression (SpO_2 < 90% is regarded as existing respiratory depression), nausea and vomiting and other adverse reactions.

2.4 Statistical Treatment

SPSS 13.0 statistical software is adopted for data analysis. The enumeration data is tested by χ^2. The measurement data is expressed by mean standard deviation ($\bar{X} \pm S$). The intra-group comparison is tested by matching t; intra-group comparison is analyzed by one-way variance, $P < 0.05$ meaning significant differences.

3 Results

The general data and inspection time of the two groups of children patients had no statistical differences ($P > 0.05$), as shown in table 1. The HR, RR and SpO_2 before and after medication had no statistical differences ($P > 0.05$), as shown in table 2. Compared with group A, the fall-asleep time of children of group B is shortened and the occurrence of crying and bucking is reduced ($P < 0.05$), are shown in table 3.

Table 1 The comparison of general data and inspection time of the two groups of children patients

Groups	Case Number	Male/Female	Age	BMI	Inspection Time
Group A	20	11/9	3.52±1.46	20.51±3.42	30.52±10.27
Group B	20	10/10	3.46±1.54	21.27±3.64	31.46±9.87

Table 2 The comparison of the HR, RR and SpO_2 before and after medication of the two groups of children patients($\bar{X} \pm S$,n=20)

Items	Groups	Before Medication(t_0)	After Medication(t_1)
HR	Group A	123.32±6.46	120.43±6.23
(/min)	Group B	121.18±6.56	116.39±4.06
RR	Group A	24.12±1.13	23.67±0.92
(/min)	Group B	23.82±0.97	22.12±1.10
SpO_2	Group A	99.23±0.38	99.07±0.65
(%)	Group B	99.11±0.47	98.32±0.46

Table 3 The comparison of fall-asleep time and adverse reactions of the two groups of children patients ($\bar{X} \pm S$,n=20)

Groups	Fall-asleep-Time	Adverse Reaction			
	(s)	Crying	Bucking	Respiratory Depression	Nausea and Vomiting
Group A	18.4±3.8	5	7	0	0
Group B	14.3±4.1*	1$^{\triangle}$	1$^{\#}$	0	0

Compared with group A,$P^{*\triangle\#}<0.05$

4 Discussion

In order to obtain a satisfying image during MRI, effective measures are usually taken on incoordinate children to make sure they are quiet in the process of scanning. As it only requires stopping movements with low analgesia, so moderate sedation, namely conscious sedation, is mainly adopted. The medication method of out-patient examination for children is mainly orally taking or enema[3-4], but due to the drug taste, operating techniques and the fear of a strange environment, orally taking or enema easily leads to crying and restlessness of children patients, so reasonable and comfortable way of medication is an important link for the sedation of children.

Administering drugs by the intranasal route is a simple and convenient way commonly used on children. It can be done by the most familiar people of children, so it is easier for children to accept. Besides, this method can also reflect the development trend of comfort and humanized medical science. Dexmedetomidine is a new-type, efficient and high-selectivity a_2 adrenoceptor agonist. Compared with other sedative drugs, it takes a quicker effect with no respiratory depression and has mild effect of analgesia. In addition, it is colorless and tasteless without any mucosa stimulation. Its concentration is higher in unit dose. Therefore, it is more suitable for nasal route[5]. The study finds that , the vital signs of the two groups of children is stable and they are calm moderately without adverse reactions such as respiratory depression, nausea, vomiting and restlessness if administrating dexmedetomidine through transnasal method.

Different with common syringe intranasal methods, MAD can make the drug liquid evenly disperse into tiny droplets and then they will adsorb on the nasal mucosa, which strengthens the absorption concentration of drugs to the utmost extent.The result of this study confirms that administering dexmedetomidine to nasal cavity by MAD can effectively shorten the fall-asleep time of children patients in comparison with ordinary nasal inhalation conducted by ordinary syringes.When dripping nose with ordinary syringes, children patients are easily to have adverse reactions such as bucking and crying[6]. The results of the study also verifies that administering dexmedetomidine to nasal cavity by MAD can effectively reduce the incidence of adverse reactions. Thus, children patients will feel more comfortable during sedation.

DOI: 10.1051/bioconf/20170801043

In conclusion, compared with traditional ways of nasal inhalation, administering dexmedetomidine to nasal cavity by MAD in MRI examination can significantly shorten the fall-asleep sleep and reduce the occurrence of adverse reactions, so it is worthy of popularization and application in clinic.

References

1.Mason KP,Robinson F,Fontaine P, Radiology, **267**,911 (2013)
2.Jia JE,Chen Y, J Clin Anesthesiol, **28**, 1172 (2012)
3.Krauss B,Green SM,Lancet,**367**,766 (2006)
4.Yuen VM,Hui TW,Irwin MG,Anesm Analg,**106,**1715 (2008)
5.Yuen VM, Irwin MG, Hui TW, Anesth Analg, **105**, 375 (2007)
6.West SK,Griffiths B,shariff Y,Br J ophthalmo,**97**,1438 (2013)

 DOI: 10.1051/bioconf/20170801044

Applicable observation of butorphanol in painless colonoscopy examination

Xuefeng Li[1], Feng Liang, and Guifeng Liu[2,a]

[1]*Department of Anesthesiology, China-Japan Union Hospital, Ji Lin University, Changchun, Jilin, China*
[2]*Department of Radiology, China-Japan Union Hospital, Ji Lin University, Changchun, Jilin, China*

Abstract. Objective:To summarize the clinical effect of butorphanol compound propofol in painless colonoscopy examination and its feasibility.Methods:100 colonoscopy examination patients (56 males and 44 females aged from 19 to 60 years old) registered between August, 2016 and September, 2016 in the endoscopy center of our hospital were randomly selected. ASA classification is I or II level. Their body weight ranged from 55 kg to 75kg. They were randomly divided into two groups and each group included 50 cases. All patients went through conventional ambrosia and liquid fasting for 8 hours before the anesthesia and they drank magnesium sulfate liquid of 2500ml to clean their gastrointestinal tracts.After patients entered the operating room, their veins of upper limb were opened so as to monitor their HR, MAP and SPO_2. After that, butorphanol of 20μg/kg was injected to patients of the experimental group while normal saline of the same amount of was injected to patients of the contrast group. After 60 seconds, propofol of 1~2 mg/kg was injected to both groups by the way of intravenous injection. The enteroscopy examination was started after patients had no eyelash reflection. Besides, actual application dose of propofol was adjusted according to clinical indications of patients and the adjusting frequency each time was controlled between 30 milligrams and 50 milligrams until the completion of the examination. SPSS 22.0 statistical software was used to analyze and handle research data of this group. Results:Anesthesia effect: The difference of inter-group comparison showed no statistical significance ($P>0.05$). The intra-group comparison and the inter-group comparison show that the difference in terms of changes of HR, MAP and SpO_2 of patients in two groups before and after the anesthesia had no statistical significance ($P>0.05$).The awakening time, VAS score, postoperative vomiting times and the occurrence rate of respiratory depression of the observation group were all significantly lower than those of the contrast group ($P<0.05$).

1 Introduction

In order to summarize the clinical effect of butorphanol compound propofol in painless colonoscopy examination and its feasibility, this research randomly selected 100 colonoscopy examination patients from the endoscopy center of our hospital and divided them into two groups. During the examination link, the contrast group only uses the propofol while the experimental group uses butorphanol compound propofol. Based on the observation results of the anesthesia effectiveness of two groups of patients, this paper studied the anesthesia analgesia effect of butorphanol in its application in painless colonoscopy. Specific contents are as follows

[a] Corresponding author: 28186282@qq.com

2 Materials and Methods

This study was approved by Ethics Committee of our hospital. In addition, our hospital has signed the informed consent form with all patients. 100 colonoscopy examination patients (56 males and 44 females aged from 19 to 60 years old) registered between August, 2016 and September, 2016 in the endoscopy center of our hospital were randomly selected. They all belonged to Level I to Level II of ASA and had no obvious heart and lung diseases. Their body weight ranged from 55 kg to 75kg. They were randomly divided into two groups and each group included 50 cases. The gender, age and body weight differences of patients in both groups had no statistical significance. All patients went through conventional ambrosia and liquid fasting for 8 hours before the anesthesia and they drank magnesium sulfate liquid of 2500ml to clean their gastrointestinal tracts. No one had the experience of preanesthetic medication. After patients entered the operating room, their veins of upper limb were opened so as to monitor their HR, MAP and SPO_2. Then, they absorbed oxygen (3 L/min). After that, butorphanol of 20μg/kg was injected to patients of the experimental group while normal saline of the same amount of was injected to patients of the contrast group. After 60 seconds, propofol of 1~2 mg/kg was injected to both groups by the way of intravenous injection. The enteroscopy examination was started after patients had no eyelash reflection. Besides, actual application dose of propofol was adjusted according to clinical indications of patients and the adjusting frequency each time was controlled between 30 milligrams and 50 milligrams until the completion of the examination. SPSS 22.0 statistical software was used to analyze and handle research data of this group. The measurement data were shown by (x±s) and inter-group comparison adopted the way of t-test. In addition, the counting data were expressed by the case number (n) and the comparison of intra-group rate (%) used the way of X 2-inspection. $P<0.05$ means that the difference was statistically significant.

3 Results

Anesthesia effect: There were 48 excellent cases and two good cases in the observation group in terms of the anesthesia effect and 45 excellent cases, 4 good cases and one bad case in the contrast group. The excellence rate of the observation group was 100.00%, which was slightly higher than that of the contrast group (98%). The difference of inter-group comparison showed no statistical significance ($P>0.05$).

Vital signs: The intra-group comparison and the inter-group comparison show that the difference in terms of changes of HR, MAP and SpO_2 of patients in two groups before and after the anesthesia had no statistical significance ($P>0.05$). Please refer to Table 1.

Table 1. Change of vital signs of two groups of patients before and after the anesthesia (x±s)

Time	Groups	HR (Times/min)	MAP (mmHg)	SpO_2 (%)
Before anesthesia	Observation group	80.23± 8.75	83.43± 9.67	98.23±1.10
	Contrast group	81.37± 8.67	82.79± 9.21	98.11±1.03
After the medicine is used for 1 min	Observation group	74.47± 7.55	72.18± 7.35	96.37±2.26
	Contrast group	5.88± 7.47	71..26± 7.21	96.06±2.35
After the inspection	Observation group	78.96± 7.98	81.57± 9.46	97.53±1.52
	Contrast group	79.87± 8.06	80.78± 9.24	97.41±1.50

Awakening time and adverse reaction: Patients of the observation group needed less awakening time than those of the contrast group. Comparison among groups showed no statistical significance ($P>0.05$). Besides, the awakening time, VAS score, postoperative vomiting times and the occurrence rate of respiratory depression of the observation group were all significantly lower than those of the contrast group ($P<0.05$). Please refer to Table 2.

Table 2 Comparison of awakening time and adverse reaction [(x±s) ,n (%)]

Group	Awakening time (min)	VAS Score	Postoperative vomiting (times)	Respiratory depression
Observation group	2.68±1.02	1.06± 0.32	1.16± 0.30	1 (2%)
Contrast group	3.06±1.39	2.10± 0.56	2.16± 0.58	7 (14%)

4 Discussion

Propofol has many advantages, such as no accumulation, fast effect and relatively short lasting period[1], and it is widely used in the field of painless gastroscopy. Even so, propofol is very limited in its analgesic effect because its anesthesia procedures can only be completed with the aid of other drugs, of which, butorphanol is the most frequently used kind. Butorphanol is a kind of mixed opioid receptor agonists-opioid receptor-antagonists and its receptor owns distinctiveness of features[2]. Thus, it can control the situation of respiratory depression and avoid the appearance of unsatisfactory situation of patients' gastrointestinal tracts. Besides, patients are not addicted to it and its drug efficacy can last relatively longer, thus, it is popular in clinical practice[3].

According to current clinical practices, propofol shall be used together with other analgesic drugs or local anesthetics for the conduction of anesthesia because this can not only improve the analgesic effect, but also reduce the adverse reactions caused by anesthesia medicines. This study uses tartaric acid butorphanol compound propofol to the anesthesia of painless colonoscopy examination[4]. The results show that its anesthesia effect is slightly higher than that of the contrast group and the pain degree (VAS) caused by it is significantly lower than that of the contrast group. This indicates that the tartaric acid butorphanol tartrate compound propofol has good anesthesia effect and it can effectively reduce the pain degree of enteroscopy examination patients. Butorphanol is a kind of opioid receptor agonists-opioid receptor excited-antagonists and it exerts analgesic effect by acting on k-receptors. Literature [5]points out that the analgesia intensity of butorphanol is five to eight times of that of morphine and it can promise quick returns because its analgesic effect can be played within one minute after the intravenous injection. The simultaneous usage of propofol and butorphanol can exert very good anesthetic effect and effectively suppress the pain of the injection site, which will eventually promote the smooth conduction of painless colonoscopy examination. Anesthesiologists always focus on the vital signs of patients and the security of anesthesia process. In this study, inter-group comparison and intra-group comparison of two groups of patients show that changes of HR, MAP and SpO2 have no obvious difference before the anesthesia, within one minute after the medicine is used and after the anesthesia. This presents that butorphanol compound propofol anesthesia do not affect vital signs of patients. Butorphanol, which is a kind of opioid receptor antagonists, exerts no respiratory depression, thus, it can effectively maintain the stability of respiration and circulation in the anesthesia process. In addition, butorphanol has weak antagonism to human μ opioid receptor and it can reduce adverse reactions of gastrointestinal tracts and the occurrence rate of nausea and vomiting after the anesthesia[6]. It can reach its peak value within four to five minutes after the intravenous injection and patients can regain consciousness fast. This research shows that patients in the observation group only need a short awakening time and the occurrence rate of their postoperative vomiting times and respiratory depression effect are all significantly lower than those of the contrast group. Domestic and foreign scholars also had similar literature reports before, showing that it is safe to use butorphanol to the anesthesia of painless colonoscopy examination.

In conclusion, butorphanol compound anesthesia is helpful to ease the pain degree, reduce the

DOI: 10.1051/bioconf/20170801044

adverse reaction, improve the anesthesia quality and promote the smooth operation of painless colonoscopy examination. Limitations of this study lie in that anesthesia time points are not subdivided and its functional mechanisms are not sufficiently analyzed. Thus, more in-depth researches still remain to be done in the future.

Conclusion

Butorphanol compound anesthesia is helpful to ease the pain degree, reduce the adverse reaction, improve the anesthesia quality and promote the smooth operation of painless colonoscopy examination.

References

1. Yingchao Xie, Jingxian Seng, Xichao Zhu, ZhiHeng Ke, Yingsheng Bo, Yingting Liang. Different dose of miss bhutto's brown composite application of propofol in painless tubal ligation [J].Practical medical journal 201329(10):1665-1667(in Chinese).
2. Bailey PL,Egan TD,Stanley TH.Intravenous opioid anesthetics[M].Miller RD.Anesia.5th ed.Philadelphia:Churchill Livingstone,2000:273-376.
3. Arora V,Bajwa SJ,Kaur S.Comparative evaluation of recovery characteristics of fentanyl and butorphanol when used as supplement to propofol anaesthesia[J].Int J Appl Basic Med Res,2012,2(2):97-101.
4. Gupta A,Kaur S,Attri JP,et al.Comparative evaluation of ketamine - propofol,fentanyl - propofol and butorphanol-propofol on haemodynamics and laryngeal mask airway insertion conditions[J].J Anaesthesiol Clin Pharmacol,2011,27(1):74-78.
5. Kaur J,Srilata M,Padmaja D,et al.Dose sparing of induction dose of propofol by fentanyl and butorphanol: A comparison based on entropy analysis[J].Saudi J Anaesth,2013,7(2):128-133.
6. Ishizuka T,Itami T,Tamura J,et al.Anesthetic and cardiorespiratory effects of propofol,medetomidine,lidocaine and butorphanol total intravenous anesthesia in horses[J].J Vet Med Sci,2013,75(2):165-172.

Experience of monitoring the recurrent laryngeal nerve in thyroid surgery with endotracheal intubation

Feng Liang[1], Guifeng Liu[2], Huiying Li[1], Yunpengfei Li[1] and Xuefeng Li[1,a]

[1]*Department of Anesthesiology, China-Japan Union Hospital, Ji Lin University, Changchun, Jilin, China*
[2]*Department of Radiology, China-Japan Union Hospital, Ji Lin University, Changchun, Jilin, China*

Abstract. Purpose:To analysis clinical experience of applying recurrent laryngeal monitoring endotracheal tube (NIM EMG Endotracheal Tube) in the surgery of thyroid.Method:A retrospective analysis of 84 patients who underwent endotracheal intubation laryngeal nerve monitoring by thyroid surgery in the Chinese-Japanese Friendship Hospital of Jilin University from March to December in 2015.To summarize the experience of intubation with NIM EMG Endotracheal Tube.Result 77 (91.7%)had initial intubation achievement in the 84 patients.FROM the 77 cases we had gotten s atisfactory nerve monitoring signal.Whereas there are 7 cases (8.3%) appear abnormal EMG or signal missing,in the 7 cases there is one which being intubated too deep ,3 cases which being intubated too shallow and 3 cases with malrotation intubation.Conclusion:We got the satisfactory signals after adjust'ing the tube by using the visual laryngoscope.

1 Instruction

Lately laryngeal nerve monitoring (EDM) technology can locate and discriminate the recurrent laryngeal nerve, which can avoid intraoperative recurrent laryngeal nerve injury [1].Recently in surgery of Thyroid intraoperative laryngeal nerve monitoring technology regards the endotracheal tube as a surface electrode EMG which can record electrodes, recording electrodes in contact with the vocal cords is good is directly related to neurophysiological monitoring effects .Whether electrode is in contact with vocal cord well relates to the effect of neurophysiological monitoring directly.Therefore, accurate intubation is an important part of the completion of the recurrent laryngeal nerve monitoring.In this paper, we analyze 84 cases that were in our hospital from March to December in 2015 with various types of thyroid surgery and summarize the experience of intubation.During the surgery we all use the recurrent laryngeal nerve monitoring endotracheal tube. Details as below

2 Material an method

2.1 General information

84 cases in our hospital for treatment of elective thyroid surgery from March to December in 2015 , their age duration is 22-65 years old, weighing duration is 48-79kg, there are 13 male and 71 female

[a]Corresponding author: 28186282@qq.com

cases.Their ASA grades are I- II level. In the 84 cases there are 53 diagnosed as papillary thyroid carcinoma.30 cases as nodular goiter, and one as lymphocytic thyroiditis.

2.2 Method

Anesthetizing 30MINS Before being anesthetized patients are given penehyclidine 1mg by intramuscular injection. After entering the operating room we take the routine testing for them as BP, HR, ECG, SpO2, open upper extremity venous access, give Ringer's solution by intravenous infusion (infusion rate 6-8ml / kg · h), after the induction of anesthetizing,we start to monitor the concentration of co2 in the end-tidal of exhaling and the concentration of sevoflurane in inhaled and exhaled process. Induction of anesthesia: intravenous inject midazolam followed by 0.05mg / kg, acid cis atracurium 0.05mg / kg, sufentanil 0.5ug / kg, propofol 2mg / kg. Ventilation mode: Chose intermittent positive pressure ventilation, make sure the respiratory rate is 12 / min, tidal volume 8-10ml / kg,inspiratory to expiratory ratio is 1: 1.5, PetCO2 maintained at 30-40mmHg.Intraoperative continuous sevoflurane - nitrous oxide - oxygen anesthesia was maintained state of the patient, to maintain the depth of anesthesia in 1.2-1.5MAC.

Before endotracheal intubation we check the package whether there are situation like damage, leaks cuff and form plastic pipe as the desired angle.After induction of anesthesia we stop giving nitrogen to give oxygen.After three minutes,we put the visual laryngoscopes in by left hand and fully reveal the glottis, Then put the NIM EMG endotracheal tube (Medtronic Xomed, USA) into the glottis by right hand, assistant to extubation guide core at this time.Continue send the tube to the subglottic, place the blue electrodes on the exposed parts of the glottis. Slow adjust the tube, ensure the tube electrode is in contact with both sides of the vocal cords fully.After intubation completing, inflate the cuff,make the tube connect to the anesthesia machine then control ventilation, so we can see the PetCO2 waveform and rise and fall movement on both sides of chest. Place the Dental pad and fix the tube to the right side of the mouth. Endotracheal intubation and anesthetizing are completed by the same anesthesiologist operation.

3 Result

77 (91.7%)had initial intubation achievement in the 84 patients.FROM the 77 cases we had gotten s atisfactory nerve monitoring signal.Whereas there are 7 cases (8.3%) appear abnormal EMG or signal missing,in the 7 cases there is one which being intubated too deep ,3 cases which being intubated too shallow and 3 cases with malrotation intubation.We got the satisfactory signals after adjusting the tube by using the visual laryngoscope.

4 Discuss

During Thyroid Surgery taking advantage of the technology of recurrent laryngeal nerve monitoring (IONM) can accurately locate and identify the recurrent laryngeal nerve, reduce the risk of recurrent laryngeal nerve injury, reduce postoperative complications. CURRENTLY Medtronic Xomed NIM-Response 3.0 EMG nerve monitor and NIM EMG endotracheal tube are the most advanced monitoring equipment. NIM EMG endotracheal tube used in this study is Medtronic Xomed NIM standard enhanced TM endotracheal tube (diameter of the catheter for men 7.0mm, women diameter of the catheter 6.0mm). The TUBE has two pairs of recording electrodes implanted in the side wall.IN the front of electrode there is one exposed portion which is about 3cm and its color is blue, only the exposed portion of the electrode is located in the middle and the recording electrode is located full access to the vocal cords (Fig. 1) , we ensure receiving the most accurate EMG signal.

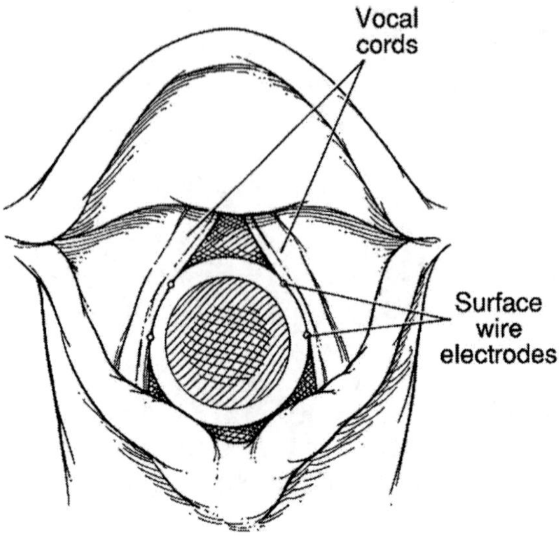

Figure 1. Recurrent laryngeal nerve monitoring

In this paper, 84 patients were given intubation through the visual laryngoscopes (Satum company, Canada). Visual laryngoscope can visually clearly exposed throat structure, reduce intubation injuries, improve the success of intubation time et al [5] Due to the contact between the recording electrode and the vocal cords is directly related to the detection effect of recurrent laryngeal nerve.Visual laryngoscope intubation operations can help us get optimum depth of intubation and visually observe contact situation between the recording electrode and the vocal cords, so that we can adjust the tube in time. Lauren et al [4] summarize the experience of apply Glidescope forNIM EMG endotracheal tube positioning for 3 cases of thyroid surgery, visual laryngoscopes provide good position for NIM EMG intubation, it is the ideal equipment in the operating room for the positing NIM EMG special endotracheal tubes, etc.

Dionigi et al [5] summarizes the 152 cases of thyroid surgery with NIM EMG intubation.There are 15 cases (10%) were not satisfied with the EMG, it is necessary to further adjust the position of the tube,and there are 93% of 15 cases whose contact between the recorded vocal and electrodes is not available.The endotracheal tube malrotation accounted for 53%, 75% of those cases are clockwise rotation to the right. Also in our summary of the 84 patients, seven initial intubation patients (8.3%) did not get a satisfactory EMG, 3 cases (43%) of them as malrotation. SO intubation malrotation is the main cause of failed intubation. According to the experience of our department to rotate the tube counterclockwise 30 ° can effectively prevent the malrotation. While to fix the endotracheal tube to the right side of the mouth is more convenient than the midline placement.

Lu [6] has makes further analysis about the depth of NIM EMG intubation of 105 cases with thyroid surgery, THE analysis showed that for the male average depth of the tube was 20.6 ± 0.97cm, and for the female the average depth of the tube was 19.6 ± 1.0cm. With the increase in height we intubate deeper. The position of the head and neck changes often leads to the displacement of the endotracheal tube.Patients with head flexion may lead the tube deep into the bronchi, the ones whose head backward may make the tube fall into the pharynx, the it can lead maximum displacement to 1.9cm, and this is particularly evident for children, the rotation of the head can also cause the displacement of the tube tip to 0.7cm [7].Thyroid surgery often require neck hyperextended, so it often leads electrode to move upward , resulting in intubation "inadequate."According to clinical observation, most cases will appear the situation of the recording electrodes move upward, so we often need to adjust the depth of intubation assisted by bronchoscopy Tsai et al gather 220 cases of thyroid surgery and divided them into two groups.The first group are the cases whose tube taped at the mouth after being intubated and then we make head move backward .IN The second group,the tube is not fixed temporarily.After getting the head backward ,adjusting and observing the position of the tube by endoscope then fix . At last two groups confirm endotracheal tube position by the endoscope.The ratio of Tube displacement is greater than 1cm groups were 12.7% for the first group,

DOI: 10.1051/bioconf/20170801045

the second group 3.6%[8].The way of intubation in second group is better than traditional way. It is recommended that before the induction of anesthesia patients should make the neck hyperextended position ,so this can avoid the displacement of head and neck posture ,avoid to adjust the depth of the endotracheal tube frequently , achieve the best effect of nerve monitoring.

Conclusion

We got the satisfactory signals after adjusting the tube by using the visual laryngoscope.

References

1. Hui Sun,Xiaoli Liu,Yantao Fu. the application of Intraoperative nerve monitoring technology in the complex thyroid surgery [J]. Chinese Journal of Practical Surgery,2010,1;66-68(in Chinese)
2. Yi Wang,Yishen Zhao,Xiaoli Liu,Hui Sun. common fault and eliminating methods of laryngeal recurrent nerve in thyroid surgery monitoring system [C].2014 CTAThe third session of thyroid surgical peak BBS and the fifth international symposium on thyroid gland cavity mirror. (in Chinese)
3. Liangyu Wang,Digui Weng,Menghua Huang.GlidescopeVisual laryngoscope clinical analysis of 80 cases of emergency tracheal intubation [J]. Fujian Medical Journal,2016,38(1);12-14(in Chinese)
4. Lauren Berkow,Alan P.B.Dackiw,Ralph P.Tufano.Use of the GlideScopefor placement of arecurrent laryngeal nerve monitoring endotracheal tube[J].Journal of Clinical Anesthesia,2011,2;81-83
5. Dionigi G,Bacuzzi A,Boni L,et al.What is the learning curve for intraoperative neuromonitoring in thyroid surgery?[J]International Journal of Surgery,2008,6;7-12
6. I-Chen Lu,Koung-Shing Chu,Cheng-Jing Tsai,et al.Optimal depth of NIM EMG endotracheal tube for intraoperative neuro monitoring of the recurrent laryngeal nerve during thyroidectomy[J].World Journal of Surgery,2008,32;1935-1939
7. Tao Zhu,Yunxia Zuo,main translation. Anesthesiology. Version 5 [M].Beijing: People's Medical Publishing House,2011;1679(in Chinese)
8. Cheng-Jing Tsai,Kuang-Yi Tseng,Fu-Yuan Wang,et al.Electromyographic endotracheal tube placement during thyroid surgery in neuromonitoring of recurrent laryngeal nerve[J].The Kaohsiung Journal of Medical Sciences,2011,27;96-101

A comparative study of sevoflurane and desflurane in the recovery of older patients undergoing thoracoscopic lobectomy

Xuefeng Li[1],Guifeng Liu[2], Yunpengfei Li[1], Huiying Li[1], and Feng Liang[1,a]

[1]*Department of Anesthesiology, China-Japan Union Hospital, Ji Lin University, Changchun, Jilin, China*
[2]*Department of Radiology, China-Japan Union Hospital, Ji Lin University, Changchun, Jilin, China*

Abstract. Purpose :To compare various recovery quality of sevoflurane and desflurane in elderly patients underwent thoracoscopic lobectomy, and to provide a reliable basis for anesthesiologists to choose anesthesia maintaining drugs.Method :Sixty elderly patients undergoing thoracoscopic lobectomy were selected from October, 2015 to May, 2016, and were divided into two groups by random number table, namely, the group of Dofetilide (group D) 30 cases, and sevoflurane (group S) 30 cases.The time of spontaneous respiration, the time of eye opening, the time of extubation and the time of awakening were recorded respectively in two groups. The incidence of nausea, vomiting and restlessness were also recorded.Result:There is no statistical difference comparing data of two groups.The time of spontaneous respiration, the duration of spontaneous breathing, the time of conscious recovery, the time of extubation and the time of leaving the room were shorter in group D than in group S, the difference was statistically significant ($P < 0.05$). (3) There were no significant differences in the incidence of adverse events between the two groups ($P > 0.05$).Conclusion :Compared with sevoflurane, desflurane can reduce the extubation and awakening time, improve the quality of awakening and speed up the turnover of surgery, and it has certain clinical application advantage for thoracoscopic lobectomy in elderly patients.

Introduction

With the development of society to the aging, more and more elderly people to face surgery problems. The elderly due to organ degeneration, heart, lung, brain and other important organs of reserve function reduced, poor tolerance to anesthetics. At the same time elderly patients with liver and kidney function decline, drug clearance cycle is extended accordingly.Therefore, the choice of anesthesia maintaining drugs in elderly patients anesthesia as the top priority. Recently,sevoflurane and desflurane as inhalation anesthesia drugs commonly used in general anesthesia, their quality of awakening after surgery is superior to traditional inhaled anesthetics. In this study, sevoflurane and desflurane used in elderly patients undergoing thoracoscopic lobectomy, compared the awakening time,extubation period and the incidences of adverse events.To provide a reasonable basis for the selection of maintaining inhalation drugs for the treatment of thoracoscopic lobectomy.

[a] Corresponding author: 295253720@qq.com

1.Materials and methods

1.1 General Information

This study was approved by our hospital ethics committee, and patients signed an informed consent form.Sixty patients aged from 65 to 82 years old, 40 males and 20 females with ASA grade II-III were selected from October 2015 to May 2016 undergoing thoracoscopic lobectomy under general anesthesia. Thirty patients in the desflurane group (group D) and 30 patients in the sevoflurane group (group S) were randomized into two groups.

1.2 Methods

1.2.1 anesthesia

30min before anesthesia, we give patients intramuscular injection of penehyclidine hydrochloride 0.01mg / kg. after patients entering the operation room, we routinely monitor their five-lead ECG, pulse oxygen saturation, BIS monitoring.Under local anesthesia we give patients radial artery puncture catheterization to monitor invasive arterial pressure. After opening the upper limb venous access, we infused lactate Ringer solution (infusion rate 5-7ml / kg · h). The patients underwent induction of anesthesia after oxygen for 5min, followed by intravenous injecting midazolam 0.03mg / kg, cisatracurium 0.3mg / kg, sufentanil 0.3mg / kg, propofol 1.2mg / Kg.After the patient's spontaneous breathing and consciousness disappearing, we administered oxygen to the patient and withdrew the nitrogen gas. After 3 minutes, the patients received double-chamber bronchial intubation (No. 35-39). After we completing intubation, we inflated the cuff and connected the endotracheal tube to the anesthesia machine. After adjusting the anesthesia machine to manual ventilation mode, we can see the continuous PetCO2 waveform and patients with good chest flutter on both sides.We identified the location of the tube with a bronchoscope,placed a dental pad and fixed the tube. Ventilation mode: single-lung intermittent positive pressure ventilation, respiratory rate 14 beats / min, tidal volume 8-10ml / kg, Inspiratory vs expiratory 1: 2, PEEP 5mmHg, PetCO2 maintained at 30-40mmHg.Patients continued inhaling sevoflurane 1.5% -3% or desflurane 4% -8%, we maintain anesthesia depth of 1.0-1.3MAC, while appropriately adjusted the depth of anesthesia according to BIS.Before surgery, we added sufentanil 5-10ug, and according to the actual situation in the additional cisatracurium 5ml / h to maintain muscle relaxation. At 15 min preoperatively, we reduced the sevoflurane and desflurane inhalation concentration according to the flow-decreasing method. At the end of the procedure, we stopped inhalation of the anesthetic and opened a large fresh gas flow (O2 5 L / min) to wash out the residual anesthetic. At the same time, in order to prevent residual effects of muscle relaxants, the two groups were given neostigmine 1mg and atropine 0.5mg antagonistic.

1.2.2 Outcome Measurement

After surgery, we recorded the time of restoring spontaneous breathing, patients' eyes opening moment, removal of endotracheal tube and leaving the recovery room. At the same time, adverse events such as restlessness, nausea, vomiting, hypotension, hypertension, arrhythmia, airway spasm were recorded.

1.2.3 Statistical Methods

All data using SPSS 22.0 statistical software for processing, measurement data using mean and standard deviation (`X ± S) to represent. And paired t test for data comparation between two groups, $P < 0.05$ considered statistically significant.

2.Results

2.1 General information

Comparing gender, age, BMI index, ASA classification and operation time of the two groups,the difference was not statistically significant (P> 0.05). Table1

Table 1 Comparison of the general data of the two groups of patients (X± S)

	S Grope	D Grope
gender(M/F)	19/11	20/10
age	71.40±4.13	70.93±4.35
weight(kg)	67.62±7.33	68.13±7.45
ASA(II/III)	17/13	20/10
Operation time(min)	125.7±15.43	127.3±16.36

2.2 Comparison of the recovery time between two groups

The time of spontaneous breathing recovery , following the command of open eyes , extubation and leaving the recovery room of Desflurane group (D group) patients is shorter than sevoflurane group (S group), and P <0.05, with statistical significance. Table 2.

Table 2 Comparison of wake-up time between the two groups (X ± S)

	S Group	D Group
Breathing recovery(min)	9.62±1.31	6.47±1.14
Open eyes time(min)	12.79±1.93	8.26±1.77
Extubation time(min)	16.57±2.61	11.63±1.85
Leaving recovery room(min)	36.73±4.69	29.18±4.02

2.3 The comparison of the incidence of adverse events between two groups

There was no significant difference in the incidence of adverse events between the two groups (P> 0.05). Table 3

Table 3 Comparison of the incidence of adverse events between the two groups(cases)

	S Group	D Group
hytertension	2	3
hypotension	1	1
arrhythmia	0	0
Nausea,vomiting	3	2
restlessness	5	3
Airway spasm	0	1

3.Discussion

For the elder there is a decline in the physiological reserve capacity and their function of important organs decreased.In the central nervous system these phenomena showed as average weight of the brain decreasing, neurons shrinking and the density reducing, reduction of cerebral blood flow, neurotransmitter and receptor,all these often lead to Waking up for a long time, poor quality of recovery, and sometimes even postoperative cognitive dysfunction [1].At the same time, most elderly patients'liver and kidney function declining leads to drugs clearance cycle extending, but also exacerbated the wake-up time delay. Therefore, looking for anesthesia maintaining drugs is

DOI: 10.1051/bioconf/20170801046

particularly important,Which can make people rapid recovery after surgery and lead less postoperative complications.

Thoracoscopic surgery has the advantages of short time, less trauma and quick recovery. Surgery often requires patients to wake up after surgery, after extubation, if the patient has a good spontaneous breathing, so as to reduce postoperative complications and can make patients early discharge [2]. Sevoflurane or desflurane is with these characteristics, is one of ideal thoracic surgery anesthetics.

Desflurane is a third-generation inhaled anesthetic with a blood / gas partition coefficient of 0.42 and a fat / blood partition coefficient of 27, whose coefficient is the smallest of existing anesthetics, and therefore it can be controlled well. Because of low tissue dissolution, desflurane can be excreted more quickly [3]. At the same time, desflurane almost has no metabolites, which is known as inhalation anesthetics which has the least biotransformation in human body.Its impact on liver and kidney function is making desflurane more suitable for elderly patients with anesthesia.

In this paper, we observed the effect of sevoflurane and desflurane on the recovery period of the elderly patients and found that the spontaneous respiration recovery time, the time to follow the command, the time to extubate the tracheal catheter and the time to leave the recovery room were shorter in sevoflurane group Group, and P values were less than 0.05, it is with statistical significance. There were no significant differences in postoperative adverse events between the two groups (P> 0.05).The two groups of patients studied in this study had similar age, body mass index, ASA classification, similar induction regimen, depth of anesthesia, and additional medication during surgery. The duration of surgery was similar and the time of discontinuation of narcotics was the same. Thus the study can present the impact of desflurane and sevoflurane on elderly patients'anesthesia recovery period objectively and correctly.Makey [4] and the other confirmed, comparing with sevoflurane, desflurane was used for anesthesia,and the patients can more rapidly recover after surgery, and their airway preventably reflexed faster, tracheal intubation was removed for a shorter time.Dexter [5] and others' meta-analysis results also show that extubation time reduced by 20% -25% in the patients who was used desflurane for anesthesia than sevoflurane. Similarly, in a recent meta-analysis, Agoliati et al. [6] demonstrated a significant advantage of desflurane in rapid extubation and reduction in wake-up delay comparing with sevoflurane. The study found that the patients who was used of desflurane for anesthesia less an average of 34% extubation time,the number of extubation delay also reduced by 95% -97%. In addition, Heavner [7] through the study found that when desflurane or sevoflurane was for older patients who undergo surgery for more than 2 hours, the patient that was used desflurane their time of eyes-open, following the command to handshake , extubation and orientation to restore was significantly better than sevoflurane group.

In thoracic surgery, one-lung ventilation may lead to alveolar damage of the ventilated lungs, this leads to pulmonary inflammatory responses that leading to lung injury . In contrast to total intravenous anesthesia, the use of sevoflurane or desflurane inhalation anesthesic can inhibit the release of inflammatory mediators in local cells, reduce the role of inflammatory factors, thereby reducing the inflammation of the ventilation lung. Modolo [8] and others summed up 20 comparative studies about anesthesia and inhalation anesthesia for single lung ventilation,then achieved a meta-analysis that pointed out there is no strong evidence making sure that anesthetic drugs will affect the prognosis of patients. Dupont [9] also reported that the impact of sevoflurane and desflurane on hemodynamics in patients with pulmonary surgery had no statistically significant difference, but desflurane group woke up faster and earlier.

HUA zhuming[10] found that the incidences of postoperative adverse events comparing sevoflurane and desflurane for elderly patients in the anesthesia through the study were not significant differences. Zhang et al found that compared with sevoflurane group, desflurane group in elderly patients had fewer postoperative complications. The two studies, as in this study, included fewer samples and the effect of sevoflurane and desflurane on the postoperative complications of elderly patients may require additional large, multicenter, randomized, controlled trials further validate.

DOI: 10.1051/bioconf/20170801046

4.Conclusion

Compared with sevoflurane, desflurane for elderly patients with thoracoscopic lobectomy can shorten the extubation and recovery time, improve the quality of recovery, speed up the turnover of surgery, it has certain clinical application advantages.

References

1. Guolin Wang, Hong Zhang, Yanlin Wang, edit the Geriatric Anesthesia [M]. Beijing: People's Medical Publishing House, 2009; 15. (In Chinese).
2. Zhonghou Cui, Chunyan Zhao. Clinical Application of Video-assisted Thoracic Surgery [J]. Chinese Journal of cardiovascular surgery, 1994, 10; 10. (In Chinese)
3. Tijun Dai, Tian Yu, edit the Anesthetic Pharmacology. Third Edition [M]. Beijing: People's Medical Publishing House 2011; 79. (In Chinese)
4. Mckay R E,Large M J,Balea M C,et al.airway reflexes return more rapidly after desflurane anesthesia than after sevoflurane anesthesia[J].Anesth Analg,2005,100(3);697-700.
5. Dexter F,Bayman E O,Epstein R H.Statistical modeling of average and variability of time to extubation for meta-analysis comparing desflurane and sevoflurane[J].International Anesth Analg,2010,110(2);570-580
6. Agoliati A,Dexter F,Lok J,et al.Meta-analysis of averageand variability of time to extubation conparing isoflurane with desflurane or isoflurane with sevoflurane[J].Anesth Analg,2010,110;1433-9.
7. Heavner J E,Kaye A D,Lin B k,et al.Recovery of eiderly patients from two or more hours of desflurane or sevoflurane anaesthesia[J].Br J Anaesth,2003,91(4);502-506.
8. Modolo N S,Modolo M P,Marton M A,et al.Intravenous versus inhalation anesthesia for one-lung ventilation[J/OL].Cochrane Database Syst Rev,2013,7;CD006313[2013-07-011].
9. Dupont J,Tavernier B,Ghosez Y,et al.Recovery after anesthesia for pulmonary surgery:desflurane,sevoflurane and isoflurane[J].Br J Anaesth,1999,82(3);355-359
10. Zhuming Hua, Shiqing Lin. Desflurane and sevoflurane for comparison of recovery after general anesthesia in elderly people [J]. Guangdong Medical, 2006, 8 (27). (In Chinese)

The effects of music intervention on the emotion, haemodynamics and endocrine hormone level of senile patients during the perianesthesia period

Feng Liang, Fang-lei Han and Shuang Qi[a]

China–Japan Union Hospital of Jilin University

Abstract. Purpose: to explore the effects of music intervention on haemodynamics and endocrine hormone level of senile patients. Methods: choosing ASA II -III level and selecting 60 senile patients performed with intraspinal anesthesia; randomly dividing them into the intervention group and control group with 30 cases in each group. Both the two groups of patients completed the SAS before entering the operating room. When the patients of the intervention group entered the operating room, selected music would be played in a continuous loop immediately till the end of the operation while the doctors of the control group just told the patient to relax without music or other special treatment. Recording and comparing the SAS scores, SBP, HR and salivary cortisol concentration of the two groups before and after intervention. Results: the SAS scores, SBP, HR and salivary cortisol concentration of the intervention group were significantly decreased compared with the value before intervention and the differences within and between the groups were significant ($P < 0.05$). Conclusion: music intervention can effectively reduce the preoperative anxiety, ease intranperative tension and maintain the stability of hemodynamics, so it is beneficial for senile parents to live through the perianesthesia period.

Due to the retrogression of organ functions caused by pathological aging, elderly people's response to stress and their adapting mechanism are decreased obviously. Especially for conscious senile patients without general anesthesia in the operation, serious preoperative anxiety and iatrogenic noise caused by psychological stress will lead to excessive excitement of sympathy - LCSA with risks of arrhythmia and sudden death[1]. And the excessive activation of hypothalamus - hypophysis cerebri - HPA will cause endocrine and immune system disorders, which will significantly affect patients' condition and prognosis [2]. Therefore, effective psychological intervention is of great significance for the maintenance of physiological and psychological stability of senile patients during the perioperative period.

Music intervention is one of the ways of mental intervention. It can relieve pain, improve mood and stabilize emotions. It is usually used to assist postoperative rehabilitation and pain therapy clinically [3~4]. But there is less research on its application on senile patients during the perioperative period.

The purpose of this study is to discuss the effects of music intervention on the emotions, haemodynamics and endocrine hormone levels of senile patients so as to provide theoretical basis for its clinical application.

[a] Corresponding author: qishuang@jlu.edu.cn

1. Materials and Methods

1. 1 General data

approved and signed by the Ethics Committee of our hospital, selecting 60 patients performed intraspinal anesthesia during January 2014 to June 2014 in China-Japan union hospital of Jilin University. The patients were 65 -75 years old with unlimited sex. The ASA was at II -III level. The operation was performed around 9 am, lasting 2-3 hours with less amount of bleeding. The patients had no heart, lung, liver, kidney or endocrine system diseases and didn't have serious visual and hearing difficulty or mental illness in the preoperative examination. The ages, income, education degrees and occupations of the two groups had no obvious differences; randomly dividing the patients into intervention group and control group with 30 cases in each group.

1.2 Research methods

consultation before the operation day could establish good communication and trust between doctors and patients. Illustrating the effects of music playing systematically and asking patients to select music from alternative music list(hypnotic music) according to their preference. Both the two groups of patients completed SAS before entering the operating room. Then their blood pressure and heart rate got routine monitoring in the operating room. The moment the intervention group entered the operating room, the selected music would be played in a continuous loop until the end of the operation. The voice was controlled in an acceptable range, generally 50 ~ 60 db. The patients of the control group were asked to relax without music playing or other special treatment. Try to keep the surroundings quiet.

1.3 Observing indexes

(1) emotional state assessment: using SAS of Zung of 1983 to have psychological assessment and then getting an initial score. The initial score multiplying 1.25 equaled to the standard score. The standard score 50 was regarded as the critical value. 50 -59 points were divided into mild degree; 60-69 points were moderate; 70 points or more were severe. The higher the score was, the heavier the anxiety symptom was [6]. Recording the SAS scores when entering the operating room (before intervention) and at the end of surgery (after intervention); (2) Hemodynamic assessment: recording the hemodynamic indexes, SBP and HR in the two time points; (3) The endocrine hormone index adopted special salivette. Pouring the sliver in the salivette to patients' mouth for 45-second chewing; then putting the sliver after chewing to the salivette tube. 0.5-1.5 ml saliva could be collected after centrifugation. The saliva was put in -84°C refrigerator for restoration and got detected intensively. Adopting ELISA to measure the concentration of salivary cortisol(kit number: HT81012; place of origin: Canada Pllabs).

1. 4 Statistical analysis

adopting SPSS 11.0 software to conduct χ^2 test to the enumeration data; the measurement data was expressed by mean ± standard deviation ($\overline{X}\pm S$); the comparison within groups was tested by matching t; the comparison between groups was analyzed by one-way variance; $P < 0.05$ meant there existed significant differences.

2 Results

2. 1 The comparison of the emotional states (SAS score) of the two groups: before intervention, the SAS scores of the two groups were both over 50 and both the two groups had one case getting more

DOI: 10.1051/bioconf/20170801047

than 60, but there was no significant difference between the two groups(P > 0.05); after intervention, the SAS score of the intervention group declined obviously, so there were significant differences between and within groups(P < 0.05); though the SAS scores of the control group had a rising trend, there was no statistical significance (P > 0.05).

2. 2 The comparison of the haemodynamics of the two groups: the SBP and HR of the two groups had no significant differences (P> 0.05); After intervention, the SBP and HR of the intervention group declined obviously and the comparison between and within groups had significant difference(P<0.05) while the control group had no significant change (P > 0. 05).

2.3 The comparison of endocrine hormone indicators: before intervention, the saliva cortisol concentration of the two groups had no significant differences (P > 0. 05); after intervention, the concentration of the intervention group decreased obviously (P < 0.05); while the salivary cortisol concentration of the control group was increased, but there was no statistical significance (P > 0. 05). See table 1.

Table 1. the comparison of the SAS scores, SBP, HR and saliva cortisol concentration of the two groups before and after intervention($\bar{X}$±S, n=30)

Groups	Item/time	Before intervention	After intervention
The intervention group	SAS scores	52.20±7.55	44.23±3.17▲△
	SBP(mmHg)	143.78±4.52	119.56±5.65•○
	HR(/min)	85.21±3.33	70.17±6.64★☆
	salivary cortisol (nmol/l)	26.11±3.45	14.75±2.12 ▼▽
The control group	SAS score	51.10±7.86	53.60±7.65
	SBP(mmHg)	144.84±4.75	145.32±5.11
	HR(/min)	83.70±3.58	82.71±3.23
	salivary cortisol (nmol/l)	21.87±3.53	22.81±3.55

After intervention: comparing the SAS scores between groups and within the intervention group, P▲△<0.05; comparing the SBP, P•○<0.05; comparing the HR, P★☆<0.05; comparing the salivary cortisol, P▼▽<0.05.

3 Discussions

As a strong psychological stressor, the fear of surgery and anesthesia has a direct impact on the emotion of perioperative patients. They usually show different degrees of anxiety [7]. Elderly people are easily affected by the recession of physiological functions, chronic diseases, changes of working and living environment, fear of therapeutic effects and other negative events, so their emotions are more likely to fluctuate. This study found that the SAS scores of the two groups before intervention were greater than 50 and both the two groups had a case reaching 60 points(moderate anxiety), which suggested the common occurrence of preoperative anxiety of senile patients; the SAS scores of the control group increased after intervention, which meant that for conscious senile patients, the continuous psychological stress caused by iatrogenic factors (such as the conversation of doctors, noise of surgical instrument and so on) might aggravate the degree of anxiety though relatively perfect anesthesia had been performed.

Intense psychological stress will stimulate hypothalamus - hypophysis cerebri - HPA and sympathetic nerve - LCSA. The excessive cortisol and epinephrine released in the two axis systems

DOI: 10.1051/bioconf/20170801047

acts on the body through blood circulation, causing the disorder of haemodynamics and immune system [8]. Elderly people's retrogression of automatic nervous system, poor cardiovascular sympathy/neuroendocrin, gradual retrogression of neurohormone and immune functions as well as intense psychological stress are more likely to cause cardio-cerebral vascular accidents and the occurrence of postoperative complications. The results of this study showed that the SBP, HR and salivary cortisol concentration of the control group increased after intervention while the salivary cortisol concentration of 2 cases reached a critical value after operation (after intervention). It suggested that the maintenance of physical steady-state of elderly people would be increasingly difficult along with the longer lasting time of psychological stress. Therefore, systematic and effective psychological intervention on senile patients during the perianesthesia period is very necessary. Music intervention is a measure of non-invasive intervention with low cost and easy operation. It avoids the influence of pharmacological intervention on patients' cognitive function after operation, so it is more acceptable for patients. The "situationality" and "infectivity" of music can bring patients to a special artistic conception.According to the results of this study, music after the intervention, the SAS scores, SBP, HR and salivary cortisol concentration of the intervention group were all significantly decreased after music intervention, which showed that music intervention could effectively relieve patients' anxiety and reduce the stress response of their bodies. Its mechanism may be: on the one hand, music intervention can alleviate excessive tension of sympathetic nerve, reduce stress reactions and promote emotional sedation [9]; on the other hand, music is acoustic vibration with certain rules. It can be transferred to neural signal via cochlea to cochlear nucleus; then it was conveyed to auditory thalamus and amygdaloid nucleus from the cochlear nucleus; it acts on hypothalamus through the neural pathway between amygdaloid nucleus and hypothalamus to reduce the activity of HPA axis so as to lower the content of cortisol released by adrenal cortex[10]. In this study, patients of the intervention group chose the music they preferred from the alternative music list. It aroused their enthusiasm of participation; in the whole operating process, patients were in beautiful melody, so their emotion was highly relaxed, which fully embodied the development trend of comfortable and humanized medicine.

In a word, music intervention can effectively reduce preoperative anxiety, relieve intranperative tension and maintain the haemodynamics stability of senile patients to help them live through the perianesthesia period safely.

References

1. Xin-Liang Zhuan, editor. Modern Anesthesiology [M]. Third Edition. Beijing: People's Medical Publishing House, 2003, 190. Chinese.
2. Price MA, Tennant CC, Buttow PN, et al. The role of psychosocial factors in the development of breast carcinoma: Part II. Life event stressor. social support, defense style, and emotional control and their interactions[J].Cancer, 2001;91(4): 686~697.
3. Krishnaswamy, Priyadharshini, et al. Effect of music therapy on pain and anxiety levels of cancer patients: A pilot study[J]Indian Journal of Palliative Care ,2016;22(3) : 307-311
4. Shu-Ming Wang, Lina Kulkarni, Jackqulin Dolev. et a1. Music and Preoperative Anxiety:A Randomized, Controlled Study[J]. Anesth Analg ,2003;94:1489~1494.
5. Juan Du. Psychological influence of music on the cesarean section and nursing experience [J]. Journal of Clinical Nursing, 2002;15(1):53-54. Chinese.
6. Nian-Feng Guo, editor. Psychologist (National Occupation Qualification Training Tutorial) [M]. Beijing: National Publishing House, 2002,4. Chinese.
7. Pin-Ling Peng, editor. General Psychology [M]. Beijing: Beijing Normal University Press, 2002,3. Chinese.
8. Natsuaki MN, Klimes–Dougan B, Ge X, Shirtcliff EA,Hastings PD, Zahn-Waxler C. Early pubertal maturation and internalizing problems in adolescence: Sex differences in the role of cortisol reactivity to interpersonalstress[J]. Journal of Clinical Child and Adolescent Psychology, 2009;38(4):513~524.

DOI: 10.1051/bioconf/20170801047

9. Wang J, Jian F, Chen J, et al. Cognitive behavioral therapy for orthodontic pain control:A randomized trial[J]. J Dent Res, 2012,91(6):580-585.
10. R. Padmanabhan AJ, Hildreth D, Laws. A prospective, randomised, controlled study examining binaural beat audio and preoperative anxiety in patients undergoing general anaesthesia for day case surgery[J].Anaesthesia,2005;60:874~877.

Randomized Trial Study on Ramosetron and Dexamethasone in the Prevention of Nausea and Vomiting after Laparoscopic Cholecystectomy under General Anesthesia

Feng Liang, Fang-lei Han and Shuang Qi[a]

China–Japan Union Hospital of Jilin University

Abstract. objective: this study aims to identify whether dexamethasone can be used in the prevention of nausea and vomiting after laparoscopic cholecystectomy (LC) under general anesthesia. Methods: 400 patients who were performed with LC under general anesthesia were chosen and divided randomly into two groups with 200 patients each, namely ramosetron group and dexamethasone group. Results: there's no significant difference in the effectiveness of complete inhibition of nausea and vomiting within 24 hours after the surgery in both groups (P> 0.050). Meanwhile, there's no significant side effects in association with the anti-emetic drugs within 24 hours after the operation in both groups, either. Conclusion: Dexamethasone plays a similar effective role as Ramosetron in the prevention of nausea and vomiting after laparoscopic cholecystectomy under general anesthesia.

1 Instruction

Postoperative nausea and vomiting (PONV) is a common and major complication of patients after surgery. Literatures show that the incidence of PONV in some operations is 25%-30%. Therefore, the control of PONV remains to be a challenge[1, 2].Even though PONV isn't a life-threatening postoperative complication, it's still a hazard that causes complications such as wound dehiscence, electrolyte disturbance and dysphoria[3,4]. Some prophylactic medications, such as antihistamines, 5-serotonin antagonist and steroid, are able to effectively reduce the incidence of PONV. These drugs, however, are limited in the widespread clinical application due to their prices and side effects. This study aims to discuss and explore whether Dexamethasone, a medicine with appropriate price and no adverse effect (or a low incidence of side effects), can achieve the same effect as ramosetron in the prevention of PONV.

1.1 Admission standard

400 patients who were about to be performed with laparoscopic cholecystectomy under general anesthesia were enrolled in this study. Before induction of anesthesia, 200 patients were treated with dexamethasone and the other 200 were treated with ramosetron. ASA grading: I-II grade, female patients aged 18-60. Obesities, pregnant women or women in lactation or menstrual period, smokers, alcoholics, drug addicts, patients that had used anti-nausea or anti-vomiting drugs within 24 hours before the surgery, patients with combined severe heart, liver, kidney, lung diseases, diabetes, patients

[a] Corresponding author: qishuang@jlu.edu.cn

DOI: 10.1051/bioconf/20170801048

with hypertension, migraine and central nervous system diseases, psychiatric patients, postoperative patients with preserved gastric tube as well as patients with PONV history were excluded. The risk score indicators were based on the simplified PONV risk scoring summarized in Apfel[5]for adult treated with inhaled general anesthesia, including women, nonsmokers, history of PONV and / or motion sickness as well as postoperative opioid use. The risk scores of patients fitting the profile of item 0,1,2,3,4 were respectively 0,1,2,3,4 with a respective incidence of PONV of 10%, 21%, 39%, 61% and 79%.

1.2 Anesthetic methods

Before the induction of anesthesia, 10 mg dexamethasone of slow intravenous injection was given to the dexamethasone group, and 0.3 mg of ramosetron intravenous injection was slowly administered to the patients in the Ramosetron group. The injection time of induced anesthesia drugs were not less than 1min. All patients underwent general anesthesia with total intravenous anesthesia after tracheal intubation. Anesthesia was induced by intravenous injection of 0.1mg/kg midazolam, 0.5μg/ kg sufentanil, 2mg/kg propofol, 0.05mg/kg cis-atracurium after rapid induction of tracheal intubation. During the operation, continuous infusion of propofol was used to maintain the target concentration of anesthesia at 4μg/ml. In addition, it's supplemented by intermittent intravenous injection of sufentanil. During the intraoperative and postoperative process, patients' ECG, cuff blood pressure and oxygen were monitored. After the surgery, it depended on patients' situation whether the mixture of 2mg neostigmine and 1mg atropine would be used to antagonize residual muscle relaxation. In terms of postoperative analgesia, they were given 0.5μg/ml sufentanil in 100ml dilution solution at the speed of 2ml/h.

1.3 Observation on curative effect in the prevention of PONV

The time of extubation was recorded and the incidence of PONV within 24 hours after extubation was recorded as well. Vomiting without gastric content, namely, retching, or vomiting with the gastric content were not strictly distinguished and were both regarded as vomiting. Patient with severe retching or vomiting might be given remedial drug intervention upon their request and the consent of the clinician.

1.4 Statistical analysis

SPSS 13.0 software was adopted to conduct data analysis. t test was used in the comparison of measurement data and the comparison of enumeration data were subjected to $\chi 2$ test. $P<0.050$ indicated significant differences.

2 Results

The basic data of the test group are as shown in table 1. There were no significant differences in age, height, weight, BMI, ASA grade, operation time (min) and bleeding volume (ml) between the two groups. The effective rates of complete inhibition of nausea and vomiting in the dexamethasone group and ramosetron group within 24 hours after the surgery were respectively 78.00% (156/200) and 84.50% (169/200). In addition, their postoperative nausea rates were respectively 22% (44/200) and 15.50% (31/200) while postoperative vomiting rates were respectively 10.50% (21/200) and 7.50% (15/200), suggesting no significant difference between the two groups ($P>0.050$). Moreover, no significant adverse effects associated with anti-emetics were observed within 24 hours postoperatively.

Table 1. The basic data of the test group

	Dexamethasone	Ramosetron	P
Average age/years	38(9)	39(10)	>0.20
Average weight /kg	56(11)	58(9)	>0.10
BMI/,kg/m2	24(9)	23(9)	>0.50
ASA□ /n(%)	91(58)	95(57)	>0.20
ASA□ /n(%)	65(42)	74(43)	>0.05
The operationg time ,min	172(5)	169(8)	>0.10
The bleeding,ml	6(2)	6(1)	>0.50

3 Discussion

Vomiting is a complex reaction mediated by the vomiting center. Nucleus tractus solitaries, dorsal nucleus and reticular structure of peripheral ambiguous nucleus in the brainstem constitute the vomiting center which is able to accept the impulses from many areas, including chemical triggering zone, gastrointestinal autonomic nervous system and vestibular organs. PONV is a common and major complication for postoperative patients. It will delay the recovery time after anesthesia in the recovery room, prolong the hospitalization time and increase the cost of treatment. PONV is a result of a variety of factors and often occurs after regular anesthesia.

A large number of prospective data show that the incidence of PONV is associated with risk factors related to patients or anesthesia, instead of the operation itself[6], which explains the impact of formulations of anesthetic drug on PONV. Sinclair and other scholars[7]found that every time the operation time was prolonged for 30min, the incidence of PONV would increase by 60%. However, the approaches of laparoscopic cholecystectomy surgeries were almost the same with similar operation time. In addition, based on the high similarity of samples, drug dosage for induction of anesthesia and anesthesia maintenance, methods and frequency were rather similar. Therefore, the sample selection criteria resulted in a greater similarity of anesthetic formula given to each patient. In this study, patients were given 10mg dexamethasone or 0.3mg ramosetron before the operation.

The effective rate of inhibition of nausea and vomiting within 24h after the surgery of dexamethasone group and ramosetron group were respectively 78.00% (156 /200) and 84.50% (169/200). The incidence of postoperative nausea were respectively 22.00% (44/200) and 15.50% (31/200) while the incidence of postoperative vomiting were respectively 10.50% (21/200) and 7.50% (15/200), indicating no significant difference between the two groups (P> 0.050). Moreover, no significant adverse effects in associated with anti- nausea and anti- vomiting medicines were observed during 24 hours postoperatively. And yet similar effects were observed in the prophylaxis of PONV in laparoscopic cholecystectomy under general anesthesia.

However, questions such as whether a smaller dose of dexamethasone will have the same antiemetic effects in the control of PONV and whether the combination use of dexamethasone and ramosetron will play a more effective role in the control of PONV still require further study.

References

1. Gan TJ, Meyer T, Apfel CC, et al. Consensus guidelines for managing postoperative nausea and vomiting [J] . Anesth Analg, 2003, 97 :62
2. Huang JC, Shieh JP, Tang CS , et al. Low-dose dexamethasone effectively prevents postoperative nausea and vomiting after ambulatory laparoscopic sur ger [J].Can J Anesth, 2001,48 (10) :73
3. Wolf gang K, Gassmayr S, Preoperative a nxiet y , stress and pre medication [J] .Baillier e' s Clin Anesth, 1998, 12 (3):2536

DOI: 10.1051/bioconf/20170801048

4. Henzi I, Walder B, Tramer MR. Dexamethas one for the prevention of postoperative nausea and vomiting: aquantitative systematic review [J] . Anesth Anal g, 2000, 90:186
5. Apfel CC , Laara E, Koivuranta M,et al.A simplified risk score for predicting postoperative nausea and vomiting: conclusion fromcross validations between two centers [J]. Anesthesiology, 1999, 91:693
6. Stadler M, Bardiau F, Seidel L, et al .Dative nausea and vomiting [J] .Anesthesiology, 2003,98 (1):46
7. Sinclair DR, Chung F,Mezei G. Can postoperative nausea andvomiting be predicted [J].Anesthesiology, 1999,91:109

DOI: 10.1051/bioconf/20170801049

The Anesthetic Effect of Dexmedetomidine on ERCP of Senile Patients

Feng Liang, Fang-lei Han and Shuang Qi[a]

China - Japan Union Hospital of Jilin University

Abstract. Objective: to discuss the effectiveness and security of Dexmedetomidine's conscious-sedation application in ERCP of senile patients. Method: selecting 80 senile patients performed with ERCP and randomly dividing them into Dexmedetomidine group (group D) and normal saline group (group C) with 40 cases in each group. Pumping 0.5 ~ 0.8 ug/kg Dexmedetomidine through the vein 15 min after surgery while pumping isodose NS to group C at the same time. Results: both the two groups completed the ERCP treatment. The tolerance of group C is poorer with more adverse reactions; the operating time of group D is shorter, but the difference has no statistical significance. During the operation, compared with group D, the MBP and HR value of group C rose obviously with significant difference ($P < 0.05$) while the MBP has no obvious change within the group; HR declined sharply after pump injection ($P < 0.05$), but the value is within the normal range. The contrast of SPO2 and RR within and between the two groups have no significant difference. Conclusion: applying Dexmedetomidine for conscioussedation in senile patients' ERCP leads to hemodynamic stability and no respiratory depression with less adverse reactions. The application during the perioperative period is effective, safe and comfortable, worthy of popularization and application in clinic.

1 Instruction

With the unceasing development of ERCP technology, ERCP has become an important diagnosis and treatment of pancreatic disease, but during the treatment, patients will suffer abdominal pain, nausea and vomiting, so the intense discomfort will make patients unable to actively cooperate with the operation. Besides, the majority of ERCP group are senile patients. The stress of operation can cause threat to the life of senile patients. Therefore, it is necessary to adopt painless technique and strengthen monitoring. But ERCP patients are in prone position with difficult management breathing. The most of the initial pain-free technology adopts trachea cannula with general anesthesia. The anesthesia is costly with complicated operation and slow postoperative recovery. In order to improve the treatment quality and guarantee the security of senile patients during the perioperative period, our hospital adopts Dexmedetomidine for conscious sedation and observes the clinical effect so as to evaluate the effectiveness and security of its application in ERCP of senile patients.

2 Materials and Methods

[a] Corresponding author: qishuang@jlu.edu.cn

2.1 General data

Selecting 80 senile patients performed with ERCP at the age from 55-75 years old and weighing 50-75 kg, including 42 males and 38 females; ASA is in II-III level. They had no heart and respiratory system disease, incomplete liver and kidney function or drug allergy history. Randomly dividing the patients into Dexmedetomidine group (group D) and NS group (group C), namely 40 cases in each group.

2.2 Methods

Preoperative patients took 10 ml 2% lidocaine mucilage orally. After the pharyngeal numbness, their ECG, NBP, SPO2 and RR would get routinely monitored. Patients added oxygen reservation during surgery by 3-min oxygen uptake through a nasal mask. Pumping 0.5~0.8ug/kg Dexmedetomidine through the vein to group D in 15 minutes and observing the patient's condition. When patients reacted slowly and went into the state of sleepiness but could still be awakened, then it is the timing to start operation. The patients were sent to the viewing room after operation. They wouldn't be sent back to the ward until complete revival; group C would be operated after pumping isodose NS at the same time.

2.3 Observing Indexes

Recording the HR, MBP, SPO_2 and RR of the two groups respectively when entering the operating room (t0), the pumping moment (t1), the end of pumping (t2), the endoscopic insertion moment (t3), the beginning of angiography (t4) and the withdrawal moment of endoscope; recording adverse reactions such as bucking, nausea, vomiting, restlessness, etc; recording the temporal summation (operating time) from endoscopic insertion to withdrawal.

2.4 Statistics processing

Using t to test measurement data and adoptingχ^2 to verify enumeration data. $P < 0.05$ means the difference has statistical significance.

3 Results

The differences of general data between groups had no statistical significance. They all completed EVL treatment. The tolerance of group C resistance was poorer with more adverse reactions; the operating time of group D was shorter than group C, but there was no statistically significant difference between the two groups (table 1). Comparing with the time point of t_0, the MBP and HR within group C and group D was significantly increased in the process of operation (t_3~ t_5), so the difference was significant ($P < 0.05$); the MBP of group D had no obvious change comparing with the values of other 5 time points; HR declined significantly after pumping (t2 ~ t5) ($P < 0.05$), but the values were all within the normal range. There was no significant difference of SPO_2 and RR value between groups and within groups (table 2).

Table 1. the comparison of adverse reactions and operation time of the two groups of patients (n = 40)

Groups	Nausea	Emesis	Bucking	Dysphoria	Operation time
Group D	5	0	0	0	9.2±2.4
Group C	30（75%）	4	8(20%)	3	11.2±1.5

DOI: 10.1051/bioconf/20170801049

Table 2. the comparison of the changes of MBP, HR, SPO_2 and RR of the two groups of patients

Indexes	Groups	t0	t1	t2	t3	t4	t5
MAP	Group D	93.2±2.4	96.7±2.1	95.2±2.1	97.2±1.9	98.2±2.4	97.2±2.9
	Group C	96.3±2.5	97.1±2.6	96.8±2.0	106.6±3.7*	103.5±3.6*	104.2±4.2*
HR	Group D	83.5±11.7	83.7±12.5	72.5±9.7$^{※}$	72.4±9.4$^{※}$	73.5±8.7$^{※}$	74.5±8.8$^{※}$
	Group C	84.5±11.9	84.3±11.9	83.5±11.9	95.6±8.2$^{\#}$	96.7±8.5$^{\#}$	97.5±8.7$^{\#}$
SPO2	Group D	96.5±2.5	96.5±2.5	96.4±2.6	96.5±2.5	95.6±2.4	96.7±2.1
	Group C	96.7±2.3	97.2±2.8	99.8±2.2	95.9±2.1	96.5±2.4	96.5±2.4
RR	Group D	13.7±6.2	13.2±6.2	11.4±7.2	12.5±6.7	12.5±6.3	12.9±6.1
	Group C	13.1±6.3	13.1±6.3	13.9±7.8	17.2±6.2$^{\triangle}$	16.8±6.6$^{\triangle}$	13.2±6.2

Comparing the data of t_3-t_5 of group C with group D: $P^{*}<0.05$; $P^{\#}<0.05$; $P^{\triangle}<0.05$;
Comparing he data of t_2-t_5 of group D with t0: $P^{※}<0.05$.

3 Discussions

Comfortable and secure medical service is the development trend of clinical medicine in recent years. Though ERCP can be performed under local anesthesia, but senile patients' tolerance of strong stress is poor, so cardiocerebral events are likely to be induced; adverse reactions such as bucking, nausea and vomiting, dysphoria and so on often interfere with the operation and increase the risk of bleeding [3]. This study found that 75% of senile patients of group C suffered different degrees of nausea and vomiting while 20% of them appeared bucking. The RR, HR and blood pressure of each time point during the operation rose obviously comparing with the moment when they entered the operating room, which suggested that performing ERCP on senile patients on the condition of topical anesthesia had certain risks.

Painless microscopy is usually completed under intravenous anesthesia. Using intravenous anesthetics or analgesia medicine often causes swallowing, bucking and other protective areflexia, so aspiration is easy to happen and difficult to deal with. In addition, the commonly used intravenous anesthetics, such as propofol and fentany, have some inhibiting effect on angiocarpy and respiratory system. The haemodynamics of most elderly people is not stable. A small amount of sedative and analgesic drugs may cause obvious inhibition of angiocarpy and respiratory functions, so such operations adopting intravenous anesthetics on senile patients should be performed particularly cautious.

Conscious sedation is a moderate level of sedation, whose level is lower than intravenous anesthesia. The patients are in the state of sleeping and they can respond to language and tactile stimulus with deglutition reflex. Their respiratory and circulatory functions are stable[4]. Dexmedetomidine is a kind of high-selectivity α^{2} adrenergic agonist, which can obviously reduce the sympathetic reactions to stress. This kind of sedation is similar to natural sleep. Unlike other sedatives, it acts on the brainstem nucleus ceruleus, so no respiratory inhibition will appear and certain swallowing function can be maintained. This study found that the patients of group D suffered little change of haemodynamics at each time point with no respiratory depression or less adverse reactions, so the sedation level reaches the level of the security, comfort and ease of operation.

Above all, applying Dexmedetomidine in ERCP for senile patients for conscious sedation is effective, secure and comfortable during the perioperative period, so it deserves popularization and application in clinic.

References

1. Schepis F,Camma C,Niceforo D,et al.Which patients with cirrhosis should undergo endoscpic screening for esophageal varices detection[J].Hepatology,2001,33:333~338.
2. Williams S GJ, Westahy D.Recent advances in the endoscopic management of variceal bleeding[J].Cut,1996,36(6):647.

DOI: 10.1051/bioconf/20170801049

3. Zigang Wu, Qiusheng Zhang, Huabin Quan. Comparison of four therapeutic methods in the treatment of esophageal varices bleeding [J]. Chinese Journal of Digestive Endoscopy, 2000, 339-341. Chinese.
4. Faigel DO,Baron TH,Goldstein JL,et a1.Guidelines for the use of deep sedation and anesthesia for GI endoscopy[J]. Gastmintest Endosc,2002,56：613~617.
5. Hao Pei. Pharmacological effects of dexmedetomidine hydrochloride and its clinical application [J]. Herald of Medicine, 2012, 12:33~34. Chinese.

DOI: 10.1051/bioconf/20170801050

Influence of Infusion of Different Liquid on Changes of Blood Gas and Electrolyte of Children during the Perioperative Period

Shuang Qi, Fanglei Han and Feng Liang[a]

China–Japan Union Hospital of Jilin University

Abstract. Objective To compare the effects of sodium potassium magnesium calcium and Glucose Injection, 5% Glucose Injection and 9% Sodium Chloride Injection on children with blood glucose, blood lactic acid and blood gas electrolyte during operation on children. Methods patients were randomly divided into A group(sodium potassium magnesium calcium and Glucose Injection group,n=20), B group(5% sodium chloride glucose injection, n=20)and C group(9% Sodium Chloride Injection group ,n=20). Blood glucose ,blood lactic acid, arterial blood Gas and electrolytes were measured and compared before infusion (T0)and on postoperative time(T1). Results blood glucose in B group increased but decreased significantly in C group($P < 0.05$).;Blood lactic acid in C group significantly decreased than the other two groups($P < 0.05$); Mg^{2+} decreased in B and C groups($P < 0.05$). Conclusion Sodium Potassium Magnesium Calcium and Glucose Injection has the same efficacy and safety as commonly used 5% sodium chloride glucose injection and 9% sodium chloride injection, when infused in pediatric patients during perioperative period.

1 Instruction

The focus of fluid management of children is the maintenance of environment stabilization during the perioperative period. Because children have their unique characteristics in terms of the dissection and physiology compared with those of adults, there have long been controversies in terms of the selection of transfusion liquid during the perioperative period [1]. Sodium Potassium Magnesium Calcium and Glucose Injection (Lejay) is a kind of compound sodium acetate electrolyte solution containing a variety of ingredients, such as sodium, potassium, magnesium, calcium, chlorine and glucose (1%). Its composition is quite similar to extracellular fluid of human body and it is widely used for adults' water complement and the maintenance of electrolyte balance of the body in clinical practice. However, few studies have focused on its effectiveness and security when it is used to children.

This study aims to study the influence of Sodium Potassium Magnesium Calcium and Glucose Injection on changes of blood glucose, blood lactic acid and blood gas and electrolyte of children during the perioperative period so as to provide theoretical basis for its clinical application to children.

2 Materials and Methods

[a] Corresponding author: 295253720@qq.com

2.1 Generally

Generally speaking, the materials shall be approved by Ethics Committee of our hospital. In addition, our hospital shall sign the informed consent form with family members of children patients. Selective operations for 60 cases of children patients (male or female) aged between 1 and 5 years old were done from January, 2014 to June, 2014 in China–Japan Union Hospital of Jilin University. Their body weight shall be between 8kg and 20kg and their ASA shall belong to level I ~ II. These children shall have no heart, lung, liver, kidney diseases and endocrine diseases before the operation. The operation was arranged between 8:30 a.m. and 9:00 a.m. and it shall not last for more than two hours. In addition, there shall be lower blood loss. The children were randomly divided into three groups: the group of Sodium Potassium Magnesium Calcium and Glucose Injection (Group A), the group of 5% dextrose injection (Group B) and the group of 9% parenteral solution (Group C). Each group includes 20 cases.

2.2 Anesthesia method

Abrosia time of children patients: They were forbidden to drink clear drinks within two hours before the anesthesia, breast milk within four hours before the anesthesia and milk or food within six hours before the anesthesia. Thirty minutes before the operation, they were injected with the atropine of 0.015mg/kg. Besides, two hours before the anesthetization, venous channels were opened in the sickroom. Children patients of Group A were injected with Sodium Potassium Magnesium Calcium and Glucose Injection (Jiangsu Hengrui Medicine co.,LTD, Lejay, State Medical Approval Number: H20051995), those of Group B were injected with 5% dextrose injection (Sichuan Kolun Pharmaceutical Co., LTD; State Medical Approval Number: H51020636) and those of Group C were injected with 9% sodium chloride injection (Sichuan Kolun Pharmaceutical Co., LTD; State Medical Approval Number: H51021158). The liquid infusion quantity can be seen in Table 1. Penehyclidine hydrochloride of 0.01mg/kg was injected to the muscles within 30 minutes before the operation. Midazolam of 0.05~0.1mg/kg, propofol of 1.5~2mg/kg, vecuronium bromide of 0.05~0.1 mg/kg and sufentanil of 0.05μg/kg were injected through the way of anesthesia induction and intravenous injection. Besides, the concentration of sevoflurane was adjusted to be between 2.5% and 5.0% during the operation so as to maintain the anesthetization effect.

2.3 Monitoring index

BP, HR, SpO_2 and $PaCO_2$ of children patients were monitored routinely in the operation. Invasive arterial blood with the total amount of 2ml was drawn from three groups of children patients at the starting point of the transfusion (T_0) and the ending time of the operation (T_1) respectively. Then, the blood-gas analyzer (i-STAT 200 of ABBOTT Company of USA) was used to detect the electrolyte (Na^+, K^+, Ca^{2+} and Mg^{2+}), the buffuer excess (BE), PCO_2, HCO_3^-, PH value, blood glucose, lactic acid concentration and other indexes. In addition, the results, operation time and adverse events were recorded.

2.4 Analyze data

The statistical software SPSS 13.0 was used to count and analyze data. Besides, SPSS 11.0 software was also used and χ^2 test was adopted to show the counting data of the observed data while mean ± standard deviation ($\bar{X} \pm S$) was used to display the measurement data. Intra-group comparison was also conducted for pairing t-testing and inter-group comparison was analyzed by the one-way variance. P $<$0.05 means that there is a significant difference.

Table 1. Supplement of demand quantity of liquid during the operation

Maintaining dosage per hour (EFR)	0~10kg 4ml/kg/h 10~20kg 2ml/kg/h >20kg 1ml/kg/h
Estimate the amount of fluid loss before the pre-operation (EFD)	3 hours×EFR Inject l/2EFD+EFR in the first hour Inject l/4EFD+EFR in the second hour
Inapparent loss amount (IL)	1ml/kg/h
The total transfusion amount during the operation	EFR+EFD+IL

3. Results

3.1 Children patients

Children patients of the three groups have no significant difference (P>0.05. refer to Table 2) in terms of their age, sex ratio, body weight and operation time. Table 2 The comparison of the general information of three groups of children patients (n=20)

3.2 Intra-group

Comparison and inter-group comparison of them show that they have no significant difference (P>0.05; Table 2) in terms of PH, PCO_2, HCO_3^- of the blood and BE value.

Table 2. Comparison of blood gas indexes of three groups of children patients ($\bar{X} \pm S$, n=20)

Group	Number of cases	Age (Years old)	Male / female	Body weight (kg)	Operation time (min)
Group A	20	3.7±1.3	11/9	13.5±5.4	111±10.8
Group B	20	3.9±1.1	13/7	13.7±5.1	110±10.6
Group C	20	3.8±1.2	12/8	13.8±5.3	112±11.1

3.3 Blood glucose

Blood glucose concentration of children patients of the three groups showed no significant difference (P>0.05) at the starting point of the transfusion (T_0). However, at the ending point of the operation (T_1), the blood glucose of Group A and Group B was all higher than T_0. The blood glucose concentration of Group B showed significant difference (P<0.05) compared with that before the transfusion and that index of Group C was lower than T_0, which displayed a significant difference (P <0.05) compared with that before the transfusion. Although the blood glucose concentration of Group A was higher than that before the transfusion, it showed no significant difference (P>0.05). The inter-group comparison of the three groups in terms of the blood glucose concentration at T_1 displayed significant difference (P<0.05), which could be seen in Table 3.

Table 3. The comparison of the blood glucose level of three groups of children patients (mmol/l, $\bar{X} \pm S$, n=20)

Group/Time	Group A	Group B	Group C
T_0	4.74±0.73	4.81±0.82	4.95±0.79
T_1	4.98±0.68$^{\#}$	6.32±1.08*$^{\#}$	3.82±0.53$^{\Delta\#}$

3.4 Lactic acid

The blood lactic acid concentration of children patients of the three groups showed no significant difference (P>0.05) at the starting point of the transfusion (T_0). However, at the ending point of the operation (T_1), the blood lactic acid concentration of Group A and Group B was all higher than T_0 but they showed no significant difference (P>0.05). The concentration value of blood lactic acid of Group C at T_1 decreased significantly (which was lower than the baseline value T_0 (P<0.05)) compared with that of the other two groups, which could be seen in Table 4.

Table 4. The comparison of the blood lactic acid level of three groups of children patients (mmol/l, $\bar{X} \pm S$,n=20)

Group/ Time	Group A	Group B	Group C
T_0	1.21±0.43	1.18±0.54	1.22±0.46
T_1	1.25±0.63	1.23±0.39	1.02±0.37*$^{\#}$

At T_1: Compared with the figure before the transfusion, the blood lactic acid level of Group C: P*<0.05; the comparison between Group C and Group A and B at T_1 $P^{\#}$<0.05.

3.5 Electrolyte

Na^+, K^+, Ca^{2+} and Mg^{2+} concentration of children patients of the three groups showed no significant difference (P>0.05) at the starting point of the transfusion (T_0). The intra-group and inter-group comparison showed that Na^+, K^+ and Ca^{2+} concentration levels of the three groups all dropped at T_1 compared with that at T_0 , but they showed no significant difference (P>0.05). The Mg^{2+} concentration level of three groups all dropped at T_1 compared with that at T_0 (P<0.05) and Group B and Group C had a more obvious decline at T_1 compared with that of Group A. The difference had statistical significance (P<0.05), which could be seen in Table 5.

Table 5. The comparison of the blood electrolyte concentration value of three groups of children patients (mmol/l, $\bar{X} \pm S$, n=20)

Group	Time	Na+	K+	Ca2+	Mg^{2+}
Group A	T_0	143.44±5.75	3.91±0.39	1.31±0.21	1.08±0.05
	T_1	141.64±4.92	3.80±0.50	1.28±0.22	1.01±0.03
Group B	T_0	142.61±5.87	3.88±0.34	1.30±0.19	1.06±0.08
	T_1	139.44±4.32	3.72±0.48	1.25±0.23	0.80±0.03*$^{\#}$
Group C	T_0	141.56±5.63	3.85±0.33	1.32±0.24	1.03±0.05
	T_1	140.27±5.28	3.79±0.27	1.27±0.26	0.86±0.05*$^{\#}$

Mg^{2+} concentration of Group B and Group C compared with that before the infusion P*<0.05; the comparison between Group B, C and Group A: $P^{\#}$<0.0

4 Discussion

People often hold the old concept that children have a relatively high basal metabolic rate and they mainly rely on glucose for their supply of energy. Thus, glucose liquid must be injected during the perioperative period regularly [2]. However, with the extension of the glucose liquid infusion time, the blood glucose concentration of children patients in Group B increases continually and there are even three cases reaching the critical value. This reminds us that the supply of glucose to children shall be handled with care and the continuous infusion of 5% glucose liquid to children during the perioperative period may cause the risk of hyperglycemia. Lactic acid, which is the substrate in the gluconeogenesis process, can partly replace some glucose to maintain the energy supply of the body in case of the short supply of glucose [3]. In this experiment, sugar solution was not injected to children of Group C. After the operation, their blood glucose level dropped but is still within the normal range. However, the lactic acid concentration of children patients is significantly lower than that of the other two groups, showing that the reason why children patients with the only infusion of sugar-free 9% sodium chloride injection have no risks of hypoglycemia during the perioperative period is more likely to be caused by more may be caused by irritable glycometabolism disorder [4] caused by anesthesia and surgical stimulation. The reduction of blood lactic acid concentration indicates that the glucose reserve in their bodies has been gradually depleted and the gluconeogenesis course has been triggered. This cannot meet energy demands of children.

Na^+ and K^+, which can maintain the normal metabolism of cells, are the most important electrolytes in the cell. Ca^{2+} in the blood can maintain the stabilization of nerves and muscles. The content of Mg^{2+} in the cell, which ranks the second, is only lower than that of K^+ and it plays an important role as the membrane stabilizer of cytomembrane and cytoplasm organelle [5]. In addition, studies have shown that Mg^{2+} is also related to the metabolism of carbohydrate and the change of its concentration can affect the steady state of the glucose and even cause the abnormal glucose metabolism [6].

Besides, Lejay adopts sodium acetate as the buffering substance. The bicarbonate radical of acetate in internal metabolism have the buffering capacity and it can be eventually transformed into carbon dioxide and water, which can reduce the metabolic burden of the liver [7]. It is the only extracellular fluid replenishing liquor which has the same PH value with that of the blood plasma at present. Thus, it can have a physiological advantage than 5% dextrose injection and 9% sodium chloride solution if it is used to the infusion of children during the perioperative period.

In conclusion, Sodium Potassium Magnesium Calcium and Glucose Injection (Lejay) is a kind of safe, effective and ideal intravenous infusion to children during the perioperative period and it is worth to be popularized and applied to clinic practices.

References

1. Xinliang Zuo, Yinming Zeng, Boluan Chen, Editor. Modern Anesthesiology [M]. Third Edition. Beijing: People's Medical Publishing House, 2003,1431～1432. (In Chinese)
2. Nakaimura K et al. Special pediatric problems in fluid and electrolyte therapy in surgery [J].Anesthesioloy, 1990, 72:1005.
3. Airu Zhou, Xiliang Cha, Editor. Biochemistry [M]. Fifth Edition. Beijing: People's Medical Publishing House, 2000,129. (In Chinese)
4. Korusi Ca, HauptmanA，BrundulaA，et al. Perioperative management with glucose solution and insulin[J].Coll antropol ,2009,33(2):653 ~657.
5. Dube L, Granry JC. The therapeutic use of magnesium in anesthesiology, intensive care and emergency medicine:a review[J].Can J Anaesth,2003,50(7):732~746.
6. Musso CG, Magnesium metabolism in health and disease [J].Int Urol Nephrol, 2009, 41(2):357~362.
7. Boluan Chen. Editor, Clinical Anesthesiology and Pharmacology [M]. First Edition. Beijing: People's Medical Publishing House, 2000, 50~584. (In Chinese)

DOI: 10.1051/bioconf/20170801051

Comparison of clinical applications of single-dose intravenous injection of mivacurium and cisatracurium in adults' vocalcordpolyps resection under self-retaining laryngoscope

Fanglei Han, Lu Zhang, Shuang Cheng and Feng Liang[a]

China–Japan Union Hospital of Jilin University

Abstract. Objective: Through comparing the clinical observation of mivacurium and cisatracurium in vocal polyp extraction, to study the advantage of mivacurium in vocal polyp extraction. Methods:Forty American Society of Anesthesiologists(ASA) physical status Ⅰ~Ⅱ patients for vocal polyp extraction,aged 18~60 years old, were randomly divided into two groups as Mivacurium injection group(Group M) and Cis-atracurium injection group(Group C) ,each group includes 20 subjects. None of the patients are allergic, has serious diseases of cardiovascular system,liver or kidney.None of them has asthma, airway high response, difficult airway or neuromuscular diseases.Those patients who use beta-blockers or calcium channel blockers for a long time were excluded .All the subjects had the same premedication, fasting and fluid fobidden time.All the subjects who get into the operating room get the routine monitoring of electrocardiogram(EEG), blood pressure(BP), heart rate(HR), pulse oxygen saturation(SpO2) and the TOF WATCH SX. Each group gives the same medicine other than the muscle relaxant during induction of general anesthesia to do the vocal polyp extraction by the same experienced operator. Two groups were recorded in each index of anesthesia induction and tracheal intubation conditions, operation conditions, anesthesia, muscle relaxation monitoring.Results: There are no statistically significant in Cormack-Lehane grading system,Cooper's grading system and operation satifaction($p>0.05$). 2.Group A have shorter intubation time than Group B($p<0.05$).3.In the TOF recovery time to 25% and 90%,extubation time, the M group was faster than the C group, and the difference was statistically significant($p<0.05$).Conclusion: 1.A single intubating dose of mivacurium can provide similar intubation and surgeon satisfaction for the vocal polyp extraction.2.Compared with cis-atracurium, mivacurium can shorten the intubation time and the recovery time of anesthesia. The adverse reactions of mivacurium is mild, and it has less Residual muscle relaxation. Therefore mivacurium is more suitable for the vocal polyp extraction than cis-atracurium.

1 Instruction

Vocalcordpolyps is a kind of benign neoplasm commonly seen in pharynx and larynx and its typical clinical manifestation is voice hoarseness of various degrees. There is an upward trend in its incidence in recent years [1]. Clinically, it is mainly treated by operations [2]. The vocalcordpolyps extracation under self-retaining laryngoscope has been widely applied in clinical practice because of its unique advantages. This surgical operation only takes a short time and surgical spots are in pharynx and larynx, which will be strongly pungent. Thus, it requires enough depth of anesthesia. During the

[a] Corresponding author: 295253720@qq.com

operation, patients shall have stable life signs and they shall be able to wake up quickly after the operation. Patients are required to be wake up smoothly and steadily. Thus, this surgical method has a high requirement to anesthesia degree and it is also extremely important to select a safe and short-acting muscle relaxing drug. This paper conducted a comparative clinical observation to the single-dose intravenous injection effect of two muscle relaxing drugs (cisatracurium and mivacurium) when they were used to vocalcordpolyps extracation under self-retaining laryngoscope so as to find a safer, more effective muscle relaxing drug with high satisfaction degrees in the surgical operation.

2. Materials and Methods

Patients (male or female) aged 18~60 years old who would undergo vocalcordpolyps extracation under self-retaining laryngoscope some time in our hospital were selected. They belonged to level Ⅰ~Ⅱ of American Society of Anesthesiologists (ASA). 40 cases were selected, including 20 male cases and 20 female cases. The patients were randomly divided into two groups (with 20 patients in each group): mivacurium induced group (Group M) and cisatracurium group (Group C).

Methods:All patients were forbidden to eat anything within 8 hours and drink anything within four hours before the operation. Before the operation, they all took penehyclidine hydrochloride of 1.0mg. After the patients came to the operating room, their venous channels of upper limb would be opened and the patient monitor would be connected. During the operation, their ECG, BP, HR and SpO2 were routinely monitored. Then, the TOF-Watch would be set and connected and the TOF stimulus mode would be used. The impulse current was 60mA, the frequency was 2Hz and the stimulus interval was 15s. After the patients had stable life signs, the anesthesia induction would be conducted. Midazolam of 0.04mg/kg, propofol of 2mg/kg, fentanyl of 4ug/kg and dexamethasone of 10mg were injected to their veins. After eyelash reflexes of patients disappeared, muscle relaxant calibration shall be done. The mode of train-of-four stimulation shall be used and the percentage of train-of-four stimulation shall be adjusted to 100% or so, which was also the baseline value. Then, mivacurium of 0.2mg/kg would be injected with the injection speed of 45s to patients of Group M while cisatracurium of 0.1mg/kg would be injected to patients of Group C. When the TOF reaction number reduced to 2 or below, oral intubation under direct vision could be done. During the operation, sevoflurane was used for anesthesia and the MAC value was maintained to be between 1.2 and 1.3. When the mouth gags were taken out, operations of the two groups would be ended and at that time, the sevoflurane shall be stopped. As long as TOP of Group C went back to more than 25 %, neostigmine of 1mg and atropine of 0.5mg could be used for antagonism. Antagonism was not done for Group M. After the patients recovered their spontaneous respiration and swallowing reflex, their deoxidation SpO2 shall be maintained to above 90% for three minutes. The endotracheal tubes could be pulled out as long as they recovered consciousness. After that, they would be moved to the post-operational care unit (PACU) for continual observation. If the patients recovered their consciousness, could complete simple actions according to instructions (such as raising their head and sticking out their tongue), returned their cough reflex to normal level, had stable life signs and Steward grade of higher than 4 points and TOF>90% and showed no attenuation in terms of the train of four stimulation, they could be sent back to the department for monitoring care and oxygen inhalation.Observe and record the intubation time, the operation time, the extubation time and the PACU time, the time used for TOF value recovering to 25% (T25), the time used for TOF value recovering from 25% to 75% (recovery index) and the time used for TOF value recovering to 90% (T90).

SPSS 13.0 software was used for statistical analysis. All measurement data were shown by Mean± standard deviation and the variance analysis and T-test method were used for inter-group comparison. $P<0.05$ meant that the difference showed statistical significance. Besides, chi-square test was used for enumeration data. $P<0.05$ meant that the difference showed statistical significance.

3 Results

The difference between Group M and Group C in their tracheal intubating conditions and operation conditions had no statistical significance (p>0.05) (refer to Table 1)

Table 1. Comparison of tracheal intubating conditions and operation conditions

	C-L grading	Irritating cough	Cooper's grading	Operation condition grading
GROUP M	19/1/0/0	2/18	17/3/0/0	2/18
GROUP C	18/2/0/0	1/19	16/3/1/1	1/19
p value	>0.05	>0.05	>0.05	>0.05

Notes: The difference of inter-group comparison had no statistical significance. p>0.05

Group M spend less time than Group C in terms of trachea anesthesia and intubation time, TOF 25% and 90% recovery time, extubation time and PACU time. Besides, the difference showed statistical significance. (Refer to Table 2 and Table 3)

Table 2. Comparison of intubation time, operation time and extubation time

	Intubation time (min)	Operation time (min)	Extubation time (min)
GROUP M	3.09±0.27	12.69±1.74	14.19±1.18
GROUP C	3.62±0.44	13.28±1.74	23.09±4.79
p value	<0.05	>0.05	<0.05

Notes: The difference of inter-group comparison had no statistical significance. p>0.05
The difference of inter-group comparison had statistical significance. p<0.05

Table 3. Comparison of TOF value

	T25(min)	T90(min)	Recovery index(min)
GROUP M	15.61±1.50	22.09±2.68	10.60±1.12
GROUP C	32.57±1.44	34.78±2.89	4.55±1.09
p value	<0.05	<0.05	<0.05

Notes: The difference of inter-group comparison had statistical significance. P<0.05

Discussion

Using your voice too much or using it improperly, upper respiratory tract infection and contacting stimulating substances can all cause vocalcordpolyps and hoarseness of various degrees. Vocalcordpolyps is a benignant hyperplasic disease of vocal cords commonly seen in the E.N.T. department. At present, its main treatment method is surgical resection. Though this operation is short in time, it is strongly irritative and requires enough anesthesia depth. During the operation, patients shall have stable life signs and shall recover themselves rapidly and smoothly after the operation. Thus, it is especially important to select and use muscle-relaxing drugs because they shall work quickly and completely relax muscles within a short acting time. Besides, the patients shall recover quickly and have a small residual amount of muscle relaxant. This research studied the two types of muscle relaxant, hoping to compare them and find one which is more suitable for this surgery way.

Cisatracurium has strong muscle-relaxing effect and weak histamine release [3] and it is widely used in clinic practice. Studies of HUANG Jiayang, etc[4] showed that the application of co-induction of 2ED95 cisatracurium and sevoflurane to vocalcordpolyps extraction could give satisfied tracheal intubation conditions and help gain satisfactory muscle relaxant effect. Some studies[5] showed that when the administration dosage is 0.1 mg/g(2ED95), the onset time was about 3.0 minutes while other

 DOI: 10.1051/bioconf/20170801051

studies [6] indicated that under the total intravenous anesthesia, if patients aged 18～75 years old were injected with twice as much ED95 cisatracurium to their veins, the duration of their muscle relaxation was (32±7) minutes on average. That was basically consistent with findings of this research. Based on the above results and comprehensively considering intubation conditions and maintenance time, twice as much ED95 cisatracurium was used for induction and tracheal intubation in this research.

Mivacurium is a kind of nondepolarizing muscle relaxant of Bis Benzylisoquinolines and is one of the most ideal short-acting muscle relaxants. It does not accumulate in the body and its action time is short. However, it can generate high plasma clearance rate without antagonism and it has no neurovirulence. Guo Chen [7] reported that mivacurium of 0.2 mg/kg could provide satisfactory intubation conditions and if the dosage increased to 0.25 mg/kg, the working time could be obviously shortened. However, when the dose was more than 0.2 mg/kg, the occurrence rate of erubescence would be raised significantly with the dosage increased. Thus, the inductive dose of mivacurium in this study was 0.2mg/kg[8]. In anesthesia and inducing period of this research, the infusion speed was strictly controlled so as to reduce the histamine release and to ensure that it can be transfused for 45s or longer. In addition, cisatracurium shall also be transfused to patients in the contrast group for 45s or longer so as to ensure that muscle relaxant infusion speed of two groups was consistent.

Experiment results show that the difference of the two groups in terms of C-L grading, irritating cough, Cooper's grading and operation condition had no statistical significance, showing that muscle relaxants used in the experimental group and the contrast group and their dosage can meet the intubation condition and operation condition for vocalcordpolyps extracation under self-retaining laryngoscope. The intubation time, extubation time, the time need for TOF value recovering to 25% (T25), the time needed for TOF value recovering to 90% time (T90), the recovery index and the PACU time of Group M are all smaller than those of Group C and their difference has statistical significance($p<0.05$), indicating that mivacurium of Group M works faster, needs shorter intubation time and its muscle relaxant recovery time is also shorter than that of Group C. Besides, its postoperative extubation time is also short, which can reduce patients' waiting time in the operating room and can be beneficial to shorten the operation cycle. The comparison difference of Steward analepsia score of patients of both groups has no statistical significance and this can ensure the anesthesia security of patients. In terms of adverse reactions of patients after their anesthesia induction and before their coming out of PACU, more patients in Group M are red in their faces and bodies than those in Group C and most flushing parts of them are in the chest. Patients in Group C have a higher occurrence rate of postoperative residual curarization, which means that, their TOF value is<90% after tubes are pulled out. This is significantly higher than that in Group M, indicating that lower risks of postoperative pulmonary complications will exist and patients are safer if mivacurium is used. In this research, when muscle relaxants are used, patients need longer recovery time than that stated in the foregoing data. This is caused by the application of sevoflurane in the anesthesia maintenance stage because the sevoflurane has a synergetic muscle relaxant effect on muscle relaxants, which can affect the measurement of values of muscle relaxant itself. However, patients of both groups all use sevoflurane for anesthetic maintenance, the synergetic muscle relaxant effect caused by it can be ignored.

During the induction process, different muscle relaxant types are used to those two groups and other drugs are all used according to their weight in fixed quantity because this can avoid the influence of other drugs on the results. During the induction process, dexamethasone of 10mg is given routinely because it can reduce the histamine release of muscle relaxants of Bis Benzylisoquinolines on the one hand and can prevent the occurrence of postoperative edema of the larynx after the operation on the other hand.

In order to ensure the safety of patients, the following requirements shall be met when patients come out of the PACU: they shall be conscious and be able to complete simple actions according to instructions (such as raising their head and sticking out their tongue); they shall have stable life signs and their Steward score shall be larger than 4 points. Their TOF shall be larger than 90% and their train-of-four stimulation shall have no attenuation.

Vocal cord impairment extracation under self-retaining laryngoscope rather than the

DOI: 10.1051/bioconf/20170801051

phonomicrosurgery technique is used as the surgical method in this research. This is a deficiency of this research in terms of its advancement. More surgical methods need to be done for researches of microsurgery anesthesia under self-retaining laryngoscope. This search also has the following shortcomings: patients of the two groups are between 18 and 60 years old and no research is done to children, the elderly, fat people or patients with internal medicine diseases.so, it has some limitations.

In conclusion, compared with cisatracurium, the application of mivacurium induction can shorten the anesthesia induction time and patients can recover quickly after the operation. Besides, they have milder adverse reactions and less muscle relaxant residuals. Thus, mivacurium induction is more applicable to vocalcordpolyps extracation under self-retaining laryngoscope.

References

1. Yingjun Shan, Jiawen Cui, Gang Fei, etal. Different surgical methods for the treatment of vocal cord polyp compared [J]. Tongji University, [Medical Edition]. 6(30):118-120(2009) (In Chinese)
2. Yingpeng Han, Chengjun Song, Xiuhui Jian, etal. Laryngeal endoscopic vocal cord polyp excision [J]. Chinese Journal of Otorhinolaryngology, 2(38):65(2009) (In Chinese)
3. Liqiu Mo, Wenqi Huang, Jiefang Tan, etal. A new effective muscle relaxant - cis - atracurium [J]. The Journal of Clinical Anesthesiology,16:170(2000) (In Chinese)
4. JiaYang Huang, Huizhang Liu, Guosheng Yan, etal. Different doses of cisatracurium and sevoflurane induced polypectomy in the application of vocal cord results [J], Practical Journal of Clinical Medicine, 189(23):187-188(2014) (In Chinese).
5. Ruijuan Wang. Comparison of the efficacy of cisapride atracurium in weightlessness and actual body weight [D], Southern Medical University, (2013) (In Chinese)
6. Yali Ma, Qingduo Guo, Rui Li, etal. Intravenous injection of cisatracurium on histamine release in patients with general anesthesia [J]. Zhonghua Anesthesiology, 29,(1):89-90(2009) (In Chinese)
7. Chen Guo, Mingjun Xu. Different doses of miracurium for gynecological laparoscopic muscle relaxation in clinical observation [J], Beijing Medical, 36(11):936-939(2014) (In Chinese)Yohitake Y,Shinohara M,Ue H,etal.Characteristics of Surface mec-hano myogran ale dependant on development of fusion of motor- units in humans[J].ApplPhysiol, **93(5)**:1744-1652 (2002).

DOI: 10.1051/bioconf/20170801052

Comparison of Various Anesthesia Induction Methods for the Anesthesia in Elderly Patients in the Abdominal Surgery

Fanglei Han, Yan Liu, Xueying Zhou and Feng Liang[a]

China - Japan Union Hospital of Jilin University

Abstract. Objective: To investigate the anesthesia induction methods in the elderly in surgeries. Methods: 60 cases of elderly patients in the Abdominal Surgery with ASA grade 1 to 3 were selected and randomly divided into three groups; intravenous injection of 4ug/kg of fentanyl and 0.1mg/kg of vecuronium bromide was conducted at first, and then group A was injected with 0.2mg/kg of midazolam, group B with 1.5mg/kg of propofol, and group C with 0.1mg/kg of midazolam and 1.0mg/kg of propofol for anesthesia induction within 1 min, and then the tracheal intubation was conducted. The effective time of anesthesia induction and the circulatory changes, as well as changes of PRA and Ang II values before anesthesia (T0), after anesthesia induction (T1), immediately after intubation (T3), and 5 min after intubation (T3) were observed. Results: The effective time of anesthesia induction was shorter in group C than in group A ($P<0.05$), and that was similar between group C and group B ($P>0.05$); compared with T0, the PRA and Ang II values in the three groups showed a decreasing trend at T1, T2 and T3, with the decrease being distinct at T1 ($P<0.05$); at T2, the PRA and Ang II values were significantly increased in group A ($P<0.05$), which were remarkably higher than those in group B and group C at the same period ($P<0.05$). The mean arterial pressure (MAP) and heart rate (HR) at T2 were notably increased in group A ($P<0.05$), while those were outstandingly reduced at T1 in group B ($P<0.05$), and group C had mild circulatory changes ($P>0.05$). The saturation of pulse oximetry (SPO2) was maintained between 97% and 100%. Conclusion: The combined use of midazolam and propofol is the ideal method for anesthesia induction in the elderly.

1 Instruction

This paper aimed to observe the effects of selecting midazolam, propofol or the combined used of the two for anesthesia induction in the elderly patients in the Abdominal Surgery on the circulatory system, so as to search for the ideal anesthesia induction method for the elderly.

1.1 General data

60 cases with ASA grade I to III were selected, including 43 male and 17 female, with the age ranging from 70 to 85 years and the body weight of 52 to 85kg. 36 cases received subtotal gastrectomy, 11 had cholecystectomy plus common bile duct exploration, 10 received laparoscopic cholecystectomy, and 13 had hemicolectomy. The patients were randomly divided into three groups, with 20 cases in each group, and different anesthesia induction drugs were selected, respectively.

[a]Corresponding author:295253720@qq.com

DOI: 10.1051/bioconf/20170801052

1.2 Anesthesia method

Intravenous injection with 0.3mg of scopolamine was conducted 30min before anesthesia, the vein was opened after entering the operating room and electrocardiogram was monitored. Intravenous injection of 4ug/kg of fentanyl and 0.1mg/kg of vecuronium bromide was conducted at first, and then group A was injected with 0.2mg/kg of midazolam, group B with 1.5mg/kg of propofol, and group C with 0.1mg/kg of midazolam and 1.0mg/kg of propofol through the vein for anesthesia induction; the drug injection in the three groups was completed within 1min, tracheal intubation was conducted after hyperventilation by pure oxygen mask, and the OHMI DA anesthesia machine was connected for breath control.

1.3 Monitoring and observation

The multifunctional circulation monitor was adopted to monitor the electrocardiogram (ECG), mean arterial pressure (MAP), heart rate (HR) and saturation of pulse oximetry (SPO2). The effective time of anesthesia induction of the three groups, which was the eyeball fixation time, was observed and recorded; and the changes of MAP, HR, SPO2, plasma rennin activity (PRA) and angiotensin II (Ang II) (radioimmunoassay) before anesthesia (T0), after anesthesia induction (T1), immediately after intubation (T2) and 5min after intubation (T3) were monitored.

2 Results

2.1 The effective time of anesthesia induction

The anesthesia induction eyeball fixation time was significantly extended in group A (8 0. 2±11 .6)s compared with that in group B (5 8. 8±1 4. 6)s and group C (5 9. 2± 1 4. 8)s, and there was no statistical difference between group B and group C (P<0.05).

2.2

Changes of PRA, Ang II and circulation Compared with T0, the PRA and Ang II in the three groups showed a decreasing trend at T1, T2 and T3, with the decrease being obvious at T1 (P<0.05); the PRA and Ang II were markedly increased in group A at T2 (P<0.05), which were significantly higher than those in group B and group C at the same period (P<0.05); and they basically recovered to the levels before anesthesia (refer to Table 1).

Compared with T0, the MAP in the three groups showed a decreasing trend after drug injection. The occurrence rate of over 20% of MAP decrease after induction was 10% (2/20), 50% (10/20) and 20% (4/20) in groups A, B and C, respectively, (P<0.05); the HR in group B was notably slowed down relative to that before anesthesia (P<0.05), and the MAP and HR were apparently increased in group A immediately after intubation (P<0.05); and the MAP and HR in the three groups basically recovered to the levels before drug injection. Refer to Table 1. SPO2 fluctuated within the range of 97% and 100%.

Table 1. Changes of PRA, Ang II and circulation in the three groups

Group	T0	T1	T2	T3
PRA（ng/ml）				
A Group	0.47±0.41	0.27±0.18$^{1)}$	0.58±0.22$^{1)}$	0.45±0.24
B Group	0.46±0.32	0.24±0.15$^{1)}$	0.38±0.21$^{2)}$	0.43±0.25
C Group	0.48±0.40	0.25±0.17$^{1)}$	0.40±0.22^{2}	0.41±0.22
Ang II (pg/ml)				
A Group	64.42±14.98	56.48±15.32$^{1)}$	75.65±21.21$^{1)}$	64.87±15.65
B Group	63.23±13.45	54.56±13.87$^{1)}$	63.21±14.68$^{2)}$	63.38±15.21
C Group	63.56±15.21	54.98±13.89$^{1)}$	64.38±13.87$^{2)}$	63.25±15.36
MAP(kPa)				
A Group	15.61±2.91	15.32±2.68	17.31±2.82$^{1)}$	16.12±2.67
B Group	15.76±2.25	12.35±2.01$^{1)}$	15.87±2.56	14.98±2.70
C Group	15.12±2.71	13.99±2.12	15.52±2.31	14.97±2.68
HR（bpm）				
A Group	78.00±18.45	80.38±12.60	90.16±19.72$^{1)}$	83.23±14.86
B Group	77.69±19.26	60.44±15.72$^{1)}$	80.53±15.28	79.99±14.65
C Group	78.51±16.96	81.63±12.92	81.62±16.71	80.89±15.84

Compared with T0: 1) P<0.05, and compared with group A: 2) P<0.05

3 Discussion

The midazolam or propofol is frequently served as the intravenous anesthesia induction drug in clinic. Midazolam has strong sedative and hypnotic effects with short elimination half life, the equivalent dose of which is only 1/10 but the effect is 1.5 to 2 times of the stabilizer. The propofol has rapid and reliable action, which can completely produce the anesthesia induction effect within a brain-arm cycle and rapidly achieve the depth of anesthesia [1]. These two drugs are adopted in this paper to implement anesthesia induction in the elderly patients, which have definite effects, and the two groups that utilize propofol have remarkably reduced anesthesia induction time that the group using midazolam.

The PRA, Ang II, blood pressure and HR in the midazolam group were markedly higher than those in the other two groups during anesthesia induction and tracheal intubation (P<0.05), which is basically identical to the reports by Kwa et al [2] that midazolam anesthesia induction can not reduce the cardiovascular response during tracheal intubation. Attention should be paid to the fact that fierce hemodynamic fluctuation in the elderly patients is likely to induce heart, brain and kidney events. Propofol is the rapid and short acting intravenous anesthetic that mainly exerts the sedative and hypnotic effects through the action of GABA [3, 4]; it has certain cardiovascular inhibitory effect, which can result in 20% drop of blood pressure, 17% decrease of cardiac output and reduced heart rate, and these may be contributed to the instable hemodynamics or hypovolemia of the patients [5]. The occurrence rate of drop of blood pressure in the propofol group in this paper reaches 50%, with concurrent decreased heart rate after drug injection. The fierce hemodynamic changes in the elderly are likely to induce cerebrovascular events; consequently, importance should be attached to the elderly with existing cerebrovascular diseases or poor functional compensation since the above changes will aggravate heart, brain and kidney function damage.

It is discovered in this research that the combined use of midazolam and propofol has short effective time of anesthesia induction and little cardiovascular response, which guarantees the anesthesia induction effect, reduces the dose of each drug and minimizes the drug adverse reaction. When conducting anesthesia induction in clinic, the combined use of two drugs can produce synergistic effect, and it is believed that the combined use of midazolam and propofol is one of the ideal methods for anesthesia induction in the elderly patients.

References

1. Chavorro C , Latorre FJ , Montero A ,et al. Comparative study of propofol versus midazolam in the sedation of critically ill patients : results of a prospectrive, randomized, multicenter trial [J] .Crit Care Med,1996,24:932
2. Kwar P ,Carson TW ,Clake RSJ ,et al. Haemodynamic changes during induction of anesthesia with midazolam and diazepam in patients undergoing coronary artery bypass surgery [J] .Anaesthesia ,1985,40:767
3. Schwieler L ,Dellbro DS ,Engberge ,et al. The anesthetic agent propofol interacts with GABA(B)-receptors : an electrophysiological study in ract [J] , Life Sei ,2003,72: 2793-2801
4. Sonner JM ,Zhany Y ,Stabernack C ,et al. GABA(A) receptor blockade antagonizes the immobilizing action of propofol but not ketamine of isoflurane in a dose-related manner[J] .Anesth Analg ,2003 ,96 :706-12
5. Brussel T ,Theissen JL , Vigfusson G ,et al. Haemodynamic and cardiody-namic effects of propofol and etomidate : negative inotropic properties of propofol[J] .Anesth Analg, 1989,69:35-40

Application of body temperature protection in patients receiving laparoscopic total gastrectomy as well as its influence

Feng Liang, Yan Liu, Xueying Zhou and Fanglei Han[a]

China–Japan Union Hospital of Jilin University

Abstract. With the extensive development of the laparoscopic total gastrectomy, increasing attention has been paid to the influence of hypothermia during the intraoperative period on patients. Therefore, effective body temperature monitoring and heating protection on patients receiving laparoscopic total gastrectomy is one of the important protective measures for guaranteeing perioperative safety of patients. This research observed the changes of the core body temperature as well as the intraoperative blood glucose and lactic acid at various time points during the perioperative period, and postoperative awakening and extubation time in patients receiving laparoscopic total gastrectomy under general anesthesia, with an aim to provide reference for clinical practice.

1 Data and methods

1.1 Object of study

The research was approved by the ethics committee of the hospital, and informed consent was obtained from all patients. 40 cases of patients that received laparoscopic total gastrectomy in our hospital were selected, among which 24 were male and 16 were female, with the age ranging from 45 to 71 years, the body weight from 45 to 66 kg, and the ASA grading of grade I to II; the cardiac, pulmonary, liver and renal functions were basically normal, the estimated duration of operation was above 2 hours, and the infusion was above 2000ml. Those with coagulation disorders, a recent history of fever or infection, hyperthyroidism or hypothyroidism were excluded. The surgeries were performed in the operating room with the room temperature of 24℃ and the humidity of 40% to 60%; the patients were randomly divided into the control group and the heating group, with 20 cases in each group, and there was no statistical difference in the general conditions between the two groups. Group A was the control group which did not take any insulation measure, and the infused liquid and the washing fluid were under room temperature. Group T was the experiment group which received temperature-adjusting blanket insulation, and the infused fluid and the washing fluid were heated to 37℃ by the infusion heating box (Mallinckrodt Medical, USA) before use.

1.2 Anesthesia method

[a] Corresponding author: 23454675@qq.com

Intramuscular injection of penehyclidine hydrochloride was conducted 30 min preoperatively; the venous access of the two groups of patients was opened through entering the heart chamber; the invasive arterial blood pressure, electrocardiography, pulse and oxygen saturation were monitored through radial artery puncture under local anesthesia; and the temperature measurement probe was inserted into the nasopharynx of the patients after entering the chamber so as to continuously monitor the nasopharynx temperature. The two groups of patients received intravenous inhalation combined with general anesthesia; 0.05mg/kg of midazolam, 0.1mg/kg of vecuronium bromide, 1mg/kg-2mg/kg of propofol, and 0.5ug/kg of sufentanil were used for post-general anesthesia induction intubation; and sevoflurane was inhaled continuously after intubation to maintain the depth of anesthesia of 1.2 to 1.4 MAC. The intraoperative fluid infusion was Lactated Ringer's solution and hydroxyethyl starch (Voluven), with the proportion of crystal liquid to colloid liquid of 1:1. Sevoflurane inhalation was ceased after the surgery, and 1mg of neostigmine and 0.5mg of atropine were administered upon the end of the surgery and the recovery of the autonomous respiration to antagonize the remaining muscular relaxation.

1.3 Observation indicators

The core body temperature at the time points of immediately after intubation (T1), as well as 30min (T2), 60min (T3), 90min (T4), 120min (T5) and 150min (T6) after anesthesia induction, and immediately after the surgery (T7) of patients in the two groups were recorded; the blood gas analysis was conducted to compare and observe the changes of blood glucose and blood lactic acid at all time points; and the duration from the closure of the anesthesia gas inhalation to extubation and to the awakening time of patients was observed and recorded.

1.4 Statistical method

The SPSS 11.0 statistical analysis software was adopted for statistical processing, all the measurement data were expressed as mean± standard deviation ($\bar{x} \pm s$), one-way analysis of variance was applied in intra-group comparison, and paired T test was adopted for inter-group comparison. The difference with $P<0.05$ was of statistical significance.

2 Results

2.1 Changes of intraoperative nasopharynx temperature between the two groups

There was no statistical significance in the nasopharynx temperature before anesthesia between the two groups ($P>0.05$). The variation trend after anesthesia: the nasopharynx temperature changes were stable at all time points in group T, which showed a decreasing trend; the nasopharynx temperature at T5, T6 and T7 in group A was remarkably decreased compared with that before anesthesia ($P<0.05$), and that was distinctly reduced at T4, T5, T6 and T7 in group A relative to group T ($P<0.05$) (refer to Table 1).

Table 1. Comparison of the intraoperative nasopharynx temperature changes between the two groups

Group	n	T1	T2	T3	T4	T5	T6	T7
A	20	36.39±0.51	36.31±0.35	36.19±0.81	35.98±0.23	35.87±0.32$^{\triangle}$	35.73±0.31$^{\triangle}$	35.61±0.68$^{\triangle}$
T	20	36.43±0.47	36.30±0.76	36.18±0.64	36.10±0.22*	36.07±0.22*	36.07±0.97	36.06±0.89*

2.2 Changes of the blood glucose at various time points between the two groups

The blood glucose of the two groups of patients showed a rising trend. There was significant difference in T6 and T7 between the two groups (refer to Table 2).

Table 2.

Project	Group	T1	T2	T3	T4	T5	T6	T7
blood sugar	A	6.30±0.64	6.88±0.71	7.74±0.78	7.60±0.94	8.00±0.95	9.50±1.05	9.95±1.15
(mmol/L)	T	6.22±0.94	6.69±1.04	7.38±1.38	7.88±1.72	8.00±1.81	8.17±1.88*	8.22±1.97*

3 Discussion

Hypothermia means that the core body temperature is lower than 35℃ by accident. Intraoperative hypothermia is not rare, but its harm has not attracted our high attention. As is demonstrated in reports, the occurrence rate of hypothermia is about 50% to 70% [1]. Sometimes, hypothermia may be beneficial to the body, but under most circumstances, it will have adverse effects. People lack the understanding towards the drawbacks of the laparoscopic surgery previously, and there are even few reports regarding the influence of hypothermia in laparoscopic surgery. In this experiment, we compare the changes of the perioperative core body temperature, blood glucose and blood lactic acid as well as their influence on the extubation and awakening time of patients receiving laparoscopic total gastrectomy under conventional and heating protection, so as to probe into the clinical significance of heating protection.

The laparoscopic total gastrectomy is relatively complicated, which has long duration of operation and CO2 pneumoperitoneum time and requires perioperative infusion of a large amount of room temperature fluid and peritoneal washing liquid, leading to progressive reduction of body temperature. This is consistent with the results in this experiment. The perioperative hypothermia can induce neuroendocrine enhancement, insulin resistance [7] and may cause blood glucose elevation. The stress response in the patients exists during the whole perioperative period as a result of the anesthesia and the surgical stimulation, in the meantime, the inhibition of anesthetics on the blood glucose metabolism also exists during the entire anesthesia and surgery process. Hypothermia can inhibit myocardial contractility, which reduces cardiac output; in addition, it can elevate the catecholamine, induce vasoconstriction, and increase peripheral resistance and blood viscosity, leading to myocardial ischemia and arrhythmia [8], peripheral vasoconstriction and reduced blood flow. All these changes will result in circulatory disturbance and being unable to completely avoid the hypoxia and increased anaerobic metabolism. Lactic acid is the product of anaerobic metabolism, which can be served as the symbol of hypoxia. For perioperative patients and those with severe trauma, blood lactic acid is an excellent indicator for evaluating disease and prognosis [9], which is verified by the changes of blood glucose and blood lactic acid at time points after 2 hours between the two groups in this experiment. The influence of longer duration of laparoscopic surgery on the internal environment changes remains to be further investigated.

The above results indicate that taking the insulation measure in patients receiving laparoscopic total gastrectomy can avoid substantial decrease in body temperature since a large amount of heat will be absorbed after the cold fluid and gas enter the body. Making the body temperature decrease becomes gentle will alleviate the perioperative stress response, and reduce the extent of elevated blood glucose. Intraoperative insulation can reduce the occurrence rate and severity of acidosis, improve the perioperative internal environment disturbance, and lower the extent of elevated lactic acid; improve the metabolism of the general anesthetics and shorten the postoperative awakening and extubation time. The core body temperature in patients receiving laparoscopic surgery for more than 2 hours may be greatly changed, thus the influence on body temperature, blood glucose, lactic acid and postoperative recovery of patients should not be neglected. The comprehensive insulation measures can obviously improve the body temperature decrease in patients and improve the internal

DOI: 10.1051/bioconf/20170801053

environment homeostasis of the perioperative patients.

References

1. Junjie Liu, Zhao Jun, ed. modern anesthesiology [M]. second edition. Beijing: People's publishing house.1997:674-676.
2. Andrzejow ski J , Hoyl e J , E apen G , et al. Ef f ect of prew arming on post induction core temperature and the incidence of inadvertent perioperet ive h ypoth ermia in pati ent sundergoin g general anaesthesia[J]. Br J Anaes th, 2008,101:627
3. Zhao Jing, Luo Luo. [J]. Chinese Journal of clinical medicine, anesthesia and perioperative period. 2003,1(10):887.
4. Puttick MI, Scott-Coombes DM, Dye J ,et al. Comparison of immunologic and physiologic effects of CO2 pneumoperitoneum at room and body temper atures[J]. Surg Endosc, 1999,13(6):572.
5. Castillo V, Gutierrez2Crespo A, Suarez F ,et al. Body temperature variations dur2ing laparoscopic cholecystectomies[J]. Rev Esp Anestesiol Reanim, 1996,43(6):201
6. Zhang Liqing, Lin Weihong, Qian Huang Jing, et al. The establishment of carbon dioxide pneumoperitoneum on the heating effect of laparoscopic hand the patient's body temperature and shivering [J]. Chinese Journal of nursing, 2007,42(10):953.
7. An Gang, Xue Fushan. Modern anesthesiology technology [M]. Beijing: Science and Technology Literature Press, 1999:804.
8. Zhang Shuyue, Zhu Junyu, Peng Yanzeng, et al. Effects of hypothermia on recovery of anesthesia and nursing intervention in patients with low temperature [J]. Chinese Journal of nursing, 2003,38(3):177.
9. Hu Xingguo. Perioperative mild hypothermia [J]. Foreign Medical Sciences – anesthesiology, 2000,21(4):225.

BIO Web of Conferences **8**, 01054 (2017) DOI: 10.1051/bioconf/20170801054
ICMSB2016

Effects of Devoflurane and Sevoflurane on Cognitive Function in Elderly Patients Undergoing General Anesthesia

Chunlu Zhang[1], Bin Wang[1], Jiali Ding[1] and Peng Chen[1,a]

[1] *Department of Anesthesiology, China-Japan Union Hospital, Ji Lin University, Changchun, China*

Abstract. Objective: To compare effects of desflurane and sevoflurane on cognitive function in elderly patients undergoing general anesthesia. Methods Eighty elderly patients undergoing general anesthesia under open surgery were enrolled in this study, ASA I ~ II, were divided into group A (desflurane group) and group B (sevoflurane group) randomly. Intravenous drugs included midazolam 0.03 mg / kg, sufentanil 0.5ug / kg, propofol 1-2mg / kg, vecuronium 0.1 mg / kg. All patients were treated with desflurane or sevoflurane to maintain anesthesia, with depth 1.2 ~ 1.4MAC. The mini-mental state examination (MMSE) scores were recorded before operation and 1h, 6h, 12h, 24h and 48h after operation. Results The MMSE scores of the two groups had no significant difference before operation but were significantly lower than those before operation ($P < 0.05$). The MMSE scores of patients in group recovered at 12h after operation($P < 0.05$). Scores of group B returned in 24h after operation. Scores of group A were higher than group B at 1 h, 6 h and 12 h after operation, the difference was statistically significant. Conclusion The recovery time of cognitive function in elderly desflurane anesthesia patients was significantly less than that of sevoflurane.

1 Background

Postoperative cognitive dysfunction(POCD) is common in elderly patients undergoing complex surgery or emergency surgery, can be expressed as abnormal cognitive function, memory impairment or decline, personality and social integration capacity changes. The current reason was divers and no clear pathogenesis. At present, it argue that which method and drugs should be used in elderly patients maintenance of anesthesia[1]. Desflurane as the current blood gas partition coefficient of the smallest inhalation anesthetics, controllability is strong. At the same time, desflurane oil / gas partition coefficient is 18.7, low blood tissue solubility[2], the clinical manifestations of desflurane anesthesia induction and recovery are more rapid. The study of effect on desflurane anesthesia on the cognitive function of elderly patients is less[3]. The aim of this study was to compare the effect of maintenance of anesthesia with desflurane or sevoflurane on the recovery of cognitive function in elderly patients, and to provide reference for clinical work.

2 Materials and methods

2.1 General Information

[a] Corresponding author: chenpeng5481@sina.com

This study was approved by the ethics committee and signed informed consent with the patient. Eighty patients underwent abdominal surgery. American Society of Anesthesiologists (ASA) graded I-II, 41 males and 39 females, aged 60-85 years, with normal body mass index. The commonly used method for the study of POCD in the elderly was the most influential method[4], the MMSE scale, to assess the degree of cognitive function change after anesthesia[5]. Inclusion criterias were education level are high school and above, no liver,no kidney dysfunction (except for the need to treat the disease), 25 points or more MMSE score. Exclusion criterias: A.have severe visual and hearing impairment B.have a history of hypoxemia or asthma C. have a history of long-term smoking or alcohol abuse. D have cognitive dysfunction before surgery E cognitive impairment. F MMSE score <25 points.The patients were divided into two groups randomly.Both group had 40 patients, group A (desflurane group) B group (sevoflurane group). There were no significant differences in gender, age, body weight and BMI between two groups. See Table 1.

Table 1. two groups of patients with gender, age, body weight, BMI ($n = 80$, $x \pm s$)

Number of groups	Male / Female (n)	Age (years)	Weight (Kg)	BMI
Desflurane group	19/21	72.13±8.59	59.40±8.10	22.55±2.27
Sevoflurane group	22/18	71.42±9.78	58.53±6.97	24.07±2.85

2.2 Anesthesia method

Heart rate(HR), blood oxygen saturation (SpO2), blood pressure and electrocardiogram (ECG) were measured and established venous access when went in the operation room. Anesthesia was induced midazolam 0.03 mg / kg, sufentanil 0.5 μg / kg, propofol 1-2 mg / kg, vecuronium 0. 1 mg / kg, mechanical ventilation after intubation, (8-10) ml / kg, maintaining the end-tidal carbon dioxide partial pressure ($PctC0_2$) between 35-45. Anesthesia maintenance used inhalation totally, groupA of inhaled desflurane,groupB inhalation of sevoflurane, MAC of 1.2 ~ 1.4 during the surgery. Adjusted the MAC of not less than 0.8 20 minutes before the end of the operation ,.At the end of surgery, stopped inhalation of anesthetic, the oxygen flow to 5L / min, when MAC was 0.2 or less and spontaneous breathing Of the 1mg plus atropine 0.5mg antagonistic residual muscle relaxation. When the patient returned to spontaneous breathing, called the patient 1 / min times, until patients clearly, removed the tracheal tube after the recovery of reflex activity. After the patient returned to the ICU, recorded MMSE scores at1, 6, 12, 24, 48 h by anesthesiologists.

2.3 Monitoring indicators

The mean arterial pressure (MAP) of T1 (anesthesia initiation), T2 (intubation), T3 (surgical incision of skin), T4 (surgical end sutured skin) and T5 (extubation) were monitored during operation. Heart rate (HR), electrocardiogram (ECG), oxygen saturation ($SP0_2$) and end-tidal carbon dioxide partial pressure ($PETC0_2$) were observed. The changes of cognitive function MMSE (Mini Mental State) were observed at 1, 6, 12, 24 and 48 h after operation. Recorded recovery period of restlessness, drowsiness, nausea, vomiting, respiratory depression and other adverse reactions.

2.4 Statistical methods

SPSS17.0 statistical software was used for data processing. Normal distribution of measurement data to mean ± standard deviation (X soil S) table is used to compare between groups t test, count data using repeated measurement of variance analysis, P <0.05 for the difference was statistically significant.

A. Recovery times when from the end of surgery to awake group A was 8.2 ± 4.6min, groupB was 12.3 ± 5.0min, two groups were significantly different (P <0.05).

B. The comparison of MAP and HR between two groups of patients at different time points T1 (anesthesia), T2 (intubation), T3 (surgical incision of the skin), T4 (surgical closure of the skin), T5 (extubation), (P> 0.05),there were no difference at same time point . MAP and HR at each time point and no significant difference compared before anesthesia (P> 0.05) Table 2 and Table 3.

Table 2. two groups of patients MAP (mmHg) comparison (n = 40, x ± s)

Group	T1	T2	T3	T4	T5
Group A	89±21	86±18	94±27	106±18	90±25
Group B	85±17	88±22	96±23	100±11	95±13

Table 3. two groups of patients HR (times / min) comparison (n = 40, x soil s)

Group	T1	T2	T3	T4	T5
Group A	78±11	93±9	90±10	91±8	85±13
Group B	80±8	88±11	92±11	93±7	90±11

C. There was no significant difference in MMSE score between the two groups before operation (P> 0.05). MMSE scores of the two groups were significantly lower than those before operation (P <0.05) at 1h, 6h and 12h after operation, The score of MMSE in group A decreased significantly (P <0.05) at 1 h and 6 h after operation, and returned to the preoperative level at 12 h after operation. The MMSE scores of group B were significantly lower than those before operation (P <0.05) at 1h, 6h and 12h after operation, and returned to preoperative level 24h after operation. The MMSE scores of group A were higher than those of group B (P <0.05). See Table 4.

Table 4. comparions of MMSE scores changes on two groups of patients before and after operation

Group	Preoperative	1h after operation	6h after operation	12h after operation	24h after operation	48h after operation
Group A	29.5 ± 0.5	26.7 ± 1.1*★	28.4 ± 1.4*★	29.1 ± 0.6	29.4 ± 1.0	29.6 ± 0.4
Group B	29.6 ± 0.6	24.7 ± 1.3*★	26.9 ± 0.7*★	27.9 ± 1.4★	29.1 ± 1.0	29.5 ± 0.3

* **Comparison between groups** P<0. 05 ★**Comparison within the group** P<0. 05

D. Comparison of adverse reactions between the two groups of nausea and vomiting and other side effects were no significant difference (P> 0.05).

3 Discussion

Postoperative cognitive dysfunction manifest as memory impairment, visual and auditory impairment, and social behavior disorder, but also for anxiety, agitation, depression and other behavior changes. The current studies show that surgery and general anesthesia will affect the cognitive function of elderly patients significantly[6], furthermore a small number of patients still exist cognitive dysfunction after a long time. At present, the etiology and pathogeny of POCD have not been clarified. However, studies at home and abroad have shown that it is clear that anesthesia drugs and aging are the cause of POCD[7].So how to avoid the occurrence of POCD in clinical work is great significancely. Desflurane is a new halogen inhalation anesthetic drug, characterized by low blood gas partition coefficient and low blood tissue solubility, which can adjust the depth of anesthesia rapidly, induction of anesthesia quickly and rapid recovery[8]. We used MMSE score criteria which widely used and most influencely at home and abroad to determine the postoperative cognitive function, the use of desflurane or sevoflurane to maintain anesthesia depth of elderly patients after surgery and evaluate the advantages and disadvantages of two inhaled anesthesia in elderly patients with postoperative cognitive function.

The results showed that there was no significant difference in MMSE scores between two groups before operation. The MMSE scores of the two groups were significantly lower than the preoperative one. The decrease was more than 2 points, this shows that the two groups of patients were effected by maintenance of inhaled anesthesia on cognitive function short-term.This may be due to the decrease in the brain's reserve capacity in the elders. Inhaled anesthetic drugs may interfere with neurons by antagonizing NMDAR (N-methyl-D-aspartate receptors) and exciting gamma-aminobutyric acid receptors (GABAAR). And synaptic development of the environment, the role of induced neurotoxicity, leading to apoptosis of brain cells, affecting synaptic plasticity, and ultimately cause cognitive impairment. At the same time hemodynamic parameters experimental were no significant differences in two groups of patients, which also ruled out hypotension, hypoxia and other factors cause POCD. Importantly, the results also showed that the MMSE scores returned to preoperative levels at 12 hours after surgery in group A, and the MMSE score returned to preoperative levels more than 12 hours in group B. At the same time, the score of MMSE in group A was significantly higher than that in group B at 1 hour and 6 hours after operation. This shows it is desflurane that the change of patient's postoperative cognitive function may be mild , and the recovery quickly. The time when all patients awaked after operations in group A were shorter than group B, which also confirmed the results indirectly. The reason may be that desflurane has a low blood gas partition coefficient and tissue dissolution rate, anesthesia induction and recovery rapid, drug uptake and release rapidly, and low pharmacokinetics in vivo. These pharmacological features include NMDAR (N-methyl- Aspartate receptor) and excitatory gamma-amino butyric acid receptor (GABAAR) receptors make it reality that have less effect on postoperative cognitive function than sevoflurane. At the same time, desflurane anesthesia wake up faster than sevoflurane anesthesia patients, it is a reason that the patient recover preoperative levels on the aspect of work and rest time confusion, the psychological and so on.

The results of this study show that the use of two inhaled anesthetics, the elders have a short period of postoperative cognitive dysfunction, but the anesthesia of desflurane anesthesia have less affect on cognitive function than sevofluran and recover faster.

References

1. Harman D. Aging:overview [J].Ann N Y Acad Sci,2001,**928**:11-21.
2. Hangzhou 200127. Pharmacology and clinical application of J. Shanghai medical journal, 2010,4:298-303.
3. Chen Bin Zheng Hui. 3, research progress J cognitive dysfunction of inhalation anesthetic and postoperative. Medical review, 2013,10:1850-1852.
4. shows. The psychiatric rating scale manual [M].2 version. Changsha: Hunan science and Technology Press, 2003,33-39.
5. Wang Huiwen, Yang Lifang, Xiong et al. Different scale assessment after general anesthesia in elderly patients with early cognitive function compared J meaning anesth, 2001,3:459-461.
6. Cao Xiaoyang J observation of cognitive dysfunction after general anesthesia in elderly patients. Journal of medical forum, 2006,3:43-44.
7. horse Ya Bing. Different ways of anesthesia research J effects on cognitive function in elderly patients after surgery. China clinical medicine, 2014,11:26-28.
8. Fu Hui. J anesthesia effect on gynecological laparoscopic intrapulmonary shunt and blood gas. China medical science, 2013,20:16-17.

A Single Intravenous Injection of Oxycodone Hydrochloride In The Treatment of Acute Post-operative Pain after Laparoscopic Rectal Tumor Resection

Chunlu Zhang[1], Jiali Ding[1], Bin Wang[1] and Peng Chen[1,a]

[1] *Department of Anesthesiology, China-Japan Union Hospital, Ji Lin University, Changchun, China*

Abstract. Objective: To investigate the efficacy of a single intravenous injection of oxycodone hydrochloride in patients undergoing laparoscopic rectal tumor resection with acute postoperative pain and evaluate the quality of Anesthesia recovery period. Methods:60 patients for laparoscopic rectal tumor resection, were randomly divided into two groups (n=30 each): oxycodone hydrochloride injection group (group I) and fentanil injection group (group II). oxycodone 0.05mg / kg (saline dilution of 1mg / ml) or fentanil 0.1ug / kg was injected intravenously at the end of surgery. Assessing the OAA / S sedation grade and Prince-Henry postoperative pain scores. Results: Demographic data and anesthesia characteristics were similar in the two groups. Group II's Prince-Henry pain score was significantly higher than the group I's (P <0.01). Conclusion: xycodone injection can reduce the incidence of acute pain after Laparoscopic rectal tumor resection, and effectively relieve the occurred acute pain, the clinical use of oxcodone is safe and effective.

1 Background

At present, the proportion of laparoscopic surgery in the surgical operation increased year by year because of its small trauma and fast recovery. The pain after laparoscopic operation has aroused great attention. The pain of visceral and radiation were more obvious after laparoscopic surgery in addition to the pain of operative incision by stretching stimulation of the diaphragmatic peritoneum and artificial carbon dioxide pneumoperitoneum and position and so on. Most clinical use of analgesics is opioid now[1]. However, it is limiting use in the recovery period because of the respiratory depression and other side effects of opioid analgesics. Oxycodone, a potent semi synthetic opioid drugs,is a type of double receptor agonist of opioid (μ receptor and κ receptor), and is widely used in moderating to severe cancer pain, postoperative acute pain and neuropathic pain[2]. The pharmacological effects of oxycodone is similar with morphine[3]. The analgesic effect of visceral pain of oxycodone is better than other opioid drug. The prominent advantage is that the sedation and respiratory depression

[a] Corresponding author: chenpeng5481@sina.com

 DOI: 10.1051/bioconf/20170801055

is mild. The study that the analgesic effect of oxycodone injection on the recovery period of acute pain after laparoscopic surgery is not too much. The control effect, the quality of recovery and other side effects after a single intravenous injection of oxycodone hydrochloride and fentanyl in patients with acute pain recovery period of laparoscopic resection of rectal tumor at the end of surgery are compared in this experiment. It can provide reference for clinical work.

2 Materials and methods

2.1 Clinical data

This study was approved by the Ethics Committee of our hospital and signed informed consents with patients. Select 60 patients who need tumor resection by laparoscopic rectal. ASA classI - class II. Male 26 cases, female 34 cases, aged 40-76 years old. No chronic pain and analgesic or sedative. No long-term history of antidepressants. No alcohol abuse history. No use of sedatives and antiemetic and anti allergic drugs for 24 h before operation. Patients were divided into two group randomly by the random number table method:Oxycodone hydrochloride injection group (group I) and fentanyl injection group (Group II).

2.2 Anesthesia methods

Opened venous pathway, monitored of blood pressure (BP) , heart rate (HR), blood oxygen saturation (SpO_2), electrocardiogram (ECG) when the patients went into the operating room routinely. Intravenous injection midazolan 0.03mg/kg, propofol 1 ~ 2 mg / kg, sufentanil 0.5ug/kg, vecuronium 0.1mg / kg was used in anesthesia induction. Inserted the endotracheal tube after aerobic nitrogen removal,and mechanical ventilation. Experimented expiration-end gas monitoring,maintaining expiration-end CO_2（PetCO2）35-45mmHg. The anesthesia was maintained with total inhalation. Inhaled sevoflurane. Maintained the depth of anesthesia in 1.2 ~ 1.5 MAC, which no additional sufentanil during the operation. Intravenous injection of vecuronium 0.05mg / kg when necessary. Stoped sevoflurane，adjusted the oxygen flow rate，eluted sevoflurane rapidly and gived oxycodone injection 0.1mg/kg or fentanyl injection 1ug/kg at the end of the operation. When the MAC were below 0.2, call patients per minutes. Pulled out the tracheal tubes after the patients were clear, the reflex activity were recovered. All the patients in each group were treated by analgesia pump after extubation.

2.3 Monitoring index

The time of anesthesia, operation , recovery , extubation and were recorded. The blood pressure, heart rate, SpO_2 of the moment of operation completing and extubation at 10min, 20min, 15min, and 5min after extubation were recorded. Each OAA/S was assessed with a 5min sedation scale after extubation. (Level 5, have a swift response to normal voice. Level 4, indifferent of normal voic. Level 3, only response to calling for name loudly or repeatedly. Level 2, only response to a slight shake on the shoulder or head. Level 1, no response to mild pushing and shaking. Level 0, to squeeze trapezius muscle. OAA/S further divided into excessive sedation score 2 or less). Postoperative pain Prince - Henry scores.(0 points, no pain when coughing. 1 points, pain when coughing. 2 points, the depth of breathing when there is pain, quiet, no pain. 3 points, the rest of the state that there is pain, but more light, can endure. 4 points, resting state that there is severe pain, unbearable. The Prince-Henry score is greater than or equal to 3 divided into the recovery period of acute pain.). The occurrence of agitation, nausea, vomiting, respiratory depression and other adverse reactions were recorded.

2.4 Statistical Analysis

Using statistical software SPSS 17.0 for data processing. The measurement data of normal distribution are expressed as mean plus or minus standard deviation($\bar{\chi}$±s). Variance analysis between groups was compared using t test. The analysis of variance of repeated measurement was used in counting data. P < 0.05 for the difference was statistically significant.

3 Result

A. There were no significant differences in sex, age, body weight and BMI status between two groups, P>0.05. See Table 1.

Table 1. Comparison of two groups of patients in general（**n=30,** χ±s）

group	The number of cases	Male / female	Age(n)	Weight（kg）	BMI
group I	30	12/18	55.17±8.59	59.40±8.10	22.55±2.27
group II	30	14/16	55.97±9.78	58.53±6.97	24.07±2.85

B. Comparison of anesthesia time and operation time between two groups，No statistical difference，P>0.05.See Table 2.

Table 2. Comparison of anesthesia time and operation time between two groups

group	anesthesia time（h）	operation time（h)
group I	4.00±0.50	4.29±0.50
group II	3.83±0.33	4.20±0.35

C. The spontaneous breathing recovery time, extubation time and recovery time of Fentanyl group was longer than oxycodone group. The difference was statistically significant (p < 0.05). See Table 3.

Table 3. The spontaneous breathing recovery time, extubation time and recovery time of Fentanyl group

group	The spontaneous breathing recovery time（min）	extubation time（min）	recovery time（min）
group I	7.57±3.22	13.00±4.09	14.80±5.22
group II	12.87±3.18	18.77±4.23	19.80±4.47

 DOI: 10.1051/bioconf/20170801055

D. Repeated measure analysis of variance indicate that the change trend of heart rate and mean arterial pressure of oxycodone group is same with the Fentanyl group at 20 minutes after extubation. See Figure 1,2.

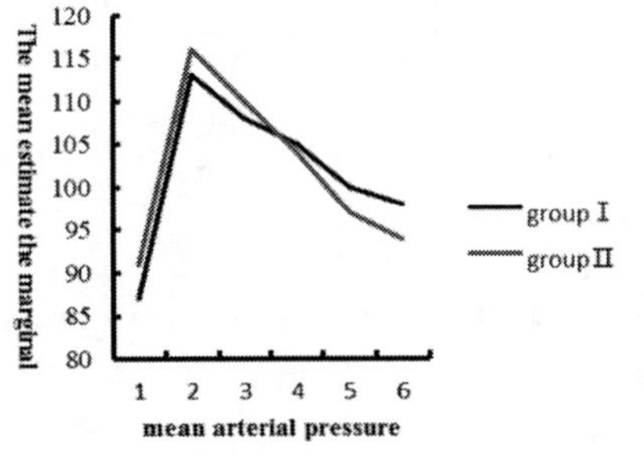

Figure 1. Contrast figure of mean arterial pressure

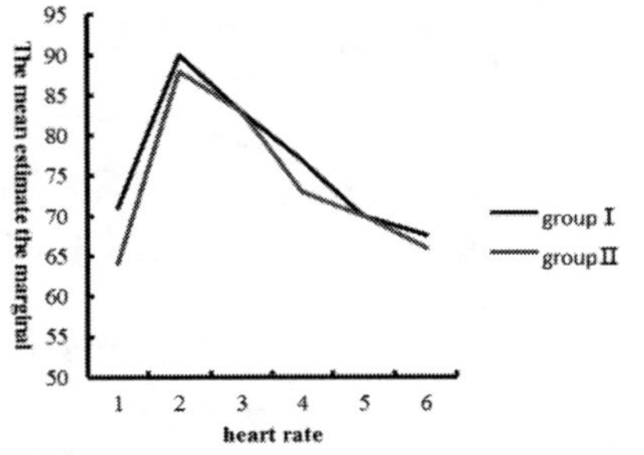

Figure 2. Contrast figure of heart rate

E. Blood oxygen saturation monitoring data displayed that oxygen saturation in fentanyl group decreased more significantly than that of oxycodone group. When the blood oxygen saturation data is lower than 90%, the oxygen mask is required. The number of cases in oxycodone group is one while the number in fentanyl group is five. See Figure 3.

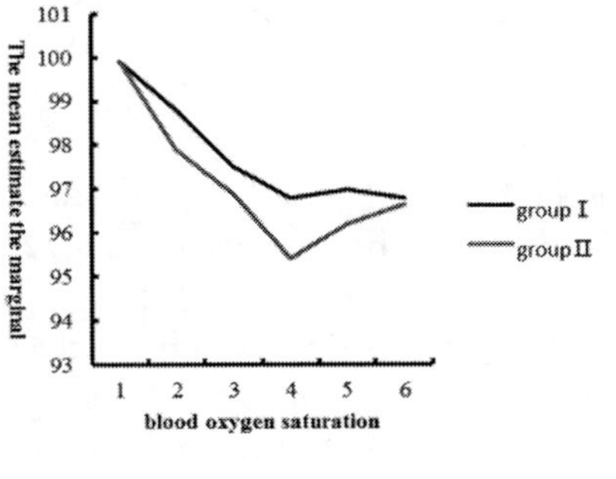

Figure 3. Contrast figure of blood oxygen saturation

F. OAA/S sedation score showed fentanyl group at each time point was lower than that of oxycodone group. The difference was statistically significant($p < 0.05$). See Figure 4.

DOI: 10.1051/bioconf/20170801055

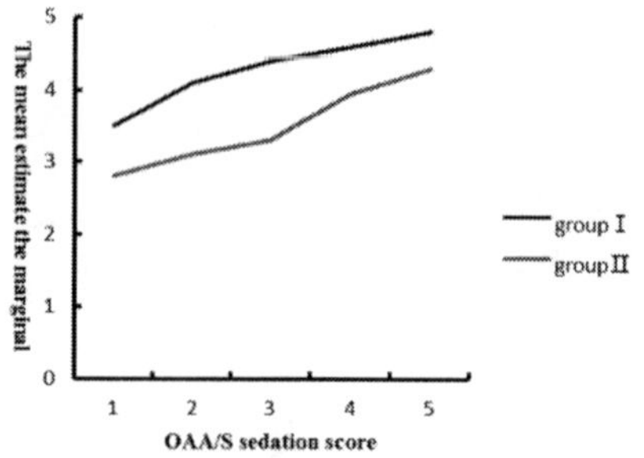

Figure 4. Contrast figure of OAA/S sedation score

G. Prince-Henry pain scores showed the score of fentanyl group was higher than oxycodone group. The difference was statistically significant($p<0.05$). See Figure 5.There were 2 patients in anesthesia recovery period having acute postoperative pain in oxycodone group of experiment. Fentanyl group had 8 patients with acute pain after surgery. When the Prince-Henry score was more than 3 points, emergency analgesia was required. The symptoms would be improved obviously after intravenous oxycodone (0.1mg/kg).

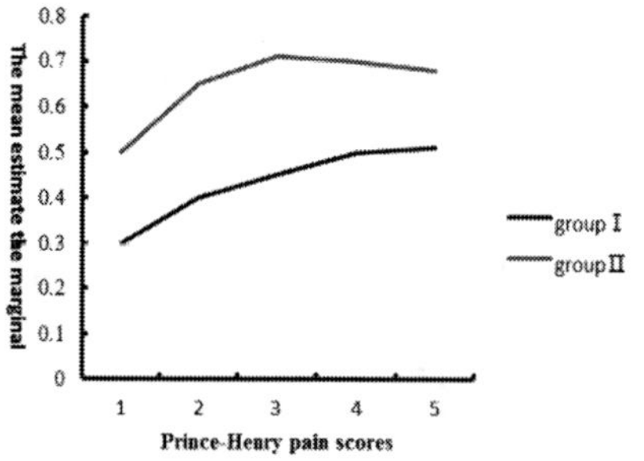

Figure 5. Contrast figure of Prince-Henry pain scores

H. There was no significant difference in the side effects of nausea and vomiting between two groups.

4 Discussion

Acute pain after operation in patients can cause a series of complications such as high blood pressure, increased heart rate, agitation, respiratory depression and so on. The characteristic of acute pain after laparoscopic operation is from visceral and radiation in addition to surgical incision. Because the pain occurs in the recovery period, the postoperative analgesia pumps and other conventional analgesia methods can not very well play the analgesic effection. It is the original intention of this study that how to do a good job of convergence and the patient is truly painless. At present, opioid drugs such as morphine, fentanyl, sufentanil is usually used in the recovery period of analgesia. However, this kind of drugs would cause respiratory depression, drowsiness and other complications with the increased of dose. Therefore, because of the quality of the patient's recovery and the fear of respiratory depression, the treatment of the recovery period is often not satisfactory. Oxycodone injection, is a new type of double opioid receptor agonist. In theory, there is a good curative effect on cutting pain and visceral pain. The effect of respiratory depression was mild and no obvious sedation.The purpose of this experiment was to compare the effects of acute pain and the quality of recovery in a single intravenous injection of oxycodone hydrochloride and fentanyl in laparoscopic resection of rectal tumor.

In the experiment, the Prince-Henry scores of oxycodone group were lower than fentanyl group at each time point. The number of cases in oxycodone group that occurred acute postoperative pain was less than fentanyl group. There are three possible reasons.

1. In the course of laparoscopic rectal tumor resection, by the stretching stimulation of the diaphragmatic peritoneum and artificial carbon dioxide pneumoperitoneum and position and other factors, visceral and radiation pain were more obvious after laparoscopic surgery in addition to the operative incision pain. Some studies have found that 63% of patients under the diaphragm and shoulder pain after laparoscopic surgery, and the degree and duration of pain is often more than operative incision pain. Patients with 73%-80% after laparoscopic surgery need to use analgesics to relieve acute pain after surgery. The onset time of fentanyl and oxycodone after intravenous injection is similar (2-3mins) and they were all strong opioids. They are mainly mediated by opioid receptors on the presynaptic membrane of nerve terminals in CNS. These opioid receptors belong to the G protein coupled receptor family（μ-，κ- and δ-）. Fentanyl is mainly through the μ receptor activation and the cutting and other sharp pain analgesic effect is good. While oxycodone is double opioid receptor agonist (μ receptor and κ receptor) and has good effect on acute postoperative pain and nerve pain. The pharmacological effects of oxycodone were similar with morphine [4] [5]. The oxycodone receptor is κ receptor. In particular, the κ2b receptor can help achieve the purpose of analgesia[6]. So the analgesic effect of oxycodone on visceral pain is better than other opioids. Animal experiments also showed that the κ receptor[7,8,9] is related to the relief of visceral pain. Therefore, the control of oxycodone on postoperative recovery period of acute pain in visceral pain may be better after laparoscopic resection of rectal tumor[10,11].

2. Because of fentanyl and oxycodone in fat-soluble difference is big, the rapid redistribution of fentanyl after intravenous injection may result in the decrease of analgesic effect.

3. In this experiment, only use the conventional dose sufentanil in induction phase, minimal residual dose at the end of the operation.

OAA/S sedation scale of oxycodone group was higher than that of fentanyl group. The number of excessive sedation was less than that of fentanyl group(1:9).The number of $SpO_2 \leq 90\%$ cases was less than fentanyl group. Spontaneous breathing recovery time, extubation time and recovery time were shorter than in the fentanyl group. There are two possible reasons.

1. Oxycodone play predominate pharmacological effectors mainly through κ receptor and the μ acceptor activation is weak. Therefore,reduce the complications such as drowsiness and respiratory depression that occur at the same time of analgesic aim realizing through the μ receptor agonist[12,13]. Due to the good analgesic effect and the patient's breathing is affected by the pain of small,oxycodone group may be greater rate.

2. Oxycodone play predominate pharmacological effects mainly through a κ r eceptor. The synergistic effect of sevoflurane of oxycodone was weaker than that of fentanyl. Therefore, the influence of the quality of recovery of patients may be more weak than fentanyl group.

In summary, a single intravenous injection of oxycodone 0.1mg/kg can prevent the occurrence of acute pain after operation better than intravenous injection of fentanyl 1ug/kg for patients of laparoscopic resection with rectal tumors at the end of the operation. Excessive sedation and respiratory depression and other side effects are smaller. The recovery of general anesthesia will be more safe.

References

1. Sun Zhengyi, Lang Jinghe, Leng Jinhua, pain after laparoscopic surgery. Chinese modern surgery magazine.2000; 11; 30 (10): 309-311.
2. Xu Jianguo Pharmacology and clinical application of oxycodone hydrochloride. Journal of Clinical Oncology, 2014,5,30 (5): 511-513.Kalso E.Oxycodone.J Pain Symptom Manage.2005;**29**:S47-56.
3. Ross FB,Smith MT.The intrinsic antinociptive effects of oxycodone appear to be kappa-opioid

receptor mediated.Pain 1997;**73**:151-157.
4. Nozaki C,Saitoh A,Tammura N,Kamei J.antinociceptive effect of oxycodone in diabetic mice.Eur J Pharmacol 2005;**524**:75-79.
5. Nielsen C,Ross FB,Lotfipour S,Saini KS,Smith MT.Oxycodone ahd morphine have distinctly different pharmacological profiles:radioligand binding and behavioural studies in two rat models of neuropathic pain.Pain 2007;**132**:289-300.
6. 2009 years of adult postoperative pain treatment expert consensus, the Chinese Medical Association Branch of anesthesiology.
7. Kalso E.Oxycodone.J Pain Symptom Manage.2005;**29**:S47-56.
8. Lugo RA,Kern SE.The pharmacokinetics of oxycodone.J Pain Palliat Care Pharmacother 2004;**18**:17-30.
9. Okura T, Hattori A, Takano Y, et al. Involvement of the pyrilamine transporter, a putative organic cation transporter, in blood–brain barrier transport of oxycodone. Drug Metab Dispos 2008; **36**: 2005– 2013.
10. M Kokki,P Valitalo,et al.central nervous system penetration of oxycodone after intravenous and epidural administration.British Journal of Anaesthesia.2014;**112**(1):133-140.
11. Xu Xing, Wu Xinmin, Xue Zhang Gang, Wang Xiangrui, Xiong, Tian Ming, Chen Yao. Oxycodone hydrochloride injection for postoperative analgesia efficacy, safety and prospective, randomized, double-blind, multicenter, positive controlled clinical research. Chinese Journal of Anesthesiology, 2013; 3; 33 (3): 269-274.
12. Sun Yang, Lei Jun, Wang Meiqing, Cui Yan. Effect of oxycodone on quality of thoracic surgery with general anesthesia recovery period. Journal of clinical anesthesiology, 2014; 10; 30 (10): 1032-1033.

BIO Web of Conferences **8**, 01056 (2017) DOI: 10.1051/bioconf/20170801056
ICMSB2016

The role of Purα in neuronal development, the relationship between Purα and epilepsy in the current researches

Shanshan Guo[1], Bingying Zhang[2], Chengmin Yuan[1], Ping Li [2], Tao Sun[1] and Jianqi Cui[1,2,a]

[1]*Ningxia Key Laboratory of Cerebrocranial Diseases, the Incubation Base of National Key Laboratory, Ningxia Medical University, Yinchuan, Ningxia, 750004, PR China*
[2]*Department of Biochemistry and Molecular Biology, School of Basic Medical Sciences, Ningxia Medical University, Yinchuan, Ningxia, 750004, PR China*

Abstract. As a ubiquitous protein in the body, Purα has been considered to possess multiple regulatory functions from DNA replication and transcription to RNA transport and translation. The recent studies have proved that Purα protein played a vital role in neuronal development with the solid and abundant evidence both in transgenic mice and human genetic DNA analysis. The Purα knocked out mice models have been successfully established in two independent research groups and their results demonstrated that lack of Purα alters postnatal brain development. Mice lacking of Purα display decreased neurogenesis and impaired neuronal development. In the mouse brain, the expression level of Purα seems to be regulated with brain development. It starts to express Purα after birth and quickly reach to highest level in the third week. The recent studies also have proved that some neurodegenerative disease also linked with Purα, such as Fragile X-associated tremor/ataxia syndrome. rCGG-mediated neuronal toxicity could be modulated by heterogeneous nuclear ribonucleoprotein A2/B1 and Purα, which is rCGG-repeat-binding proteins. The 5q31.3 microdeletion syndrome is another disorders that is associated with the mutation of Purα gene, it has been demonstrated that all the symptoms such as the neurocognitive impairment, neurodevelopmental delay and learning disability are caused by the mutation of Purα gene. In this mini review, we will retrospectively review the current progress of Purα in neuronal development and try to figure out the connections between these observed phenomena and the biological function of Purα.

1 Introduction

Purα [1], encoded by PURA gene, could bind to single-stranded DNA and RNA [2, 3] in a sequence-specific way. Purα is a highly conserved protein throughout the whole evolution; it has been reported that between mouse and human beings, there are only 2 amino acid residues differences in the sequence of all the 322 amino acids of Purα [3]. Purα has a broad potential functions, the increasing research works have demonstrated that the initiation of DNA replication, gene transcription, and the compartmentalized mRNA translation [4-6]. In addition, Purα is also involved in cell cycle regulation, oncogenic transformations [1, 7-9], and DNA repair [10, 11]. Purα is though expressed in every metazoan tissue [5], it predominantly plays a regulatory function in the developmental stage [12, 13] in a tissue-specific way. Although the precise roles of Purα in the brain are still debatable, most of the current research works demonstrated that Purα possibly participates both in neuronal proliferation and in the maturation of dendrites. In Purα knock out mice, the animals look normal at birth, but soon

[a] Corresponding author: Jianqi@gmail.com

develop neurological deficit at 2 weeks of age, and die within 4 weeks [14]. Another recently discovered neurodegenerative disorder is Fragile X-associated tremor/ataxia syndrome [15], the adult carriers of permutation alleles (55-200 CGG repeats) of the fragile X mental retardation 1 [16] gene [17, 18] would be affected. In addition, another neuronal developmental disorder linked with Purα is 5q31.3 Microdeletion Syndrome reported by Lalani et al [19] and Hunt et al [20]. For more detailed information about Purα please refer to the review written by Johnson EM [5, 21], White MK [22] and Chai J et al [40]. In this review, we will mainly discuss the functions of Purα protein on the neuronal development, the transgenic model of Purα as well as the functions of Purα on neurological disorders.

2 The functions of purα on neuronal development

Purα is a ubiquitous nucleic acid-binding protein originally purified from mouse brain. Purα has an ability to bind to a DNA sequence derived from the promoter of the mouse myelin basic protein gene [12, 23]. Human Purα possesses an ability to bind to a DNA sequence presented upstream of the human c-Myc gene. The Purα cDNA was cloned from HeLa cells and its sequence was also analysed [3]. Purα has a very high conservation in their sequence and there is only 2 out of 322 amino acid residues different between the mouse [24] and human beings [2]. Same as other transcriptional factors, Purα also has a modular structure with a central-binding domain containing 3 class I repeats and 2 class 2 repeats. There is a glycine-rich domain in its N-terminal and a "psycho" domain and glutamine-rich domain and glutamate-rich domain in its C-terminal. Its DNA binding domain is intensely conserved throughout evolution [22]. (Figure 1)

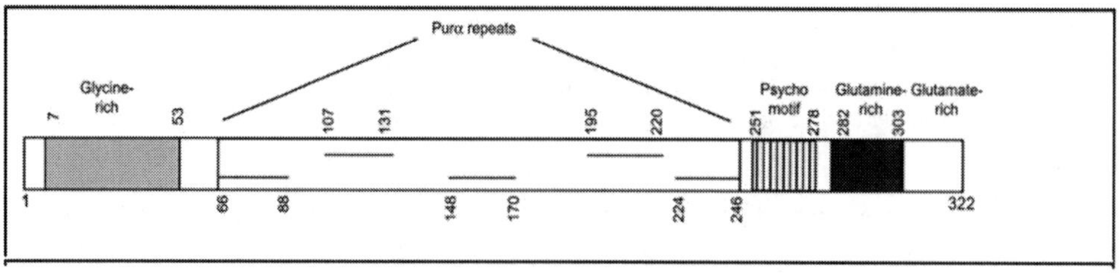

Figure 1. Structure of Purα protein

Pur family includes three different proteins, Purα, Purβ [2] and Purγ. Since they use the alternative polyadenylation sites [25], there are two isoforms existed, and in almost every metazoan tissue can detect the expression of Pur proteins [5]. As a multifunctional protein, Purα [5] can bind to both DNA and RNA and because of this property, it has been reported that the initiation of DNA replication, control of transcription and mRNA translation [2, 4] are all associated with Purα. It has been reported that the DNA sequences close to viral and cellular origins of replication could be associates with Purα, Purα might have an ability to destabilize DNA helix [26], as we know that the initiation of transcription and replication requires unwinding of duplex DNA, the critical role of Purα in neuronal development and synaptogenesis is undoubted. Being a multifunctional protein, Purα binds to single-stranded nucleic acids in a sequence-specific way to fulfil its regulatory functions in many critical life activities, which could be the reason why its structure is strongly conserved through the evolution [5]. The extensive characteristics of Purα is its ability to bind to single-stranded DNA and RNA in a sequence specific way, therefore, it has been originally cloned from HeLa cell according to its affinity for a single-stranded DNA element with the sequence GGNGGN [2]. Since Purα plays critical roles in DNA replication and gene transcription [4, 5], the increasing research works reported that it also acted as a critical component in the compartmentalized translation of RNA [6]. In addition, as a comprehensively expressed protein, Purα plays developmental and tissue-specific regulatory roles in the central nervous system [14, 23]. Khalili et al reported that during embryogenesis period Purα protein levels kept in a quite low level in mouse brain, but after the first postnatal week, it quickly begin to increase and reach their peak at 18 days postnatal. From then on, Purα is retained at comparatively high levels in the adult brain [14]. Consequently, these data suggest that Purα levels inversely correlate during mouse brain development.

The distribution of Purα in neurons is specific and it has been reported that Purα is mainly expressed in the cytoplasm, particularly in dendrites at branch points [6]. Interestingly, Purα is also to be found in a complex form with polyribosomes and heterogeneous nuclear ribonucleoproteins (hnRNP) proteins existed in the neuron. Purα knockout mice die shortly after birth with a variety of organ defects [14]. In particular, mice lacking Purα show neurological dysfunction, and histopathological examination of knockout mouse brains has shown defects in neuronal cell development and differentiation which leads to neuronal dropout as well as abnormal and delayed corticogenesis. Defects in dendrite development throughout the cortex, the hippocampus, and the cerebellum were observed [14, 27]. Hippocampal neurons fail to form synaptic connections in the absence of Purα, as visualized by a significant lack of post-synaptic density-95 (Psd95) foci. Similar, a possible role in regulation of immune cell function and response to hypoxia has also been attributed to Purα [27].

3 Postnatal brain development was affected when purα has been disrupted in knock out mice model

Previous experiments have demonstrated that Purα is essential for postnatal development and developmentally coupled cellular proliferation in the mouse. Khalili et al [14] has reported that when both alleles of PURA gene have been disrupted, the mice displayed special features. The mice seem normal at birth, but gradually developed neurological problems at 2 weeks of age, also companied with some abnormal signs such as severe tremor and spontaneous seizures. The mice were bound to die by 4 weeks.

Histological examination demonstrated that the numbers of neurons in regions of the hippocampus and cerebellum in PURA knock-out mice was much less than that in the age-matched wild type littermates, and also, lamination of these regions is deviant at time of death. MCF7 is a protein marker for DNA replication, the PURA knock-out mice demonstrated that the proliferation of precursor cells is deficient in these regions when MCF7 levels were immunohistochemically analysed. These phenomenon were also observed in several other tissues, such as those of myeloid linage, but in heterozygous mice, these aberrant phenomenon is relatively less severe than that in homozygous mice. Histological examination also demonstrated that the level of myelin and glial fibrillary acidic protein were reduced both in oligodendrocytes and astrocytes, which implies that these cells undergo a pathological development. A crucial time for cerebellar development appears at the 5th day after birth, as a signal protein marker for these is the Purα and Cdk5, the expression level of them were both in their highest levels in bodies and dendrites of Purkinje cells of wild type mice, in normal condition, Purα and Cdk5 can be coimmunoprecipitated from brain lysates of wild type mice, but unfortunately, both of them were missing in dendrites of PURA knock-out mice. In accompany with these observation, the level of both phosphorylated and nonphosphorylated neurofilaments in dendrites of the Purkinje cell layer and of synapse formation in the hippocampus were all remarkably reduced with immunohistochemical studies. All above discoveries indicate that Purα participates in the developmentally timed DNA replication in specific cell types, this might be a new arising role of Purα in compartmentalized RNA transport and translation in neuronal dendrites [14].

Interestingly, another comparable work reported by Hokkanen et al reported that deficiency of Purα changes postnatal brain development and gives rise to megalencephaly [28]. In their generated Purα-deficiency mice model, they found that the mice were born with no obvious pathological condition and could be raised to 6 months but not died at week 4 as reported by Khalili's work. Hokkanen thought that the postnatal proliferation of neuronal precursor cells both in the hippocampus and in the cerebellum were only extended when Purα was disrupted, but the overall number of post mitotic neurons were not affected. In previous studies, the transition of dendritic protein MAP2 is thought to be Purα-dependent, Hokkanen's work also found the changes of this protein's expression and distribution. Same as Khalili's work, they also found that the PURA knock-out mice developed a continuous tremor at the age of 2 weeks and this symptom continued throughout lifetime; this observation is consistent with Khalili's report, it indicates that disruption the endogenous level of Purα expression would be associated with the onset of epilepsy, this phenomenon is also confirmed by other researches related to the changes of Purα [19, 29], about this we will discuss later in the review.

In contrast to Khalili's work, some other new findings such as a megalencephaly, axonal swellings and hyperphosphorylation of neurofilaments were observed in PURA knock-out mice. Nevertheless, both of these two research works confirmed the effects of Purα on the brain development, but there are still some differences on the observed phenotypes, especially whether the Purα knock-out mice die at the 4th week after the birth or can live to the adulthood, the reason for this is not very clear right now. Hokkanen's work did not show the western blotting results of the of expression level of Purα in their knock out mice, the efficiency of knock out of Purα is complete or not, or it might be only a truncated form of Purα being knocked out, all of these has to be considered as a reason for this differences.

4 Pura mutation is associated with 5q31.3 microdeletion syndrome

5q31.3 microdeletion syndrome is a rare disorder and firstly reported by Shimojima et al [30], which is characterized by severe developmental delays, distinctive facial features, and delayed myelination. The subsequent researches continually reported this new discovered disorder associated with neuronal abnormalities, including neonatal hypotonia, encephalopathy, and severe developmental delay, with or without seizure. All the researches demonstrated that the disorder is linked to the minimal critical deletion interval harbours three genes [30-33].

Recently, Lalani et al has reported 11 individuals with clinical features of 5q31.3 microdeletion syndrome and they found de novo mutations in PURA within the critical region [19]. And their data demonstrated that the severe neurological abnormalities observed in this syndrome is caused by the mutation of PURA gene. Among 2908 successive subjects, 11 individuals were found to have de novo heterozygous mutations in the single-exon gene PURA, in which there were four truncating (three nonsense and one frameshift), five missense and two in-frame deletions. All 11 individuals appeared significant hypotonia at birth, other symptoms included respiratory insufficiency (central and obstructive sleep apnea and recurrent pulmonary aspiration). Since the patients often encountered early-onset feeding difficulties with moderate dysphagia and evidence of tracheal aspiration, nasogastric or gastric-tube feeding were often required. Another common features observed in the subjects with 5q31.3 microdeletion syndrome is their myopathic facies, which were characterized as an open mouth and high arched palate. There were 4 out of 11 individuals presented with hypomyelination or myelin-maturation delay when checked by brain-imaging studies. Another frequently observed symptoms in infancy were myoclonic jerks which caused 6 out of 11 children to progress to develop epilepsy. Most individuals had abnormal EEG and two individuals were diagnosed Lennox-Gastaut syndrome. It has to be pointed out that the exome sequencing is a valuable method for the diagnosis of this disorder and it is helpful distinguish this disease with other disorders. All the above mentioned results indicate that severe neurodevelopmental profile of 5q31.3 microdeletion syndrome is closely associated with PURA mutations since all the individuals with this mutation share the common pattern of clinical manifestations.

A similar results was reported by Hunt et al, they found de novo mutations of PURA in family trios, which causes severe, sporadic, undiagnosed neurodevelopmental delay and learning disability through a whole exome sequencing [28]. There were four subjects has been identified with this de novo mutation which causes the structural changes of the PURA encoded protein Purα, usually it is considered as a transcriptional activator. The four mutations including two different frameshifts, one in-frame deletion and one missense mutation had been identified through a whole exome sequencing. Although the phenotype of the mutations in this gene is variable, the clinical manifestations are common to all individuals, which included moderate to severe neurodevelopmental delay and learning disability, neonatal hypotonia, early feeding difficulties and seizures, or 'seizure-like' movements. In addition to the above findings, this disorder might also be associated with anterior pituitary dysregulation. When Purα repeat III was disrupted, a more severe phenotype appeared, so it is reasonable to think there might be a possible genotype–phenotype correlation existed in the brain though psychomotor developmental outcomes seem variable among patients. All these findings provide the solid evidence and confirmed that PURA mutations cam cause 5q31.3 microdeletion syndrome, the clinical manifestations of this disorder include neurodevelopmental delay, learning disability, and neonatal hypotonia, feeding difficulties, abnormal movements and epilepsy. It is also helpful to elucidate the role of PURA in other microdeletion phenotype.

5 Purα and fragile x tremor/ataxia syndrome

Fragile X syndrome is previously described by Mareni et al [31] in 1981 and he reported that individuals with the form of X-linked mental retardation is associated with the presence of a cytologically variant X chromosome in which there is a secondary constriction or "fragile site" at Xq 27-28 (Fra X) existed. From then on, the increasingly research works focused on this project and confirmed that this disorder is caused primarily by expansion of the CGG trinucleotide repeat in the 50 untranslated region (50 UTR) of the fragile X mental retardation 1 (FMR1) gene [38,39]. It has been demonstrated that fully affected individuals have more than 200 CGG repeats that are referred to as full mutation alleles, whereas the normal individuals generally only possess between 5 and 54 repeats. The FMR1 gene is considered as precursor of full mutation alleles when they expanded, they can change from a permutation alleles into full mutation form through genetic instability during maternal germline transmission [39]. The premutation carriers of Fragile X-associated tremor/ ataxia syndrome (FXTAS), which is considered as a recently identified neurodegenerative disorder, were all found in male and female individual carriers [32, 33]. Although the incidence to develop FXTAS is much higher in male carriers, female carriers also can develop FXTAS [33]. There is the ubiquitin-positive intranuclear inclusion present in both neurons and astrocytes throughout the brain and it has been considered as the remarkable neuropathological features of FXTAS [32] and the progressive intention tremor, gait ataxia, parkinsonism, autonomic dysfunction, and cognitive decline were comprised of the most common clinical manifestations of FXTAS [34]. Additionally, the marked dropout of Purkinje cells, Purkinje axonal torpedoes, and Bergmann gliosis were also observed in the cerebellum of FXTAS patients. Immunohistochemical examination demonstrated that there was no intranuclear inclusions present in Purkinje cells, and they appeared in the dentate nucleus in a small number of neurons and diffusely distributed in cerebellar astrocytes [35].

The research work from Jin et al demonstrated the mechanism of Purα in the FXTAS syndrome, they successfully established a Drosophila model of fragile X tremor/ataxia syndrome in which they found that Purα could bind to rGGG repeats and modulate repeat-mediated neurodegeneration [16]. They suggested that rCGG-repeat-binding proteins (RBPs) (in which Purα and hnRNP A2/B1 has been identified as RBPs) may be sequestered from their normal function by rCGG binding since transcribed premutation repeats alone are sufficient to cause neurodegeneration. They also show that Purα and rCGG repeats interact in a sequence-specific way and it is strongly conserved through evolution. rCGG-mediated neurodegeneration could be suppressed with overexpression of Purα in Drosophila in a dose-dependent way. Additionally, it also has been illustrated that Purα existed in the inclusions of FXTAS patient brains. Above all, it is not difficult to conclude that Purα plays an important role in the pathogenesis of FXTAS since rCGG repeat sequestration of specific RBPs and result in neuronal cell death.

Interestingly, the research work from another group supports Jin's work. Tan et al [36] also found that in an FXTAS Drosophila model, gypsy, a specific retrotransposon named gypsy played an important role in the regulation of rCGG-mediated neurodegeneration when it was activated. On the other hand, the neuronal toxicity caused by rCGG repeats could be suppressed when gypsy was knockdown. They also found that hnRNP A2/B1, one of the rCGG-repeat-binding protein is involved in this process through interaction with heterochromatin protein 1. All these findings indicate an important role for retrotransposition in neurodegeneration [36].

The role of Purα in rCGG–mediated neurodegeneration was further investigated by Qurashi et al [37]. They identified more than 100 proteins that interact with Purα through a proteomic approach. It is interesting to notice that rCGG–mediated neurodegeneration could be regulated by Rm62, which is a kind of p68 RNA helicase existed in Drosophila. The expression of Rm62 could be reduced posttranscriptionally by rCGG repeats and result in the Hsp70 transcripts accumulated in the nucleus and, which will act as additional mRNA to participate in stress and immune response. All these data might point to a new mechanism of FXTAS pathogenesis in which, the nuclear exports were impaired by these mRNA abnormally accumulated inside the nucleus.

6 Conclusions

Purα is an important transcriptional factor in different cellular processes. Purα plays a critical role in the developing and mature brain and targeted disruption of the PURA gene in both alleles has also confirmed its functions in postnatal brain development. Purα-deficient mice appear a serial pathologic phenotypes in neuronal system manifested by severe tremor and spontaneous seizures. The neuronal cells were found very low in number in hippocampus and cerebellar regions in addition to many other abnormalities. Although other similar work demonstrated that the PURA-/- mouse can be raised to 6 months without dying, but the brain development was altered when Purα is knocked out. Both model displayed the phenomenon that disruption the Purα expression will lead the mouse to induce seizures. All these results indicate some connections between Purα and epilepsy.

Purα is also associated with 5q31.3 microdeletion syndrome and the mutation of PURA would be the cause of this disorder, which has been approved by different individual research groups. The patient with 5q31.3 microdeletion syndrome displayed severe neurological conditions, including neonatal hypotonia, feeding difficulties, an abnormal electroencephalogram(EEG), hypomyelination and severe psychomotor delay, the spontaneous seizure or tremor is also observed in the patient. With whole exome sequencing analysis, it has been confirmed that the mutation of Purα can cause this disorder.

Purα is also involved in the Fragile X tremor/ataxia syndrome since it can bind to rCGG repeats and modulate repeat-mediated neurodegeneration. Overexpression of Purα in Drosophila could suppress rCGG-mediated neurodegeneration in a dose-dependent manner.

Above all, Purα is an important transcriptional factor and play a critical role in the neuronal development. Targeted disruption the expression of Purα would lead severe neurological conditions, especially, it might be involved in the onset of seizure.

Acknowledgments

This work was supported by grants from NFCS to Dr. Jianqi Cui (81260197) and National 973 Project to Dr. Jianguo Niu (2014BC560711) and Grant of Dominant Subject Group for Tao Sun (XY201511) supported by Ningxia Medical University scientific research project.

References

1. S.M. Barr, E.M Johnson. Journal of cellular biochemistry. 2001; 81(4):621-38.
2. A.D. Bergemann, Johnson EM. Molecular and cellular biology. 1992; 12(3):1257-65.
3. A.D. Bergemann, Z.W. Ma, E.M. Johnson. Molecular and cellular biology. 1992; 12(12):5673-82.
4. G.L. Gallia, E.M. Johnson, and Khalili K. Nucleic Acids Res. 2000; 28(17):3197-205.
5. E.M. Johnson. Anticancer Res. 2003; 23(3A):2093-100.
6. E.M Johnson, Y. Kinoshita, D.B. Weinreb, et al. J Neurosci Res. 2006; 83(6):929-43.
7. D.W. Stacey, M. Hitomi, M. Kanovsky, et al. Oncogene. 1999; 18(29):4254-61.
8. N. Darbinian, G.L. Gallia, J. King, et al. J Cell Physiol. 2001; 189(3):334-40.
9. K. Lezon-Geyda, V. Najfeld, E.M. Johnson. Leukemia. 2001; 15(6):954-62.

DOI: 10.1051/bioconf/20170801056

10. H. Wang, M.K. White, R. Kaminski, et al. Anticancer research. 2008; 28(3A):1441-7.
11. H. Wang, M. Wang, K. Reiss, et al. Cancer biology & therapy. 2007; 6(4):596-602.
12. S. Haas, J. Gordon, K. Khalili. Molecular and cellular biology. 1993; 13(5):3103-12.
13. A. Tretiakova, G.L. Gallia, N. Shcherbik, et al. The Journal of biological chemistry. 1998; 273(35):22241-7.
14. K. Khalili, L. Del Valle, V. Muralidharan, et al. Mol Cell Biol. 2003; 23(19):6857-75.
15. J.G. Daigle, K. Krishnamurthy, N. Ramesh, et al. Acta Neuropathol. 2016; 131(4):605-20.
16. P. Jin, R. Duan, A. Qurashi, et al. Neuron. 2007; 55(4):556-64.
17. R.J. Hagerman, M. Leehey, W. Heinrichs, et al. Neurology. 2001; 57(1):127-30.
18. S. Jacquemont, R.J. Hagerman, M. Leehey, et al. American journal of human genetics. 2003; 72(4):869-78.
19. S.R. Lalani, J. Zhang, C.P. Schaaf, et al. American journal of human genetics. 2014; 95(5):579-83.
20. D. Hunt, R.J. Leventer, C. Simons, et al. Journal of medical genetics. 2014; 51(12):806-13.
21. E.M. Johnson, D.C. Daniel, J. Gordon. Journal of cellular physiology. 2013; 228(5):930-7.
22. M.K. White, E.M. Johnson, K. Khalili. Cell Cycle. 2009; 8(3):1-7.
23. S. Haas, P. Thatikunta, A. Steplewski, et al. J Cell Biol. 1995; 130(5):1171-9.
24. Z.W. Ma, A.D. Bergemann, E.M. Johnson. Gene. 1994; 149(2):311-4.
25. H. Liu, E.M. Johnson. Nucleic Acids Res. 2002; 30(11):2417-26.
26. N. Darbinian, G.L. Gallia, K. Khalili. J Cell Biochem. 2001; 80(4):589-95.
27. T. Kong, M. Scully, C.S. Shelley, et al. J Immunol. 2007; 179(3):1934-41.
28. S. Hokkanen, H.M. Feldmann, H. Ding, et al. Hum Mol Genet. 2012; 21(3):473-84.
29. D. Hunt, R.J. Leventer, C. Simons, et al. J Med Genet. 2014; 51(12):806-13.
30. K. Shimojima, B. Isidor, C. Le Caignec, et al. American journal of medical genetics Part A. 2011; 155A (4):732-6.
31. C. Mareni, B.R. Migeon. American journal of human genetics. 1981; 33(5):752-61.
32. P.J. Hagerman, R.J. Hagerman. Mental retardation and developmental disabilities research reviews. 2004; 10(1):25-30.
33. R.J. Hagerman, B.R. Leavitt, F. Farzin, et al. American journal of human genetics. 2004; 74(5):1051-6.
34. R.J. Hagerman, M.Y. Ono, P.J. Hagerman. Current opinion in psychiatry. 2005; 18(5):490-6.
35. C.M. Greco, R.J. Hagerman, F. Tassone, et al. Brain: a journal of neurology. 2002; 125(Pt 8):1760-71.
36. H. Tan, A. Qurashi, M. Poidevin, et al. Human molecular genetics. 2012; 21(1):57-65.
37. A. Qurashi, W. Li, J.Y. Zhou, et al. PLoS genetics. 2011; 7(6):e1002102.
38. S.T. Warren, and S. L. Sherman, (2001). The fragile X syndrome. In The Metabolic & Molecular Bases of Inherited Disease, C.R. Scriver, A.L. Beaudet, D. Valle, B. Childs, K.W. Kinzler, and B. Vogelstein, eds. (New York: McGraw-Hill Companies), pp. 1257–1290.
39. S.Sherman, (2002). Epidemiology. In Fragile X Syndrome: Diagnosis, Treatment and Research, R.J. Hagerman, ed. (Baltimore, MD: The Johns Hopkins University Press), pp. 136–168.
40. J. Chai, Y. Li, H. Wang, and J. Cui, Archives of medicine, 2015, Vol. 7 No. 4:5.

 DOI: 10.1051/bioconf/20170801057

Precision medicine and traditional chinese medicine of dialogue

Xin Lou[1], Ming Bai[1] and Mingsan Miao[1,a]

[1]*University of Traditional Chinese Medicine, yaolixueke, 450046 zhengzhou, China*

Abstract. The precision medicine is more precise individualized medicine, based on the patient's genes or physiological to formulate the specific treatment plan, for the realization of individualized treatment of various diseases to provide valuable information.But with the progress of modern science and technology, modern medicine dependence on medical instruments are too serious, traditional ways are gradually forgotten.If the machine depends on the instrument test results too serious which don't combined with the actual diagnosis, the cause of misdiagnosis, so we should pay attention to the overall analysis of diseases and systematic diagnosis and examination, use of the overall treatment concept traced back to find the cause of Traditional Chinese Medicine, finally decide to select a best treatment plan.We should use the dialectical attitude to look at the precise medical.Not blindly requirements according to the road of precision medicine of Traditional Chinese Medicine to go, to shine in himself field, form of self characteristic of Traditional Chinese Medicine.Can learn some of the advantages of accurate concept, the good and rejecting the bad, hope the Traditional Chinese Medicine in the modern environment more walk more far.

The concept of precision medical is drugs in the future will be customized for each individual or a small group of people, At present using mass production of a class of drugs, and to all the people of the same kind of disease take the same medicine situation will be gradually eliminated.It is more precise individualized medicine, based on a genetic or biological treatment to customize.As follow is the Specific steps:First, in patients with a large number of individuals and genome sequencing to build a huge medical data information database, and then through the analysis of the genetic information of the comparison of different individuals, to further understand the common cause and individual cause of various diseases, in the last developed for particular patients with specific diseases pathogenic gene targeted drugs and therapies, of course, also for personalized preventive care for healthy people.Each patient has a "card", Doctor "swipe" identification information, symptomatic to people;To see a doctor and treating according to each person's genome, take medication as the main body of the existing medical reform, thus will influence and change the future of medical care, drug development and use.Will also find more and more treatments for cancer and for the realization of individualized treatment of various diseases to provide valuable information.

With the concept of precision medical widely promoted and in clinical application,the effect of the play is also more and more obvious, especially in the cancer therapy has made big contribution, but has some concrete problems to be solved on the clinical application, precision medical has some far-fetched in the face of the whole concept and theory of Traditional Chinese Medicinal and the theory of property of chinese herbs, medicinal herbs of warm, hot, cold, cool, cannot be explained by the

[a] Corresponding author: miaomingsan @163.com.

concept of precision medicine, in the face of these problems we need to correct view precision medicine, reasonable applied to clinical therapy.

1 The application of precision medical

Precision medical plays a significant role in the treatment of diseases, such as heart cerebrovascular disease, cancer, neurodegenerative diseasesand so on.

1.1 Precise application in cardiovascular medicine

In recent years, on the basis of the gene sequencing of precision medical get rapid development, heart vascular disease, clinical go up to with hypertrophic cardiomyopathy, sudden cardiac death, such as genetic hypertension monogenic disease genetic cardiovascular disease has through genetic testing for early diagnosis, early warning and targeted therapy, pathogenic gene mutation carriers and their families for genetic screening, genetic block, predict curative effect, individualized treatment, cardiovascular field will become the new direction of precision medical.

1.2 Precision medical applications in cancer

Presently, precision medical in the treatment of cancer get good progress and results.For example, there are more and more patients with breast cancer, lung cancer, colorectal cancer, and melanoma are applied precise treatment to cure the disease and tested for genome in treatment, doctors can according to each person's genome differences make the best treatment plan.Breast cancer after decades of research, for example, has been divided into 10 categories, according to a new system of classification and diagnosis, clinicians can according to different category, take accurate treatment plan to make possible a variety of breast cancer to cure.

1.3 Precision medical applications in neurodegenerative diseases

Neurodegenerative disease is a kind of complex diseases, including Alzheimer's disease, Parkinson's disease, Amyotrophic Amyotrophic lateral sclerosis, such as its pathogenesis, early diagnosis difficult, complex causes the lack of treatment, the serious influence the patient's quality of life and work.Currently, neurodegenerative diseases "precision medical" just getting started, there are a lot of work to be carried out.For example, the use of genomics, proteomics, metabolomics technology such as finding more disease related gene, molecular Marker, and by adopting the technology of pharmacogenomics, the development of individualized targeted drugs, "tailored" to use drugs, to achieve the ultimate neurodegenerative diseases, to promote the "precision medical treatment", better service for patients with medical treatment.In early diagnosis, need to be patient of genomics, proteomics, metabolomics etc information analysis, to fully assess risk of patients;In the treatment of disease, patients with combined drug genome information needed, personalized treatment options.

1.4 Precision medical problems in the treatment of cancer

In tumor treatment, for example, has now been confirmed in cancerous cells exist in the process of a large number of genetic mutations, suggesting that most of the mutations, there is no significant effect on tumor occurrence and evolution, but how to determine and accurately interpret these key mutations will not be easy;And the tumor cells in the process of treatment is not completely passive, tumor cells can evolve through natural selection, or evolution, single drug treatment all cannot kill tumor cells is also one of the reasons. This is because the tumor is the result of the interaction between genes and environment, environmental factors caused by tumor cell gene structure and function of change, and ultimately make tumor cells to better adapt to the environment. Only from the tumor on the one hand, accurate treatment unsolved problem existed, is not to remove a lesion can be cured.Therefore, the precise treatment in clinical medical treatment still has a long way to go.

1.5 for the treatment of using precise idea of thinking

With the implementation of the concept of "precision medicine", it should be optimistic about the concept, but also should be questioned.Questioning the idea comes from the understanding of the disease itself, That is, whether the problem can be accurate diagnosis and treatment.Until today in development of medicine, It can cure diseases are numbered, Most chronic diseases are usually not just one of the body's problems, but one or more of the system's problems,and may be the whole body of one or more of the system is out of the question.What is more important to human health are the result of complex multi-factor comprehensive effect, at the same time, because each person's physical quality is different, each kind of disease according to the specific circumstances of individual patients have different manifestations. So you should see, most of the complexity of the disease is caused by a single factor. Therefore, because the complexity of the disease itself, which makes the precise concept more stay at the ideal level.

2 Precision medicine in the application of traditional chinese medicine

Figures and tables, as originals of good quality and well contrasted, are to be in their final form, ready for reproduction, pasted in the appropriate place in the text. Try to ensure that the size of the text in your figures is approximately the same size as the main text (10 point). Try to ensure that lines are no thinner than 0.25 point.

2.1 Characteristics of traditional chinese medicine thinking - the whole

The whole concept has a pivotal position in the Traditional Chinese Medicine, to observe and explore the relationship between the human body and the human body and the external environment and the clinical diagnosis and treatment of diseases has an important guiding significance, believes that the human body is an organic whole, constitute the various parts of the human body in the physiological, pathological are inseparable, mutual influence; At the same time we also recognizes the unity of body and the environment. The unity of internal and external environment, the body's own integrity of the idea, referred to as the overall concept. In some respects, Human beings live in nature, there exist necessary conditions for the survival of human beings. At the same time, the changes of nature will directly or indirectly affect the physiological or pathological state of the human body, and the body will react accordingly. Under the influence of the overall concept, Traditional Chinese Medicine only consider the overall accuracy is the true precision!

2.2 Characteristics of traditional chinese medicine treatment-syndrome differentiation

Analysis, identification of disease syndrome is a syndrome differentiation. Identify the cause of the disease, the nature of the site, and then determine where to judge; The disease syndromes performance varied, pathological changes is very complex, lesions with different time, space and different individuals have different influence on the changes of disease. By syndrome differentiation, distinguish between the phenomenon and the essence of the disease, you can seek the treatment. According to the different time of onset, different individual characteristics, location and time to be patient, because the treatment of the person. According to different causes, illness, disease, pathogenesis to determine the specific treatment methods, and then choose an effective drug, which can also be said to be a precise expression.

 DOI: 10.1051/bioconf/20170801057

2.3 Characteristics of Traditional Chinese Medicine - compatibility

Etiology is not unique, the only precise is the narrow sense of precision, For the cause of the compound is accurate. Zhongjing formula, such as less of its ingredients, but fujin made excellent compatibility, if we change the amount of a medicine in the formula, it is possible to result in great changes of prescription effect.

In the case of Guizhi decoction, the party by 90 g cassia twig, peony, licorice, ginger 90 g, 90 g of 12 pieces, 60 g, big jujube efficacy for muscle functions, to reconcile camp guard, can also be used as a basic prescription for the treatment of other diseases and other diseases. such as the following: (1) Add Houpu 60g, almond fifty, jieji functions of dispelling wind, gasp for;(2) Increase the amount of the single twig, cassia twig and 150g, namely Gui Gui Tang, reuse of cassia twig, intended to Wen Tong Xinyang, in order to make kidney water, flush down the inverse;(3)The prescription of cassia twig to peony and Aconite Decoction on the sun card table mistakenly and evil injury Yang invaginations of the pathogenesis and the addition and subtraction, Yang if the injury is not heavy, can be removed if the injury is heavy Yang peony, resulting in "cold" disease, need to add temperature by Fuyang aconite;(4) Based on the reuse of Guizhi Decoction plus peony Xiaojianzhong decoction is composed of maltose, sweet and sour sweet combination prescription can regulate Yin and Yang.

The application of Guizhi Decoction to maltose, licorice, jujube sweet to health priorities, with Guizhi, Xin Yang Laitong regulating Wei ginger, with peony acidity and camp is pain, replenishing spleen progenitor, provides an important basis for the clinical medical treatment of offspring spleen and stomach diseases. The above party is based on square chmical process, total herbs are not many, but reflects the subtleties of the classic prescription, fully demonstrated the characteristics of a person's diagnosis and treatment of one party.

2.4 Traditional Chinese Medicine application features - páo zhì

The drug is according to the condition to use, improve the efficacy, reduce toxicity must be considered is the focus of medication,so drug processing in this action can not be ignored.At the same time, taking drug treatment regimens for efficacy of increase or decrease also have an important influence.

If the drug dosage is two to six times the clothing and a variety of methods. The specific choice of drug medication, and according to the specific characteristics of pathogenesis and disease. The meal served drug concentration, rapid effect; to decline Yang, Yin Han Sheng syndrome; for old pain, is a virtual reality of the card to see the three directions; four doses were found only with serious and complicated illness; five times, six times those seen in Angelica four inverse plus Evodia Decoction and ginger pig skin soup; take a small amount of service frequency, mainly to the drug continued constantly. From the above we can see that the application of traditional Chinese medicine is also a certain precision at all.

2.5 The similarities between Chinese medicine and Precision medicine

Zhang zhongjing of the han dynasty, study of huangdi neijing, the concept of appropriate due to illness better applied to specific diseases,dissolve into the specific prescription. In TCM clinical syndrome differentiation process, also need to grasp the pathogenesis in order to establish the core of treatment. Zhongjingfang spread very long. Secondly, since ancient times are in use, it compared to the updating speed of the western medicine fast, inspire us to think about Traditional Chinese Medicine formula "precision medicine". The ancient one, due to illness condition from a certain perspective, which belongs to the category of medicine treatment of individual.

According to the characteristics of TCM syndrome differentiation treatment, The sick people are different, and the reaction to the disease must be different. Disease have different time, place and it will be different; Disease no matter occurrence, development, recovery and so on are not immutable and frozen, such as the SARS virus, its infectious and biological characteristics will change with the change of the condition such as time, space, thereby increasing the treatment difficulties, we according to the characteristics of the disease drug use, so that is the precisely reflected.

See the application from the overall concept of Western medicine, and can not be completely out of the whole concept to grasp, to comprehensive clinical examination was performed only after an antidote against the disease, at this point, medicine also need to use the whole idea as a foundation, comprehensive inspection of an antidote against the disease in order to better treatment of diseases.

2.6 the difference between TCM and Precise medical

Traditional Chinese Medicine treatment is more dependent on the method of inspection, auscultation and olfaction, inquiry, and pulse-taking and palpation, to an overall understanding of the symptoms of the patients with causes, and symptomatic treatment.Such as pulse condition, The formation of pulse sequence, and the heart, blood and body fluid are closely related.Its different change reflects the heart strong or weak, relaxation, qi-blood virtual lag changes in the three aspects.Only will be the heart of the fixed, context and the flow of qi-blood, body fluid together analysis, to reveal the essence of the change of the pulse, What letter what Maicai reasoned, is not like water without source.Just from this aspect of pulse, different pulse condition implicating the body all aspects of function, it is only through the overall analysis to suit the remedy to the case to achieve the purpose of treating diseases.In this aspect Traditional Chinese Medicine treatment notice or as a whole, precise prefer to individualized treatment.

2.7 The problems encountered in the application of Traditional Chinese Medicine

Precise medical treatment concept in a person of Traditional Chinese Medicine, syndrome differentiation and treatment aspects are reflected,also play an important role, but in the face of the overall content of property theory in the treatment of Traditional Chinese Medicine and Traditional Chinese Medicine, you can't explain it, so we should dialectical view of medicine application in Traditional Chinese Medicine, can not produce a plausible theory of Traditional Chinese Tedicine treatment; for example, the Traditional Chinese Medicine about four, refers to the cold temperature Liangsi different resistance, also known as four.It reflects the vicissitudes of drugs on the body of yin and Yang, cold and heat changes the tendency. It's not just to show that medicine is to make people feel cold or warm, and modern instruments can't solve the mystery, it not because of which a specific substance to determine the temperature of drug and cold, so that the precise idea is not all powerful.

Treatment of disease of TCM should pay attention to avoid the "no overall plan for a fundamental transformation".Such as the common cold,common have a bad cold, fever, head and pain and a series of symptoms, However, due to the different pathogenic factors and physical, it is often characterized by cold, the wind hot cold, heat, and internal heat cold and other different types of evidence, only through the through overall analysis of the patient, to distinguish clearly what kind the cold, can determine the corresponding treatment methods, to give appropriate treatment, rather than just "hot" fever of local treatment.So the treatment should start from the overall concept, with the disease treatment, flexible.

3 Discussion

Precision treatment has many advantages in clinical application, such as accurate positioning of the lesions, reduce the patient's pain and solve some of the complicated diseases that are difficult to be solved now. at the same time there are a lot of problems need us to solve, to better provide methods for clinical treatment.

With the progress of modern science and technology, precision medicine in the application process of the instrument is too serious, the traditional diagnosis method was gradually forgotten, although the inspection instrument is accurate and objective, the use of modern science and technology in the diagnosis and treatment has certain advantages, but clinical experience is more important, if the machine depends on the instrument test results, and can not combine the actual condition, it can lead to misdiagnosis, it is not uncommon in clinical.This tells us that if we rely solely on the instrument data and the results of the examination, the temporary removal of a lesion site, the temporary relief of the patient suffering, but the follow-up related diseases caused by this can not be without thinking.So in the face of the accurate treatment of such a deficiency, we should reflect on whether you would have a long and Traditional Chinese Medicine diagnosis and treatment technology without hesitation abandon and into the precise treatment arms.

The biggest characteristic of Traditional Chinese Medicine treatment is to grasp the patient's overall treatment, find the source for treatment.It is the wisdom of the spread of thousands of crystals, but also apply to modern disease treatment.Such as the classic "the headache medicine foot" said,this is from the meridian theory. relevant organs lesions occurred.According to the meridian courses route, starting from the overall concept, to find out the source of the disease, which can treat a range of diseases.So we should start from the whole, find the source, combination with the idea of the precise treatment to achieve twice the result with half the effort to disease treatment, better services for patients.But the doctor of Traditional Chinese Medicine treatment there is a problem, such as the treatment cycle is long, the interaction mechanism between the drug problems such as unclear, it is also need we dialectical view.

From the above discussion we can see that in the concept of treatment of TCM due to illness condition, dialectical treatment and Chinese medicine taking method in some degree belongs to the medicine category, we can see that they are common in medical research in the future, the two complementary treatment may become a trend, but can not be a requirement for Tratidional Chinese medicine according toWestern medicine routine to go. each concept all have their own characteristics, and has their own areas of expertise, in their own field to shine, to form the self characteristic of Chinese medicine,, in order to go out of the country, so that more people recognized, and it is not a simple precise medical treatment concept can be carried forward in traditional Chinese medicine, advanced the concept of medicine does have merit, you can learn from Traditional Chinese Medicine, its essence, to its dregs, not blindly copy,we should use dialectical attitude to treat the precise medicine. Hope that the Traditional Chinese Medicine in the modern environment, the more far away from the ancients retained the wisdom of a better use of our lives.

The articles will be quoted as follows:

project team of outstanding scientific and technological innovation in Henan province (TCJ2014-391), the natural science foundation of Henan province (132300410019), science and technology innovation team of Zhengzhou city (131PCXTD612)

References

1. J. Zheng, W. Li, The reconsideration of precision medical. Journal of medical informatics. J. **8**, (2016)
2. X.L. Ru, X.H. Chen, Zhang zhongjing for quantitative identification and accurate treatment of disease and analysis. World - the modernization of traditional Chinese medicine science and technology. J. **3**,10(2015)
3. F. Liu, Y.S. Yang, The characteristics of TCM syndrome differentiation, value and significance of the author . Journal of traditional Chinese medicin.. J. **12** (2009)
4. Y.L. Wu, Accurate cancer medicine: the way into the future.Journal of evidence-based medicine.J. **1**(2015)
5. J. Wang, China's basic medical journal of traditional Chinese medicine, J. (2012)
6. X.H. Yang, Y.N. Wen, Treatment based on syndrome differentiation of traditional Chinese medicine research progress review. Journal of traditional Chinese medicine information. J. **12**(2012)

DOI: 10.1051/bioconf/20170801057

7. Y.L. Jiao, J.C. Wang, Q. Zhang, G.X. He, China facing the opportunities and challenges in the field of precision medical. ournal of public health management in China. J. (2015)
8. F. Xiao, From the evidence-based medicine to the precise medical thinking. Chinese kidney disease research on electronic magazine. J. **1** (2014)
9. Q.M. Zhan, China the strategic needs and key tasks for the development of precision medical. Chinese nerve trauma surgery electronic journal. J. **3** (2015)
10. M.Y. He, J.L. Xia, X.D. Wang, Accurate medical research progress. Journal of clinical medicine in the world. J. (2015)
11. A.W. Wu, J.F. Ji, Implementation of precision medical opportunities and challenges. Chinese journal of clinical laboratory management electronic. J. **1** (2015)
12. X.Q. Guo, W.R. Huang, Z.M. Cai, Francis choi. Cancer precision medicine science. J. **5** (2015)
13. F. Jiao, S.Y. Xie, Y.J. Li, Z.L. Li, Y.F. Yan, Accurate medicine and uncertainty. Journal of medicine and philosophy. J. **1** (2015)
14. Z.F. Jiang, Breast cancer treatment decisions: from individualized treatment to precision medical. Chinese journal of practical surgery. J. (2015)
15. H.X. Zhang, Q.Y. He, Based on the accurate medical targeted anti-tumor drugs sensitivity prediction and its application research and development. Chinese journal of new drugs. J. (2015)
16. M.C. Gong, Accurate medical challenge to medical informatics and task. Journal of medical informatics. J. **1** (2016)
17. X. Zhang, X.G. Mao, W. Zhang, Big data era of precision medical. Chinese journal of neurosurgery disease. J. **11** (2016)

Preparation of 5-fluorouracil loaded chitosan microparticle and its drug release properties

Mingming Li[a], Shiping Zhan[b]

Department of Environment and Chemical Engineering, Dalian University, Dalian, China

Abstract. Chitosan is one kind of good biocompatible polymer and is suitble for drug carriers. Preparation of 5-fluorouracil (5-Fu) loaded chitosan (CS) particles and *in vitro* release experiment were performed using ionic crosslinking method with sodium tripolyphosphate (TPP) as crosslinker. The optimal preparing parameters were verified by 5-Fu release experiments. The drug loading, and release behavior of drug loaded microparticles in vitro were investigated. The optimal preparation conditions were：the temperature 25℃, the ratio of CS to TPP 5:1， the CS concentration 1.5g/L, stirring speed 650rpm. Under these conditions, the drug loading of particles was up to 45%.

1 Introduction

5-Fluorouracil (5-Fu), an analogue of the natural pyrimidine uracil, is an anticancer drug that has been widely used in the management of solid malignancies arising from breast, gastrointestinal tract, head and neck, and colorectal cancer in particular [1-4]. 5-Fu is having lack of site specificity and poor therapeutic effect. Also toxic effects to healthy cells and unavailability of major proportion of drug at the colon region remain as limitations [5]. To reduce its adverse effects, several efforts have been made to improve the delivery and the bioavailability of 5-Fu. The controlled drug delivery system of the drug-loaded polymer provides an alternative nonsurgical treatment for cancer. Moreover, the advantage of the drug-loaded micropaticles as drug delivery system is that this system can be maintained effective dosage longer than that with simple solution of the drug in lesion location [6,7]. Chitosan (CS), a kind of natural polysaccharide, is widely used as drug carrier, due to its good biocompatibility and biodegradability. Drug loaded with chitosan microspheres can get to the Tumor targeted area and control the slow-release of the drug. Thereby, it can reduce the toxic and side effects of drugs on normal tissues.

This work was aimed to prepare 5-Fu loaded chitosan microparticles for sustained delivery application with ionic crosslinking method. The effects of preparing parameters on the drug loading of particles and release characteristic were studied.

2 Materials and methods

2.1 Materials

[a] Corresponding author: limingming@dlu.edu.cn
[b] Corresponding author: zhanshiping@dlu.edu.cn

DOI: 10.1051/bioconf/20170801058

5-Fu (purity >98%), used as active pharmaceutical ingredient, was purchased from Zhuorui Chem. (Shanghai, China). CS (purified), used as polymer was obtained from Haiyun Chem. (Dalian, China). Sodium tripolyphosphate(TPP, purified), used as crosslinker was obtained from Tianjin Kemiou Chemical Reagent Co.(Tianjin, China). Phosphate buffer saline (PBS) tablets (pH 6.8), which were used to prepare the drug release medium, were obtained from Tianjin Damao Chemical reagent Factory (Tianjin, China). Dialysis bag was perchused from Beijing JingKeHongDa (Beijing, China). All of the reagents were used without any further purification.

2.2 Preparation of 5-Fu loaded chitosan microparticles

The method used for the preparation of chitosan microparticles containing 5-FU is ionic crosslinking technique. Different CS concentrations (0.6g/L, 0.8g/L, 1.0g/L, 1.5g/L, 2.0g/L, 2.5g/L) was selected in order to optimize the best one and also to observe the effect of polymer on the formulation. Acetic acid was used as organic solvent with the concentration of 0.1% (w/v). Drug was dissolved in acetic acid with varying polymer concentrations. Then it was followed by addition of TPP with varying ratio of CS to TPP (3:1, 4:1, 5:1, 6:1, 8:1), and was stirred continuously for 1.5h with different stirring rates(650rpm, 750rpm, 850rpm, 950rpm, 1050rpm). Then the suspension was subjected to centrifugation. Supernatant was removed and washed repeatedly three times. Microparticles were obtained after drying in a vacuum oven.

2.3 Drug loading

Microparticles after drying were added to their specific solvent to facilitate the coat of the microparticles to get dissolved. The resultant suspension was subjected to evaporation for further removal of the solvent prior to filtration. Then the residue was washed and diluted appropriately with phosphate buffer of pH6.8 to determine drug content and entrapment efficiency. Samples were measured at an absorbance of 265 nm in ultraviolet-spectrophotometer (UV; 756PC, Shanghai Spectrum Instument Co., Ltd. Shanghai, China). Drug content loading of 5-Fu in CS microparticles was determined by the following equation [8]:

$$\textit{Drug loading content}\ (\%\ w/w) = \frac{\textit{Weight of drug in microparticle}}{\textit{Weight of microparticle recovered}} \times 100 \tag{1}$$

2.4 In Vitro experiment of drug release

In drug release test, approximately 20mg sample was weighed and filled into a dialysis bag which was placed into a conical flask containing 30 mL of PBS (pH 6.8). Then, the conical flask was placed in a shaking incubator (SHZ-82; FeiPu Experimental Instrument Factory, Changzhou, China) at100rpm and 37°C to simulate physiological conditions. At sampling points, 3mL of the PBS/drug solution was removed from the flask and a fresh PBS solution with an identical volume was added to the maintain sink conditions. The 5-Fu in the sample was measured at 265nm with an ultraviolet-spectrophotometer. The release amount of 5-Fu was calculated with a predetermined standard curve (C = 18.6916A-0.6093, $R^2 = 0.9996$).

3 Results and discussions

3.1 Drug loading studies

Figure 1 shows the effect of CS concentration on the drug loading of 5-Fu loaded CS microparticles. The drug loading increases with the increasing CS concentration, and achieves the highest amount of

DOI: 10.1051/bioconf/20170801058

nearly 50% at 1.5g/L CS concentration. Then the drug loading decreases with the rising CS concentration. As long as the CS concentration increase, the reaction time gets longer, which leads to the more adsorption of 5-Fu onto the microparticles.

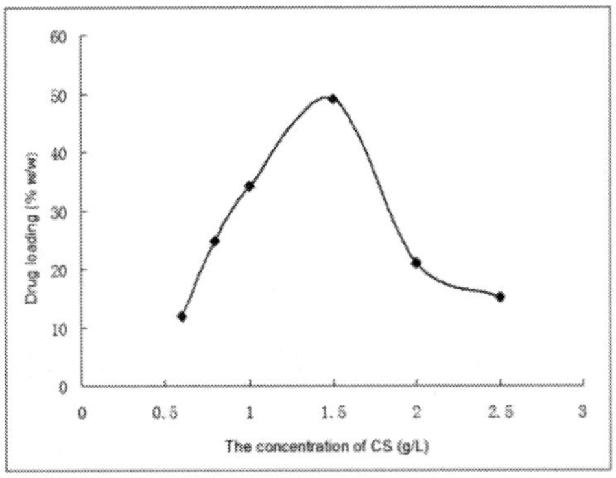

Figure 1. The effect of CS concentration on the drug loading of 5-Fu loaded CS microparticles

Figure 2 shows the effect of CS to TPP ratio on the drug loading of 5-Fu loaded CS microparticles. As shown in Figure 2, the amount of drug loading changes more evidently with the increase of ratio of CS to TPP. When the ratio is 5:1, the drug loading is about 34(%, w/w)

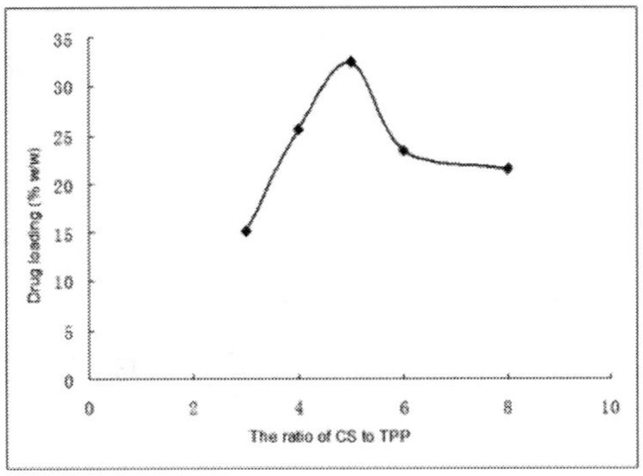

Figure 2. The effect of CS to TPP ratio on the drug loading of 5-Fu loaded CS microparticles

Figure 3 shows the effect of stirring rate on the drug loading of 5-Fu loaded CS microparticles. As shown in Figure 3, the drug loading presents a obviously decreases with the increase of stirring rate. When the rate is 650rpm, the drug loading is about 25(% w/w). When the stirring rate is too fast, the vigorous stirring leads to the agglomeration of the dispersed microparticles, hence to the decreasing of 5-Fu drug loading.

3.2 Drug release studies

Figure 4 shows the 5-Fu release profiles from the CS microparticles with different CS concentrations. The release is completed in 12h under PBS solution (pH6.8) condition, although the test continued for 24h in order to ensure that no more drug released at the end of the test. Total amount of impregnation mass is assumed to correspond to the accumulated drug released at the end of the test. The release studies show that, in the case with 1.0g/L CS concentration, the cumulative release rate is highest.

 DOI: 10.1051/bioconf/20170801058

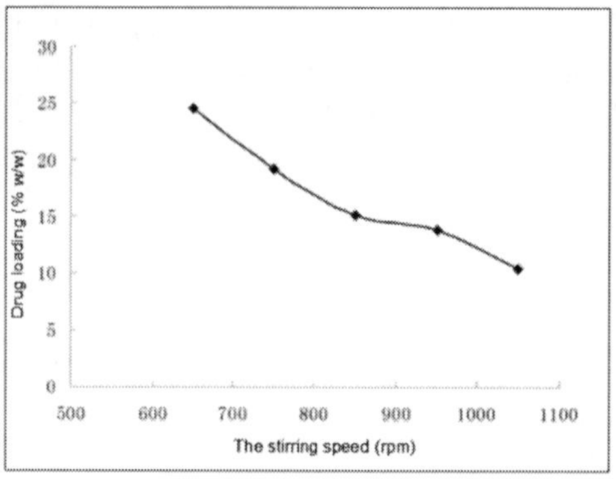

Figure 3. The effect of stirring rate on the drug loading of 5-Fu loaded CS microparticles

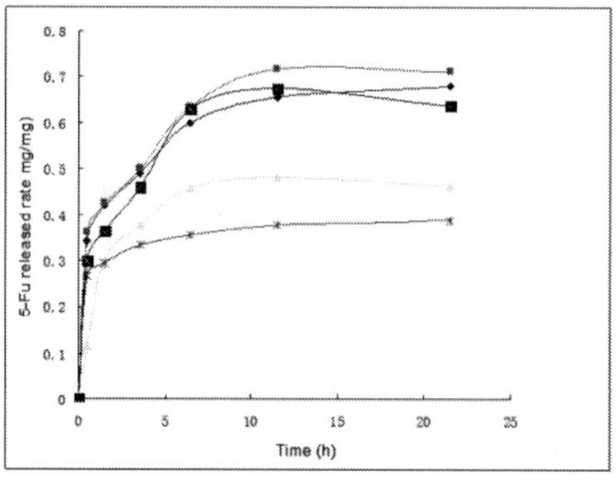

Figure 4. 5-Fu released rate from CS microparticles, with different CS concentrations(g/L): (◆) 0.8; (▪) 1.0; (▲) 1.5; (×)2; (*) 2.5

Figures 5 show the effects of ration of CS to TPP on the impregnation efficiencies. As shown in the curves, the ratio of CS to TPP plays an importent role in the 5-Fu release rate. In all the range of ratios, the drug release shows the same variation trend. In the first 6h, the release rate increase dramatically while the curve increase slowly in the next 6h. Then, the release rate remains smoothly during the last 12h. When the ratio of CS to TPP is 8 :1, the 5-Fu release rate is highest with the amount of about 0.7mg/mg.

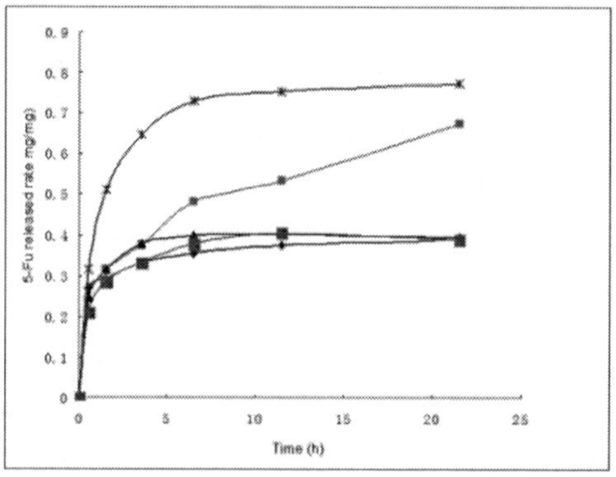

Figure 5. 5-Fu released rate from CS microparticles, with different ratio of CS to TPP: (◆) 3 :1; (▪) 4 :1; (▲) 5 :1; (×)6 :1; (*) 8 :1

DOI: 10.1051/bioconf/20170801058

Figures 6 show the effects of the stirring rate on the 5-Fu release rate. As shown in the Figures 6, when the stirring rate is 950rpm, the drug release rate can achieve the highest level, which is about 0.9mg/mg in the first 12h.

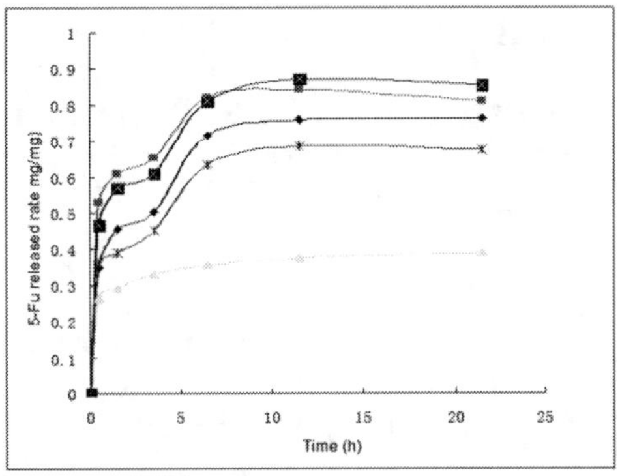

Figure 6. 5-Fu released rate from CS microparticles, with different stirring rate(rpm): (◆)650; (■)750; (▲)850; (×)950; (*)1050

4 Conclusions

In this work, the CS microparticles are impregnated with 5-Fu by the ionic crosslinking method using TPP as crosslinker in order to develop hydrophilic drug release systems. The drug loading and release behavior of the 5-Fu loaded CS microparticles are investigated. The rusults show that, this method is simple and reproducible while drug release performance *in vitro* show that the microparticles have obviously sustained release effect.

Acknowledgments

This research is supported by the Doctoral Fund of Dalian University.

References

1. F. Casale, R. Canaparo, L. Serpe, E. Muntoni, et al, Pharm. Res. **50**, 173−179(2004)
2. M. Simeonova, R. Velichkova, G. Ivanova, V. Enchev, I. Abrahams, Int. J. Pharm. **263**, 133-140(2003)
3. M.V. Natu, M. Gil, H.C. Sousa, J. Supercrit. Fluids. **47**,93-102(2008)
4. G.C. Pisanelli, C.L. Wilt, H.M. Pinedo, F. Franchi, P. Noordhuis, B.J.M. Braakhuis, J.A.M. Laar, G.J. Peters, Eur. J. Cancer, **31**, 1517-1525(1995)
5. T. Shashank, M.N. Satish Kumar, P. Ashwari, Saudi. Pharm. J. **23**,308-314(2015)
6. S.P. Zhan, C. Chen, Q.C. Zhao, W.J. Wang, Z.J. Liu, Ind. Eng. Chem. Res. **52**, 2852-2857(2013)
7. Y.L. Wang, Y.P. Wang, J. Yang, R. Preffer. R. Dave, B. Michniak, Power Technol. **164**, 94-102(2006)
8. B. Avinash, J.S. Steven, L, Karen, Int. J. Pharm. **51**, 87–92(2007)

Study on the expression of MMP-9 and NF-κB proteins in epithelial ovarian cancer tissue and their clinical value

Shen Wei[1,a], Chen Juan[2], Li Xiurong[1] and Yin Jie[1]

[1]*Department of Obstetrics and Gynecology, the Forth Hospital of Hebei Medical University, Shijiazhuang, 050011, Hebei, China*
[2]*Health Examination Center, the Fourth Hospital of Hebei Medical University, Shijiazhuang, 050011, Hebei, China*

Abstract. Objective: To investigate the expression of matrix metalloproteinase-9 (MMP-9) and nuclear factor κappa-B (NF-κB) in different ovarian tissue and explore the relationship between their expression and clinicopathological features of epithelial ovarian cancer. Methods: The expression of NF-κB and MMP-9 in 15 cases of normal ovarian tissue and 80 cases of epithelial ovarian cancer were detected by mmunohistochemistry PV method. Results: Expression of MMP-9 and NF-κB in epithelial ovarian cancer was significantly higher than those in normal ovarian, and the difference was statistically significant (P<0.05). The expression of MMP-9 protein was not related to histology classification and differentiation degree, but it was significantly related with lymphatic metastasis and clinical stage (P<0.05). The expression of NF-κB protein was significantly associated with histology classification, lymphatic metastasis, clinical stage, and differentiation degree (P<0.05). In epithelial ovarian cancer tissue, the expression of MMP-9 and NF-κB proteins showed a significant positive correlation (P<0.01). Conclusion: The expression MMP-9 and NF-κB proteins are closely related to pathological features of epithelial ovarian cancer and there are important significance to the development, invasion and metastasis of epithelial ovarian cancer. So they might become useful prognostic indicator for diagnosis and prognosis of epithelial ovarian cancer.

1 Introduction

Ovarian cancer was one of the common malignant tumors for women. Its incidence rate ranked third next to the cervical cancer and uterine cancer, but its mortality rate was of being first on the above three kinds of malignant tumors of female genital tract. Its occurrence and development was a complicated process involved in many ways and many genes [1]. Matrix metalloproteinase-9 (MMP-9), a key enzyme involved in the degradation of extracellular matrix, was closely related to tumor invasion and metastasis [2]. Nuclear factor kappa B (NF-κB) was a eukaryotic transcription factor, which existed in almost all cells. NF-κB was usually the form of homologous or heterologous dimer, and its most common form was P50/P65 dimer, which played a passing information role in the process of cell differentiation, apoptosis and tumor growth [3]. In this study, the expression of MMP-9 and NF-κB proteins in normal ovarian tissue and epithelial ovarian carcinoma was detected by immunohistochemical method. The relationship between MMP-9 and NF-κB protein expression was

[a] Corresponding author: shenwei1980112@163.com

analyzed, in order to provide valuable theoretical basis for the mechanism of invasion and metastasis of ovarian cancer and prognosis effect.

2 Material and method

2.1 General data

95 cases of epithelial ovarian cancer specimens, which were from ovarian cancer tissue of gynecologic surgical resection, were collected from the fourth hospital of Hebei Medical University during January 2010-January 2012 year. A total of 80 patients with epithelial ovarian cancer were selected, whose age were from 43 to 75 years, with a median age of 55.8 years, including 41 cases of serous cystadenocarcinoma, 28 cases of mucinous cystadenocarcinoma, and 11 cases of endometrioid adenocarcinoma. Histological staging was performed according to NCCN2012. In 80 cases of specimens, there were 38 cases in stage I and II, 42 cases in stage III and IV, 39 cases of well-differentiated, 26 cases of moderately differentiated, and 15 cases of poorly differentiated, 39 cases with lymphatic metastasis, and 41 cases without lymphatic metastasis. 15 cases of normal ovarian tissue were collected, which were obtained by double adnexectomy for uterine myoma. The median age was from 36 to 58 years, and the median age was 42.3 years. All the specimens were confirmed by pathological examination, and the data were complete. All patients had not received chemotherapy or radiotherapy before operation.

2.2 Main reagents

Rabbit anti human NF- kappa b/P65 polyclonal antibody and Rabbit anti human MMP-9 polyclonal antibody were purchased from Santa Cruz Company. PV-6001 and PV-6002 detection reagents and DAB kit were purchased from Beijing Zhongshan-jinqiao Reagent Company. PBS buffer solution was purchased from GIBCO Company.

2.3 Immunohistochemical staining method

Immunohistochemistry PV method was used in experiment. Surgical specimens were fixed in 10 % formalin, embedded in paraffin. Paraffin specimens were separated into 5 mm sections. After conventional dewaxing, they were immersed in 3 % hydrogen peroxide methanol solution at room temperature for 10-15 min to eliminate endogenous peroxidase activity. After the antigen restore used autoclave for 2 min, they were sealed with 1 % BSA (bovine serum albumin), incubated at room temperature for 10-15 min (closed tissue charge), and poured to serum. Adding primary antibody NF-ϰb/P65 and MMP-9 (titer 1: 50) , they were placed in a refrigerator at 4 °Cfor the night, washed with 0.01 M PBS, and incubated for 5 minute for 3 times, and then adding the secondary antibody PV-6001 and PV-6002 reagent, they were incubated at 37 °C for 60 min. They were vibrated wash in 0.01M PBS for 5 min for 3 times, and stained (PBS 4 ml, DAB 5 g, H_2O 25 ul) around 5 min (with a microscopically moderate as well). They were washed 5 times by the distilled water, redyed with hematoxylin for 1 min, dehydrated, hyalinized with xylene, sealed with neutral gum, and observed under the light microscope.

2.4 Result determination

MMP-9 and NF-ϰB were detected positive, if brown-yellow granules were discovered in cytoplasm. Comprehensive evaluation methods: The score of the degree of staining was as follows: Colourless for 0, light yellow for 1, claybank for 2, brown for 3. The counting of the ratio of positive cell, which was positive cells accounted for the percentage of tumour cells, was as follows: less or equal to 5 % for 0, 6-25 % for 1, 26-50 % for 2, >50 % for 3. Assessment: Multiplication of the above two score value

DOI: 10.1051/bioconf/20170801059

was adopted, and the result was 0 for the negative (-), 1- 2 for weak positive (+), 3- 4 for moderate positive (+ +), and > 4 for strong positive (+ + +).

2.5 Statistical methods

The SPSS13.0 statistical software was used, and measuring data was expressed by ($\bar{x} \pm s$). The correlation among the positive rates was analyzed by χ^2 test. The correlation analysis of protein expression was analyzed by Spearman correlation analysis. $P<0.05$ was statistically significant.

3 Results

3.1 The expression of MMP-9 and NF-κB proteins in different ovarian tissue

The positive expression rate of MMP-9 and NF-κB proteins in epithelial ovarian carcinoma tissue, was significantly higher than that in normal ovarian tissue, and the difference was statistically significant ($P<0.05$, Table 1). The results of MMP-9 and NF-κB proteins in epithelial ovarian carcinoma and normal ovarian tissue were detected by PV method, and the results were shown in Figure 1.

Table 1. Expression of MMP-9 and NF-κB proteins in different ovarian tissue

Group	n	MMP-9 expression		NF-κB expression	
		n, positive rate (%)	P-value	n, positive rate (%)	P-value
Normal ovarian	15	0(0)	<0.01	1(6.7)	<0.01
Epithelial ovarian carcinoma	80	54(67.5)		62(77.5)	

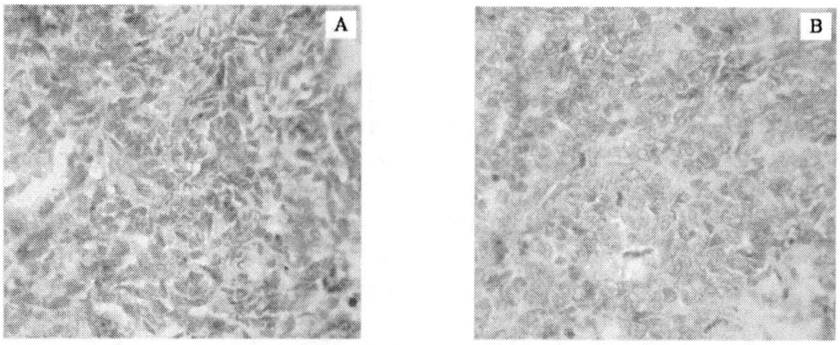

Figure 1. Immunohistochemical results of MMP-9 and NF-κB proteins in epithelial ovarian carcinoma. (A) Positive expression of MMP-9 protein; (B) Positive expression of NF- kB protein.

3.2 Correlation between MMP-9 protein expression and clinicopathological parameters in epithelial ovarian cancer tissue

The positive expression rate of MMP-9 protein among serous cystadenocarcinoma, mucinous cystadenocarcinoma and endometrioid adenocarcinoma had no significant difference ($P>0.05$). The positive expression rate of ovarian cancer tissue in stage I - II was lower than that of ovarian cancer in stage III - IV, and the difference was statistically significant ($P<0.01$). In ovarian cancer tissue of the

DOI: 10.1051/bioconf/20170801059

high differentiation, middle differentiation and low differentiation, the positive expression rate gradually increased with the decrease of the differentiation degree, but there was no significant difference (*P*>0.05). In ovarian cancer tissue, the positive expression rate of ovarian cancer tissue with lymphatic metastasis was higher than that without metastasis (*P*<0.05, Table 2).

3.3 Correlation between NF-κB protein expression and clinicopathological parameters in epithelial ovarian cancer tissue

The positive expression rate of NF-κB protein in serous cystadenocarcinoma, mucinous cystadenocarcinoma and endometrioid adenocarcinoma gradually increased, and there was significant difference (*P*<0.05). The positive expression rate of stage I - II of ovarian cancer tissue was lower than that of stage III - IV, the difference was statistically significant (*P*<0.01). With the reduced of differentiation, the positive expression rate of ovarian cancer tissue significantly increased, and the difference was statistically significant (*P*<0.05). The positive expression rate of ovarian cancer tissue with lymphatic metastasis was higher than that without metastasis, and the difference was statistically significant (*P*<0.01, Table 2).

Table 2. Expression of MMP-9 and NF-κB proteins and their correlation with clinicpathological features of epithelial ovarian cancer

Characteristics	n	MMP-9 expression		NF-κB expression	
		n, positive rate (%)	P-value	n, positive rate (%)	P-value
Histology classification					
Serous cystadenocarcinoma	41	26(63.4)	>0.05	27(65.9)	<0.05
Mucinous cystadenocarcinoma	28	20(71.4)		24(85.7)	
Endometrioid adenocarcinoma	11	8(72.7)		11(100)	
Clinical stage					
I-II	38	18(47.4)	<0.01	22(57.9)	<0.01
III-IV	42	36(85.7)		40(95.2)	
Differentiation degree					
High	39	25(64.1)	>0.05	25(64.1)	<0.05
Middle	26	18(69.2)		23(88.5)	
Low	15	11(73.3)		14(93.3)	
Lymphatic metastasis					
Positive	39	32(82.1)	<0.05	36(92.3)	<0.01
Negative	41	22(53.7)		26(63.4)	

3.4 The relationship between MMP-9 and NF- κB protein expression in epithelial ovarian cancer

The analysis of 80 cases of samples of ovarian cancer tissue by Spearman rank correlation showed that the MMP-9 protein expression was positively correlated with NF-κB protein expression in ovarian cancer tissue (P<0.01).

4 Discussions

The formation and development of tumor were related to the activation of oncogenes or gene inactivation of antioncogene for cell gene mutation, and they were also influenced by the changes of the microenvironment of the tumor cells. Matrix metalloproteinases (MMPs) were a class of proteins with very close relationship with the invasion and metastasis of tumor. MMP-9 was an important member of the MMPs family, also known as type IV collagenase and gelatinase B, and could degrade the main components in the cell membrane matrix, namely type IV and V collagen and gelatin, which resulted in the tumor cell invasion to the surrounding tissue through the basement membrane loss, thereby promoted tumor invasion and metastasis. In recent years, the studies found that MMP-9 played an important role in the evolution of malignant tumor. In thyroid papillary carcinoma, non-small cell lung cancer, primary carcinoma of liver and other tissue, MMP-9 showed high expression, and was closely related with tumor progression. It was a sensitive biological indicator by used as preoperative diagnosis, judging postoperative prognosis and determining with and without lymph node metastasis [4-5]. This paper studies found that MMP-9 protein was high expression in ovarian tumor tissue, and the positive expression rate was closely related to clinical staging and lymph node metastasis. It suggested that the expression of MMP-9 was involved in the invasion and metastasis of ovarian cancer, so it could be used as one of the judgment index of prognosis of ovarian cancer patients.

NF-κB was a nuclear protein found in B cells that binded to the B site of the immunoglobulin kappa light chain enhancer and could regulate the transcription of the immunoglobulin kappa light chain, so it was named nuclear transcription factor κB. NF -κB, as a kind multidirectional and multifunctional nuclear transcription factor, could inhibit apoptosis by inducing the expression of anti-apoptotic genes. It had a close relationship with occurrence and development of tumor and chemotherapy drug resistance, and acted a key role in the anti apoptosis mechanism of tumor. Zhang Jian et al by transfection of IκBα gene demonstrated that NF-κB activation in A549 cells could inhibit their invasion ability, and abnormal expression and activation of NF-κB was also detected in esophageal squamous cell carcinoma, gastric carcinoma, liver cancer and other tissue [6-8]. This paper studies showed that NF-κB in ovarian cancer tissue was significantly higher than that in normal ovarian tissue, and was significantly correlated with histological type, lymph node metastasis, clinical stage, degree of differentiation, further suggesting that the expression of NF-κB protein up-regulation could be an important mechanism of the occurrence and development of ovarian cancer.

Kim Y et al found that MMP-9 gene promoter region contained multiple binding sites of NF-κB transcription factor, and inhibition of NF-κB activity could significantly inhibit the expression of MMP-9 gene [9]. Shishodia S et al found that NF-κB could promote MMP-9 expression in lung cancer, and they in cervical cancer, thyroid papillary cancer, lung cancer were positively correlated [10]. The above results showed that MMP-9 and NF-κB protein expression in epithelial ovarian cancer was positively correlated, and NF-κB activation could regulate the transcription and expression of MMP-9 in ovarian cancer cells, and promote the invasion and migration of ovarian cancer cells.

References

1. A. Jemal, R. Siegel, E. Ward, J. Xu, M.J. Thun, CA-Cancer J. Clin., **58,** 71(2008)
2. L. Liu, J. Wu, Z. Ying, M. Li, Cancer Res., **70**, 3750(2010)
3. J. Liu, G. Yang, Cancer Res., **64**, 1655(2004)

DOI: 10.1051/bioconf/20170801059

4. K. Zarrabi, A. Dufour, J. Li, J. Cao, J. Biol. Chem., **286**, 33167(2011)
5. Y. Koyama, H. Naruo, Y. Yoshitomi, K. Oguri, J. Biochem., **143**, 581(2008)
6. J. Zhang, Y.J. Xu, W.N. Xiong, Z.X. Zhang, C.L. Du, L.F. Qiao, W. Ni, S.X. Chen, Chin. J. Cancer, **27**, 710(2008)
7. B.B. Aggarwal, Cancer Cell, **6**, 203(2004)
8. A.C. Bharti, B.B. Aggarwal, Biochem. Pharmacol., **64**, 883(2002)
9. Y. Kim, H. Kang, S.W. Jang, J. Ko, Cell Physiol. Biochem., **28**, 175(2011)
10. S. Shishodia, P. Potdar, C.G. Gairola, B.B. Aggarwal, Carcinogenesis, **24**, 1269(2003)

An improve RCB method based on microwave induced thermo acoustic tomography

Jinsong Du, Wei Wang, Jie Gao and Yang Gao[a]

Shenyang Institute of Automation, Chinese Academy of Science, Key Laboratory on Radar System Research and Application Technology of Liaoning Province, Shenyang 110179, China

Abstract. Thermoacoustic imaging (TAI) is a non-ionizing and non-invasive imaging method, which combines the merits of high ultrasound imaging resolution with high microwave imaging contrast. In TAI, a short non-ionizing microwave pulse irradiates tissues to induce a small temperature rise, which consequently causes thermoelastic expansion to generate TA signals. By using an image reconstruction algorithm, TAI can then recover the microwave absorption distribution inside the tissue and further distinguish abnormal areas from background normal tissues. TAI for breast cancer detection is the main purpose of this study, the basic theory of TAI was introduced at first, especially the RCB reconstruction algorithm for TAI. After that, in this dissertation, two sets of thermoacoustic imaging systems were developed, which named thermoacoustic tomography and ultra-short pulse based high resolution TAI. According to experiments and theoretical studies carried out in this dissertation, the feasibility of thermoacoustic imaging method by Robust Capon Beam-former (RCB) for breast cancer detection before its clinical investigation is fully validated. Due to high imaging performance needs for the early detection of breast cancer. Based on the advantages of TAI, the potential applications of TAI for finger joints and brain diseases diagnosing are explored, which is opening up a new field for TAI.

1 Introduction

Recently, there has been increasing interest in using the thermoacoustic (TA) effect for biomedical applications [1,2]. TA imaging takes advantage of the excellent resolution of ultrasound imaging while keeping the merit of high contrast and good tissue penetration of microwave[3]. The TA signal strength is proportional to the microwave energy density and microwave-to-acoustic conversion efficiency[2,4]. Further enhancement of the conversion efficiency is critical for non-invasive imaging. Extensive efforts have been made to improve the efficiency of TA conversion by using exogenous contrast agents; a significant improvement is yet to be achieved[5].

Till now, X-ray mammography is the most widely used modality for breast tumour screening in high risk groups[6], but it also results in unnecessary surgery and anxiety, along with the risks of more frequent X-ray exposure causing a small but significant increase in breast cancer induced by ionizing radiation[7]. MRI provides the good tissue structure imaging in detail but still cannot provide electromagnetic parameters or acoustic parameters distribution of breast environment[8], beside its

[a] Corresponding author: gaoyang@sia.cn

unacceptable cost for screening purpose[1,2]. Thermoacoustic imaging (TAI) is a non-ionizing and non-invasive imaging method, which combines the merits of high ultrasound imaging resolution with high microwave imaging contrast. In TAI, a short non-ionizing microwave pulse irradiates tissues to induce a small temperature rise, which consequently causes thermoelastic expansion to generate TA signals.

In recent years, various MITAT demonstrations have been reported. In 1984, Guo et al. introduced thermodynamical process. Due to the contrast in the absorption coefficient between tumor and background medium[3,5], the consequent thermal expansion caused by heating effect induces an acoustic wave at the position of the tumour. Lin and Chan built a prototype of 2-D imaging system for biological tissue. Kruger et al. performed the thermoacoustic tomography with 434-MHz microwave in five patients with documented breast cancer to show the contrast enhancement of tumors[7,8]. The microwave field energy distribution is not yet studied in-depth. LV Wang has studied the system experimental platform and qualitative image reconstruction[4,6]. The density distribution of microwave energy has been carried out by Xing Da et al. Lou et al. recently proposed a new approach of ultra-short microwave source to enhance the induced acoustic signal, which is shown to be very helpful to increase the signal-to-noise ratio (SNR) of MITAT applications.

2 Theory Analysis

In the thermoelastic regime the thermoacoustic pulse waveform with be determined by both the microwave absorption as well as the pulse duration. Irradiation of the absorbing layer with a microwave pulse will result in an instantaneous heat deposition inside this layer and excites TA waves[15]. As in the case of linear acoustic system, the equation of thermoelastic sound excitation in a heat-conducting medium can be given as[2,6]

$$\alpha = \omega\sqrt{\mu\varepsilon}\left[\frac{1}{2}\left(\sqrt{1+\frac{\sigma^2}{\varepsilon^2\omega^2}}-1\right)\right] \tag{1}$$

Where ω is the angular frequency, μ is the permeability, ε is relative permittivity, σ is the conductivity[5,6]. From (1), it is found that the absorption coefficient α is related with ε and σ under the fix frequency condition[5,8].

The TA signal strength is thus decided by the microwave radiation distribution. On the other hand, the propagation process of TA is related to the tissue characteristics.

$$\left(\nabla^2-\frac{1}{v_s^2}\frac{\partial^2}{\partial t^2}\right)p(r,t)=-\beta\rho\frac{\partial^2 T(r',t')}{\partial t^2} \tag{2}$$

where $p(\boldsymbol{r},t)$ is the induced acoustic signal, c_0 is the speed of sound, ρ_0 is the equilibrium density, C_p is the constant pressure heat capacity per unit mass and β is the coefficient of thermal expansion, and $H(\boldsymbol{r},t)$ is the heating function[5,7].

As shown in Figure 1, there are N ultrasonic sensors in the platform, which is used to receive the TA signal. The received signal is shown as[4,7]

$$x(t)=s(t)+n(t) \tag{3}$$

In (3), $x(t)$ is the signal from the sensors, $s(t)$ is TA signal from the sample, and $n(t)$ is noise and interference from the system. It is assumed that $s(t)$ and $n(t)$ is uncorrelated, and (4) can be deduced:

$$R_{xx} = R_{ss} + R_{nn} \tag{4}$$

Where R_{xx} is the self-covariance matrix about $x(t)$, and so on.

The key of RCB is to find out the optimal weighting vector W [1,2], when the gain is constant in the desired direction θ. The minization of array output energy can reduce interference and enhance the SNR by the adaptive diagonal loading method. ξ is the error constant[3,4].

$$\min_a a^* R^{-1} a \quad subject\, to \|a-\bar{a}\|^2 = \xi \tag{5}$$

In (5), $\bar{a} = [1,1...1]$ is the direction vector preseted. Through Lagrange multiplier method

$$a = \bar{a} - (I + \lambda R)^{-1} \bar{a} \tag{6}$$

Where I is the unit diagonal matrix, and λ can be calculated[6,8].

$$\left\|(I - \lambda R)^{-1} \bar{a}\right\|^2 = \xi \tag{7}$$

$\lambda \geq 0$ is the monotone decreasing function, which can be deduced by newton iteration or bisection method. From the equation (6), a and the final result can be calculated.

$$\hat{W} = \frac{R_{xx}^{-1} a(\theta)}{a(\theta) R_{xx}^{-1} a(\theta)} \tag{8}$$

This paper has presented an improvement RCB method, which can achieve dynamic range path variation, and combine the synthetic aperture image and self-adaption beam-former. Comparing with the traditional TA imaging algorithm, the method can enhance SNR and increase the image resolution.

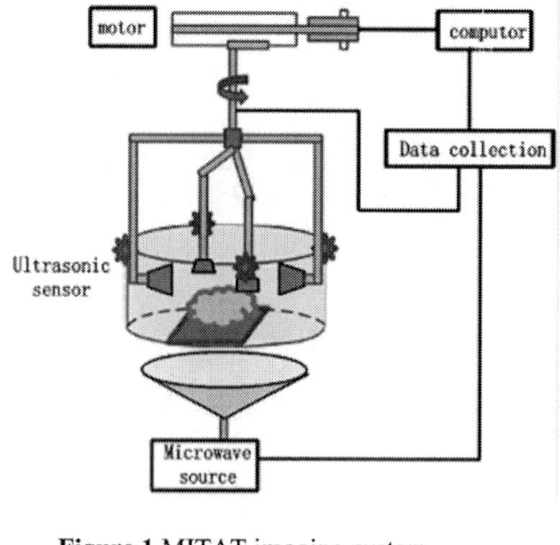

Figure 1.MITAT imaging system.

3 Experimental Research

In order to verify the imaging performance of the TA system, and evaluate the image reconstruction method, we used the analogy absorber as the sample which is similar with true biological tissue in the physical characteristics. The absorber is composed of agar, water and salt which were mixed by a certain percentage. Figure 2 shows TA image reconstruction result for the signal sample. In the figure, coinciding with the actual size of the absorber, the reconstruction size of the sample is 5mm×5mm. (a)

DOI: 10.1051/bioconf/20170802001

used the traditional Daley and Sum, and (b) did the improved RCB. From the results comparison, the result had the better SNR and resolution by improved RCB.

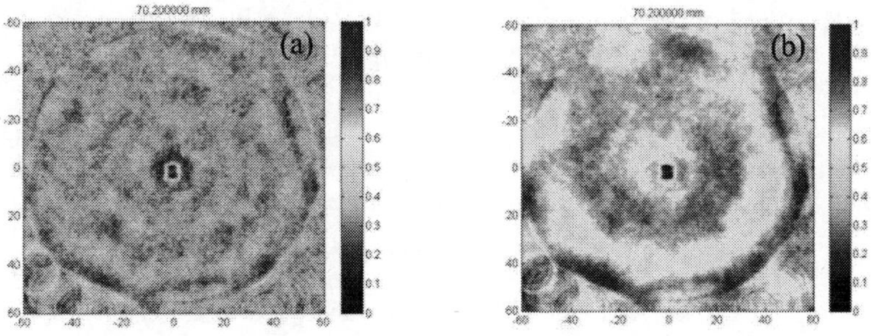

Figure 2. Recovered TA images for single absorbers.

Figure 3 showed the TA reconstruction result of both absorbers with the same physical characteristics. The distance of the both samples is 3mm. (a) used the traditional Daley and Sum method, and (b) did the improved RCB method. From the figure (b), the image edge is clearer, and has a better resolution and SNR. It is import to introduce the MITAT technology into the breast cancer clinical detection.

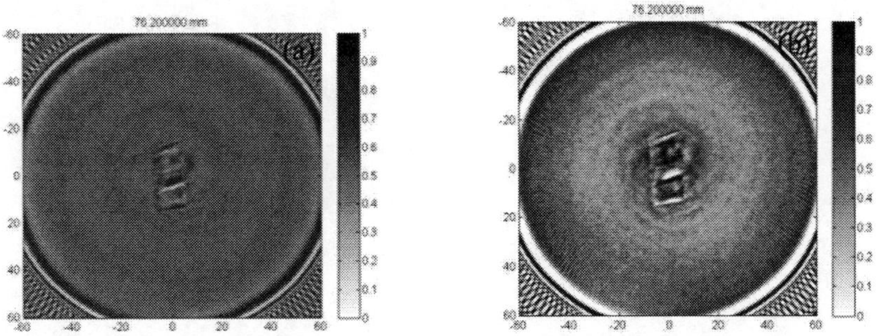

Figure 3. Recovered TA images for single absorbers.

4 Results and Discussion

TA signal is the ultrasound which can reflect the energy absorption distribution of biological tissue, and the reconstruction algorithm is important for reconstructing the internal structure of tissue. This paper has presented an improvement RCB method, which can achieve dynamic range path variation, and combine the synthetic aperture image and self-adaption beam-former. Comparing with the traditional TA imaging algorithm, the method can enhance SNR and increase the image resolution. Combine the theoretical analysis and experimental data, which comes from the S-band microwave induced thermo-acoustic imaging system, we have demonstrated that the image algorithm is important.

References

1. Cunguang Lou, Sihua Yang, Zhong Ji, Qun Chen, and Da Xing. Ultra-short Microwave-Induced Thermoacoustic Imaging: A Breakthrough in Excitation Efficiency and Spatial Resolution. PHYSICAL REVIEW LETTERS, pp. 208101-1~208101-5 (2012).
2. Pramanikm KUG, Changhui Li, et al. Designand evaluation of a novel breast cancer detection system combining both thermoacoustic andphotoacoustic (PA) tomography[J]. Medical Physics, 35(6) :2218-2223 (2008).
3. Nie Liming, Zhou Quan, Yang Diwu. Microwave-induced thermoacoustic scanning CT for high-contrast and noninvasive breast cancer imaging[J]. Medical Physics, 35(9): 4026-4032 (2008).

4. KELLNBERGERS, HAJIABOLIA, RAZANSKYD, et al. Near-field thermoacoustic tomography of small animals[J]. Physics inMedicine andBiology, 56(11): 3433-3444 (2011).
5. Xin Bi, Lin Huang, Jinsong Du, Weizhi Qi, Yang Gao. Pulsed Microwave Energy Spatial Distribution Imaging by Means of Thermoacoustic Tomography, Acta Phys. Sin., 64(1), pp. 014301-1~014301-6 (2015).
6. Jinsong Du, Yang Gao, Xin Bi, Weizhi Qi, Lin Huang, Jian Rong. S-band microwave induced thermo-acoustic tomography system, Acta Phys. Sin. 64(3), pp. 034301-1~034301-6 (2015)
7. Huabei Jiang, Yong Xu, Nicusor Iftimia Julie Eggert and etc., “Three Dimensional Optical Tomographic Imaging of Breast in A Human Subject”, IEEE Trans on Med., Vol. 20, No. 12, pp: 1334-1340 (2010).
8. Zeng L M, Xing D, Gu H M, Fast microwave-induced thermoacoustic tomography based on multi-element phase-controlled focus technique. Chin Phys Lett, 23(5), pp. 12125-1218 (2006).

Construction of risk prediction model of type 2 diabetes mellitus based on logistic regression

Jian Li[1,#], Qin Huang[2], Minghua Dong[3,#], Wei Qiu[3], Lixia Jiang[4], Xiaoting Luo[5], Zhengchun Huang[1], Shuiqin Chen[5], Qinfeng Wu[3], Lu Ou-Yang[3], Qin Wu[4], Lihua Liu[5] and *Shumei Li*[3, a]
[1]*Department of Anatomy Gannan Medical College, Ganzhou, Jiangxi, China*
[2]*Department of General Medicine, St.Michael hospital, Shanghai, China*
[3]*Department of Epidemiology in Preventive Medicine, Gannan Medical College, Ganzhou, Jiangxi, China*
[4]*Department of Clinical laboratory of 1'th affiliate hospital, Gannan Medical College, Ganzhou, Jiangxi, China*
[5]*Department of Biochemistry and Molecular Biology, Gannan Medical College, Ganzhou, Jiangxi, China*

Abstract. Objective: to construct multi factor prediction model for the individual risk of T2DM, and to explore new ideas for early warning, prevention and personalized health services for T2DM. Methods: using logistic regression techniques to screen the risk factors for T2DM and construct the risk prediction model of T2DM. Results: Male's risk prediction model logistic regression equation: logit(P)=BMI × 0.735+ vegetables × (-0.671) + age × 0.838+ diastolic pressure × 0.296+ physical activity× (-2.287) + sleep ×(-0.009) +smoking ×0.214; Female's risk prediction model logistic regression equation: logit(P)=BMI ×1.979+ vegetables× (-0.292) + age × 1.355+ diastolic pressure× 0.522+ physical activity × (-2.287) + sleep × (-0.010).The area under the ROC curve of male was 0.83, the sensitivity was 0.72, the specificity was 0.86, the area under the ROC curve of female was 0.84, the sensitivity was 0.75, the specificity was 0.90. Conclusion: This study model data is from a compared study of nested case, the risk prediction model has been established by using the more mature logistic regression techniques, and the model is higher predictive sensitivity, specificity and stability.

1 Introduction

With the rapid development of social economy, people's life style and dietary structure have changed and population aging intensified, diabetes prevalence rate rose rapidly, especially in our country [1], it has brought heavy economic burden to society and family. In recent years, foreign countries have tended to use a risk assessment tool to the risk of type 2 diabetes mellitus(T2DM) for prediction and risk score, for finding the early identification of high-risk groups, to control risk factors by carrying out the health education and lifestyle intervention , and then reduce the T2DM [2]. Some domestic scholars also constructed T2DM risk model with risk factors scoring method and OR data score and so on, to predict high-risk T2DM;and, in this aspect, they did some propaganda and education to make people accept the T2DM risk individual prediction model [3-4] in the identification of high-risk

[a] Corresponding author: Shumei Li, gnyxylsm@163.com. Phone: 008615083787928.
[#]These authors contributed equally to this study and share first authorship.
This study was supported by scientific fund from National Natural Science Fund in China (No.81360445), Science and technology support program of Jiangxi Province (NO.20132BBG70086).

groups gradually .

2 Research design and program

2.1 Research objects

Researchers chose the sample from T2DM baseline survey for the project study in 19 districts and counties in Ganzhou City in 2009.The study subject is a total of 8086 copies of data, age 35-64 years old. The baseline survey and follow-up data content include general demographic characteristics, smoking and drinking history, personal health history, diabetes family history, physical activity and physical exercise and so on; human body measurements included height, weight, BMI, waist circumference, hip circumference, waist hip ratio (WHR), blood pressure, lung capacity and so on ; biochemical metabolic index included fasting blood glucose (FBG), total cholesterol (TC), triglyceride (TG), high density lipoprotein cholesterol (HDL-C), low density lipoprotein cholesterol (LDL-C), blood uric acid (UA) (save all to MSSQL database).

2.2 Research methods

Researchers randomly selected 2 / 3 samples (5391persons) from MSSQL database as the training group, the pre-selected variables is age, sex, BMI, waist circumference and waist to hip ratio (WHR), blood pressure (BP), T2DM family history, physical activity, smoking and drinking, FBG, TC, TG, HDL-C, LDL-C; and then made single factor analysis for the pre-seleced variables . Multivariate analysis mainly adopted mature logistic regression model, as to whether diabetes as the dependent variable (diabetic patient with the value 1, non diabetic patient with the value 0), other variables (suspected risk factor variables) as independent variables were multivariate logistic regression, $\text{logistic}(p)= \ln(\frac{p}{1-p})$,to construct the T2DM risk prediction model. In order to increase the prediction accuracy, prediction model of gender were fitted.Another 1/3 sample (2695 people) in the study cohort was used as the test group, the area under the receiver operating characteristic curve (ROC) was used to analyze the sensitivity and specificity of the model, to evaluate the predictive effect of the model .

3 Results

3.1 Relationship between age, sex and T2DM

Research data showed the prevalence of T2DM was 7.7%, with group of every 5 years , people will be divided into 35-, 40-, 45-, 50-, 55-, 60-64, T2DM prevalence was compared between six groups by gender , the differences of T2DM prevalence between each group were not statistically significant ($p > 0.05$).(see Table 1).

Table 1. Different gender and age in T2DM prevalence rate.

age(year)	male			female		
	n	T2DM	%	n	T2DM	%
35 -	288	8	2.8	368	13	3.5
40-	362	20	5.5	439	12	2.7
45-	375	22	5.9	402	21	5.2
50-	273	22	8.1	395	28	7.1
55-	390	50	12.8	475	41	8.6
60-64	378	53	14.0	363	57	15.7
total	2066	175	8.5	2442	172	7.0

3.2 Relationship between the body mass index (kg / M^2) and T2DM

Body mass index (BMI) were divided into three groups: the normal group (18.5-23.9),T2DM prevalence was 6.7% (88/1315);hyper height group(≥24.0), T2DM prevalence was 9.6% (86/897); obesity group(≥28), T2DM prevalence was 14.9 % (35/235). The three groups were compared with the increase of BMI and the prevalence of T2DM, the difference was statistically significant (x^2=19.15, p<0.001).
BMI group by gender: Male, T2DM prevalence of normal group was 7.3% (42/578);T2DM prevalence of overweight group was 9.9% (41/414); T2DM prevalence of obesity group was 17.2% (16/93); There are differences between these groups and these differences were statistical significant (x^2=10.03, $p < 0.05$). Female, T2DM prevalence of the normal group was 6.2% (46/737); T2DM prevalence of overweight group was 9.3% (45/483); T2DM prevalence of obesity group was 13.4% (19/142); There are differences between these groups and these differences was statistical significant (x^2=9.72, $p < 0.05$).(see Table 2).

Table 2. T2DM prevalence rate of different BMI In different gender.

BMI	male			female		
(kg/m^2)	n	T2DM	%	n	T2DM	%
18.5-23.9	578	42	7.3	737	46	6.2
≥24.0	414	41	9.9	483	45	9.3
≥28	93	16	17.2	142	19	13.4
total	1085	99	9.1	1362	110	8.1

3.3 Relationship between waist circumference and waist to hip ratio of T2DM

Waist samples were divided into two groups: normal group (male <85cm, female <80cm), T2DM prevalence was 5.7% (79/1377); abnormal group (male ≥85cm, the female≥80cm), T2DM prevalence was12.2% (129/1060). Waist circumference increased, the prevalence of T2DM increased, the difference was statistically significant (x^2=31.74, p<0.001).Male, T2DM prevalence of normal group was 7.0% (45/640);T2DM prevalence of abnormal group was 12.2% (54/443); There are differences between these groups and these differences were statistical significant (x^2=8.39, $p < 0.05$). Female, T2DM prevalence of the normal group was 4.6% (34/737); T2DM prevalence of abnormal group was 12.2% (75/617); There are differences between these groups and these differences was statistical significant ($p < 0.05$).(see Table 3).

Table 3. Relationship between waist circumference and T2DM

waist (cm)	male			female		
	n	T2DM	%	n	T2DM	%
male <85, female <80	640	45	7.0	737	34	4.6
male ≥85, female≥80	443	54	12.2	617	75	12.2
total	1083	99	9.1	1354	109	8.1

The samples was divided into three groups by the ratio of waist to hip(W-H), the low group (male WH<0.82, female WH<0.78), the prevalence of T2DM was 3.9% (16/410); the normal group (male 0.82≤WH< 0.91, female 0.78≤WH<0.87) ,the prevalence of T2DM was 6.6% (83/1267); the high group (male WH≥0.91, female≥0.87), the prevalence of T2DM was 14.2%, 107/754). Compared with the three groups between male and female, WH increased, the prevalence of T2DM increased, the difference was statistically significant (p<0.05).(see Table 4).

Table 4. Waist hip ratio and the prevalence of T2DM relations.

W-H	male			female		
	n	T2DM	%	n	T2DM	%
male <0.82, female <0.78	207	13	6.3	203	3	1.5
male 0.82-0.91, female 0.78-0.87	588	44	7.5	679	39	5.7
male ≥0.91, female ≥0.87	286	40	14.0	468	67	14.3
total	1081	97	9.0	1350	109	8.1

3.4 Relationship between blood pressure and the prevalence of T2DM

The samples were divided into two groups by the blood pressure,the normal group (systolic blood pressure SBP<140 mmHg and diastolic blood pressure DBP<90 mmHg), and hypertension group (SBP≥140 mmHg, DBP ≥90 mmHg) .
Normal group of Systolic blood pressure(SBP)'s T2DM prevalence was 6.9% (132/1920). Hypertension group of SBP 's T2DM prevalence was 14.8% (76/512); the prevalence of T2DM in hypertension group increased, and the difference is statistically significant (p <0.05). The normal group of diastolic blood pressure(DBP)'s T2DM prevalence was 7.6% (141/1853) ;hypertension group of DBMS's T2DM prevalence was 11.7% (67/575), the prevalence of hypertension group was different in T2DM,for male ,the difference has no statistical significance ($p > 0.05$);for female, the difference has statistical significance ($p < 0.05$).(see Table 5,6).

Table 5. SBP and T2DM prevalence rate.

SBP	male			female		
(mmHg)	n	T2DM	%	n	T2DM	%
<140	842	63	7.5	1078	69	6.4
≥140	237	35	14.8	275	41	14.9
total	1079	98	9.1	1353	110	8.1

Table 6. DBP and T2DM prevalence rate.

DBP	male			female		
(mmHg)	n	T2DM	%	n	T2DM	%
<90	805	66	8.2	1048	75	7.2
≥90	273	32	11.7	302	35	11.6
total	1078	98	9.1	1350	110	8.1

3.5 Relationship between total cholesterol (TC) and T2DM

The total cholesterol index sample was divided into three groups: the low group ,TC < 5.18 mmol / L, T2DM prevalence was 6.2% (130/2105); intermediate group ,5.18 mmol/L≤TC<6.22 mmol/L, T2DM prevalence was 10.2% (80/783); high group ,TC≥6.22 mmol/L, T2DM prevalence was 18.6% (49/264). The prevalence of hypertension group was different in T2DM:for male and female ,the difference of group has no statistical significance ($p > 0.05$);for whole sample the difference has statistical significance ($p < 0.05$).(see Table 7)

Table 7. Regardless of gender, the relationship between cholesterol and prevalence rate of T2DM.

TC	male			female		
(mmol / L)	n	T2DM	%	n	T2DM	%
< 5.18	927	74	8.0	1178	56	4.8
5.18-6.22	328	33	10.1	455	47	10.3
≥6.22	101	14	13.9	163	35	21.5
total	1356	121	8.9	1350	138	7.7

3.6 Relationship between triglyceride (TG) and T2DM

Triglyceride sample was divided into three groups: low group ,TG < 1.70 mmol / L, the prevalence of T2DM was 6.0% (132/2209); intermediate group, 1.70 mmol/L≤TG<2.26 mmol/L, the prevalence of T2DM is 9.5% (46/482); high group , TG≥2.26 mmol/L), the prevalence of T2DM was 17.9% (82/459). Compared with the three groups between male and female, TG increased , the prevalence of T2DM increased, the difference was statistically significant ($p < 0.001$).(see Table 8).

Table 8. Relationship between triglyceride index and the prevalence rate of T2DM.

TG	male			female		
(mmol / L)	n	T2DM	%	n	T2DM	%
< 1.70	918	64	7.0	1291	68	5.3
1.70-2.26	210	18	8.1	272	28	10.3
≥2.26	225	39	17.3	234	43	18.4
total	1353	121	8.9	1797	139	7.7

3.7 Relationship between smoking and T2DM

The subjects were divided into two groups, low exposure group (no smoking) and high exposure group (smoking): there was no significant difference in the prevalence of low exposure group and high exposure group (x^2=0.831, p>0.05).

3.8 The Relationship between dietary factors and T2DM

Compared the people who often eat coarse grain with the people who don't eat the coarse grain, male's OR was 0.41, 95% confidence interval (0.25, 0.67), The differences of prevalence have statistical significance (x^2=13.12. p< 0.001), coarse grain is the protective factor; Female's OR is 0.73, 95% confidence interval for (0.47, 1.50), the differences of prevalence have statistical significance (x^2=13.18, $p < 0.05$); coarse grains is the protective factor.
Compared the people who often edible meat with the people who don't eat meat,male's OR is 2.37, 95% confidence interval (1.46, 3.84), the differences in prevalence have statistical significance (x^2=12.92, $p < 0.001$), meat is the risk factor ; The difference of female prevalence have no statistical significance (x^2= 0.26, p> 0.05).
Compared the people who often edible vegetable with the people who don't edible the vegetables, male's OR is 0.59, 95% confidence interval for (0.37, 0.93), the differences in prevalence have statistical significance (x^2=5.16, $p < 0.05$),vegetable is the protective factor; Female's OR is 0.50, 95% confidence interval for (0.32, 0.77),the differences in prevalence have statistical significance (x^2=9.92, $p < 0.05$); vegetable is the protective factor.

3.9 Multiple factors analysis of logistic regression

The analysis of Significant predictive variables had showed : male subjects in the study taking BMI, age, diastolic blood pressure, smoking as the risk factor, and vegetable intake, physical activity, sleep as the protective factor (β<0) ,and risk prediction model regression equation logit (P) = BMI ×0.735+ vegetables× (-0.671) + age ×0.838+ diastolic pressure × 0.296+ physical activity×(-2.287) sleep ×(-0.009) +smoking ×0.214. The female subjects in the study taking BMI, age, diastolic pressure as the risk factor, and vegetable intake, physical activity, sleep as the protective factor, risk prediction model regression equation logit (P) = BMI ×1.979+ vegetables × (-0.292) + age× 1.355+ diastolic pressure × 0.522+ physical activity × (-2.287) + sleep × (-0.010). (see Table 9,10)

Table 9. The Results of logistic regression analysis of male relative indexes.

Variable beta	β	S.E	x^2	P
Body mass index	0.735	0.905	0.659	0.417
Vegetable	-0.671	0.292	5.291	0.021
Age	0.838	0.497	2.843	0.092
Diastolic blood pressure	0.296	0.303	0.956	0.328
Physical activity	-2.287	0.106	465.380	0.000
Sleep	-0.009	0.166	0.003	0.956
Smoking	0.214	0.235	0.828	0.363

Table 10. The Results of logistic regression analysis of female related indexes.

Variable beta	β	S.E	x^2	P
Body mass index	1.979	1.056	3.512	0.061
Vegetable	-0.292	0.316	0.855	0.355
Age	1.355	0.588	5.318	0.021
Diastolic blood pressure	0.522	0.333	2.460	0.117
Physical activity	-2.287	0.106	465.380	0.000
sleep	-0.010	0.123	0.007	0.956

3.10 The Internal authenticity of the model and its evaluation

Male' s area under the ROC curve area was 0.83, sensitivity was 0.72, specificity was 0.86; female's area under ROC curve was 0.84, sensitivity was 0.75, specificity was 0.90, two sets of data showed that model had a high predictive value.(see Figure 1).

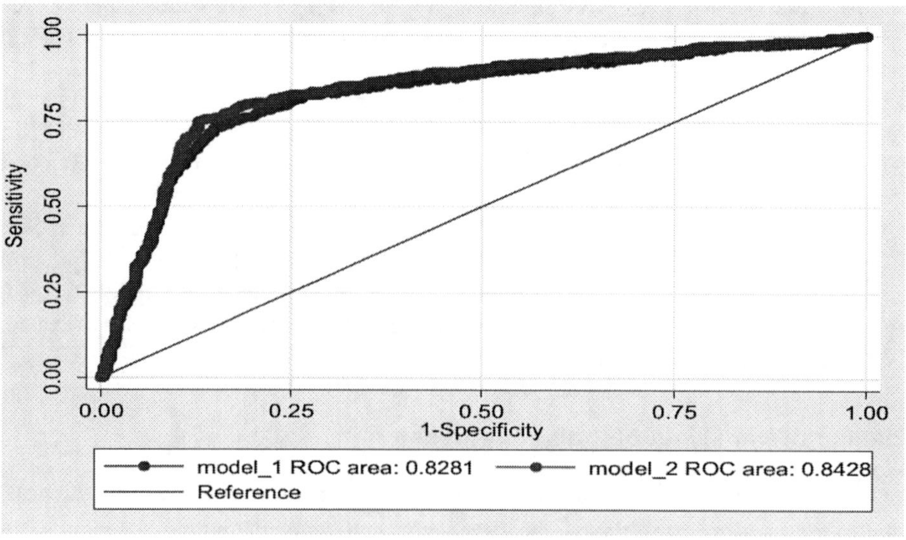

Figure 1. Sensitivity and specificity of the model.

4 Discussion

The modeling data was from compared study of prophase nested case, and used the less error and more mature logistic regression techniques in theory to deal with the variables. The risk model that had been established on this base has higher sensitivity, specificity and stability. The development of T2DM risk model started late. Since 2003, some type 2 diabetes risk assessment tools had been

 DOI: 10.1051/bioconf/20170802002

gradually developed and applicated in foreign country [5-9], but due to the difference in race, behavior and lifestyle, diet, medical and health services, economic and cultural factors, foreign risk assessment tool was not fully applicable to Asian and / or Chinese [10-11].Therefore, since 2006, scholars in our country [12-14] have concluded that the main risk factors of diabetes mellitus and its relative risk to establish evaluation methods and evaluation model of adult individual diabetes risk in our country ,on the basis of epidemiological survey of diabetes ,by risk factors scoring method, disease risk score, the analysis of diabetes risk factors and prevalence data. These models mostly based on the compared study cases , prospective study was insufficient; Our study team has fully considered the conditions which required for robust risk model, include stable and representative large samples , a prospective observation, multivariate estimation and mature computer technology platform .But the research is still in initial stage with many remaining problems: the first is the study did not exclude the influence of the linear of predictive variables in model to the stability of the model; researchers only evaluate the predictive model by the similar people , part of the data is not so complete; We will evaluate and verificate this model by screening return visitors in the latter part of the study .

References

1. Yang WY, Lu JM, Weng Jp, N Engl J Med ,*Prevalence of diabetes among men and women in China*,**362**,1090-1101(2010).
2. Douglas N, Rohini M,Tom D, BMJ,*Risk models and scores for type 2 diabetes: systematic review*, **343**, d7163(2011).
3. Mann DM, Bertoni AG, Shimbo D, Am J Epidemiol, *Comparative Validity of 3 Diabetes Mellitus Risk prediction Scoring Models in a Multiethnic US Cohort The Multi-Ethnic Study of Atherosclerosis*, **171**,980–988(2010).
4. Almeda-Valdes p, Cuevas-Ramos D, Mehta R,Curr Diabetes Rev,*UKpDS Risk Engine, decode and diabetespHD models for the estimation of cardiovascular risk in patients with diabetes,* **6**,1-8(2010).
5. Buijsse B, Simmons RK, Griffin SJ, Schulze MB,Epidemiol Rev, *Risk Assessment Tools for Identifying Individuals at Risk of Developing Type 2 Diabetes,* **33**,46-62(2011).
6. Lindström J, Tuomilehto J, Diabetes Care,*The diabetes risk score: a practical tool to predict type 2 diabetes risk*,**26**,725–731(2003).
7. Schulze MB, Weikert C, pischon T, Diabetes Care,*Use of multiple metabolic and genetic markers to improve the prediction of type 2 diabetes: the EpIC-potsdam Study*,**32**,2116–2119(2009).
8. Glmer,Charlotte,Carstensen,Bendix,Sandbk,Annelli,Lauritzen,Torsten,Jrgensen,Torben,Borch-Johnsen,Knut, Diabetes Care ,*A Danish Diabetes Risk Score for Targeted Screening* ,**3**,727-733(2004).
9. Heikes KE, Eddy DM, Arondekar B, Diabetes Care, *Diabetes Risk Calculator: a simple tool for detecting undiagnosed diabetes and pre-diabetes*,**31**,1040-1045(2008).
10. Liu M,Pan CY, Jing MM, Su HY,Lu JM, Chinese Journal of diabetes, *Evaluation of diabetes risk score in the evaluation of new onset diabetes*,**17**,201 -204(2009).
11. 12.Calvin Woon-Loong Chin, Elia n Hui San Chia, Stefan Ma, Derrick Heng, Maud rene Tan, Jean ette Lee, E Shyong Tai, Agus Salim, BMC,*The ARIC predictive model reliably predicted risk of type II diabetes in Asian populations*,**1** -13(2012).
12. Li XY, Li R, Zhang SN, Chinese public health, *Effect evaluation of different screening methods for asymptomatic diabetes*,**22**,687-689(2006).
13. Wu HY, Pan P, He Y, Chinese Journal of health management, *The risk assessment method for adults with diabetes in China*,**1**,95-98(2007).
14. Zhang L, Shi K, Yi D, Chongqing medical, *Study on risk assessment of diabetes in community residents in Chongqing City*,**40**,1885-1888(2011).

Meta-Analysis of the association of *IGF2BP2* gene rs4402960 polymorphisms with T2DM in Asia

Zhengchun Huang[1], Minghua Dong[2], Jian Li[1], Wei Qiu[2] and Shumei Li[2,a]

[1]*Department of Human Anatomy, Gannan Medical College, Ganzhou, Jiangxi, China*
[2]*Department of Epidemiology in Preventive Medicine, Gannan Medical College, Ganzhou, Jiangxi, China*

Abstract. In order to evaluate the association of IGF2BP2 (SNPS:rs4402960) gene polymorphism and type 2 diabetes mellitus (T2DM) susceptibility, we searched for all the related research literature by CNKI, Wanfan, Pubmed and Springer link database to collect data.Meta-analysis software(RevMan5.0) was applied for heterogeneity tests in genotype level, two methods(fixed-effects and random-effects model)were performed to pool the odds ratio(OR). Random-effects model was adopted if there were obvious heterogeneity among studies. On the contrary, a fixed effect model was used. At the same time, publication bias was examined by Funnel plot. The studies collected 28 articles(including 35 studies), and included 52277cases of T2DM and 54168 controls.The pooled ORs of allele (T vs G)of rs4402960 polymorphic loci in IGF2BP2 was significant association with T2DM (OR=1.163 95%CI=[1.138,1.189] P<0.00001).

1 Introduction

Type 2 diabetes mellitus (type 2 diabetes, T2DM), characterized by significant genetic heterogeneity, and are influenced by various environmental factors, resulted in peripheral tissue insulin resistance (IR) and insulin secretion deficiency, which lead to absolute or relative deficiency of insulin and high blood glucose by reducing uptake glucose [1]. With the new technologies like DNA sequencing, gene chip technique and other techniques application [2], they provide the new methods for researching T2DM susceptibility genes.Three different GWAS research groups in 2007 found that Insulin growth factor 2 mRNA binding protein 2 (insulin-like growth factor 2 mRNA binding protein 2, IGF2BP2) gene polymorphism and T2DM is associated[3-5]. Subsequently, multiple research groups carried out a lot of research about the association between T2DM and IGF2BP2, but research conclusions are not consistent in different populations from racial, ethnic, national, regional groups. Therefore, this research has comparative and cohort studies on associated cases published in Asia about relationships between the IGF2BP2 gene rs4402960 polymorphic loci and T2DM, with comprehensive evaluation the relationships between them, it offers valuable advice for high-risk T2DM screening and early prevention.

[a] Corresponding author: Shumei Li, gnyxylsm@163.com.
This study was supported by scientific fund from National Natural Science Fund in China (No.81360445).

2 Data and methods

2.1 Data sources

The title Resources are from Joint search of CNKI, Wanfang database, and PubMed and Springer link database. The searching key words includes "T2DM, IGF2BP2, rs4402960 polymorphic loci, case-control study, chort study". The deadline for retrieval is May 31th, 2016.

2.2 literature inclusion and exclusion criteria

The literature included in the study was a case control study or cohort study; the study involved a number of areas of the world and the populations of races; Type 2 diabetes diagnosis were consistent with 1999 WHO diagnostic criteria and 2010 the American Diabetes Association (ADA) diagnostic criteria, patients were randomly selected regardless of age, sex, combined disease history, family disease history, and there were no special limit about years of research conducted or papers published. There are specific provisions on the content of sample; the original literature data provide OR value and 95% CI or can be transformed into the OR value and 95% CI; literature contains enough information for comparison between T2DM patients and controls the IGF2BP2 gene rs4402960. Patients with type 1 diabetes, gestational diabetes, other special type of diabetes (insulin β cell function of gene defects, insulin action gene defects, endocrine disease, etc.) are exclusive.

2.3 Literature information extraction and statistical analysis

2.3.1 Data extraction and sorting

The research related information are independently extracted from selected literature, and after data collection, they are filled in meta analysis of data abstraction, the information of the IGF2BP2 gene rs4402960 polymorphic loci and T2DM tables are formed, finally, the corresponding database are established according to the data from the original literature (OR and 95% CI) .

2.3.2 Heterogeneity test

In order to explore whether there is a clear heterogeneity between the studies, the Q test based on the chi square test was used for quantitative analysis, that is, the use of I2 to measure the size of the heterogeneity [6]. If the study shows no statistical heterogeneity (P>0.1, I2<=25%), it will use the mantel Haenszel fixed effects model (Peto method) to merge data. If there is heterogeneity (P<0.1, 25%<I2<50%), the Dersimonian-Laird random effect model (D-L method) will be used to combine the data; if I2 greater than 50%, it is thought that there is a obvious heterogeneity, subsequently, subgroup analyses exploring sources of heterogeneity is going to happen.

2.3.3 The significance test of combined OR value

Using the the Z value and P value in the part "test for the overall effect" in the RevMan5.20 software forest plots to do the significant test , if the calibration standard P≤0.05, we can consider it significant [7].

DOI: 10.1051/bioconf/20170802003

2.3.4 Bias analysis

Through the funnel plots to examine into the literature publication bias, if the funnel plot asymmetry may indicate the presence of publication bias, the funnel plot symmetry shows that there was no evidence of publication bias, the analysis is completed by statistical software RevMan5.20.

3 Results

3.1 Basic information

By searching, 28 articles conformed to associated cases control study about the IGF2BP2 gene rs4402960 polymorphic loci and T2DM are screened out, with 35 studies which contains 106445 people in seven Asian countries[6-33], including 52277 T2DM and 54168 controls .

3.2 Meta analysis of the association between T2DM and IGF2BP2 gene rs4402960 polymorphism

Figure 1 is the type of genes in the *IGF2BP2* gene rs4402960 polymorphic loci (allele gene T vs G) associated with T2DM and a meta analysis forest graph.

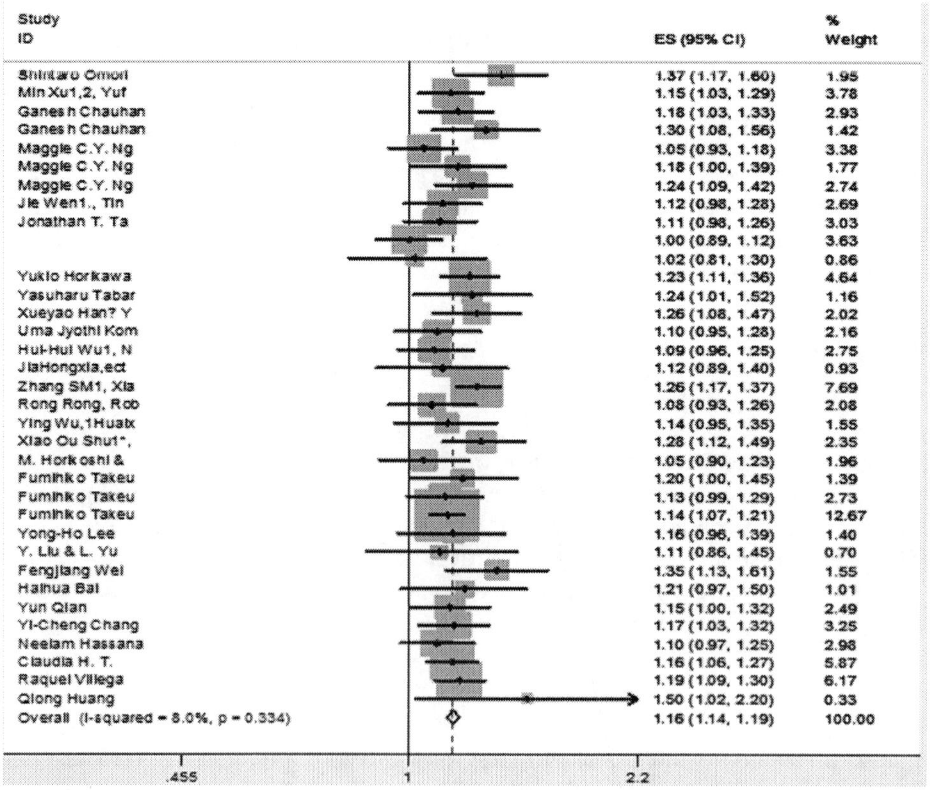

Figure 1. Meta analysis forest graph of rs4402960 IGF2BP2 gene polymorphism loci of genotype (T vs G) associated with T2DM.

3.3 Heterogeneity analysis

Heterogeneity is an important link in the Meta analysis test, the purpose is to examine whether the results of independent research has can be merged. Rs4402960 gene in IGF2BP2 polymorphism loci genotypes (T vs G) has no obvious heterogeneity (Q=36.95,P=0.334,I 2=8.0%) , it uses the fixed effect model to merger the OR values and 95%CI. Results showed,the mergered OR values and 95%CI were 1.163(1.138,1.189).So it can be thought that rs4402960 gene in IGF2BP2 polymorphism loci of the allele T is the susceptibility genes of T2DM.

3.4 The evaluation of publication bias

From Figure 2, it can be seen that Meta analysis of inverted funnel figure of rs4402960 gene in IGF2BP2 polymorphism loci of the allele (T vs G) is basic symmetric, it is less likely to exist publication bias. And the Begg's Test and Egger's Test for Publication Bias also showed that both P-values were greater than 0.1,it can be thought there is no publication bias.

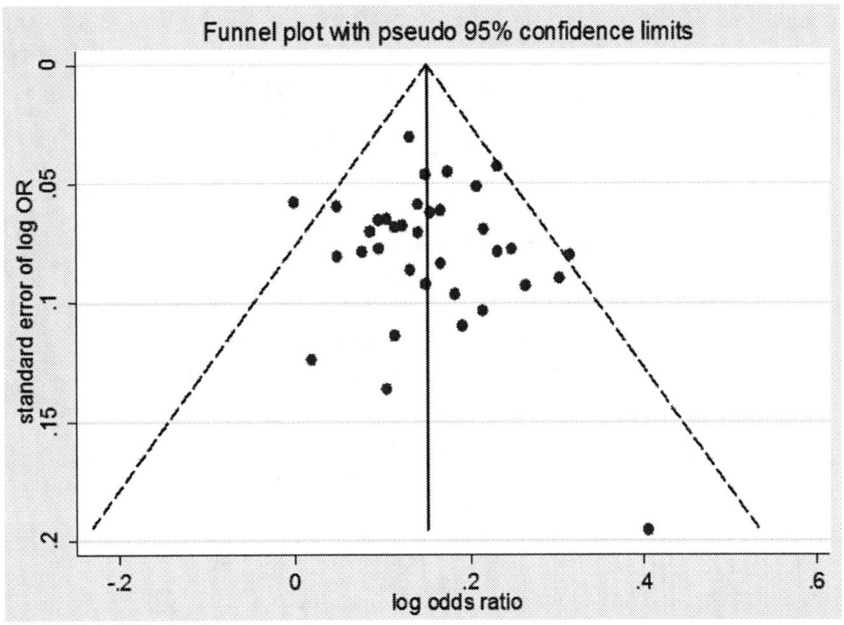

Figure 2. Bias evaluation funnel figure of rs4402960 polymorphic loci alleles (T vs G) associated with T2DM.

4 DISCUSSION

From gene level and with the theory and methods of evidence-based medicine, this study evaluates the relevance between rs4402960 IGF2BP2 gene polymorphism loci and T2DM. The mergered OR value and 95%CI of the allele (T vs G) were 1.163(1.138,1.189), it showed there are relevance between the rs4402960 IGF2BP2 gene polymorphism loci and T2DM. This study screening at home and abroad nearly all correlation research literature concerned with rs4402960 IGF2BP2 gene polymorphism loci and T2DM, and evaluate and deal with the quality of literature with scientific methods; then to these obvious heterogeneity between rs4402960 polymorphic loci alleles (T vs G) .Results shows that there is no obvious heterogeneity in rs4402960 polymorphic loci alleles (T vs G) in Asia population. At the same time, bias may exist in every step of Meta analysis, so the study use the funnel figure to evaluate bias, it turns out that inverted funnel figures of IGF2BP2 rs4402960 polymorphic loci of genotypes and alleles are basic symmetric, indicating there is less likely to have a publication bias.

The test results of this research may have bias to a extent, due to different detection technologies, different levels of detection and other inevitable errors within the study of the Meta analysis. The occurrence of T2DM has something with genetic, environmental and many other factors, this article only discusses the research of the correlation between rs4402960 IGF2BP2 gene polymorphism loci and the T2DM, however, other IGF2BP2 gene polymorphism loci are not involved, at the same time, such factors as diet, lifestyle, climate, environment may have interaction effect on T2DM, it is still necessary to further explore the molecular mechanism of IGF2BP2 result in T2DM.

References

1. Zai-ying Lu, Nan-shan Zhong editor, *Internal medicine*(People's Medical Publishing House, 2008).
2. Yan Chen, Guifeng Liu, Mei Wu.Chinese Journal of Gerontology **24**,6346-6347(2013).
3. Saxena R, Voight BF, Lyssenko V, et al. Science **316**,1331-1336(2007).
4. Scott LJ, Mohlke KL, Bonnycastle LL, et al. Science **316**,1341-1345(2007).
5. Zeggini E, Weedon MN, Lindgren CM, et al. Science **316**,1336-1341(2007).
6. Qiong Huang, Ji-ye Yin, Xing-ping Dai, et al.Acta Pharmacol Sin **31**,709–717(2010).
7. Wu H-H, Liu N-J, Yang Z, et al. European Journal of Medical Research **19**,40(2014).
8. Omori S, Tanaka Y, Takahashi A, et al. Diabetes **57**,791-795(2008).
9. Xu M, Bi Y, Xu Y, et al. PLoS One **5**(2010).
10. Chauhan G, Spurgeon CJ, Tabassum R, et al. Diabetes **59**,2068-2074(2010).
11. Ng MCY, Park KS, Oh B , et al. Diabetes **57**,2226-2233(2008).
12. Wen J, Rönn T, Olsson A, et al. PLoS One **5**(2010).
13. Tan JT, Ng DP, Nurbaya S, et al.J Clin Endocrinol Metab **95**,390-397(2010).
14. Horikawa Y, Miyake K, Yasuda K, et al. J Clin Endocrinol Metab **93**,3136-3141(2008).
15. Tabara Y, Osawa H, Kawamoto R, et al.Diabetes 58,493-498(2009) .
16. Han X, Luo Y, Ren Q, et al. BMC Medical Genetics **11**,81(2010).
17. Kommoju UJ, Maruda J, Kadarkarai S, et al. Meta Gene **1**,15-23(2013).
18. Hongxia Jia,Bin Gao,Yu Jie.The Chinese journal of clinical physicians **6**,7528-7531(2012).
19. Zhang Si-min, Xiao Jian-zhong,et al. ChinMed J **126**,4013-4018(2013).
20. Rong R, Hanson RL, Ortiz D, et al. Diabetes **58**,478-488(2009).
21. Wu Y,Li H, Loos RJF, et al. Diabetes **57**,2834-2842(2008).
22. Shu XO, Long J, Cai Q, et al. PLoS Genetics **6**(2010).
23. Horikoshi M, Hara K, Ito C, et al. Diabetologia **50**,2461-2466(2007).
24. Takeuchi F, Serizawa M, Yamamoto K, et al. Diabetes **58**,1690-1699(2009) .
25. Lee YH, Kang ES, Kim SH, et al. J Hum Genet **53**,991-998(2008).
26. Liu Y, Yu L, Zhang D, et al. Diabetologia **51**,2134-2137(2008).
27. Fengjiang Wei, Chunyou Cai, Shuzhi Feng,et al.Sci Rep **5**(2015).
28. Haihua Bai, Haiping Liu, Suyalatu Suyalatu,et al.J Diabetes Res **6**(2015).
29. Yun Qian, Feng Lu, Meihua Dong, et al.PLoS One **10**(2015).
30. Yi-Cheng Chang, Pi-Hua Liu, Yu-Hsiang Yu, et al.PLoS One **9**(2014).
31. Neelam Hassanali, N. Maneka G. De Silva, Neil Robertson, et al.PLoS One **9**(2014).
32. Claudia H. T. Tam, Janice S. K. Ho, Ying Wang,et al.PLoS One **8**(2013).
33. Raquel Villegas, Ryan Delahanty, Yu-Tang Gao,et al.PLoS One **7**(2012).

BIO Web of Conferences **8**, 02004 (2017) DOI: 10.1051/bioconf/20170802004
ICMSB2016

Research on community elderly quality of life and its influencing factors in Changchun City of Jilin Province

Fengge Dong, Xuejiao Zheng, Wenjing Sun, Xiufeng Zhao, Bing Yu and Yujin Liu[a]

College of Humanities & Sciences of Northeast Normal University, Changchun City 130117 Jilin Province, China

Abstract. Objective: To understand the status and influencing factors of the quality of life of the elderly patients in the community of Changchun City, and provide the basis for drawing up the interventions about improving the quality of life of the elderly patients in the community. Methods: 498 elderly patients were investigated by using random sampling method and QOL - BREF and self-designed questionnaire. Results: The scores of the quality of life were: PSYCH (13.00±1.91), ENVIR (13.39±2.23), PHYS (13.47±1.71) and SOCIL (14.21±2.56). The scores of quality of life self-evaluation and health condition self-evaluation were respectively (72.01±17.14) and (66.30±14.23). The results showed that gender, age, degree of education, marriage situation, vocation, income level can affect four aspects of the quality of life. The main factors influenced the scores of QOL were: leisure opportunities, satisfaction with health care services, satisfaction with sleep, interpersonal relationship, self-appreciation in appearance and satisfaction with the degree of the ability of daily life. Conclusion: The effective measures should be taken to improve the quality of the elderly patients, such as improving the healthy care function, establishing great supporting system in community.

1 Introduction

The elderly population of China is first in the world[1]. The fifth population census showed that Jilin Province entered the aging society since 2003, there were about 3.095 million old people who were more than 65 years old in Jilin Province, accounting for 11.24% of the total population by 2015[2]. As the aging process of Jilin Province deepening, the aging problem widely attracts the focus of the people, especially the old people's quality of life. Because of affecting by economy and medical level, the elderly of Jilin Province have their own special life form, quality of life is not optimistic. Taking Changchun City as an example, to investigate the quality of life and influence factors of old people in community, to provide evidence for making the intervention measures of the quality of life.

2 The respondents and methods

2.1 The respondents

[a] Corresponding author: 543055940 @qq.com

By using simple random sampling method to extract six communities in Changchun City from January to May in 2015. With the help of community staff and the cooperation of the old people who are more than 65 years , we finished 500 questionnaires, of which 498 is efficient, the effective rate is 99% .

2.2 Research method and content

Under the status that the respondents knew all the details, with the guidance of the investigators, the old people filled questionnaires honestly. The survey content includes General Situation Questionnaire and the QOL-BREF. The General Situation Questionnaire includes the basic information such as gender, age, marriage, income, education, vocation and so on. And QOL-BREF[3] includes 26 items, which covers PHYS, PSYCH, SOCIL and ENVIR and another two independent analysis items. According to the positive score (the higher the score, the higher the quality of life), the average score of subordinate entries plus 4 times is the scores of the related field.

2.3 Statistical methods

Established a database with SPSS16.0 software, making data input, t test, variance analysis and multiple linear regression statistical analysis.

3 Results

3.1 The result of QOL-BREF

PSYCH (13.00±1.91) scored lowest, ENVIR (13.39±2.23), PHYS (13.47±1.71), SOCIL (14.21±2.56) scored highest. The score≥60 of self-evaluation score of quality of life and self-evaluation score of health condition that accounted for 77.25% and 72.62%, and the scores were respectively (72.01±17.14)and(66.30±14.23).

3.2 The result of single factor variance analysis

Gender, age, education, marriage, vocation, income level all can affect the quality of life. The male scored more than female in the field of physiological($p < 0.05$). Age was negatively related to the scores of all areas in the quality of life($p < 0.05$). The old people of high culture scored also higher in PSYCH and PHYS($p<0.05$). The elderly who were in the state of marriage scores higher in PSYCH and PHYS($p<0.05$). Workers scored low in all area of quality of life except SOCIL, but the medical staff scored higher than other people in ENVIR, PSYCH and PHYS($p<0.05$).The higher the income, the higher the score in ENVIR and PSYCH($p<0.05$). As shown in Table 1 and Table 2.

Table 1. Single factor variance analysis result ($\bar{x} \pm s$)

Items	Category	Quantity	ENVIR	SOCIL	PSYCH	PHYS
Gender	Male	203	13.53±2.16	14.39±2.51	13.12±2.68	13.72±1.51*
	Female	295	13.29±2.26	14.09±2.59	13.00±1.97	13.29±1.81*
Age	65~75	290	13.63±2.28*	14.61±2.50*	13.51±2.54*	13.78±1.65*
	76~85	190	13.25±2.51*	14.07±2.67*	12.38±1.63*	13.05±1.66*
	≥85	18	13.03±2.06*	13.61±2.53*	12.66±2.16*	12.73±2.02*

Education Level	Primary school and below	173	13.37±2.49	13.97±2.87	12.94±1.98	13.22±1.71*
	Middle school	275	13.36±2.12	14.31±2.41	13.06±2.50	13.53±1.66*
	College degree or above	50	13.59±1.71	14.50±2.16	13.37±2.04	13.95±1.88*

Note: *means p<0.005

Table 2. Single factor variance analysis result $(\bar{x} \pm s)$

Items	Category	Quantity	ENVIR	SOCIL	PSYCH	PHYS
Marital status	Unmarried	1	13.50±0.00	10.67±0.00	12.00±0.00	13.54±0.00*
	Married	437	13.40±2.20	14.28±2.56	13.08±2.32*	14.29±0.00*
	Divorced	1	12.74±0.00	13.71±0.00	12.41±0.00	12.86±0.00*
	Widow or Widower	59	14.06±2.17	13.81±2.33	12.38±2.40	12.93±2.72*
vocation	Teacher	40	13.35±1.71*	13.66±2.53	13.75±4.80*	13.40±1.50*
	Worker	303	13.05±2.21*	14.05±2.70	12.65±1.77*	13.27±1.72*
	Self-employed	86	14.00±2.55*	14.65±2.33	13.00±1.68*	13.58±1.51*
	Medical staff	20	14.52±1.57*	14.46±2.25	14.33±2.30*	14.74±1.81*
Income (Yuan)	≦1500	70	12.50±2.03*	13.91±2.83	12.64±1.96	13.26±1.75*
	1501~2500	232	13.22±2.42*	14.43±2.21	13.28±1.95	13.33±1.74*
	2501~3500	169	13.89±1.89*	14.48±2.26	13.40±2.70	13.72±1.52*
	≥3501	27	14.00±1.96*	14.51±2.59	13.72±2.31	13.99±2.13*

Note: *means p<0.005

DOI: 10.1051/bioconf/20170802004

3.3 The result of multiple linear regression analysis for quality of life

Table 3. The result of multiple linear regression analysis for quality of life

Factors		Regression coefficient	Standard error	t	p
ENVIR	Leisure entertainment opportunities	1.053	0.060	17.510	0.000
	Satisfaction with health care service	0.821	0.069	11.977	0.000
	Satisfaction with sleep	0.330	0.074	4.457	0.000
	Satisfaction with interpersonal relationship	0.314	0.078	4.005	0.000
SOCIL	Satisfaction with interpersonal relationship	1.646	0.076	21.589	0.000
	Self appreciation in appearance	0.410	0.085	4.800	0.000
	Satisfaction with sleep	0.263	0.090	2.928	0.004
PSYCH	Self appreciation in appearance	0.772	0.090	8.619	0.000
	Satisfaction with daily life ability	0.603	0.100	6.044	0.000
	Satisfaction with health care service	0.185	0.077	2.391	0.018
PHYS	Satisfaction with daily life ability	1.174	0.086	13.689	0.000
	Satisfaction with sleep	0.639	0.083	7.745	0.000
	Leisure entertainment opportunities	0.335	0.054	6.143	0.000

The factors that influenced ENVIR in the descending order were: leisure opportunity, satisfaction with health care services, satisfaction with sleep, satisfaction with interpersonal relationship. The factors that influenced SOCIL in the descending order were: satisfaction with interpersonal relationship, self-appreciation in appearance, satisfaction with sleep. The factors that influenced PSYCH in the descending order were: self-appreciation in appearance, satisfaction with daily life ability, satisfaction with health care services. The factors that influenced PHYS in the descending order were: satisfaction with daily life ability, satisfaction with sleep, leisure opportunity. As shown in Table 3.

4 Discussions

The survey results show that the quality of life of old people in Changchun City is in medium level. The scores of self-evaluation of Quality of life and health condition are both more than the four fields(PHYS,PSYCH,SOCIL,ENVIR) in quality of life, and this is basically identical with Hong Xu's research results[4].

4.1 The result of multiple linear regression analysis for quality of life

4.1.1 Gender, age and marriage

The male scored higher than female in the field of physiological ($p < 0.05$), due to differences in physiological structure between male and female, women are aging faster than men. After retirement women take more family responsibility than men, so compared with men, women are easy to produce physical problem. Due to immunity of the organism and athletic ability declined, the older the lower that the scores of quality of life. The elders who were in marriage score higher, it is because that the couple could take care of and company with each other.

4.1.2 Education, vocation and income level

The higher level of education, the more opportunity to obtain health information that the old people have, the stronger self-health care consciousness, so it could improve the elderly living conditions, optimize the physical and mental state, improve the scores of mental and physical fields. This is basically identical with Hanli Zhan's research result[5]. In different vocations, the more opportunity that contact with pharmaceutical and health care knowledge, the more score were obtained in ENVIR PSYCH and PHYS. The higher the income level, the more score were obtained in ENVIR and PHYS, It is basically identical with Yuli Wang's research result[6].

4.1.3 Leisure entertainment opportunities

Because it is not convenient for the elderly do physical activity and their visual, acoustic, and expression ability gradually decline, most old people will choose to stay in the room to kill time, this will definitely reduce the old people's leisure entertainment opportunities, at the same time they give up the opportunity to alleviate physical fatigue, restore physiological balance, expand the scope of social activities and get comfort in mind, this is the reason that leisure entertainment opportunities was positive correlation to ENVIR, PSYCH and PHYS.

4.1.4 Satisfaction with sleep

Sleep is the main indicator that measures old people' physiological function, it almost takes up one third of the whole life, so it is closely related to people's health, sleep determines the quality of life in a sense. This study shows that the elderly who were satisfaction with their sleep, their score were higher in PHYS. Poor quality of sleep will react on the PHYS, PSYCH, SOCIL and ENVIR. The research result is basically identical with scholars both in China and abroad[7,8].

4.1.5 Satisfaction with the ability of daily life

The ability of daily life is an important indicator that evaluates quality of life of the old people, it is closely related to physical and psychological health, the elderly who are not satisfied with their daily life ability will have a lower level of quality of life. Once the old people lose self-care ability, it will not only restrict their freedom, lower their living standards, but also bring a heavy load to family and society. It is basically identical with Jinli Yang's [9] and Xueyuan Liu's[10] research results.

4.1.6 Satisfaction with interpersonal relationship

Interpersonal relationship is the basic social needing. This study showed that interpersonal relationship was not only affect SOCIL, but also had a certain effect on the ENVIR. This shows that a good interpersonal relationship can help the old people obtain a good quality of life, the old people's main relationships is family relationships. Relationships between mother-in-law and daughter-in-law, parents and children, husband and wife is good or not also have effect on the quality of life.

4.1.7 Satisfaction with health care service

Preventing and treating disease is the main function of health care service system. As the growth of the age and the spread of the epidemic disease, the health care services that the elderly demand for is also growing, health care service has become a part of the life of the elderly, convenient health care service system will help to improve the life quality of old people.

4.1.8 Self-appreciation in appearance

As the growth of age, the old people have great changes in appearance, they lose their youth and beauty, became doddering and ill, walk unsteadily, some old people are afraid to look into the mirror and unwilling to accept aging. Survey results show that self-appreciation in appearance directly affect the score in PSYCH, it also have certain influence in SOCIL. Because those who appreciate themselves are more likely to please themselves, have more confidence, they're more likely to have better chance and social relationships.

4.2 Suggestions to improve the quality of elderly life

Improve the community medical and health care function, establish system of home visit, physical examination and personal health education. The scientific and effective intervention of physical and mental health for the elderly can improve the elderly's self-care ability, remove obstacles of obtaining health knowledge, and then improve the elderly's quality of life. It was reported that we can improve the ability in health knowledge, beliefs, behaviours and skills to improve the elderly health level[11].

To establish perfect supporting system in community, offer more care social support for the elderly, take the elderly who are unable to move freely especially those who lost companion to the crowd, to participate in the activities in community and have an enjoyable life etc.

The society should improve the insurance coverage of the elderly health, establish the elderly care insurance system, make it easy for the elderly to go to see a doctor, improve the elderly health care system, make sure that the elderly can obtain adequate treatment and care in order to improve the quality of survival.

Acknowledgements

This work was supported by Soft Project of Health and Family Planning Commission of Jilin Province under Project 2014R029.

References

1. Yushao Wu. The development of Chinese aging report(2013)[M].Jing Bei: Social Sciences Academic Press(China), (2013):2-3.
2. The national economic and social development statistical bulletin of Jilin province in 2015. http://tjj.jl.gov.cn/tjgb/ndgb/201603/t20160321_2179875.html.
3. Yuantao Hao, Jiqian Fang. The introduction and instructions to the world health organization quality of life measurement scale for Chinese version[J]. *Modern Rehabilitation*,(2008),**08**:1127-1129-1145.
4. Hong Xu, Jing Xiao, Xun Zhuang, Shaoying Lan, Xiguo Xu, Yuexia Gao. Quality of life and its influencing factors of elders in Nantong[J]. *Chinese Journal of Gerontology*,(2012),**07**:1450-1452.
5. Hanli Zhan, Qingxia Zeng, Xiaoying Bai, Peiming Liu. The community elderly quality of life and the investigation and analysis of community care needs[J]. *Journal of Traditional Chinese Medicine Management*,(2016),**12**:11-12.
6. Yuli Wang. The elderly life quality and health quotient study in Changchun[D]. Changchun University of Chinese Medicine,(2012).
7. Yue Dai, Baoquan Zhang, Yinglan Li, Yi Shu, Caixiao Jiang, Hong Sun. Meta analysis of correlation of Chinese old people sleep quality and depression, anxiety[J]. Chinese Journal of Nursing, (2016),**04**:488-493.
8. Huang HC, Chou CT, Lin KC, et al. The relationships between disability level, health-promoting lifestyle, and quality of life in outpatients with sys-temic lupus erythematosus[J]. J *Nurs Res*(2007),**15**:21-32.
9. Jinli Yang, Changxiang Chen, Shuxing Li, Miaomiao Tian. The correlation of daily life ability and the elderly mental health in community[J]. *Hebei Medical Journal*,(2014),**14**:2185-2186.
10. Xueyuan Liu, Lei Zhao, Yueyun Xie, etc. The influence between elderly daily life ability and mental health. *China Journal of Health Psychology*,(2009),**17**:201-202.
11. Xiaoyun Hu, Yukai Du. Health literacy among residents and its impact on health in Hubei province[D]. Collected papers of graduate student in Huazhong University of Science and Technology, (2009):32-35.

Chaos based blood glucose noninvasive measurement: new concept and custom study

Li Cui[1], Yuhang Dai[2], Wenshi Li[2,a] and Yejia Feng[2]

[1]*China Tower, Suzhou Branch Company, 215011 Suzhou, China*
[2]*Soochow University, Microelectronics Department, 215006 Suzhou, China*

Abstract. Background. Non invasive monitoring of Blood Glucose (BG) has been a challenge calling for new accurate and fast measurement methods. Objective. To propose new concept of chaos based BG non invasive test aiming at personal customization requirements. Methods. First to build the compact RC model of tissue BG through impedance precision measuring Kit, then to simulate and soft-test BG by Boolean chaotic Codec circuits in soft tool Multisim 13.0, The third to capture the chaotic decoding outputs with the Kit plus PC in calculated signatures of resistor and phase of the tested impedance at the subjects' left wrist in synchronous test by Bayer BG meter. Results. All in controlled trials of Bayer BG meter, the chaotic BG modelling had gained three new compared formulae in merits of errors less than 1mmol/L and latency less than 1minute. Conclusion. During further verification of this chaotic test paradigm, the opened logic route of above methods will boost measurement experts' confidence in overcoming future problems of blood glucose monitoring in vivo.

1 Introduction

Four kinds of methods on Blood Glucose non invasive measurement are in chemistry, electrical, optical, and acoustic principles [1,2]. Till now there have been puzzles on non invasive measurement paradigm as multi variables effects. However the BG test accuracy and latency problems are two more prominent issues. So entering personalized customization era, how to understand easily creative application of classic biological impedance will undoubtedly open new window of old obstacles.

Chaos based Blood Glucose non invasive measurement built by the corresponding author was inspired by three events: (1) With white noise excitation to measure and compute BG in 2006 contributed by Master student Xiaobing Hou at Beijing University of Chinese Medicine; (2) Modulated +5V DC-DC supply power in chaotic waves in auricular NIRS BCI in 2012 contributed by Brainchildren Lab of our research group; (3) Water conductivity test. Human body can be modelled into a warm water bag as embedded into a single branch voltage divider circuit under test [3,4]!

Next we report the personalized customization journey of chaotic BG impedance test modelling.

Section 2 starts with digging out the RC model of BG tissue impedance (innovation point 1) [5,6].

[a] Corresponding authors: lwshi@suda.edu.cn

This work was supported by the Natural Science Foundation of Jiangsu Province, China in 2014 (No. BK20141196). This work was supported by the 2015 Annual General College Graduate Research and Innovation Programs (No. KYLX15_1244) in Jiangsu Province, China.

Section 3 focuses on RC-based chaotic Codec BG sensing (innovation point 2) [7,8].
Section 4 extracts three custom formulae in contrast test of Bayer BG meter (innovation point 3).
Section 5 throws up the conclusions caring the illustration of test accuracy and latency indexes.

2 Wrist blood glucose modelling in RC net

The idea searches for BG model in RC net. Figure 1 and 2 show needed instruments. Wherein the chip AD5933 is high speed biological impedance acquisition system in key parameters of 12bit, 1MSPS.

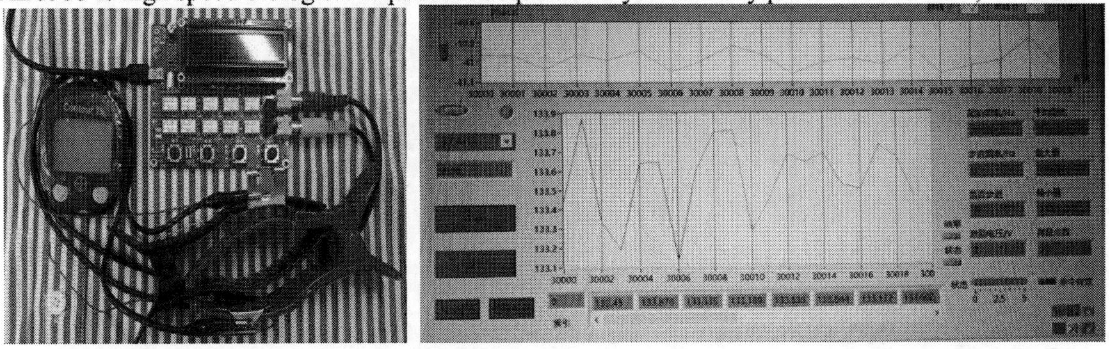

Figure 1. Bayer BG meter and Impedance precision measuring Kit (left) and PC terminal Interface (right).

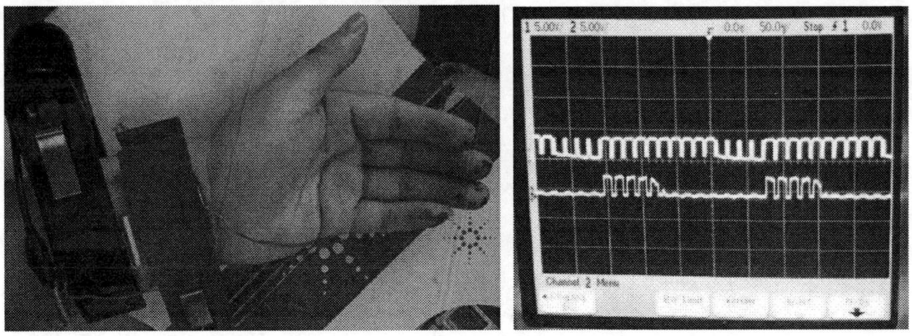

Figure 2. IEC standard ECG clamp and test site (left) and Agilent 54264 A Oscilloscope display of chaos (right).

Table 1. Compact RC model of tissue BG

Tested BG / mmol/L	R/Ω // C/nF
6.6	102//11
7.3	104//15
7.7	104//33
8.8	104//5
7.3	106//328
7.6	106//510

6.8	105//75
6.7	109//78

The accuracy of Bayer BG meter is less than 0.6 mmol/L with test time 8 of seconds.

Subjective: Male, 52 years old in healthy state @ room temperature of 28 degrees Celsius, Humidity of 60%.

Step-1: BG test value sampling invasively after Alcohol disinfection and filling Tab. 1.

Step-2: During 5 minutes interval, to uniformly capture 5- ~ 10-group data (20-trial per group set in one second). To anti-compute the RC model from the wrist impedances and then fill Tab. 1.

Step-3: To use the RC model in Table 1 as the "wrist in vivo" under test by Multisim 13.0 exploring new chaotic Codec BG test method.

3 RC-based chaotic Codec BG sensing

Figure 3 introduces this section's idea flowchart in detail. Boolean chaotic circuit is with coding core of XOR gate (U2A) and with decoding core of XOR gate (U5A), the known devided voltage branch includes the RC model (R1//C1) and sampling resistor R2 and the chaotic decoding output load R3. The key waveforms (right) cares ports of IO1 ~ IO4. Based on Table 1 we run 8 times for constructing Figure 4 (left) with the right mapping under the standard of Point 5 (7.3mmol/L).

Summary: The new formula y = *f*(x) (Equation 0) is with error less than 1mmol/L.

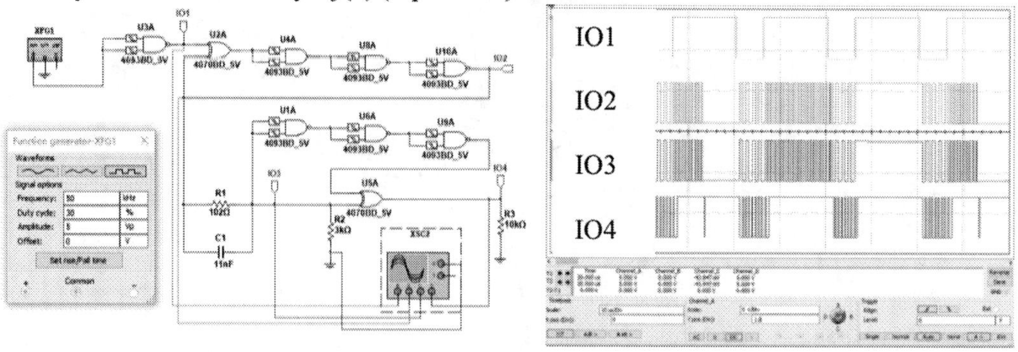

Figure 3. RC model based new chaotic coding and decoding BG soft-test (left) and key waveforms (right).

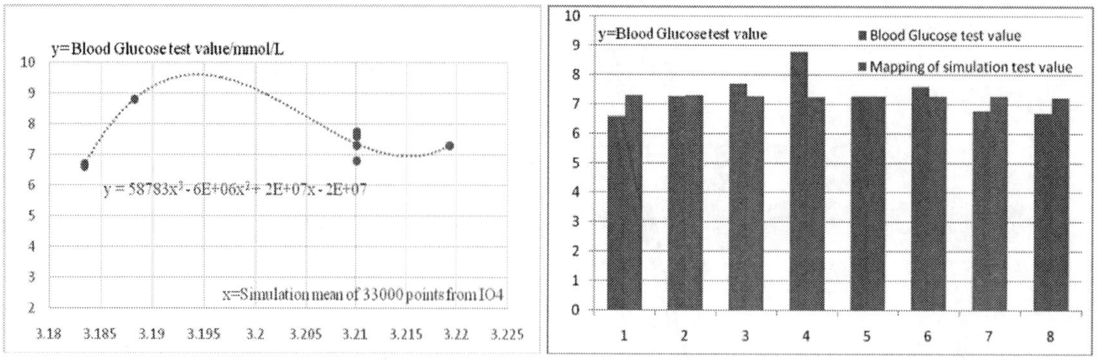

Figure 4. Soft-test of BG with simulation data (left) and their mapping "true" values (right).

4 Custom Chaos BG glucose non-invasive monitoring

In fact like in Figure 3 we set generator frequency of 50kHz, 30 percents of Duty in chaotic coding block with MCU of MSP430G2553. And the acquisition methods are also kept the same style in Section 2.

Figure 5 ~ Figure7 illustrate the mappings of wrist BG Resistors and Phases respectively with three kinds of extracting mean constrains: of 200R (P), of 30% largest ones and of 30% smallest ones.

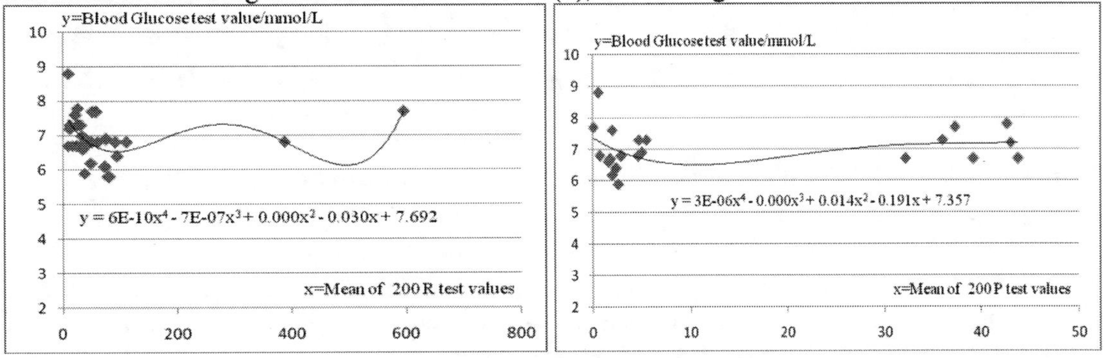

Figure 5. Chaotic impedance test based BG mapping (left R: mean) and (right P: mean).

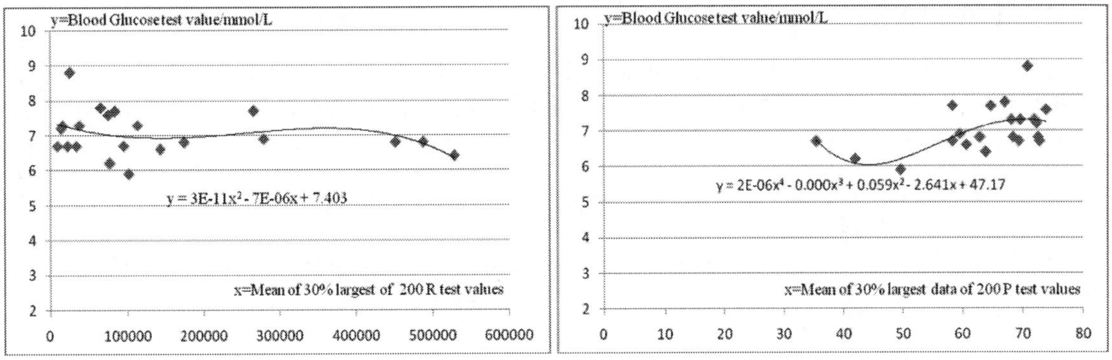

Figure 6. Chaotic impedance test based BG mapping (left R: mean of 30% Max) and (right P: mean of 30% Max).

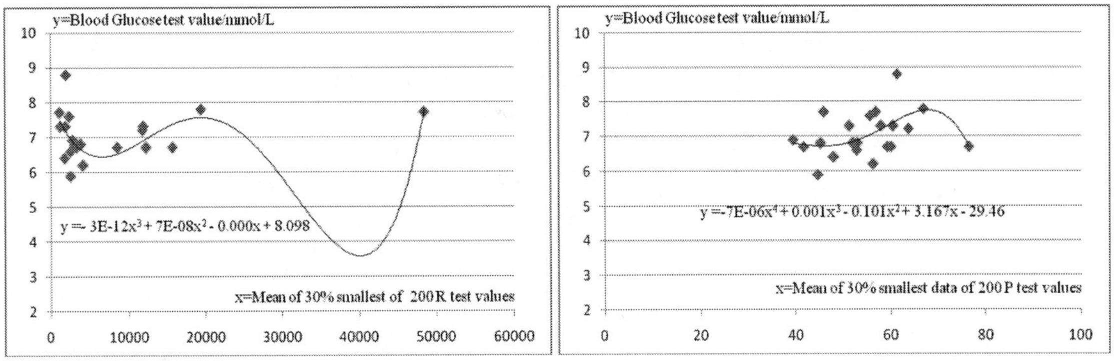

Figure 7. Chaotic impedance test based BG mapping (left R: mean of 30% Min) and (right P: mean of 30% Min).

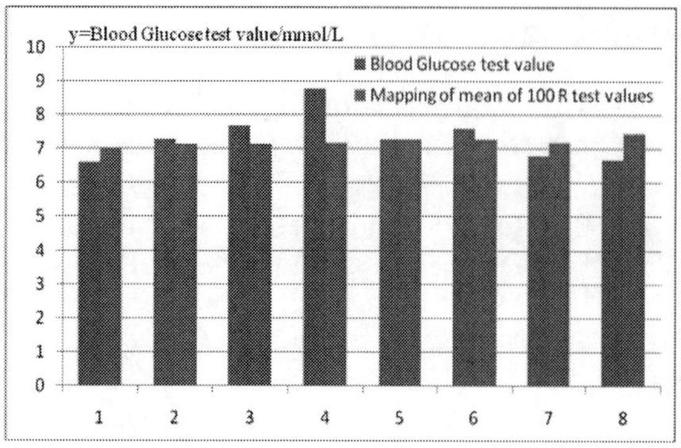

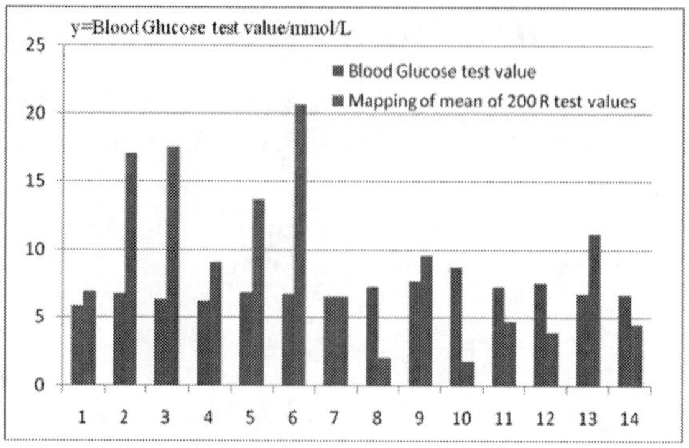

Figure 8. Impedance test based BG mapping (left: Sine wave excitation) and (right: Chaotic Codec)

To analyze Figure 5 ~ Figure 7 we must point out that all the derailed dots are for reason of 30kHz stimulus in Kit in Fig.1 (All data is met before and after the continues four lunches).

To compare Figure 8 @ left and right we believe that classic inputs will result in basic accuracy of BG test mapping while in about 5 minutes delay times, fortunately the right results in Figure 8 with all 50kHz in Kit and in Codec circuits met minute-level latency.

5 Conclusions

To help build new RC model of BG non invasive measurement our co-works had shown the personalized customization method from test to simulation to test all in contract of Bayer BG meter monitoring for Subject One.

The grasped optimized experiments' condition includes 30% Duty 40 kHz ~ 50kHz excitation, and the digital Boolean Codec tech with 1 ~ 3 stages delay chain @ room temperature of 28 degrees Celsius, Humidity of 60%.

Further works will inhibit the effect from the temperature variable in order to stabilize this accuracy of about 0.5mmol/L in chaotic measurement for Blood Glucose monitoring in vivo.

References

1. J. Yadava, A. Rania, V. Singha, Biomedical Signal Processing and Control, **18** (2015).
2. M. Goodarzi, S. Sharma, H. Ramon, Trends in Analytical Chemistry, **67** (2015).
3. Li Wenshi, Chinese Journal of Electronics, **25** (2016).
4. R. Brown, L. O. Chua, B. Popp, Int. J. of Bifurcation and Chaos, **2** (1992).
5. A. Caduff, E. Hirt, Yu Feldman, Biosensors and Bioelectronics, **19** (2003).
6. K. Song, U. Ha, S. Park, J. Bae, Hoi-Jun Yoo, IEEE J. of Solid-State Circuits, **50** (2015).
7. B. Muthuswamy, L. O. Chua, Int. J. of Bifurcation and Chaos, **20** (2010).
8. Rui Zhang. Physical Review E, **80** (2009).

DOI: 10.1051/bioconf/20170802006

The pairwise phase consistency in cortical network and its relationship with neuronal activation

Daming Wang and Yaoru Sun[a]

Department of Computer Science and Technology, Tongji University, Shanghai, China

Abstract. Gamma-band neuronal oscillation and synchronization with the range of 30-90 Hz are ubiquitous phenomenon across numerous brain areas and various species, and correlated with plenty of cognitive functions. The phase of the oscillation, as one aspect of CTC (Communication through Coherence) hypothesis, underlies various functions for feature coding, memory processing and behaviour performing. The PPC (Pairwise Phase Consistency), an improved coherence measure, statistically quantifies the strength of phase synchronization. In order to evaluate the PPC and its relationships with input stimulus, neuronal activation and firing rate, a simplified spiking neuronal network is constructed to simulate orientation columns in primary visual cortex. If the input orientation stimulus is preferred for a certain orientation column, neurons within this corresponding column will obtain higher firing rate and stronger neuronal activation, which consequently engender higher PPC values, with higher PPC corresponding to higher firing rate. In addition, we investigate the PPC in time resolved analysis with a sliding window.

1 Introduction

Neuronal gamma-band oscillation and synchronization (30–100 Hz) have been observed in several cortical and subcortical areas [1-4] and numerous species [1, 5, 6], and suggested to be associated with a variety of cognitive functions, including signals routing [7-9], feature integration [10, 11], selective attention [12, 13], memory [14, 15] and so on. The properties of gamma band oscillation and its role for cognitive functions are extensively investigated [16-18]. A hypothesis, referred to as CTC (Communication through Coherence), often manifests itself as a plausible mechanism for neuronal oscillation [8, 19].

The CTC hypothesis concerns two aspects for two signals, if their amplitudes correlate (power correlation) or if their peaks and valleys align (phase coherence) [20]. With regard to amplitude aspect, increasing the amplitude of the neuronal activity can implement attention selection to a certain sensory stimulus [21]. Besides, switching one of several convergent pathways from asynchronous to oscillatory state can accurately establish signal routing, because the spatial pattern of amplitude of firing rate can reproduce the spatial pattern of firing rates of neuronal population [22]. As for phase aspect, the oscillatory phase, describing the temporal relationship relative to the background

[a] Corresponding author: yaoru@tongji.edu.cn
This work was supported by the National Natural Science Foundation of China (61173116).

oscillation, is a vital oscillatory parameter and proposed to underlie information processing and transferring in brain cortex [23].

The phase of oscillation is indicated to support multiple cognitive functions, for instance feature coding [24, 25]. The phenomenon of theta-phase precession occurs in the hippocampus and the spatial information can be encoded by theta phase. when the animal moves across the place field of a particular neuron, the phase gradually advances towards the peak of the theta cycle [26]. The gamma cycle hypothesis proposes that the gamma cycle is determined by synchronized inhibitory interneurons, and the excitatory input to a pyramidal neuron can be transformed into a temporal code [27]. The physiological experiment in visual cortex of awaken monkey suggests that the phase can be coded systematically as a function of neuronal activation strength in the gamma cycle [28]. Hippocampal theta has also been proposed to implement information encoding and retrieving in different phases of the theta cycle [4, 15, 23]. The oscillatory phase for other cognitive functions is elucidated in information integration [10, 29], phase-dependent coding of memorized objects [30] and letter recognition [31].

Several computational simulations and theoretical analyses also confirm the general functions of oscillatory phase. The phase of oscillation carries information about stimulus identity and can be shifted with stimulus features, with spike phase coding for stimulus strength [32]. The interaction strength between two neuronal groups can also be established by their relative phase [5, 16, 33].

Two aspects correlated with oscillatory phase are taken into account: the phase of oscillation, representing input stimulus or neuronal activation, and the phase synchronization, serving as an instance of CTC approach for signal routing, information communication and selective attention. The spike-LFP phase, a kind of phase representation, relates spike activity to its background LFP (Local field potential) oscillation [34].

With respect to phase synchronization, PLV (Phase Locking Value) [35], general coherence [5] and PPC (Pairwise Phase Consistency) [34, 36] are three major measures, each of which can quantify phase consistency or the strength of rhythmic synchronization. However, both PLV and general coherence approaches are strongly influenced by the variable and uncontrollable sample size, such as different spike and trial numbers. In order to overcome this bias influence, the PPC measure is introduced, which executes the computation of vector dot product instead of the operation of vector addition across pairs of oscillatory phases [36]. Unfortunately, there still exists further obstacles for PPC approach and a revised version of point-field PPC $\hat{P}_1$, with respect to spike-LFP phase, is proposed. The point-field PPC $\hat{P}_1$ measure only implements dot product computations across different trials in order to eliminate the phase dependence within the same trial, which is induced by non-Poisson history effects, such as refractoriness and bursting [34].

In this experiment, we construct a spiking neuronal network to simulate activities of orientation columns in primary visual cortex, which is inspired by the physiological experiment in visual cortex of awaken monkeys [28]. The properties of PPC and its relationships with input stimulus, orientation preference, firing rate and neuronal activation are extensively investigated and thoroughly demonstrated. The PPC, statistically quantifying the consistency of oscillatory phases, shows significant values all within the frequency of gamma-band. When the input orientation is similar to or as same as the preferred orientation of a certain column, higher firing rate and stronger neuronal activation will be brought about for this corresponding column, which consequently leads to higher PPC values.

2 Methods

2.1 Neuronal model

A leaky integrate-and-fire neuronal model is adopted in this paper, which evolves according to the following equation [37]:

$$C_m \frac{dV}{dt} = -g_L(V - V_{rest}) + I_{AMPA} + I_{GABA} + I_{bg} \quad (1)$$

, where C_m is a membrane capacitance, g_L a membrane leak conductance, and I_{bg} a constant tonic background current. When the membrane potential V surpasses the threshold potential of V_{thres}, an action potential is engendered, then the membrane potential V is reset to the resting potential V_{rest}, and it remains clamped for a refractory period of T_{ref}. The excitatory (AMPA) and inhibitory (GABA) synaptic currents I_{AMPA} and I_{GABA} obey (2) and (3)

$$I_{AMPA} = g_{AMPA}(V_E - V) \quad (2)$$

$$I_{GABA} = g_{GABA}(V_I - V) \quad (3)$$

, where V_E and V_I are excitatory and inhibitory reversal potential respectively. The dynamics of the excitatory and inhibitory conductance g_{AMPA} and g_{GABA} comply with the equations as below:

$$\frac{dg_{AMPA}}{dt} = -\frac{g_{AMPA}}{\tau_{AMPA}} \quad (4)$$

$$\frac{dg_{GABA}}{dt} = -\frac{g_{GABA}}{\tau_{GABA}} \quad (5)$$

, where τ_{AMPA} and τ_{GABA} are excitatory and inhibitory synaptic decay time, and the default values of all parameters can be referred to the Table 1 in the following section.

2.2 Network Architecture

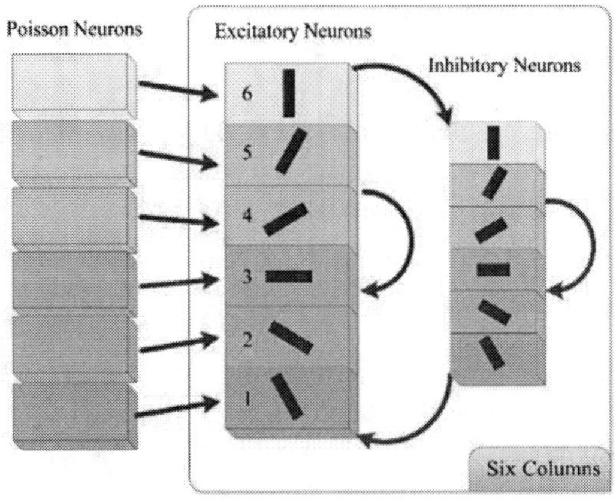

Figure 1. Network architecture

The architecture of neuronal network is elucidated in Figure 1, the left part of which is six groups of Poisson neurons, providing external spike inputs, and the right is six columns, simulating orientation columns in the primary visual cortex. There are 1200 Poisson neurons, 200 for each Poisson group, and totally 1200 excitatory neurons and 300 inhibitory neurons in the six orientation columns. Each of the six orientation columns, decorated with same color, is comprised of 200 excitatory and 50 inhibitory neurons. The ratio of excitatory neurons is 80%, which is in accordance with the experimental evidence [38]. The red and blue arrows represent excitatory and inhibitory synaptic connections respectively, with the parameter $\varepsilon = 0.2$ as their connection probability.

Owing to orientation selectivity and tuning curve of firing rate with regard to stimulus orientation in visual cortex [39], both excitatory and inhibitory neurons in each column are specified with a preferred orientation, with $-\frac{\pi}{3} + \frac{\pi}{6}i$ for the ith column. Therefore the vector of preferred orientation for six columns is $\left\{-\frac{\pi}{3}, -\frac{\pi}{6}, 0, \frac{\pi}{6}, \frac{\pi}{3}, \frac{\pi}{2}\right\}$. Besides, the same orientation preference is also assigned to the neurons of Poisson groups as their corresponding orientation columns, and Poisson rate is determined according to (6):

$$R_{input} = \left[\cos\left(2(\theta_{pref} - \theta_{stim})\right) + 1\right] F_{max} \tag{6}$$

, where θ_{pref} and θ_{stim} are the preferred orientation of Poisson neuron and the stimulus orientation respectively, F_{max} the maximal Poisson firing rate.

There are forward connections from Poisson groups to their corresponding columns, the weights (W_f) of which are all the same and constant. Whereas the connection weights within six orientation columns depend on the difference of preferred orientation between presynaptic and postsynaptic neurons.

$$W_{ij} = \text{Wexp}\left(\beta\left[\cos\left(2(\theta_{pre} - \theta_{post})\right) - 1\right]\right) \tag{7}$$

, where θ_{pre} and θ_{post} denote the preferred orientations for presynaptic and postsynaptic neurons respectively.

The LFP data is simulated as the sum of the absolute values of constant background current, excitatory and inhibitory synaptic currents from all excitatory neurons within a nearby neuronal group [40].

$$\text{LFP} = \text{R}\left(\sum_{i=1}^{n}\left(|I_{AMPA}| + |I_{GABA}| + |I_{bg}|\right)\right) \tag{8}$$

, where the parameter R represents the impedance of electrodes [41].

2.3 Experiment simulating

In the middle of each orientation column, we continuously select 10 excitatory neurons and record their spike activities (action potentials) and synaptic currents, which bring about one LFP channel data. In consequence, there are totally 6 LFP and 60 neuron channels to simulate 6 recording sites by means of intracranial electrodes.

We implement 15 trials altogether for this experiment. For each trial, there is a pre-stimulus baseline period of 500 ms, during which only a tonic background current I_{bg} and low Poisson spike inputs with a rate of F_{bg} are supplied, which simulates the spontaneous activity in visual cortex. Then a constant orientation of 30 degrees is presented for a duration of 1500 ms, in order to mimic the drifting gratings. The input orientation is as same as the preferred orientation of the fourth column, which renders neurons within this column have higher firing rate, stronger neuronal activation and higher phase consistency.

For each spike from a certain neuron channel, six LFP segments except the one within the same column are cut out, all with the spike time as their segments center. Through Discrete Fourier Transform with a Hanning window, all LFP segments are decomposed into spike-triggered LFP spectrum by (9)

$$\bar{X}_i(f) = \frac{1}{5}\sum_{j=1}^{5}\frac{X_i^j(f)}{|X_i^j(f)|} \tag{9}$$

, where $X_i^j(f)$ denotes the spike-triggered LFP spectrum from the jth LFP channel and ith spike of a certain neuron. Then the spike-LFP phase can be easily obtained through (10) [28].

$$\Theta_i = \text{angle}\left(\bar{X}_i(f)\right) \tag{10}$$

, the angle function obtains the phase angle of the input spectrum. The term $\boldsymbol{U}_{k,m}$ is a vector representation of spike-LFP phase Θ_i. Then we can determine point-field PPC $\hat{P}_1$ with the help of the following equation [34]:

$$\hat{P}_1 = \frac{\sum_{m\in M}\sum_{l\in L}\sum_{k=1}^{N_m}\sum_{j=1}^{N_l}(U_{k,m}\cdot U_{j,l})}{\sum_{m\in M}\sum_{l\in L}(N_m N_l)} \tag{11}$$

, where M and L represent the trial number, N_m and N_l denote the spike number for trial m and trial l.

Finally, we demonstrate time resolved analyses with a sliding window, and the window length and time step are 75 ms and 10 ms respectively. All the neuronal model simulation and network construction are conducted on the basis of an open source Python package of Brian simulator, with a time step 0.1 ms. Then the following spectral analyses are implemented based on a MATLAB toolbox FieldTrip. In addition, the parameters encountered in previous sections are set according to the Table 1:

Table 1. Parameter Settings.

Parameter	Value	Parameter	Value	Parameter	Value	Parameter	Value
C_m	250 pF	g_L	10 nS	V_{rest}	-65 mV	V_{thres}	-45 mV
I_{bg}	300 pA	V_E	0 mV	V_I	-75 mV	τ_{AMPA}	5 ms
τ_{GABA}	10 ms	T_{ref}	5 ms	F_{max}	40 Hz	ε	0.2
β	2	W_f	0.05	R	1.0 MΩ	F_{bg}	3 Hz

3 Results

3.1 Neuronal activities

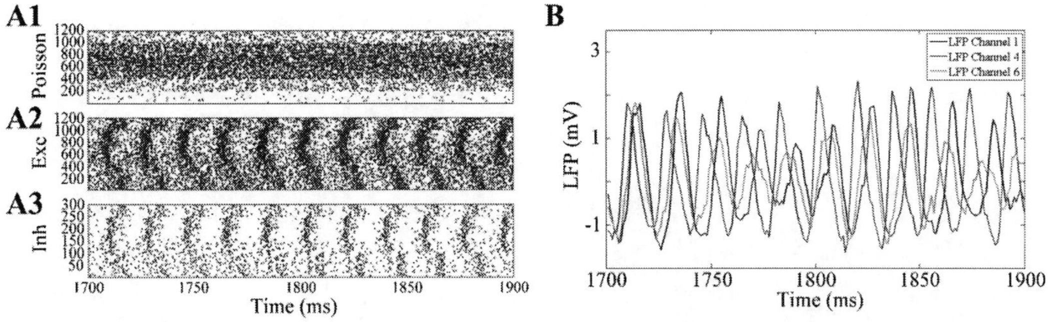

Figure 2. Spike raster and LFP oscillation.
(A) Spike raster of 200 ms period for (A1) Poisson neurons, (A2) excitatory neurons, (A3) inhibitory neurons. (B) Three time sequences of LFP oscillations for the first, fourth and sixth orientation columns.

As demonstrated in Figure 2A, there are three spike raster plots of 200 ms period, representing for Poisson, excitatory and inhibitory neurons respectively. Each interval between index numbers in the vertical coordinate corresponds to one orientation column. Owing to the Poisson rate vector of {3 Hz, 23 Hz, 63 Hz, 83 Hz, 63 Hz, 23 Hz} for six Poisson groups, determined by the differences between preferred orientation of Poisson groups and stimulus orientation, the spike raster of Poisson neurons from the fourth column is densest and gradually declines upwards and downwards to other columns

(Figure 2A1). According to experimental evidences [12], individual neurons fire irregularly, whereas population neurons fluctuate approximately at a certain frequency, giving rise to appropriate oscillation and synchronization. It has been observed that there are strong oscillations in both pyramidal and inhibitory neurons, both around the frequency of 55 Hz (Figures 2A2 and 2A3). If not mentioned otherwise, the subsequent spectral analyses, i.e. PPC computing, are all restricted to this frequency.

We record three LFP oscillation traces from the first, fourth and sixth orientation columns, the peaks of which are entirely different, with the channel four, six and one as a peak time sequence (Figure 2B). Because spikes are more likely to discharge in the trough of the LFP oscillation [41], neurons in the fourth column will fire early with greater likelihood, and then the sixth and first columns, which accounts for the generation of different oscillatory phases and distinct PPC values.

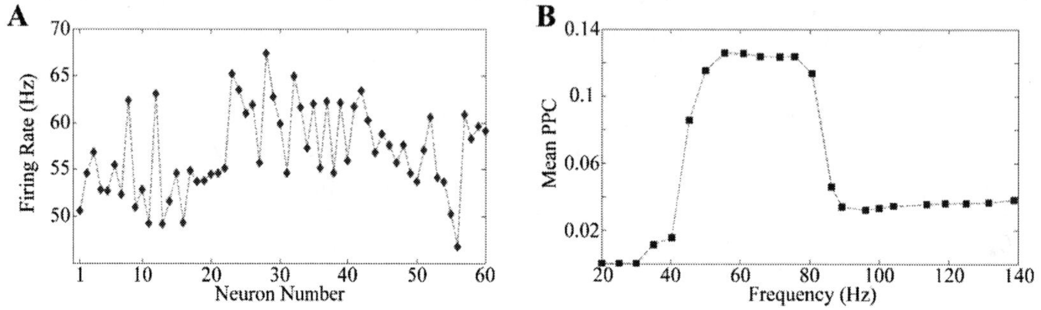

Figure 3. Neuronal firing rate and mean PPC.
(A) Neuronal firing rates of 60 neurons across 15 trials. (B) Mean PPC averaged across 60 neuron channels as a function of frequency

The neuronal firing rates of all 60 neurons across 15 trials are precisely computed and exhibited in Figure 3A. Though to a certain degree do the firing rates of all neurons fluctuate, there are higher firing rates for the third, fourth and fifth columns, and lower firing rates for other columns, because of different orientation preference and the tuning curves of Poisson rates.
Through the measure of point-field PPC $\hat{P}_1$, we calculate PPC for each neuron and then average them to obtain the mean PPC as a function of frequency (Figure 3B). The mean PPC value, with a peak at 55 Hz, demonstrates significant phase consistency all within the frequency of gamma range, which is consistent with the monkey physiological experiment [28].

3.2 Pairwise phase consistency

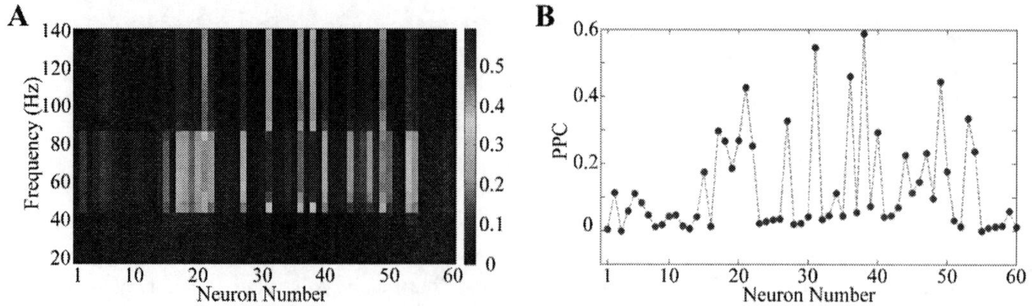

Figure 4. PPC of 60 neurons.
(A) PPC of 60 neurons for the frequency range of 20-140 Hz. (B) PPC of 60 neurons at 55 Hz

The PPC values of all 60 neurons, with the frequency range of 20-140 Hz, are thoroughly elucidated in Figure 4A, which are derived by means of point-field PPC $\hat{P}_1$ approach. The significant phase consistency values of all neurons are all within the frequency of gamma-band. Furthermore, the PPC values for the fourth orientation column are the most powerful and gradually decrease towards left and right to other columns. We select the PPC at the frequency of 55 Hz, the same frequency of LFP oscillation, and illustrate it as a function of neuron number (Figure 4B).

The input orientation of 30 degrees is the same as the preferred orientation of the fourth column, which gives rise to stronger Poisson inputs, higher firing rates and neuronal activation. Thus the spikes are more likely to be concentrated on a certain part of LFP oscillations and therefore lower phase dispersions are engendered. Consequently, the PPC measure, quantifying the consistency of oscillatory phases, obtains higher values in the fourth column, and declines gradually to its nearby columns.

3.3 Time resolved analysis

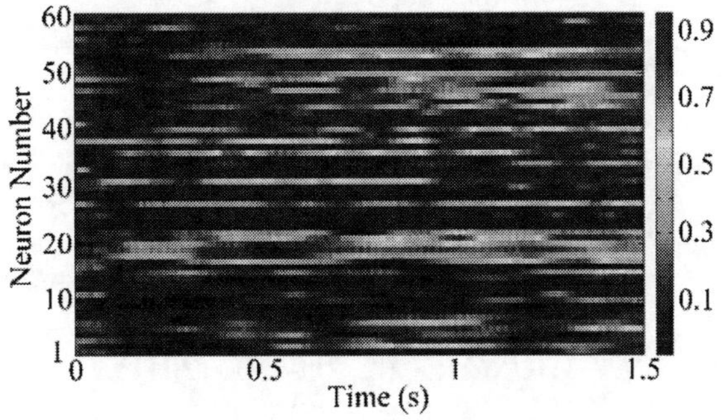

Figure 5. Time resolved analysis of PPC for 60 neurons.

With a sliding window, we analyze the PPC values of all 60 neurons in a time resolved way (Figure 5). The PPC manifests higher values for neurons from the fourth orientation column, and then decreases gradually upwards and downwards to other columns. Though neurons within the same orientation column receive the same Poisson input and obtain similar neuronal activation, the PPC of which are not always the same, but slightly different across different neurons for the same column. Additionally, the PPC for the beginning period is considerably small and comparatively similar across all 60 neurons because of the transient stimulus onset effect.

4 Conclusions

In order to investigate the oscillatory phase, PPC and their relationships with input orientation, firing rate and neuronal activation, we construct a spiking neuronal network according to the physiological experiment of awaken monkeys [28]. Six Poisson groups, with different Poisson rates, provide external spike inputs and six columns, with distinct orientation preference, simulate orientation columns in primary visual cortex. Then we conduct data recording and spectral analyses, which include spike-LFP phase calculation [28] and point-field PPC $\hat{P}_1$ computation [34]. The most powerful PPC values of neurons are all within the gamma-band frequency, which is consistent with physiological experiments [28]. If the preferred orientation of one column is close or equivalent to the input orientation stimulus, the neurons within this column will obtain higher firing rate and stronger activation, thereby giving rise to stronger synchronization and higher PPC values. However, the PPC will degrade if the input orientation is far from the preferred orientation of a certain column. Furthermore, we adopt the leaky integrate-and-fire neuronal model because of its biological plausibility and computational efficiency, however other models such as Hodgkin–Huxley model and SRM model are feasible.

References

1. C. M. Gray, P. Konig, A. K. Engel, W. Singer. *NATURE* **338**, 334-337 (1989).
2. J. M. Schoffelen, R. Oostenveld, P. Fries. *SCIENCE* **308**, 111-113 (2005).
3. T. Spellman, M. Rigotti, S. E. Ahmari, S. Fusi, J. A. Gogos, J. A. Gordon. *NATURE* **522**, 309 (2015).
4. K. M. Igarashi, L. Lu, L. L. Colgin, M. Moser, E. I. Moser. *NATURE* **510**, 143 (2014).
5. T. Womelsdorf, J. Schoffelen, R. Oostenveld, W. Singer, R. Desimone, A. K. Engel, P. Fries. *SCIENCE* **316**, 1609-1612 (2007).
6. R. T. Canolty, E. Edwards, S. S. Dalal, M. Soltani, S. S. Nagarajan, H. E. Kirsch, M. S. Berger, N. M. Barbaro, R. T. Knight. *SCIENCE* **313**, 1626-1628 (2006).
7. E. Salinas, T. J. Sejnowski. *NAT REV NEUROSCI* **2**, 539-550 (2001).
8. P. Fries. *TRENDS COGN SCI* **9**, 474-480 (2005).
9. T. Akam, D. M. Kullmann. *NAT REV NEUROSCI* **15**, 111-122 (2014).
10. W. Singer, C. M. Gray. *ANNU REV NEUROSCI* **18**, 555-586 (1995).
11. N. Brunet, C. A. Bosman, M. Roberts, R. Oostenveld, T. Womelsdorf, P. De Weerd, P. Fries. *CEREB CORTEX* **25**, 918-926 (2015).
12. P. Fries, J. H. Reynolds, A. E. Rorie, R. Desimone. *SCIENCE* **291**, 1560-1563 (2001).
13. D. Baldauf, R. Desimone. *SCIENCE* **344**, 424-427 (2014).
14. B. Pesaran, J. S. Pezaris, M. Sahani, P. P. Mitra, R. A. Andersen. *NAT NEUROSCI* **5**, 805-811 (2002).
15. J. Fell, N. Axmacher. *NAT REV NEUROSCI* **12**, 105-1500 (2011).
16. P. Fries. *ANNU REV NEUROSCI* **32**, 209-224 (2009).
17. G. Buzsaki, X. J. Wang. *ANNU REV NEUROSCI* **35**, 203-225 (2012).
18. C. A. Bosman, C. S. Lansink, C. M. A. Pennartz. *EUR J NEUROSCI* **39**, 1982-1999 (2014).
19. P. Fries. *NEURON* **88**, 220-235 (2015).
20. A. Z. Harris, J. A. Gordon. *ANNU REV NEUROSCI* **38**, 171-194 (2015).
21. M. Zeitler, P. Fries, S. Gielen. *J COMPUT NEUROSCI* **25**, 89-107 (2008).
22. T. Akam, D. M. Kullmann. *NEURON* **67**, 308-320 (2010).
23. M. A. Wilson, C. Varela, M. Remondes. *CURR OPIN NEUROBIOL* **31**, 250-253 (2015).
24. M. A. Montemurro, M. J. Rasch, Y. Murayama, N. K. Logothetis, S. Panzeri. *CURR BIOL* **18**, 375-380 (2008).
25. C. Kayser, M. A. Montemurro, N. K. Logothetis, S. Panzeri. *NEURON* **61**, 597-608 (2009).
26. J. O'Keefe, M. L. Recce. *HIPPOCAMPUS* **3**, 317-330 (1993).
27. P. Fries, D. Nikolic, W. Singer. *TRENDS NEUROSCI* **30**, 309-316 (2007).
28. M. Vinck, B. Lima, T. Womelsdorf, R. Oostenveld, W. Singer, S. Neuenschwander, P. Fries. *J NEUROSCI* **30**, 1250-1257 (2010).
29. T. Womelsdorf, P. Fries. *J PHYSIOL-PARIS* **100**, 182-193 (2006).
30. M. Siegel, M. R. Warden, E. K. Miller. *P NATL ACAD SCI USA* **106**, 21341-21346 (2009).
31. R. Polania, M. A. Nitsche, C. Korman, G. Batsikadze, W. Paulus. *CURR BIOL* **22**, 1314-1318 (2012).
32. P. H. Tiesinga, T. J. Sejnowski. *FRONT HUM NEUROSCI* **4** (2010).
33. A. Buehlmann, G. Deco. *PLOS COMPUT BIOL* **6**, e1000934 (2010).
34. M. Vinck, F. P. Battaglia, T. Womelsdorf, C. Pennartz. *J COMPUT NEUROSCI* **33**, 53-75 (2012).
35. J. P. Lachaux, E. Rodriguez, J. Martinerie, F. J. Varela. *HUM BRAIN MAPP* **8**, 194-208 (1999).
36. M. Vinck, M. van Wingerden, T. Womelsdorf, P. Fries, C. M. A. Pennartz. *NEUROIMAGE* **51**, 112-122 (2010).
37. T. P. Vogels, H. Sprekeler, F. Zenke, C. Clopath, W. Gerstner. *SCIENCE* **334**, 1569-1573 (2011).
38. H. Hu, J. Gan, P. Jonas. *SCIENCE* **345**, 529 (2014).
39. D. Ferster, K. D. Miller. *ANNU REV NEUROSCI* **23**, 441-471 (2000).
40. A. Mazzoni, S. Panzeri, N. K. Logothetis, N. Brunel. *PLOS COMPUT BIOL* **4**, e1000239 (2008).
41. A. Barardi, B. Sancristobal, J. Garcia-Ojalvo. *PLOS COMPUT BIOL* **10**, e1003723 (2014).

The simulation study of three typical time frequency analysis methods

Yifeng Li[a], Lihui Zhang, Baohui Li, Yan Xu, Sanyuan Wu, Xiaoyang Wei, Xiaoyan Liu, Rong Lin and Quan Wang

Institute of Aviation Medicine, Air Force, 100142, Beijing, China

Abstract. The principals and characteristics of three typical time frequency analysis methods that Short Time Furious transformation, wavelet transformation and Hilbert-Huang transformation are introduced, and the mathematical definition, characteristics and application ranges of these analysis methods and so on are pointed out, then their time-frequency local performance is made analysis and comparison through computer programming and simulation.

1 Introduction

In the field of Aeronautics and Astronautics, in order to prevent pilots from air syncope in flying and to get a better flight, the human physiological signals under +Gz acceleration need to be monitored and studied. The recognition, feature extraction and monitoring analysis and other technologies of variety of physiological signals including EEG signal has become a hot topic of research in recent years [1]. All kinds of analysis methods such as Fourier transformation [2-4], Short Time Fourier transformation[5-7],wavelet and wavelet package transformation[8-12], Hilbert-Huang Transformation (HHT) [13-17] and other technologies have been applied to analysis of signal. Fourier transformation is the basic of analysis of time domain and frequency domain for signal, and is very valid to the analysis of periodic stable signal. But in nature and engineering areas there are a large number of non periodic non stationary signal, and EEG is typical non periodic signal, so the classic analysis method based on Fourier transformation can not accurately reflect the local time-varying frequency spectral characteristic of signal, unable to obtain many of the key information. So a series of new signal time frequency analysis methods are put forward and developed. The basic idea of time frequency analysis is to design the joint function of time and frequency, and at the same time, the density and intensity of energy of signal at different time and frequency is described. Before the analysis and comparison on time-frequency performance of various of methods has been seen, but the research on the comparison and analysis on Short Time Fourier transformation, wavelet transformation, and HHT, this three kinds of methods is rare through the designed simulation signal. This study first briefly introduces of the principles and characteristics of this several methods, Short Time Fourier transformation, wavelet transformation and HHT, then EEG signal is simulated by the designed suitable simulation signal through programming, so the time frequency graphs are made out through simulation signal, and time frequency performance of three methods is made a contrast and analysis, to explore the application prospect of them in processing and analysis of dynamic EEG signal.

[a] Corresponding author: liyf8886@sina.com

DOI: 10.1051/bioconf/20170802007

2 The introduction of basic principle and method

2.1 Short Time Furious transformation (STFT)

STFT assumes non stationary signal as that every small segment is smooth, intercepts signal through using a sliding window, and makes Fourier transformation for the intercepted signal again and again, so as to obtain the frequency spectrum of signal at any time. Its definition is as:

$$STFT\ (t,f)=\int x(t)\ h(-t)e^{-j\pi f}dt \tag{1}$$

The advantage of STFT is that its physical meaning is clear and there is no cross term appearing. It is a time frequency analysis method which application is the most. According to the Heisenberg uncertainty principle, the product of time width and bandwidth of signal should be a constant relevant to sampling rate. In other words, when time resolution is needed to improve, the resolution of frequency is often needed to sacrifice, and vice versa. When the analyzed signal contains many kinds of types of scale components with large differences, STFT is powerless. Although STFT there is the defects such as resolution ratio is not high and so on, its algorithm is simple, and easy to implement, so in a long time it is a powerful tool of non stationary signal analysis is, and still is widely used.

2.2 Wavelet transformation

Wavelet transform, as a kind of new linear time frequency analysis method, is developed in the Middle and late period of the 1980s of 20th century. It makes multi-scales analysis of signal by the operation function of dilation and translation and others. It is a milestone development in reconcile analysis development history. The wavelet transformation definition of $x(t)$ is as:

$$\begin{aligned} WT_x(\alpha,\tau) &= \frac{1}{\sqrt{a}}\int x(t)\phi^*\left(\frac{t-\tau}{\alpha}\right)dt \\ &= \langle x(t),\phi_{\alpha\tau}(t)\rangle \end{aligned} \tag{2}$$

It is the inner product result of function and wavelet basis ,and among this wavelet basis:

$$\varphi_{\alpha\tau}=\frac{1}{\sqrt{\alpha}}\varphi\left(\frac{t-\tau}{\alpha}\right) \tag{3}$$

α is the scale parameter, and τ is the translation parameter. The wavelet basis is different with the difference of α and τ , and it can describe the frequency band characteristic of signal, that is it can show time-frequency characteristic of signal at the same time, and can very sensitively detect mutation of signal. Wavelet transformation is also subjected to the Heisenberg uncertainty principle, but the window of wavelet transformation is adjustable time-frequency window, using a narrow window in high frequency, and using a wide window in low frequency, fully embodying the idea of multi-resolutions analysis, and this is consistent with the feature of time varying non stationary signal is. The characteristic of multi scales and multi resolutions makes wavelet transformation be known as "mathematical microscope". But wavelet transformation does not possess characteristic of adaptive decomposition, and the finite length of wavelet basic can also cause energy leakage, and its selection principle is still a difficult problem. After development of about 30 years, wavelet transform has made breakthrough progress, and formed the multi resolutions analysis, framework and filter group-this three big complete and abundant wavelet transformation theory system, and now it has been widely used in signal elimination noise, medicine and many other areas.

2.3 Hilbert-Huang transformation (HHT)

Hilbert Huang transformation is a new signal processing method proposed by N.E.Huang et al. in 1998, and it is considered as a major breakthrough of linear and steady-state spectral analysis based on Furious transformation in recent years. Essentially it is the process that making smoothing processing for signal, and decomposing stepwise the wave of different scales in signal, formatting intrinsic mode function component, then making Hilbert Huang transformation for these components. HHT method consists of two main steps:

(1)The original data is made preprocess, and through empirical mode decomposition (EMD) method, the data is decomposed into intrinsic mode function (IMF) of n order, that is:

$$x(t) = \sum_{i=1}^{n} c_i + r_n \tag{4}$$

(2)The decomposed each order IMF component is made Hilbert Huang transformation, and the respective instantaneous frequency is got, then the time frequency diagram is made.

Hilbert Huang transformation is an effective tool for nonlinear and non stationary data analysis, and it is a self-adaptive treatment method. It's the biggest characteristic is that by EMD decomposition of signal, making non stationary signal smooth, resulting in instantaneous frequency has physical meaning, and then exporting the meaningful Hilbert time frequency spectrum. The analysis basic of HHT is: the data structure of the modulation mechanism of intra wave and no fixed basement (that is IMF characteristic). It doesn't take that time and frequency is absolutely two space variables, and no longer believes that the fundamental component part of signal is complete sine function, namely in the basic component part , instantaneous frequency is also changing with time, but it is worth noting, now the proof of HHT obviously falls after its application. The further development of HHT also requires that the envelope problem, EMD decomposition theory verifying, the end point effect suppression in boundary treatment and other issues have a better way to solve. Its theoretical perfection will be a long process. While this does not affect its application. In just a few years from the method of Hilbert Huang transformation publishes up to now, HHT method has always subjected an extensive attention of scholars at home and abroad and it has been widely used in biomedicine, equipment diagnosis and other various of scientific research and engineering field.

3 Simulation signal contrast and analysis study

In order to better contrast the treatment effects of Short Time Fourier transformation, wavelet transformation and Hilbert Huang transformation, this several methods on EEG data, and taking into account the basic rhythms signals in EEG [18], that is EEG is composed of low frequency and high frequency signals, so the segmented two sine functions are used to simulate the dynamic EEG data, frequency is respectively taken as 10 Hz and 50 Hz, and sampling frequency is taken as 1000Hz, taking data of 2s. The time point of mutation is taken at the time of about 1s. The simulation signal is as:

$$x(t) = \begin{cases} \sin(20\pi t) & 0 \le t \le 1s \\ \sin(100\pi t) & 1 < t \le 2s \end{cases}$$

STFT, wavelet transformation, and HHT,this several kinds of transformation treatment are respectively made for this simulation signal, and the time frequency graphs of this several kinds of transformation are made out. The simulation graph and time frequency graphs of this several kinds of transformation are shown in the following graphs.

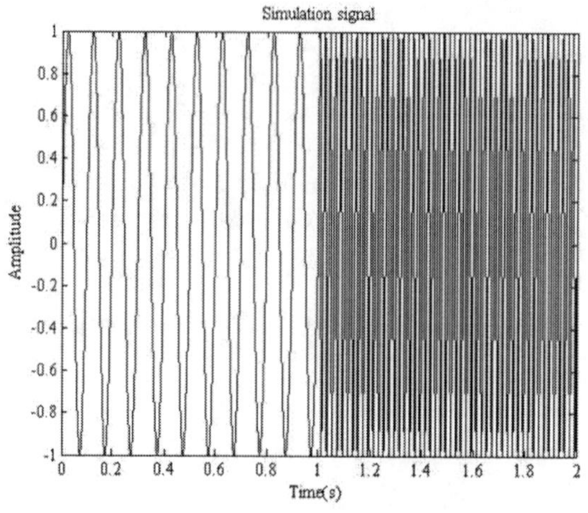

Figure 1. The simulation signal.

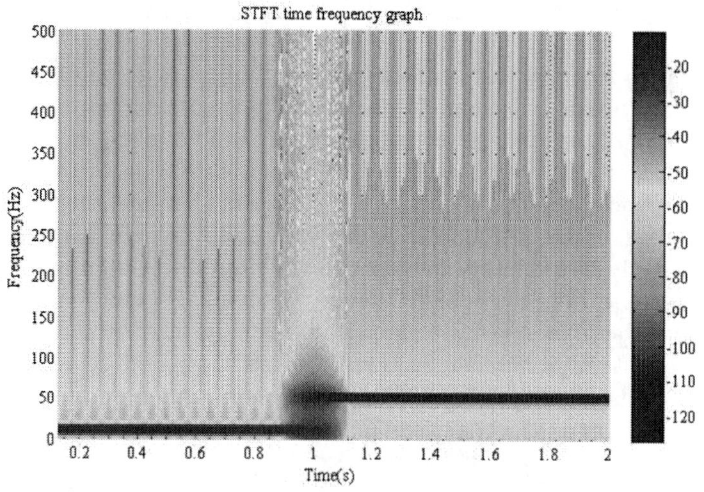

Figure 2. STFT time frequency graph.

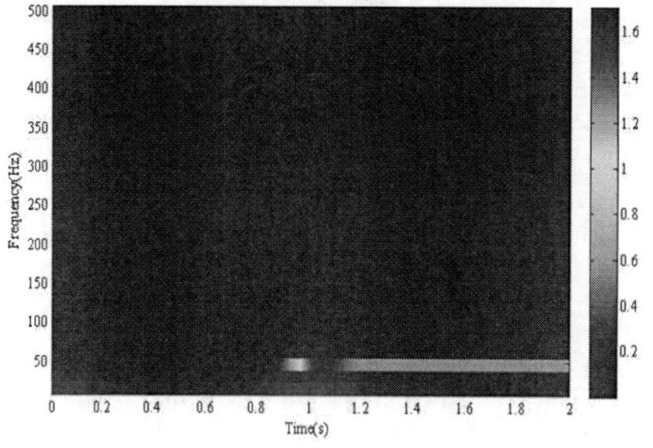

Figure 3. Wavelet time frequency graph.

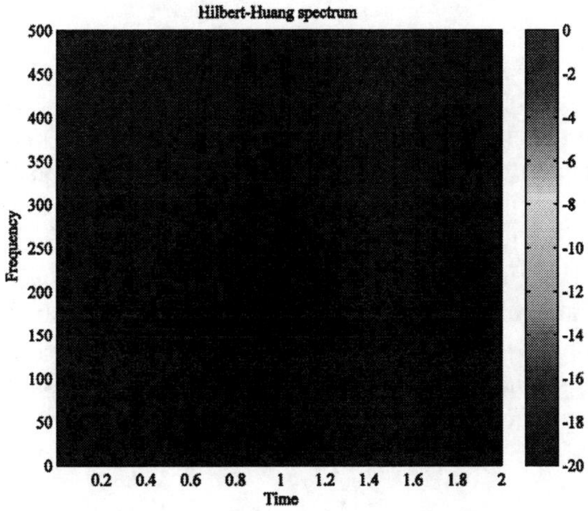

Figure 4. HHT spectrum graph.

Three methods all can distinguish the change of main frequency. From Figure 2, spectrum graph of Short Time Furious transformation it can be seen that, although this figure can distinguish the moment of two main frequency components and mutation occurrence, the resolution is not high. According to the Heisenberg uncertainty principle, the product of time width and bandwidth of signal should be a constant relevant to sampling rate. In other words, when time resolution is needed to improve, the resolution of frequency is often needed to sacrifice, and vice versa. For EEG signal, when the frequency resolution is required, the time window length is likely to appear too large, and at this time FFT may be difficult to meet the requirement. The treatment effect is not satisfactory for non-stationary nonlinear signal such as EEG. Figure 3 is a wavelet spectrum. Although the time-frequency resolution of wavelet transformation are both very good, for the smaller frequency component, the time point of frequency hopping is described not enough accurately. Figure 4 is the HHT spectrum. It can be seen from the figure that processing effect of HHT on nonlinear unstable signal has a unique advantage, because this method does not exist the steps of partition and translation of Windows when using, adopting that separately making analysis for the short time series. Therefore, it makes a good balance between the requirements of time resolution and frequency resolution, that is to say, as long as it meets the Shannon theorem (i.e., sampling law), this method can improve time domain resolution as much as possible under the situation that without sacrificing frequency domain resolution. HHT spectrum displays the distribution of signal energy in time frequency domain, and can help more clearly understand the distribution situation of low and high frequency part in signal. Moreover, HHT method structures basis function from the signal itself, directly reflects the characteristic of signal itself, and eliminate the generated virtual volume due to the mismatch of base function and signal in transformation process. so the obtained HHT spectrum can not only locate the position of frequency mutation point, but also can accurately distinguish two main frequency components of 10 Hz and 50 Hz, and contains no crossing interference term, with more superior ability of detecting signal singularity than FFT and wavelet transformation .In addition, HHT can use some unique forms, such as instantaneous energy spectrum, time frequency spectrum and others to display new characteristic. But relatively speaking, this method quite consumes computer memory and calculation time, that is, relatively calculation is slower, and this has limitation in practical application. Moreover, envelope rationality, EMD decomposition theory verifying, and end point effect restrain in boundary treatment and other issues still need further to be proofed, namely relatively speaking the theoretical development of this method is not as perfect as other two, and its theory perfection will be a longer process relatively.

4 Conclusions

This study introduces the principal and method of STFT, wavelet transformation, and HHT, three typical time-frequency analysis methods, and the time frequency local performances of this several methods are made analyzed and compared through computer simulation, Then the advantages and disadvantages of this several methods are discussed. EEG is a typical time varying, non stationary signal, time-frequency union analysis technology is created and developed according to the need of real science application. Three kinds of methods all can better distinguish the main frequency, but relatively speaking, HHT method, has a good time-frequency and adaptability for signal. But this method also has its limitations. It is believed that with the continuous perfection and development of modern time frequency analysis theory and technology, in the field of EEG monitoring and analysis of aeromedicine, time frequency union analysis technology will have a more broad application prospect and greater development potential.

References

1. Koji Morikawa, Akinori Matsumoto, Shrishail Patki, Bernard Grundlehner,Auryn Verwegen, Jiawei Xu, Srinjoy Mitra and Julien Penders, Compact Wireless EEG Sys tem with Active Electrodes for Daily Healthcare Monitoring, IEEE International Conference on Consumer Electronics (ICCE).2013:204-205.
2. SN. Tang, FC. Jan , HW. Cheng ,CK. Lin , G Z. Wu, Multimode Memory-Based FFT Processor for Wireless Display FD-OCT Medical Systems, IEEE Transactions on Circuits and Systems I: Regular Papers.2014; **61** (12): 3394 – 3406.
3. Jing Han, Lingling Zhang ,G. Leus, Partial FFT Demodulation for MIMO-OFDM Over Time-Varying Underwater Acoustic Channels, IEEE Signal Processing Letters.2016: 23 (2): 282 – 286.
4. T. Yang,H. Pen , Z. Wang , C. S. Chang, Feature Knowledge Based Fault Detection of Induction Motors Through the Analysis of Stator Current Data, IEEE Transactions on Instrumentation and Measurement.2016 ;**65** (3): 549 – 558.
5. W. Liu, X. Zhang, Signal modulation characteristic analysis based on SFFT-Haar algorithm, IET International Radar Conference 2013, Xi'an, 2013: 1 – 4.
6. B. Kim, S.H. Kong, S. Kim, Low Computational Enhancement of STFT-Based Parameter Estimation, IEEE Journal of Selected Topics in Signal Processing, 2015, **9**(8):1610 - 1619 .
7. K. Jaganathan, Y. C. Eldar, B. Hassibi, STFT Phase Retrieval: Uniqueness Guarantees and Recovery Algorithms, IEEE Journal of Selected Topics in Signal Processing, 2016,**10** (4) :770 - 781 .
8. R. Mothi, M. Karthikeyan, A wavelet packet and fuzzy based digital image watermarking, 2013 IEEE International Conference on Computational Intelligence and Computing Research (ICCIC), Enathi, 2013:1 – 5.
9. J.S Pan, and L. Yue, "A ridge extraction algorithm based on partial differential equations of the wavelet transform, "2014 IEEE Symposium on Computational Intelligence for Multimedia, Signal and Vision Processing (CIMSIVP), Orlando, FL, pp.1 – 5, 2014.
10. S.A.R. Naqvi, I. Touqir, and S.A. Raza, "Adaptive geometric wavelet transform based two dimensional data compression, "2013 7th IEEE GCC Conference and Exhibition (GCC), Doha , pp.583 – 588, 2013.
11. G. Serbes, N. Aydin, and H.O. Gulcur, "Directional dual-tree complex wavelet packet transform, "2013 35th Annual International Conference of the IEEE , Engineering in Medicine and Biology Society (EMBC), Osaka, pp.3046-3049, 2013.
12. S. Bharkad, and M. Kokare, "Fingerprint matching using discreet wavelet packet transform , "2013 IEEE 3rd International Advance Computing Conference (IACC), Ghaziabad, pp.1183-1188, 2013.

DOI: 10.1051/bioconf/20170802007

13. K. Rai , V. Bajaj ,A. Kumar, Hilbert-Huang transform based classification of sleep and wake EEG signals using fuzzy c-means algorithm, 2015 International Conference on Communications and Signal Processing (ICCSP), IEEE, Melmaruvathur, 2015: 0460 – 0464.
14. Y. Wang ,C. Chen , H. He, Large Eddy Simulation and Hilbert Huang Transform for fluctuation pressure of high speed train, 2015 IEEE International Instrumentation and Measurement Technology Conference (I2MTC), Pisa, 2015: 284 – 288.
15. Rahul Kumar Chaurasiya , K. Jain , S. Goutam ,Manisha, Epileptic seizure detection using HHT and SVM, 2015 International Conference on Electrical, Electronics, Signals, Communication and Optimization (EESCO), Visakhapatnam, 2015: 1 – 6.
16. K. Y. Lin , D. Y. Chen , W. J. Tsai, Face-Based Heart Rate Signal Decomposition and Evaluation Using Multiple Linear Regression, IEEE Sensors Journal ,2016; **16** (5): 1351 – 1360.
17. P. Y. Chen ,Y. C. Lai , J. Y. Zheng, Hardware Design and Implementation for Empirical Mode Decomposition, IEEE Transactions on Industrial Electronics,2016; PP (99): 1.
18. M. Nakamura, Q. Chen, T. Sugi, A. Ikeda, and H. Shibasaki, "Technical quality evaluation of EEG recording based on electroencephalographers' knowledge,"Medical Engineering & Physics, vol. (27),2005, pp.93-100.

BIO Web of Conferences **8**, 02008 (2017) DOI: 10.1051/bioconf/20170802008
ICMSB2016

Study on the identification of *Polygonatum kingianum* and its adulterants by PCR amplification of specific alleles

Yanying Li[1,2], Xinyan Zhu[1], Yiguo Li[3], yan Yang[1], Yana Shi[1], Xiaobin Hu[1], Jiajing Wang[1], Pengzhang JI[1a]

[1]*Institute of Medical Plant, Yunnan Academy of Agricultural science, Kunming 650223, China*
[2]*College of Food Sciences and Technology, Yunnan Agricultural University, Kunming 650092, China*
[3]*Facalty of Sciences and Technology, Kunming University of Sciences and Technology, Kunming 650500, China*

Abstract. Using the allele-specific diagnostic PCR for the rapid and efficient identification of Chinese herbal medicines and their adulterants. In this paper, we analyze the research object of *P. kingianum* and its adulterants. The total DNA of the samples to be identified was extracted, and the PCR was amplified by using universal primers. After the homology comparison, the specific sites were found out by using BioEdit software, and the specific PCR primers were designed by Premier primer 5.0 software, and the samples were amplified by the primers. The 331bp bands were amplified by *P. kingianum* of the specific primers, but the results of their adulterants were not amplified. The results show that the method can provide guidance for the allele-specific diagnostic PCR for quick and easy to select *P. kingianum*, which is not affected by environmental factors, affecting the growth period of the plant, were identified directly from the molecular level of DNA. This method having important application value in accurate introduction and clinical application can provide a model for other identification.

1 Introduction

According to the flora of Chinese records, *Polygonatum kingianum Coll. et Hemsl.* is Polygonatum plants grown in the forest, shrub or damp grass, sometimes in the rock, elevation of 700-3600 meters. It is distributed mainly over Yunnan, Sichuan, Guangxi and Guizhou provinces and other places [1]. According to Pharmacopoeia of People's Republic of China, *P. kingianum* has beneficial effects on the spleen, lungs and kidney. And it can be used for the treatment of spleen qi deficiency, body tired fatigue, stomach yin deficiency, dry food less, lung deficiency cough dry cough, hemoptysis, blood deficiency, Yaoxisuanruan, premature graying, heat diabetes and other symptoms [2]. In the dictionary of traditional Chinese medicine, there are three original plants included in *P. sibiricum*, and they are classified as *P. sibiricum* in the flora of Chinese records and the Pharmacopoeia[1,3]. But there are significant differences between the different producing areas for the same kind of Chinese medicinal herbs in the contents of main chemical components, such as polysaccharides and saponins. At present, in order to guarantee the quality and clinical effects, people rely mainly on the

[a] Corresponding author: pengzhji@126.com
This work was supported by the Nature Science Foundation of Yunnan Province (2015FB161), the Environmental Protection Foundation of Yunnan Province (2014BI006) and the BioIndustry Project of Kunming City (2016-17).

morphological and physiological methods, however *P. kingianum* and its adulterants were highly similar, it was not easy to identify, and traditional morphological and chemical identification is easily affected by man-made and environmental factors. In recent years, with the continuous development of molecular identification technology, more and more researchers will begin to apply molecular identification technology to the identification of Chinese herbal medicines. However, the molecular identification methods reported in the present are often complicated and long time, such as identify *ruthenicum L.* used by ITS [4], using NCBI nucleotide database analysis DNA identification of Zijingpi's adulterant species *Schisandra sphenanthera* [5], using cpDNA RBCL and translated amino acid sequences to distinguish *Dryopteris crassirhizoma* [6]. Site specific PCR requires only a single PCR reaction, and then we can judge the genuine from whether there is amplification, is very simple and fast. Today, the site-specific PCR technology has been applied in *Panax ginseng* [7], *Humulus lupulus L.* [8], *Sarcandra glabra* and *Chloranthus* [9], *Amomum villosum Lour.* [10], *Mentha haplocalyx* and *Mentha spicata* [11].

Today, people often mix *P. kingianum* up with *P. odoratum* and *P. uncinatum* on the market. While, the method can overcome obstacles of Large sample size and Strong subjectivity, provide a set of PCR primers and identification of Polygonatum and its adulterants, only PCR amplification reaction in a reaction system. One can quickly and accurately identify *P. kingianum* and its adulterants.

2 Materials and methods

2.1 Plant materials and extraction of the total dna

The experimental materials include *P. kingianum*, *P. odoratum*, Ku Huang Jing, *P. uncinatum Diels*, *Solanum tuberosum*, *Ipomoea batatas*, *Dioscorea polystachya Turcz.*, *Zingiber officinale* and *Colocasia esculenta.* Which was collected and acquired from Yunnan, China in 2015 (Table 1). *P. kingianum*, *P. odoratum*, Ku Huang Jing and *P. uncinatum Diels* were wild, the remaining samples for sale in the market. Wild samples were taken from leaves 1g while the samples for sale were taken from the tuber 1g, using the modified CTAB method to extract the total DNA, and the DNA was dissolved into the sterile ddH$_2$O, which was preserved in -20°C.

Table 1. Samples used in this study.

Chinese name	Amounts	Location	Time	Scientific name
Dian Huang Jing	21	Yanjin, Yunnan, China	Mar-15	*Polygonatum kingianum*
Yu Zhu	7	Kunming, Yunnan, China	Mar-15	*Polygonatum odoratum*
Ku Huang Jing	5	Kunming, Yunnan, China	Mar-15	
Xiao Huang Jing	8	Kunming, Yunnan, China	Mar-15	*Polygonatum uncinatum Diels*
Yang yu	6	Kunming, Yunnan, China	Jul-15	*Solanum tuberosum*
Fan shu	6	Kunming, Yunnan, China	Jul-15	*Ipomoea batatas*
Shu Yu	6	Kunming, Yunnan, China	Jul-15	*Dioscorea polystachya Turcz.*
Jiang	6	Kunming, Yunnan, China	Jul-15	*Zingiber officinale*
Yu	6	Kunming, Yunnan, China	Jul-15	*Colocasia esculenta*

2.2 PCR amplification

The amplification of universal primers (psbA-trnH-F/R) was amplified by PreMix (2×PCR MasterMix, Sangon Biotech, Shanghai, China). The simple PCR cycling parameters implemented using one cycle of 5 min at 94 °C, followed by 35 cycles of 1min at 94 °C, 1 min at 56 °C, and 1 min extension at 72 °C with the final extension at 72 °C for 7 min. PCR products were analyzed via 1.5% agarose gel electrophoresis (Promega, Madison, Wisconsin, USA) and visualized by Nucleic acid dye (Dingguo Changsheng, Beijing, China) under UV. (Figure 1(B))

2.3 Propagating sequence analysis and primer design

BioEdit software was used for sequence analysis, collation and comparison, to find out the specific mutation sites in the *P. kingianum.* Using premier 5.0 primer software to design a pair of specific PCR primers (DHJ-F, DHJ-R). (Table.2)

2.4 Specific pcr for authentication of *p. kingianum* and assay

The total volume of the PCR reaction system was 25μL, the reaction system including 2.5μL10 * PCR Taq buffer (Mg2+plus 20mM), 0.5μL (10mM) dNTP, primers and 1μL (10μM), DNA Taq polymerase (Dingguo Changsheng, Beijing, China) 0.5μL (2U/μL), DNA template 1μL (10-100ng), distilled water 18.5μL. Using oscillators to shake evenly after the preparation of PCR reaction solution, put PCR tube into the instrument of PCR reaction, which is in the 4min at 94°C of pre-degeneration, and degeneration of 94°C for 1min, annealing at 56°C for 45s, and 1 min extension at 72 °C with the final extension at 72 °C for 10 min in total 32 cycles, with 4°C to save. After the amplification, the products were amplified by 5μL, and 40min was observed by 1.5% agarose gel electrophoresis, 100V voltage electrophoresis and staining with nucleic acid dyes. (Figure 1(C))

3 Results

3.1 Assay of the total DNA extraction

In experiments, extraction of different samples DNA detected by the concentration of 1% agarose gel electrophoresis, mix sample with the 8L DNA solution and bromophenol blue solution, 0.5 times TBE, 120V DC voltage in the gel electrophoresis of 30min, automatic camera observation imaging system. As show in Figure 1 (A).

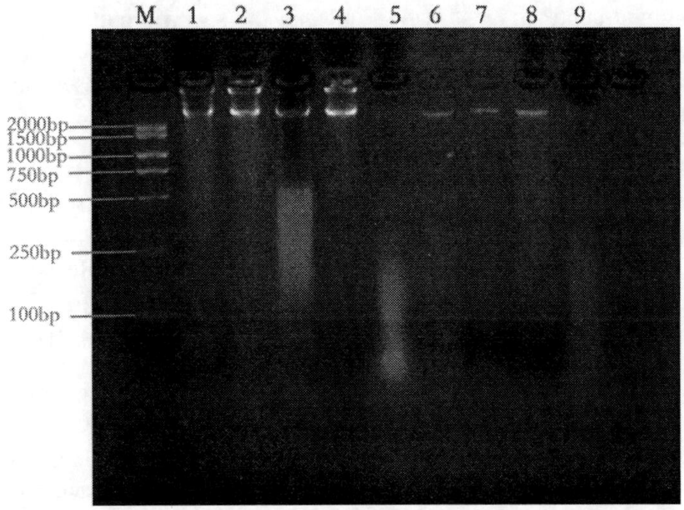

Figure 1(A). Gel electrophoresis of total DNA to *P. kingianum* and its adulterants.

1. *P. kingianum* 2. *P. odoratum* 3. Ku Huang Jing 4. *P. uncinatum Diels* 5. *Solanum tuberosum* 6. *Ipomoea batatas* 7. *Dioscorea polystachya Turcz.* 8. *Zingiber officinale* 9. *Colocasia esculenta*

3.2 Assay of psbA-trnH amplification

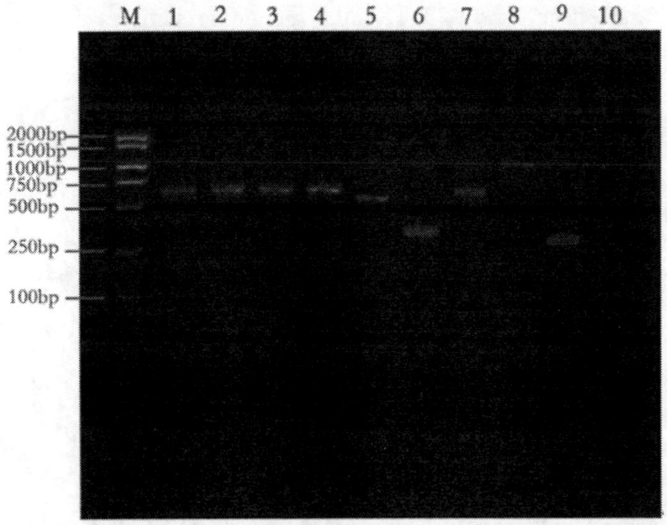

Figure 1(B). Gel electrophoresis of psbA-trnH amplification to *P. kingianum* and its adulterants.
1. *P. kingianum* 2. *P. odoratum* 3. Ku Huang Jing 4. *P. uncinatum Diels.* 5. *Solanum tuberosum* 6. *Ipomoea batatas* 7. *Dioscorea polystachya Turcz.* 8. *Zingiber officinale* 9. *Colocasia esculenta* 10. the negative control

The selection of universal primer psbA-trnH for the amplification of *P. kingianum* and its adulterants, and electrophoresis detection, as shown in the figure 1 (B).

3.3 Specific primer design

Selection of universal primers psbA-trnH for the amplification of *P. kingianum* and its adulterants, the amplified samples were sent to the Shanghai biotechnology company to sequence. The results of psbA-trnH sequencing to *P. kingianum* and its adulterants using BioEdit software for sequence analysis, proofreading and contrasting. It is find out that the specific mutation of *P. kingianum* is A, while is T of *P. odoratum* and *P. uncinatum Diels* in 71bp, and other adulterants is missing. At 402bp of *P. kingianum* is C, Ku Huang Jing is A, *Colocasia esculenta* is T. Based on the difference, it is designed a couple of specific PCR primers (DHJ-F, DHJ-R) using premier 5.0 primer software, the nucleic acid sequence as shown in Table 2.

Table 2. Primers and PCR reaction conditions used in this study

Primer name	Sequence	Reaction condition
psbA-trnH-F	5'-GTTATGCATGAACGTAA TGCTC-3'	94℃5min, 94℃1min, 56℃1min, 72℃1min, 72℃7min, 32cycles
psbA-trnH-R	5'-CGCGCATGGTGGATTCA CAATCC-3'	94℃5min, 94℃1min, 56℃1min, 72℃1min, 72℃7min, 32cycles
DHJ-F	5'-ATGTATTAAGAATCGTT GAAGGAGC-3'	94℃4min, 94℃1min, 56℃45s, 72℃1min, 72℃10min, 32cycles
DHJ-R	5'-AGCTAATCATTTATCGA GAAAAATG-3'	94℃4min, 94℃1min, 56℃45s, 72℃1min, 72℃10min, 32cycles

3.4 Specific PCR amplification

Put 5μL PCR amplification products in 1.5% agarose gel with nucleic acid dyes, 100V stable voltage electrophoresis 40min, observe the UV Gel imaging system. The results are shown in Figure 1 (C), it reflects that the sample produce band at 331bp, the remaining samples did not produce bands. The 331bp bands were amplified of *P. kingianum* by the specific primers, while that were not amplified of their adulterants.

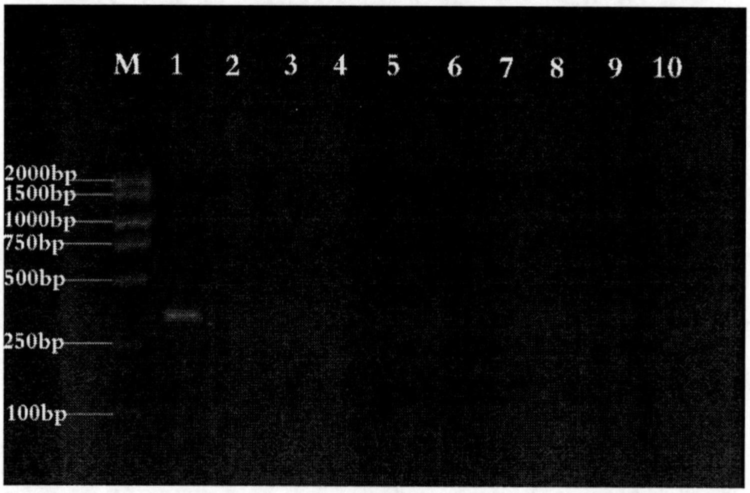

Figure 1(C). Gel electrophoresis of specific primer amplification to *P. kingianum* and its adulterants .
1. *P. kingianum* 2. *P. odoratum* 3. Ku Huang Jing 4. *P. uncinatum Diels* 5. *Solanum tuberosum* 6. *Ipomoea batatas* 7. *Dioscorea polystachya Turcz.* 8. *Zingiber officinale* 9. *Colocasia esculenta* 10. the negative control

4 Discussions

As part of the confusion of Chinese herbal medicines caused great danger to the safe use of traditional Chinese medicine, but the limitations of traditional identification methods affect the safe and effective use of traditional Chinese medicine. The technology of traditional Chinese medicine identification method runs through every link of the planting, processing and production, with the development of Chinese medicine industry, the traditional identification methods have been difficult to meet the rapid identification and normal needs of hospitals, pharmacies, enterprises, customs and other industries, thus a new methods of identification becoming urgent.

The characteristic of AS-PCR is just should consider in the 3 'end sites when designed primer sequence, while a primer for site specific identification of PCR and ARMS can improve the specificity of amplification not only 3' end sites, but also can be in the 3 'end of the primer close to the second to sixth base position mutation base. This method is widely used in identification of Chinese medicinal materials [12-15].

Allele-specific diagnostic PCR is a type of genotyping based on SNP gene loci, because this method only need a PCR, and PCR products were gel electrophoresis, and then distinguish between wild and mutation type according to the electrophoresis strip size and whether they have, fast and brief, has a unique advantage in the authenticity of products for rapid detection of medicinal materials, be good use of rapid identification of traditional Chinese medicine. There are no reports that identification of *P. kingianum*, *P. odoratum* and *P. uncinatum Diels* by allele-specific diagnostic PCR, for the future interrelated researches or identifications it can play a reference role [16].

5 Conclusions

Early studies indicated that molecular identification of Polygonatum was limited within DNA sequence information, and that the lack of a focus and simplicity needed a great deal of preliminary work of DNA sequence information in Polygonatum cultivars. Overcoming this limitation or improving effected on identification be use of cultivar-specific SNP marker, finding the specific sites and establishing allele specific PCR system to the rapid identification of *P. kingianum*. *P. kingianum* and its adulterants were not easy to identify, and traditional morphological and chemical identification is easily affected by man-made and environmental factors. In this study, the results show that the method can provide guidance for the allele-specific diagnostic PCR for quick and easy to select *P. kingianum*, which is not affected by environmental factors, affecting the growth period of the plant, were identified directly from the molecular level of DNA. This method having important application value in accurate introduction and clinical application can provide a model for other identification.

References

1. Chinese Flora Commission, Science Press and Missouri Botanical Garden Press, *Flora of China*, **24** (2000) (http://foc.eflora.cn/content.aspx?TaxonId=200027858)
2. Chinese Pharmacopoeia Commission, *Pharmacopoeia of the People`s Republic of China,* 306 (China Medical Science Press, 2015)
3. Z. Guoping, D. Shen, C. Renshou, etal, SSTP, *Dictionary of Chinese Traditional Medicine*, 2828(2005)
4. Gu X, Zhang XQ, Song XN, Zang YM, Li Yan-peng, Ma CH, Zhao BX, Liu CS, Chin J Chin Mater Med, *A new herbs traceability method based on DNA barcoding-origin-morphology analysis--an example from an adulterant of 'Heiguogouqi'*, **39**, 4759-62 (2014)
5. Cheng XL, Liao CL, Liu CS, Bai ZF, Yang YJ, Zheng J, Zhang J. Chin J Chin Mater Med, *DNA identification of Zijingpi's adulterant species Schisandra sphenanthera based on NCBI nucleotide database analysis*, **37**, 2534-7 (2012)
6. Zhao ZL, Leng CH, Wang ZT. Planta Med, *Identification of Dryopteris crassirhizoma and the adulterant species based on cpDNA rbcL and translated amino acid sequences*,**73**,1230-3 (2007)
7. Wang H, Li G, Kwon WS, Yang DC, Int J Mol Sci, *Development of EST Intron-Targeting SNP Markers for Panax ginseng and Their Application to Cultivar Authentication*, **17**, 884-895 (2016)
8. Henning JA, Coggins J, Peterson M, BMC Res Notes, *Simple SNP-based minimal marker genotyping for Humulus lupulus L. identification and variety validation*, 542-553 (2015)

9. Wci Y, Chen Y, Huang Y, Liu J, Liang Y, Mitochondrial DNA A DNA Mapp Seq Anal, *Molecular authentication and quantitative analysis of Sarcandra glabra and adulterated chloranthus products using SNP markers*, **27**, 3618-25 (2016)
10. Huang Q, Duan Z, Yang J, Ma X, Zhan R, Xu H, Chen W, PLoS One, *SNP typing for germplasm identification of Amomum villosum Lour. Based on DNA barcoding markers*, **9**, e114940 (2014)
11. Cao L, Qin SS, Yuan Y, Zhu XQ, Chin Med Mat, *Molecular identification of Mentha haplocalyx and Mentha spicata with specific primers multi-PCR system*, **37**, 41-5 (2014)
12. Wang C, Liang Y, Guo B, Xiao P, Chin J Chin Mater Med, *Molecular authentication of medicinal Epimedium species--a case of diagnostic PCR authentication of Epimedium koreanum*, **35**, 1799-802 (2010)
13. Gajić B, Stevanović J, Radulović Ž, Kulišić Z, Vejnović B, Glavinić U, Stanimirović Z, Exp Appl Acarol, *Haplotype identification and detection of mitochondrial DNA heteroplasmy in Varroa destructor mites using ARMS and PCR-RFLP methods*, (2016)
14. Suhda S, Paramita DK, Fachiroh J, Asian Pac J Cancer Prev, *Tetra Primer ARMS PCR Optimization to Detect Single Nucleotide Polymorphisms of the CYP2E1 Gene*, **17**, 3065-9 (2016)
15. Wang H, Wang J, Li G, Genet Mol Res, *A simple real-time polymerase chain reaction (PCR)-based assay for authentication of the Chinese Panax ginseng cultivar Damaya from a local ginseng population*, **15** (2016)
16. Oueslati A, Ollitrault F, Baraket G, Salhi-Hannachi A, Navarro L, Ollitrault P, BMC Genet, *Towards a molecular taxonomic key of the Aurantioideae subfamily using chloroplastic SNP diagnostic markers of the main clades genotyped by competitive allele-specific PCR*, **17**, 118 (2016)

DOI: 10.1051/bioconf/20170802009

Toxicology evaluation of selenium protein powder

Shunyi Qin[1], Lina Liu[1], Jingshen Wang[2], Fanghong Zhao[1] and Jifei Ma[1,a]

[1]*College of Animal Science and Veterinary Medicine, Tianjin Agricultural University, Tianjin 300384, China*
[2]*Tianjin Selenium Egg Biological Technology Co., Ltd., Tianjin 300100, China*

Abstract. The experiment was conducted to evaluate the safety of selenium protein powder, a novel organic selenium nutritional supplement, and reported corresponding data and results based on a series of toxicological tests. It was examined to evaluate oral acute toxicity by median lethal dose test and mutagenic potential by bone marrow cell micronucleus test and sperm abnormality test using Kun-Ming mice. The results showed that the oral LD_{50} of selenium protein powder exceeded 31.25 g/kg body weight in mice. No mutagenicity was found by mouse bone marrow cell micronucleus test and mouse sperm abnormality test. The results suggested greater safety of selenium protein powder as a nutritional selenium supplement, and selenium protein powder has the potential for development and application in food systems or functional foods.

1 Introduction

As an essential trace element for human and animal, selenium is actively involved in the antioxidant defense systems [1], immune function [2], thyroid function and reproduction [3]. The bioavailability of Selenium varies by sources. It is generally accepted that the organic Selenium sources are more bioavailable than the inorganic Selenium sources[4]. Recently, a novel organic selenium nutritional supplement, selenium protein powder is attracting increasing attention due to its high bioavailability and nutrition[5]. Although safety evaluation of different types of selenium compounds had been widely studied [6-8], there was no available data about the safety evaluation of selenium protein powder in the literatures. Therefore, this paper evaluated oral acute toxicity and mutagenicity of selenium protein powder by median lethal dose (LD_{50}) test, bone marrow cell micronucleus test and sperm abnormality test.

2 Materials and methods

2.1 Selenium protein powder samples

Selenium protein powder (Tianjin Selenium Egg Biological Technology Co., Ltd) was made from selenium-enriched eggs. Briefly, hens were fed a diet supplemented with selenized yeast (Alltech Inc., Nicholasville, KY) so that the selenium-enriched eggs were obtained, and then, selenium protein

[a]Corresponding author: hbmjfts@126.com

 DOI: 10.1051/bioconf/20170802009

powder were obtained from selenium-enriched eggs by means of low temperature baking. The total Selenium concentrations in selenium protein powder determined by hydride generation–atomic fluorescence spectrometry method was 38.76 μg/g.

2.2 Oral acute toxicity test

The median lethal dose (LD_{50}) test was conducted to assess the acute toxicity of selenium protein powder. The test was performed by the Spearman-Karber method (Finney, 1978). In brief, one hundred Kun-Ming mice, approximately 4 weeks old, were divided into five groups with equal number of both sexes and equal initial body weight. The mice were administered (via gavage) with selenium protein powder suspended in distilled water at the doses of 12.80, 16.00, 20.00, 25.00 or 31.25 g/kg body weight, respectively. Following dosing, the mice were observed for one week for mortality, clinical observations and gross necropsy findings.

2.3 Bone marrow micronucleus test

Fifty 8-week-old Kun-Ming mice were divided into five groups with equal number of both sexes and equal initial body weight. Distilled water was given by oral gavage as a negative control, and cyclophosphamide (40 mg/kg) in physiological saline was given by intraperitoneal injection as a positive control. Selenium protein powder (suspended in distilled water) were given by oral gavage for 2 days (24 h between doses) at doses of 580, 1150 or 2300 mg/kg body weight to mice in other three groups, respectively. The mice were anesthetized and sacrificed 6 h after the last administration. Bone marrow from the sternum was collected, sectioned and stained by routine clinical protocol, and then observed under the microscope. The numbers of polychromatic erythrocytes (PCE) and normochromatic erythrocyte (NCE), the numbers of PCE with micronuclei were record, and then micronucleus rates were calculate and analyzed by U-test of Poisson distribution.

2.4 Sperm aberration test

The sperm aberration test was performed by the standard methods [9]. Twenty-five 8-week-old Kun-Ming male mice, with an average body weight of 30 g, were randomly divided into five groups with equal number and equal initial body weight. Distilled water was given by oral gavage as a negative control, and 40 mg/kg cyclophosphamide in physiological saline was given by intraperitoneal injection as a positive control.Selenium protein powder were suspended in distilled water, and given by oral gavage for 5 days, at doses of 2250, 5000 or 10000 mg/kg body weight to mice in other three groups, respectively. The mice were anesthetized and sacrificed 35 days after the first administration. The bilateral epididymides were excised and placed in physiological saline and minced with ophthalmic scissors. Smears were prepared on clean slides, fixed with methanol and stained with 1% eosin. The slides were air-dried and coded for subsequent examination. Morphological evaluation for sperm was carried out using a high-magnification microscope. One thousand sperm were observed for each mouse to record the numbers of the aberrated sperm and aberration ratios were analyzed by Chi-square (χ^2)test .

3 Results

3.1 Oral acute toxicity assay

The results showed that mice administered selenium protein powder did not develop any clinical signs of toxicity either immediately or during the post-treatment period even at the highest dose of 31.25 g/kg body weight. The general conditions of all mice were normal. No mortality occurred either immediately or during the one-week observation period (Table 1). Therefore, the oral LD_{50} for selenium protein powder is considered to be greater than 31.25 g/kg for mice.

Table 1. Results of oral median lethal dose test.

Dosages (g/kg)	Number of Animals	Number of Animal death	Mortality (%)
12.80	20	0	0
16.00	20	0	0
20.00	20	0	0
25.00	20	0	0
31.25	20	0	0

3.2 Micronucleus test

The results showed that the incidence of micronucleus ratios for the male animals were within the range of 0.85-1.40‰ except for the positive control (24.00‰)(Table 2), and that the incidence of micronucleus ratios for the female animals were within the range of 1.3-1.9‰ except for the positive control (25.10‰) (Table 3). The data showed that there were no statistically significant differences between any of the selenium protein powder dosage groups and the negative control group ($P>0.05$), whereas the positive control group was significantly different ($P<0.05$) from others. The results also showed that there were no statistically significant differences of PCE/NCE in all groups ($P>0.05$). The results indicated that within the dose range of 580–2300 mg/kg body weight, selenium protein powder did not trigger an increase of micronucleus ratios *in vivo*.

Table 2. Results of micronucleus test of bone marrow PCE cells in male mice

Dosages(mg/kg)	Animals	Number of PCE	MNF(‰)	PCE/NCE
Control	5	5×2000	1.40±0.46	0.88±0.05
580	5	5×2000	0.80±0.30	0.93±0.03
1150	5	5×2000	1.70±0.44	0.85±0.06
2300	5	5×2000	1.20±0.44	0.83±0.03
CP 40 mg/kg	5	5×2000	24.00±2.30*	0.93±0.05

Table 3. Results of micronucleus test of bone marrow PCE cells in female mice

Dosages (mg/kg)	Animals	Number of PCE	MNF(‰)	PCE/NCE
Control	5	5×2000	1.30±0.25	0.83±0.06
580	5	5×2000	1.50±0.42	0.88±0.07
1150	5	5×2000	1.90±0.33	0.90±0.07
2300	5	5×2000	1.50±0.35	0.87±0.05
CP 40 mg/kg	5	5×2000	25.10±1.73*	0.90±0.07

* $P<0.05$ compared with control group.

3.3 Sperm aberration test

The results showed that the aberration rates for animals given selenium protein powder were within the range of 1.82% to 2.46% while that for the positive control was 12.14% (Table 4). The data showed that there was no statistically significant difference ($P>0.05$) between any of three dosage groups and the negative control group, whereas the positive control group was significantly different ($P<0.01$) from the negative control group. The results indicated that, within the dose range of 2250-10000 mg/kg body weight, the selenium protein powder did not have any detectable mutagenic effect on sperm cells in mice.

Table 4. Results of sperm abnormality test of mice

Dose (mg/kg)	Animals	Numbers of Sperms	Numbers of Aberrated sperms	Frequencies of abnormality (%)
Control	5	5 ×1000	24.60±3.56	2.46±0.36
2250	5	5 ×1000	18.20±3.12	1.82±0.31
5000	5	5 ×1000	24.20±6.39	2.42±0.64
10000	5	5 ×1000	18.40±3.17	1.84±0.32
CP 40	5	5 ×1000	121.40±8.49	12.14±0.85**

** $P<0.01$ compared with control group.

DOI: 10.1051/bioconf/20170802009

4 Discussion

Our results showed that the oral LD_{50} for selenium protein powder is considered to be greater than 31.25 g/kg body weight for mice, indicating that selenium protein powder can be considered a non-toxic material on the basis of the criteria of acute toxic classifications by Ministry of Health, China[9]. Our results also indicated that, as organic selenium, selenium protein powder was shown to have a significantly lower acute oral toxicity than sodium selenite in mouse. Based on our results and previous results[10-12] the toxicity order of Selenium compounds in mice determined as: sodium selenate > sodium selenite > elemental nanoSe > nutritional additive Sel-Plex (selenized yeast) > probiotic LactoMicroSe > selenium protein powder.

As a well-known genotoxic assay, the mouse bone marrow micronucleus test has proven suitable for the evaluation of genotoxic and antigenotoxic actions of compounds [13]. And the frequency of micronucleated polychromatic erythrocytes is a reliable measure of both chromosome loss and breakage [14]. In the micronucleus test, our results showed that within the dose range of 580-2300 mg/kg body weight, selenium protein powder did not trigger an increase of micronucleus ratios in mce, which indicated that selenium protein powder was not mutagenic to somatic chromosomes in mice.

The sperm aberration test is a method of evaluating genetic injury to male reproductive cells of mammals (including man) *in vivo*. Our results showed that, within the dose range of 2250-10000 mg/kg body weight, the selenium protein powder did not have any detectable mutagenic effect on sperm cells, and suggested that it would not trigger genetic injury to male reproductive cells of mammals *in vivo*. According to the National Research Council, the recommended dietary allowance of selenium is 55μg/day for adult men and women[15], and the recommended daily oral dose of selenium protein powder is 500-1000 mg/day. In sperm aberration test, the highest dose of selenium protein powder was settled to 10000 mg/kg because the dose had reached 700 times of the recommended daily oral dose (1000 mg) for a person of 70 kg body weight.

Base on the results of above studies, we could conclude that selenium protein powder is safe at the dose of daily-recommended consumption and has the potential for development and application in food systems or functional foods.

5 Conclusions

The oral LD_{50} for selenium protein powder exceeded 31.25 g/kg body weight in mice. Selenium protein powder were found to be non-mutagenic and non-genotoxic in the micronucleus test and mouse sperm aberration test, and it has greater safety as a nutritional selenium supplement.

Acknowledgements

This research was supported by Research Program of Application Foundation and Advanced Technology of Tianjin (grant number: 15JCYBJC30600) and One-three-one Program of Innovative Talents Training of Tianjin.

References

1. B.R. Cardoso, B.R. Roberts, A.I. Bush, D.J. Hare, Metallomics **7**, 1213-1228 (2015)
2. W. Baowei, H. Guoqing, W. Qiaoli; Y. Bin, J. Anim. Physiol. Anim. Nutr. **95**, 440-448 (2011)
3. M.P. Rayman, Lancet **365**, 233-241(2000)
4. M. Kieliszek, S. Blazejak, Nutrition **29**, 713-718 (2013)
5. X. Wang, Y. Wang, S. Qin; H. Wang, J. Wang, China Animal Husbandry and Veterinary Medicine **41**, 148-152 (2014)
6. R.J. Shamberger, Mutat. Res. **154**, 29-48 (1985)
7. J.C. Griffiths, R.A. Matulka, R. Power, Int. J. Toxicol. **25**, 477-485 (2006)

DOI: 10.1051/bioconf/20170802009

8. F.R. Moore, G.A. Urda, G. Krishna, J. C. Theiss, Mutat. Res. **367**, 33-41 (1996)
9. Ministry of Health, PR China, National Standards of PR China, GB15193.1–15193.21.(2003)
10. I. Benko, G. Nagy, B. Tanczos, E. Ungvari, A. Sztrik, P. Eszenyi, J. Prokisch, G. Banfalvi, Environ. Toxicol. Chem. **31**, 2812-2820 (2012)
11. J.C. Griffiths, R.A. Matulka, R. Power, Int. J. Toxicol. **25**, 465-476 (2006)
12. M. Shakibaie, A.R. Shahverdi, M.A. Faramarzi, G.R. Hassanzadeh, H.R. Rahimi, O. Sabzevari, Pharm. Biol. **51**, 58-63 (2013)
13. B.H. Narayan, N. Tatewaki, V.V. Giridharan, H. Nishida, T. Konishi, Food Funct. **1**, 174-179 (2010)
14. M.A. Furtado, L.C.F Almeida, R.A. Furtado, W.R. Cunha, D.C. Tavares, Mutat. Res. Genet. Toxicol. Environ. Mutagen. **657,** 150-154 (2008)
15. National Research Council. *Dietary Reference Intakes*, 300-307 (Washington, DC: National Academy Press, 2000)

Analysis of genotype effects for the immunosuppression via two-step method

Xiaona Sheng[1], Wanqiu Xie[2] and Ying Zhou[2, a]

[1]*School of Information Engineering, Harbin University, Harbin 150086, China*
[2]*School of Mathematical Sciences, Heilongjiang University, Harbin 150080, China*

Abstract. This paper studies the main effects and interactive effects between genes on immunosuppression susceptibility caused by ultraviolet radiation in population of mice. We present a two-step strategy, i.e., we first establish one full linear model based on all main effects and interactive effects, and use the Dantzig selector method to screen the genotype effects preliminary; then via the idea of stepwise regression, under the other model we further detect the significant main effects and interactive effects for the UV-induced immunosuppression susceptibility. The most significant main effect site that we identified is D10Mit170, and the most significant interactive sites are D6Mit389 and D16Mit131.

1 Introduction

The main effects and interactive effects of genotypes play important role on the expression of the trait of biology [1]. Previously, researchers mainly focus on the detection of main effects, however, more and more studies have shown that main effects only explain part of genetic variation of the trait of biology, and the interactions between loci are the important genetic foundation which causes some complex traits [2, 3]. Therefore, in current genome-wide association analysis (GWAS), it is necessary to recognize the loci with interactions, although it is considered as open difficulty.

Millions of SNPs are the research object in the GWAS. However, when detecting interactions for the high-dimension data, traditional methods face unprecedented challenge on the aspects of complex degree of algorithm, computing speed, etc. Currently, some new statistical methods have been applied to the detection of interactive effects, such as the machine learning method [4, 5], the data mining method [6], variable selection [7, 8], two-step method[3, 9], and so on. The Dantzig selector method proposed by Candes and Tao can deal with the problem that the number of variables is mcuh larger than that of observations [7].

In this paper, we performed deep statistical analysis on the genotype effects for immunosuppression in Mice. The data is from the literature [10]. Clemens et al. [10] collected the genotype data on 64 SNPs and immunosuppression data of 134 backcross individuals, and they detected some main effects and interactions among these loci. Here we proposed a new strategy and found different genotype effects for the immunosuppression.

[a] Corresponding author: yzhou@aliyun.com

2 Theory and method

From a model selection view in multivariate regression, Candes and Tao [7] proposed a famous and effective method called Dantzig selector (DS), i.e., for linear model $y=X_{n\times p}\beta+z$, where parameter vector $\beta \in R^p$ and $p>>n$, X is a data matrix, and z is error vector. The DS estimator is solution to the l_1 regularization problem

$$\min_\beta \|\beta\|_{l_1} \quad \text{subject to} \quad \|X^T(y-X\beta)\|_{l_\infty} \le (1+t^{-1})\sqrt{2\log p}\cdot\sigma$$

This estimator can control the loss within a reasonable region of mean squared error, and it can well deal with the situation where the number of variables or parameters is much larger than the number of observations n in usual linear model.

In the genetical problem we considered in this paper, the true parameter vector is high-dimension and sufficiently sparse in general, therefore we can take advantage of the DS method to estimate the effect parameters by building linear model. Next, we presented our new two-step strategy of effect estimating.

Step I: Detecting possible main effects and interacting loci

First, we consider the following full linear model composed of all main effects and interactive effects

$$E[Y \mid G] = \gamma + \sum_{i=1}^{p} a_i g_i + \sum_{j<k} b_{jk}(g_j g_k) \qquad (1)$$

After computation with the DS method, we obtain the estimates of QTL effects, many of which are zero in fact, and some significant main effects are same with those given in the literature [10]. Owning to this fact, we choose the most significant six main effects and the possible interacting loci to build a new statistical model to further detect the significant interacting loci.

Step II: Further searching significant interactive loci

The new statistical model built in Step II is

$$E[Y \mid G'] = \mu + a_1 g_1' + a_2 g_2' + a_3 g_3' + a_4 g_4' + a_5 g_5' + a_6 g_6' + \sum_{i<j} b_{ij}(g_{ij}') \qquad (2)$$

where $g_1', g_2', g_3', g_4', g_5'$ and g_6' denote the genotypes of the 6 main-effect loci associated with the immunosuppression (i.e., D1Mit411, D6Mit389, D10Mit170, D14Mit260, D17Mit49 and D19Mit19), and g_{ij}' ($i<j$) denote the 89 genotype pairs of the possible interactive loci detected in Step I. Here we applied the stepwise regression to deeply estimate the genotype effects via model (2).

3 Results

The most significant main-effect locus detected by the proposed two-step method is D10Mit170, with the effect estimate -36.847 and the P-value is 0.0188. Meanwhile, we further detected some interactive loci which were not reported in existing literatures. The information of the most significant 11 interactive effects obtained by the two-step method was listed in the following Table 1. The results obtained from the new strategies can well supplement the existing results in current research field. The last column denotes the P-values of testing interactive effects, and the smaller the values are, the more significant the corresponding interactive effects are. Meanwhile, we considered the correlation coefficients of each interactive pairs, and correlation coefficient of D14Mit266 and D19Mit34 was the largest, which further verified that their interaction was significant.

Table 1. Information of interactive effects obtained by the two-step method.

Locus 1		Locus 2		Test results		
Chr.	Position	Chr.	Position	C(P) value	F-value	P-
6	D6Mit389	16	D16Mit131	-32.787	8.18	0.0049
7	D7Mit238	18	D18Mit110	-28.840	10.59	0.0014
14	D14Mit266	19	D19Mit34	-39.554	3.31	0.0071
17	D17Mit187	17	D17Mit123	-39.076	6.53	0.0118
7	D7Mit362	9	D9Mit18	-35.022	6.05	0.0152
1	D1Mit411	5	D5Mit239	-40.832	6.03	0.0155
8	D8Mit14	15	D15Mit226	-40.742	3.12	0.0800
8	D8Mit4	18	D18Mit110	-39.052	2.58	0.1111
11	D11Mit339	19	D19Mit34	-39.605	2.49	0.1169
4	D4Mit16	9	D9Mit297	-40.109	2.25	0.1361
5	D5Mit31	18	D18Mit49	-39.155	2.13	0.1474

4 Simulation studies

Simulation studies are performed to illustrate and evaluate the proposed two-step algorithm of detecting genotype effects.

For illustration, we consider the situation that a quantitative trait is contributed by four SNPs on a single chromosome. Loci 3 and 4 have interactive effects, and the trait value is generated by the following model

$$Y = b_0 + \sum_{i=1}^{4} b_i G_i + b_{34} G_3 G_4 + \varepsilon$$

where $\varepsilon \sim N(0, \sigma^2)$. We choose sample size n = 300, 500, 750 and 1000, respectively. The simulation is performed 1000 times and the power that the interactive effects are correctly detected is used to measure the precision of the new two-step strategy.

To further examine the effect of different possible factors on the performance of the proposed method, we considered two scenarios: (i) Loci 3 and 4 only have interactive effect, i.e., $b_3 = b_4 = 0$; (ii) Loci 3 and 4 have both interactive effect and main effects. Different values of heritability are taken so that the coefficients in model (3) and σ^2 can be determined, and the simulated data can be generated correspondingly.

The simulation results under the first scenario were presented in Figure 1. It can be seen from the simulation results: (1) In each case of sample size, the detecting powers increase with the increase of heritability, for example, the detecting power increases from 0.316 to 0.674 when the heritability changes from 0.1 to 0.5 for the sample size n = 300; (2) As expected, the detecting power increases as the sample size increases. From the character of the curves in Figure 1, there is no interaction between the two factors of the sample size and the heritability. The simulation results under the second scenario were similar.

5 Discussions

In this paper, we have developed an efficient two-step method to estimate genotype effects for a real data set of mice. Since the number of parameters is much larger than that of observations, under the framework of linear model, we adopted the DS method to decrease the dimension of parameters and obtained some candidate main-effect loci and interactions in the first step, and then we search deeply among these loci by the stepwise regression in the second step. By analyzing the mice data, we found some existing genotype effects that have been reported; meanwhile we also detected some new main effects and interactions (Additional results were not shown in this paper, limited by the length of the paper).

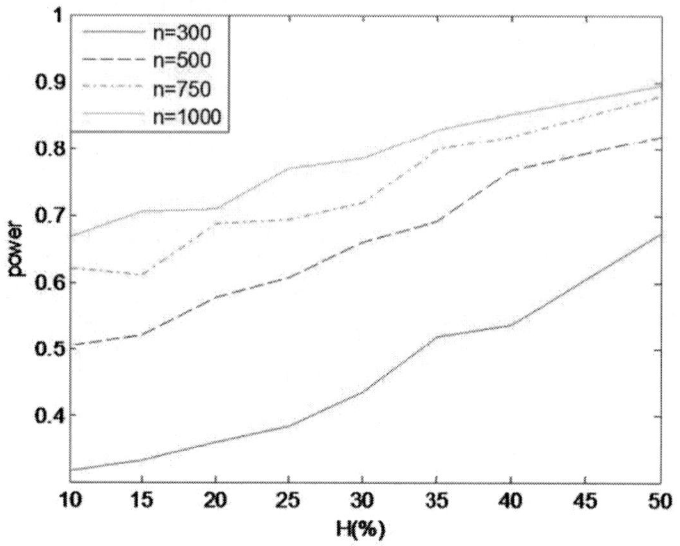

Figure 1. Detecting powers of the new method under different conditions

From the simulations we found all detecting powers of are reasonable and satisfactory in each simulation scenario, which shows the performance and advantage of the new strategy.

Although we describe our methods in the context of a mice population, it can be extended straightforwardly to the case of other populations including human population. Aiming at the detections of genotype effects for complex traits (or diseases) in human population, however, the strategy of effect detecting needs further research.

Acknowledgments

This research was supported by the Scientific Research Foundation of Department of Education of Heilongjiang Province of China (No. 1253G044) and the Science and Technology Innovation Team in Higher Education Institutions of Heilongjiang Province (No. 2014TD005).

References

1. S.K. Musani, D. Shriner, N. Liu, R. Feng, C.S. Coffey, N. Yi, H.K. Tiwari, D.B. Allison, Detection of gene-gene interactions in genome-wide association studies of human population data. Hum. Hered. **63**, 67-84 (2007)
2. J.Y. Dai, C. Kooperberg, M. Leblanc, R.L. Prentice, Two-stage testing procedures with independent filtering for genome-wide gene–environment interaction. Biometrika **99**, 929-944 (2012)
3. W. Ma, C.F. Yuan, H. Liu, W. Zheng, Y. Zhou, Genome-wide interaction analysis of quantitative traits in outbred mice. Genet. Res. **97**, 1-8 (2015)
4. B.A. McKinney, D.M. Reif, M.D. Ritchie, J.H. Moore, Machine learning for detecting gene-gene interactions: a review. Appl. Bioinformatics **5**, 77-88 (2006)

DOI: 10.1051/bioconf/20170802010

5. S.J. Winham, C.L. Colby, R.R. Freimuth, X. Wang, M. de Andrade, M. Huebner, J.M. Biernacka, SNP interaction detection with random forests in high-dimensional genetic data. BMC Bioinformatics **13**, 164 (2012)
6. M.D. Ritchie, L.W. Hahn, N. Roodi, L.R. Bailey, W.D. Dupont, F.F. Parl, J.H. Moore, Multifactor-dimensionality reduction reveals high-order interactions among estrogen-metabolism genes in sporadic breast cancer. Am. J. Hum. Genet. **69**, 138-147 (2001)
7. E. Candes and T. Tao, the Dantzig selector: statistical estimation when p is much larger than n. Ann. Statist. **35**, 2313-2351 (2007)
8. R. Tibshirani, Regression shrinkage and selection via the lasso. J. Royal Statist. Soc. B **58**, 267-288 (1996)
9. P. Zhang, J.P. Lewinger, D.V. Conti, J.L. Morrison, & W. J. Gauderman, Detecting gene-environment interactions for a quantitative trait in a genome-wide association study. Genetic Epidemiology **40**, 5 (2016)
10. K.E. Clemens, G. Churchill, N. Bhatt, K. Richardson, F.P. Noonan, Genetic control of susceptibility to UV-induced immunosuppression by interacting quantitative trait loci. Genes and Immunity **1**, 251-259 (2000)

Study on treatment effect of hearing aids on bilateral prolonged tinnitus in Chinese patients

Feng Liang[1], Fanglei Han[1] and Lin Li[2,a]

[1]*Department of Anesthesiology, China- Japan Union Hospital, Jilin University, 130033, China*
[2]*Dept. Otolaryngolgy and head & Neck, China- Japan Union Hospital, Jilin University, 130033, China*

Abstract. Purpose: This study aims to discuss the treatment effect of hearing aids on tinnitus. Methods Send questionnaires to 48 cases of bilateral prolonged tinnitus patients with course of more than 1 year to evaluate the subjective perception state of the tinnitus after 24 months. At the same time, follow up 10 cases of bilateral prolonged tinnitus patients without receiving the treatment of hearing aids and use them as the contrast group. Results: Through the treatment of 24 months, the treatment effects on tinnitus are as follows: 23 cases were cured (47.92%), 14 patients were improved significantly (29.17%), 6 cases were improved (12.5%), and 6 cases were invalid (12.5%). The overall effective rate was 87.5%.Conclusion: The hearing aids have treatment effect on the bilateral prolonged tinnitus patients in China.

1 Instruction

The tinnitus usually occurs with deafness and can be a primary central disease (which may be caused by central blood supply efficiency, brain tumor, cranial injury and operation and neurasthenia), and long-term abnormal stimulation for the auditory pathway will also induce a memory in the auditory center to generate long-term tinnitus. The drug treatment effect on the tinnitus is poor. Questionnaires for the bilateral prolonged tinnitus patients who were suggested to wear the hearing aids were used in this study to evaluate the treatment effect of the hearing aids on tinnitus.

2 Materials and Methods

2.1 Patient data

There were 48 patients who wore the hearing aids for deafness associated with bilateral prolonged tinnitus in Otorhinolaryngology Head and Neck Surgery Department of China-Japan Union Hospital of Jilin University from January to December, 2013, and their courses of disease were more than 12 months. 20 cases were males, 28 cases were females. They were 56-76 years old and the median age was 68. The contrast group had 10 cases, with 5 males and 5 females, they were 46-70 years old, and the median age was 65.

The study was approved by the institutional review board of China-Japan Union Hospital. All the participants provided written informed consent before the study.

[a] Corresponding author: lilin01@jlu.edu.cn

The diagnostic assessment of patient with tinnitus should include a case history, audiologic assessment (audiometry, stapedial reflexes, otoacoustic emissions [OAEs]), psychoacoustic tinnitus measures (minimum masking level, loudness, pitch) before trying on the hearing aids. Assessment of tinnitus severity and psychiatric comorbidity (depression, anxiety) using standardized questionnaires before the treatment and at the end of two years follow up.

According to the severity of tinnitus and tinnitus with and without concomitant symptoms, the degree of tinnitus was divided into 6 grades [1]: grade 0: the patient is free of tinnitus; 1 grade: the patient has tinnitus occasionally, but he does not feel pain; grade 2: the patient has prolonged tinnitus, and the tinnitus aggravates when the environment is quiet; grade 3: the patient still has prolonged tinnitus even in noisy environment; grade 4: the patient has prolonged tinnitus associated with attention and sleep disorders; grade 5: the patient has prolonged severe tinnitus and cannot work properly; grade 6: the patient has the suicidal tendency because of the severe tinnitus. Both ears of each of 48 patients in the treatment group wore the hearing aids.

2.2 Study design

Digital hearing aids are selected according to the degree of hearing loss, after the patient wears the hearing aids for more than 24 months, subjective changes of the tinnitus are observed, and the grade of tinnitus is assessed according to the grading methods of tinnitus. The patient uses the hearing aids for more than 10 hours every day. After 2 years, the patient were asked whether they accept any other drugs to treat the tinnitus again, and if there was a history of treatment , the patient would withdrawn from the study.

2.3 Criteria of curative effect and follow-up

All the 48 patients finish the study, at the end of the 24months follow-up, all of them filled in a questionnaire to give the result of the treatment and also whether they were satisfied with the result. Curative effect: It is effective if the tinnitus symptom improves to grade 2; it is remarkably effective if the tinnitus symptom improves to grade 2 or above; and the patient is cured if the tinnitus symptom disappears. The patients do not use drugs for tinnitus treatment in this period.

Results

All the 42 patients finished the study. 42 patients who wore the hearing aids on their two ears clearly stated that the treatment was effective. 23 cases of patients were cured (47.92%), 14 patients were improved significantly (29.17%), 6 cases were improved (12.5%), and 6 cases were invalid (12.5%). The overall effective rate was 87.5%. While in the contrast group, no patient stated that the tinnitus was relieved. The hearing aids have treatment effect on the tinnitus through the chi-square test ($P< 0.05$).

Table 1. the effect on the treatment of tinnitus compared with hearing aids group and control group.

	Effective	invalid
Hearing aids group(n=42)	36	6
Control group (n=10)	0	10
P<0.05		

Discussions

Tinnitus is a distressing problem involved in auditory perception, and affects the quality of life of the suffered. Until now, there was no effective or specific drug therapy available, although much research work into mechanism and possible treatment is underway. Many clinic studies found that the drug treatment on subjective tinnitus is uncertain [2]. The tinnitus retraining therapy (TRT) was now the suggested treatment for tinnitus [3], the mechanisms is that TRT could reduce the sensitivity of a central system to the tinnitus until it disappears, to rebuild the filtering function of the auditory system, to stop the auditory feeling on tinnitus and to enable the auditory center to forget the memory for tinnitus gradually. TRT consists of counseling and sound therapy, both based on the neurophysiological model of tinnitus [4]. Although TRT is efficacious in most patients, but some patients could not accepted the treatment because of the economical burden and district limited.

The tinnitus happened together with the hearing loss in many cases. Hearing aid was used to improve the hearing level and language reorganization, and nowadays the hearing aid was already accepted by the doctor and patients in China. In some case the tinnitus was found immediately dismissed even when the patient was trying on the hearing aid. But sometimes the patients felt no changing on their tinnitus. But more and more researches pointed out that the treatment of hearing aids may have certain treatment effect on the tinnitus associated with hearing loss[5], and some methods, for example, both tinnitus handicap inventory (THI) and the international outcome inventory for hearing aids could decrease the tinnitus significantly, both of them improved the patients' satisfactions[6]. After the patients wore the hearing aids, background noise was amplified properly, and the hearing aids could have the effect of the tinnitus retraining therapy when the surrounding environment was quiet. Our study found that if the patients wore the hearing aids for more than 10 hours each day continuously for a long time (2 years), they would have a good treatment effect on most bilateral prolonged tinnitus. In addition, the psychological states of the patients may also be adjusted by the hearing aids, so that the subjective feeling of the tinnitus of the patients can be alleviated [7-8].

In conclusion, in Chinese group, we could use the hearing aid to treat the prolonged tinnitus even the patients who found it was no effect at the beginning of the hearing aids trying. And patient's education might be also important to encourage the patients to have confident on the longtime treatment.

Acknowledgement

All authors declared no conflict of interest in the research. The work presented in this report is the subject of a pending patent filed by the China-Japan Union Hospital of Jilin University (201510382781.X). This study was supported by the National Natural Science Foundation of China (811007020), the Health and Family Planning Foundation of Jilin Province (20152046), the Science and Technology Development Plan Foundation of Jilin Province (20160101070JC), and the Norman Bethune Program of Jilin University (2015305).

References

1. Trevis KJ, McLachlan NM, Wilson SJ. Front Psychol. **7** 1262 (eCollection 2016)
2. Michiels S, Van de Heyning P, Truijen S, Hallemans A, De Hertogh W. Man Ther. **26** 125-131 (2016)
3. Kim SH, Byun JY, Yeo SG, Park MS. J Int Adv Otol. **12(1)** 72-76 (2016)
4. Jastreboff PJ. HNO.**63(4)** 307-311(2015)
5. Sereda M, Hoare DJ, Nicholson R, Smith S, Hall DA.Ear Hear. **36(4)**:417-29(2015)
6. Zarenoe R, Söderlund LL, Andersson G, Ledin T. J Am Acad Audiol. **27(8)** 669-76 (2016)
7. Cabral J, Tonocchi R, Ribas Â, Almeida G, Rosa M, Massi G, Berberian AP. Int Tinnitus J. **20(1)** 54-58 (2016)
8. Derek J Hoare, Phillip E Gander, Luke Collins, Sandra Smith, and Deborah A Hall. J Eval Clin Pract. **18(2)** 326–334(2012)

Effect of Cornus officinalis glycosides on the myelin basic protein in blood serum and the nerve function in Balb/c mice after sciatic nerve injury

Feng Liang [1], Suming Wu [2] and Shuang Qi [1,a]

[1]*China–Japan Union Hospital of Jilin University*
[2]*The 4th Hospital of Jilin University(The General Hospital of FAW)*

Abstract Objective: to explore the effect of cornus officinalis glycosides(COG) on the myelin basic protein(MBP) in blood serum andthe nerve function in Balb/c mice after sciatic nerve injury.Methods : Adult male Balb/c mice were randomly divided into high-dose, middle-dose, low-dose COG groups,.Transecting and inosculating the right sciatic nerve; testing the concentration of MBP by ELISA in 3, 5 days, 1 , 2 , 4 , 8 weeks and having electroneurophysiological detection in 4, 8 and 12 weeks after surgery respectively; making statistical analysis of the results. .Results : The MBP concentration and results of lectroneurophysiological detection have clear dose-effect relationship with the COG dose. Conclusion: Through inhibiting the demyelination of the sciatic nerve , COG can lower the degree of nerve damage so that accelerating the nerve regeneration.

1 Introduction

The thesis applies COG to Balb/c mice with sciatic nerve injury and intends to provide theoretical foundation for the effects of COG on peripheral nerve injury through the fuctional determination of MBP and nerve electrophysiology.

2 Materials and Methods

2.1 Animal Grouping and Molding

84 8-week healthy male Balb/c mice weighing (20 + 2) g were raised in the same condition. Injecting with anesthesia in abdominal cavity by 1% pentothal sodium 100 mg/kg. The operating field was disinfected routinely. Oblique incision was performed on the right side of the hip; the muscle got blunt separation; the sciatic nerve was exposed. Completely cutting off the nerve trunk 0.5cm under the ischial tuberosity; under 12-times surgical microscope, using 11-0 atraumatic suture needle to anastomose the lacerated ends; then closing the wound layer by layer. Randomly dividing the mice into high-dose COG group (320 mg/kg/d), middle-dose COG (160 mg/kg/d), low-dose COG group (80 mg/kg/d) and normal saline control group with 16 cases in each group. Each group was

[a] Corresponding author: qishuang@jlu.edu.cn

administrated the drug by gavage continuously till death. Observing the mice at 7 time points, namely 3 days, 5 days, 1 week, 2 weeks, 4 weeks, 8 weeks and 12 weeks after surgery.

2.2 Detection of Related Indexes

Using Medtronic Keypoint myoelectricity/evoked potential machine to test the bilateral sciatic nerves of mice 4 weeks, 8 weeks and 12 weeks after surgery. The indoor temperature was 24 °C; using 1% pentothal sodium for intraperitoneal anesthesia; making routine disinfection on the surgical field; the prone position showed the sciatic nerve; stabbing the soleus of mice with pin needle electrode; the muscle belly was a recording spot (M); the grounding electrode was placed at the tails of mice. Performing parallel stimulating electrode (the fixed space between the two electric poles was 2 mm) on the anastomotic stoma of the sciatic nerve near the ischial tuberosity level (P) and the branch of distal sciatic nerve (D) with strong stimulation(10mA electric current). Using coliper to measure and input the distance between the stimulating electrodes; then the instrument would calculate MNCV automatically. The formula is as follows:

MNCV = the distance between the two stimulating electrodes/latency difference value of the action potential.

Measuring the MBP of mice with ELISA method :According to time sequence, getting 3 mice from each group; drawing about 0.5ml eyeball blood; holding the blood with an Ep tube in 4°C refrigerator overnight after 2-hour standing in normal temperature. In the morrow, centrifuging the blood at the speed of 5000 r/min for 3 min; taking the upper serum and storing it in -20 °C refrigerator for standby application; having ELISA determination after the animal materials were obtained.

2.3 Statistical Analysis

Applying statistical software SPSS 13.0 for data analysis. The results of each group of samples were expressed by mean±standard deviation ($\bar{X} \pm S$) ;using t to verify the contrast between administered groups as well as the contrast between administered groups and the control group; if $P < 0.05$, then the difference has statistical significance.

3. Results

comparing the MNCV of each COG group with the control group 4 weeks, 8 weeks, 12 weeks after surgery, the data got from each group has difference. Comparing different time points of high-dose group, middle-dose group and low-dose group respectively, significant differences exist while there was no significant difference between high-dose group and middle-dose group. The MNCV which was obtained 4 weeks, 8 weeks and 12 weeks after surgery is shown in table 1.

The contrasting result of the MBP concentration in serum at each time point is shown in table 2. Using t to check the results of high-dose, middle-dose and low dose groups with the control group respectively. If $P < 0.05$, then there exist significant differences.

Table 1 Testing results of each group 4, 8 and 12 weeks after surgery ($\bar{X} \pm S$ n=3), m/s

Groups	4 weeks after surgery	8 weeks after surgery	12 weeks after surgery
High-dose group	41.1 ± 0.72^{abc}	63.9 ± 1.64^{abc}	64.8 ± 1.52^{abc}
Middle-dose group	37.4 ± 0.93^{ab}	58.4 ± 1.17^{ab}	59.7 ± 0.83^{ab}
Low-dose group	33.8 ± 0.79^{a}	52.6 ± 0.79^{a}	54.1 ± 0.39^{a}
Control	28.6±0.55	50.4±0.46	47.7±3.36

group

Comparing each COG groups with the control group, [a]0.01 < P < 0.05; comparing high-dose and middle-dose groups with low-dose group, [b]0.01 < P < 0.05; comparing high-dose group with middle-dose group, [c]P > 0.05

Table 2 The comparison results of the MBP concentration in serum at different time points ($\bar{X} \pm S$ n=3), ng/l

Time points	High-dose group	Middle-dose group	Low-dose group	Control group
3d after surgery	6984±330.2[a]	7625±209.5[b]	8329±207.4[c]	9657±323.5
5d after surgery	6119±75.9[a]	6550±199.2[b]	7441±175.2[c]	8752±206.1
1w after surgery	4812±142.7[a]	5402±124.6[b]	6035±114.3[c]	7821±245.2
2w after surgery	3084±122.3[a]	3815±110.9[a]	4723±109.1[b]	6836±122.4
4w after surgery	1665±140.8[a]	1987±76.5[a]	2196±96.1[b]	2456±82.6
8w after surgery	1422±66.0a	1596±63.2a	1822±72.3b	2108±77.6

Comparing the data of each COG groups with the control group, [a]P<0.01, [b]0.01<P<0.05, [c]P>0.05.

4 Discussion

COG is a kind of traditional Chinese medicine with multiple effects such as immunosuppression, antidiabetic, anti-inflammatory effect, oxidation resistance, antiarrhythmic effect, etc[1]. In recent years, many studies show that COG can significantly increase the number of newborn mature neurons and blood vessels of mice with cerebral injury and improve their nervous system functions[2]; but there has no report about its impact on nerve regeneration and repairing after peripheral nerve injury.

The majority of peripheral nerves are medullated nerves. The regeneration and repairing of myelin sheath has close relationship with the regeneration of nerve fibers. MBP is the main protein of myelin sheath and it has neural specificity. Under normal physiological condition, its content in serum is extremely low[3-4]. After peripheral nerve injury, the parclose between nervus vascularis is damaged and the expression quantity of MBP in blood volume increases rapidly, so the different expressions of MBP can dynamically observe the situation of myelin damage and repairing[5].

This experiment adopted ELISA method to detect the MBP concentration in the sciatic nerve and serum. The experimental results show that in the early and middle stage of nerve repairing, the MBP concentration of COG groups is lower than the control group and there exist significant differences, which means that COG can effectively reduce the demyelination degree of sciatic nerve in nerve regeneration. The testing results of nerve electrophysiology suggest that the MNCV of each group increases continuously over time and the MNCV of COG groups has great advantage over the control group in each time point. It shows that COG can lower the expression of MBP in serum to reduce the disintegration of myelin sheath and axon so as to protect the myelin sheath and promote the repairing of injured sciatic nerve.

In brief, COG can accelerate the regeneration myelin sheath and axon of the injured sciatic nerve. Its mechanism may be inhibiting the expression of MBP in serum to reduce partial damage level of injured nerves so as to accelerate the nerve regeneration.

References

1. Miao Y, Xiaoxian W, Military Medical Journal of Southeast China,12,240(2010)
2. Yao RQ,Zhang L,Wang W, Brain Res Bull, 79,69, (2009)
3. Min Y,Kristiansen K,Boggs J M, Proceedings of the National Academy of Sciences,106,315, (2009)

4. Tompa P,FEBS letters,579,3346 (2005)
5. Brettschneider J, Petzold A, Sussmuth S,Journal of neurology, 256,3(2009)

BIO Web of Conferences **8**, 02013 (2017) DOI: 10.1051/bioconf/20170802013
ICMSB2016

Research on Algorithm for Feature Extraction and Classification of Motor Imagery EEG Signals

Juan Tian[1,a] and Zhaochen Zhang[1]

[1]*College of Medical Information and Engineering, Taishan Medical University, Taian 271016, China*

Abstract. This paper made a research on the feature extraction and pattern recognition of left and right hands motor imagery EEG signals. In combination with the data from BCI Competition III, denoising preprocessing is carried out for EEG signals firstly; and then, the relative wavelet energy is extracted as a feature vector from the Channels C_3 and C_4 by use of the algorithm for relative wavelet energy, and pattern recognition is carried out by use of the radial basis function neural network (RBFNN). Simulation results show that the proposed method achieves good classification results.

1 Introduction

Brain-Computer Interface (BCI) is a short form of the interface between brain and computer. It reflects the consciousness of people through the EEG signals but not the normal physiological output pathways constituted by peripheral nerves and muscles, so it is a new way of communication and control [1]. There are many kinds of EEG signals generally used in BCI. As the modes of generation are different, EEG signals are mainly divided into two types, i.e., induced EEG signals and spontaneous EEG signals [2]. Motor imagery EEG signal is a kind of spontaneous EEG signal which is the most widely used. The research on BCI projects based on motor imagery EEG signals was firstly carried out by Pfurtscheller et al. The result showed that, during unilateral limb movement or motor imagery, Event related desynchronization(ERD) was produced in the contralateral brain areas and event related synchronization (ERS)[2] was produced in the ipsilateral brain area. Specifically, during the experiment, the cerebral cortex where channels C_3 and C_4 were located represented the area of right and left hands motor imagery in the brain; during the motor imagery, ERD/ERS in the areas on both sides of the brain was symmetrical, i.e., during the right hand motor imagery, ERD occurred on the cerebral cortex for motor imagery in the left area of the brain while ERS occurred on the cerebral cortex in the right area of the brain; and vice versa. Further research showed that during the motor imagery, obvious ERD/ERS phenomenon occurred mainly within the frequencies of Rhythm μ (8-12Hz) and Rhythm β (18-24Hz) of the corresponding cerebral cortex for motor imagery in the contralateral brain areas, and at these time points, the energy in the frequencies of Rhythm μ and Rhythm β decreased (or increased)[3-6].

A complete BCI system mainly includes acquisition, preprocessing, feature extraction, pattern recognition, control equipment and peripheral equipment for EEG signals [7], among which feature extraction and pattern recognition are the most critical parts. The main contents of this paper include feature extraction and pattern recognition of EEG signals by use of wavelet transformation and fuzzy

[a] Corresponding author: tianjuan0001@163.com

neural network, taking the ERD/ERS phenomena corresponding to left and right hands motor imagery as the differentiating criteria in combination with the characteristics.

2 Feature Extraction

The purpose of feature extraction is to transform the preprocessed EEG signals into different feature vectors representing different consciousness tasks. Feature extraction is the most critical step in the identification of BCI signals, directly impacting on the design for the later classifier. In this paper, the relative energy extracted by use of the wavelet transformation technology from the characteristic frequency band is mainly used as the characteristic value.

2.1 Source of Experimental Data

The experimental data used in this paper comes from Dataset IIIb of BCI Competition III held in 2005. It was provided by Graz University of Technology in Austria and divided into two categories: left hand motor imagery and right hand motor imagery. The EEG data was collected from three healthy subjects, denoted as O3, S4 and X11. Among them, O3 had 640 pieces of data, and the experimental task was to adopt the method of virtual reality to control the feedback items on the screen by use of left and right hands motor imagery; S4 and X11 had 1,080 pieces of data respectively, and the experimental task was to control the virtual ball through motor imagery, moving the ball to the left or right end of the screen. All the data were collected from the EEG signals recorded in two-lead mode at electrodes C_3, C_Z and C_4, among which the electrodes C_3 and C_4 were located in the functional area of the primary sensorimotor cortex of the brain, and they could reflect the most effective information about the changes in brain status during the left and right hands motor imagery of subjects; as a reference electrode, the sampling frequency of C_4 is 125Hz after 0.5-30Hz filtering. In order to reduce the amount of calculation, the experimental data was selected from S4 for this time, and there was a total of 1171677 x 2 pieces of data[8].

The complete experimental procedure is described as follows: the duration of each experiment is 8s, and it is divided into three stages. During the first 2s, the subjects have a rest to keep a relaxed state. When T=2s, they begin to enter the stage of imagery; when T=3s, a short beep sounds and it suggests that the subjects' consciousness is being concentrated and the motor imagery is beginning; meanwhile, the cursor on the screen moves left or right, which will last for the whole process of the experiment, while the subjects imagine the movements as planned according to the movement of the cursor. When T=4s, the stage of information feedback begins, and the duration is 4s (within 4s-8s). The classification results of the feedback stage are given to the subjects from the classifier. The experimental process is shown in Figure 1.

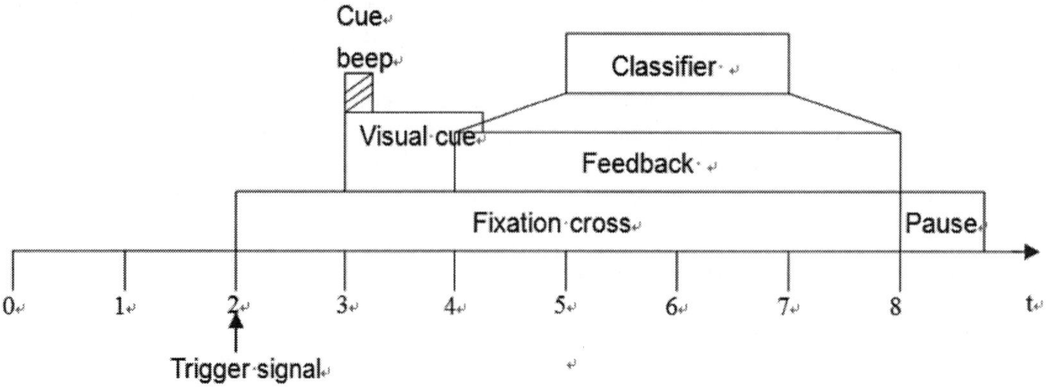

Figure 1. Experiment Process of Left and Right Hands Motor Imagery

2.2 Preprocessing of Experimental Data

The research shows that, during the acquisition process of EEG signals, false information such as ECG, EMG, REM, blink and power frequency interference will interfere with the observation and analysis against the EEG signals by people. Therefore, before the feature extraction, in this paper, denoising is carried out for the original EEG data by use of the wavelet transformation technology and collect the total a trigger 8 seconds in the data. The comparison of EEG signals before and after denoising is shown in Figure 2.

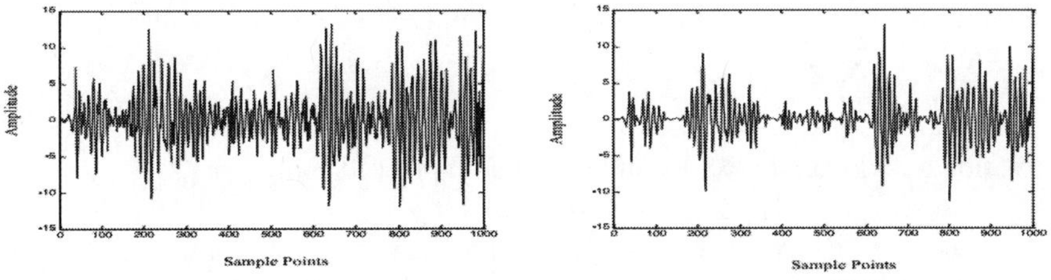

(a) (b)

Figure 2. Comparison of EEG Signals Before and After Denoising (a) Before Denoising (b) After Denoising

2.3 Feature Extraction Based on Wavelet Transformation

2.3.1 Basic Theory of Wavelet Transformation

Wavelet transformation is a multi-scale signal analysis method. In both time and frequency domains, it represents the local signal characteristics and is used for multi-resolution analysis, so it is applicable to analyze the transient and time-varying characteristics of non-stationary signals. EEG signals are a kind of complex nonlinear and non-stationary signal. Therefore, in this paper, wavelet transformation is used for the analysis and feature extraction of EEG signals [9].

It is assumed that the wavelet basis function is $\psi_{a,b}(t)$, then,

$$\psi_{a,b}(t)=\frac{1}{\sqrt{|a|}}\psi(\frac{t-b}{a}) \tag{1}$$

where, *a* represents the contraction-expansion factor or scale factor, and *b* represents the shift factor.

Discrete binary method is adopted for the calculation of "*a*" and "*b*", and the following discrete binary wavelet function can be obtained:

$$\psi_{j,k}(t)=2^{-\frac{j}{2}}\psi(2^{-j}t-k) \tag{2}$$

The discrete wavelet transformation coefficient:

$$DWT_x(j,k)=2^{-\frac{j}{2}}\int_{-\infty}^{\infty}f(t)\psi^*(2^{-\frac{j}{2}}t-k)dt \tag{3}$$

The inverse transformation:

$$f(t)=\sum_{j=-\infty}^{\infty}\sum_{k=-\infty}^{\infty}DWT_x(j,k)\psi(2^{-\frac{j}{2}}t-k) \tag{4}$$

It is assumed that the length of EEG data segment used for feature extraction is m x 2^N points, then, Formula (4) is converted into:

$$f(t)=\sum_{j=0}^{m\times 2^{j-1}}\sum_{k=0}^{m\times 2^{N-j}-1}DWT_x(j,k)\psi(2^{-\frac{j}{2}}t-k) \tag{5}$$

According to the Mallat decomposition algorithm, the finite layer decomposition for $f(t)$ is

DOI: 10.1051/bioconf/20170802013

obtained:

$$f(t) = A_J + \sum_{j=1}^{J} D_j \tag{6}$$

where, J represents the decomposition layer, A_J represents the low frequency approximate component coefficient, and D_j represents the high frequency detail coefficient at different scales.

Under the premise that the wavelet basis function is orthogonal basis function, according to the energy conservation property of wavelet transformation, the formula for wavelet energy is obtained:

$$E_{tot} = \sum_{t} |f(t)|^2 \tag{7}$$

In combination with Formula (6), it can be further expressed as:

$$E_{tot} = A_J^{\ 2} + \sum_{j=1}^{J} D_j^{\ 2} = \sum_{j=1}^{J+1} |D_j|^2 \tag{8}$$

Then, the relative wavelet energy[10] is:

$$p = \frac{E_j}{E_{tot}} = \frac{|D_j|^2}{\sum_{j=1}^{J+1} |D_j|^2} \tag{9}$$

2.3.2 Analysis of Simulation Results

As the wavelet db possesses excellent local characteristics in both time and frequency domains[11], in this paper, wavelet db4 in the time length of 1s is selected as the wavelet function to carry out 4-layer decomposition for the preprocessed EEG signals; the obtained wavelet coefficients of the detailed part are D_1, D_2, D_3, and D_4, and that of the approximate part is A_4. The corresponding frequency ranges are: 8-13.5Hz, 13.5-19Hz, 19-24.5Hz, and 24.5-30Hz, respectively. And then, for the EEG signals of Channels C_3 and C_4, the relative wavelet energies within 4-7s are selected as the characteristic values of left and right hands motor imagery. The simulation results are shown in Figures3 and 4.

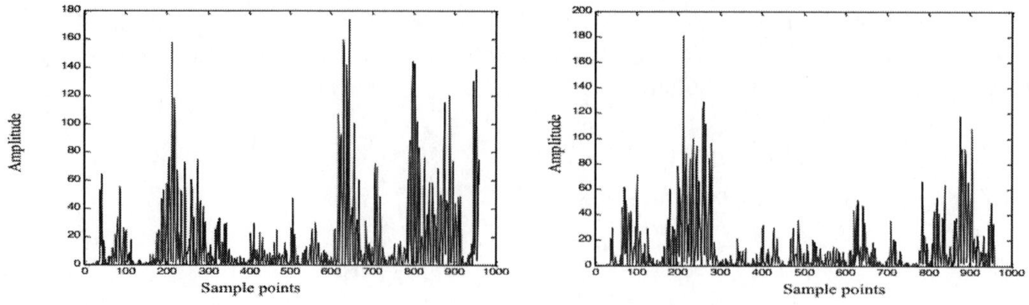

(a) (b)

Figure 3. Relative Energies in Channels C3 and C4 during Left Hand Motor Imagery(a) Relative Energy in Channel C3 (b) Relative Energy in Channel C4

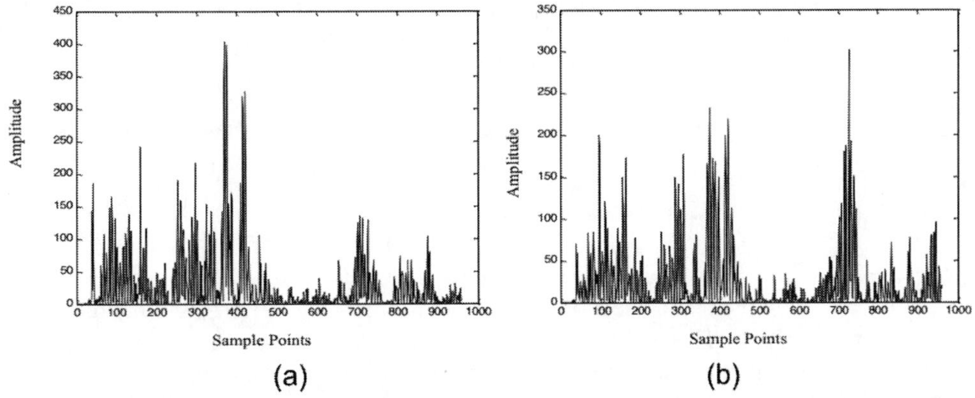

Figure 4. Relative Energies in Channels C3 and C4 during Right Hand Motor Imagery(a) Relative Energy in Channel C3 (b) Relative Energy in Channel C4

From Figure3, it can be seen that during left hand motor imagination, obvious ERD/ERS occurs at the sample points of 600-700 in Channels C_3 and C_4; from Figure 4, it can be seen that during right hand motor imagination, obvious ERD/ERS occurs at the sample points of 700-800 in Channels C_3 and C_4.

3 Pattern Recognition

Pattern recognition is a very important part of BCI system, and the core part of pattern recognition is design of classifier. The common classification algorithms mainly include linear discriminant method, support vector machine-based method, and artificial neural network method. The artificial neural network method is widely used because of its simple application, convenient parameter selection, and high accuracy of classification results. According to the different neural network models, artificial neural network is divided into feed-forward neural network, feedback neural network, and self-organizing neural network, among which feed-forward neural network includes perceptron network, BP network, linear neural network, and radial basis function neural network. For radial basis function neural network, the nearest neighbor clustering algorithm is adopted and it has such advantages as short learning time, small amount of calculation and good performance, so in this paper, RBF neural network is selected for the classification of left and right hands motor imagery EEG signals.

3.1 Basic Theory of RBF Neural Network

RBF neural network is composed of input layer, hidden layer and output layer. For the hidden layer, radial basis function is adopted as the excitation function.

Gauss's function is taken as the radial basis function, and is recorded as:

$$g_i = \exp(-\frac{\|x - c_i\|^2}{2\sigma_i^2}) \tag{10}$$

The output form is:

$$y = w_0 + \sum_{i=1}^{m} w_i g_i \tag{11}$$

where w_i is the weight from the hidden layer to the output layer. c_i and σ_i^2 are the center and the variance of the radial basis function, respectively. m is the number of the centers. According to the empirical formula, c_i can be 0, σ_i can be 1, and w_i can be 0-0.5. In order to simplify the calculation, in this paper, c_i is taken as 0, σ_i is taken as 1, and w_i is taken as 0.5.

3.2 Analysis of Classification Result Analysis

In order to reduce the amount of calculation, in this paper, 300 sets of characteristic data from Channels C_3 and C_4, among which 200 sets of data is taken as training samples and the other 100 sets of data is taken as test samples. The features of the training samples are used as the input data of RBF neural network, and the thinking task categories of the training sample are used as the output data of RBF neural network, with the category number recorded as "0" and "1". The rate of correct identification of the test samples (i.e., the ratio of the number of correctly identified samples to the total sample number) is used as the evaluation index. According to the above algorithm, this paper uses Matlab to compile the corresponding program to realize recognition of left and right hands motor. Test for 100 times and the overall recognition rate approaches 91%.

4 Conclusion

Feature extraction and feature classification are one of the key problems of BCI system. In this paper, through the research on left and right hands motor imagery EEG signals, ERD/ERS phenomenon is taken as the differentiating criteria, relative wavelet energy in the special frequencies of Channels C_3 and C_4 is used as feature vectors of left and right hands motor imagery EEG signals, RBF neural network is used for features classification, and simulation is carried out in combination with the data from BCI Competition III held in 2005. The results show that good classification effects are achieved. The research on BCI, as a new human-computer interface (HCI) mode, needs multidisciplinary cooperation. At present, the BCI system is not perfect and the corresponding theory and algorithm are not mature, so further research is required. With the continuous improvement and perfection of technology, the BCI system will eventually come out of the laboratory and be widely used in real life[12].

Acknowledgments

The work is supported by Project of Shandong Province Higher Educational Science and Technology Program, China(Grant No. J13LN04)

References

1. 1.J. R. Wolpaw, N. Birbaumer, W. J. Heetderks,et al. Brain-computer interface technology: a review of the first international meeting[J],*IEEE Trans.Rehab.Eng.*,2000,**8(2):**164-173
2. P.Missonnier, G.Gold, F.R.Herrmann, et al.Decreased theta event-related synchronization during working memory activation is associated with progressive mild cognitive impairment.[J]. *Dementia & Geriatric Cognitive Disorders*, 2006, **22(22)**:250-259.
3. 3.B.Akinci,N.G.Gencer,Online cue-based discrimination of left/right hand movement imagination[J],*Biomedical Engineering Meeting*,2010,1-4
4. S. Shahid, R. K. Sinha, A. G. Prasad, Mu and beta rhythm modulations in motor imagery related post-stroke EEG: a study under BCI framework for post-stroke rehabilitation[J],*Bmc Neuroscience*,2010,**11(1)**:1-2
5. E. Thomas, J, Fruitet, M, Clerc, Investigating brief motor imagery for an ERD/ERS based

BCI[J],*International Conference of the IEEE Engineering Medicine*,2012,**(4)**:2929-2932

6. E. Thomas, J, Fruitet, M. Clerc, Combining ERD and ERS features to create a system-paced BCI[J],*Journal of Neuroscience Methods*,2013,**216(2)**:96-103
7. J. R. Wolpaw, N. Birbaumer, D. J. McFarland, et al. Brain–computer interfaces for communication and control[J]. *Clinical neurophysiology*, 2002, **113(6)**: 767-791
8. BCI Competition III:2005, URL<http://www.bbci.de/competition/iii >
9. Jasmin Kevric, Abdulhamit Subasi, Comparison of signal decomposition methods in classification of EEG signals for motor-imagery BCI system[J],*Biomedical Signal Processing and Control*,2017,**31**:398-406
10. 10.A.S.Sankar, S. S. Nair, V. S. Dharan, P. Sankaran, Wavelet Sub Band Entropy Based Feature Extraction Method for BCI[J], *Procedia Computer Science*,2015,**46**:1476-1482
11. 11.M.H.Alomari, E. A. Awada, O. Younis, Subject-Independent EEG-Based Discrimination Between Imagined and Executed, Right and Left Fists Movements[J], *European Journal of Scientific Research*, 2014,18(3): 364-373
12. A. Ortizrosario, A. Adeli, Brain-computer interface technologies: from signal to action[J], *Reviews in the Neuroscience*, 2013, 24(5): 537-552

 DOI: 10.1051/bioconf/20170802014

Reduction of Scan Time and Echo Time in 3D Time-of-Flight (TOF) MR Angiography, Using Reverse Elliptical Centric Dual-Echo Acquisition

Mingliang Chen[1], Zhaoyang Jin[1], Zhixing Wang[4], and Yiping P. Du[2, 3, a]

[1]*Department of Automation, Hangzhou Dianzi University*
[2]*Department of Biomedical Engineering, Shanghai Jiao Tong University*
[3]*Shanghai United Imaging Healthcare co.,ltd*
[4]*Department of Biomedical Engineering, Zhejiang University*

Abstract. In the conventional single-echo three dimension (3D) time-of-flight (TOF) magnetic resonance angiography (MRA), it always takes long scan time to obtain the good images. In addition, high-resolution MRA with flow compensation requires long echo time (TE). Here, a reverse elliptical centric dual-echo 3D TOF MRA pulse sequence was developed to reduce the scan time and minimize the effective TE. Dual-echo and reverse elliptical centric acquisition in the phase encoding and slice phase encoding direction were used to reduce the scan time by 61%. Furthermore, the effective echo time (TE) was also minimized from 4.8ms to 4.1ms. The developed pulse sequence was tested on clinical MRI systems and the contrast to noise ratio (CNR) was measured to evaluate the results. The results show that the CNR only decrease by 2.5% with the scan time reduced to 39% when compared with the conventional single-echo 3D TOF. Compared with the conventional dual-echo 3D TOF acquisition, the developed pulse sequence brings advantages in three aspects: (1) reduce scan time by 22%; (2) increase CNR by 10%; (3) reduce the artifacts induced by dual-echo effects. In addition, while keeping the similar scan time as the conventional single-echo 3D TOF acquisition, the results show that reverse elliptical centric dual-echo 3D TOF acquisition increase CNR by 54.3%.

1 Introduction

In magnetic resonance imaging (MRI), the image degradation due to the signal loss remains a significant problem. And TOF MRA is sensitive to signal loss in two basic mechanisms: (1) saturation of flowing spins (2) intra-voxel spin phase dispersion. Here we mainly discuss about the second mechanism. In TOF MRA, it can be caused by two factors: (1) the inhomogeneity of the main magnetic field, the influence of eddy current and magnetic susceptibility; (2) the complex flow of blood, such as acceleration flow, pulsation[1]. Previous studies have indicated that the signal loss caused by the intra-voxel spin phase dispersion is greatly improved in short echo time (TE). Thus, on the previous basis using partial RF pulse[2], the maximum amplitude of the flow compensation gradients, as well as partial echo acquisition technology[34] and so on, the novel design for 3D TOF in this article reduce the TE further to improve the blood vessel imaging.

In the conventional 3D TOF single-echo imaging, the scan time often requires more than 10 minutes due to the phase encoding in two directions and the use of multiple overlapped thin slab

[a] Corresponding author: yipingdu@sjtu.edu.cn

(multi-slab)[5]. Therefore, it's difficult for the patients to tolerate it, which will absolutely affect the final imaging. To solve this, someone proposed the conventional dual-echo 3D TOF[6] to reduce the scan time by 50% when compared with the conventional single-echo 3D TOF. In this paper, a reverse elliptical centric dual-echo 3D TOF pulse sequence is developed, and the scan time can be decreased to 39% of the conventional single-echo 3D TOF.

2 Principle and method

2.1 First order flow-compensation

For the 3D TOF pulse sequence, the first order flow compensation is applied on the readout and slice phase encoding directions to null the first-order moment at the peak of the echo.

Figure 1 shows a first order flow-compensation diagram. G_{max} and T_r are the maximum gradient amplitude and the corresponding shortest ramp time, respectively. The widths of the flow-compensation gradient pairs along k_z direction are derived as:[78]

$$W_1^z = \frac{1}{2}\left\{-3T_r + \sqrt{T_r^2 + 4\left[\frac{A_z - M_0}{G_{max}}\left(\beta + \frac{T_r}{2} + \frac{A_z - M_0}{2G_{max}}\right) - \frac{M_1}{G_{max}}\right]}\right\} \tag{1}$$

$$W_2^z = \frac{A_z - M_0}{G_{max}} + \frac{1}{2}\left\{-3T_r \pm \sqrt{T_r^2 + 4\left[\frac{A_z - M_0}{G_{max}}\left(\beta + \frac{T_r}{2} + \frac{A_z - M_0}{2G_{max}}\right) - \frac{M_1}{G_{max}}\right]}\right\} \tag{2}$$

where A_z are the net encoding areas along the phase and slice directions, respectively, and they must be incremented with constant steps with the differences being between two consecutive steps proportional to Δk_z. β is the time between the end of the second flow compensation pulse with W_2^z. W_0 , T_r^0, and G_0 are the widths of the plateau, the ramp time, and the amplitude of the plateau, respectively. M_0^0 and M_1^0 are the zeroth and first moments of the slab-selection gradient.

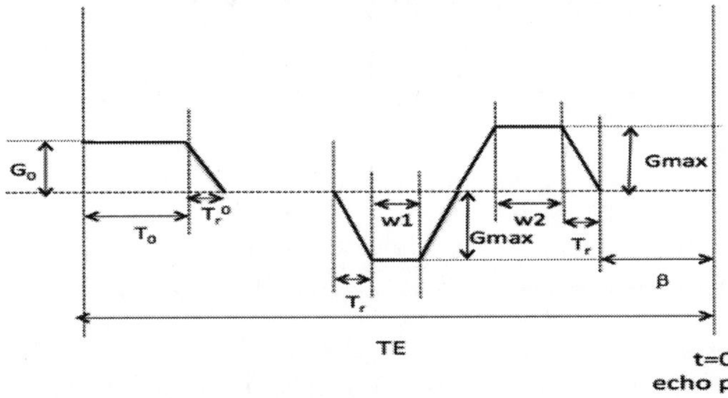

Figure 1. Gradient diagram of the flow-compensations along the slice phase-encoding directions for 3D TOF imaging. It starts from the center of RF pulse and ends in the echo peak.

2.2 Revers elliptical centric dual-echo view order

To realize this, each of the $(N_y \times N_z)$ views in the $k_y - k_z$ plane is initially sorted according to their distance from the center of k-space. The views that are outside of the ellipse:

$$\frac{k_y^2}{k_{ymax}^2} + \frac{k_z^2}{k_{zmx}^2} = 1, \tag{3}$$

but inside the standard $k_y - k_z$ rectangle are skipped. As shown in the Fig.2, the open dots in the four corners are the omitted views, which are not acquired and substituted with zeros during reconstruction. And in the previous work, it has been shown that the elliptical views can reduce the scan time of 3D acquisition by 21% with a slight penalty on the image quality[9]. Next, the dots located in the ellipse are further divided into two equal parts according to the distance from the origin of k-space. The red dots are the views close to the center of k-space and are acquired in the first echo with a relatively short echo time (TE), determining the signal-noise ratio (SNR) and contrast-noise ratio (CNR) of the images. The black dots are the views far away from the origin of k-space and are acquired in the second echo with a long TE. For dual-echo, two echoes are acquired within one repetition time (TR). The SNR and CNR of images are mainly determined by the first echo, which is also called the effective TE. Starting from their respective edge, both red dots and black dots with the views of the decreasing distance from the origin of k-space are acquired subsequently.

Such design has brought some advantages. (1) Starting from the periphery of k-space views, it can reduce the use of dummy scans (i.e., scan with the data acquisition disabled), which are used to reach a steady state[8]. (2) By collecting the views close to the center of k-space centrally, the artifacts can be reduced. Because 3D TOF acquisitions are commonly phase encoded in two orthogonal directions, there is considerable flexibility in the $k_y - k_z$ view order. Using the reverse elliptical centric view order, the conspicuous ghosts can be reduced by evenly spreading the ghosts in two dimensions (phase and slice)[10]. (3) Combining the dual echo with reverse elliptical centric view order techniques, the scan time can be reduced to about 39% of the conventional single-echo 3D TOF. In addition, owing to the decrease of the slice phase encoding area in the first echo, from the calculation equations (1) and (2) in the flow compensation part above, the effective TE can be reduced. The degree of reduction is determined by the maximum gradient and corresponding ramp time used in the design and the number of phase encodings in the slice direction.

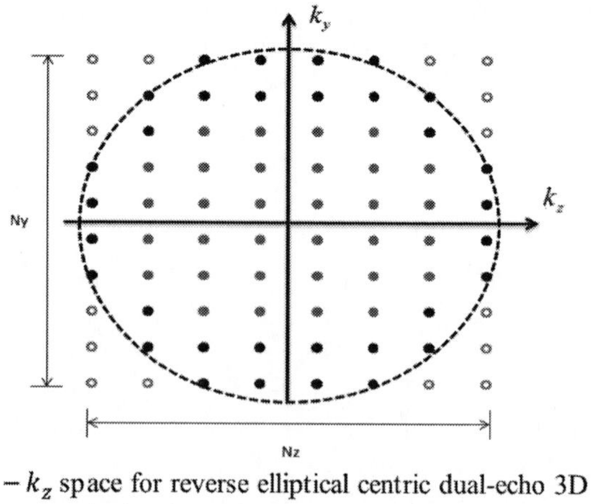

Figure 2. The views order in $k_y - k_z$ space for reverse elliptical centric dual-echo 3D TOF acquisition. The data for the views in the corners are not acquired and set to zeros during image reconstruction. The remaining views (the solid black dots and red dots) are further divided into two equal parts according to the distance from the origin of k-space. The solid red dots and the black dots are acquired in the first and second echo, respectively.

2.3 MRI

To demonstrate the potential advantages of reverse elliptical dual-echo 3D TOF acquisition, an initial comparison was made with the conventional single-echo 3D TOF and the conventional dual-echo 3D TOF acquisition, respectively.

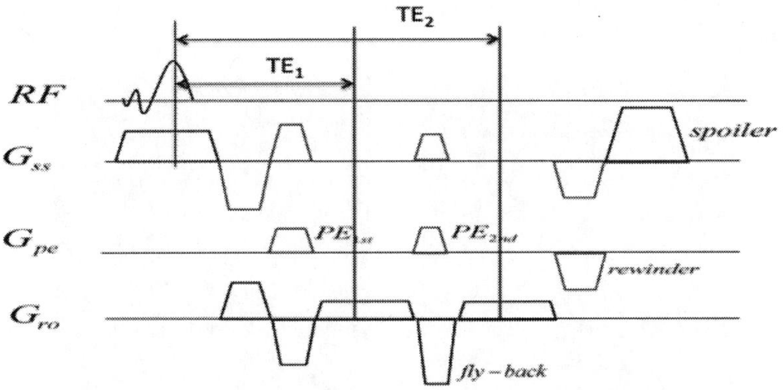

Figure 3. Diagram of the reverse elliptical centric dual-echo 3D TOF pulse sequence. To achieve the short TE, the asymmetric RF pulse, partial-echo acquisitions and flow-compensated gradients with minimum duration are applied in the sequence. In addition, by reducing the slice phase-encoding areas of the first echo, the effective TE (TE_1) is shortened. A fly-back gradient is located in the middle of two echoes to restore flow compensation in the second echo. And the gradients in phase-encoding and slice phase-encoding directions are rearranged to accomplish the reverse elliptical centric dual-echo view order. The red gradient means that the gradients will vary according to the specific view order.

Figure 3 shows the reverse elliptical centric dual-echo 3D TOF sequence. The acquisition of a second echo with the extra phase encoding and slice encoding is added to a conventional multiple overlapped thin slab acquisition of a 3D TOF pulse sequence. Both the view orders of the two echoes are rearranged to cover the elliptical $k_y - k_z$ space.

Five healthy volunteers were scanned on a 1.5T MR scanner (United Imaging Healthcare, Shanghai, China) using a 16-channel phased array head coil, the maximal gradient field adopted here was the 15.5mT/m, and the corresponding ramp time was 270us. For the three pulse sequences, an asymmetric RF pulse and 71% of partial echoes acquisition were used to reduce TE. The effective TE of conventional single-echo 3D TOF sequence and conventional dual-echo 3D TOF were both limited to 4.8ms, but was reduced to 4.1ms in the reverse elliptical centric dual-echo 3D TOF. In addition, the second TE in the conventional dual-echo 3D TOF and reverse elliptical centric dual-echo 3D TOF acquisition were 11.8ms and 11.1ms, respectively. The other parameters for these scans were same, i.e., 22cm field-of-view (FOV), 256*256 matrix, 1mm slice thickness, 25° flip angle, 26ms repetition time (TR), ±30kHZ bandwidth (BW). To reduce the inflow saturation[11], four slabs were acquired with 24 slices per slab and a 25% overlap between two slabs. In order to avoid the signal wrap-around along the slice-selection direction, the imaged slab width was set as 28mm corresponding to 24mm wide for the excited slab.

Images reconstruction and post-processing were performed by using MATLAB. In order to reduce the partial volume effect[12], zero-filling interpolation[13] was applied to k-space data to reconstruct 3D complex images with a total 256*256*156 matrix. Then CNR was measured on the final MRA images, including different anatomic locations. To quantify the CNR, the method developed by Du et al and refined by Chapman et al[14], was used.

3 Results

Figure 4 shows the projections of the MRA images acquired by the conventional single-echo 3D TOF acquisition in the left column and the reverse elliptical centric dual-echo 3D TOF acquisition in the right column, respectively. The three rows correspond to the MIP (maximum-intensity projection) of MRA in the transverse, sagittal and coronal plane. The CNR of the four vessels indicated by the arrows in the Fig.4 was measured to evaluate the performance. To distinguish them, the four blood vessels were marked in four different colours, and they are red (vessel one), blue (vessel two), green (vessel three) and yellow (vessel four).

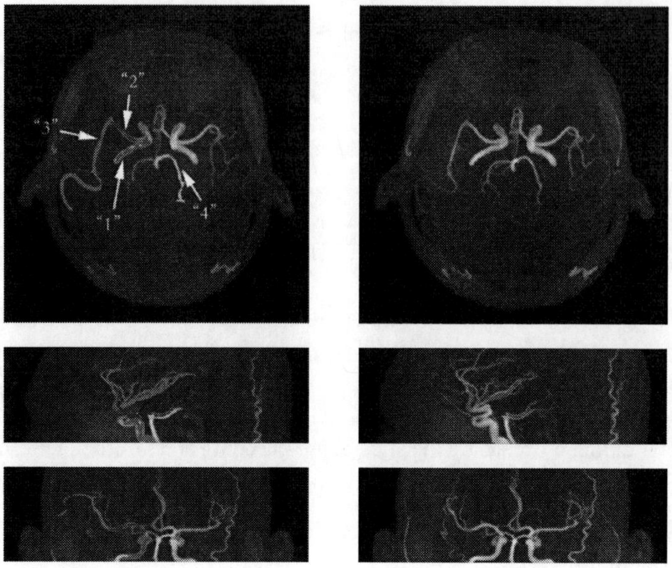

(a) (b)

Figure 4. The MRA images acquired by reverse elliptical centric dual-echo 3D TOF (column a) and conventional single-echo 3D TOF (column b) acquisition. From top to bottom, the MIP of MRA in the transverse, sagittal, coronal plane were shown to observe the difference. After reducing the scan time by 61% and the effective TE about 0.7ms, the MRA image of reverse elliptical centric dual-echo 3D TOF method maintains the similar vascular visualization with the conventional single-echo 3D TOF method.

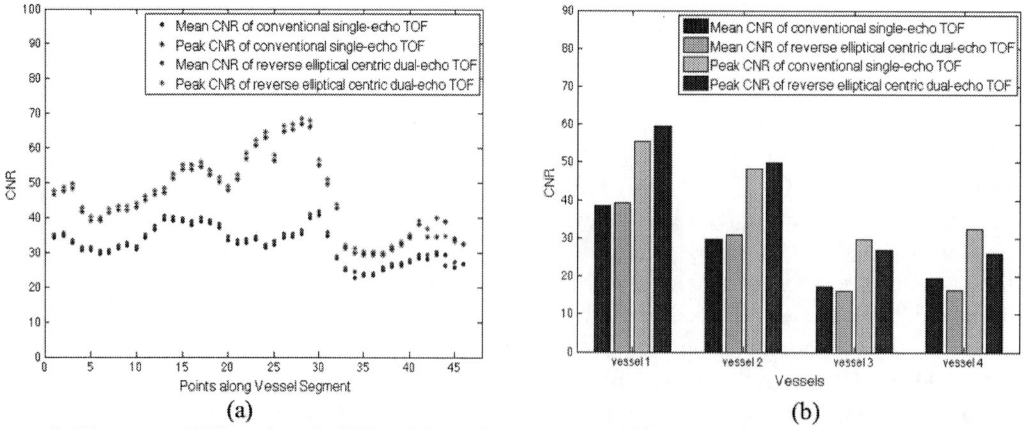

(a) (b)

Figure 5. The mean CNR and peak CNR of the points along the big vessel labeled "1" in conventional single-echo and reverse elliptical centric dual-echo 3D TOF acquisition are shown in (a). A histogram shown in (b) is used to graphically summarize and display the mean CNR and peak CNR of four vessel segments in Fig.4.

Fig.5 (a) shows the representative local vessel CNRs of the big vessel "1". The solid points designate the mean local vessel CNR and the asterisks designate the peak local vessel CNR. The histogram in Fig.5 (b) shows the mean CNR and peak CNR of the four vessel segments averaged from five volunteers. From analysis of the results, it can be found that for the big vessels like vessel one, two, the CNR of reverse elliptical centric dual-echo 3D TOF MRA is a little higher than that in the conventional single-echo 3D TOF MRA. And the CNR for small vessels, such as vessel three, four, decrease a little because of the long echo time of the acquisitions for high frequency parts.

DOI: 10.1051/bioconf/20170802014

Furthermore, the blurring and ghosting due to off-resonance effects that can be predicted theoretically by point spread function (PSF) analysis[15] may also degrade the CNR.

In conclusion, compared with the conventional single-echo 3D TOF, only 39 percentage of the scan time is taken, but the MRA with similar CNRs is obtained in the reverse elliptical centric dual-echo 3D TOF pulse sequence.

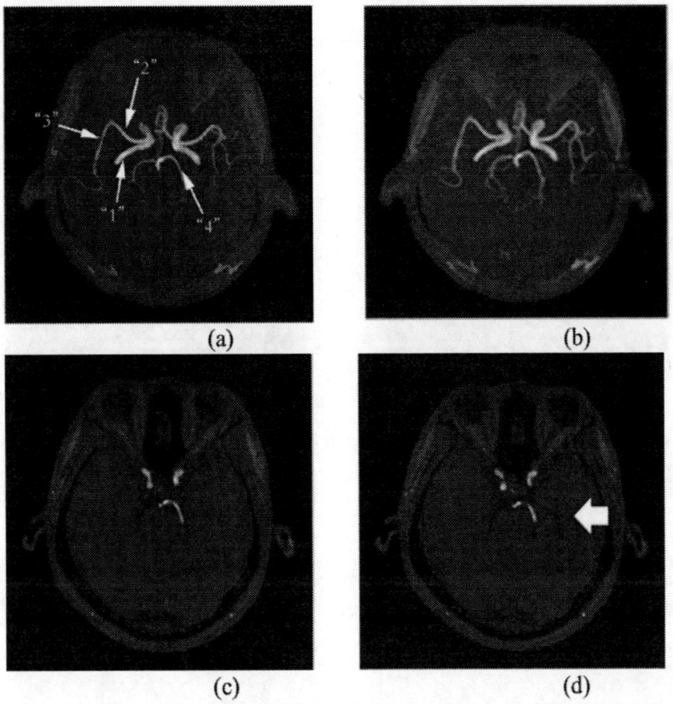

Figure 6. The MRA images acquired by reverse elliptical centric dual-echo 3D TOF (a) and conventional dual-echo 3D TOF (b) acquisitions (c) and (d) are one of the images acquired by reverse elliptical centric dual-echo 3D TOF pulse sequence and conventional dual-echo 3D TOF pulse sequence, respectively.

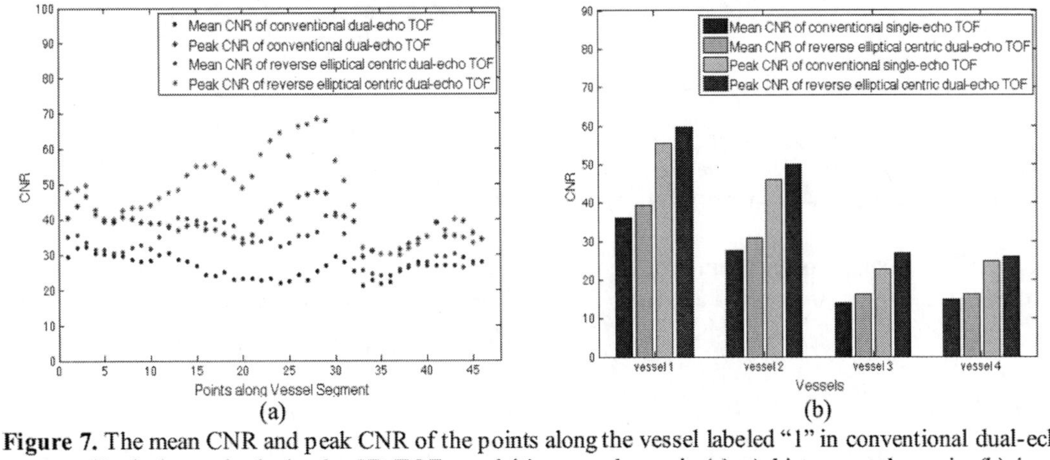

Figure 7. The mean CNR and peak CNR of the points along the vessel labeled "1" in conventional dual-echo and reverse elliptical centric dual-echo 3D TOF acquisition are shown in (a). A histogram shown in (b) is used to graphically summarize and display the mean CNR and peak CNR of four vessels averaged from five volunteers.

Fig.6 shows the MRA data acquired by the reverse elliptical centric dual-echo 3D TOF (the left column) and the conventional dual-echo 3D TOF (the right column) acquisition, respectively. The top row shows the MIP of MRA, and the bottom row shows the specific slice of the same patient. In the bottom row, the artifacts indicated by the arrow in Fig.6 (d) were reduced much in the reverse elliptical centric dual-echo 3D TOF. The artifacts due to the off-resonance effects were predicted theoretically by PSF analysis[15]. Beyond that, the local vessels CNR of the four vessels were measured

 DOI: 10.1051/bioconf/20170802014

to compare in Fig.7 (a) and (b). Hence, compared with the conventional dual-echo 3D TOF MRA, the current method not only reduce the scan time by 22%, but also decrease the artifacts and improve the CNR of images.

To further compare the reverse elliptical centric dual-echo 3D TOF and the conventional single-echo 3D TOF pulse sequence, Fig 8.(a) and Fig 8.(b) show the MRA data acquired by reverse elliptical centric dual-echo 3D TOF pulse sequence after averaging and conventional single-echo 3D TOF pulse sequence, respectively.

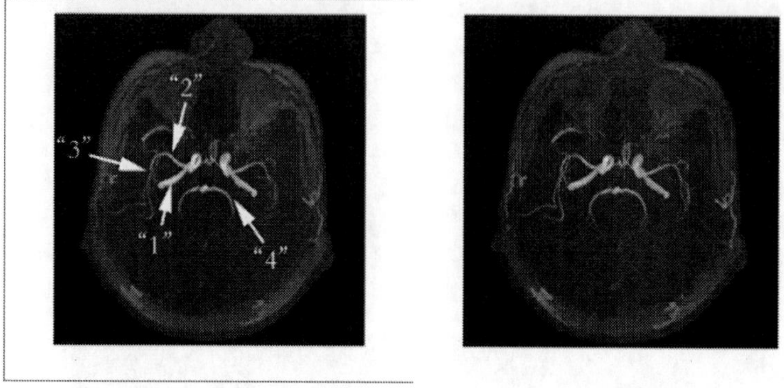

(a) (b)

Figure 8. The MRA images acquired by reverse elliptical centric dual-echo 3D TOF after averaging (a) and conventional single-echo 3D TOF (b) acquisitions.

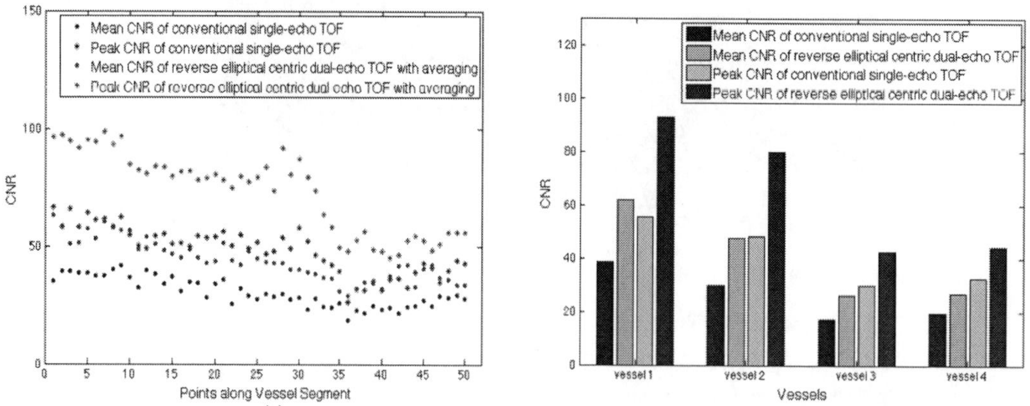

(a) (b)

Figure 9. The mean CNR and peak CNR of the points along the big vessel labeled "2" in conventional single-echo and reverse elliptical centric dual-echo 3D TOF with two-average acquisition are shown in (a). A histogram shown in (b) is used to graphically summarize and display the mean CNR and peak CNR of four vessels averaged from five volunteers.

And the CNR measurements were made in the Fig.9. For the big vessels like "1" and "2", the CNRs of reverse elliptical centric dual-echo 3D TOF MRA with averaging are much higher. In addition, the CNRs of small vessels such as "3" and "4", are also higher.

4 Discussion

The applications of MRA always confront two problems. (1) Due to the long scan time, it may bring the motion artifacts and also occupy the time required for other scans, affecting the effective diagnosis of the patient's condition. (2) In vascular MRI, the phase dispersion caused by complex flow might lead to the signal loss, which can be avoided by reducing TE. The developed pulse sequence can resolve the two problems by reducing the scan time to about 39% of the conventional method and decreasing the effective TE from 4.8ms to 4.1ms.

DOI: 10.1051/bioconf/20170802014

Table 1. Mean and peak CNR of the vessels

	Conventional single-echo TOF	Conventional dual-echo TOF	Reverse elliptical centric dual-echo TOF	Reverse elliptical centric dual-echo TOF after averaging
Mean/Peak	26.29/41.47	23.11/37.29	25.63/40.64	40.56/64.85

TABLE I shows the mean and peak CNR of the MRA of the vessels averaged from the five volunteers acquired by different acquisition methods. Compared with the conventional single-echo 3D TOF, the mean and peak CNR of reverse elliptical centric dual-echo 3D TOF MRA are decreased by 2.5% and 2%, respectively. Compared with the conventional dual-echo 3D TOF, the mean and peak CNR of reverse elliptical centric dual-echo 3D TOF MRA are increased by 10.9% and 8.9%, respectively. In addition, the comparison of CNR between the conventional single-echo TOF MRA and the reverse elliptical centric dual-echo TOF MRA with two-average shows that the mean CNR and peak CNR are increased by 54.3% and 56.4%, respectively. In conclusion, the results reveal the CNR advantage of reverse elliptical centric dual-echo 3D TOF pulse sequence.

Acknowledgment

The author thanks JinWang, Lele Zhao, Suhang Mao, Xiaoqian Huang, Ke Meng and Tianyi Zeng, for their assistance in the preparation of this paper. This study was greatly supported by the National Natural Science Foundation of China (61372024, 81371518).

References

1. S.N. Urchuk, D.B. Plewes, JMRI, Mechanisms of flow-induced signal loss in MR angiography, **2**(4): 453-462 (1992)
2. H.T.C. Nielsen, G.E. Gold, E.W. Olcott, et al, MRM, Ultra-short echo 2D time-of-flight MR angiography using a half-pulse excitation, **41**(3): 591-9 (1999)
3. A.J. Evans, D.B. Richardson, R. Tien, et al. AJNR, Poststenotic signal loss in MR angiography: effects of echo time, flow compensation, and fractional echo, **14**(3): 721-9 (1993)
4. P. Schmalbrock, C. Yuan, D.W. Chakeres, et al, Radiology, *Volume MR angiography: methods to achieve very short echo times*, **175**(3): 861-5 (1990)
5. D.L. Parker, C. Yuan, D.D. Blatter, MRM, *MR angiography by multiple thin slab 3D acquisition,* **17**(2): 434-51 (1991)
6. W. Fang, Y.P. Du, BMEI, Optimization of 3D time-of-flight MR angiography with shortened echo time and dual-echo acquisition, 53-58 (2013)
7. K. Ying, P. Schmalbrock, B. Clymer. MRM, Echo-Time Reduction for Submillimeter Resolution Imaging with a 3D Phase Encode Time Reduced Acquisition Method, **33**(1): 82-87 (1995)
8. EM. Haacke, RW. Brown, MR. Thomson, R. Venkatesan. Wiley-Liss, *Magnetic resonance imaging: physical principles and sequence design* (1999)
9. M.A. Bernstein, S.B. Fain, S. J. Riederer, JMRI, Effect of windowing and zero-filled reconstruction of MRI data on spatial resolution and acquisition strategy, **14**(3): 270-80 (2001)
10. King, K. Franklin, Elsevier, *Handbook of MRI Pulse Sequences*, 687 (2004)
11. F. R. Korosec, MRA, Basic Principles of MRI and MR Angiography, 3-38 (2004)
12. T. Doi, T. Okamoto, K. Nakagawa, Japanese Journal of Radiological Technology, *The partial volume effect in MRI*, 51 (1995)

13. Yiping P. Du MS ME, D.L. Parker, W.L. Davis, et al, JMRI, Reduction of partial-volume artifacts with zero-filled interpolation in three-dimensional MR angiography, **4**(5): 733–741 (1994)
14. Y. Du, D.L. Parker, W.L. Davis, et al, Investigative Radiology, Contrast-to-noise-ratio measurements in three-dimensional magnetic resonance angiography, **28**(11): 1004-9 (1993)
15. E. K. Jeong, D.L. Parker, J. S. Tsuruda, et al, MRM, Reduction of flow-related signal loss in flow-compensated 3D TOF MR angiography, using variable echo time (3D TOF-VTE), **48**(4): 667-76 (2002)

TrnH-psbA sequence analyses of asparagus cochinchinensis from different geographical origin in China

Yingzi Ma[a], Jialing Zhang and Jianghai Liu

College of Life Science and Technology, Central South University of Forestry and Technology, Changsha, Hunan 410004, China

Abstract. This template explains and demonstrates how to prepare your camera-ready paper for The trnH-psbA sequences of 13 Asparagus cochinchinensis populations from 6 provinces of China were studied. The results showed that length of trnH-psbA change and the mutation of GC content were small. The length of trnH-psbA sequences were from 619 bp to 632 bp, and the GC content was about 36%. The total variation rates of 13 populations were from 2.21% to 3.47%, when the missing sites were considered as variation sites. A. cochinchinensis from different sources had 10 information sites in trnH-psbA sequence, accounting for 1.58% of the total sequence. The information sites were located in the sites 8, 9, 120, 457, 458, 486, 487, 491, 492, and 593, respectively. Clustering analysis showed that the Qianxi and Hengshan populations clustered together; Dushan, Yuqing, and Guangzhou populations were grouped; Nanning and Xinning populations formed another cluster. trnH-psbA sequences could identify different A. cochinchinensis populations. Clustering of different A. cochinchinensis populations related primarily to latitude and had little relationship with longitude.

1 Introduction

Asparagus cochinchinensis (Lour.) Merr. belongs to Liliaceae –(or lily family). It is mainly produced in Guizhou, Yunnan, and Hunan provinces in China. Its roots have been used as a traditional Chinese herbal medicine to nourish'YIN', clear lung, alleviate constipation, and cure antibacterial infection and cancer[1]. Current researches of A. cochinchinensis focused on its chemical composition, pharmacological effect, and clinical application. The previous studies showed that main components were carbohydrate[2,3], amino acids [4], and saponins [5]. Clinical applications have been conducted on hyperplasia of mammary glands [6], malignant lymphosarcoma [7,8], antitussive and expectorant[9], immunity[10], and anti-radiation [11]. While genomic sequence and genuine of this plant were few reported.

The external influence to cytoplasm is smaller than to the nucleus, and the cytoplasm is relatively conservative and reflects the evolutionary trend. TrnH-psbA fragment is one of chloroplast intergenic regions. It has 75 bp conserved sequence at both ends of fragment which is easy to design primerswith

[a] Corresponding author : ma_yingzi@163.com
This work was financially supported by Hunan Provincial Science and Technology department (2013TZ2024), and the Research Program from Education Department of Hunan Province of China(2014, NO.199).

good universality and high amplification success. Previous research showed trnH-psbA sequence could be used as DNA barcoding to identify plant [12,13,14]. In this study, the genetic polymorphisms and relationships of 39 individuals from 13 populations were evaluated by using trnH-psbA fragment in order to provide accurate identification information for germplasm resources of A. cochinchinensis.

2 Materials and methods

Total 13 populations, 3 individuals per populations, were collected from Guangdong, Guangxi, Guizhou, Hunan, Zhejiang, and Qinghaiprovince (Table 1). DNA was extracted according to the method described by Rogers and Bendic (1988).

Table 1. Source and location of A. cochinchinensis

Location	# of Sample	Source		
			Latitude(°)	Longitude(°)
Xining, Qinghai	3	Wild	36.57	101.81
Hangzhou, Zhejiang	3	Cultivated	29.18	118.35
Yanhe, Guizhou	3	Wild	28.21	108.01
Hengshan, Hunan	3	Wild	27.28	112.71
Qianxi, Guizhou	3	Wild	26.90	105.68
Huaxi, Guizhou	3	Wild	26.43	106.68
Xinning, Hunan	3	Wild	26.25	110.30
Yuqing, Guizhou	3	Wild	25.32	107.28
Dushan, Guizhou	3	Wild	25.01	108.31
Yongzhou, Hunan	3	Wild	24.65	111.00
Xingyi, Guizhou	3	Wild	24.63	104.85
Guangzhou, Guangdong	3	Cultivated	23.115	113.25
Nanning, Guangxi	3	Cultivated	22.22	107.90

The polymerase chain reactions (PCR) using trnH-psbA primerswere conducted according to Peterson [15]. The PCR products were fractionated on 1 % agarose geland visualized using GelLogic 100 image system. The target fragments were isolated from the agarose gel under UV radiation, purified with the reagent kit (Tiangel midi purification kit), and then directly sequenced. Sequence analysis was performed by DNAMAN and MEGA 5.0.

3 Results and discussion

The trnH-psbA length of 39 samples was from 619 to 632 bp, and GC content was from 35.0% to 36.2% (Table 2). trnH-psbA sequences of 3 samples from a same population were almost the same and they only had few variable sites. Different populations had different variable sites.

Table 2. Length and GC content of trnH-psbA sequences

Location	Length bp	G C content %	Location	Length bp	G C content %
Xining	621	35.6	Yuqing	620	36.0
Hangzhou	620	36.0	Dushan	620	36.0

Yanhe	621	35.8	Yongzhou	619	36.2
Hengshan	621	35.9	Xingyi	621	36.1
Qianxi	621	35.7	Guangzhou	620	35.8
Huaxi	632	35.0	Nanning	621	36.1
Xinning	621	36.1			

The loss rates were about 2.00% except Huaxi population that was 0.32%. Mutational rates were from 0 to 2.52%.Yuqing population had no mutations, while Huaxi population had the highest mutational rate, 2.52%. The total mutational rates ranged from 2.21% to 3.47% when a vacancy was counted as a mutational site. Xingyi population had the hightest mutational rate of 3.47%, and Yuqing population had the lowest mutational rate of 2.21% (Table 3).

Table 3. The loss and mutational rate of different populations

Populations	Loss rate	Mutational rate	Total	Populations	Loss rate	Mutational rate	Total
Xining	2.05	0.95	3.00	Yuqing	2.21	0.00	2.21
Hangzhou	2.21	0.63	2.84	Dushan	2.21	0.32	2.53
Yanhe	2.05	0.32	2.37	Yongzhou	2.37	0.32	2.69
Hengshan	2.05	0.47	3.52	Xingyi	2.05	1.42	3.47
Qianxi	2.05	0.32	2.37	Guangzhou	2.21	0.16	2.37
Huaxi	0.32	2.52	2.84	Nanning	2.05	0.32	2.37
Xinning	2.05	0.32	2.37				

There were 10 information sites which were single nucleotide sites (SNPs). When a vacancy as an absence in different populations from different provinces were at positions 8, 9, 120, 457, 458, 486, 487, 491, 492 and 593. They were transition at positions 492 and 593, transversion at positions 8,120, 486,487 and 491 (Table 4).

Table 4. trnH-psbA sequences information sites of different provinces

Origin	Information site									
	8	9	120	457	458	486	487	491	492	593
Guizhou Province	C	A	A	A	-	A	G	G	C	A
Guangxi Province	C	A	G	A	-	A	G	G	C	T
Guangdong Province	A	-	A	A	-	A	G	G	C	A
Hunan Province	C	A	G	A	-	A	G	G	C	T
Zhejiang Provinc	C	A	G	-	-	A	G	G	C	T
Qinghai Province	A	-	A	A	A	G	A	A	G	A

Note: "-":absence

UPGMA (unweighted pair-group method with arithmetic means) dendrogram based trnH-psbA sequences showed that the populations were divided into two branches. One included Yuqing, Guangzhou, Dushan, Xining, Qianxi, Hengshan, Yongzhou, and Huaxi populations, the other consisted of Hangzhou, Xingyi, Yanhe, Nanning, and Xinning. Yuqing, Guangzhou, and Dushan populations made a subgroup, while Xining, Qianxi, and Hengshan populations in another subgroup Xingyi and Yanhe population clustered, and Nanning and Xinning populations firstly clustered in the second branch. Guangzhou and Dushan populations were closer clustering although their latitudes had large distance (Figure 1).

There was a certain relationship between phylogenetic distribution of plants and the general geographic location (especially the latitude) [16,17]. This paper showed interspecific variation and phylogenetic distances of A. cochinchinensis mainly affected by latitude based the populations had closer relatives those had closer latitude with the analyse of trnH-psbA sequences that was consistent the results using ITS sequences [18]. Such as the similarity of the Qianxi and Hengshan populations was 99.8% with the latitude of the Qianxi populations was 27.28 N and Hengshan population was 27.05 N, the similarity of the Huaxi and Yongzhou populations was 99.2% with the latitude of theHuaxi populations was 26.43 N and Yongzhou population was 26.13 N. But there are some exceptions, the similarity of the Yuqing and Dushan populations was 99.7% while the latitude of the Yuqing populations was 22.50 N and Dushan population was 35.80 N, this showed that the latitude may be factors on affecting the distribution of plant, but not the decisive factor. The populations from Guangzhou were firstly clustering with Dushan and Yuqing, Hangzhou were firstly clustering with Xingyi and Nanning were firstly clustering with Xinning, and they were cultivated populations, it indicated these populations might be introduced from Guizhou, Guizhou and Hunan Province, respectively

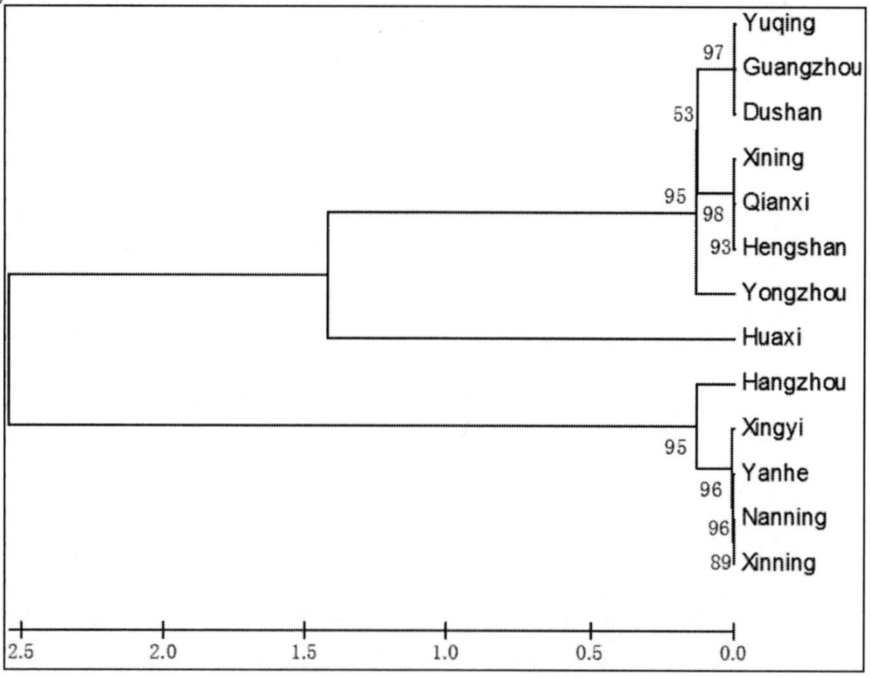

Figure 1. UPGMA homologous tree of A. cochinchinensis based trnH-psbA sequences

4 Conclusion

Wild *A. cochinchinensis* were mainly used and wild resources were gradually depleted with demand growing in recent years, we must rely on artificial cultivation and cultivated varieties mainly introduced by the off-site. An important feature of Chinese medicine was genuine, different local production of medicines had different active ingredients and content, so the genuineness of *A. cochinchinensis* medicine was important prerequisite to guide the introduction and selective improved variety. DNA sequence analysis was effective means to identify genuineness of medicinal materials. Our results showed the populations from Guizhou province had more genetic polymorphism based their trnH-psbA had more mutation rate, it suggested Guizhou province might be a genuine producing area, trnH-psbA sequence might clarify its genetic background from the molecular level to overcome

 DOI: 10.1051/bioconf/20170803001

varietal complexity induced the decline of yield and quality degradation and guide introduction and cultivation of *A. cochinchinensis*.

References

1. Yao NH, Kong NY. Developments of chemical compositions and biological action of asparagus L. Nat.l Prod. Res. Deve. Vol.11(1999), p. 67-71.
2. Du XH, Guo YZ. Studies on the isolation of antitumor active constituents of polysaccharides from Chinese herb asaragus cochinchinensis(Lour.) merr. Shenyang Phar. Univ. Vol.7(1990), p.197-201.
3. Li ZX, Huang CG, Cai YJ, Chen XM, Wang F, Chen YZ. The chemical structure and anttoxidative activity of polysaccharide from Asparagus cochinchinensis. Acta Phar. Sin. Vol.35(2000), p. 358-362.
4. Ni JM, Zhao NN, Wang R. Comparison of amino acid of Asparagus before and after processing. Chin. Trad. Herb. Drug. Vol.23(1992), p. 182.
5. Liang ZZ, Aquino R, De Simone F, Dini A, Schettino O, Pizza C. Oligofurostanosides from Asparagus cochinchinensis L. Planta Med., Vol. 54(1988), p. 344-346.
6. Shen ZY, Ping JL. 513 cases on hyperplasia of mammary glands with tamoxifen and asparagine. Clin. Med. Chin. Vol.13(1997), p. 138-139.
7. Xu L, Lao LX, Ge A, Yu S, Li J, Mansky PJ. Chinese herbal medicine for cancer pain. Integr. Cancer Ther. Vol.6 (2007), p. 208-234.
8. Zhang HJ, Sydara K, Tan G-T, Ma C-Y, Southavong B, Soejarto DD, Pezzuto JM, Fong HYS. Bioactive constituents from Asparagus cochinensis L. J Nat Prod. Vol.67(2004), p. 194-200.
9. Luo J, Long QD, Li CX, Li L, Huang NH, Nie M. Comparison of antitussive, expectorant and anti-asthmatic effect between ALWB and ACM. J. Guiyang Med. Coll. Vol.23(1998), p. 132-134.
10. Kumari R, Tiwary BK, Prasad A, Ganguly S.. Study on the immuno- modulatory effect of herbal extract of Asparagus racemosus Willd. in broiler chicks. Global J. Res. Medi. Pl. & Indigenous Medi, Vol. 1(2011), p. 1-5.
11. Li QS, Li YJ, Liu Y. Protective effect and mechanism of Chinese Asparagus on damaged rat's heart induced by chronic radiation. Chin. J. Trad. Chin. Med. Pharm . Vol. 26(2011), p. 375-377.
12. Kress WJ, Wurdack KJ, Zimmer EA, Weigt LA, Janzen DH. Use of DNA barcodes to identify flowering plants. PNAS, Vol.(2005)102, p. 8369-8374.
13. Nomura N, Takaso T, Peng CI, Kono Y, Oginuma K, Mitsui Y, Setoguchi H .Molecular phylogeny and habitat diversification of the genus Farfugium (Asteraceae) based on nuclear rDNA and plastid DNA. Ann. Bot., Vol.106(2010), p, 467 – 482.
14. Shaw J, Lickey EB, Schilling EE, Small RL. Comparison of whole chloroplast genome sequences to choose noncoding regions for phylogenetic studies in angiosperms: the tortoise and the hare III . Amer. J.Bot., Vol.94(2007), p. 275-288.
15. Peterson, John H, Koch E, Peterson J. . A molecular phylogeny of the genus Gagea (Liliaceae) in Germany inferred from non- coding chloroplast and nuclear DNA sequences . Plant Syst. Evol., Vol.245(2004), p. 145-162.
16. Zhu SH, Zhang QF, Wang MQ. Ribosomal DNA polymorphisms of the common wild rice from China. Acta Gene. Sin. Vol.25(1998), p. 531-537.
17. Wang YH, Wang H, Gao LZ. Study on genetic diversity of Oryza rufipogon Griff. by simple seqence repeat. Acta Bot. boreali-occidentalia Sin. Vol.23(2003), p. 1750-1754.
18. Ou LJ, Zhang RW, Tan ZW, et al.. Nuclear DNA ITS sequence analysis of Asparagus cochinchinensis from different geographical regions in China. Chin. Trad. Herb. Drugs. Vol.42(2011), p. 1402-1406.

Screening and identification of lactic acid bacteria strains with high acid-producing from traditional fermented yak yogurt

Xiaoyong Chen[1,2,3,#], Xin Zhao[3,4,#], Yu Qian[3,4], Jian Li[5,6], Lianhong Chen[5,6], Juan Chen[6], Yu Zhang[1,2] and Huayi Suo[1,2,a]

[1]*College of Food Science, Southwest University, 400715, Chongqing, China*
[2]*Chongqing Engineering Research Center of Regional Food, 400715, Chongqing, China*
[3]*Chongqing Collaborative Innovation Center for Functional Food, Chongqing University of Education, 400067, Chongqing, China*
[4]*Department of Biological and Chemical Engineering, Chongqing University of Education, 400067, Chongqing, China*
[5]*Institute of Qinghai-Tibetan Plateau, Southwest University for Nationalities, 610041, Chengdu, China*
[6]*College of Life Science and Technology, Southwest University for Nationalities, 610041, Chengdu, China*

Abstract. A total of 57 strains of lactic acid bacteria (LAB) were isolated and purified from traditional fermented Yak Yogurt in Hongyuan-Sichuan and Yangbajing-Tibet. The strains with high acid-produced were screened by soluble calcium circle and titratable acidity determination. The five strains, 7-1, 22-1, 28-1, 34-1 and 62-1, possessed the high acid-producing and the value of titratable acidity is 196.2, 191.1, 192.2, 194.8 and 200.2 T respectively. Based on 16S rDNA sequence analysis, 22-1 was identified as *Lactococcus lactis* subsp. *lactis*, 28-1 as *Lactobacillus casei*, 34-1 as *Lactobacillus fermentium*, 7-1 and 62-1 as *Enterococcus durans*. This study could provide the evidence for researching fermentation strains to improve yogurt quality.

1 Introduction

Lactic acid bacteria refer to a group of bacteria with the capacity to ferment carbohydrate to generate lactic acid. It is a main fermentative strain to ferment traditional fermented foods like dairy product, pickle and fermented sausage etc. Lactic acid bacteria generates flavor substances such as micromolecule aldehyde, acid and ester etc. through fermenting saccharides to promote the formation of traditional fermented food's flavor, which plays a great importance in fermented food[1, 2]. In the fermentation process of dairy product, the acid-producing ability of lactic acid bacteria is a key factor to measure the quality of dairy product [3]. Bian Lihong and others [4] researched and found that the acid-producing ability of lactic acid bacteria in different brands of plain yogurt is closely related to the

[a] Corresponding author: birget@swu.edu.cn, [#]Co-first author
This work was partly supported by Special Fund for Agro-scientific Research in the Public Interest (201203009), Special Fund of Chongqing People's Livelihood and Technology Innovation (cstc2015shmszx80021) and Fundamental Research Funds for the Central Universities(XDJK2016E109).

separation of curd and whey in yogurt. In addition, the lactic acid bacteria with a high acid-producing ability can not only shorten curd time and enhance productivity but also perfect yoghourt's texture and improve the quality of yoghourt etc. Consequently, to select the lactic acid bacteria with a high acid-producing ability has an important significance for improving the quality of yoghourt. At present, to select the lactic acid bacteria with a high acid-producing ability among traditional fermented foods has gotten a wide attention from many domestic and overseas scholars. Zhang Heping etc. [5] researched the characteristics of acid production for the lactic acid bacteria separated from natural fermented dairy product, from which a *Streptococcus thermophilus* with short curd time, high rate of acid production and weak post-acidification ability is obtained. Ma Yuxiao etc. [6] get three lactic acid bacteria with high curd speed, high acid-producing ability and fine curd quality through the screening from dzo yoghourt. In addition, Carafa etc. [7] researched the experiment on glycometabolism, acidification ability and acid and bile salt tolerance etc. of lactic acid bacteria in natural fermented cheese of Alps, the selected lactic acid bacteria from which can be applied in the production of fermented dairy product. The good curd quality in traditional yak yoghourt has shown that there are lactic acid bacteria with a high performance. The special growing environment and the narrow distribution range of yak have limited the research of lactic acid bacteria in traditional fermented yak yogurt.

Our country is the headstream of yak in the world. Yak resource is very rich in Qinghai-Tibet Plateau, where herdsman adopts traditional handicraft and natural fermentation method to make yak yogurt [8, 9]. It is an indispensable necessity in herdsman's daily life, as well as a traditional featured delicious food in this area. The unique geographical environment, climatic condition, Tibetan herdsman's scattered community and nomadic pasture pattern have together formed the special resource pool of lactic acid bacteria in Qinghai-Tibet Plateau, which offers a guarantee for selecting the lactic acid bacteria with high fermentative capability. This research screens the lactic acid bacteria in traditional fermented yak yogurt and evaluates its acid-producing ability to provide the reference for better using fine strains and improving the quality research of traditional fermented yak yogurt.

2 Material and method

2.1 Reagents

2.1.1 Bacterial source

Fifty seven lactic acid bacteria strains separated from the traditional yak yogurt in Hongyuan of Sichuan and Yangpachen of Tibet are kept in store by College of Food Science, Southwest University.

2.1.2 Reagent

MRS broth, agar from Beijing Solarbio Science and Technology Ltd. Sodium hydroxide, calcium carbonate analytically pure from Chengdu Kelong Chemical Reagent Factor. λDNA/HindIII, 6×DNA Loading Buffer,100 bp DNA Ladder, 2×Taq PCR MasterMix, bacterial genome DNA extraction kit (DP302) from Tianjin Biotech (Beijing) Co., Ltd. forward primer 27F (5-AGAGTTTGATCCTGGCTCA-3), reverse primer 1495R (5-CTACGGCTACCTTGTTACGA-3) compounded by Sangon Biotech (Shanghai) Co., Ltd.

2.2 Instrument and equipment

Storm Gradient PCR, S1000 Thermal Cycler and Mini-Sub Cell GT from American Bio-Rad Company, Centrifugal machine 8510 from Germany Eppendorf Company, Water-Jacket Thermostatic Constant Incubator, GHP-9160 from Shanghai Keelrein Scientific Instrument Co., Ltd. Clean bench,

SW-CJ-2F from Suzhou Antai Airtech Co., Ltd. pH meter,PHS-3C from INESA Analytical Instrument Co., Ltd.

2.3 Methodology

2.3.1 Preliminary screening of lactic acid bacteria with high acid-producing ability

In clean bench, add 1.5% sterile $CaCO_3$ into the sterile MRS medium. Then shake to make $CaCO_3$ well mixed. In this process, bubble should be avoided. When it is cooled to about 46°C, pour it into the plate immediately. After solidification, it is as standby application. Select an appropriate dilution of bacterial liquid to make spread plate cultivation at 37°C for 24h. Finally, make the preliminary judgment for the acid-producing ability of experimental bacteria through the size of $CaCO_3$ lysis zone and bacterial colony to select the bacterial strain with an obvious transparent zone as the alternative bacterium with high acid-producing ability.

2.3.2 Determination of acidity for lactic acid bacteria with high acid-producing ability

Inoculate the cultivated bacterial liquid into MRS fluid medium as per 2% inoculum size. Measure the titration acidity after cultivating for 48h at 37°C. Take 10 mL nutrient solution into 150 mL erlenmeyer flask. Then add 20 mL water after boiled and cooled and 2 drops of phenolphthalein indicators to make well mixed. Use acidometer to indicate end point and 0.1 mol/L NaOH standard solution to titrage to pH, 8.2. Record the volume of NaOH standard solution consumed. Finally calculate the acidity of nutrient solution, which will be indicated by Thorner degrees (°T) [10].

2.3.3 Analysis on identification and phylogenetic tree of lactic acid bacteria with high acid-producing ability

Preparation of DNA template: extract genome DNA (template) of lactic acid bacteria with high acid-producing ability by the method in the specification of DNA extraction kit. Use 0.7% agarose gel electrophoresis to make the test.

PCR amplification: PCR reaction system is 25 μL. Template is 1 μL. Primers, 27F and 1495R are 1μL, 2×Taq PCR MasterMix 12.5 μL respectively, complemented to 25 μL through adding sterile ultrapure water. PCR reaction conditions include 94°C 5min, 94°C 40s, 55°C 40s and 72°C 1min, totaling 35 cycles as well as 72°C 10 min. After the reaction is finished, take 5 μL PCR products to make the test through 1.5% agarose gel electrophoresis.

Sequencing comparison: send PCR products to BGI Science and Technology Ltd. to make sequencing. The sequencing data is submitted to GenBank of NCBI. Use Blast program to search the sequence with maximal homology, and finally determine the species of lactic acid bacteria with high acid-producing ability.

Analysis on phylogenetic tree: take 10 kinds of 10 lactic acid bacteria's gene sequences in the same segments from GeneBank database. Use Alignment program in Clustalx1.83 software to make multi-sequencing matching comparison for the tested sequences and the reference sequences obtained from BLAST searching. Then use Kimura 2-parameter mode in MEGA5.0 software to calculate genetic distance. The phylogenetic tree of homologous sequence to separate lactic acid bacteria is constructed by neighbor-joining method. Each branch of phylogenetic tree adopts bootstrap method to make confidence coefficient test. The bootstrap data set is 1000[11-13].

3 Results and analysis

3.1 Preliminary screening results of high acid produced lactic acid bacteria

Table 1 and Figure 1 show that the calcium-dissolving zones of 17 lactic acid bacteria are found not obvious. 30 lactic acid bacteria can generate calcium-dissolving zone. But the transparent zones are different to some extent for different bacterial strains.

Table 1. The situation of lactic acid bacteria produced transparent circle.

Serial No.	No.	Transparent circle size	Serial No.	No.	Transparent circle size	Serial No.	No.	Transparent circle size
1	27	+	20	4-1	+	39	25-1-1	++
2	31	-	21	4-2-2	++	40	25-1-2	++
3	32-2	++	22	5-1	++	41	25-2	+
4	37	-	23	6-1	+	42	26-1	-
5	38-2	-	24	6-2	+	43	27	++
6	41-1	-	25	7-1	++	44	27-2	++
7	43-2	-	26	7-2	++	45	28-1	++
8	54	++	27	12-1	++	46	32-1-1	++
9	55	-	28	16	+	47	32-1-2	++
10	57-2	+	29	17-2	++	48	34-1	++
11	58	-	30	18-1-1	++	49	40	++
12	60	-	31	18-1-2	++	50	41-1	++
13	64	+	32	18-2	+	51	41-2	+
14	71	-	33	18-3	++	52	42-1	+
15	77	++	34	19-1	++	53	53-1	++
16	88	++	35	21-1	+	54	53-2	+
17	89	-	36	21-2	++	55	53-3	+
18	3-1-1	++	37	22-1	++	56	62-1	++
19	3-1-2	+	38	23-1	++	57	67-1	+

Note: "++", strong transparent circle, "+", weak transparent circle, "-", no transparent circle.

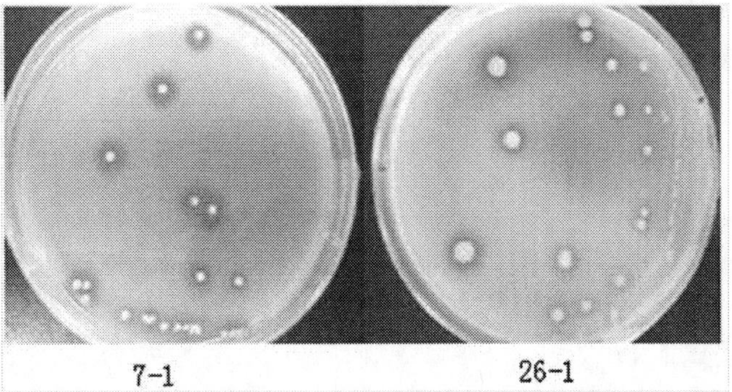

Figure 1. The lactic acid bacteria produced transparent circle in MRS calcium carbonate medium.

3.2 Acidity determination results of high acid produced lactic acid bacteria

Table 2 shows that the acid-producing range of experimental bacterial strain is from 69.2 to 200.2°T, among which the acid-producing contents of No.7-1, 22-1, 28-1, 34-1 and 62-1 bacterial strains are higher. The titration acidities are 196.2, 191.1, 192.2, 194.8 and 200.2 o T respectively. The acid-producing contents of experimental bacterial strains researched by Liu Zhenming etc. [14] are from 64.4°T to 294.2°T, and the ones researched by Zhang Lanwei etc. [15] from 44.35°T to 57.65°T. Compared with the screening results made by other scholars, the lactic acid bacteria with bacterial strains No. 7-1, 22-1, 28-1, 34-1 and 62-1 have a higher acid-producing ability. Under the same culture medium and condition, the diffident acid-producing ability of different strains shows that difference exists between the acid-producing ability of bacterial strain metabolism's carbohydrate and its acid resistance.

Table 2. Acidity determination results of high acid produced lactic acid bacteria after screening.

Serial No.	No.	Acidity (°T)	Serial No.	No.	Acidity (°T)	Serial No.	No.	Transparent circle size
1	32-2	91.9±0.36	11	17-2	128.8±1.08	21	27	103.3±0.53
2	54	91.8±0.53	12	18-1-1	104.8±0.26	22	27-2	69.2±0.79
3	77	98.0±0.20	13	18-1-2	105.9±0.35	23	28-1	192.2±0.30
4	88	87.8±0.26	14	18-3	104.0±0.26	24	32-1-1	129.7±0.36
5	3-1-1	86.8±0.46	15	19-1	103.2±0.30	25	32-1-2	104.0±0.26
6	4-2-2	102.9±0.36	16	21-2	110.2±0.36	26	34-1	194.8±0.79
7	5-1	134.2±0.89	17	22-1	191.1±0.20	27	40	100.4±0.44
8	7-1	196.2±0.26	18	23-1	177.8±0.23	28	41-1	128.0±0.26
9	7-2	145.3±0.36	19	25-1-1	103.0±0.70	29	53-1	94.2±0.44
10	12-1	122.4±0.30	20	25-1-2	153.8±0.26	30	62-1	200.2±0.46

3.3 Identification of high acid produced lactic acid bacteria

Figure 2A shows that genome DNA of bacterial strain is in around 23130bp, strip neat and luminosity obvious, which means the extraction effect of genome DNA is good. The target fragment after PCR amplification gets close to 1500bp (Figure 2B), no tailing phenomena in strip, clear and bright, and no strip in negative control, which means PCR amplification succeeds with a better effect.

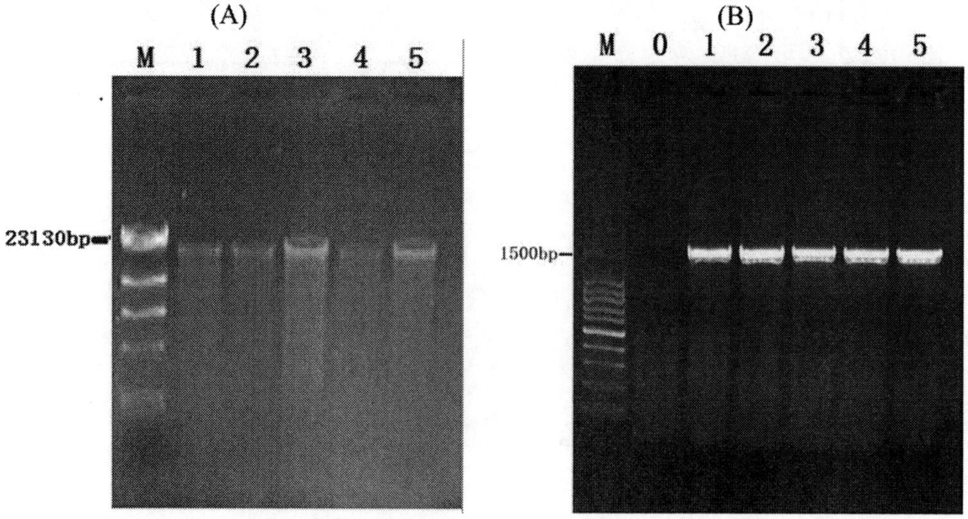

Figure 2. Agarose gel electrophoresis of genomic DNA (A) and PCR products (B).

Traditional fermented yak yogurt is a dairy product with unique flavor, co-fermented by complex microflora taking actic acid bacteria and saccharomycetes as main bacteria and microorganism. At present, it is found through research that its fermentation microorganism [16] is mainly composed by *Enterococcus faecium* and *Enterococcus durans* in *Enterococcus*, *Lactobacillus casei*, *Lactobacillus plantarum*, *Lactobacillus bulgaricus*, *Lactobacillus fermentium* and *Lactobacillus helveticus* in *Lactobacillus* and *Leuconostoc mesenteroides* subsp. *mesenteroides* as well as *Streptococcus thermophilus* etc. Traditional fermented yak yogurt is characterized by high acidity. The acid generally [17, 18] can reach to 198.7°T. The research finds various lactic acid bacteria including *Enterococcus faecium*, *Enterococcus durans*, *Lactobacillus casei*, *Lactobacillus paracasei*, *Lactobacillus fermentium* and *Lactococcus lactis* as well as *Lactococcus lactis* subsp. *lactis* and *Streptococcus thermophilus* etc. contribute to the formation of yak yogurt's acid and flavor. Table 3 shows the experiment also has verified that *Enterococcus durans*, *Lactococcus lactis* subsp. *lactis*, *Lactobacillus casei* and *Lactobacillus fermenti* play a significance role in the formation of yak yogurt's acid. Although enterococcus durans have not been recorded in the list of bacteria used in our country's food industry, it is an important composition in microflora of traditional fermented yak yogurt. It has a high acid-producing ability, so its effect is more significant in the quality formation of traditional fermented yak yogurt.

Table 3. Identification results of high acid produced lactic acid bacteria

No.	Similarity(%)	Latin name
7-1	100	*Enterococcus durans*
22-1	100	*Lactococcus lactis* subsp. *lactis*
28-1	100	*Lactobacillus casei*
34-1	100	*Lactobacillus fermentium*
62-1	100	*Enterococcus durans*

3.4 Phylogenetic tree analysis

Figure 3 shows that different genera of lactic acid bacteria are in their own branch. The branch length represents the degree of species evolution. In cladogram, *Lactococcus*, *Leuconostoc*, *Galactococcus* and *Lactobacillus* are included. The bacterial strains with No.7-1 and 62-1 are determined as *Enterococcus durans* to their 100% bootstrap value. In phylogenetic tree, they are determined as *Enterococcus durans*. Both of them have a close genetic relationship. Meanwhile, the bacterial strain with No. 22-1 and *Lactococcus lactis* subsp *lactis* are classified into a group due to their 100% bootstrap value, determined as *Lactococcus lactis* subsp. *lactis*. It is known through the topological structure of cladogram that the genetic relationship between *Lactococcus* and *Enterococcus* is closer. But it has a distant genetic relationship with *Lactobacillus* and *Leuconostoc*. The bacterial strains with No. 28-1 and 34-1 are in *Lactobacillus* branch, having a closer genetic relationship. In phylogenetic tree, they are classified into a group with *Lactobacillus paracasei*, *Lactobacillus casei*, and *Lactobacillus plantarum* as well as *Lactobacillus fermentium*. Among them, the bacterial strain with No. 28-1 has the same bootstrap value as *Lactobacillus casei* and *Lactobacillus paracasei*. This is because *Lactobacillus casei* and *Lactobacillus paracasei* have a very close generic relationship. No. 34-1 bacterial strain and *Lactobacillus fermentium* are classified into a group due to their 100% bootstrap value, determined as *Lactobacillus fermentium*. The analysis result of phylogenetic tree and the result identified by 16S rDNA method show that 16S rDNA method is feasible, and the result, reliable.

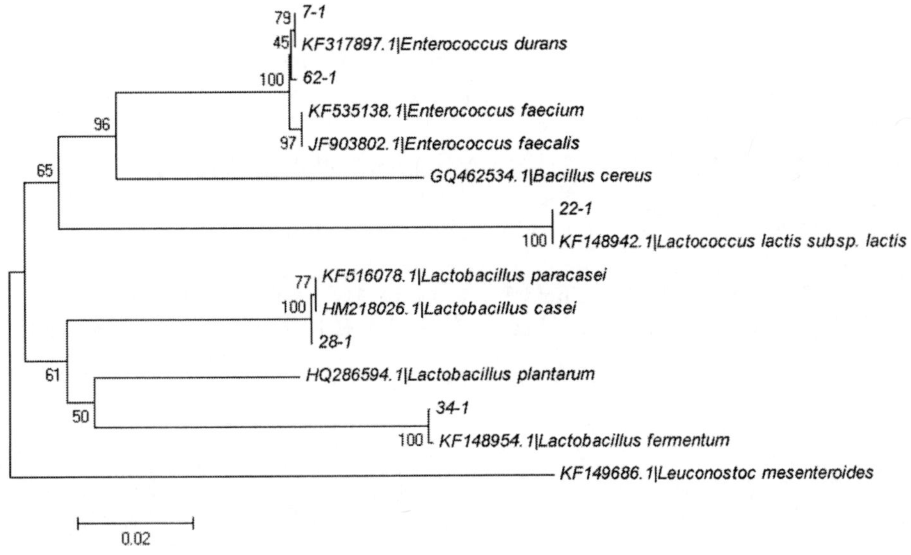

Figure 3. Phylogenetic tree of lactic acid bacteria rDNA 16S sequence.

4 Conclusion

In this research, the lactic acid bacteria with high acid-producing ability in traditional fermented yak yogurt is screened and selected by calcium-dissolving zone method and through determination of titration acidity. The research shows the abilities of different bacterial strains in yak yogurt to generate calcium-dissolving zone are different. The acid-producing abilities are also different. The acidity is between 69.2 and 200.2 o T. Among them, the bacterial strains with No. 7-1, 22-1, 28-1, 34-1 and the bacterial strain with No. 62-1 have a higher acid-producing ability. After cultivating for 48h at 37^{oC}, titration acidities are 196.2, 191.1, 192.2, 194.8 and 200.2 o T respectively. No. 22-1 bacterial strain is identified as *Lactococcus lactis* subsp. *lactis*, No. 28-1 bacterial strain as *Lactobacillus fermentium*, and the bacterial strains with No. 7-1 and 62-1 as *Enterococcus durans* through 16S rDNA sequence analysis. Although *Enterococcus durans* have not been recorded in the list of bacteria used in our country's food industry, it is an important composition in microflora of traditional fermented yak yogurt. The experimental result shows that it has a high acid-producing ability, and playing an important role in the formation of flavor and quality of traditional fermented yak yogurt.

References

1. T.Q. Zhang, Z.N. Yang, B.H. Kong, Food Sci 637-642, **29** (2008).
2. X. Xin, G. Song, X.H. Zhou, J.W. Zhao, L. Chen, Q.P. Ge, W.X. Ping, J. Chinese Inst. Food Sci. Technol. 202-207, **14** (2014).
3. H. Qin, X. Liang, W.B. Zhang, Y. Zhang, L. Mi, Food Sci. 241-246, **34** (2013).
4. L.H. Bian, Y. Wang, Y.T. Zhang, L.N. Qu, Guangdong Agri. Sci. 128-130, **39** (2012).
5. H.X. Chen, L.L. De, Y. Ren, D.L. Zhang, Y.R. Yang, W.J. Liu, H.P. Zhang, J. Dairy Sci. Technol. 1-5, **2015** (2015).
6. H. Tian, X.P. Zhang, Y.L. Yan, Food Sci. 152-156, **31** (2010).
7. I. Carafa, T. Nardin, R. Larcher, R. Viola, K. Tuohy, E. Franciosi, Food Microbiol. 123-132, **48** (2015).
8. C.S. Wu, C. Shu, J. Li, Y. Qian, H.Y. Suo, Food Ind. 129-133, **33** (2012).
9. Y.W. Wang, C.M. Zhang, H.Y. Suo, H. Yue, J. Li, C. Tang, Sci. Technol. Food Ind. 184-187, **35** (2014)
10. C.X. Liu, Jiangsu J. Prev. Med. 69-70, **19** (208).
11. N.A. El-Naggar, S.A. Haroun, E.A. Oweis, A.A. Sherief, Prep. Biochem. Biotechnol. 712-729, **45** (2015).
12. K. Tamura, D. Peterson, N. Peterson, G. Stecher, M. Nei, S. Kumar, Mol Biol Evol. 2731-2739, **28** (2011).
13. S. Ennahar, Y. Cai, Y. Fujita, Appl. Environ. Microbiol. 444-451, **69** (2003).
14. Z.M. Liu, Y.Y. Wang, J. Dairy Sci. Technol. 169-172, **33** (2010).
15. C.L. Ma, L.W. Zhang, Sci. Technol. Food Ind. 189-190, **31** (2010).
16. L.V. McFarland, Clin. Infect. Dis. 85-90, **60** (2015).
17. Z. Luo, Y.G. Shi, L. Yu, B.Z. Han, China Brew. 40-41, **2005** (2005).
18. J. Xu, Y.Y. Yun, W.Y. Zhang, Y.D. Shao, B. Menghe, H.P. Zhang, China Dairy Ind. 23-27, **34** (2006).

 DOI: 10.1051/bioconf/20170803003

Cognition investigation and analysis of hand hygiene of nursing staff in pension institutions

Bing Yu, Yujin Liu [a], Wenjing Sun, Xiufeng Zhao, Zhuang Cong and Wei Meng

Welfare College of Nursing of College of Humanities&Northeast Normal University, 130117, Changchun, China

Abstract. Objective: We investigate and analyze the cognition of hand hygiene of nursing staff in nursing institutions. This method provides reference for improving the hand hygiene condition of nursing staff in nursing homes, and we put forward the feasible solution to this situation. Methods: 100 nurses from five nursing institutions in Changchun were investigated using a self-designed questionnaire, and the results were analyzed. Results: due to the most old-age care institutions personnel are less educated, rapid flow of talents, and the health care training from endowment organization are low frequency or not organized, so most of nursing staff think hand hygiene has nothing to do with the health of the elderly. They grasp the knowledge of some common hand washing, and hand washing before and after the elderly care is not ideal. The effects of hand washing have a number of factors, the main factor is that the pension institutions hand washing facilities incomplete, and in fact the nursing staff work very busy. Discussion: from the government to pension institutions should strengthen and pay more attention to the training of pension agency front-line care staff hand hygiene knowledge, increase the training of professional nursing practitioners, focus on bringing in Colleges and universities social science or medical science related, similar professional personnel and institutional strengthening management of washing hands, according to the actual conditions to improve the hand washing facilities, adhere to wash their hands properly, so as to avoid occurrence of handling of disease spread.

1 Introduction

With the increasing of the aging of the population in China, nearly one million employees in China more than 40000 social welfare institutions, nursing staff is the largest group of [1], constitute the main force in the industry. Nursing staff during routine care for the elderly, often to help the old man eating drinking water and the old man's body surface skin, waste and old man around various objects surface contact, resulting in nursing staff hand carrying a variety of pathogenic bacteria. Nurses and contact with the elderly opportunity most longest contact time, the old resistance strength is low, if the before and after nursing activities without enough opponent cleaning or disinfection, it is easy to cause hand the spread of pathogens to the next An old man and infection diseases affect health, even the old man one of the causes of death. In the hospital, the hospital environment and the surfaces of various

[a]Corresponding author: Liu Yujin, E-mail:1257819211@ qq.com
This work was supported by a Social Sciences planning project of Department of Education of Jilin Province(2014.12.1).

DOI: 10.1051/bioconf/20170803003

objects exist lots of microorganisms, the hands of medical workers in a variety of clinical operation inevitably often contact the microbial contamination and become the potential factors of nosocomial infection. Studies have shown that the hands of medical personnel in the spread of bacteria and the occurrence of hospital infection accounted for about 30%[2], therefore, hand contamination of medical staff is an important route of transmission of the hospital infection caused by. Cleaning and disinfection of hand skin is very important, and wash it the most simple and feasible to control the hospital infection measures. Public places Hand hygiene is to prevent infection diseases of important means, in the developing world hand hygiene can significantly reduce the outbreak of the disease in children. Good hand hygiene habits can effectively prevent diarrhea and respiratory diseases, such as infectious diseases. However, the residents of China proper hand washing rate is only 4%[3]. based on the above, in access to a large number of documents at the same time, there are no about nursing home care personnel hand hygiene cognition survey report, this paper on nursing home care hand hygiene knowledge were investigated.

2 Objects and methods

2.1 Research objects

A questionnaire survey was conducted on 100 nurses in eight nursing homes in Changchun city.

2.2 Survey methods

According to the World Health Organization of < medical and health institutions guideline for hand hygiene > and China's 2009 < medical staff hand hygiene practices > file based, designed the "endowment nursing staff hand hygiene cognition questionnaire. The questionnaire is divided into four parts: the first part is the basic information survey, including unit, position, seniority, sex, marriage, education. The second part is nursing staff opponents hygiene knowledge to grasp the situation. The third part is the nursing staff in the correct way to wash your hands to grasp the situation. The fourth part is factor of the effects of hand washing[4].

2.3 Statistical methods

All data were collected and analyzed by SPSS17.0 statistical software. The descriptive statistical method was used.

3 Results and analysis

3.1 analysis of the general situation of nursing staff

100 nurses, including 12 male and 88 female people; More than 2 years over the age of 45; 10 58 29 1 100 nursing, male 12, female 88 people, 30 - 40, 22, 41 - 50 years of age 45, 51 years old and above 33 people, and the seniority in the following 32 people, length of 1 - 2 years 43 people, age 2 years above 25 people; cultural degree: elementary school, junior high school, high school, college and above 3 people ': urban laid-off workers, migrant workers 30 people. The results of the survey show that pension institutions nursing personnel age mainly concentrated in 41, urban laid-off workers and migrant workers. Personnel flow, length of service is only 25 people, so to him The training needs to be targeted, training frequency should be repeated, so that it changes the concept.

3.2 The analysis of nursing staff on hand indications

From Table 1 can see nursing staff to contact with the old man's blood, body fluids, secretions, mucous membrane, incomplete skin wash to the indications for 100% is much higher than the other hand washing indications, showing that self protection consciousness is strong. Wash your hands after contact with the elderly around objects to indications minimum was 23%, and protection of the elderly, to prevent cross infection awareness is weak, and nurses cultural degree low, ID thought, nor did he accept any hand hygiene knowledge, there are many relationships.

Table 1. The analysis of nursing staff on hand indications

Choice of indications for hand washing or hand disinfection	Need	Unwanted	Correct rate
Visible blood stains or by hands, body fluids contaminated dirt	100	0	100%
Direct contact with an old man's body before	45	55	45%
After each contact an old man intact skin (such as the measurement of vital signs, patient handling, help)	65	35	65%
To assist the elderly taking oral medicine, eating, drinking water before and after	54	46	54%
Exposed to the old man's mucous membrane, after the skin is not complete	100	0	100%
After the body fluid, the secretion, the secretion or the secretion after the dressing of the old man	100	0	100%
In the process of nursing for the elderly, the hand from the old man's body to clean parts before the Department of pollution	76	24	76%
After each contact with an old man around the object	23	77	23%
Do you need to wash your hands before and after wearing gloves (including disposable gloves and non disposable gloves)	32	68	32%
After meals	80	20	80%

3.3 Analysis of nursing staff's health related knowledge mastery

Department of nursing staff opponents of bacteria related knowledge mastery is 10%, the six step hand washing method is 4%. See Table 2, which shows that the nursing staff opponents of health related knowledge master level rate is low, opponents of health knowledge, there is no complete system learning.

Table 2. Analysis of nursing staff's health related knowledge mastery

Project	Correct	Not correct	Correct rate
Bacteria parasitic on the hand are mainly divided into temporary and resident bacteria.	4	96	4%

The main hand bacteria causing nosocomial infection were resident bacteria.	10	90	10%
Hand washing is mainly in order to remove the hand of the resident bacteria	9	91	9%
If the gloves are not obviously contaminated, do not need to be replaced for other patient services	35	65	35%
October 15th is the international day of washing"	2	98	2%
Do you usually wash your hands in accordance with the six steps of washing	4	96	4%
Does soap need to be kept dry	24	76	24%
Care for more than one elderly, whether the need to replace gloves	34	66	34%
Disposable gloves may be reused	45	55	45%
Before using the "fast hand", you need to keep your hands dry.	17	83	17%

3.4 Factors influencing hand washing of nursing staff

The main factors that affect the hand washing of nursing staff are busy, not enough time, the rate is 96%, followed by the forgotten in what situations should wash their hands, the rate is 94%, the specific Table 3.

Table 3. Factors influencing hand washing of nursing staff

Project	Yes	No	Influence rate
Cleaning agent, disinfectant to stimulate the skin, causing dry skin	91	9	91%
The number of hand washing pool is too small in the working environment.	87	13	87%
The position of hand washing pool is not convenient in working environment.	90	10	90
Hand wash basin is not equipped with enough cleaning agent	67	33	67%
Hand wash basins are not equipped with dry hand facilities (such as disposable hand towels, sterile towels or dryer, etc.)	82	18	82%
Liquid soap, hand disinfection, dry toilet paper costs too high	65	35	65%
Work too busy, not enough time	96	4	96%
The risk of being contaminated by a nurse is very low.	76	24	76%

Person to hand washing or hand disinfection in the work is not enough attention	88	12	88%
Forgot to wash hands under what conditions	94	6	94%

4 Discussion

The nurses completed, pension institutions service quality depends on the nursing staff to take care of the elderly is special. Hand hygiene as nurses for the elderly to implement each service operation premise, has very important significance. The investigation shows, nursing staff on hand to hand hygiene knowledge and related sign is not clear, the understanding is superficial, to wash the finger and hand hygiene related knowledge has a lot of mistakes, influence of nursing staff and the wash factors affect the rate was above 65%. Hand washing behavior is a complex phenomenon, hindering handwashing behavior factors is multi-level, there are internal and external causes in the pension institutions all care activities for the elderly, involving the environment, cognitive, interactions between many of the emotional level factors. Reason analysis for China's pension institutions started late, for the elderly health management lacking, for the fine service to do does not reach the designated position. Pension institutions may did not do a good job of nursing personnel training and health propaganda work, the local government while carrying out the pre service system, but the training is not standardized, as far as the training requirements. From the point of view of management and there is no understanding in hand hygiene of importance and neglect the opponents of health management. In institution endowment nursing staff general cultural quality low, high school degree or above 13% of nursing staff in the loss of serious, length of service 2 years above is only a community service center in Beijing 25%[5]. 2005 had free training more than 100 laid-off personnel, but after a lapse of one year has no people in nursing jobs. High turnover rate, affecting the normal work of old-age care institutions, and even the some pension agencies did not dare to implement strict management of nursing staff. Age generally are relatively large, than younger people have more experience of life and experience, work conscientiously, steadfast. But in this age of cultural quality of personnel is low, has formed inherent mode of thinking and living habits, difficult to change the mode of thinking and bring personal lifestyle and health habits to work, in an Hand hygiene related theoretical knowledge of poor ability. In the wash factors, mainly pension institutions hand washing facilities are not complete, the nursing staff workload. Factors of hand hygiene behavior is multifaceted, it is result of personnel and the surrounding environment interact continuously, even including psychology, habits, and from the perspective of behavioral science to analysis, knowledge is the premise that belief structures, but faith will guide a person's behavior[6]. Based on the above reasons, first of all, should accelerate the old-age nursing team construction, along with the development of the aging of the population and the old social welfare undertakings, China needs a large number of high-quality nursing staff , enhancing the nursing staff treatment, which is to speed up the building of the contingent of the key. Vigorously develop professional workers, should pay attention to the introduction of special talent for pension institutions, nursing staff and the introduction should focus on Colleges and universities social science or medical science related, similar professional personnel. Secondly, increased government support and investment in infrastructure, implement preferential policies and help pension institutions to reduce construction and operating costs, improve the environment and facilities, and satisfy the basic needs of the front-line staff. Again, a strong proponent of nursing staff in nursing homes in the health of continuing education, the government of the nursing personnel implement free training to endowment organization training or care Professionals themselves to spend money training - at least for now, reality, the implementation of government funded free training, improve the quality of personnel, pension institutions overall level of health services can be improved[7]. Finally, the government should strengthen supervision and management, who hand washing facilities non-compliance, should warn rectification; pension institutions should also be to strengthen their own management, improve hand washing facilities,

DOI: 10.1051/bioconf/20170803003

urging nursing staff to establish a habit of hand hygiene, convenient wash, dry hands, hand measures is to ensure that all effective hand washing, a prerequisite to enhance awareness of nursing staff hand washing; pension institutions can post corresponding publicity column or a photograph in the old room or corridor, to promote nursing staff to develop a good hand Health habits, while obtaining the understanding of the elderly[8].

In short, hand washing is one of the most simple and easy way to prevent and control the spread of pathogens is an important measure to prevent the elderly from being infected with infectious diseases, but also the effective means of two-way protection of elderly nursing staff and the elderly.

References

1. The Ministry of civil affairs. China Civil Affairs statistical yearbook. Beijing: China Statistics Press(2009).
2. Yang Shuping. Investigation and Analysis on cognitive behavior of nursing students to wash hands and Countermeasures. nursing practice and research,**6**, 21(2009).
3. Leaf Daxun. Wash your hands. A simple but meaningful measures, the British Medical Journal (Chinese Edition),**3**(2000).
4. A white paper on hand washing status of residents in the Ministry of health of the Ministry of healt,**10**(2011).
5. WHO advisory panel and WHO Advisory Committee and hand hygiene program drafting group - hand hygiene guidelines for medical and health institutions - WHO - hand hygiene guidelines in medical activities(2005).
6. Huang Juru, Zhong Chunjiao, Xiao Fenglan et al. Investigation of hand washing behavior of nurses. Journal of nosocomial infection in China,**14**,12(2004).
7. Chen Zhuoyi, Chen Wei, present situation and Countermeasures of the construction of nursing staff in China. Journal of Changsha District Vocational Technical College,**12**,16(2009).
8. Yuan Xiaoning, Wang Shaoli, Zhao Xinmao et al. Investigation and Analysis on the knowledge and behavior of the hand hygiene of the hospital assistant staff. The nursing management in China,**4**,8(2008).

DOI: 10.1051/bioconf/20170803004

Plant regeneration and *Agrobacterium*-mediated transformation of Vacuolar H^+-ATPase c Subunit Gene in hybrid poplar *populus davidiana* Dode × *P. bollena* Lauche

Xue Han[1] and Bing Han[2,a]

[1]*College of Accountancy, Harbin University of Commerce, Harbin 150028, China*
[2]*College of Food Science and Engineering, Harbin University of Commerce, Harbin 150076, China*

Abstract. An efficient regeneration and transformation system was developed for hybrid poplar *populus davidiana* Dode × *P. bollena* Lauche. Several factors, such as a competent regeneration protocol, antibiotics, bacterial concentration, infection time have been shown to affect transformation efficiency. It is the first report describing an efficient protocol for Vacuolar H^+-ATPase c Subunit Gene from salt tolerant plant *Puccinellia chinampoensis*, which transformed into *populus*. Stable transgene integration was confirmed by the expression of GFP fusion proteins in leaves, stem and root for populus.

1 Introduction

Development of an efficient, competent plant regeneration system is essential for transformation. Poplar regeneration has been reported in the past decades. Different genotype of poplar have been used to perform the regeneration experiment [1-3]. The leaf disc transformation method established by Horsch et al. [4]. The technology has been applied to various *populus* species for transformation effeiency [5-7], The systematic analysis of several factors to improve the transformation efficiency, including poplar genotype, *Agrobacterium tumefaciens* strain, bacterial concentration and so on [8-10].

Populus davidiana Dode × *P. bollena* Lauche is the hybrid cultivar of poplar that using *Populus bolleana* as male parent and *Populus davidiana* Dode as female parent, which is planted on a significant scale in northeast of China due to its characteristics as graceful shape, cold resistance, fast growth, and suitability to develop transgenic.

Puccinellia chinampoensis is a special gramineous plant, which can survive well in the saline and alkali environment with severe drought and salt stress. *Puccinellia chinampoensis* has resistance mechanism of adaptation in the saline and alkali soil. Vacuolar H^+-ATPase c Subunit Gene (VHA-c) was isolated from *Puccinellia chinampoensis*, which was used to transformation in the study. Several researchs and informations about VHA have been published including structure, function and regulation mechanism[11-14]. Vacuolar H^+-ATPase c Subunit Gene has been studied in *Mesembryanthemum crystallinum* [15], sugar beet [16], *Arabidosis thaliana*[17,18] and tobacco [19].

[a] Corresponding author: hb790118@163.com

 DOI: 10.1051/bioconf/20170803004

In the present study, we developed a high efficient regeneration system for *Populus davidiana* Dode × *P. bollena* Lauche. In the basis of the regeneration system, a simple and efficient protocol for transformation, selection and regeneration of *popolus* was developed. Although VHA-c has been successfully used to transfer into a wide range of gramineae plant species, it has received little attention for transformation of woody plant. To our knowledge, this is the first report describing an efficient protocol for Vacuolar H^+-ATPase c Subunit Gene from salt tolerant plant *Puccinellia chinampoensis*, which transformed into *populus*. Stable transgene integration was confirmed by the PCR analysis and the expression of GFP fusion proteins.

2 Materials and methods

2.1 Plant material, *Agrobacterium* strain and plasmid

Populus davidiana Dode × *P. bollena* Lauche leaf explants were obtained from shoot cultures. The *Agrobacterium tumefaciens* EHA105, containing the plasmid pBI121-VHA-c-GFP. This plasmid driven by the CaMV 35S promoter.

2.2 Regeneration system

MS medium was supplemented with 6-BA and NAA such as Table 1. The leaves were cut into 0.5*0.5 cm^2 blade and were cultured in flasks. They were transferred to fresh regeneration medium at 10-day intervals.

Regenerated shoots, 3–5 cm in length were excised and rooted on MS medium containing various concentrations NAA (0.1, 0.2, 0.3 or 0.4 mg/L). All the media included 3% (w/v) sucrose and 0.8% (w/v) agar as gelling agent. The pH value of the media was adjusted to 5.8 after adding the hormone and autoclaved at 121°C for 20 min. AS and all antibiotics such as Kan and Cef were filter-sterilized and added to media cooled to 50–60°C after autoclaving.

After 1-2 months the root was lignified, and the well-developed plantlets were taken out from media. They were washed under gentle stream of running tap water to remove the adhering media, and then transferred to plastic cups containing a mixture of sterile compost. The plastic cups were covered with transparent polyethylene bags to maintain the internal humidity and placed in growth room. Plantlets in plastic cups were watered every 2 days with regular water and acclimatized. After 2 weeks in the culture room, opening the plastic bags progressively until the plants were ready for transfer to the greenhouse.

Regeneration experiments were designed with 10 Petri dishes per treatment, each dish contained at least 6 leaves for shoot regeneration and 3 shoots for the rooting of shoots. Experiments were replicated three times. These cultures were incubated at 25±2°C temperature under 16 h photoperiod and were observed for regeneration.

2.3 Assessing antibiotics for transgenic shoot selection and bacterial elimination

For the experiments testing Kan as a selective agent, untransformed leaf explants were cultured on induction media containing 0, 5, 10, 20, 30 or 40 mg /L Kan, aiming to identify the most suitable concentration for selection of transgenic plants. The experiment was carried out with 30 explants per treatment and replicated three times. Growth and differentiation were observed for 1 month.

Sensitivity of rooting of shoots to Kan was determined by culturing the elongated adventitious shoots (2–3 cm) on the rooting medium with 0, 5, 10, 20, 30, and 40 mg/L Kan for 1 months. The experiment was carried out with 30 shoots per treatment and replicated three times. Rooting was observed over 1 month.

For the experiments testing Cef as a bacterial elimination, untransformed leaf explants were cultured on induction media containing 0, 100, 200, 300, 400, 500, 600, 700, 800 mg /L Cef, aiming to identify the most suitable concentration for control bacterial growth and the plant growth well. The experiment was carried out with 30 shoots per treatment and replicated three times. Growth and differentiation were observed over two-month period.

2.4 *Agrobacterium* concentration and infection time

A single colony was inoculated into 3 mL of liquid YEP medium with 100 mg/L Rif and 50mg/L Kan, incubated at 28°C for 24 h with constant agitation (175 rpm). An additional 30 mL YEP was then added and the culture was grown overnight at the same culture conditions upto OD_{600} =1.2. The cells were then centrifuged at 5,000 rpm for 5 min and the pellet was diluted to OD_{600} = 0.5, 0.8, 1.0, 1.2 for transformation. The leaf explants were incubated in the bacterial suspension with different times such as 10min, 20min, 30min, 40min.

2.5 Transformation and plant regeneration

The bacterial suspension containing 100μmol/L AS and 0.1% TritonX-100 was poured into petri dish. Leaves were cut into 0.5 cm*0.5 cm blade in the bacterial suspension of petri dish, and inoculated for different time in the bacterial suspension . Leaf explants were then dry on sterile filter paper and cultured in the dark on the appropriate tissue culture medium without Kan and Cef. After 48h co-cultivation, leaf explants were washed 3 times (20-30 min each) in liquid co-cultivation medium plus 200 mg/L Cef, and then blotted dry on sterile filter paper. The explants were transferred to regeneration medium containing Kan and Cef for selection and cultured in the light at 25±2°C for 4-6 weeks. To inhibit *Agrobacterium* overgrowth, cultures were transferred to fresh selective medium at 15-day intervals until the leaves differentiation. Shoots were transferred to rooting medium containing Kan and Cef after the transformation leaves differentiation shoots . Approximately 100 leaf explants were used for each transformation experiment and replicated three times. Transformation efficiency were calculated from the number of genetically independent transgenic polar plants divided by the number of explants that were infected as to optimize parameters affecting poplar transformation.

2.6 DNA extraction and PCR analysis

After selection, leaves from regenerated transgenic shoots were harvested for PCR analysis. Genomic DNA was extracted from transformed and non-transformed leaves according to the method of CTAB. The primers 5 - ATGTCGTCGGTGTTCAGCG -3 and 5 - ATCTGCGCGGGATTGGCCG -3 that amplifies a 498-bp fragment corresponding to the coding region of VHA-c gene were used. Reactions were performed in a 20 μL volume containing 1 μL (100–500 ng) of genomic DNA, 1μL (20 μmol/L) of each primer, 10 units GoTaq DNA polymerase. Reactions were first heated at 94°C for 5 min followed by 30 cycles of 94°C for 30 s , 58°C for 30s , 72°C for 1 min, and followed by a final extension at 72°C for 10 min. Amplified DNA was detected under UV light, after electrophoresis of the amplification reaction mixture for each sample in 1% (w/v) agarose gels and staining with ethidium bromide.

2.7 Expression of GFP fusion proteins in populus of tissue culture

The leaf, shoot and root were taken away from *populus* of tissue culture and observed in confocal microscopy. Confocal microscopy was performed with a TCS SP2 laser-scanning confocal imaging system (Leica). GFP fluorescent signals were detected with excitations at 488 nm and 543 nm, respectively.

3 Results and Discussion

3.1 Effects of media composition on plant regeneration

The adventitious shoot induction is influenced by the growth regulators concentration and combination. After 10 days of culture, the leaves were curled and expanded. Shoot buds differentiated directly from both the cut blades of leaves explants on some growth regulators concentrations and it took 15–20 days for bud differentiation. All combinations of different concentrations of cytokinins (BA) and auxins (NAA) induced direct shoot buds from leaves explants, especially in the main vein of leaves blades. A combination of BA (0.1 mg/L) and NAA (0.01mg/L) regenerated minimum leaf explants differention frequency (0%) (Table 1). The combination of BA (0.1 mg/L) , NAA (0.05, 0.08, 0.1 mg/L) and BA (0.5 mg/L) , NAA (0.01, 0.05, 0.08, 0.1 mg/L) also regenerated lower shoots than others. BA has vital effect on shoots differentiation, lower or higher concentration lead to the lower percent explant regeneration directly. A combination of BA (0.3 mg/L) and NAA (0.08 mg/L) regenerated 91% shoots. (Table1, Figure 2ABC).

Table1. Effect of different combinations and concentrations of cytokinins (BA) and auxins (NAA) in MS medium on shoot regeneration from leaf explants of *populus davidiana* Dode × *P. bollena* Lauche

Medium	BA(mg/L)	NAA(mg/L)	Per cent explant regeneration(%)
1	0.1	0.01	0
2	0.1	0.05	15.8±1.37c
3	0.1	0.08	6.56±1.37c
4	0.1	0.1	3.12±1.62c
5	0.3	0.01	78.7±6.47ab
6	0.3	0.05	18.6±4.17c
7	0.3	0.08	91.0±7.98a
8	0.3	0.1	58.6±3.09b
9	0.5	0.01	17.4±2.85c
10	0.5	0.05	41.0±10.8bc
11	0.5	0.08	3.83±3.54c
12	0.5	0.1	6.56±2.85c

a Denotes highly significant value

Rooting experiment results showed that formation and growth of the new roots were promoted by NAA at any concentrations that have been setting, the percents of root regeneration were 100%. All the tissue culture derived shoots (2–4 cm in length) developed roots within 10–15 days of MS medium supplemented with various concentrations of NAA. The growth of roots was rapid within 30 days. Of the two concentrations of NAA (0.1 mg/L , 0.2 mg/L) evaluated in the present study, the root of growth and development status were fine root , the number of the roots reduces and the root of length shorter. The status of the rooting were the length was smaller, the number of the roots reduces and thick root at the concentrations of NAA (0.3 mg/L, 0.4mg/L). NAA at 0.3 mg/L was the best for rooting as the growth status of roots were length and polyrhizal (Figure 2D).

Plantlets were taken out of the culture tubes and hardened after two months when the root induction. 90% of plants survived after the hardening process(Figure2 EFGK).

3.2 Assessing antibiotics for transgenic explants selection and bacterial elimination

At 20 mg/L Kan, leaves chlorosis were observed in all explants by the end of the 2 months. Concentrations above 30 mg/L Kan completely inhibited regeneration. At 30 mg/L Kan, non-rooting were observed in all explants by the end of the 2 months. Concentrations above 40 mg/L Kan completely inhibited the roots regeneration. Based on these datas, 30 mg/L Kan is the most suitable concentration for use to shoots regeneration and 40 mg /L Kan is the most suitable concentration for use to roots regeneration in genetic transformation experiments of *populus davidiana* Dode × P.

bollena Lauche(data not shown).

The percentage of shoot regeneration on culture medium containing 200 mg/L Cef did not significantly differ from that obtained on the control treatment (culture medium devoid of antibiotics). The percentage of shoot regeneration on culture medium containing 200–500 mg /L Cef reduced. Cef at 500 mg /L or more gave the worst responses, the status of having little shoots regeneration. (data not shown)

3.3 Effects of Agrobacterium concentration and infection time on plant transformation

The higher transformation frequency were observed at OD_{600} = 0.8 with little contamination(Figure1). Lower transformation frequency were obtained in lower *Agrobacterium* concentrations(OD_{600}=0.5). The same with infection time, the frequency of transformation and contamination increased by the time longer, the higher contamination frequency can lead lower frequency of transformation. The explant inoculation time were reasonable at 20-30min (Figure 1).

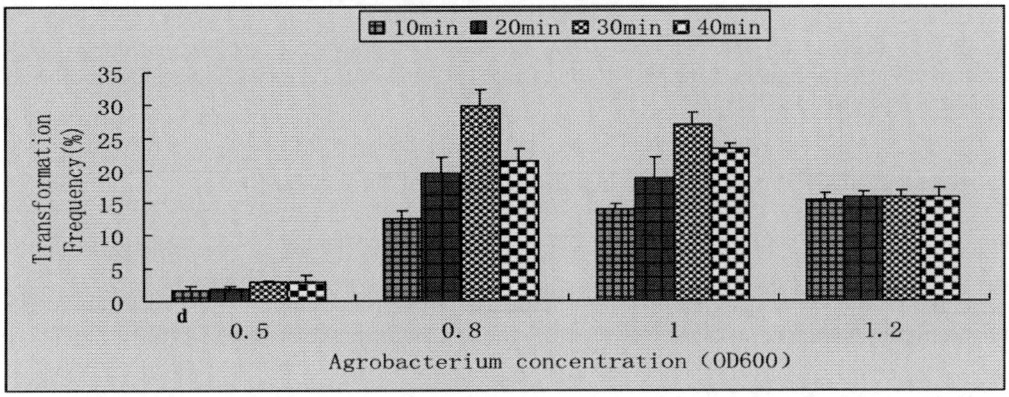

Figure1. Frequency of the transformation of *populus davidiana* Dode × *P. bollena* Lauche with different *Agrobacterium* concentration and the infection time.

Figure 2. High frequency plant regeneration, shoot induction, rooting, micropropagation and high frequency transformation of *populus davidiana* Dode × P. *bollena* Lauche. AB. Regenerated shoots from leaf explants; C. Elongation of regenerated shoots; D. Roots regenerated; E. Regenerated plantlets with medium from flask; FG. Growth of regenerated plantlets in pots after 20 d of transfer. HIJ. Transgenic shoots differentiation. K. Growth of regenerated plantlets in pots after six months of transfer.

 DOI: 10.1051/bioconf/20170803004

3.4 DNA extraction PCR analysis

The successful incorporation of the transgene was proven by the genomic PCR. No positive result occurred with DNA isolated from control plants (non-transformed). Sixteen transformed plants were produced. The data of Figure 3 shown that Vacuolar H^+-ATPase c Subunit Gene has been transferred into hybrid poplar *populus davidiana* Dode × *P. bollena* Lauche.

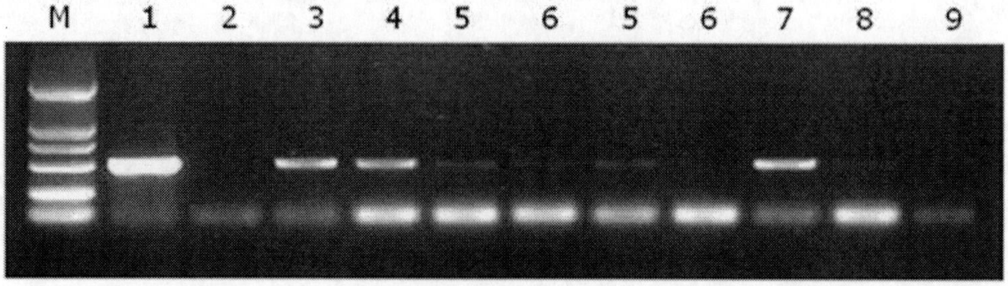

Figure 3. PCR analysis of total genomic DNA isolated from transgenic and untransformed plants by amplification of the VHA-c genes. Lane M, DL2000; Lane1, plasmid pBI121; Lane2, untransformed plant; Lanes 3–8, transformed plants; Lane 9, H_2O.

3.5 Expression of GFP fusion proteins in *populus* of tissue culture

A plant transformation vector designed to express the GFP gene was used to detect expression of GFP fusion proteins in *populus* of tissue culture. From the Figure 4, GFP fluorescence was detected in leaves, stem and root, which proved that the Vacuolar H^+-ATPase c Subunit Gene has been transferred into hybrid poplar *populus davidiana Dode* × *P. bollena* Lauche and expression exactly.

DOI: 10.1051/bioconf/20170803004

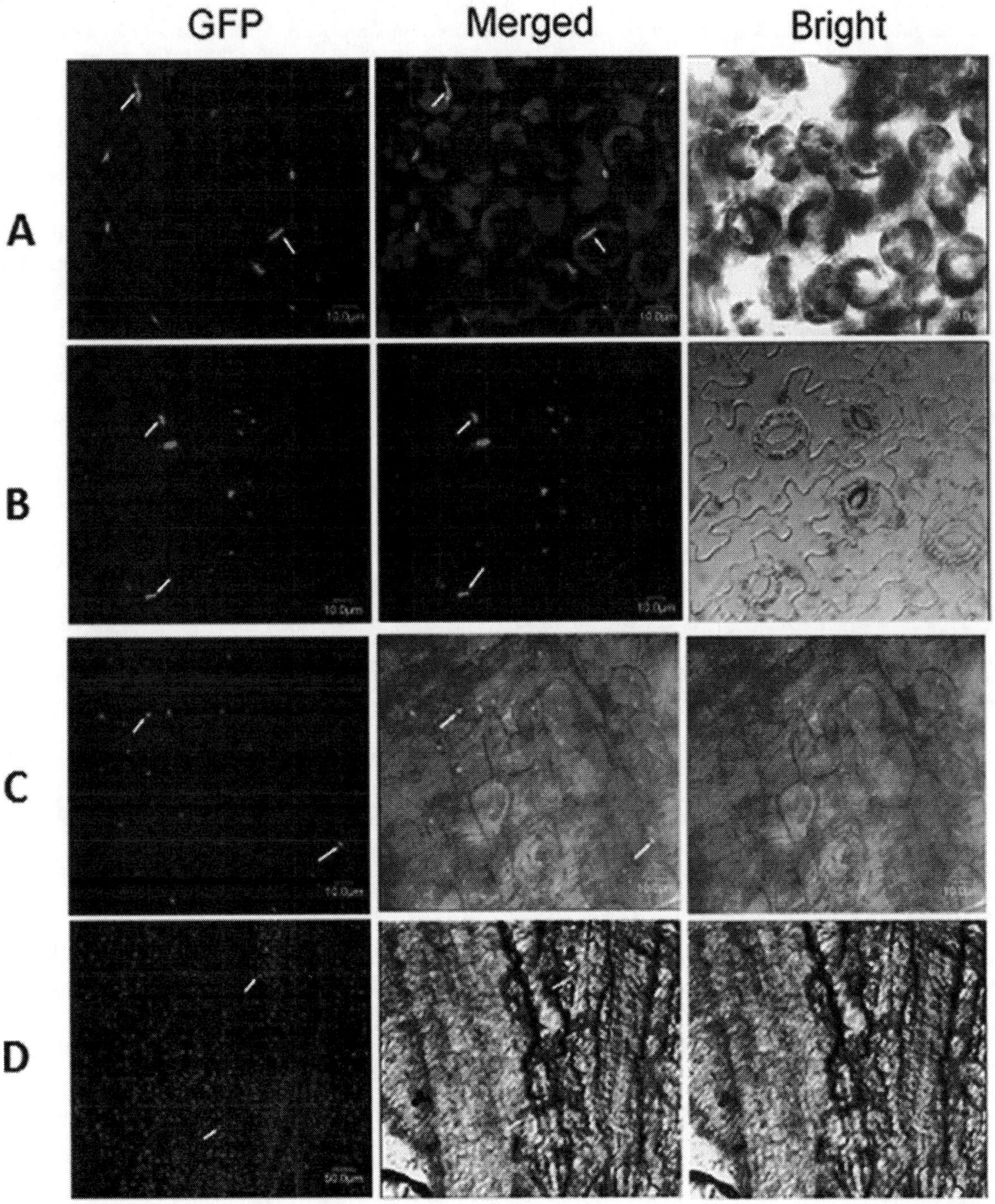

Figure 4. Expression of MD-GFP fusion protein in populous
A Expression of MD-GFP in mesophylls (white arrows). **B** Expression of MD-GFP in epidermal of leaf (white arrows). **C** Expression of MD-GFP in stem (white arrows). **D** Expression of MD-GFP in root (white arrows). Scale bar (A, B, C) 10 μm; (D) 50 μm.

4 Discussion

The regenerated adventitious shoots along with petiole explants of popular oftenly turned brown, woody plants and especially angiosperms, oftenly secrete polyphenols and tannins, which can inhibit development of explants/organogenesis and cause necrosis, this problem can be partially overcome by frequent sub-culturing or variation in medium nutrients and hormone composition[20]. In this study, the higher concentrations of 6-BA(0.8,1.0) combination with any concentrations of NAA(0.01-0.1) lead the leaf explants turning brown(data not shown). Our data show that the appropriate hormone concentrations and ratio can overcome this problem.

Higher contamination rates can be obtained by increasing the *Agrobacterium* concentration, although transformation frequency increased with increased *Agrobacterium* concentration. It is universally acknowledged that leaf explants were washed five to six times (20-30 min each) in liquid co-cultivation medium plus 200 mg/L Cef may induce leaf explants albinic or low transformation frequency. But our date show that the method was used to remove the excess *Agrobacterium* ahering to leaf explants, which was valuable to inhibit the number of *Agrobacterium*. The result showed that lower *Agrobacterium* contamination rates had a rise in transformation.

References

1. M.D. Block, Plant Physiol, 1110-1116 (1990).
2. R. Yadav, P. Arora, D. Kumar, D. Katyal, Plant Biotechnol Rep, 3:175-182 (2009).
3. D.Q. Huang, W.H. Dai, Plant Cell Tiss Organ Cult, 107:169-174 (2011).
4. R.B. Horsh, Science, 217:1229-1231(1985).
5. K.H. Han, M.P. Gordon, S.H. Strauss, Can J For Res, 27:464–470 (1997).
6. M. Nishiguchi, K. Yoshida, T. Mohri, T. Igasaki, K. Shinohara, J For Res, 11:175–180 (2006).
7. D.P. Yevtushenko, Plant Cell Rep, 29:211–221(2010).
8. M. Confalonieri, A. Balestrazzi, S. Bisoffi, Plant Cell Reports, 13:256-261 (1994).
9. M. Confalonieri, A. Balestrazzi, S. Bisoffi, R. Cella, Plant Cell, tissue and Organ Culture, 43: 215-222 (1995).
10. M. Confalonieri1, A. Balestrazzi, R. Cella, Plant Cell, Tissue and Organ Culture, 48: 53–61 (1997).
11. M. Hasenfratz, C. Tsou, T.A.Wilkins, Plant Physiol, 108: 1395-1404 (1995).
12. R. Viereck, M. Kirsch, R. Lrw, T. Ransch, FEBS Lett, 384:285-288 (1996).
13. R. Ratajczak, Biochim Biophys Acta, 1465:17–36 (2000).
14. Y. Arata, T. Nishi , Biochim Biophys Acta, 1555: 71–74 (2002).
15. M. Tsiantis, D.M. Bartholomew, J.A.C. Smith, Plant J, 9: 729-726 (1996).
16. A. Lehr, M. Kirsch, R. Viereck, J. Schiemann, T. Rausch, Plant Mol. Biol., 39: 463– 475 (1999).
17. S. Padmanaban, X. Lin, I. Perera, Y. Kawamura, H. Sze, Plant Physiol, 134: 1514–1526 (2004).
18. T. Seidel, C. Kluge, M. Hanitzsch, J. Rob, M. Sauer, K. Dietz, D. Golldack, Journal of Biotechnology, 112: 165–175 (2004).
19. C.X. Xu, L. Zheng, C. Gao, C. Wang, G.F. Liu, J. Jiang, Y.C. Wang, Plant Mol Biol Rep, 4:1-7 (2010).
20. A.K. Thakur, A. Saraswat, D.K. Srivastava, J. Plant Biochem. Biotechnol, 9: 1-7 (2011).

The establishment of insulin resistance model

Lanlan Liu, Xian Zhang, Fei Chen, Jianwen Hu and Bin Zeng[a]

School of Life Science, Jiangxi Science & Technology Normal University, Nanchang, China

Abstract. Establishment of insulin resistance model of HepG2 cells in vitro is aimed at studying puerarin responsible for the hypoglycemic effect. HepG2 cells induced by palmitate and glucosamine were used to establish the model of insulin resistance. Simultaneously, the effect of the two inducing methods was compared. From the test results of MTT(3-(4,5-Dimethylthiazol-2-yl)-2,5-diphenyltetrazoliumbromide) and glucose uptake, it was significantly found insulin resistance of HepG2 cells when it was treated with palmitate under the concentration of 0.125mM. The hypoglycemic effect of puerarin testing by insulin resistance model of HepG2 cells indicated that HepG2 cells induced by low concentration of Palmitate can be used to produce stable and reliable insulin resistance. The establishment of insulin resistance model have a certain value for in vitro screening hypoglycaemic herbal medicine.

1 Introduction

Type 2 diabetes accounts for over 90% of diabetic cases, the occurrence of which are closely associated with the development of insulin resistance, followed by the progressive decline of pancreatic-β cell function that results in relative deficiency of insulin secretion [1]. Type 2 diabetes is primarily featured by insulin resistance ,which occurs when the insulin-responsive tissues, mainly skeletal muscle, adipose tissue, and the liver, cannot respond to insulin properly [2]. Insulin resistance plays a key role in the pathophysiology of metabolic syndrome which is a constellation of common metabolic disorders that is associated with diabetes mellitus [3-4]. Simultaneously, insulin resistance is also correlated with several other chronic diseases, such as dyslipidaemias, cardiovascular diseases, neurodegenerative disorders and cancers [5-6], also including obesity, atherosclerosis, hypertension, and hepatic steatosis [7-8]. HepG2 cell is a table hepatocellular carcinoma cell line derived from human fetal liver tumor cells, which is very similar to liver cells [9]. Insulin resistance is manifested by free fatty acid production, increased hepatic glucose production, and decreased skeletal muscle uptake of glucose. Therefore, this experiment was conducted using the HepG2 cells to establish hepatic insulin resistance model by Palmitate and for the preliminary screening of herbal medicine.

[a] Corresponding author: Zengtx001@aliyun.com
This work was supported by these projects in China(31171731, 31460447, 20142BDH80003, 2013-CXTD002,"555 talent project" of Jiangxi Province, Jiangxi Province Key Laboratory of Bioprocess Engineering and Co-Innovation Center for In Vitro Diagnostic Reagents and Devices of Jiangxi Province).

2 Materials and methods

2.1 Main materials

The Hepg2 cell line were bought from Procell of Wuhan city, DMEM(Dulbecco's modified Eagle's medium):(HyClone), MTT (Biotopped), DMSO (Sangon Biotech), palmitic acid (PA, Sigma), Insulin from bovine pancreas (Sigma), fetal bovine serum (FBS, Gibco), Glucosamine (Beyotime Biotechnology), BSA (Roche), Puerarin (Nanchang Wante).

2.2 Main instruments

IX-71 type inverted optical microscope (Olympus, Janpan), carbon dioxide incubator (Thermo scientific forma, USA), forma class-II/A2 biological safety cabinet (Forma, USA), 5804R type centrifuge (Eppendorf, Germany), multiscan MK3 type full automatic microplate reader (Thermo lab systerms, USA), milli-Q water purification systems (Millipore, Shanghai).

2.3 Methods

2.3.1 Cell culture

HepG2 cells were cultured in DMEM low glucose medium containing 10% fetal bovine serum, 100 units/mL penicillin, and 100 μg / mL streptomycin. Cells were incubated at 37°C in a humidified incubator supplied with 5% CO_2. When the cells grew to 80% confluence, passing the cells with 0.25% trypsin.

2.3.2 Determination of cell viability in different concentrations of different inducing agents

For the development of insulin resistance by high glucose and palmitate, HepG2 cells were seeded in 96-well plates at a density of 2×10^4 cells/well [1] and grown for 24 hours to reach 80%-100% confluence. The cells in 96-well plates were divided into control group and model group. The model group was incubated for 24 hours with low glucose complete medium containing 0.125 mM, 0.25 mM, 0.5 mM, 1 mM, 2 mM concentrations of palmitate, the control group an equal volume of the vehicle for 24 hours. Another model group of cells were inducing by 20 mM, 30 mM, 40 mM, 50 mM, 60 mM, 70 mM concentrations of high glucose medium for 24 hours. After 24 hours, the cells were washed twice with PBS, each well was added fresh complete medium with 20μl MTT and incubated for 4 hours to add. After in the incubator 4 h, each well had 150 ul DMSO, by MTT method detection of cell growth.

2.3.3 Optimization time, concentration and detection of cell viability

A low concentration inducer was beneficial to the growth of the cell. Through many experiments, the cell density was adjusted accordingly and the cell growth status is better under the density of 5×10^4 cells/well. HepG2 cells were seeded in 96-well plates at a density of 5×10^4 cells/well and grown for 24 hours to reach 60% confluence. Palmitate was added to the medium to reach 0.125 mM, 0.25 mM, 0.50 mM. High glucose was added to the medium to reach 20 mM, 30 mM, 40 mM. Each of the 96-well plates was corresponding to the corresponding inducer and the cell viability was detected by

DOI: 10.1051/bioconf/20170803005

MTT in 6, 12, 24 and 36 hours respectively.

2.3.4 Repeated experiments to determine the induction time and concentration

When HepG2 cells were in the logarithmic phase, they were seeded in 96-well plates at a density of 5 × 10^4 cells/well and grown for 24 hours to reach 50% confluence. Cells incubated in 0.05 mM, 0.1 mM, 0.125 mM, 0.25 mM, 0.5 mM, 1 mM, 2 mM by Palmitate for 24, 48, 72 hours. The cell viability was detected by MTT.

2.3.5 Establishment of insulin resistance model and the hypoglycemic effect of Puerarin

HepG2 were maintained in DMEM low glucose medium. When subcultured at 80% confluence, HepG2 cells were seeded in 96-well plates at a density of 5 × 10^4 cells/well and grown for 24 hours to reach 70%-80% confluence. Cells were induced by palmitate in the concentration of 0.125 mM for 24 hours, with PBS washed twice, then respectively set blank group, control group, model group. Each control group and model group were added to puerarin, rosiglitazone, metformin. Blank groups were divided into without DMSO and DMSO in the same medium. Absorbance at 550 nm was then determined using a microplate reader. Puerarin concentration was 25 ug/ml, rosiglitazone20 ug/ml, metformin was 5 ug/ml. It was indicated that insulin resistance model was established successfully from glucose consumption difference between the control group and model group. The concentration of glucose is in direct proportion to the absorbance. Five microliters of the resulting supernatant was mixed with 195 μL of Glucose kit in a 96-well plate and incubated at 37 °C for 20min. Absorbance at 550 nm was then measured using a microplate reader.

3 Results

3.1 The results of absorbance and cell viability of HepG2 with palmitate and high glucose

The concentration of palmitate is 0.125-2 mM and high glucose is 20-70 mM effecting on cell proliferation after 24 hours. As shown in Table 1 and Table 2, Figure 1, Palmitate(PA) at low concentration has little effect on the cells, while high glucose concentration on cell survival rate obviously higher inhibited.

Table 1. The absorbance and Cell viability of HepG2 at A492 with palmitate for 24h($p<0.05$).

Concentration(mM)	Absorbance	Cell viability
0	0.847±0.033	1
0.125	0.893 ±0.002	1.054 ±0.002
0.25	0.865 ±0.017	1.021 ±0.020
0.50	0.660 ±0.005	0.780 ±0.005
1	0.485 ±0.002	0.573 ±0.002
2	0.232 ± 0.006	0.274±0.006

 DOI: 10.1051/bioconf/20170803005

Table 2. The absorbance and Cell viability of HepG2 cells at A492 with high glucose for 24h($p<0.05$).

Concentration (mM)	Absorbance	Cell viability
0	1.277±0.02	1
20	0.805±0.008	0.630±0.008
30	0.796±0.002	0.623±0.002
40	0.735±0.014	0.576±0.014
50	0.750±0.017	0.587±0.017
60	0.662±0.013	0.519±0.013
70	0.575±0.027	0.451±0.027

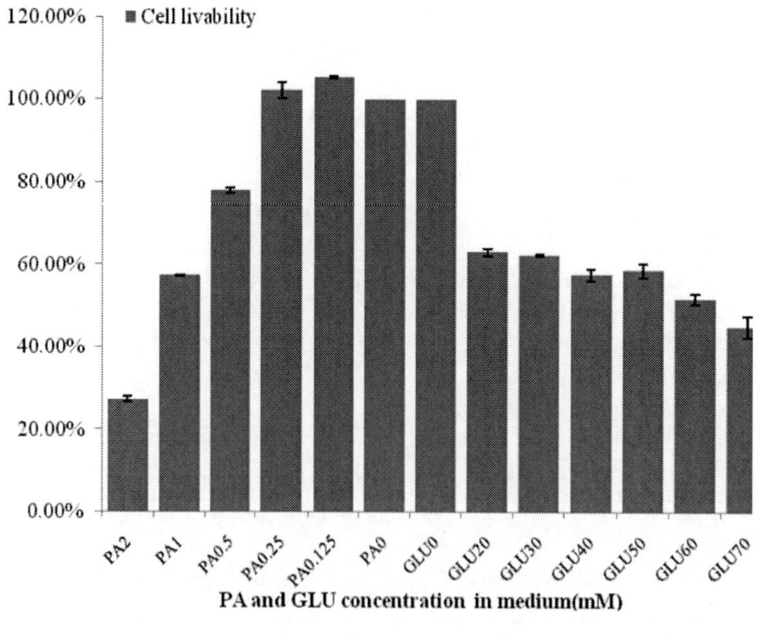

Figure 1. Effects of cell viability with different concentration of High glucose and palmitate induced HepG2 cells. (PA0 was without palmitate, GLU 0 was without glucose, $p<0.05$).

3.2 The results of optimization time and concentration and detection of cell viability

Figure 2 shows that the cell viability was gradually decreased with the increase of inducer concentration and induction time. The inhibitory effect of Palm Acid on cells was less than that in high glucose group(judged by both p between groups<0.05). Subsequent experiments with the palm acid as inducer, while the concentration of high glucose still need to make appropriate adjustments.

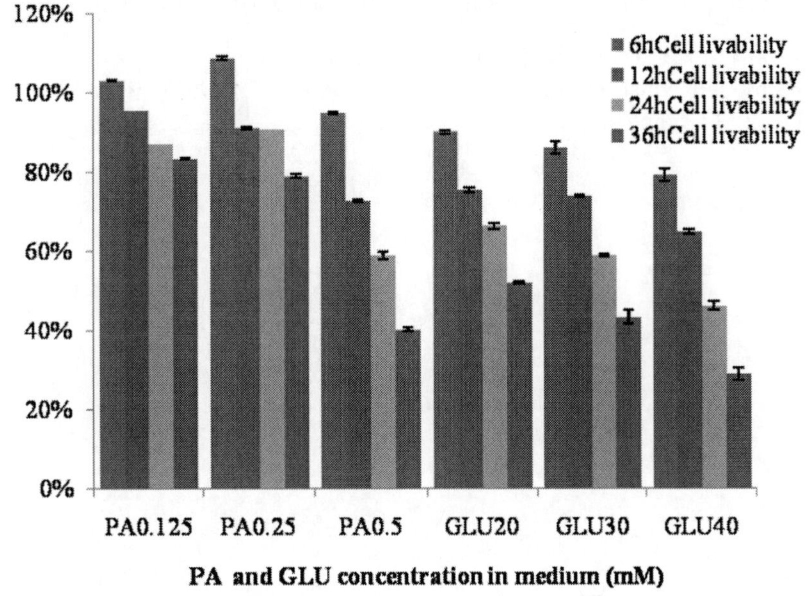

Figure 2. 6h, 12h, 24h, 36h, cell viability (Data were represented as means ±S.D. $p<0.05$).

3.3 The results of the induction time and concentration

From the chart below(Figure 3), cells in low concentration 0.05-0.125 mM were less inhibitory effect on cell growth, while in 0.25mM-2mM cell viability decreased significantly. High concentration is not conducive to the subsequent experiment. Under comprehensive consideration, subsequent induction agent was in low concentration of palmitate in the concentration of 0.125 mM for 24 hours.

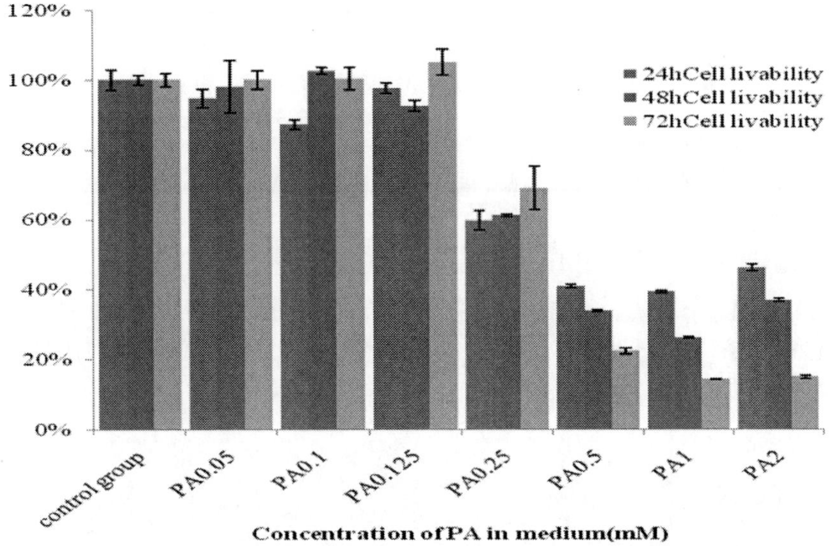

Figure 3. Detection of 24h, 48h, 72h, cell viability(Data were represented as means ±S.D. $p<0.05$).

 DOI: 10.1051/bioconf/20170803005

3.4 The standard curve of glucose(presented as the absorbance at 500 nm)

The linear regression of the concentration X was carried out in the light absorption value Y, and the standard curve regression equation was Y=0.0552x+0.0565, R^2 =0.9997. The linear range of the standard curve is 0.02~20 mmol/L.

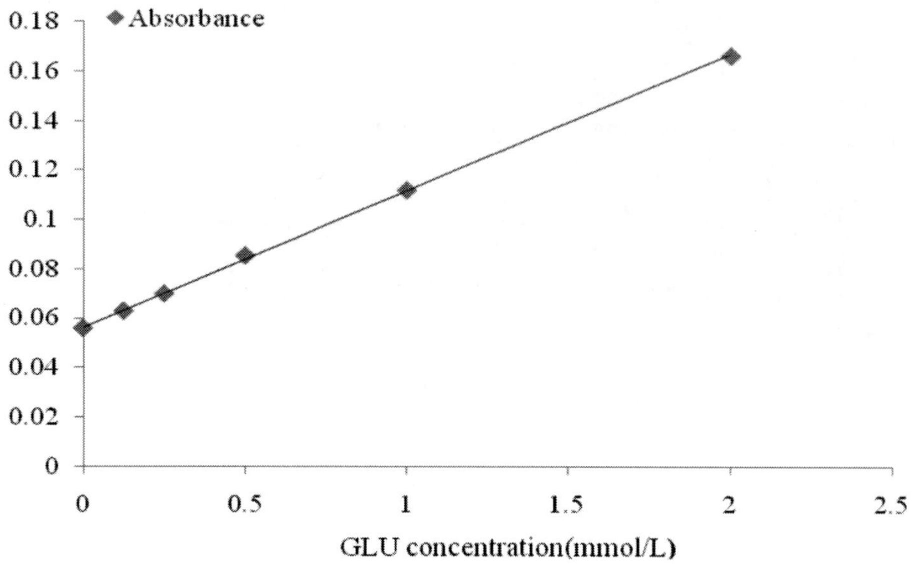

Figure 4. The standard curve of different concentration of glucose.

Table 3. Establishment of insulin resistance model and effect of Puerarin on hypoglycemia(Data were represented as means ±S.D.p<0.01).

Group	Glucose consumption(mM)	Glucose consumption difference (%)
control+ puerarin	1.585 ± 0.002	
Model+ puerarin	1.295 ± 0.002	18.297
control+ rosiglitazone	2.672 ±0.003	
Model+ rosiglitazone	2.473 ±0.001	7.478
control+ Metformin	2.129 ±0.005	
Model+ Metformin	2.092 ±0.004	1.378

3.5 The results of glucose uptake

The concentration of glucose was in direct proportion to the absorbance. Five microliters of the resulting supernatant was mixed with 195 μL of Glucose kit in a 96-well plate and incubated at 37 °C for 20 min. Absorbance at 550 nm was then measured using a microplate reader .

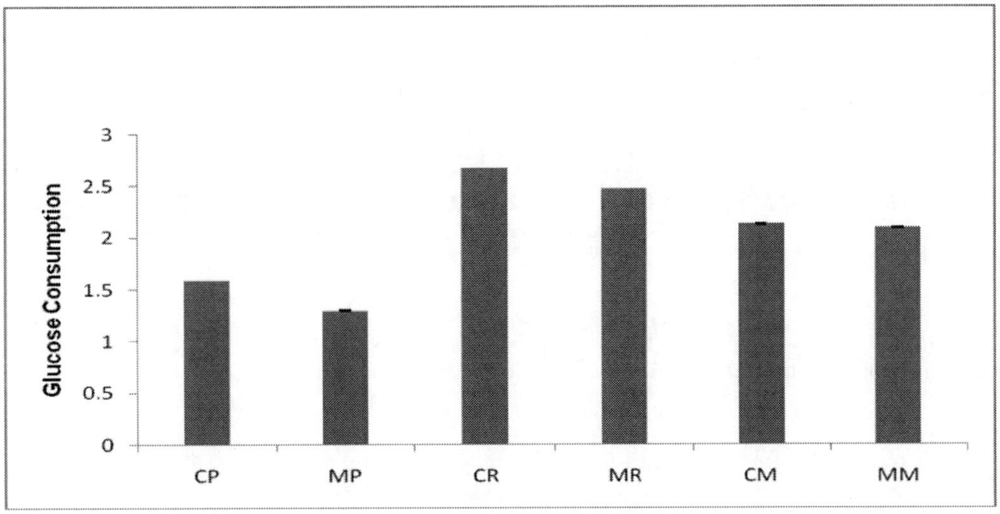

Figure 5. Insulin resistance model glucose consumption(mmol/L). Control+ Puerarin(CP), Model + Puerarin (MP), Model+ Rosiglitazone(MR), Control+ Metformin(CM), Model + Metformin(MM).

From the charts above, the results showed that glucose consumption of model groups under the same concentration and time of puerarin and rosiglitazone were significantly lower than that of the control group. To explain the resistance of cells induced by 24hour in the post - HepG2 cells.

4 Summary

It was found that low consentration of palmitate achieved better results on establishing insulin resistance model using HepG2 cells induced by palmitate and glucosamine. The insulin resistance model was used to verify hypoglycemic effect of puerarin, which showed that an effective and stable insulin resistance model was established successfully. Treatment of insulin resistance is a major strategy in the prevention and management of type2 diabetes[10-12]. The underlying mechanism about the phenomena of palmitate - treated HepG2 cells needs further research.

References

1. H.L. Cheng, H. K. Huang, C. I. Chang, C. P.Tsai, C. H. Chou, J. Agric. Food **56**, 6835 (2008).
2. C. I. Chang, C. H. Chou, M. H. Liao, T. M. Chen, C. H. Cheng, R. Anggriani, H. L. Cheng, J. Funct. Foods **13**, 214 (2015).
3. L. Guariguata, D. R. Whiting, I. Hambleton, J. Beagley, U. Linnenkamp, J. E. Shaw, Diabetes. Res. Clin. Pr **103**, 137 (2014).
4. R.H. Eckel , S.M. Grundy, P.Z. Zimmet, Lancet **365,** 1415 (2015).
5. B. Mlinar, J. Marc, A. Janež, M. Pfeifer, Clin. Chim. Acta. **375**, 20 (2007).
6. S. M. Hirabara, R. Gorjao, M. A. Vinolo, A. C. Rodrigues , R. T. Nachbar, R. Curi, Bio-Med. Res. Int (2012).
7. K. G. M. Alberti, P. Zimmet, J. Shaw, IDF Epidemiology Task Force Consensus Group, Lancet **366**, 1059 (2005).
8. B. B. Kahn, J. S. Flier, J. Clin, Invest **106**, 473 (2000).
9. J. R. Levy, M. Belsky, Am. J. Med. Sci **299**, 302 (1990).
10. W.Y. Fujimoto, Am J Med **108**,9 (2000).
11. H.E. Lebovitz, M.A. Banerji, Eur J Pharmac **490**, 135 (2004).
12. M. Kasuga, J. Clin. Invest ***116***, 1756 (2006).

Recombinant expression and purification of an Oxysterol Binding Protein from *Aspergillus oryzae* 3.042

Xian Zhang, Lanlan Liu, Jianwen Hu, Fei Chen and Bin Zeng[a]

School of Life Science, Jiangxi Science & Technology Normal University, Nanchang, China

Abstract. A full-length cDNA encoding a candidate Oxysterol-binding protein(OSBP) from *Aspergillus oryzae* (AoOSBP) was cloned and expressed in Escherichia coli as a maltose-binding protein (MBP) fusion protein. The MBP-AoOSBP protein from the importantly industrial fungus *A. oryzae* was purified by amylose resin and chromatography column. SDS-PAGE showed that MBP-AoOSBP has an estimated molecular weight of 182 kDa. OSBP and its homologues (ORPs) own the affinity for oxysterols, cholesterol and glycerophospholipids. According to the superiority of *A. oryzae* in the fermented foods and also in food-grade productions pharmaceutical enzyme manufacture, it is meaningful to identify the biochemical properties of OSBP in *A. oryzae*.

1 Introduction

Aspergillus oryzae, as a kind of GRAS (generally recognized as safe) strain, has been widely used for the production of certain traditional fermented Asian foods and alcoholic beverages [1] .This safe strain is also wildly being used for production of various industrial enzymes including amylases, proteases, lipases and phytases etc. [2]. The genome of *A. oryzae* has been sequenced in 2005 [3], which has facilitated research efforts to understand its basic biology and develop better industrial strains of this significant fungus [4] .

The study of *A. oryzae* genome database reveals the presence of a presumptive Oxysterol-binding protein (AoOSBP) coding gene which has a predicted open reading frame (ORF) 3792 bp capable of encoding a polypeptide of 1263 amino acids. Oxysterol-binding protein (OSBP), a cytoplasmic protein, is widely distributed in eukaryotes. OSBP and its homologues (ORPs) have been confirmed owning affinity for oxysterols, cholesterol and glycerophospholipids, due to their properties of the lipid-binding/transfer [5-9]. In yeast *Saccharomyces cerevisiae* are gene/protein family consists of seven members (designated Osh)[10], while in mammals including humans there are 12 ORP genes/proteins [11-13]. Some researchers studied the vertebrate model to illustrante the function of conservative OSBP-related proteins, known as lipid binding/transfer proteins, between zebrafish and

[a] Corresponding author: Zengtx001@aliyun.com
This work was supported by these projects in China(31171731, 31460447, 20142BDH80003, 2013-CXTD002, "555 talent project" of Jiangxi Province, Jiangxi Province Key Laboratory of Bioprocess Engineering and Co-Innovation Center for In Vitro Diagnostic Reagents and Devices of Jiangxi Province).

human [14].

Until now, no investigation concerning OSBP from filamentous fungi has been documented. According to the superiority of *A. oryzae* in the fermented foods and also in food-grade productions pharmaceutical enzyme manufacture, it is meaningful to identify the biochemical properties of OSBP in *A. oryzae* as well as its expression level in the process of growth. In this study, the AoOSBP gene was cloned, recombinantly expressed in *E. coli* cells and the MBP-AoOSBP fusion protein were purified by mylose resin and chromatography column. purified. It is beneficial to the further research with the combined effect of phytosterol and its application in the fermentation process.

2 Materials and methods

2.1 Strains, plasmids and biochemical reagents

Escherichia coli JM109 and *Aspergillus oryzae* 3.042 were stored in Tianjin University of Science & Technology. *E. coli* OverExpress C43 (DE3) was bought from Shanghai Microgene Biotech Co.,Ltd. Vector PMAL-c4x (Invitrogen, USA) was used for fusion expression of AoOSBP protein from *A. oryzae*. Amylose resin was purchased from NEB (USA) and the chromatography column was from QIAGEN (Germany). Reagents and chemicals used for DNA manipulation and RNA extraction were purchased from Takara (Dalian, China). All the other reagents were of the analytical grade.

2.2 Total RNA extraction and reverse transcription

In order to acquire fungal mycelia for total RNA isolation, an approximately 1×10^7 spore suspension was inoculated into 50 mL Yeast Extract Peptone Dextrose Medium in a 250 mL flask and allowed to grow at 30 °C for 24 h with shaking and then fungal mycelia were harvested and washed with deionized water. To isolate total RNA, mycelia were ground in liquid nitrogen in an ice-cold mortar and 0.1 g of the resultant mycelium powder was used for total RNA extraction with TRIzol® reagent (Promega, USA) according to the manufacturer's instructions. The quality and integrity of RNA was determined by 1% agarose gel electrophoresis. Then 3μg of total RNA was subjected to reverse transcription using random heptamer primers with the manufacturer's instructions. Next, the resultant cDNA was then used for PCR amplification of AoOSBP cDNA.

2.3 Construction of expression vector

To express AoOSBP as a maltose-binding protein (MBP) fusion, the pMAL-c4x vector (New England Biolabs) was used. Briefly, the full open reading frame (ORF) of AoOSBP was amplified from cDNA synthesized from mycelium total RNA using primer pairs with a *Sal*I restriction site incorporated at 5' end (5'-gtcgac TGGCCTGGTCGCGATAGTGATTCG-3') and a *Pst*I restriction site at the 3' end (3'-ctgcagCCTGGAGCAGGTACGCTCCAACCA-5'). High-fidelity Probest DNA polymerase (TaKaRa) was used to minimize potential mutations introduced during amplification. The thermocycler conditions were 95 °C for 2 min; 30 cycles of 95 °C -1 min, 55 °C -1 min, 72 °C -1 min; and afinal extension at 72 °C for 10 min. The PCR product was first cloned into T vector and verified by sequencing, followed by digestion with *Sal*I and *Pst*I to release the OSBP gene fragment. pMAL-c4x vector (New England Biolabs) was linearized with *Sal*I and *Pst*I, and the OSBP insert was ligated into it following digestion with the same enzymes. The ligation product was transformed into *E. coli* JM109 cells (Invitrogen, USA) and transformants were selected on LB plates supplemented with 100 μg/mL ampicillin. The resultant recombinant expression plasmid, designated pMAL-AoOSBP, was confirmed by digestion of appropriate restriction enzymes and subsequent sequencing.

 DOI: 10.1051/bioconf/20170803006

2.4 Expression of recombinant OSBP in *E. coli*

The expression plasmid pMAL-AoOSBP was transformed into *E. coli* OverExpress C43 (DE3) cells for AoOSBP expression. Briefly, a single colony of *E. coli* transformant was inoculated into 50 ml LB broth containing 100 μg/mL ampicillin in a 250 mL flask. After overnight cultures at 37 °C with shaking at 220 rpm, 4 mL cells were inoculated into 400 mL fresh LB medium containing 100 μg/mL ampicillin in a 1 L flask and grown at 37 °C with shaking. When cell OD_{600} reached 0.6-0.8 putting the flask on the ice 15 min, IPTG was added to a final concentration of 0.5 mM and the speed was slowed down to 160 rpm. The cells were harvested after a further 20 h incubation at 13 °C.

2.5 Purification of AoOSBP

Crude protein preparations and all further work with OSBP were carried out at 4°C. After cultivation, cells were collected by centrifugation at 5,000 × g for 15 min and pellets resuspended in 20 mL binding buffer (20 mM Tris-HCl, 200 mM NaCl, 1mM EDTA, pH 7.4). Cells were lysed by sonication(15 min cycle of 3 s on, 3 s off, at 45% amplitude) on the ice and cell debris removed by centrifugation (40,000 × g, 4 °C, 30 min). After centrifugation, the supernatant was filtered and mixed gently for 1 h at 4 °C with 2 mL of amylose resin (NEB) that was equilibrated in binding buffer. Then the mixture was flowed through the chromatography column, and repeated this step for two or three times. The non-adsorbed protein fraction was eluted from the column with binding buffer and the remaining protein eluted with a step gradient of maltose . Target protein fractions were pooled, concentrated using an Amicon Ultra-4 unit (MWCO 30 kDa), and stored at -80 °C for subsequent analysis.

3 Results and Discussion

3.1 Cloning of the AoOSBP gene from A. oryzae

A large eukaryotic gene family with homology to OSBP has been revealed by analysis of genomic and cDNA databases [15-17]. OSBP is characterized for two domains, the pleckstrin homology (PH) domain at the N-terminus, and the ligand-binding (LB) domain at the C-terminus. The PH domain, which is found in many membrane-binding proteins, mediates membrane association of the OSBP through phosphoinositide interaction [18,19], and The C-terminal domain that in some cases has been shown to bind oxysterols, cholesterol and ergosterol [20,21].

For expression of AoOSBP protein in *E. coli* cells, the full-length 3792 bp cDNA was amplified by RT-PCR from *A. oryzae* mycelium total RNA (Figure 1). The sequence of the amplified AoOSBP cDNA fragment was confirmed in the GENEWIZ. Then the amino acid sequence of AoOSBP was carried on the comparison in the NCBI BLAST, and the result is an exact match with *Aspergillus oryzae* 3.042, as shown in Figure 2. The gene of AoOSBP was connected to the expression vector pMAL-c4x to express the MBP-AoOSBP fusion protein in *E. coli* cells. The Figure 3 showed that pMAL-AoOSBP expression vector was constructed successfully.

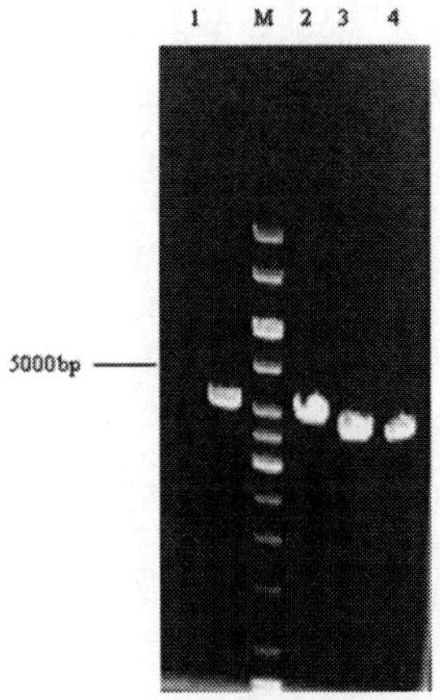

Figure 1. The agarose gel of the genome and cDNA of AoOSBP. M: KB Ladder; Lane1 and Lane 2: The genomic size of AoOSBP (~ 4000bp); Lane 3 and Lane 4: The cDNA size of AoOSBP (~3800bp).

Sequences producing significant alignments:

Select: All None Selected:0

Alignments Download GenPept Graphics Distance tree of results Multiple alignment

Description	Max score	Total score	Query cover	E value	Ident	Accession
oxysterol-binding protein [Aspergillus oryzae 3.042]	2606	2606	100%	0.0	100%	EIT77350.1
oxysterol binding protein (Osh1) [Aspergillus oryzae RIB40]	2603	2603	100%	0.0	99%	XP_001825518.1
oxysterol binding protein [Aspergillus flavus AF70]	2601	2601	100%	0.0	99%	KOC17205.1
hypothetical protein ANOM_001793 [Aspergillus nomius NRRL 13137]	2373	2373	99%	0.0	94%	XP_015411210.1
oxysterol binding protein (Osh1) [Aspergillus fumigatus Z5]	2001	2001	100%	0.0	78%	KMK55954.1
oxysterol-binding protein homolog C2F12.05c [Aspergillus udagawae]	1981	1981	100%	0.0	78%	GAO85745.1
oxysterol binding protein [Aspergillus kawachii IFO 4308]	1975	1975	100%	0.0	78%	GAA91102.1
unnamed protein product [Aspergillus niger]	1956	1956	100%	0.0	78%	CAK40942.1
oxysterol binding protein [Aspergillus niger]	1954	1954	99%	0.0	78%	GAQ46371.1

Figure 2. The results of Comparing in the NCBI BLAST shows the gene of AoASBP is extracted from the *Aspergillus oryzae* 3.042.

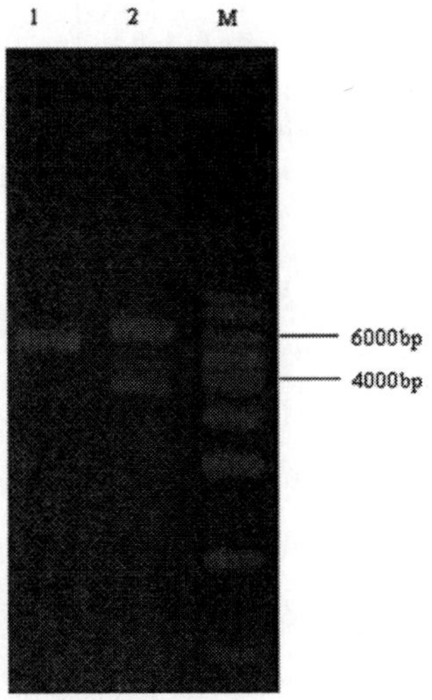

Figure 3. The agarose gel of expression vector pMAL-AoOSBP. M: KB Ladder; Lane 1: The plasmid of pMAL-AoOSBP; Lane 2:The pMAL-AoOSBP digested by SalI and PstI (The size of vector is 6645bp and the gene is 3792bp).

3.2 Heterologous expression and purification of AoOSBP

To determine the induction conditions for optimal AoOSBP expression in OverExpress C43 (DE3) cells, various IPTG concentrations and incubation times were examined. It was found that under the condition of 0.5 mM IPTG and incubation at 13°C for 20 h was suitable for AoOSBP expression (data not shown). And under the same condition , added inducers and without it the fusion protein expression of the target protein have obvious differences(Figure 4 A). To purify MBP-AoOSBP protein, the crude protein preparation was combined with amylose resin at 4 °C and the mixed was throughed chromatography column, then eluted with the washing buffer. The purified MBP-AoOSBP was analyzed by SDS-PAGE (Figure 4 B). Based on the SDS-PAGE, the gene of OSBP from *A. oryzae* was successfully expressed in *E. Coli.*

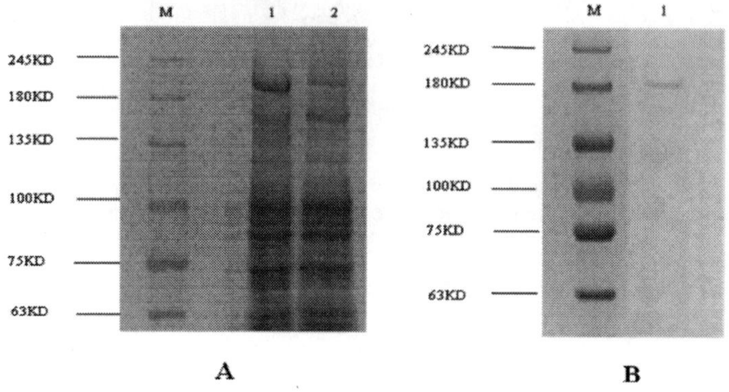

Figure 4. 8% SDS-PAGE analysis of the MBP-AoOSBP fusion protein. (A) Lane M: protein molecular weight marker. Lane 1: The expression of the MBP-AoOSBP fusion protein after adding inducers; Lane 2: The broken supernatant that without inducers. (B) lane M is the protein marker; lane 2 shows purification of AoOSBP by the amylose resin and chromatography column.

4 Summary

Strains of *A. oryzae* are used for the production of a number of popular traditional Asian fermented foods, in particular for large-scale manufacture of soy sauce whose unique flavors are favored by consumers. It is well known that *A. oryzae* plays a significant role in the taste and smell of its fermented final product and lipid metabolism contributes to the flavor formation during fermentation. OSBP is a lipid-binding protein that has been implicated in the regulation of various cellular processes, including nonvesicular cholesterol transport, signaling, lipid metabolism, and vesicular trafficking. In this study, we successfully obtained target gene of AoOSBP through reverse transcription-polymerase chain reaction. The sequence of the amplified AoOSBP gene were confirmed by sequencing. Then the AoOSBP gene were connected to the expression vector pMAL-c4x to express the MBP-AoOSBP fusion protein in *E. coli* cells. The MBP-AoOSBP fusion protein were purified by amylose resin and chromatography column.

References

1. M. Machida, O. Yamada, K. Gomi, DNA Res. **15**, 173 (2008).
2. L. Liu, A. Feizi, T. Österlund, C. Hjort, J. Nielsen, BMC Syst. Biol. **8** , 73 (2014).
3. M. Machida, K. Asai, M. Sano, T. Tanaka, T. Kumagai, G. Terai, K. Abe, Nat. **438**, 1157 (2005).
4. Q. Hao, X. Liu, G. Zhao, L. Jiang, M. Li, B. Zeng, Biotechnol. Lett. **38**, 519 (2016).
5. P. A. Dawson, N. D. Ridgway, C. A. Slaughter, M. S. Brown, J. L. Goldstein, J. Biol.Chem. **264**, 16798 (1989).
6. F. R. Taylor, S. E. Saucier, E. P. Shown, E. J. Parish, A. A. Kandutsch, J. Biol.Chem. **259**, 12382 (1984).

7. P. Y. Wang, J. Weng, R. G. Anderson, Sci. **307**, 1472 (2005).
8. B. Mesmin, J. Bigay, J. M. von Filseck, S. Lacas Gervais, G. Drin, B. Antonny, Cell **155**, 830 (2013).
9. H. Kentala, S. G. Pfisterer, V. M. Olkkonen, M. Weber Boyvat, Steroids **99**, 248 (2015).
10. C. J. Jaworski, E. Moreira, A. Li, R. Lee, I. R. Rodriguez, Genomics **78**, 185 (2001).
11. A. M. Anniss, J. Apostolopoulos, S. Dworkin, L. E. Purton, R. L. Sparrow, DNA Cell Biol. **21**, 571 (2002).
12. M. Lehto, S. Laitinen, G. Chinetti, M. Johansson, C. Ehnholm, B. Staels,V. M. Olkkonen, J. Lipid Res. **42**, 1203 (2001).
13. C. T. Beh, L. Cool, J. Phillips, J. Rine, Genet. **157**, 1117 (2001).
14. Y. Zhou, G. Wohlfahrt, J. Paavola, V. M. Olkkonen, Bioche. Biophys. Res. Commun. **446**, 675 (2014).
15. M. Lehto, S. Laitinen, G. Chinetti, M. Johansson, C. Ehnholm, B. Staels, E. Ikonen, V.M. Olkkonen, J. Lipid Res. **42**, 1203 (2001).
16. J. Jaworski, E. Moreira, A. Li, R. Lee, I. R. Rodriguez, Genomics **78** 185 (2001).
17. C. T. Beh, L. Cool, J. Phillips, J. Rine, Genetics **157**, 1117 (2001).
18. A. A. Kandutsch, E. B. Thompson, J. Biol. Chem. **255**, 10813(1980).
19. F. R.Taylor, A. A.Kandutsch, Chem. Phys. Lipids **38**, 187 (1985).
20. Y. J. Im, S. Raychaudhuri, W. A. Prinz, J. H. Hurley, Nature **437**, 154(2005).
21. N. D. Ridgway, P. A. Dawson, Y. K. Ho, M. S. Brown, J. L. Goldstein, J. Cell Biol. **116**, 307 (1992).

Value measurement and case study of development of traditional Chinese medicine resources and the protection of biodiversity

Xin Wang, Yanxia Lv, Chenting Zhou and Junlong Shen[a]

School of Health Economics and Management, Nanjing University of Chinese Medicine, Xianlin Road 138#, 210023, Nanjing, China

Abstract. China has rich traditional medicine resources. However, with economic development, the excessive exploitation and utilization of traditional Chinese medicine resources has driven many species to the verge of extinction. However, in recent years, the protection of such resources through advocacy, research, and law enforcement has increased, and populations of some important species have been recovered. Nevertheless, the overall diversity of traditional Chinese medicine resources has faced a notable crisis. Therefore, the development of diverse traditional Chinese medicine resources can improve people's understanding of the importance of these resources, and also play a role in the rational allocation and utilization of such resources, thereby protecting the biological diversity of traditional Chinese medicine resources.

1 Over exploitation of traditional Chinese medicine resources hinders conservation of biodiversity

Due to the rapid development of China's economy and the improvement of the health awareness of its population, traditional Chinese medicine resources are in demand in healthcare, and in prevention and treatment of diseases due to their effectiveness . Both the domestic, and foreign demand for traditional Chinese medicine has rapidly increased. Driven by economic interests, enterprises have sprung up that excessively utilize traditional Chinese medicine resources, causing destruction of the ecological environment and reducing and even depleting these resources in the wild. It has been reported that more than 100 such resources, such as liquorices, notopterygium root, Vitex trifolia, Cistanche, Pinellia ternate, shikonin, etc., are now too scarce to meet medicinal needs. Due to the scarcity of wild resources, the availability of more than 30 types of plants, such as crescent wild lotus, Dysosma versipellis, Magnolia officinalis, Eucommia, wild ginseng, etc., cannot be guaranteed to meet the demand for commodities[1]. The biological diversity of traditional Chinese medicine is decreasing, and the harmonious relationship between humans and nature has been reduced or even lost, causing a loss of biological diversity in the traditional Chinese medicine.

[a] Corresponding author:jlshen2005@126.com

2 Evaluation and measurement of the ecological value of traditional Chinese medicine resources

2.1 International experience

Traditional Chinese medicine resources form an aspect of animal and plant resources in general; thus, we can use methods for evaluating the ecological values of animal and plant resources to evaluate the ecological values of traditional Chinese animal and plant medicinal resources.

There are a number of studies on the measurement of the ecological value of animal and plant resources in general. Drumarx used the cost of expenditure in the assessment of the value of wildlife over several decades; travel costs, conditions, and other parameters were measured in the assessment of the application and development of ecosystem services[2]. Costanza[3] used a physical quality evaluation method, energy analysis method, market value method, and opportunity cost method, calculated for global ecosystem services and capital asset value; this study formed the basis for many further studies. In the measurement and evaluation of the ecological value of traditional Chinese medicine resources, Han Song and Lin Wei-wei established index systems and measurement methods for the measurement of the ecological value of wild animal and wild plant resources; these methods are referenced in this paper. Han Song considered the ecological value of wildlife resources based on wild animals that are involved in maintaining the integrity of the food chain value as the value of biodiversity. He designed a specific evaluation index system, using the opportunity cost method and an alternative cost method to calculate the ecological value of wild animal resources[4]. Lin Wei-wei divided the ecological value of resources into water conservation, solid soil, carbon sequestration, and storage nutrition substance-specific indicators, and used the shadow price method to calculate the ecological value of wild plant resources [5].

2.2 System for classification and evaluation of ecological value of traditional Chinese medicine resources

Traditional Chinese medicine resources form an organic part of the natural ecosystem, and its ecological value is mainly reflected in the maintenance of ecological balance[6]. the ecological value of traditional Chinese medicine resources is mainly reflected in three factors:

First, traditional Chinese medicine animal resources that maintain the integrity of the food chain, the smooth progress of the nutrient cycle. Second, as part of a natural ecological system, traditional Chinese medicine plant resources play a role in natural ecological functions. Third,Traditional Chinese medicine resources contribute to this biological diversity, and thus maintaining biodiversity is one of the ecological values of traditional Chinese medicine resources.

Therefore, the ecological value of traditional Chinese medicine resources is equal to sum of the ecological value of the animal resources, the ecological value of the plant resources and the maintenance of biological diversity of traditional Chinese medicine resources, which can be expressed as:

$$VTCM = VMB + VFB + VDB \tag{1}$$

Where V_{TCM} is the ecological value of traditional Chinese medicine resources; VFB is the value of traditional Chinese medicine animal resources in maintaining the integrity of the food chain; VAB is the ecological service function value of traditional Chinese medicine plant resources, and VDB is the value of traditional Chinese medicine resources in maintaining biodiversity.

2.3 Measurement of ecological 1value of traditional Chinese medicine resources

2.3.1. Ecological value of animal resources

2.3.1.1 Value of animal resources in maintaining the integrity of the food chain

Wild animal resources in Chinese herbal medicine maintain the integrity of the food chain, thereby maintaining the smooth circulation of nutrients. This value includes the biological control value, seed transmission value, improving the soil value, and purifying the environment value. It can be measured as follows:

$$VFB = VFB_1+VFB_2+VFB_3+VFB_4 \tag{2}$$

Where VFB_1 is the biological control value; VFB_2 is the seed dispersal value; VFB_3 is the improving soil value; and VFB_4 is the purification of the environmental value.

(1)The biological control value:

Wildlife resources add benefit to human life; certain birds and mammals, and even certain pests and rodents are the main food items to humans.Thus, the ecological value of wild animal resources can be used to indicate the cost of prevention and control of pests, and can be calculated as follows:

$$VFB_{1i} = \sum K_j \cdot Q_{ij}/q_j \tag{3}$$

where VFB_{1i} is the i species of wild animals in biological control income; K_j is the costs of control in a unit area of J species of insect pests; Q_{ij} is the number of J species of harmful animals eaten by i species of wild animals on average every year; and q_j is the j types of harmful animals each year, indicating the density of harm.

(2)The spread of seeds value

Some wild animals seeds for food, and thus the wild animals have a significant effect on seed propagation. Animals such as squirrels, jays, crossbills, rodents, and other animals who store pine nuts for food, and as a food storage habit, are conducive to seed dispersal and planting. This value can be calculated as follows:

$$VFB_{2i} = H_i \cdot M \tag{4}$$

Where VFB_{2i} is the seed transmission yield of wild animal resources; H_i is the area of i species of wild animals; and M is the annual planting cost per unit area.

(3)Improving the value of the soil

Wild animals improve the value of the soil, as mainly reflected in two aspects. First, wild animal manure can improve soil fertility, promote soil fertility (which assumes a moderate amount of feces from wild animals, and no soil fertility damage). Second, they improve soil permeability and ventilation, and is calculated as follows:

$$VFB_{3i} = VFB_{31i} + VFB_{32i} = (NI_i \cdot P_{NI} + PH_i \cdot P_{ph} + PO_i \cdot P_{po})\, Q_{di} + FL \cdot Q_{hi}/q_{hi} \tag{5}$$

Where VFB_{3i} is the i type of wild animal that improve soil yield; VFB31i is the i type of wild animal that improves the monetary value of soil fertility; VFB32i is the i type of wild animal that improves the monetary value of the soil permeability and gas permeability; NIi is the nitrogen content per kilogram of feces of i type of wild animal; PNI is the average cost of industrial nitrogen; PHi is the i type of phosphorus content per kilogram of wild animal feces; Pph is the average cost of industrial production of elemental sulfur; POi is the i species of wild animal feces content per kilogram of potassium; Ppo is the average cost of the industrial production of potassium; Qdi is the i type of the amount of manure produced annually by wild animals; FL is the cost per unit area of arable land; Qhi is the i species of wild animal in the number of mining caves; and qhi is the density of i species of wild animals excavated in the cave.

2.3.1.2 Animal resources for maintaining the value of biodiversity

A review of the literature showed that the opportunity cost method was the best method for evaluating forest biodiversity; in this study, cost measurement was chosen to evaluate the Chinese medicine resources that maintain the value of biodiversity.

Wildlife survival depends on habitat; once habitat is destroyed, wildlife resources will also cease to exist. The opportunity cost is the same as habitat resource opportunity costs, and consists of two parts: the habitat resources for the construction projects, requiring the community to give up their benefits, and for construction projects leaving the community to increase wear and tear. This includes income that depends on wildlife resources and the protection of other wildlife. Such investments are measured using the following formula:

$$VBD_{1i} = \sum Q_{Bij} P_{Bj} + INV_i \tag{6}$$

Where VBD1i is the i species of wild fauna that maintain the value of biodiversity; QBij is the i type of wild animals and plants that occupy j types of resources annually; PBj is j type of average market prices of resources; INVi is i type of investment in wildlife conservation.

2.3.2 Plant resources ecosystem roll functions

Plant resources with ecological roles, including water conservation value, carbon sequestration value, oxygen value, and soil conservation value, can be expressed by the formula:

$$VFB = VFB_1 + VFB_2 + VFB_3 + VFB_4 \tag{7}$$

Water conservation value

The water conservation value of Chinese medicinal plant resources can be divided into the storing precipitation value and the purifying water quality value. The specific calculation method is as follows:

1) Storing precipitation value: Wild plant resources play the important role of capturing precipitation in ecosystems, which increases the effective amount of water and regulates runoff. Storing precipitation is a method for maintaining water balance, as calculated by the formula below:

$$W = (P - E) \cdot A \tag{8}$$

Where W is the stored precipitation (m^3/a); P is the average annual rainfall (mm/a); E is the average annual evaporation (mm/a); and A is the area of Chinese medicine plant resources (hm^2).

2) Water purification value: Plant resources are involved in absorption, adsorption, precipitation decomposition of harmful substances, and improvement of water quality. The value of water purification plant resources can be calculated using the alternative engineering method as follows:

$$VS = C \cdot K \tag{9}$$

Where VS is the water purification value (yuan); C is the intercepted precipitation (m^3/a); K is the unit per volume of water purification costs, which is 0.9885 yuan/m^3.

Therefore, the value of water conservation by plant resources is expressed as:

$$VFB_1 = W + VS = (P - E) \cdot A + C \cdot K \tag{10}$$

Carbon fixation value

Through photosynthesis, plants convert solar energy into chemical energy and synthetic organic materials. The fixation of absorbed CO_2 in the plant body can be calculated using the equation of photosynthesis—$6CO_2 + 12H_2O \rightarrow (CH_2O) + 6O_2 + 6H_2O$—as 1.639 g CO_2 per gram of dry matter. The calculation of pure carbon fixed in plants can be calculated by the following formula:

$$C_c = 0.2729 C_g \tag{11}$$

Where C_c is the fixed carbon content (t/a), and C_g is the fixed CO_2 (t/a).
Oxygen production value

O_2 and CO_2 emissions of a fixed amount are calculated in the same manner; per 1 g of dry matter 1.19 g O_2 can be released, the economic value of O_2 released can be calculated using the reforestation cost method and the price of industrial oxygen substitution method.

Soil conservation value

There are many methods for assessing soil conservation by plants. Taking into account the feasibility of the operation, we selected an alternative engineering method. The basic steps for assessment are as follows: first, we calculated the amount of plant and soil conserved and converted this into land area multiplied by the appropriate engineering reclamation costs, to obtain the value of plant and soil conservation. Second, we calculated the loss of N, P, K, organic matter content in the soil taken away; then, if this is equated to the appropriate amount of commercial fertilizer, and multiplied by the average market price of fertilizer, it is possible to calculate the value of fertilizer plants [7].

2.3.3 Special value of the biological diversity of traditional Chinese medicine resources

The ecological value of traditional Chinese medicine resources and the ecological value of plant and animal resources in general, also have its own value, such as pest control, providing natural medicines for other organisms, and the number of species needed to maintain stability.

Ginkgo is a good example; its fruit used not only for treating phlegm, cough, lungs, and is used as a diuretic, and the ginkgo tree plays a role in ecological water conservation, soil conservation, etc., but also prevents pests and diseases, and plays a role in crop protection[8]. Studies have found that ginkgo leaves are able to protect against the invasion of pathogens and can secrete an antibacterial insecticidal substance[9]. In addition, more importantly, the value of traditional Chinese medicine resources in biodiversity is reflected in its role in human health, disease prevention, strategic resources, public health, and as a human resource library for overcoming health disorders. For instance, extracts from the annual stems and leaves of the Artemisia plants have anti-malarial effects. Extracts from Ginkgo biloba can be used to treat cardiovascular disease.

3 Case analysis of biodiversity conservation and development of traditional Chinese medicine

Bletilla is a typical traditional Chinese medicine plant from the western region; this orchid's dry tubers stops bleeding, reduces swelling, and promotes tissue regeneration. However, Bletilla seeds are very small and have no endosperm[10]; it is difficult to germinate and grow this plant under natural conditions, seedling cultivation is more difficult, and is only achieved by plant breeding[11]. Moreover, the plant has a long breeding cycle and low reproductive efficiency. It is difficult to meet the high demand for cultivation of this plant, and thus the current market supply mainly depends on wild Bletilla. In 2006, the dosage of Bletilla medicinal goods reached 1000 tons; the total market demand broke through 1500 tons in 2007. Because the market demand for Bletilla has increased markedly, the wild resources decline year by year. According to reports, the Bletilla output for 2007 was about 300 tonnes in Guizhou, 500 tons in Yunnan, about 100 tons of Guangxi, and the total output of the western region was about 900 tons[12]. In terms of the market price, in the 1980s, the basic price of Bletilla was ¥ 4~6 per kg. The price began to rise in 1991, and reached ¥ 35~42 per kg in 2005~2006, and in 2014~2015, the market price had risen to ¥ 450~680 per kg. Westerners played a role in stimulating economic interests in the over-exploitation of wild Bletilla. It is predicted that wild Bletilla resource reserves is declining at a rate of 10%~15%. In 2007, Guizhou resources declined by about 40% or more, so that it is now difficult to find wild mountain Bletilla.

DOI: 10.1051/bioconf/20170803007

References

1. Ch. H. Chang. Protection of rare and endangered species of wild medicinal plant[J]. *Gansu Agriculture*, 3, 99 (2006).
2. M. Faber, R. Winkler. Heterogeneity and Time: From Austrian Capital Theory to Ecological Economics[J].*The American Journal of Economics and Sociology*, 65, 803-825, (2006).
3. R. Costanza, R. Arge, R. Groot, et al. The Value of the World's Ecosystem Services and Natural Capital[J]*Nature*, 386, 253-260, (1997).
4. S. Han. The Study of Application and Estimation on Wildlife Resource Value in China[D].Beijing Forestry University, (2008).
5. W. W. Lin. Study on Economic Valuation of Wild Plant Resources. Beijing Forestry University, (2005).
6. L. Qu, G.H. Liu, W. X. Nan, et al. Causes of Biological Species Reduction and Related Countermeasures[J]. *Tianjin Science & Technology*, 6, 50-51(2015).
7. W. Sh. Luo. A Discuss on forest ecological service function evaluation and compensation problem of National nature reserve of dayaoshan in Guangxi Province[D].Guangxi University, (2014).
8. L. Liu. Study on Forest Plant Germplasm Resource Value Evalution[D]. Northeast Forestry University, (2013).
9. Zh. J. Fu, Sh. Y. Xing, X. J. Liu, etal. The growth characteristics analysis and Investigation on the Ancient ginkgo resources in Hubei Province[J]. Journal of Forestry Engineering, 27, 50-54, (2013).
10. J. H. Zhang. Legal protection of wild herbs resources research in western national areas[D]. Mingzu University of China, (2013).
11. T. Zhou, W. K. Jiang, Sh. H. Wei, et al. The investigation and utilization of resources of wild bletilla present situation analysis[C]. Academic Conference on identification of traditional Chinese medicine of the Chinese Academy of traditional Chinese Medicine, (2008).
12. J. Hu, J. Qiao, Sh. Liu, et al. The research on bletilla resource[C]. Seminar on the teaching reform and teaching material construction of the Chinese traditional medicine product academic conference and the identification of Chinese Medicine, (2015).

DOI: 10.1051/bioconf/20170803008

Microwave assisted extraction of flavonoids from pomegranate peel and its antioxidant activity

Ju Huang [a], Weiping He [2], Chen Yan[1], Xihua Du[1] and Xiaoqin Shi[1]

[1]*School of Chemistry & Chemical Engineering, Xuzhou Institute of Technology, Xuzhou 221111, Jiangsu, China*
[2]*School of Chemical Engineering, Xuzhou College of Industrial Technology, Xuzhou 221140, Jiangsu, China*

Abstract. Pomegranate peel, a waste generated from fruit processing industry, which contains maximum amounts of bioactive compounds than the juice, is a potential source of active ingredient such as flavonoids that are known for their antioxidant properties. Microwave-assisted extraction (MAE) of flavonoids from pomegranate peel and its antioxidant activity were investigated. Three independent and main variables, including ethanol concentration(%), solvent/solid ratio(mL/g) and MAE time(min), which were of significance for the yields of flavonoids were studied. The optimal conditions were determined by the orthogonal design. The maximum extraction yield of 4.260% was achieved at ethanol concentration of 60%, ratio of solvent to material of 40:1 and MAE time of 3min. In vitro antioxidant assays showed that the extract exhibited a strong DPPH radical-scavenging ability with IC_{50} value of 0.187 mg/mL. Results indicated that MAE was a suitable approach for the extraction of flavonoids from pomegranate peel.

1 Introduction

Pomegranate (Punica granatum L.) is a shrub or a small tree widely distributed in both tropical and subtropical areas. It is an important source of bioactive compounds and has been used for folk medicine for many centuries [1]. It has nutritional and medical benefits such as antioxidantive, anticancer, and antimutagenic efficacy. Many researches indicate that the Pomegranate peels contain much more bioactive compounds than the juice, which are one of the most valuable by-products of the food industry [2-4].

Compared with the traditional methods, MAE has many advantages, such as higher extraction rate, shorter extraction time, use of less solvent, better productivity and higher quality products. It is a widely used technique for the extraction of valuable bioactive components from plant materials. In microwave assisted solid-liquid extraction, the extraction is accelerated due to the changes in cell anatomy caused by the impact of microwave radiation. The accelerated extraction may be attributed to synergistic action of temperature and concentration gradients [5-7].

The objective of this study was to optimise the MAE conditions for the extraction of flavonoids from Pomegranate peel. The orthogonal design was employed to study the optimal ethanol

[a] Corresponding author: 281770895@qq.com.
This work was supported by the research projects of college(No. XKY2013317), Xuzhou institute of technology, China.

DOI: 10.1051/bioconf/20170803008

concentration (%), solvent/solid ratio (mL/g) and MAE time(min) which could maximise the yield of flavonoids from Pomegranate peel. Then, the colorimetric method with $AlCl_3$/ethanol system was used to quantify total flavonoids. In addition, the antioxidant activity of flavonoids from Pomegranate peel, obtained under optimised MAE extraction conditions was determined by 2,2-diphenyl-1-picrylhydrazyl (DPPH) radical-scavenging assay.

2 Materials and methods

2.1 Raw materials and reagents

Pomegranate peels were obtained from the fresh fruit purchased from the local market in Xuzhou City, Jiangsu Province,China. The peels were dried in an oven with air circulation at 60°C and then ground into a fine powder(40-60 mesh).

Rutin was purchased from Aladdin reagents, 2,2-diphenyl-1-picrylhydrazyl (DPPH) and ascorbid acid(Vc) were purchased from sigma. All the other reagents were of analytical grade.

2.2 Instruments

UV-VIS spectrophotometer T6, Beijing PERSEE General Instrument Co., Ltd. A microwave oven P70F20L-DG(S0), Galanz microwave life Electric Appliance Co., Ltd. Electronic balance JA2003, Shanghai Precision Scientific Instrument Co., Ltd. Blower drying box DHG-9070A, Suzhou Jiangdong Precision Instrument Co. Ltd.

2.3 Microwave assisted extraction and experimental design

A modified microwave oven (P70F20L-DG(S0), Galanz, China) was used for microwave assisted extraction. The system was equipped with a temperature sensor for monitoring and regulating the internal temperature of the extraction vessels. Dry samples (1 g) were placed into 50 mL beaker extraction vessels and the solvent added. The container was capped, and the system was started-up. The exact temperature was detected by a sensor. When the desired temperature was reached, the heating device would automatically shut down for a while until the temperature dropped. The extraction temperature and time were set at different degrees according to different conditions. After extraction, the vessels were left for several minutes to cool down to room temperature.

In order to optimize microwave assisted extraction (MAE) of antioxidant compounds from Pomegranate peels, especially flavonoids, different ethanol concentration (40-80%), solvent/solid ratio(10-50mL/g) and MAE time(1-5min) were evaluated individually. The results of the single factor analysis were combined together in order to maximize the extraction in terms of flavonoids yield and antioxidant activity with orthogonal experiments.

2.4 Determination of flavonoids content and yield of flavonoids

The content of flavonoids was determined by spectrophotometry using the aluminium chloride colorimetric method with some modifications [8]. Briefly, 2 mL diluted solution containing flavonoids was added to a 25mL volumetric flask, and 1 mL of 5% (w/w) $NaNO_2$ was mixed for 6 min, and then 1 mL of 10% $AlCl_3$ (w/w) was added and mixed, 6 min later, 10mL of 1 mol/L NaOH was added. After 15 min, the absorbance of the solution at 400 nm was measured with uv/vis spectrophotometer (T6, PGENENAL, Beijing, China) against the same mixture, without the sample as a blank. The calibration curve was prepared by preparing Rutin solutions at concentrations from 0 to 0.1 mg/mL in ethanol. It was determined to be as the following: $A=7.7750c+0.0176$, where A is absorbance value of sample, c is sample concentration ($R^2=0.9945$).

The extraction yield of flavonoids was expressed according to Eq. (1):

Yield of flavonoids

$$Y = \frac{C \times N \times V}{m} \times 100\% \quad (1)$$

Where C is mass concentration of flavonoids,N is dilute times,V is the volume of the extraction liquid, m is sample mass(Pomegranate peels).

2.5 Determination of antioxidant activity

The scavenging activity of flavonoid from Pomegranate peels towards DPPH-radical was measured according to the method by Xie[7], with some modifications. Briefly, 0.2 mmol/L solution of DPPH in ethanol was prepared daily before measurements, 2mL of various concentrations (0.1, 0.2, 0.3,0.4, 0.5, 0.6, 0.7, 0.8, 0.9 and 1.0 mg/g) of the sample solutions were thoroughly mixed with 2mL of freshly prepared DPPH and 1mL ethanol. The mixture was shaken vigorously and allowed to stand for 30 min in the dark, and the absorbance at 518 nm was then measured against a blank with uv/vis spectrophotometer (T6, PGENENAL, Beijing, China). A lower absorbance value of the reaction mixture indicates a higher free radical scavenging activity. Ethanol was used as the blank control, ascorbic acid was used as positive controls. All tests were carried out in triplicate. The capability to scavenge the free radical DPPH in percentage of sample (I%) was calculated according to the following equation.

$$I\% = \left(1 - \frac{A_i - A_j}{A_o}\right) \times 100\% \quad (2)$$

Where A_0 is the absorbance of the incubation DPPH solution without addition of the sample or positive controls, A_i is the absorbance of the incubation mixture containing both the test sample and DPPH solution and A_j is the absorbance of the sample without DPPH solution. The concentration of sample producing a 50% reduction of the radical absorbance (IC_{50}) was used as an index to compare the antioxidant activity.

3 Result and discussion

3.1 Effect of ethanol concentration, solvent/solid ratio and MAE time

Experiments were conducted to study the effect of ethanol concentration on the yield of flavonoids from Pomegranate peels. The extraction was performed at 40, 50, 60, 70and 80% respectively, the extraction time was fixed to 3 min, the ratio of solvent to material was fixed to 20:1, and the other conditions were kept the same. Figure 1(a) shows the effect of different ethanol concentration on extraction of total flavonoids. The results showed that the yield increased rapidly at first, then decreased when ethanol concentration exceeds 60%. So the optimized ethanol concentration is 60% for further experiments.

The effect of the ratio of solvent to material on the MAE process was carried out at 10:1, 20:1, 30:1, 40:1 and 50:1,respectively, while the other experimental conditions were as follows: extraction time of 3 min, and solvent of 60% (v/v) ethanol. As shown in Figure 1(b) the yield of flavonoids first increased then decreased with increasing the ratio of solvent to raw material, where an optimized ratio is 30:1~40:1 mL/g. In general, a higher ratio of solvent to raw material could result in larger concentration gradient during the diffusion from solid into the solution, and a igher extraction yield. On the other hand, too much solvent would decrease the microwave adsorption of material, since more energy was needed to be absorbed by the solvent. Therefore, the breakage of material cell wall and mass transmission might be negatively influenced.Thus, the ratio of solvent to raw material of 30:1~40:1 mL/g was adopted in the present work.

The extraction yield of flavonoids affected by different extraction time is shown in Figure 1(c) when other extraction conditions were as follows: ethanol concentration of 60% and ratio of solvent to material weight of 30:1 mL/g, respectively. The extraction yield of flavonoids shows the trend of first increase then decrease with increasing the extraction time, where a maximum yield (4.1%) was obtained at the extraction time of 3 min. MAE is a fast extraction process where microwave energy is delivered efficiently to materials through molecular interaction under the electromagnetic field, offering a rapid transfer of energy to the extraction solvent and plant materials. However, long extraction time induced the degradation of flavonoids, and the yield of extraction decreased. Therefore, 3 min was selected as the center point for further experiments.

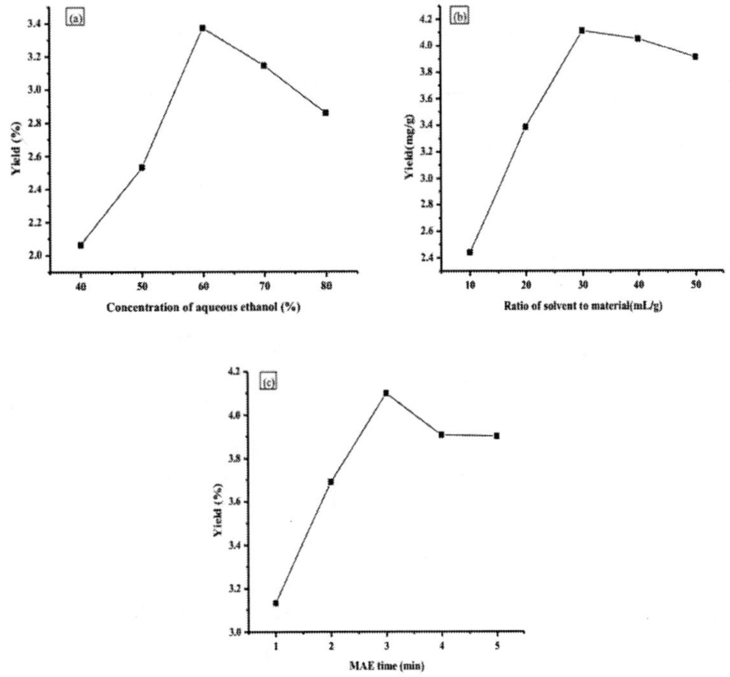

Figure 1. The effects of (a) concentration of aqueous ethanol, (b) Ratio of solvent to material and (c) MAE time on the extraction yields.

According to the single-factor experiments, we adopted ethanol concentration 50-60-70 %, the ratio of solvent to material 20:1-30:1-40:1 mL/g and extract time 2-3-4 min for orthogonal experiments.

3.2 Optimization of MAE operating parameters by orthogonal experiment

The yield of flavonoids at different expermental combinations were given in Table 1. The results showed that the effect on the yield of total flavonoids was decreased in the following order when MAE was used to extract total flavonoids from Pomegranate peels: ethanol concentration, the ratio of solvent to material and MAE time, that is A>B>C. The optimal conditions for flavonoids extract from Pomegranate peels are ethanol concentration of 60%, ratio of solvent to material of 40:1 and MAE time of 3min.

DOI: 10.1051/bioconf/20170803008

Table 1. Experimental design runs in orthogonal method $L_9(3^4)$ and the observed value of extraction yield.

Experiment number	A	B	C	D	yield(%)
	ethanol concentration(%)	solvent/solid ratio (mL/g)	MAE time(min)	blank column	
1	50	20	2	1	2.946
2	50	30	3	2	3.256
3	50	40	4	3	3.171
4	60	20	3	3	3.586
5	60	30	4	1	3.978
6	60	40	2	2	4.186
7	70	20	4	2	3.590
8	70	30	2	3	3.684
9	70	40	3	1	4.126
k_1	1.039	1.125	1.202	1.228	
k_2	1.306	1.211	1.216	1.224	
k_3	1.267	1.276	1.193	1.160	
R	0.799	0.476	0.069	0.203	

3.3 Analysis of variance and verification

The interactions and coefficients on the orthogonal experiments were obtained by analysis of variance (ANOVA) (Table 2). The results show that factor A is very significant, while B and C are not.

Table 2. ANOVA for the MAE extraction conditions.

factor	the sum of variances	degrees of freedom	variance	F-value	significant
A	1.118	2	0.559	14.333	*
B	0.311	2	0.1555	3.987	
C	0.007	2	0.0035	0.09	
D	0.078	2	0.039	1	
error	0.08	8			

In order to validate the reliability of the optimum MAE conditions, five experiments were executed at the optimal conditions. The experimental yield of flavonoids from Pomegranate peels was 4.260% (n=5, RSD <0.17).

DOI: 10.1051/bioconf/20170803008

3.4 Antioxidant activity

The antioxidant activity of flavonoids from Pomegranate peels was assessed with DPPH-scavenging assay, in comparison with known antioxidants ascorbic acid (Figure 2). It was observed that the flavonoids from Pomegranate peels exhibited notable DPPH radical-scavenging activity, and the DPPH radical scavenging effects were increased with increasing flavonoids concentrations. It increased from 31.75% to 92.85% when the concentrations of the flavonoids increased from 0.109 mg/mL to 0.886 mg/mL. However, DPPH free radical-scavenging of flavonoids from Pomegranate peels was less than that of VC, but the results indicated that flavonoids extracted by MAE had a strong DPPH radical-scavenging activity, with an IC_{50} value of 0.187 mg/mL.

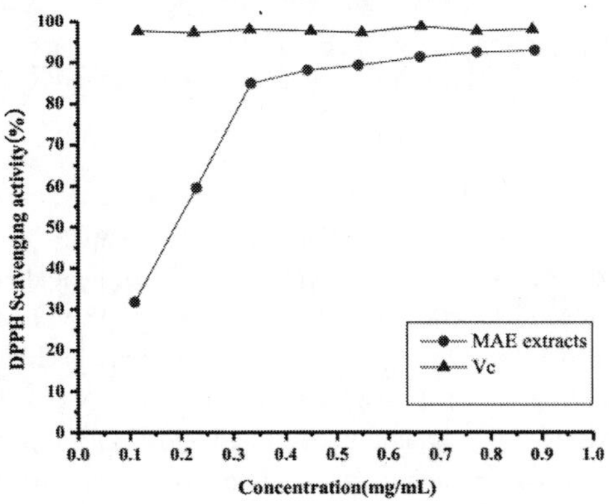

Figure 2. Antioxidant activity of flavonoids from pomegranate peel assessed by DPPH radical scavenging assay.

4 Conclusion

Microwave irradiation, which has proved to be a clean, efficient and convenient energy source, has been utilised in the extraction of flavonoids from Pomegranate peels. Orthogonal experiment was successfully applied for optimisation of MAE of flavonoids from Pomegranate peels. The optimal flavonoids yield of 4.26% was obtained when the optimum conditions (ethanol concentration of 60%, ratio of solvent to material of 40:1 and MAE time of 3min.)were used. Application of MAE in the extraction of flavonoids from Pomegranate peels dramatically reduced extraction time. The flavonoids obtained by MAE exhibited good antioxidant activity with an IC_{50} value of 0.187 mg/mL. Thus, MAE was proposed as an alternative for the extraction of flavonoids from Pomegranate peels, and orthogonal design was an useful tool for the optimisation of MAE process of flavonoids from Pomegranate peels.

References

1. Wang C, Shi L, Fan L, Ding Y, Zhao S, Liu Y, et al. Ind Crops Prod.**42**,587(2013).
2. Sood A, Gupta M. Food Bios. **12**,100(2015).
3. Zhu C, Liu X. Carbohydr Polym. **92**(2),1197(2013).
4. Amyrgialaki E, Makris DP, Mauromoustakos A, Kefalas P. Ind Crops Prod.**59**,216 (2014).
5. Thirugnanasambandham K, Sivakumar V, Maran JP. Int J Biol Macromol.**72**,1(2015).
6. Bouras M, Chadni M, Barba FJ, et al. Ind Crops Prod. **77**,590 (2015)
7. Xie JH, Dong CJ, Nie SP, Li F, Wang ZJ, Shen MY, et al. Food Chem. **186**,97-105(2015).
8. Amir, M., Mujeeb, M., Khan, A., Ashraf, K., Sharma, D., & Aqil, M. International Journal of Green Pharmacy, **6**, 67–72 (2012).

DOI: 10.1051/bioconf/20170803009

Screening ssDNA aptamers against HIV P24 antigen using agarose beads as carriers

Jiayu Zeng[1], Xing Li[1], Hongxia Yuan[2], Meilan Ma[1], Dongdong Li[2], Jing Ma[2] and Shiqi Liao [2,a]

[1]*College of Life Sciences, Northwest Normal University, Lanzhou, 730070, China*
[2]*Laboratory of Molecular Biology, Gansu Academy of Medical Sciences, Lanzhou, 730050, China*

Abstract. In this study, carboxylated agarose beads were used as screening carriers combined with the subtractive systematic evolution of ligands by exponential enrichment (SELEX) technology to obtain an ssDNA secondary library, which could specifically bind to HIV P24 antigen after nine screening rounds from a random ssDNA library. An electrophoretic mobility shift assay (EMSA) and quantitative real-time PCR assay (RT-qPCR) was performed to monitor in real-time the binding specificity of the ssDNA secondary library to the HIV P24 antigen. The obtained ssDNA secondary library was converted to the corresponding dsDNA library, ligated to a PMD18-T vector, and transformed into *E. coli* DH5α and sequenced. The results demonstrated that after nine rounds of screening, two aptamer sequences specifically bound to the HIV P24 antigen were obtained.

1 Introduction

Currently, the diagnosis and treatment of human AIDS virus (HIV)-caused acquired immune deficiency syndrome (AIDS) is still a medical issue. Early diagnosis of AIDS is an important part of treatment. Predicting the spread possibility of AIDS and evaluating treatment effects both require highly sensitive and specific diagnostic methods. There are many ways to measure HIV infection; detecting the HIV P24 antigen and its antibody is one of the most commonly used methods in HIV clinical screenings.

The HIV P24 antigen is a major structural component of the HIV envelope, with a molecular weight of 24 KD. Each complete virus particle contains approximately 2,000 HIV P24 antigen molecules. During early HIV infection, the HIV P24 antigen expresses in the human body and can be detected in the blood. However, the induced HIV P24 antibody subsequently binds to the HIV P24 antigen, which makes it difficult to detect the HIV P24 antigen in the blood.

Systematic evolution of ligands by exponential enrichment technique (SELEX) is a novel *in vitro* screening technique that has emerged in the last 20 years. It combines library technology and screening technology to screen out ssDNA oligonucleotide molecules, the so-called aptamers from a random ssDNA library. Aptamers are similar to antibodies because they can bind to the corresponding targets with high affinity and specificity. Currently, screening aptamers with SELEX is developing quickly. Aptamers can be widely applied in various fields, including medical biology. In terms of virus research and treatment, SELEX has been used to screen out various relevant aptamers of viruses,

[a] Corresponding author
This work was supported by Chinese National Science Foundation (81360333; 81560346).

DOI: 10.1051/bioconf/20170803009

providing a new direction for preventing and treating related diseases.

In this study, SELEX was applied to screen aptamers for the HIV P24 antigen, laying the foundation for new methods of AIDS diagnosis and treatment.

2 Materials and Methods

2.1 Materials

Carboxylated agarose beads were provided by Gansu Academy of Medical Sciences; the HIV P24 antigen (2 mg/ml) was from Beijing Wanyumeilan Reagent Company; Streptavidin was synthesized by TaKaRa; the PMD18-T vector system was from Shanghai Sangon Biotech Co., Ltd.; the *E. coli* DH5α was from Gansu Academy of Medical Sciences; the random ssDNA library was synthesized by Shanghai Sangon Biotech Co., Ltd. with the following sequence: 5'-CTATAGCAATGGTACGGGTAC-TTCC-(40N)-CAAAAGTGCACGCTACTTTGCTAA-3' (N = A, G, T, C); primer P7 with the sequence 5'-CTATAGCAATGGTACGGTACT-TCC-3' and primer P11 with the sequence 5'-TTAGCA-AAGTAGCGTGG-ACTTTTG-3' was synthesized by TaKaRa; the anti- HbsAg was from Beijing Wanyumeilan Reagent Company; the PCR-related reagents were from Shanghai Sangon Biotech Co., Ltd. and TaKaRa; all other reagents were analytical grade.

The GeneAmp PCR System 2700PCR instrument was from AppliedBiosystems (USA); The MVT-28 gel imager was from Herolab (Germany); the ROTOR-GENE 3000 Real Time PCR System was from Rotor-Gene (USA); the mini double vertical electrophoresis tank DYCZ-24DN was from Beijing 61 Instrument Factory; the SHZ-82 thermostat shaker was from Changzhou Guohua instruments Co., Ltd.; the TGL-16A high-speed refrigerated centrifuge was from Changsha Pingfan Instruments Co., Ltd.

2.2 Methods

2.2.1 Screening aptamers for the HIV P24 antigen

1) Screening Round 1

Coating: The carboxylated agarose beads were placed in a 5 ml Eppendorf tube and washed with 5×beads volume deionized water three times. A volume of 500 μl 100-fold diluted HIV P24 antigen was added and incubated at 37 °C for 1 h.

Blocking: The supernatant was discarded and the precipitant was washed three times with 0.01 mol/L pH7.4 PBS. Then, a 1 ml blocking solution (0.1g sucrose, 12.5 mg bovine serum albumin, and 12.5 mg casein dissolved in 10 ml PBS) was added and incubated at 37 °C for 1 h. Again, the supernatant was discarded and the precipitant was rinsed with 0.01 mol/L pH 7.4 PBS three times. Then, it was transferred to a new Eppendorf tube.

Pretreatment of the ssDNA library: The new synthesized random ssDNA library was dissolved in a 300 ml binding buffer, heated in a boiling water bath for 5 min, and was then rapidly cooled for 10 min.

Binding: The pretreated ssDNA library was added to the abovementioned Eppendorf tube and incubated at 37 °C for 2 h. The supernatant was discarded and the precipitant was washed with a 2×saline-sodium citrate (SSC) solution 5 times, and rinsed with pure water after replacing the tube.

Eluting: The supernatant was discarded. Then, 100 μl of water was added to the Eppendorf tube and boiled in a boiling water bath for 5 min. After it was centrifuged at 12,000 rmp for 5 min, the supernatant was collected. This step was repeated three times and the supernatants were combined.

Preparation of the dsDNA library [12]: A sample of 80 μl of recovered supernatant was used as a template to prepare dsDNA in a 100 μl PCR reaction. Besides the template, the PCR solution included a 10 μl 10×Buffer, fluorescent dye supergreen I 1.5 μl, dNTP mixture (25 mmol/L of each dNTP) 0.8 μl, primer P7 1.0 μl, 5'-biotinylated primer P11 1.0 μl, and TapDNA polymerase 0.5 μl, added with deionized water to a total volume of 100 μl. The 100 μl solution was aliquoted into PCR tubes with 24 μl per tube to perform PCR amplification at the following conditions: 95 °C for 6 s, then 60 °C for 34 s; after reaching the S-curve peak, the PCR amplification was terminated. Thus, the biotinylated dsDNA was obtained.

Preparation of the ssDNA library [12]: The prepared streptavidin-biotin magnetic beads were rinsed three times in 0.01 mol/L pH7.4 PBS. The biotin-labeled dsDNA from the last step was added to the beads and incubated in a rotating shaker for 30 min at 37 °C. The supernatant was discarded, and then a sample of 0.4 ml 5% formamide was added and incubated in a 40 °C water bath for 3 min. After it was rinsed 4 times with deionized water, the sample was transferred into a new Eppendorf tube. A sample of a 200 μl binding buffer was added into the tube and incubated in a 100 °C dry block heater for 5 min. The Eppendorf tube was removed and the supernatant was transferred into a new tube. Thus, the ssDNA secondary library was obtained and to be used for the next round of screening.

2) Screening Round 2

Coating: The carboxylated agarose beads were washed with deionized water with 5×beads volume deionized water three times and evenly allocated into two Eppendorf tubes, labelled as the positive and negative selection tube. In the positive selection tube, a volume of 500 μl 100-fold diluted HIV P24 antigen was added; in the negative tube, the same volume of 50-fold diluted Anti-HBsAg was added, and both tubes were incubated at 37 °C for 1 h.

Blocking: For both of the positive and negative selection tubes, the supernatants were discarded and the precipitants were washed three times with 0.01 mol/L pH7.4 PBS. Then, a sample of a 1 ml blocking solution was added into each tube and incubated in a 37 °C rotating shaker for 1h.

Binding: The supernatants in both tubes were discarded and the precipitants were each washed with 0.01 mol/L pH7.4 PBS three times. The ssDNA secondary library obtained from the last round was added to the negative selection tube and incubated at 37 °C for 2 h. Then, the supernatant from this tube was transferred into the positive selection tube and incubated at 37 °C for another 2 h. The supernatant from the positive selection tube was discarded. Then, the precipitant from each tube was washed five times with a 2×SSC solution and rinsed with pure water after replacing the tubes.

Eluting: A sample of 100 μl pure water was added to each tube and boiled in a boiling water bath for 5 min. After it was centrifuged at 12,000 rmp for 5 min, the supernatant from each tube was collected. This step was repeated three times and the supernatant was combined from each tube, respectively.

The preparation of the ssDNA secondary library was the same as in the first round of screening.

3) Re-screening

In subsequent screening process rounds, the HIV P24 antigen was always used as the target protein for the positive selection, and the anti-HbsAg and goat polyclonal HIV-P24 antibody were used alternatively as the non-target protein for the negative selection. In each screening round, the obtained ssDNA secondary library was used as the starting library for the next round, whereas all other operations were the same as the second round. Meanwhile, in each round, RT-qPCR was performed to monitor the enrichment of target ssDNA until it reached the maximum. This completed the screening process.

4) Determining the ssDNA library enrichment

RT-qPCR was used to determine the ssDNA library enrichment. The PCR solution included the ssDNA library 6.0 μl as a template, 10×Buffer 10 μl, dNTP mixture (25 mmol/L of each dNTP) 0.8 μl, primer P7 0.4 μl, 5'-biotinylated primer P11 0.4 μl,TapDNA polymerase 2.0 μl, and fluorescent dye 1.0 μl, with added deionized water to a total volume of 100 μl. After it was rapidly and thoroughly mixed, the solution was aliquoted into PCR tubes at 24 μl per tube. The PCR reaction condition was 95 °C for 6 s and 60 °C for 4 s, for a total of 40 cycles. The negative control was prepared with deionized water as a template, and the positive one was prepared with the ssDNA starting library as a template, with all other ingredients and steps the same as above.

2.2.2 Determining the specific binding of ssDNA with the target protein

1) The electrophoretic mobility shift assay (EMSA) was performed to determine the specificity of HIV P24 antigen-ssDNA complexes formed by ssDNA and the target protein. A sample of 5 μl HIV P24 antigen was mixed with 10 μl ssDNA and incubated at 37 °C for 1 h. Then, the agarose gel electrophoresis was conducted with the following condition: 1.5 % W/V agarose, 1×TAE buffer, a constant voltage of 80 V and 40 min electrophoresis. At the same time, two controls, the HIV P24 antigen and ssDNA, were set up and the gel was stained with Coomassie blue.

2) RT-qPCR was conducted to determine the binding specificity between the ssDNA and the target protein, the HIV P24 antigen. Two samples of the same amount of carboxylated agarose beads were coated with the HIV P24 antigen and the non-target protein, Anti-HbsAg, respectively. According to the screening method for round 1 in section 1.2.1, the ssDNA libraries bound with the target protein and the non-target protein were prepared respectively. Afterwards, the prepared ssDNA libraries were used as templates for RT-qPCR according to the method for determining the ssDNA library enrichment in section 1.2.1.

2.2.3 Sequencing

The ssDNA aptamers library which specifically bound to the target protein HIV P24 antigen that was used to prepare the corresponding dsDNA library (with the downstream primer of 5'-biotinylated P11 instead and all other operations the same as the screening method for round 1 in section 1.2.1). Then, the dsDNA library was ligated to the PMD TM 18-T vector and transformed into *E. coli* DH5α. The positive clones were sent for sequencing at the Shanghai Personalbio Biological Technology Co., Ltd. The ligation and transformation followed conventional methods.

3 Results and Analysis

3.1 Screening HIV P24 antigen aptamers

In the newly synthesized ssDNA library, the abundance of ssDNA specifically bound to the target protein was low. Therefore, the negative selection process was skipped in the first round. The ssDNA enrichments of the first round determined by RT-qPCR demonstrated clearly separated PCR curves for the positive control, the screening sample, and the negative control (Figure 1), with a significantly higher DNA abundance in the screening tube than in the negative control. This indicated that the target protein in the screening tube bound a certain amount of ssDNA. Subsequently, the ssDNA served as a template for preparing the ssDNA secondary library, the starting library for the next round.

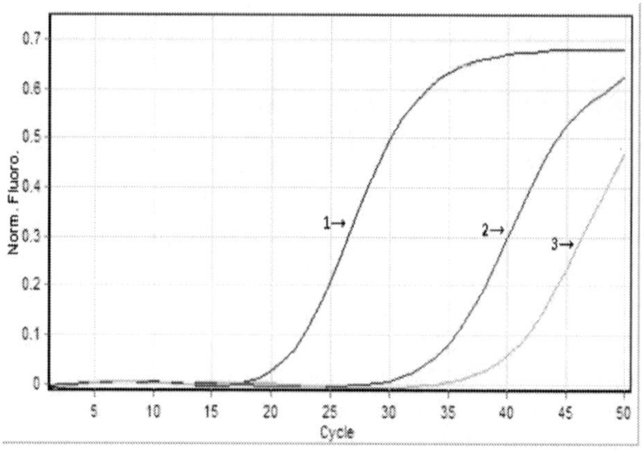

Figure 1. RT-qPCR determination of the ssDNA enrichment for the first round of screening. 1. The positive control; 2. The screening tube; 3. The negative control. The vertical axis is the fluorescence intensity, and the horizontal axis is the number of PCR cycles. After 18 cycles, the ssDNA in the positive control started to enrich; after 27 cycles, the ssDNA in the screening tube started to enrich; after 33 cycles, the ssDNA in the negative control started to enrich.

After the first round of screening, a large amount of non-target ssDNA was removed and the abundance of the target ssDNA was greatly improved. Hence, to more accurately evaluate the target ssDNA abundance, the negative selection was established from the second round of screening. From the RT-qPCR results for rounds 2-4 of screening, the PCR curves were distinct for the positive selection tube, and the positive and negative controls; the PCR curves were distinct for the negative selection tube, and the positive and negative controls. The PCR curves for the positive and negative selection tubes were not distinct from each other. Figure 2 demonstrates the RT-qPCR results for round 4 of screening. The results indicated that during screening rounds 2-4, both the positive and negative selections enriched ssDNA. However, the largely enriched ssDNA in the negative selection indicated high abundant non-target ssDNA in the starting libraries, and a low abundance of the target ssDNA.

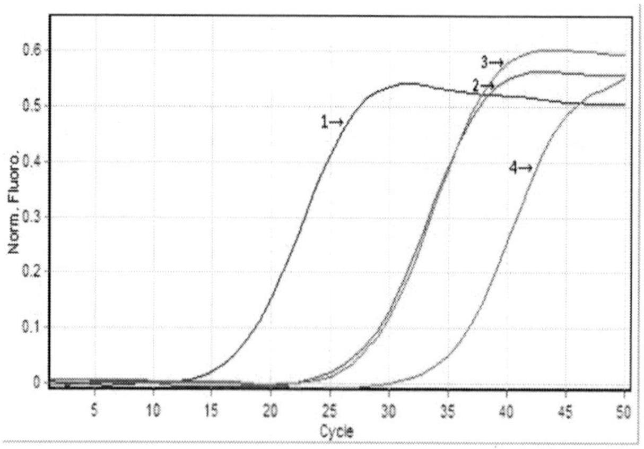

Figure 2. RT-qPCR determination of the ssDNA enrichment for the fourth round of screening. 1. The positive control; 2. The positive selection tube; 3. The negative selection tube; 4. The negative control. The vertical axis is the fluorescence intensity, and the horizontal axis is the number of PCR cycles. After 14 cycles, the ssDNA in the positive control started to enrich; after 23 cycles, the ssDNA in the positive and negative selection started to enrich; after 26 cycles, the ssDNA in the negative control started to enrich.

Using the ssDNA secondary library from round 4 as the starting library, the 5th round of screening was conducted. Figure 3 shows the RT-qPCR results for the 5th round. The results demonstrated distinct PCR curves for the positive selection tube, and the positive and negative controls; it also showed distinct PCR curves for the negative selection tube, and the positive and negative controls. Meanwhile, the PCR curves for the positive and negative selection tubes started to separate, indicating that the ssDNA abundance was different between them, with a one cycle difference. This suggested the presence of non-target ssDNA in the starting library of round 5, only with a low abundance. At the same time, the ssDNA that was specifically bound with the HIV P24 antigen started to enrich.

Subsequently, a continuous five rounds of screening was conducted. As shown in the RT-qPCR results for rounds 6-10, the PCR curves for the positive and negative selection tubes separated further and further. The curves for the negative selection became closer to the negative control, and the difference was only one cycle for round 9. There were no significant differences between screening rounds 9 and 10, as shown from the PCR results. These results indicated that during rounds 6-10, the target ssDNA was enriching, and the non-target ssDNA was disappearing. At round 9, the abundance of the non-target ssDNA reached the lowest point, while the target ssDNA reached the highest point. Therefore, the obtained ssDNA secondary library was the target ssDNA that bound specifically to the target HIV P24 antigen, the so-called ssDNA aptamers for the HIV P24 antigen. Figure 4 shows the RT-qPCR results for the 9th round of screening.

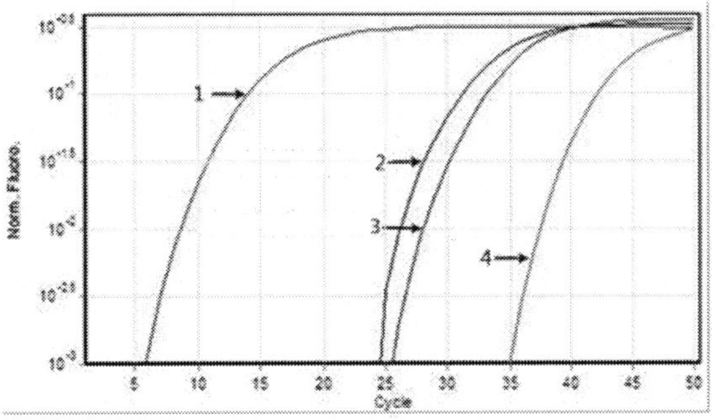

Figure 3. RT-qPCR determination of the ssDNA enrichment for the fifth round of screening. 1. The positive control; 2. The positive selection tube; 3. The negative selection tube; 4. The negative control. The vertical axis is the fluorescence intensity, and the horizontal axis is the number of PCR cycles. After 6 cycles, the ssDNA in the positive control started to enrich; after 25 cycles, the ssDNA in the positive selection started to enrich; after 26 cycles, the ssDNA in the negative selection started to enrich; after 35 cycles, the ssDNA in the negative control started to enrich.

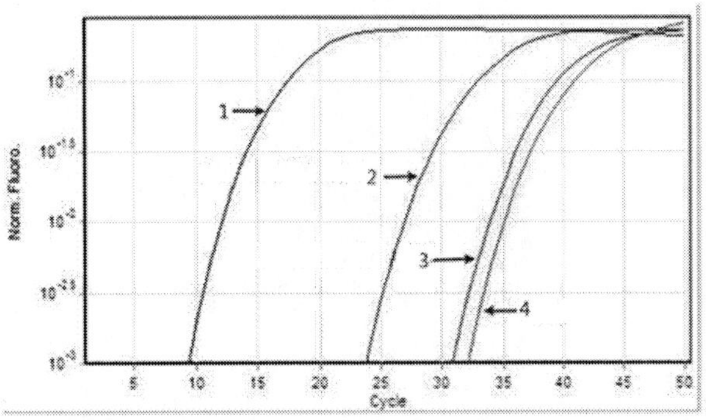

Figure 4. RT-qPCR determination of the ssDNA enrichment for the ninth round of screening. 1. The positive control; 2. The positive selection tube; 3. The negative selection tube; 4. The negative control. The vertical axis is the fluorescence intensity, and the horizontal axis is the number of PCR cycles. After 9 cycles, the ssDNA in the positive control started to enrich; after 24 cycles, the ssDNA in the positive selection started to enrich; after 31 cycles, the ssDNA in the negative selection started to enrich; after 32 cycles, the ssDNA in the negative control started to enrich.

 DOI: 10.1051/bioconf/20170803009

3.2 Preparation of the ssDNA secondary Library

The ssDNA secondary library obtained after the 9th round of screening was the target of ssDNA aptamers that could specifically bind to the HIV P24 antigen. The preparation of the ssDNA secondary library included two steps. First, the supernatant ssDNA product of the 9th round of screening was PCR amplified to prepare the biotinylated dsDNA library. Second, the streptavidin-biotin beads were used to prepare the corresponding ssDNA library from the above biotinylated dsDNA library. Thus the ssDNA aptamers were obtained, and the methods were as described for the first round of screening.

Figure 5 shows the RT-qPCR results of the preparation of the biotinylated dsDNA library. It was demonstrated that after about 15 cycles, the DNA started to enrich, and reached the S curve peak after 35 cycles for all four parallel PCR tubes. Figure 6 demonstrates the polyacrylamide gel electrophoresis (PAGE) results of the prepared dsDNA and ssDNA libraries. There were clear bands of approximately 80 bp at lanes 1 and 2, consistent with the molecular weight of the synthesized random library, which indicated the acquisition of the dsDNA and ssDNA libraries. In addition, the dsDNA had a higher mobility rate due to the formation of a compact double helix structure.

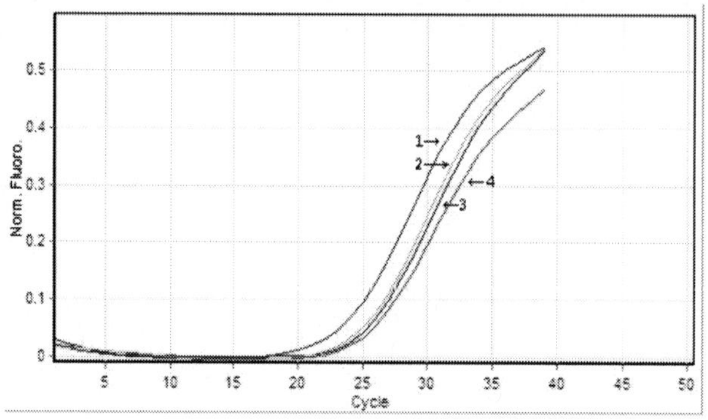

Figure 5. RT-qPCR determination of the preparation of the dsDNA library. 1, 2, 3, and 4 are four parallel PCR tubes, all templated with the ssDNA secondary library from the ninth round of screening. The vertical axis is the fluorescence intensity, and the horizontal axis is the number of PCR cycles.

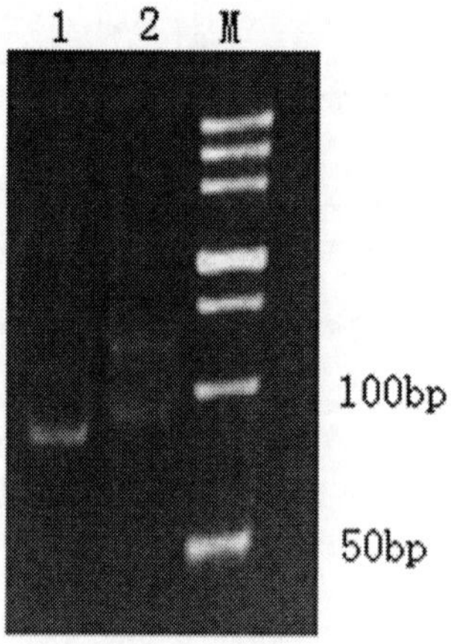

Figure 6. PAGE detection of the prepared dsDNA and ssDNA libraries. Lane M. DNA markers; Lane 1. dsDNA sample; Lane 2. ssDNA sample. A 10% non-denaturing polyacrylamide gel was used with a loading volume of 1 μL.

3.3 EMSA identification of the aptamers

The ssDNA library of the aptamers of the 9th round of screening should be able to recognize and bind to the target protein HIV P24 antigen specifically to form HIV P24 antigen-ssDNA complexes. The molecular weight and charge properties of the complexes were different from the HIV P24 antigen and ssDNA, causing different electrophoretic mobility rates. Thus, EMSA was conducted to identify the HIV P24 antigen-ssDNA complexes. Figure 7 shows the PAGE results. As shown in this figure, there was a clear band in lanes 1 and 2, with lane 2 showing a higher mobility rate than lane 1. This result suggested that the HIV P24 antigen bound with ssDNA aptamers and formed HIV P24 antigen-ssDNA complexes. Since the complexes had higher molecular weights than the HIV P24 antigen, they encountered large gel retardation and resulted in decreased mobility. Lane 3 was ssDNA and would not be stained by Coomassie blue.

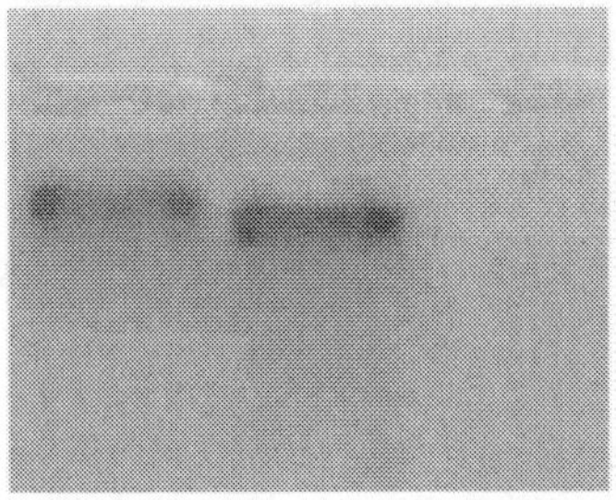

Figure 7. The 1% agarose gel results of the aptamer complexes identified with EMSA. Lane 1. HIV P24 antigen-ssDNA complexes; Lane 2. HIV P24 antigen; Lane 3. ssDNA. A 10% non-denaturing polyacrylamide gel was used with a loading volume of 10 μl.

3.4 Identification of binding specificity of ssDNA and HIV P24 antigen by RT-qPCR

The ssDNA library of the aptamers obtained from the 9th round of screening was bound to the beads coated with the target protein HIV P24 antigen and the non-target protein Anti-HbsAg, respectively, to prepare the specific binding ssDNA secondary libraries. RT-qPCR was performed to monitor the ssDNA abundance in the secondary libraries; Figure 8 shows the results.

The results showed distinct enrichment curves for the HIV P24 antigen-coated tube and the negative control, indicating a high abundance of ssDNA in the HIV P24 antigen-coated tube. Which meant that the beads-coated with the HIV P24 antigen captured large amounts of ssDNA. The enrichment curves for the Anti-HbsAg-coated tube and the negative control nearly superimposed each other, indicating little to no ssDNA in the Anti-HbsAg-coated tube. This meant that the beads-coated with Anti-HbsAg failed to capture ssDNA. The above results further confirmed that the ssDNA obtained from the 9th round of screening could specifically bind to the HIV P24 antigen, and it could serve as HIV P24 antigen aptamers.

Moreover, from the experimental results, it can be seen that the enrichment curves for the Anti-HbsAg-coated tube and the negative control were not identical. This might be due to the non-specific binding between ssDNA and Anti-HbsAg, or an incomplete wash.

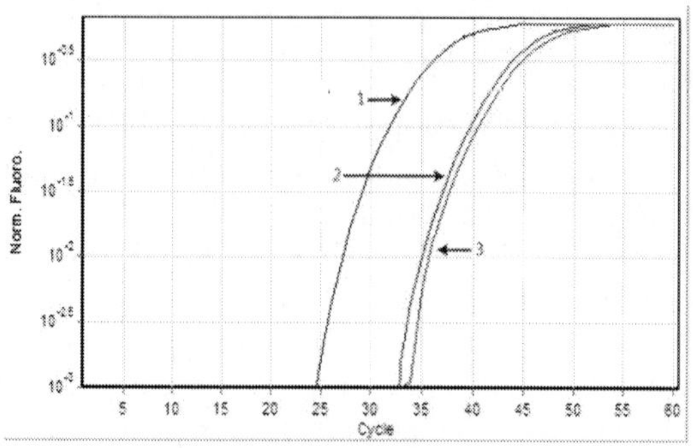

Figure 8. The binding specificity between ssDNA and HIV P24 antigen identified by RT-qPCR. 1. HIV P24 antigen-coated tube; 2. Anti-HbsAg-coated tube; 3. The negative control. The vertical axis is the fluorescence intensity, and the horizontal axis is the number of PCR cycles. After 25 cycles, the ssDNA in the HIV P24 antigen-coated tube started to enrich; after 33 cycles the ssDNA in the Anti-HbsAg-coated tube started to enrich; after 34 cycles, the ssDNA in the negative control started to enrich.

3.5 Sequencing

The obtained ssDNA library of aptamers for HIV P24 antigen was converted into a dsDNA library and transformed into *E. coli* DH5α. Three positive clones were picked and sent for sequencing and chemical synthesis at Shanghai Personalbio Biological Technology Co., Ltd. The three synthesized ssDNA were amplified by PCR and the HIV P24 antigen binding specificity was confirmed by EMSA and RT-qPCR. Eventually, two HIV P24 antigen-specific binding aptamers were obtained, with sequences as follows:

Chain 1:

5'-TCGGCACGTTCTCAGTAGCGCTCGCTGGTCATCCCACAGCTACGT-3';

Chain 2:

5'-GGACGCGGTTAGGGGAGAATTTCCATACCTTTGGGTTGTA-3'.

4 Discussion

Due to its unique advantages, aptamer plays a very important role in molecular diagnostics, molecular imaging, targeting delivery, and treatment of cancer. Aptamers have structural diversity, demonstrating a high specificity and a strong affinity upon binding to a variety of target proteins. In terms of protein detection, aptamers are superior to the conventional antigen-antibody means [11-12], and has great potential for broad applications.

This study combined subtractive SELEX and RT-qPCR, using carboxylated agarose beads as the carrier and HIV-P24 antigen as the target molecule, to screen specific oligonucleotide aptamers. Compared to conventional screening techniques, this method is inexpensive, simple, fast, and can maximize the target ssDNA enrichment with high screening efficiency and specificity. In addition, the use of carboxylated agarose beads could largely avoid contamination and hard separation caused by other traditional carriers. Carboxylated agarose beads could improve the coupling efficiency between the target protein and the carrier. The streptavidin-biotin magnetic beads would improve the preparation efficiency and the purity of ssDNA because of the coupling effect for dsDNA and the carrier due to high affinity between streptavidin and biotin. Our next step would be to further explore and establish new approaches for screening and preparing ssDNA aptamers. Improving the screening efficiency would be an important part of our future work.

After 9 rounds of screening, two aptamer sequences specifically binding to HIV P24 antigens were obtained. Our study lays the foundation for new approaches to quickly, accurately, and efficiently

 DOI: 10.1051/bioconf/20170803009

detect HIV P24 antigens with aptamers.

References

1. Cai D, Xu D, Zhang Q, *et al.* Classification of lung cancer using ensemble-based feature selection and machine learning methods [J]. *Molecular biolystems*, 2014, **464**(7291): 1071-1076.
2. ZENG Jia-yu, YANG Nan, WAN Bo, *et al.* The preparation and characterization of epoxy-modified magnetic microsphere [J]. *Journal of Northwest Normal University (Natural Science)*, 2014, **50**(2): 93-99.
3. Kulbachinskiy AV. Methods for selection of aptamers to protein targets [J]. *Biochemistry*, 2007, 72: 1505-1518.
4. Ireson CR, Kelland LR. Discovery and development of anti-cancer aptamers [J]. *Mol Cancer Ther*, 2006, 5: 2957-2962.
5. Barbas AS, Mi J, Clary BM, *et al.* Aptamer applications for targeted cancer therapy [J]. Future Oncol, 2010, 6: 1117- 1126.
6. Elyse D. Bernard, Kathy C. Nguyen, Maria C. DeRosa, Azam F. Tayabali, Rocio Aranda Rodriguez. Development of a bead-based aptamer/antibody detection system for Creactive protein. Analytical Biochemistry. 2015: 67-74.
7. Fang X, Tan W. Aptamers generated from cell-SELEX for molecular medicine: a chemical biology approach [J]. *Acc Chem Res*, 2010, 43: 48-57.
8. Sturgeon JP, Shawcross DL. Recent insights into the pathogenesis of hepatic encephalopathy and treatments [J]. Expert Rev Gastroenterol Hepatol, 2014, **8**(1): 83-100.
9. Sefah K, Meng L, Lopez-Colon D, *et al.* DNA aptamers as molecular probes for colorectal cancer study [J]. PLoS One, 2010, 5: e14269. doi: 10.1371/journal. pone. 0014269.
10. Keefe AD, Pai S, Ellington AD. Aptamers as therapeutics [J]. *Nat Rev Drug Discovery*, 2010, 9: 537-550.
11. Liao SQ, Liu YQ, Zeng JY, *et al.* Aptamer-based sensitive detection of target molecules via RT-PCR signal amplication [J]. *Bioconjugate Chem*, 2010, 21: 2183-2189.
12. Yoshida Y, Horii K, Sakai N, *et al.* Antibody-specificaptamer-based PCR analysis for sensitive protein detection [J]. Anal Bioanal Chem, 2009, 395: 1089-1096.
13. Constantine NT, Zink H. HIV testing technologies after two decades of evolution [J]. *Indian J Med Res*, 2005, 121: 519-538.
14. Willner I, Zayats M. Electronic aptamer-based sensors [J]. *Angew Chem Int Ed*, 2007, **46** (27): 6408-6418.

Purification of total flavonoids from loquat leaves by macroporous resin and corresponding antioxidant capacity

Suhua Huang, Biaosheng Lin, Binbin Li, Bi Tan and Yanping Hong[a]

College of Life Science, Longyan University, Longyan, Fujian 364012, China

Abstract. Flavonoids is one of the major active compounds in loquat leaves. In this study, the purification process of total flavonoids in loquat leaves by macroporous resin was researched and the antioxidant activity of total flavonoids was determined. The active compounds were extracted by 95% ethanol, and the total flavonoids was purified by macroporous resin. Comparing the static and dynamic adsorption and desorption characters of 6 macroporous resin, the best type of macroporous resin was determined. The desorption-solvent was determined by the gradient elution. Moreover, DPPH and FRAP analysis methods were used to determine the antioxidant activity of total flavonoids in this study. The result showed that X-5 was the optimum macroporous resin to purify the total flavonoids of loquat leaf. After adsorbed by X-5 macroporous resin, removing impurity with 15% ethanol and desorbed with 65% ethanol, the total flavonoids purity of the eluate was 82.20%, the purification factor was 2.14 times and the recovery was 81.08%, and the macroporous resin should be regeneration after using 3 times. As the flavonoids content of extraction increase, its antioxidant activity increase. In summary, the X-5 macroporous resin was suitable to purify the total flavonoids of loquat for higher purification fold and higher recovery. Purified extract had higher content of flavonoid and better antioxidant activity, it suggested the flavonoids played an important role in antioxidant activity.

1 Introduction

Loquat (Eriobotrya japonica (Thunb1.) Lindl.), a kind of subtropical fruit tree in China, has nutritious and delicious fruits with high medical value. Its leaves, fruits, seeds and flowers can be used as medicine, and are usually used to prepare loquat cream and loquat dew [1]. In addition, loquat leave is a traditional Chinese herb with cough-suppressing and phlegm-transforming functions [2-3]. In Japan, loquat leaves are often prepared as loquat tea [4]. The triterpenoids and flavonoids extracted from leaves, flowers and fruits have anti-inflammatory, antioxidant, anticancer, and hypoglycemic effects [5-7] The purification and activity of triterpenoids have been extensively explored. Similarly, its flavonoids and polyphenols also have excellent medicinal efficacy [8-13]. Macroporous resin adsorption method with multiple advantages of removing impurities, high adsorption capacity and repeated utilization during the purification process of natural products has been gradually recognized

[a] Corresponding author: kk9096@126.com
This study was supported by Fujian Provincial Key Science and Technology Project (Grant No. 2012N0019), Fujian Provincial Natural Science Foundation (Grant No. 2012D100), Fujian Provincial Department of Education Project (Grant No. JB12204) and The Undergraduate Scientific and Technology Innovation Project of Fujian Province, China (201511312058).

and applied[14-17]. Many experiments show that macroporous resin has excellent separation efficiency for flavonoids [18-20]. The purification and enrichment of flavonoids through macroporous resin can provide a basis for the further exploitation and utilization of loquat leaves.

In the present study, ethanol extraction method was used for obtaining the crude extract of total flavonoids from loquat leaves. The purification efficiency of 6 kinds of macroporous resins for the crude extract of total flavonoids was compared to realize the screening of optimal resin and purification process. The antioxidant activities of purified total flavonoids were evaluated through DPPH and FRAP analysis, which provided a reference for industrial production of total flavonoids from loquat leaves.

2 Material and methods

2.1 Materials and reagents

Fresh and mature loquat leaves collected from Putian, Fujian, were subjected to wiping and natural drying as well as pulverizing to obtain the dried powder.

The specific parameters of macroporous resins including ADS-7, ADS-17, AB-8, D-101 and X-5 (Nankai University), as well as DM-130 (Shanxing Resin Co., Ltd.) were shown in Table 1. Ursolic acid, 1,1-diphenyl-2-picryl hydrazine (DPPH), terpyridine triazine (TPTZ), Folin-ciocalteu reagent and gallic acid were purchased from Sigma Company. Rutin was purchased from Sinopharm Chemical Reagent Co., Ltd.. Other reagents were of analytical grade.

Table 1. The table of specific parameters of 6 kinds macroporous resins.

Resin type	Polar	Colour	Particle diameter(mm)	Specific surface area(m^2/g)	Average pore diameter(nm)
ADS-7	Polar	Pale yellow	0.3—1.25	100-120	25-30
ADS-17	Medium polarity	Milky white	0.3—1.25	90—150	25—30
DM-130	Weak polarity	Milky white	0.3—1.25	500-550	90—100
AB-8	Weak polarity	Milky white	0.3—1.25	480—520	13—14
D101	Non polarity	Milky white	0.3—1.25	600—700	10—12
X-5	Non polarity	Milky white	0.3—1.25	500—600	29—30

2.2 Methods

2.2.1 Preparation of loquat leaf extract

Approximately 1 kg of loquat leaf dried powder was subjected to the extraction by 95% industrial alcohol with the material-liquid ratio of 1:10 for three repeats. The extract was condensed until no taste of alcohol under the reduced pressure condition. The appropriate amount of distilled water was added to the crude extract for standing and the formation of a large amount of precipitate. The mixture was subjected to filtration to obtain the supernatant and precipitate. The supernatant was condensed to a certain concentration as the sample solution for the purification of flavonoids, and the sample solution was stored at 4 °C for use.

2.2.2 Establishment of standard curve and determination of total flavonoids in samples

The standard curve of total flavonoid content was established according to the previous methods from Hong *et al* [8]. Rutin was used as the standard sample. The sodium nitrite-aluminum nitrate method was applied. The stock standard solution of rutin at the appropriate amount was diluted by 30% ethanol to 1 mL. Then, 60 μL of 5% sodium nitrite was added to the diluted rutin standard solution with the standing for 5 min. Similarly, 60 μL of 10% aluminum nitrate was added with the standing for 6 min, and 400 μL of 1.0 mol/L sodium hydroxide and 480 μL of 30% ethanol were sequentially added with the standing for 15-20 min. Finally, the optical density (OD) values of rutin standard solutions with various concentrations at 510 nm were determined and the standard curve was established according to the OD values from gradient concentrations. The tested samples were diluted to a certain concentration, and their absorbance was measured with three repeats in the same manner. The contents of total flavonoids in tested samples were calculated according the standard curve of rutin.

2.2.3 Screening of optimal macroporous resin

Static adsorption and desorption of macroporous resins: Totally 1 g of pretreated ADS-7, ADS-17, DM-130, AB-8, D-101 and X-5 resins was mixed with 20 mL of sample solution, respectively. The adsorption was conducted at 25 °C water bath for 6 h. Then, the contents of total flavonoids were determined to calculate the absorption amount of flavonoids. The resin was washed by water and subjected to desorption in 95% ethanol. The contents of flavonoids in desorption solution was determined to calculate the desorption rate. All experiments are in triplicate.

Dynamic adsorption and desorption of resins: Totally 14 mL of resin suspension was packed into the column (10 mm * 200 mm) through wet-packing method. The sample loading and absorption were conducted with sample solution of 50 mL and flow rate of 2 BV/h, and column washing with column-washing water of 3 BV and flow rate of 2 BV/h. The contents of total flavonoids in effluent and distilled water eluent were determined to calculate the adsorption rate. The desorption was conducted using 95% ethanol for washing column at the flow rate of 1.5 BV/h. The contents of total flavonoids in the recycled 95% ethanol was determined to calculate the desorption rate.

2.2.4 Observation of static behavior from selected resin

A total of 1 g of selected resin was sequentially subjected to pre-treatment, the addition of sample solution (20 mL) at various concentrations. The absorption was conducted in 25 °C water bath for 6 h. Then, the contents of total flavonoids were determined to calculate the absorption amount and establish an isotherm curve of static absorption at 25 °C.

Similarly, a total of 1 g of selected resin was sequentially subjected to pre-treatment, the addition of sample solution (20 mL). The absorption was conducted in 25 °C water bath, and the sample collection was conducted with the time interval of 1 h. Then, the contents of total flavonoids were determined to calculate the absorption amount and establish an absorption dynamic curve.

Moreover, 1 g of resin with the adsorption of flavonoids was subjected to the gradient desorption through 20 mL of ethanol at various concentrations. The contents of total flavonoids in desorption solution were determined to calculate the desorption rate.

2.2.5 Observation of dynamic behavior from selected resin

Totally 20 mL of resin suspension was packed into the column (10 mm*300 mm) through wet-packing method. The sample solution was loaded to the column at the flow rate of 0.5 BV/h. The effluent was collected at each fraction of 10 mL and the contents of total flavonoids in effluent solutions were determined to draw a leakage curve.

Totally 20 mL of resin suspension was packed into the column (10 mm*300 mm) through wet-packing method. The sample solution was loaded to the column at the flow rate of 0.5 BV/h until the termination of leakage. Then, the elution was conducted by 5-70% ethanol in a gradient elution manner. The contents of total flavonoids in eluent solutions were determined to draw an elution curve.

2.2.6 Determination of purification efficiency and stability of selected resin

DOI: 10.1051/bioconf/20170803010

Totally 20 mL of resin suspension was packed into the column (10 mm * 300 mm) through wet-packing method. According to the optimal sample loading, impurity cleaning and elution process, the eluent was collected, concentrated and dried to obtain a purified product. The appropriate amount of the purified product was dissolved in methanol to determine the content of total flavonoids and to calculate the purity and purification factor of purified total flavonoids.

In addition, the selected resin without regeneration was used to repeat above elution process to determine the recovery rate of total flavonoids, which can be used for evaluating the number of repeated use.

2.2.7 Antioxidant activity of total flavonoids from loquat leaves

DPPH method: According to the method from Benzuie and Strain [21], the sample solutions before and after purification by macroporous resin were diluted as 5 concentration gradients. Then, 0.2 mL of each diluted sample was mixed with 1.98 mL of DPPH ethanol solution for 30 min reaction. The OD values of the tested samples with various concentrations after reaction were determined to calculate DPPH radical scavenging rate, as expressed in EC_{50} (mg powder/mL extract, scavenging rate at the extract concentration of 50%).

FRAP method: According to the method from Benzuie and Strain [21], the purified flavonoids solution was subjected to the 50-80 fold dilution. Then, 0.2 mL of diluted flavonoids solution was mixed with 1.8 mL of TPTZ working solution for the reaction at 37 °C for 10 min. The OD value of the mixture after reaction at 593 nm was determined to calculate the total reducing power, as expressed in the equivalent $FeSO_4$ μmoL/g dried power.

3 Results and discussion

3.1 Standard curve of total flavonoids

The sodium nitrite-aluminum nitrate method was applied to establish the standard curve using rutin as the standard sample, the absorbance value from rutin at various concentrations as the Y-axis, and absolute content of rutin (mg) as the X-axis. The linear regression equation was determined as $y = 5.82x - 0.0056$ ($R = 0.9993$).

3.2 Screening for resin type

3.2.1 Static adsorption and desorption of resins

When the initial concentration of total flavonoids from loquat leaves was 8.33 mg/mL, the results for dynamic absorption and desorption of 6 kinds of resins were shown in Table 2. Among these resins, resin ADS-7 revealed the highest absorption rate, which reached up to 79.5%; however, the desorption rate of resin ADS-7 was only 10.09%, and it is difficult to complete the recovery of flavonoids. Therefore, resin ADS-7 is not suitable for the purification of flavonoids. Based on the analysis of static absorption data, the absorption rates of resins X-5 and DM-130 were just a little lower than that of resin ADS-7, but resin X-5 revealed the significantly higher desorption rate than resin DM-130. Taking the comprehensive consideration, resin X-5 should be the best one.

3.2.2 Dynamic adsorption and desorption of resins

The dynamic and static absorption and desorption of 6 kinds of resins exhibited the similar results, and the absorption and desorption rates were shown in Table 2. The highest absorption rate and the lowest desorption rate of resin ADS-7 indicated its large dead absorption and low recovery for flavonoids so that it is not suitable for the purification of flavonoids from loquat leaves. On the other hand, resin X-5 revealed a higher adsorption rate and desorption rate. Based on the comprehensive analysis of dynamic and static absorption and desorption efficiency, resin X-5 among these 6 kinds of resins is the best resin for the purification of flavonoids from loquat leaves.

Table 2. The results of static/dynamic absorption and desorption of 6 kinds of resins.

Resin type	Static adsorption rate (%)	Static resolution ratio (%)	Dynamic adsorption rate (%)	Dynamic resolution ratio (%)
ADS-7	79.53±0.74 [a]	10.09±0.21 [d]	68.54	29.65
ADS-17	24.73±0.95 [d]	63.05±2.93 [c]	50.07	93.98
DM-130	35.85±0.43 [b]	70.46±2.62 [b]	45.25	109.81
AB-8	26.96±0.58 [c]	68.09±0.75 [b]	56.05	105.61
D101	27.27±0.29 [c]	70.72±1.74 [b]	60.63	105.24
X-5	36.90±0.47 [b]	82.48±00.81 [a]	66.37	104.21

Note: The value in the table was were presented as mean ± standard error, and the different superscript letters indicate that there had a significant different in the same column ($p<0.05$).

3.3 Performance determination of resin X-5

3.3.1 Static behavior of resin X-5

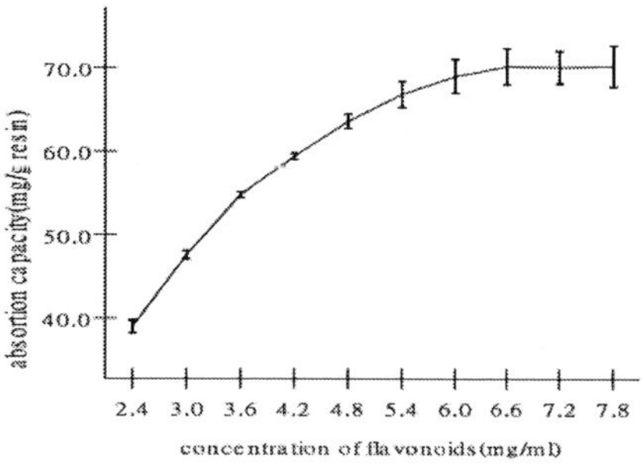

Figure 1. Adsorption isotherm of X-5 resin.

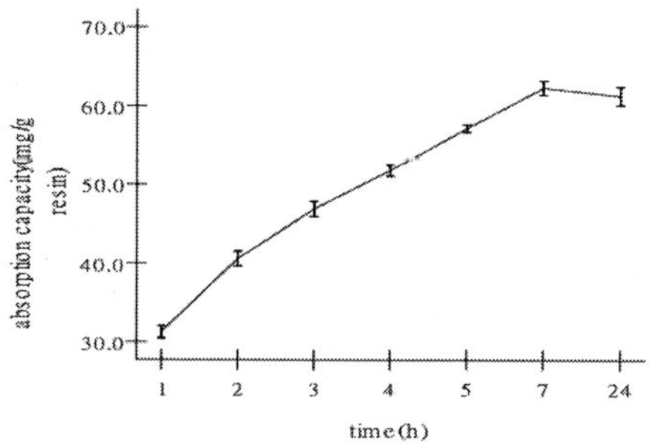

Figure 2. Dynamic curve of X-5 static adsorption.

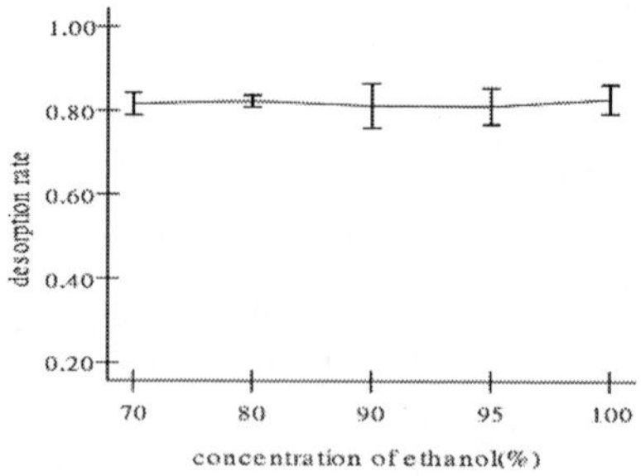

Figure 3. Static desorption curve of X-5 resin.

The static absorption performance of resin X-5 at 25 °C was determined. As shown in Figure 1-3, the concentration of total flavonoids was 5.4 mg/mL, static adsorption of resin X-5 reached up to equilibrium; meanwhile, the adsorption reached up to the saturation status at the absorption time of 6-7 h. In contrast, the full desorption of flavonoids could be achieved at the ethanol concentration of 70%.

3.3.2 Dynamic behavior of resin X-5

The flavonoids solution at the concentration of 5.45 mg/mL was loaded on resin column. The effluents were collected at the each fraction of 10 mL. The dynamic absorption performance of resin X-5 was determined, as shown in Figure 4. The leakage point was reached when sample solution was loaded at the volume of 30 mL, namely 1.5 BV. In addition, the saturation was reached when sample solution was loaded at the volume of 300 mL, namely 15 BV. Figure 5 demonstrated the gradient elution curve of ethanol. An elution peak was observed at ethanol concentration of 35-40%, and a complete elution was achieved at ethanol concentration of 65%. In contrast, no elution of flavonoids was observed if ethanol concentration was lower than 15%. Therefore, 15% ethanol was chosen for the clearance of impurities and 65% ethanol was chosen for the elution of flavonoids.

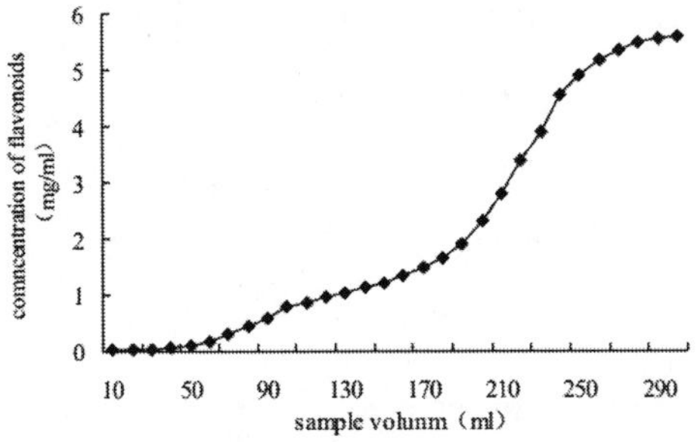

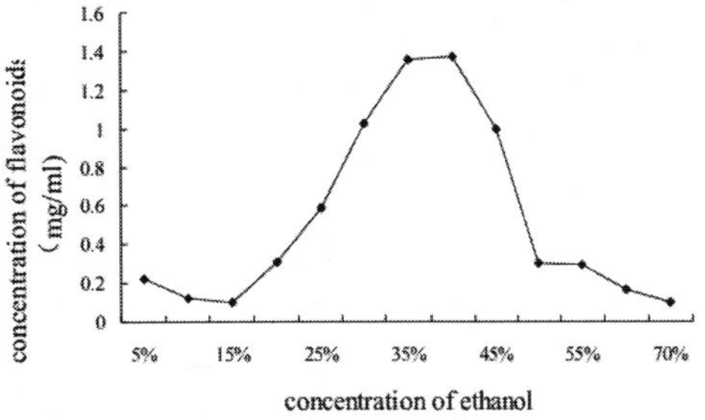

Figure 4. Leakage curve of X-5 dynamic adsorption **Figure 5.** Gradient elution curve of X-5 dynamic adsorption.

3.4 Determination of purification efficiency and column efficiency for optimal process

Totally 20 mL of X-5 resin suspension was pocked into the column through wet-packing method. The sample solution with the initial flavonoids concentration of 5.47 mg/mL was loaded, and the loaded volume was 1.5 BV (30 mL). After the loading of sample solution was completed, 15% ethanol was used for the clearance of impurities and 65% ethanol was used for the elution of flavonoids. The purified product was obtained through elution, condensation and drying, which was used for the determination of purification efficiency. Results indicated that the recovery rate of total flavonoids was 81.08%, the purity of total flavonoids was 82.20% and the purification factor was up to 2.14.

Table 3. Effect of repetition loading on the stability of the column.

Loading times	1	2	3	4	5	6
Adsorption rate(%)	95.28	95.09	93.94	90.70	87.86	84.36
Loss of flavonoids(%)	9.26	9.82	10.23	11.43	12.69	13.99
Recovery rate of flavonoids(%)	79.96	73.25	68.41	62.07	55.73	46.78

Continuous sample loading, impurity cleaning and elution were conducted to evaluate the column efficiency. As shown in Table 3, the repeated use of resin column could result in a gradual decrease in absorption capacity and recovery rate of flavonoids. If the number of continuous sample loading was 3, the recovery rate of flavonoids revealed the reduction by 10%. Therefore, the repeat use of resin column was more than 3 times, the clearance of impurities and the regeneration of resin column are required to improve the absorption capacity of resin to total flavonoids.

3.5 Antioxidant activity of total flavonoids from loquat leaves

The ethanol solution of DPPH as the standard sample was diluted to a series of concentrations. The OD values of diluted DPPH standard solutions were determined at 517 nm. The DPPH standard equation was obtained to be $y = 30.951x + 0.0030$ ($R^2 = 0.9993$) using DPPH concentration (μg/mL) as the X-axis and absorbance value as the Y-axis. In addition, 0.2 mL of $FeSO_4$ solution with various concentrations was added to 1.8 mL of TPTZ working solution for determining the absorbance at 593 nm. The FRAP standard curve was obtained to be y = 16.212x - 0.0608 (R^2 = 0.9940) using ferrous sulfate content (μmol) as the X-axis and absorbance as the Y-axis.

The contents and antioxidant performance of crude flavonoids and purified flavonoids were determined, as shown in Table 4. Subjected to the purification by macroporous resin, the contents of purified flavonoids exhibited a significant increase. Meanwhile, the FRAP value also revealed a significant increase, suggesting the significant enhancement of total reducing power. In addition, the EC_{50} of DPPH clearance revealed a significant decrease, suggesting the enhanced scavenging capacity to free radicals. All of these results indicated that the content of total flavonoids in loquat leaves was consistent with its antioxidant capacity; thereby flavonoids play important roles in anti-oxidation.

Table 4. The contents and antioxidant performance of crude flavonoids and purified flavonoids.

Items	Flavonoid content (g/g DW)	FRAP ($FeSO_4$ mmol/g)	DPPH (EC_{50} mg/g)
purified flavonoids	0.839 ± 0.012^{aA}	4.559 ± 0.009^{aA}	0.360 ± 0.02^{a}
Crude flavonoids	0.385 ± 0.005^{bB}	2.771 ± 0.002^{bB}	0.782 ± 0.12^{b}

Note: The same column of different lowercase letters indicate significant difference ($p<0.05$), and capital letters indicate extremely significant difference($p<0.01$).

4 Discussions

In the present study, the preparation of crude extract from loquat leaves was subjected to cold extraction through 95% industrial ethanol with three repeats, condensation under reduced pressure, and precipitation through sitting down. The precipitate and filtrate has high contents of total terpenes and flavonoids. The yields of flavonoids and total terpenes in precipitate are 1.22% and 3.57%, respectively; the yields of flavonoids and total terpenes in extract are 3.28% and 1.90%, respectively. In addition, the yields of flavonoids and total terpenes in filtrate before condensation are 4.05% and 1.90%, respectively. However, the content of flavonoids in extract after drying was decreased, and the content of total terpenes was no change, suggesting that flavonoids may be unstable due to oxidation. Therefore, the water-soluble filtrate after extraction and condensation should be the major sample solution for purifying total flavonoids from loquat leaves.

The flavonoids from loquat leaves not only have antioxidant activity, but also have anti-inflammatory activity [22]. In this study, the extract from loquat leaves has the enrichment of flavonoids after cold ethanol extraction. The purification process of the extract with enriched flavonoids through macroporous resin was optimized. Based on the experiments of static and dynamic absorption and desorption, resin X-5 was the optimal one; meanwhile, the optimal extraction process for total flavonoids from loquat leaves has been established through analyzing the static and dynamic absorption and desorption behaviors of resin X-5. Experimental results indicated that resin X-5 has better absorption and desorption capacity for total flavonoids from loquat leaves when compared with other 5 kinds of macropourous resins. When the concentration of total flavonoids reached up to 5.4 mg/mL at 25 °C, the maximum absorption amount of resin X-5 for total flavonoids was up to 62.32 mg/g. Subjected to the impurity clearance by 15% ethanol and elution by 65% ethanol during the optimal purification process of macroporous resin, the purity of total flavonoids from loquat leaves can reach up to 82.20% with the purification factor of 2.14, which reveals the higher purity than other reported purification methods[23-24]. Moreover, the purification through macroporous resin could result in the obvious enrichment of flavonoids, thereby greatly improving the antioxidant capacity, and revealing the important roles of flavonoids in anti-oxidation.

References

1. S.M. Lu, W.H.Wang, M.Y. Zheng, Chin. J. Biol. Progr. **6**, 453-458 (2013).
2. S.Q. Lin, *Acta. Hortic. 750*, 37-44 (2007).
3. J.Y. Lin, C.Y.Tang, *Food Chem*. **107**, 1587–1596(2008).
4. H. Tan, S.Furuta, T. Nagata, K.Ohnuki, T.Akasaka, B.Shirouchi, M.Sato, R. Kondo, K. Shimizu, *J. Agric. Food Chem*. **4**, 836-41 (2014).
5. D.Dufour, A.Pichette, V. Mshvildadze, M.E. Bradette-Hébert, S.Lavoiem, A. Longtin, C. Laprise, J.Legault, *J. Ethnopharmacol*. **111**, 22–28 (2007).
6. J.B.Wu, Y.H.Kuo, C.H.Lin, H.Y.Ho, C.C. Shih, *J. Agric. Food Chem*. **62**, 10717-10726 (2014).
7. J.H.Ju, L. Zhou, G. Lin, D. Liu, L.W. Wang, J.S. Yang, *Chin. J. Chin. Pharm. J*. **38**, 752-757 (2003).
8. Y. P.Hong , S. Q. Lin, Y. M. Jiang, M. Ashraf, *Plant. Food Hum. Nutr*. **63**, 200–204(2008).
9. Y. Huang, J. Li, Q. Cao, S.C. Yu, X.W. Lv, Y.Jin, L. Zhang, Y.H. Zou, J.F. Ge, *Life Sci*. **78**, 2749–2757 (2006).
10. H.A.Jung, J.C.Park, H.Y.Chung, J.Kim, J.S. Choi, *Arch. Pharm. Res*. **22**, 213–218(1999).
11. G.I.Nonaka, I.Nishioka, *Chem. Pharm. Bull*. **30**, 4268–4276 (1982).
12. H.J. You, C.Y. Choi, J. Y. Kim, S.J. Parkk, K.S. Hahm, H.G. Jeong, *FEBS Lett*. **509**, 156-160 (2001).
13. Y.P. Hong, S.H. Huang, H.Y. Cao, S.Q. Lin, Z.L. Liu, *Chin. J. Acta Horticulturae Sinica*. **36**, 898 – 904 (2009).
14. D.H. Li, Y.Wang, Y.S. Lv, J.H.Liu, L. Yang, S.K.Zhang, Y.Z. Zhuo, *Biomed. Res. Int*. **2015**, 1-9 (2015).
15. R.Yang, D. Meng, Y. Song, J. Li, Y. Zhang, X. Hu, Y. Ni, Q.Li, *J. Agric. Food Chem*. **60**, 8450-8456 (2012).
16. Z. Yu, L. Liu, Y. Xu, L. Wang, X. Teng, X. Li, J. Dai, *Carbohydr Polym*. **132**, 180-186 (2015).
17. M. Zhuang, M. Zhao, L. Lin, Y. Dong, H. Chen, M. Feng, W. D. Sun, G. Su, *Food Chem*. **190**, 338-44 (2016).
18. Y. Dong, M. Zhao, W. D. Sun, M. Zhuang, H. Chen, M. Feng, L. Lin, *Food Chem*. **168**, 538-545 (2015).
19. J. Kim, M. Yoon, H. Yang, J. Jo, D. Han, Y.J. Jeon, S. Cho, *Food Chem*. **162**, 135-142 (2014).
20. P. Zhao, C. G. Qi, Wang, X. Dai, X. Hou, *J. Chromatogr B Analyt Technol. Biomed. Life Sci*. **1007**, 8-17 (2015).
21. I.F. Benzuie, J.J. Strain, *Anal. Biochem*. **239**, 70-76 (1996).
22. K. Maher, B.A. Yassine, B. Sofiane, *Afr. Health. Sci*. **15**, 613-20 (2015).
23. Y. Liu, X.T. Zhu, *Chin. J. Exp. Traditional Med. Formulae*. **18**, 55-57 (2012).
24. H. Lv, J.H.Teng, J. Chen, L. Ma, B.R. Ren, W.L. Li, *Chin. J. Mod. Appl. Pharm*. **31**, 40-44 (2014).

BIO Web of Conferences **8**, 03011 (2017)
ICMSB2016

DOI: 10.1051/bioconf/20170803011

Modification and characterization of an aptamer-based surface plasmon resonance sensor chip

Junpeng Tan[1,2,3], Bin Hao[1,2], Zhicheng Liu[1,2], Fengbo Bai[4], Ruiqin Yang[3,a] and Hongxia Hao[1,2,b]

[1]*Collaborative Innovation Center of Judicial Civilization, China*
[2]*Key Laboratory of Evidence Science, China University of Political Science and Law, Ministry of Education, Beijing 100088, China*
[3]*People's Public Security University of China, Beijing 100038, China*
[4]*Hong TeCH(Beijing)Co.,Ltd, Beijing, 100021,China*

Abstract. Recently, aptamer-based surface plasmon resonance (SPR) sensors have become increasingly popular due to their high specificity, high sensitivity, real-time detection capabilities, and label-free features. The core component of an aptamer-based SPR sensor is a chip. This paper presents the modification steps and the characterization results of a sensor chip for the construction of a 2, 4, 6-trinitrotoluene-targeted, aptamer-based, SPR sensor. After cleaning the aptamer-based SPR sensor chip, polyethylene glycol (PEG) with functional thiol groups at one end was added to the chip surface by Au-S covalent bonds to form a self-assembled film. Then, the carboxyl groups at the other end of PEG and the carboxyl groups of trinitrophenyl-glycine (TNP-Gly) were activated and connected via ethylenediamine (EDA). This effectively completed the chip's modification. During the modification process, relevant experimental conditions were optimized. The chip's surface elements, as well as their chemical states, were characterized by X-ray photoelectron spectroscopy (XPS). The results, outlined in the following study, demonstrate that this modification of an aptamer-based SPR sensor chip adhered to normative expectations. Thus, the modification process proposed here establishes an important foundation for subsequent study of TNT detection.

1 Introduction

As early as 1990, Tuerk and Ellington [1] independently screened RNA oligonucleotide aptamers out from oligonucleotide libraries. Aptamers can be defined as randomly obtained, single-stranded

[a] Corresponding author: Rqyang1971@yahoo.com
[b] Corresponding author: haohx@126.com
We are grateful to Humanities and social science research program of China university of politics and law (10ZFQ82009) and Academician Foundation of the Ministry of Public Security of the People's Republic of China (No. 2011-23214203, 2011-23215243) , Beijing Municipal Education Commission University Science and Technology Park Construction Project (2011-08111603) and Program for Young Innovative Research Team in China University of Political Science and Law (14CXTD04, 16CXTD05) for their financial supports.

oligonucleotides that can bind specifically to target proteins and other small molecules through the systematic evolution of ligands by exponential enrichment (SELEX) technique [2, 3]. Aptamers can be classified as DNA, RNA, or even XNA, and their molecular weights are relatively small compared to common antibodies [4]. Aptamers have been applied and developed in biochemistry, sensor development, protein research, drug detection, and numerous other areas for many years [5-10].

Similar to the operation of antibodies, the screening targets of aptamers include all kinds of areas, from inorganic ions, amino acids, and organic dyes to proteins, drugs, and explosives. Based on SELEX technology, we can screen aptamers for any target molecules [11]. The affinity between an aptamer and its target molecule is exquisitely high, with a usual dissociation constant somewhere between 10^{-12} mol/L and 10^{-9} mol/L [12], much higher than common ligands. Moreover, aptamers are capable of distinguishing subtle differences between different ligands, and can even distinguish a single protein molecule's different configurations and/or functional states. Compared to the synthesis of antibodies, the preparation of aptamers only relies on in vitro screening of SELEX technology. Not only is this more cost-effective, but it also has other advantages, such as smaller molecule weights than antibodies. For this reason, it can be used for specific recognition of small molecules. In addition, aptamers show high stability during repeated screening, and they can be recycled in various external environments.

Due to their high affinity with target molecules, as well as their strong specificity, aptamers are potentially highly usable in fields such as analytical chemistry, gene detection, and drug development. However, since they cannot generate detection signals when bound to target substances, aptamers usually need the help of surface plasmon resonance (SPR), fluorescence detection [13], electrochemical detection [14], Raman spectroscopy [15], or colorimetry [16] to carry out signal conversion and detection. Among these technologies, a combination of an SPR sensor with aptamers for the detection of targets has recently become a research priority.

SPR is a physical and optical phenomenon. By measuring a metal surface's refractive index fluctuations, SPR can be used to study the optical properties of different materials [17]. Figure 1 shows the SPR sensor's basic principles [18]. When a polarized light beam hits the surface of a prism in a specific range of angles, an interface plasmon wave is generated between the prism's surface and the metal film (gold or silver). When the propagation constant of the incident light wave matches the propagation constant of the interface plasmon wave, the metal film's free electrons resonate (in other words, the surface plasmon resonates). During our analysis, we fixed a layer of a self-assembled film to the sensor chip's surface; the sample then flowed over the chip's surface. If there are molecules in the sample that can potentially bind to the self-assembled film, the refractive index of the gold film surface changes, leading to additional changes in the SPR resonance angle. The detection of these changes provides particular information, such as the analytes' concentration, affinity, kinetic constants, and specificity [19-23].

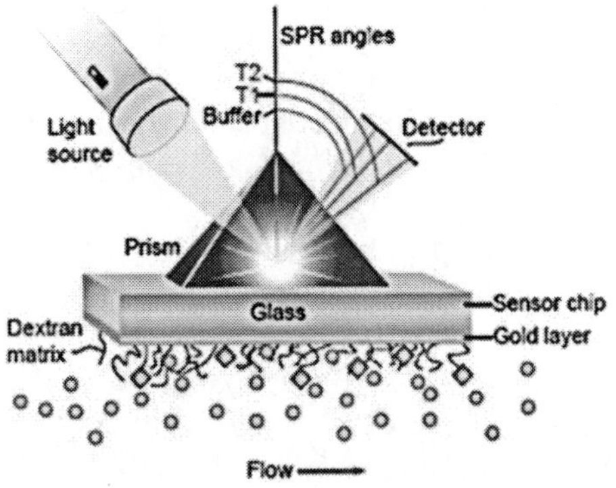

Figure 1. Schematic view for the principle of SPR sensor.

Based on the methodology of detecting 2, 4, 6-trinitrotoluene (TNT) with an aptamer-based, SPR sensor chip, our study focuses on the sensor's core, which was the modification and characterization of a trinitrophenyl-glycine (TNP-Gly) chip. In our process, we took advantage of the high specificity of aptamers and the high sensitivity of SPR, constructing an SPR sensor that could detect TNT directly and rapidly. Due to the stability of Au-S bonds [24, 25], TNP-Gly with carboxyl groups effectively and stably bound to the gold film surface with the help of modified polyethylene glycol (mPEG-SH) and the activation of 1-(3-dimethylaminopropyl)-3-ethylcarbodiimide hydrochloride/ N-hydroxylsuccinimide (EDC/NHS). Since polyethylene glycol (PEG) contains bifunctional groups (HS-PEG7-OH and HS-PEG7-COOH), the functional –SH groups were coupled to the SPR chip surface; NHS/EDC was used to activate the –COOH at the other end, and via amide reactions with ethylenediamine (EDA), the TNP-Gly with carboxyl groups was connected to the chip's surface. In this way, we were able to complete TNP-Gly chip synthesis. For the characterization process, an X-ray photoelectron spectroscopy (XPS) was used to perform quantification analysis during each stage. Figure 2 presents the basic principles and synthesis methods.

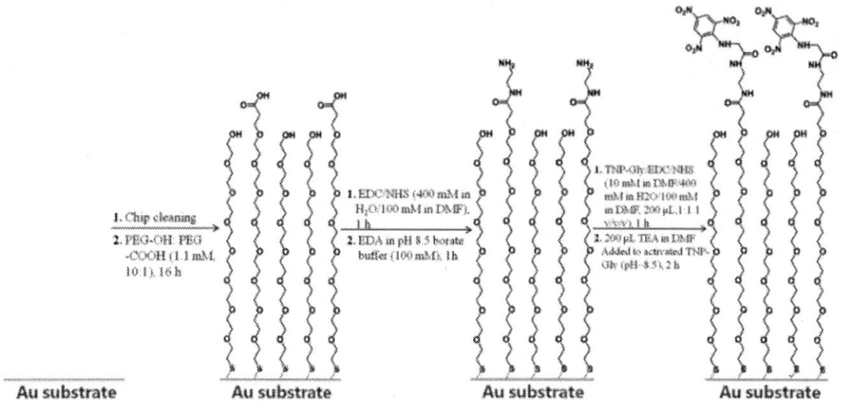

Figure 2. The basic principles and synthesis methods of the SPR sensor chip.

2 Experiment

2.1 Materials and instruments

The TNP-Gly we used was obtained from Research Plus (US). Thio PEG Hydroxyl (HS-PEG$_7$-OH) and thio PEG carboxylic acid (HS-PEG$_7$-COOH) were obtained from Ploypure (Norway). N-hydroxysuccinimide (NHS) and 1-(3-dimethylaminopropyl)-3-ethylcarbodiimide hydrochloride (EDC•HCl) were provided by from Sinopharm Chemical Reagent Co., Ltd. Sodium dodecyl sulfate (SDS) and triethylamine (TEA) were obtained from Shanghai Anpel Laboratory Technologies Inc. Ethylenediamine (EDA) and N, N-dimethylformamide (DMF) were obtained from Tianjin Fuchen Chemical Reagents Factory. Methanol, ethanol, and acetone were provided by Beijing Chemical Reagent Company.

Our X-ray photoelectron spectroscopy analyzer (XPS, ESCALAB 250Xi) was produced by Thermo Fisher Scientific Inc. The plasma cleaner (PDC-32G-2) was obtained from Mycro Technologies Co., Ltd. The shaker (cyclotron oscillator HY-5A) was from Ronghua Instrument Manufacturing Co., Ltd. The vortex (2E-161240) was manufactured by Scientific Industries Inc., US. The base for our aptamer-based, SPR sensor chip was made of BK7 optical glass (2.5 cm × 2.0 cm × 0.1 cm), with one side coated with 2 nm of chromium film and (afterwards) with 45 nm of gold film using a high vacuum coating method.

DOI: 10.1051/bioconf/20170803011

2.2 Chip cleaning

Our method for the preparation of 1% SDS washing solution was as follows. a sample of 10 g SDS was added to a 100 mL beaker and diluted to 1 L in a volumetric flask. Cleaning the chip with 1% SDS solution: the gold-coated chip (by vacuum deposition) was placed in a glass dish, with the gold film side facing up. Then an appropriate amount of 1% SDS solution was added to soak and wash the chip three times by hand-shaking. The chip was then soaked in the SDS solution and shaken in a shaker at 45 r/min for 20 min. Afterwards the chip was removed with a tweezer and rinsed off any residual SDS solution with deionized water, and dried with nitrogen. The chip was then placed in a clean glass dish with the gold film facing up, soaked with acetone, and washed by hand-shaking for another three times. Soaked in acetone, the chip was shaken at 45 r/min for 3 min. Afterwards acetone was removed and ethanol was added to soak and wash the chip by hand-shaking for an additional three times. Soaked in ethanol, the chip was shaken at 45 r/min for 3 min. Then ethanol was removed and methanol was added to soak and wash the chip by hand-shaking for another three times. Soaked in methanol, the chip was shaken at 45 r/min for 3 min. After cleaning, the chip was removed with a tweezer and dried with nitrogen. The dried chip was cleaned further in a plasma cleaner at medium-low level for 60 s. Finally it was sealed and stored.

2.3 Modification of the chip

A sample of 10 μL HS-PEG-OH and 10 μL HS-PEG-COOH were respectively added to a 2 mL microtube. A volume of 1 mL deionized water was added into each tube and properly shaken, thus the stock solutions were prepared. A volume of 3.529 mL ethanol was added to a glass dish; 421 μL HS-PEG-OH stock solution and 50 μL HS-PEG-COOH stock solution were added subsequently to make a molar ratio of 10:1 for PEG-OH: PEG-COOH. The stored chip was soaked into the above solution with the gold film facing up. After sealed, it was shaken at 45 r/min for 16 h to allow complete contact and reaction between the chip and PEG.

During the reaction, a sample of 0.6901 g EDC was dissolved in 3 mL deionized water and 0.1036 g NHS was dissolved in 3 mL DMF, to make a molar ratio of EDC: NHS = 4:1 later on. After 16 h, the SPR chip in the glass dish was removed and first rinsed with ethanol, then cleaned with an ultrasonic cleaner for 5 s, and finally dried with nitrogen. In a clean glass dish, 1.333 mL deionized water, 1.333 mL EDC stock solution and 1.333 mL NHS stock solution were added, then the dried chip was soaked in and shaken at 45 r/min for 1 h.

During the EDC/NHS activation period, EDA stock solution was prepared: in 3.737 mL borate buffer (pH 8.5), 13-14 drops of EDA reagent were added dropwise. After the activation reaction, the SPR chip was removed, rinsed with deionized water, and dried with nitrogen. In a new glass dish, a 500 μL EDA stock solution and 3.5 mL borate buffer (pH 8.5) were combined, then the above dried chip was soaked in this solution and shaken at 45 r/min for 1 h. Afterwards, the SPR chip was once again removed, rinsed with deionized water, and dried with nitrogen.

During the reaction between the chip and the EDA solution, a TNP-Gly stock solution was prepared: a sample of 0.0056 g TNP-Gly was dissolved in 400 μL DMF. Then 200 μL of the above TNP-Gly solution was transferred to a clean glass dish, and combined with 200 μL EDC stock solution, 100 μL NHS stock solution, and 100 μL DMF successively. After mixed uniformly, it was shaken at 45 r/min for 1 h. The TEA solution was prepared as well: 16 drops of TEA were added to 400 μL DMF. After the TNP-Gly solution was shaken for 1 h, it was added with 200 μL TEA solution and mixed well to prepare the TNP-Gly stock solution. The 600 μL TNP-Gly stock solution was then transferred to cover the gold film surface of the prepared chip with a 1 mL transfer pipette. After kept still for 2 h, a plastic head dropper was used to remove all the TNP-Gly stock solution along the chip's edge. The chip was rinsed with deionized water, dried with nitrogen, and stored.

 DOI: 10.1051/bioconf/20170803011

3 Results and discussion

3.1 Chosen of chip modifiers

PEG is soluble in water, methanol, and benzene, while it is insoluble in ether and n-hexane. Due to its special structure, both ends of PEG can be modified by a variety of functional group types to meet different needs. Additionally, as a polymer, PEG's relative molecular weight varies depending on its polymerization degree. A variety of PEGs with different molecular weights have been widely applied to excipients in the pharmaceutical industry, stabilizers in the chemical industry, matrix in the textile and cosmetics industry, among others. In laboratory studies on the surface modification of gold nanoparticles, PEG is normally used to stabilize the electrons in the nanoparticle surface.

After an extensive review of the relevant literature, our lab decided to use the mPEG-SH type of PEG as our gold film surface modifier [26-29]. The primary reasons for this decision were as follows. (1) The gold film deposited by our group was a naked gold film with only three elements on its surface: C, O, and Au. In order to make the chip's surface TNP-Gly modifiable, it need to be modified and activated first. It was then more conducive to further modification with compounds such as EDC/NHS and EDA. (2) PEG is antigen-free, immunogenic-free, non-toxic, and biocompatible, qualities that are particularly important for subsequent binding with aptamers. (3) In physiological environments, PEG has no adsorption of or interactions with other biological molecules. Therefore, we estimated that it would not cause interference of elements from external environments when coupled to the gold film surface. This was crucial for the XPS analysis, since it ensured the high stability of gold film. (4) Based on PEG'S advantage of having two ends that can be linked to functional groups, our study selected mPEG-SH-type PEG (HS-PEG_7-OH and HS-PEG_7-COOH) with strong activity and with average molecular weights within the PEG400 category. Due to Au's high affinity with S on one end, it is coupled to the chip surface stably, leaving the other end with a functional group for stabilizing further modification. After testing HS-PEG_7-OH: HS-PEG_7-COOH with ratios of 5:2, 4:1, 6:1, 10:1, and 20:1, as Figure 3 shows, the 10:1 molar ratio was chosen to modify the gold film chip..

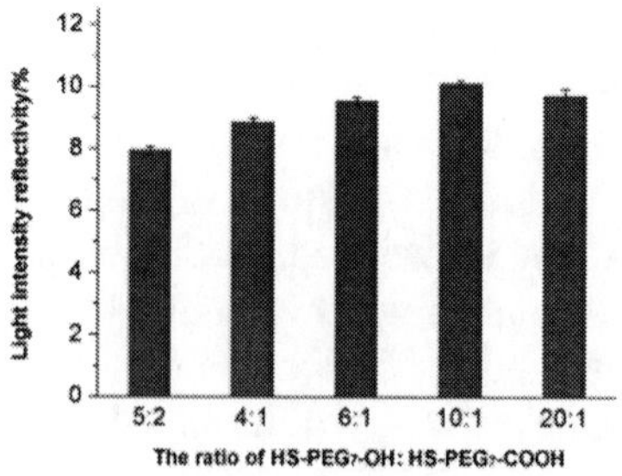

Figure 3. The SPR response signal of five ratios of HS-PEG_7-OH: HS-PEG_7-COOH.

3.2 Chosen of chip activator

EDC•HCl is soluble in water and often serves as a dehydrating agent to react with a carboxyl group. NHS is soluble in water and DMF, and it is usually used as a condensing agent to improve condensation efficiency in EDC reactions. These two elements are often used together to activate carboxyl groups to form acyl groups, and then to replace the hydrogens of amino groups to produce amides in amide reactions.

During the modification process of the chip's gold film, PEG with -COOH was used. Therefore, as Figure 2 shows, there were -COOH groups in the chip's surface after modification. The TNP-Gly to be modified, on the other hand, carried amino groups. Therefore, EDC/NHS was chosen to serve as

 DOI: 10.1051/bioconf/20170803011

the activator for the coupling reaction [30-33]. As shown in Figure 4, consulting the test results of different EDC/NHS ratios, the ratio of EDC: NHS = 4:1 was chosen for further experiments.

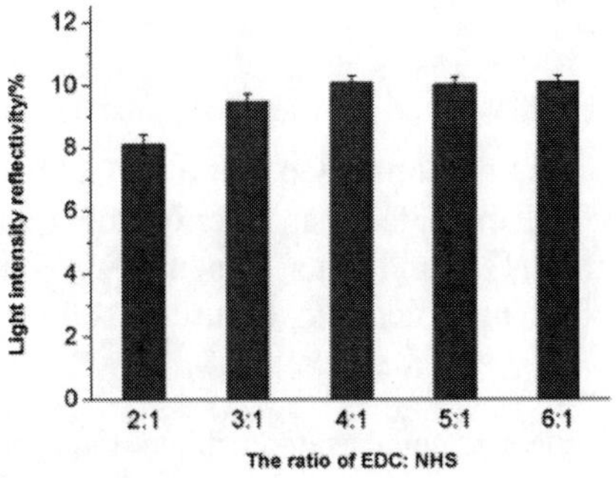

Figure 4. The SPR response signal of different ratios of EDC: NHS.

3.3 Chosen of reaction time and pH

Since the whole synthesis process was broken down into four stages, the chosen of reaction time and pH for each stage was very important. The reaction did not require a special temperature, therefore, room temperature was used throughout the whole experiment.

Stage I, as Figure 2 shows below, consisted of coating the chip with HS-PEG-OH and HS-PEG-COOH. Since the gold film had a strong S affinity, the formation of Au-S bonds did not require special external conditions except for a thorough reaction time. This study used a ratio of HS-PEG-OH: HS-PEG-COOH = 10:1 to bind to the naked chip. Due to the large area of the gold film (20 mm × 15 mm × 1 mm), there were less -SH groups from PEG than Au elements on the chip's surface. Thus, after further literature review the reaction time was adjusted from the usual 24 h to 16 h [34]. Moreover, to avoid high pH interference on the activation reaction during the introduction of EDC/NHS later on, the reaction between PEG and the chip was in an ethanol solution rather than the alkaline SDS solution usually used for PEG modification. In this experiment, SDS was only used during cleaning and deionization of the naked chip.

Stage II consisted of EDC/NHS activation and EDA introduction. In carboxyl-modified experiments, EDC usually achieves the best result between pH 4 and pH 6, due to its own structural characteristics. However, for the modified HS-PEG_7-OH and HS-PEG_7-COOH, a weak alkaline environment was required to keep PEG activity. Later, during the TNP-Gly linking process, a weak alkaline environment was also required to avoid the non-coupling of TNP-Gly. Therefore, the carboxyl activation for the gold film was performed under a neutral pH, with a reaction time of 1 h (slightly longer than usual). To avoid any external interference during TNP-Gly coupling, EDA was introduced. During the PEG modification and EDC/NHS activation process, the acyl groups were significantly excessive compared to the resulting TNP-Gly. Therefore, to avoid the interference of external compounds and groups due to excessive acryl, EDA with double amino groups were used to neutralize the acryl groups. To keep the reaction in a weak alkaline environment, $_{p}$H 8.5 borate buffer was used, with a reaction time of 1 h.

Stages III and IV included the preparation of the TNP-Gly stock solution and its binding to the modified chip. To ensure complete binding between TNP-Gly and the chip, EDC/NHS was added into the TNP-Gly stock solution. Following both the literature references and our own tests, the reaction time for the binding of TNP-Gly to the chip was set to 2 h.

3.4 Characterization and analysis of the sensor chip

During the synthesis of the chip and TNP-Gly, the unmodified chip, the PEG-modified chip, the EDC/NHS-activated chip, and the fully modified chip were all analyzed with XPS to record the binding of the modified process at each stage. Results are shown below.

3.4.1 Unmodified chip

The XPS analysis of the unmodified chip demonstrated that only C, O, and Au elements were detected on the chip's surface (see Table 1). This indicated that, after cleaning with SDS, organic solvents, and plasma cleaner, any interferences from N, S, and other elements were successfully removed. The chip was therefore completely clean before modification. Figure 5 shows the XPS spectra of the three elements mentioned above.

Table 1. Element contents of the unmodified chip.

Element	Peak Value	Content (%)
C_{1s}	285.03	30.27
O_{1s}	532.33	14.40
Au_{4f}	84.07	53.64

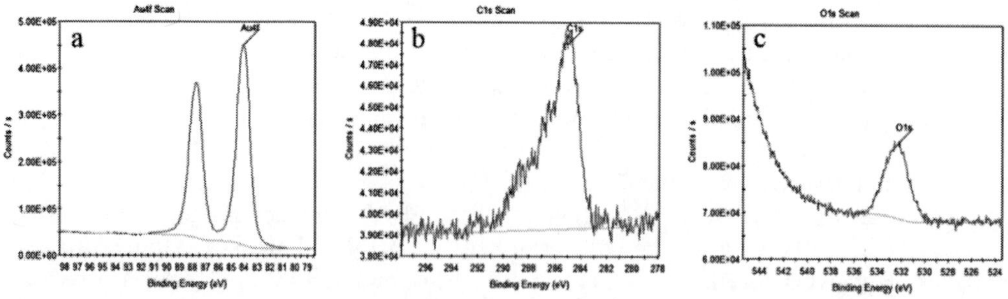

Figure 5. The XPS spectra of three elements of the unmodified chip (a:Au-SCAN, b: C-SCAN, c: O-SCAN).

3.4.2 PEG-modified chip

Table 2 shows the element contents from the XPS analysis of the PEG-modified chip. These results demonstrated that S appeared on the chip's surface, indicating the coupling of S elements to the gold film. Since the PEG we used contained -SH groups, PEG had formed stable Au-S bonds on the gold film surface (see Figure 6a). In Figure 6b it is clear that the element content of C was elevated compared to the unmodified chip, and that there was an O-C=O absorption peak at 288.5 eV and a C-O-C absorption peak at 286.7 eV. As Figure 6c shows, the element content of O was also increased compared to the unmodified chip, and there was a C=O absorption peak at 533 eV. Our results suggested the existence of -COOH groups on the gold film surface. Therefore, it can be concluded that HS-PEG_7-COOH and HS-PEG_7-OH were coupled to the chip's surface through Au-S bonds, and the surface of the gold film was successfully modified with PEG with -OH and –COOH groups, forming an effective Au-OEG_7OH:OEG_7COOH structure.

 DOI: 10.1051/bioconf/20170803011

Table 2. Element contents of the PEG-modified chip.

Element	Peak Value	Content (%)
C_{1s}	286.70	42.96
O_{1s}	532.91	23.11
Au_{4f}	84.01	31.05
S_{2p}	162.23	2.44
Zn_{2p3}	1022.34	0.45

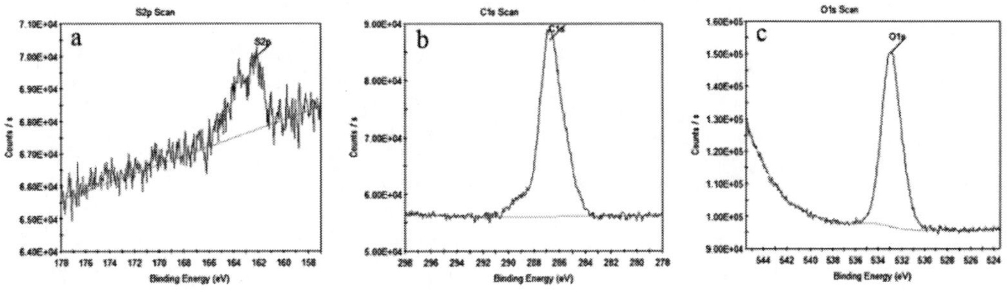

Figure 6. The XPS spectra of three elements of the PEG-modified chip (a:S-SCAN, b: C-SCAN, c: O-SCAN).

3.4.3 EDC/NHS-activated chip

The XPS analysis has been performed for the EDC/NHS-activated chip. Table 3 shows the element contents, while Figure 7 presents the XPS spectra of the S, N, C, and O. A new element N appeared during detection on the chip's surface, indicating successful coupling of N elements to the chip's surface. Moreover, there was a C-NH2 absorption peak at 400 eV but no O-C=O absorption peak at 288.5 eV, suggesting that the original –COOH on the chip underwent amino reactions with EDA after EDC/NHS activation. Thus, the original –COOH disappeared and only C-NH2 groups could be found on the chip's surface, forming an Au-OEG7CO-NH (CH2CH2) NH2 structure.

DOI: 10.1051/bioconf/20170803011

Table 3. Element contents of the EDC/NHS-activated chip.

Element	Peak Value	Content (%)
C_{1s}	286.76	36.99
O_{1s}	533.02	18.53
Au_{4f}	84.09	38.04
S_{2p}	162.29	2.76
N_{1s}	400.49	3.69

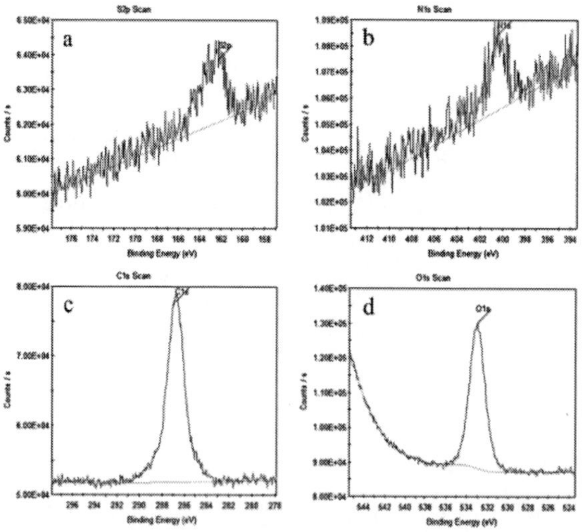

Figure 7. The XPS spectra of four elements of the EDC/NHS-activated chip (a:S-SCAN, b: N-SCAN, c: C-SCAN, d:O- SCAN).

3.4.4 Fully modified chip

The XPS analysis has been performed for the fully modified chip. Table 4 shows the element contents, which exhibited no significant change from the EDC/NHS-activated chip. Figure 8 presents the XPS spectra of the S, N, C, and O. There was a $-NO_2$ absorption peak at 406 eV and a $C-NH_2$ peak at 400 eV. The $-NO_2$ came from TNP-Gly and the C-NH2 came from both TNP-Gly and EDA. Our results indicated that $Au-OEG_7CO-NH(CH_2CH_2)N$-Gly-TNP formed on the chip's surface.

Table 4. Element contents of the fully modified chip.

Element	Peak Value	Content (%)
C_{1s}	286.68	38.26
O_{1s}	532.99	18.60
Au_{4f}	84.04	37.01
N_{1s}	400.54	3.41
S_{2p}	162.31	2.71

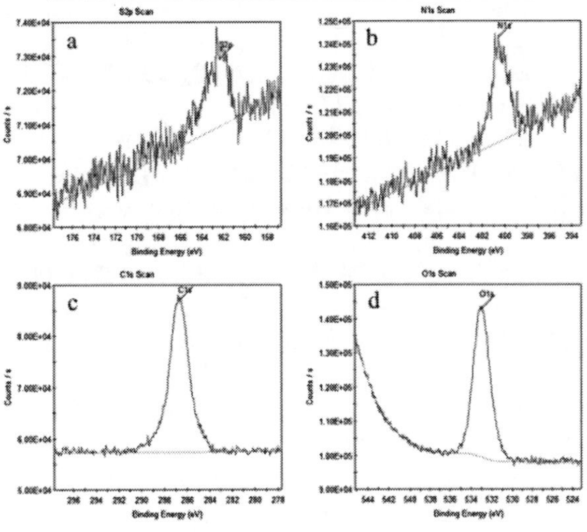

Figure 8. The XPS spectra of four elements of the fully modified chip (a:S-SCAN, b: N-SCAN, c: C-SCAN, d:O- SCAN).

4 Conclusion

Taking advantage of the stable binding between -SH and Au, as well as the PEG quality of having two ends that can bind to functional groups, this study used the mPEG-SH type of PEG that contains both -COOH and -OH groups to bind to a chip. Following EDC/NHS activation, this type of PEG connected to the amino-containing TNP-Gly to assemble the TNP-Gly chip. During our experimental design, the main difficulty we encountered was in selecting the appropriate ratio between HS-PEG-OH and HS-PEG-COOH. To solve this difficulty, the ratios of 4:1, 5:2, 7:3, 10:1, and 20:1 were tested with the XPS spectra at different ratios for each stage. Based on our own results and a comprehensive literature review, the ratio of 10:1 was chosen for further modification, due to its superior performance. After coupling with TNP-Gly, the XPS spectra showed an expected $-NO_2$ absorbance peak. The successful chip synthesis we present in this study establishes a solid foundation for further investigation of TNT detection.

References

1. A.D. Ellington, J.W. Szostak. Nature, **346**(6872), 8818-8822 (1990).
2. S Song, L Wang, J Li, et al. Trac-Trend Anal Chen, **27**(2), 108-117(2008).
3. W Sun, L.M. Du. Curr. Pharm. Design, **16**(20), 2269-2278(2010).
4. C. Tuerk, L. Gold. Science, **249**(4968), 505-510 (1990).
5. H.R. Jang, A.W. Wark, S.H. Baek, et al. Anal Chen, **86**(1), 814-819(2013).
6. P. Subramanian, A. Lesniewski, I. Kaminska, et al. Biosens Bioelectron, **50C**(24), 239-243(2013).
7. H. Bai, R. Wang, B. Hargis, et al. Sensors, **12**(9), 12506-12518(2011).
8. M.Y. Kim, S. Jeong. Nucleic Acid Therapeutics, **21**(3), 173-178(2011).
9. P. Janardhanan, C.M. Mello, B.R. Singh, et al. Talanta, **117**(24), 273-280(2013).
10. B. Prieto-Simón, J. Samitier. Anal Chen, **86**(3), 1437-1444(2014).
11. S. Tombelli, M. Minunni, M. Mascini, Biosens Bioelectron, **20**(12), 2424-2434(2005).
12. N. Chelyapov. Biochemistry, **45**(7), 2461-2466 (2006).
13. R. Nutiu, Y.F. Li. J Am Chen Soc, **125**(16), 4771-4778 (2003).
14. XL Zuo, SP Song, J Zhang, et al. JACS, **129**(5), 1042-1043 (2007).
15. J.W. Chen, J.H. Jiang, X. Gao, et al. Chem. Eur. J, **14**(27), 8374-8382(2008).
16. R. Jin, G. Wu, Z. Li, et al. JACS, **125**(6), 1643-1654(2003).
17. R.W. Wood. Proceedings of the Physical Society of London, **18**(1), 269-275(1902).
18. P. Englebienne, A.V. Hoonacker, M Verhas. Spectrosc-Int J, **17**(2-3), 255-273(2003).
19. S. Sonezaki, S Yagi, E Ogawa, et al. J Immunol Methods, **238**(238), 99-106(2000).
20. W. Lee, D.B. Lee, B.K. Oh, et al. Enzyme Microb Tech, **35**(6-7), 678-682(2004).
21. J. Homola, J. Dostálek, S. Chen, et al. Int J Food Microbiol, **75**(1-2), 61-69(2002).
22. J. Wei, Y. Mu, D. Song, et al. Anal Biochem, **321**(2), 209-216(2003).
23. D.R. Shankaran, K.V. Gobi, T. Sakai, et al. Biosens Bioelectron, **20**(9), 1750-1756(2005).
24. T.M.H. And, M.J. Tarlov. J Am Chen Soc, **119**(38), 8916-8920(1997).
25. J. Wang, R. Lv, J. Xu, et al. Anal and Bioanal Chem, **390**(4), 1059-1065(2008).
26. M. Yasuura, K. Toko, T. Onodera. Sensor Mater, **23**(1), 21-37(2011).
27. T. Onodera, Y. Mizuta, K. Horikawa, et al. Sensor Mater, **23**(1), 39-52(2011).
28. K. Nagatomo, T. Kawaguchi, N. Miura, et al. Talanta, **79**, 1142-1148(2009).
29. T. Kawaguchi, D.R. Shankaran, S.J. Kim, et al. Talanta, **72**, 554-560(2007).
30. I. Coille, G. Gauglitz, J. Hoebeke. Anal Bioanal Chem, **372**, 293-300(2002).
31. I. Rault, V. Frei, D. Herbage, et al. J Mater Sci Mater Med, **7**(7), 215-221(1996).
32. J.M. Lee, H.H.L. Edwards, C.A. Pereira, et al. J Mater Sci Mater Med, **7**(9), 531-541(1996).
33. L.H.H.O. Damink, P.J. Dijkstra, M.J.A.V. Luyn, et al. Biomaterials, **17**(7), 679-684(1996).
34. J. He, F. Zhao, C. Wu, et al. J Mater Chem B, **1**(40), 5398-5402(2013).

Analysis of major triterpene acids and total polysaccharides in the leaves of 11 species of *Eriobotrya*

Yanping Hong[1], Biaosheng Lin[1], Hongyun Cao[1], Yongshun Gao[1] and Shunquan Lin[2,a]

[1]*College of Life Science, Longyan University, Longyan, Fujian 364012, China*
[2]*College of Horticulture, South China Agricultural University, Guangzhou 510642, China*

Abstract. Loquat (*Eriobotrya japonica*) is a subtropical tree with commercially important fruit and with leaves that have medicinal uses against cough and asthma. Leaves contain significant amounts of phenols, flavonoids, triterpene acids, and polysaccharides. However, information of the triterpene acids and polysaccharides in leaves is limited. In this study, the contents of five major triterpene acids and total polysaccharides in the leaves of 11 species of Eriobotrya were determined using RP-HPLC and phenol-sulfuric acid method, respectively. Total concentration of triterpene acids varied from 8.38 to 22.35 mg g-1 DW. The concentration of triterpene acids was greater than 20 mg g-1 DW in *E. bengalensis*, *E. prinoides* and *E. fragrans*, and it was less than 10.0 mg g-1 DW in *E. elliptica*. Different drying temperature (40-80°C) of leaves did not affect the content of polysaccharides with highest content (80.8 mg g-1 DW) from *E. prinoides*. The contents of triterpene acid in six wild species and polysaccharides in 5 wild species were higher than that in cultivated loquat.

1 Introduction

Dried leaves of loquat (*Eriobotrya japonica* (Thunb) Lindl.) have been used as a traditional Chinese medicine for the treatments of cough and asthma [1-2]. Most investigations on functional compounds in loquat leaves have been concentrated on only one species *Eriobotrya japonica* [3-9]There are a few studies on triterpene acids and polysaccharides [10].

Polysaccharides, which play an important role in the structural composition of living systems, are being identified via the processes of biological recognition, regulation, and information storage. In the last decades, some polysaccharides extracted from natural sources showed important biological activities, such as antitumor, immunomodulatory, hypoglycemic and anti-inflammatory and are attracting increasing attention in medicine. There are many investigations related to the viability of using ursolic acid as an effective medicine in anti-tumor treatment [11] and antifatigue agents [12].

2 Material and methods

[a] Corresponding author: kk9096@126.com
This work was supported by Guangdong Provincial Natural Science Foundation (Grant No. 07118121), Fujian Provincial Key Science and Technology Project (Grant No. 2012N0019), Fujian Provincial Natural Science Foundation (Grant No. 2012D100), Fujian Provincial Department of Education Project (Grant No. JB12204).

2.1 Materials

Mature leaves were obtained from 11 species of *Eriobotrya* trees grown in the orchard in South China Agriculture University. Leaves were dried for 3-4 days at about 40 °C, and then crushed into powder (80 mesh). The ground powder was stored at about -20 °C prior to analysis. The 12 samples of 11 species used in this study are listed in Table 1.

Table 1. The contents of triterpene acid (mg/g powder) and polysaccharide (amount to glucose mg/g DW) in various species of Eriobotrya.

Species	English name	A	B	C	D	E	The content of triterpene acid	The content of polysaccharide
E. japonica cv. Zaozhong No. 6	Common loquat cv. Zaozhong 6	3.13	1.47	1.02	1.90	7.35	14.88	42.83±2.05e
E. japonica Lindl.	Common loquat (wild tree)	2.50	0.64	0.50	1.69	5.57	10.90	44.97±1.66e
E. elliptica Lindl.	Tibet loquat	3.34	1.23	0.81	—	3.01	8.38	45.39±1.95e
E. prinoides var. *dadunensis*	Daduhe loquat	3.81	1.13	0.83	1.44	9.20	16.41	60.24±3.00c
E. prinoides Rehd & Wils	Oak leaf loquat	2.04	5.25	1.27	2.31	9.53	20.40	80.79±0.72a
E. deflexa Nakai	Taiwan loquat	2.11	1.52	0.61	1.30	5.86	11.40	45.23±0.33e
E. deflexa var. *koshunensis*	Hengchun loquat	1.51	1.79	0.86	1.36	9.73	15.25	42.88±0.55e
E. fragrans Champ	Fragrant loquat	1.53	5.76	0.98	2.81	9.13	20.21	72.68±3.23b
E. kwangsiensis Chun	Guangxi loquat	2.25	4.14	0.41	0.96	5.17	12.93	41.78±1.51ef
E. bengalensis Hook. f.	Bengal loquat	1.89	3.20	0.80	2.46	14.01	22.35	54.28±2.70d
E. obovata W. W. Smith	Obovata leaf loquat	2.91	5.64	0.51	—	3.57	12.63	41.94±1.95ef
E. cavaleriei Rehd	Big flower loquat	2.97	4.87	1.04	1.42	8.69	18.99	55.90±1.19d

Note: 1. A: tormentic acid; B: 2α,19α-dihydroxy-3-oxo-urs-12-en-28-oic acid;
C: 2α-hydroxyoleanolic acid; D: oleanolic acid; E: ursolic acid
2. Meaning with different letters within the same column are significantly different at the level of 5%.

2.2 Sample Preparation and HPLC Analysis of triterpene acid

Sample preparation and HPLC analysis were carried out according to Ju *et al* [4]. with some modifications: 0.8g dried powder was suspended in 20 mL chloroform and exposed to ultrasonic twice, each for 20 min, with an interval of 10 min. The suspension was kept over night (about 12 h, 25°C), and then recovered to its original volume followed by filtration; 5 mL filtrate was dried and dissolved in 5 mL of methanol. The solution was filtered through a membrane with a pore size of 0.22 μm.

Filtrate (20μL) was injected into the HPLC apparatus, which is consist of an Aglien HP1100 quaternary HPLC pump system with VWD system, a SK5200H (200W) ultrasonic launder and a Laborata 4000 (Heidolph) evaporator. The column was 5μm hypersil ODS2 (250 mm * 4.6 mm I.D.) in 25 °C. The mobile phase was a solution composed of acetonitrile, methanol and 21 mg mL^{-1} aqueous solution of ammonium acetate (v:v:v=60: 14: 26), with a flow rate at 1.5 ml min^{-1}.The detector wavelength was 215 nm.

Ursolic acid and oleanolic acid were provided by the National Institute for the Control of Pharmacerticals and Biological Products (Beijing, China), and tormentic acid, 2α,19α-dihydroxy-3-oxo-urs-12-en-28-oic acid, 2α-hydroxyoleanolic acid (maslinic acid) were isolated from Fragrent loquat[8], all these were used as standard. Methanol and acetonitrile were of chromatographic purity and all the other chemicals used were of analytical purity. All the aqueous solutions were prepared with double distilled water and filtered through 0.45μm nylon filters before use. Data was acquired and analyzed using Aglien Chem Stations.

2.3 Sample preparation and determination of polysaccharides contents

2.3.1 Orthogonal assay of polysaccharides extraction

Three factors and three-level orthogonal test were carried out to determine the optimum extraction method of polysaccharides. The dried powder of 'Zaozhong 6' loquat were firstly defatted with ethanol solution and the resulting residues were suspended in deionized water of different volumn (Solid-liquid ratio were 1:10, 1:15, 1:20), and then treated for 3/6/9 hours by warm water bath (temperature were 60 °C, 80 °C, 100 °C), after that, the supernatant were collected for the analysis of polysaccharides by phenol - vitriolic colorimetry.

2.3.2 Dried temperature

Leaves of 'Zaozhong 6' loquat were dried in three temperatures (40°, 60° and 80°C) for 3-4 days.

2.3.3 Determination of polysaccharides contents

The dried powder (0.5 g) of leaves from each species was firstly defatted in 20 mL of 50 % ethanol solution by ultrasonic treatment, with 20 min, using an ultrasonic instrument (SK5200H Ultrasonic Launder, 200 W, Kedao Supersonic Equipment Co. Ltd., Shanghai, China). The residues were suspended in 10 mL of deionized water, and then treated for 3 hours by boiling water bath, after that, the suspension were centrifuged at 12000 g for 15 min.

The contents of polysaccharides of the supernatant were determined by phenol-vitriolic colorimetry. Firstly, 975 μL deionized water was added to 25 μL supernatant of each samples for diluting extraction, then 0.4 mL of 5% phenol was added to 0.2 mL of each diluted samples, after that, 2 mL of sulfuric acid was added immediately. The mixed solution was shaken vigorously under room temperature (about 25 °C) for 30 min. The absorption was measured at 490 nm, using distilled water as a blank. The amount of polysaccharides was measured and expressed as glucose equivalent (GE) on dry weight basis.

2.3.4 Data analysis

Each experiment was repeated by three times. The data were analyzed using the statistical package SPSS 10.0. Significant differences were calculated according to the Duncan's Multiple range test. Differences at the level of 5% were considered as statistically significant.

3 Results and discussion

3.1 Quantitative determination of triterpene acid

The samples of 'Zaozhong No.6' and ursolic acid were used for precision test, reproducibility test, recovery test and stability test. The results showed the method of HPLC was credible[6].

DOI: 10.1051/bioconf/20170803012

A series of solutions of tormentic acid, 2α,19α-dihydroxy-3-oxo-urs-12-en-28-oic acid, 2α-hydroxyoleanolic acid (maslinic acid), oleanolic acid and ursolic acid at gradient concentrations were prepared in methanol. The standard curve of concentration vs peak area was established. All the standard curves fitted perfect linear equation.

tormentic acid: Y=5.46827479X+10.15579 (r=0.99954);

2α,19α-dihydroxy-3-oxo-urs-12-en-28-oic acid: Y=3.99517844X+ 4.5830234 (r=0.99967);

2α-hydroxyoleanolic acid: Y=5.855837X+40.010592 (r=0.99954);

oleanolic acid: Y=5.3474258X+9.7835004 (r=0.99955);

ursolic acid: Y =4.33280277X+19.362927 (r= 0.99969).

The concentration of triterpene acids in different species of *Eriobotrya* varied from 8.38 to 22.35 mg g^{-1} DW (table 1). Triterpene acids concentration was higher than 20.0 mg g^{-1} DW in Bengal loquat, Oak leaf loquat and Fragrant loquat, while it was lower than 10.0 mg g^{-1} DW in Tibet loquat. The content of triterpene acids in 6 species were higher than the normal loquat, it suggested that these wild species could be better candidate materials for utilizing triterpene acids.

3.2 Quantitative Determination of polysaccharides

3.2.1 Standard curve of glucose

A series of solutions of glucose which were dried to constant weight at gradient concentrations (40, 80, 120, 160 and 200 μg/mL) were prepared in deionized water. The OD490nm of those solutions were determined using phenol-vitriolic colorimetry. The standard curve of concentration vs OD490 nm was established. The standard curve fitted a perfect linear equation: Y=0.0051X-0.0047 (r= 0.999).

3.2.2 Orthogonal test of extraction

Polysaccharides were determined for 9 different treatments (Table 2). The analysis of variance results showed that the temperature of water bath and the Solid-liquid ratio influenced the yield of polysaccharides, the optimum condition was boiling water bath (100 °C) for 3 hours using a Solid: liquid ratio of 1:20.

Table 2. The result of orthogonal test of polysaccharid extraction.

No.	Temperature of water bath (°C)	Solid-liquid ratio	Time (hours)	Content of polysaccharide (mg GE /g DW)
1	60	1:10	3	10.448
2	60	1:15	6	17.176
3	60	1:20	9	21.776
4	80	1:10	6	18.089
5	80	1:15	9	23.343
6	80	1:20	3	26.759
7	100	1:10	9	23.670
8	100	1:15	3	34.410
9	100	1:20	6	36.895

3.2.3 Dried temperature and the contents of polysaccharides

The leaves of 'Zaozhong 6' loquat were dried in three temperatures (40°, 60° and 80°C) and the content of polyssachrids was 42.83±2.05 mg GE/g DW, 45.97±2.23 mg GE/g DW and 40.95±1.65 mg

GE/g DW respectively, the analysis of variance showed that the dried temperature had no apparent affect on the content of polysaccharides.

3.2.4 Determination of the contents of polysaccharides

Oak leaf loquat had the highest content of polysacharaides (80.79mg GE/g DW, p=0.05), followed by Fragrant loquat, Daduhe loquat, Big flower loquat and Bengal loquat (Table 1). The lowest amount of polysaccharides (p=0.05) was observed in Guangxi loquat and Obovata leaf loquat. The contents of the polysaccharides of 5 species were higher than those of the normal loquat. This indicates that there is a significant difference in polysaccharides content among the 11 Eriobotrya species. Therefore, rather than the normal loquat, some wild species could be identified as better material for utilizing the polysccharides from the leaves.

4 Conclusion

The contents of triterpene acid and polysaccharides in the leaves of Oak leaf loquat, Fragrant loquat, Daduhe loquat, Big flower loquat and Bengal loquat were higher than those of the normal loquat while Oak leaf loquat, Daduhe loquat, Big flower loquat and Bengal loquat contained significantly higher level of total phenolics and flavonoids, and had stronger antioxidant activities compared with those of the cultivated species'Zaozhong 6' loquat [7]. It is reported that, the contents of polysaccharides and flavones in old leaves were quiet different between different loquat cultivars [13]. Our research also showed that the contents of polysaccharides, triterpene acid and flavones in leaves differed significantly among different species of *Eriobotyra*. Our study suggests that high level of functional components may be found in wild species of loquat and could be useful for medicinal use.

References

1. S.Q. Lin, *Acta. Hort.* **750,** 37-44 (2007).
2. J.Y. Lin, C.Y.Tang, *Food Chem.* **107,** 1587–1596 (2008).
3. D.O. Gray, *Phytochem.* **11,** 751–756 (1972).
4. J.H. Ju, L. Zhou, G. Lin, D.Liu, L.W. Wang, J.S. Yang, *Chin. Pharm. J.* **38,** 752-757 (2003).
5. T.H. Lee, S.S.Lee, Y.C.Kuo, C.H. Chou, *J. Nat. Prod.* **64,** 865-869 (2001).
6. Y.P. Hong, S.Q.Lin, X.M. Huang, *Acta. Hort.* **750,** 225-232 (2006).
7. Y.P. Hong, S.Q.Lin, Y.M.Jiang, M. Ashraf, *Plant Foods Hum. Nutr.* **63,**200-204 (2008a).
8. Y.P. Hong, Y.C. Qiao, S.Q.Lin, Y.M Jiang, F.Chen, *Sci. Hort.* **118**, 288-292(2008b).
9. K. Kobaa, A.Matsuokab, K.Osadac, Y.S. Huangd, *Food Chem.* **104,** 308–316 (2007).
10. G. F.Duan, B.Y. Yu, R. B.Wu, M.Q. Chen, B. J.Li, W. Chen, *Chin. J. Stra. Pharm. J.* **21,** 50-53 (2009).
11. J. Huang, Y. Sun, *China New Med. J.* **6,**101-104 (1997).
12. A. P. Huang, H.Y.Gao, X.P.Chen, T. Li, S.Q. Zhen, *Chin. J. Fujian Fruit* **137,** 13-14(2006).
13. F. Jiang, S.Q.Zheng, H.Y.Gao, A.P. Huang, X.P.Chen, Acta. Hort., **750**, 321-323 (2007).

DOI: 10.1051/bioconf/20170803013

Preparation and performance optimization of TPBISi green-light organic luminescent material devices

Huajing Zheng[a], Quan Jiang, Yadong Jiang and Ruan Zheng

State Key Laboratory of Electronic Thin Films &Integrated Devices, School of Optoelectronic, University of Electronic Science and Technology of China, Chengdu 610054, China

Abstract. The Study analyzed and tested the absorption spectrum, photoluminescence spectrum, and device's electroluminescence spectrum of a new silole material. The device with Silol as an emitting layer, emitted green-light whose structure is ITO/NPB/2,2,3,3-tetraphenyl-4,4-bisthienylsilole(TPBTSi)/Alq3/Mg: A by improvement of preparation technology and optimization of thin film. It reaches the maximum luminescence of 11290.2 cd/m^2, the maximum luminous efficiency of 0.84 lm/W, luminescence spectrum of 516 nm, chromaticity diagram CIE coordinate of(0.275, 0.4568) when voltage is 15V. All of the above is the green characteristic spectrum of TPBTSi.

1 Introduction

In recent years, OLED has attracted people's wide attention [1-5]. The materials of OLED are divided into two parts such as conjugate polymer materials and small molecule materials. The small molecule materials have definite chemical structure and better color purity, but are easy to crystallize which has become the main factor to shorten the life expectancy of the device. The polymer molecule materials can be processed easily but are hard to purify. In order to solve this problem, scientists have developed star shaped helical molecules containing heteroatom [6]. Compared to the frequently-used heterocyclic compounds containing N, the heterocyclic compound containing Si(Silole) has lower LUMO energy level, attributed to Silol's special electronic structure. The energy level of LUMO is mixed with σ^* orbit on Silicon atom and π^* orbit on part of butadiene[7]. In 1996, 2,5-diarylsiloles 20a-f was first applied to electronic transport materials, and thereafter came the silole derivatives (21a and 21b, 22a-c) in succession. The 20d film precipitated under vacuum performs fast and non-dispersion electron transport (μ_e≈2×10-4/in comparison to the environment of 6.4×105 V/cm). Even in the presence of oxygen, the 20d film produces very little electron trap. The electron transport speed rate of 20d film is 100 times more than that of Alq3, due to the fact that the Alq3 produces electron trap and leads to dispersion electron transport. Compered to Alq3, to a certain extent, the 20d has a larger LUMO value (3.3V), that results in better characteristics on electron transport. However, in electrochemical research, most Silole derivatives are irreversibly oxidized and reduced. Inserting 20d between LiF/Al cathode and Alq3 light emitting layer or 20d can improve device EL efficiency (from 1.9 to2.6 lm/W). Due to the fact that neither the Tm (175 °C) or Tg of 20d are high, the surface morphology is damaged

[a] Corresponding author : zhj12@163.com
This work was supported by Fund project: National Natural Science Fund (20674049).

by crystalizing while the device is working. The method to improve the Tg (the highest to 81°C) is to increase the number of substituent aromatic, such as compounds 20c-e. However, since the Tg of 20c-e is lower than Alq3, the device prepared by 20d for ETM tested under air environment has a better stability. This is probably due to the fact that 20d has a better stability under air. Apart from being used as an ETM, Silole (20a and 20f) can also be used as a luminescent material [8-10].

Luminescence mechanism is discussed in this study through four types of Silole complexes 2,2,3,3-tetraphenyl-4,4-bisthienylsilole(TPBTSi) being used as emitting layer, by choosing hole transport layer material NPB and electronic transport layer material Alq_3 used as hole and electronic transports layers, preparing high brightness electroluminescence device with simple structure by vacuum thermal evaporation method, charactering device's photoelectric properties and by the analysis of energy level structure.

2 Test

2.1 TPBTSi film surface analysis

Figure1, 2 show the absorption and emission spectra tested on quartz glass. Figure 3 outlines the surface morphology. From Figure1, the maximum absorption of TPBTSi solid film is 246 nm, the PL peak is 498 nm, which are the characteristics of green light. .

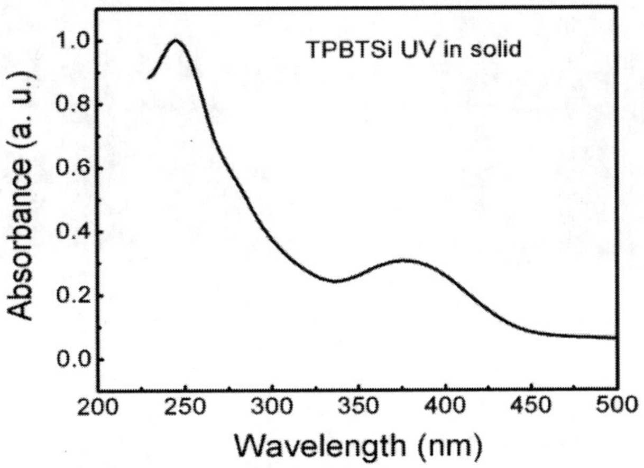

Figure 1. TPBTSi film absorption spectrogram.

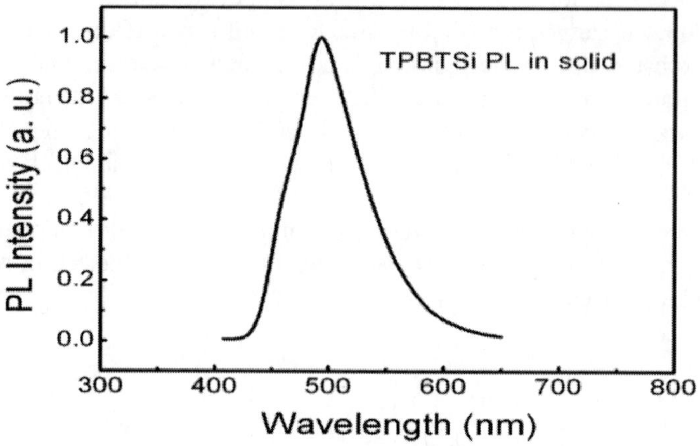

Figure 2. TPBTSi film luminescence spectrogram.

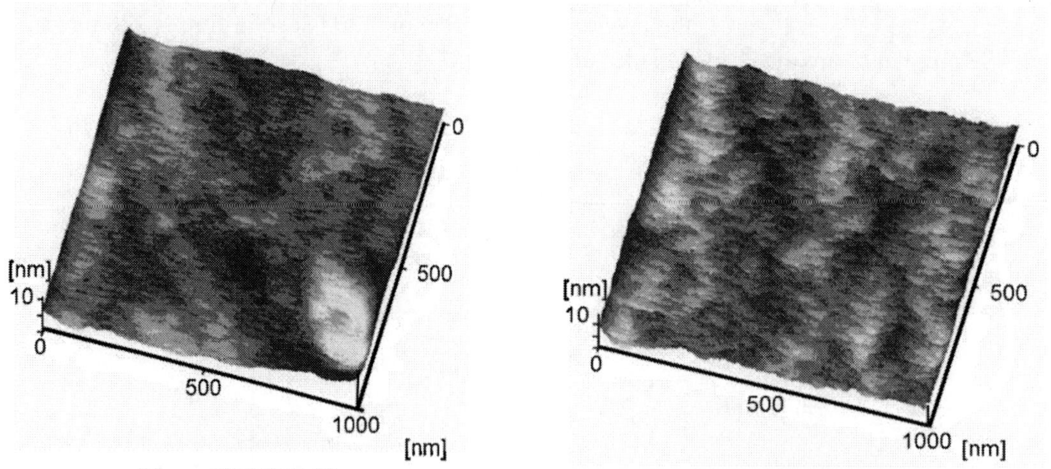

Figure 3. TPBTSi film surface morphology.

Table 1 outlines the surface morphology parameters of two substrates and the data shows clearly that the $R_{P\text{-}V}$ of 20nm film is obviously higher than that of 30nm. Figure 3 shows that the film with such thickness has pinhole. Both the average roughness (R_a) and the range from peak to trough ($R_{P\text{-}V}$ decrease significantly, meanwhile surface particle radium decreases in at a different degree. The decreas of average roughness and surface particle radium are obviously beneficial to increasing the contact area for the film with other organic layers, which is beneficial to them in terms of attaching closely. In order to use TPBTSi film worked as emitting layer, the electronic and the hole must be controlled in this layer to compound luminescence. In order to avoid the carrier loss caused by the increase of turn-on voltage and the film disadvantage, the thickness of emitting layer TPBTSi should be 30 nm.

Table 1. TPBTSi film surface morphology parameters.

substrate	Surface morphology parameters		
	R_a(nm)	$R_{P\text{-}V}$(nm)	R_{rms}(nm)
1# 20 nm	0.74	5.67	1.01
2# 30 nm	0.49	1.06	1.00

2.2 OLED device prepared by TPBTSi

Table 2. Experiment Medicine.

	Manufacturer	Purity
NPB	Aldrich	99 %
Alq_3	Aldrich	99.9 %
CBP	Aldrich	99 %
BCP	Aldrich	98 %
Mg	Sinopharm Chemical Reagent Co., Ltd	99.9%
Ag	Sinopharm Chemical Reagent Co., Ltd	99.5%
ITO Glass	Shenzhen Nanpo.,Co., Ltd	11 Ω/ filtered 80%
Acetone	Chengdu Kelong Chemical Reagent Co., Ltd	99.9 %
Ethanol	Chengdu Kelong Chemical Reagent Co., Ltd	99.9 %

Organic multifunctional film-forming equipment OLED-V was sourced from Shenyang Vacuum Technology Institute, vertical manifold of clean bench was sourced from Nantong Cleaning Equipment Co., Ltd, spin coater KW-4A was sourced from Shanghai Chemat Function Ceramics Technology Co., Ltd. Sartorious BS electronic balance was purchased from Beijing Sartorious Mechatronics T&H Co., Ltd, luminance meter ST-86LA and spectrometer OPT-2000 was purchased from Beijing Normal University, semiconductor test system Keithley 4200 was sourced from Keithely Company and ultrasonic cleaner KQ-200DB was bought form Nantong Ultrasonic Cleaning Equipment Co., Ltd.

Place the cleaned substrate into preprocessing vacuum chamber of the film-forming equipment Sonicel plus200 (Sonic, Korea). Prepare the oxygen plasma treatment, cleaning the surface dirt furthermore, improving surface oxygen content of ITO to increase function. Then, with no system vacuum damaging, make the vapor plating of organic materials and the electrode continuously and successively. The vacuums of vapor plating organic materials and mental electrodes is 2.1×10^{-4} Pa and 3.8×10^{-3} Pa, the speed is 0.3-1 nm/ and 1.1-1.3 nm/s in order. Figure 4 shows the material's formula. Device structure is as follows: ITO/NPB (20 nm)/TPBTSi (30 nm)/Alq_3(x)/Mg:Ag (200 nm), (Device A: x=0 nm, Device B: x=10 nm).

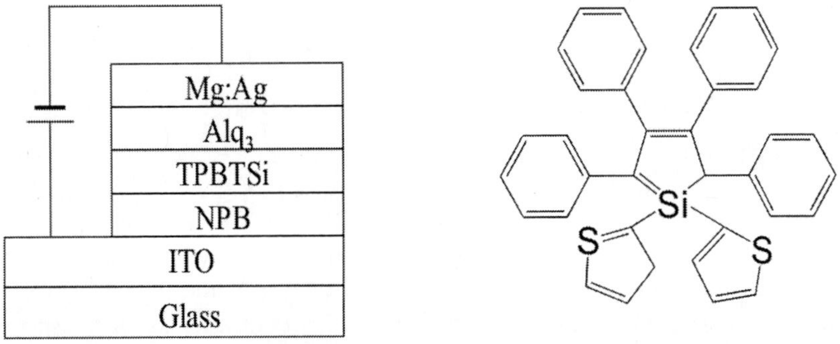

a. Device Structure Diagram b. TPBTSi

Figure 4. Adopted Device Structure(a)and Materials Molecular Structural Formula(b).

DOI: 10.1051/bioconf/20170803013

3 Results and discussions

3.1 TPBTSi Green OLED device's performance test and analysis

The relationship between current density and Bias voltage and the relationship between luminance and Bias voltage about double-layer (Device A) and tree-layer (Device B) TPBTSi Devices are represented in Figure 5 and Figure 6:

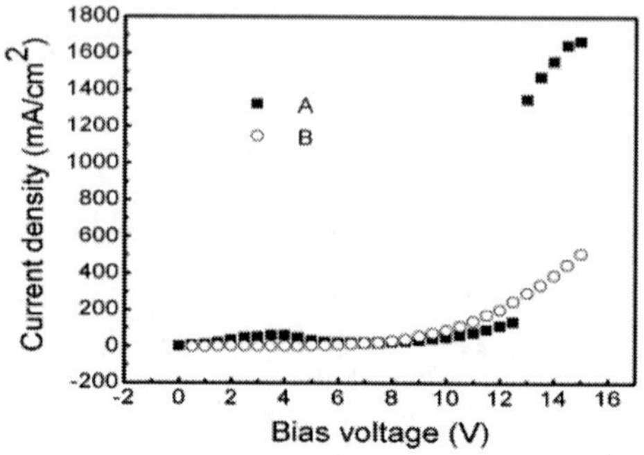

Figure 5. TPBTSi Device's *J-V* characteristic curve.

From Figure 5 and Figure 6, it is evident that with the same voltage, when the luminance of double-layer device is low, the maximum luminance is only262 cd/m^2, at the same time the current density is only134 mA/cm^2. But the luminance of tree-layer device is higher than double-layer's/. Under the voltage of 15V, the maximum luminance reaches 7032 cd/m^2 with a current density of50.7 mA/cm^2.

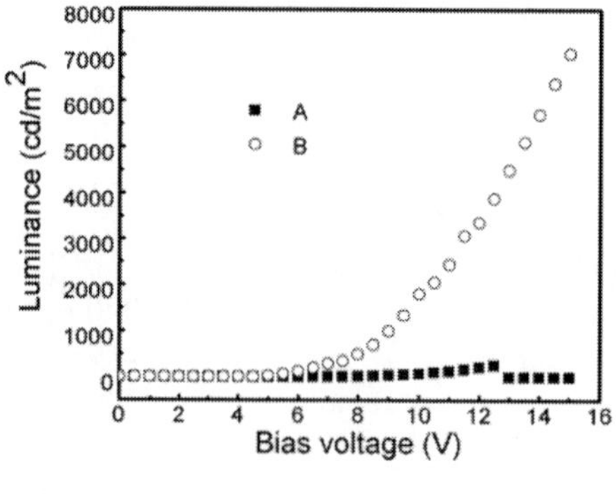

Figure 6. TPBTSi Device's *L-V* characteristic curve.

From the characteristic curve of current density and voltage, it is evident that with the voltage of 12V, the current of Device A increases sharply which is caused by the number imbalance of hole and electronic current-carrying. The compound probability of carriers decreases leading to ample thermal current due to the fact that the number of the hole is greater than the electronic. With the decrease in distance between recombination zone and metal electrode /organic layer , the related exciton quenching process increases resulting in the probability of emitting photons, especially when the exciton make the radiative transition. Therefore, the transmitting capability of TPBTSi carriers is not good. In order to improve balance ability of carriers, a good film-forming electron transport material must be introduced. In addition, the current density change of Device B in which added electron transport material Alq3 is slow indicates that the effect is obvious. The electronic transports well from cathode to emitting layer through Alq3 layer and the barrier of hole from anode trough NPB/TPBTSi enters emitting layer too. This leads to a balance of carrier numbers, an obvious improvement of device efficiency and the device's luminous efficiency of0.43 lm/W when the luminance reaches 300cd/m^2.

Figure 7 shows the EL spectra test result of device. The EL spectral peaks of two devices are located in the same wavelength where both are 516nm. Compared to NPB/TPBTSi fluorescent spectra (PL), it is verified that this is TPBTSi characteristic spectra. From the structures of two devices, the recombination zone of hole and electronic is located in the same emitting layer and such a phenomenon is explained by energy level structure diagram of Figure 8.

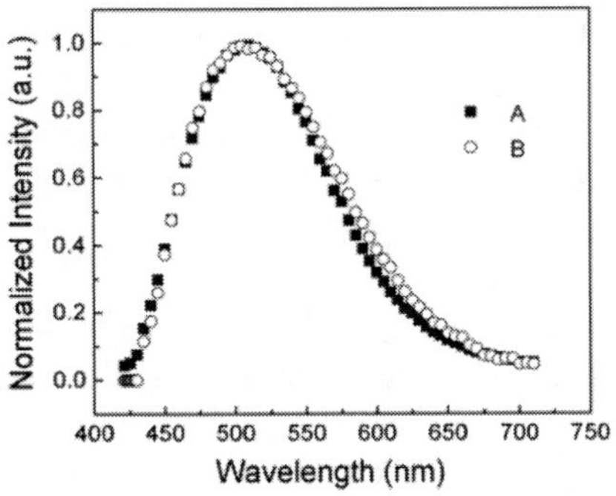

Figure 7. Device's EL spectrogram.

In the double-layer device structure, NPB plays a role of hole transport layer and the TPBTSi are both emitting layer and electron transport layer. The luminescence of device is green which is the characteristic spectrum of TPBTSi, it indicates that carrier recombination zone is completely located in TPBTSi layer. The electronic implanted into emitting layer form cathode, parts of hole implanted from anode is injected into emitting layer by overcoming the high hole barrier (10 eV) of interface and is injected into emitting layer by Coulomb electrostatic force of space electronic charge, which forms luminescent exciton with accumulated electron from emitting layer and then sends out TPBTSi characteristic spectrum by radiative decay.

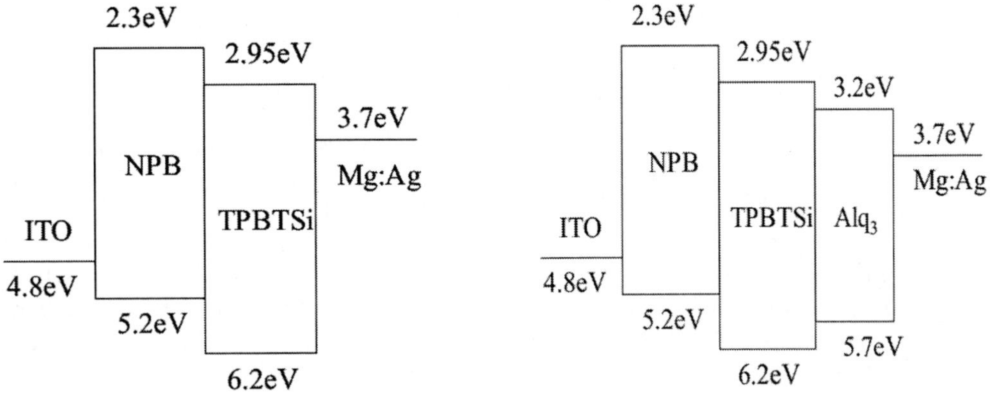

Figure 8. Device's energy level structure diagram.

In the three-layer device, the device is still with TPBTSi characteristic spectrum of green luminescence, which indicates that the compound position of carrier is in the same organic layer too. The reason for such a phenomenon is due to the fact that on the one hand, most holes can't get into Alq_3 layer being blocked by highly ionization potential of TOBTISi. The Alq_3 only plays a part of transporting but unsuccessful in participating in luminescing. On the other hand, the electron implanted from cathode can get into TPBTSi layer by easily penetrating into the low electronic barrier height, but with the bad electronic transmitting capacity of TPBTSi itself, the electron can't get into

 DOI: 10.1051/bioconf/20170803013

NPB layer. The NPB therefore does not participate in luminescing but only remains a part of hole transport layer. The evidence suggests that, the electron and hole are both bound in the TPBTSi layer. For the high fluorescence efficiency of this layer itself, the LUMO energy level of TPBTSi ground state molecules can take electrons easily. Furthermore the HOMO energy level loses electrons easily that results in molecules migrating into exited state and loses energy by radiative decay meanwhile gives off characteristic luminescence of TPBTSi.

3.2 Thickness optimization of three-layer TPBTSi device

Without the optimization of tree layer device, in order to increase the number of electronic and improve the property of device, the electronic layer Alq_3 should be added into two layer device TPBTSi. To improve the luminance and efficiency, the two kinds of carriers of the device have to be kept in the best balanced state. There are two ways to keep the carriers balanced, one is to increase minority carrier, and the other is to decrease majority carrier.

(1) Increase of minority carrier

From the test of the double-layer and three-layer devices, the luminance increases fast after adding electron transport layer. But for most OLED, the hole is the major carrier. Considering the fact that the conductive capacity of NPB is relatively good, in emitting layer TPBTSi, the hole is mostly the major carrier and the electron is minor carrier. It is likely that with little number of minority carrier, there is few number of exiton which is formed by recombination resulting in low luminescence of the device. To improve the luminescence of the device, it is necessary to increase the number of electronic carrier of NPB emitting layer. On the other hand, as the conductivity of Alq_3 is not as good as that of the NPB's, the change of thickness of Alq_3 layer therefore influences device's electronic filed distribution and current more obviously. This in turn influences the luminescent properties of the device. For this change to be the thickness of Alq_3 which is electron transport layer and preparer the light-emitting device with following structure: ITO/NPB (20 nm)/TPBTSi(30 nm)/Alq(x nm, x=10,20,30,40)/Mg:Ag(200 nm).

The luminance-voltage (*L-V*) diagram is showed on Figure 9. From the Figure it is evident that with the increase of the voltage the luminance of each device increases too, which presents the obvious characteristics of diode. But under the same voltage, with different thickness of Alq_3 the luminance is different. When the voltage is 10V, the luminance of each device is as follows: (A)1800 cd/m^2, (B)1840.6 cd/m^2, (C)344.6 cd/m^2, (D)384.5 cd/m^2. Under voltage of 15V, the maximum luminance of each device is as follows: (A)7032.3 cd/m^2, (B)9368.2 cd/m^2, (C)3068.8 cd/m^2, (D)3658.3 cd/m^2. It is obvious that when the thickness of Alq_3 is larger than 20nm, under the same voltage, the luminance decreases obviously with the increase in thickness due to the fact that the electron injected from electrode of thinner device can easily get through recombination zone under high filed. If the Alq_3 is too thick, the traps of capturing electron will get more resulting in the reduction of the compounding and colliding probabilities. And there will be a big loss of electron's capacity with a long distance transmission [10] which will reduce the number of electron reaching the recombination zone so that it will decrease the brightness (luminance) and the current. In addition, the voltages to test the device are (A)3.5 V, (B)4.0 V, (C)5.0 V, (D)5.5 V. From the above it is evident that the thickness of Alq_3 plays a part in turn-on voltage and the voltage goes to the lowest when the thickness is 10nm.

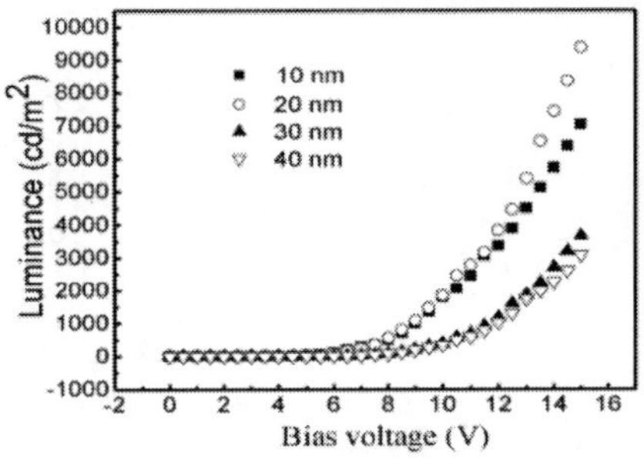

Figure 9. TPBTSi *L-V* diagram with different Alq_3 thickness.

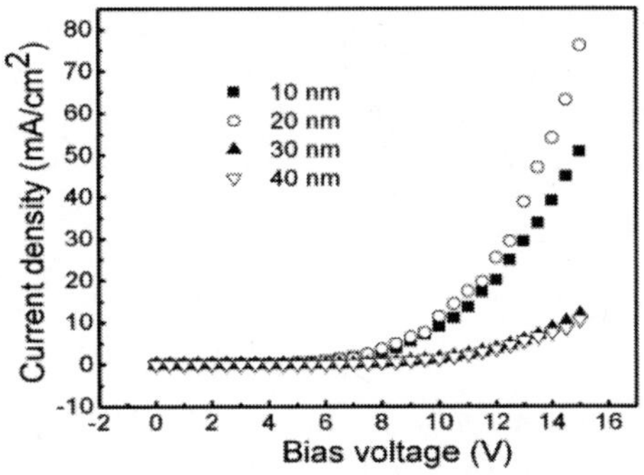

Figure 10. TPBTSi *J-V* diagram with different Alq_3 thickness.

From Figure 10, it is evident that the current density and luminance of the device has the same effect tendency even with different Alq_3 thicknesses. When the Alq_3 thickness is more than 20nm, the device current decreases obviously with the Alq_3 thickness increases and when the voltage is 15V, the current of each device are respectively 50.7(10 nm), 76.3 (20 nm), 11.8 (30 nm), 10.6(40 nm) mA/cm^2 ,which has a proportional relation with the luminance (brightness) of device. It is found from the normalized EL spectrogram of device, the differences of the first three devices are little and it is hardly to observe the migrations of spectral peaks which are all kept at 516 nm, and that is the typical TPBTSi spectral peak value. But when the thickness of Alq_3 layer increased into 40nm, the spectral peak of the device displaced to the direction of long wave (522nm) and made a redshift of 6nm which means changes to the light-emitting Alq_3layer. This may be attributable to the thickness of Alq_3 layer that is too big and at the same the mobility of the electron is as low as two orders of magnitude so that parts of electrons restricted in Alq_3 layer lead to a recombination luminescence with the hole came from anode. Hence it is proved that the thickness of Alq_3 layer used as the electron transport should be controlled within 30 nm, and considering the current should be low (low joule while device working) and the luminescent properties should be good, the Alq_3 layer is chosen of 20nm.

 DOI: 10.1051/bioconf/20170803013

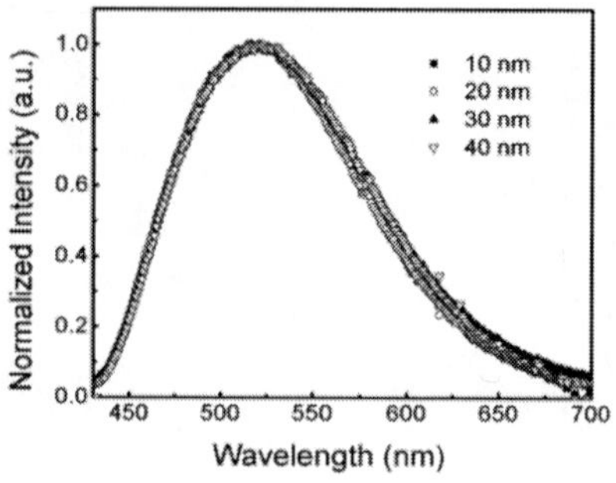

Figure 11. TPBTSi EL diagram of different Alq_3 layer thicknesses.

(2) Decrease of majority carrier

After the electron transport layer is determined, to improve the property of the device by making a balance of two kinds of carriers by decreasing the number of majority carrier hole. With no thickness changes of emitting layer and Alq_3 layer, prepare the devices bellowed by changing the thickness of hole transport layer: ITO/NPB(x nm, x=10,15,20,25)/TPBTSi(30 nm)/Alq_3(20 nm)/Mg:Ag.

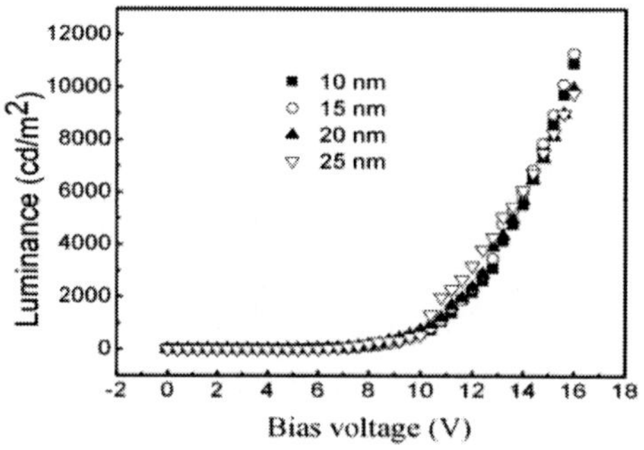

Figure 12. TPBTSi *L-V* diagram with different NPB thicknesses.

From Figure 12 and Figure 13 it is evident that the NPB thickness change of the hole transport layer has little effect on the luminance (brightness) and current density. With the voltage of 16V, the luminance (brightness) and current density of the devices are respectively: 10902.8 cd/cm^2 (56.8 mA/cm^2), 11290.2 cd/cm^2 (54.4 mA/cm^2), 9960.7 cd/cm^2(57.8 mA/cm^2), 9780.7 cd/cm^2(42.6 mA/cm^2), that won't affect the efficiency and the property of the device which indicates that the hole has a high migration (shift) rate in NPB. To choose a proper thickness of the whole device, the bellowed are chosen to the best optimized structures: ITO/NPB(20 nm)/TPBTSi(30 nm)/Alq_3(20 nm)/Mg:Ag(200 nm).

DOI: 10.1051/bioconf/20170803013

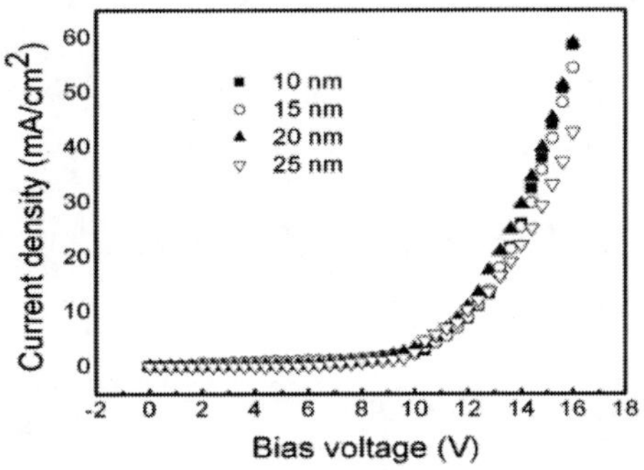

Figure 13. TPBTSi *J-V* diagram with different TPBTSi thicknesses.

4 Conclusions

Using new green luminescent materials TPBTSi as emitting layer, by adding organic material Alq_3 as a transport layer, to make a three layer device with structure of ITO/NPB/TPBTSi/Alq_3/Mg:Ag. The device has the maximum luminance of11290.2 cd/m^2, CIE coordinate of (0.275, 0.4568) and the maximum luminous efficiency of0.84 lm/W. Compared to the two layer device, the TPBTSi is a kind of luminescent material with good luminescent property but lack of the performance of electron transport. The optimized green light-emitting device can be made by adding an electron transport layer and optimizing the device by changing the organic layer's thicknesses.

References

1. T. H. Kim, H. K. Lee, O. O. Park, et al. White-light-emitting diodes based on iridium complexes via efficient energy transfer from a conjugated polymer. Adv. Funct. Mater., 2006, 15: 611-617.
2. J. Wang, Y. D. Jiang, J. S. Yu, et al. Low operating voltage bright organic light-emitting diode using iridium complex doped in 4,4′-bis[N-1-napthyl-N-phenyl-amino] biphenyl. Appl. Phys. Lett., 2007, 91: 131105.
3. X. Gong, M. R. Robinson, J. C. Ostrowski, et al. High-efficiency polymer-based electrophosphorescencet devices. Adv. Mater., 2002, 14: 581-585.
4. C. H. Lee, G. W. Kang, J. W. Jeon, et al. Blue electroluminescence and dynamics of charge carrier recombination in a vacuum-deposited poly(p-phenylene) thin film. Thin solid films, 2000, 363: 306-308.
5. C. W. Tang, S. A. VanSlyke, C. H. Chen. Electroluminescence of doped organic thin films . J. *Appl. Phys*., 1989, 65: 3610-3616.
6. T. Mori, K. Obata, T. Mizutani, et al. Electroluminescence of organic light emitting diodes with alternately deposited dye-doped aluminium quinoline and diamine derivative. J. *Appl. Phys*., 1999, 32: 1198-1203.
7. Q. Li, J. S. Yu, L. Li. High bright blue organic light emitting diodes based on a novel silole derivative. SPIE, 2007, Chengdu, 2-170, inpress.
8. A. K. Kulkarni, K. H. Schulz, T. S. Lim, et al. Dependence of the sheet resistance of indium-tin-oxide thin films on grain size and grain orientation determined from X-ray diffraction techniques. Thin Solid Films, 1999, 345: 273-277.
9. H. Antoniadis, M. A. Abkowitz, B. R. Hsieh. Carrier deep-trapping mobility-lifetime products in poly(p-phenylene vinylene). Appl. Phys. Lett., 1994, 65: 2030-2032.
10. J. H. Burroughes, D. C. Bradley, A.R. Brown, et al. Light-emitting diodes based on conjugated polymers. Nature, 1990, 347: 539-541.

DOI: 10.1051/bioconf/20170803014

Sound speed optimization based on fuzzy sets using image texture as quality factors

Xingwu He

Network & Information Management Center, Chengdu Normal University, Chengdu, Sichuan, China

Abstract. A constant sound speed of 1540 m/s is usually used by ultrasound imaging systems for calculating delay curves in beamforming. A mismatch of this constant system sound speed and the real one in human tissue could result in significant defocusing and degrade the image resolution in the modern ultrasound imaging system. This paper has investigated characteristics and potential relationship of image texture parameters to image focusing quality due to unmatched speed of sound. Algorithms for sound speed optimization (SSO) system based on fuzzy process and image texture parameters are presented. The proposed method has been investigated and verified from phantom imaging including speckle, point targets and cyst images. Test results show that individual texture feature may help to determine appropriate sound speed used for beamforming; however, only the proposed fuzzy process based system can achieve the best sound speed detection with consistent and robust output.

1 Introduction

In medical ultrasound imaging systems, the sound speed is an important parameter for beamforming. It is usually set as a constant in human tissue, typically 1540 m/s and his parameter is used to calculate electronic time delay [1]. These delay curves ensure that the ultrasound signals of all the array elements are in phase at the focal point. The actual propagation velocity varies over a very large range in human tissues, for instance, 1450 m/s in fat and 1560 m/s in kidney [2]; moreover, the same tissue may have different sound speeds in different patients.

The mismatched sound speed would lead to a shift of structure position, widening the point spread function and increasing side lobe levels of the image and then degrades image detail and contrast resolution [3]. Therefore, finding a way to select the best global sound speed used in system's delay profiles could be seen as a first-order phase aberration correction method. Zhao and Trahey [4] developed a method which could correct phase aberration by maximizing either mean brightness of diffuse or point targets. Napolitano et. al. [5] described an algorithm to find out the sound speed that produces the best overall lateral image quality by analyzing the spatial frequency content of the envelope data using various trial sound speeds. He and Liu [6] developed methods with raw radio frequency (RF) data and experimental results show more accurate sound speed than image based methods, while it needs much more computational requirement. Due to the shift of the focusing away from its theoretical acoustic focusing caused by the mismatched sound speed, Li and Liu [7] proposed a method to find the best sound speed by one of image texture parameters.

On account of significant defocusing and some geometric distortions leading to the effect of aberration detected in the reconstructed image as opposed to the aperture data. Authors [24] proposed an image registration-based sound speed estimation method that can be implemented using two images with different steer angles. However, for accurate registration, good image qualities in terms of detail and contrast resolution, as well as sophisticated registration procedures are all necessary. Moreover, the varying local echo phase when changing the transmit beam steering angle directly reflects the varying arrival time of the transmit wave front. Michael Jaeger [21] have described a method by sensing the angle-dependent aberration delay in a spatially resolved way and thus aberration correction for a spatially distributed volume aberrator, which achieved location-independent diffraction-limited resolution as well as accurate display of echo location based on reconstructing the speed-of-sound spatially resolved in phantoms. Clinical research in [22] shows sound speed correction can be used as an initial diagnosis in the disagree of homogeneous fatty liver. Guenther [23] proposed a method based on a modified inverse Radon transform for speed-of-sound correction, and the reconstruction using the modified inverse Radon transform shows a small but distinguishable improvement.

Quantitative results from [8] show that the sound speed error degrades the contrast of the point target, while the texture of the speckle does not appear significantly changed. Meanwhile, isolated point reflectors couldn't be present in clinical ultrasound imaging [9]. Therefore, to deal with complex clinical cases, this paper presents a sound speed optimization system based on fuzzy sets with image texture as quality factors. It not only integrated the sound speed correction method proposed by [4] and the focusing quality factor based on image texture parameters described by [7] but also built a decision-making system with fuzzy logic in order to make the texture analysis be more robust. This is a practical approach for sound speed optimization. Our experimental results on the phantom proved that this sound speed optimization system is accurate and robust on simulated tissue for speckle, focusing (from point targets) and cyst.

2 Method

2.1 Image texture as quality factors

Image texture has been used as a tool in image analysis; in [11], there are nine commonly used parameters for texture classification. In this paper, we classify them into four groups with properties related to image quality.

Group 1 (mean): represents the intensity of local region. The better the beam focusing, the greater it is.

Group 2 (variance, correlation, cluster shade and cluster prominence): describes the degree of image difference or linearity. Since the defocusing from mismatched sound speed wouldn't definitely produce the change of image difference or linearity for nonlinear tissue structures; and from our experiments, all of them are not quite associated with the focus quality. Thus, this paper doesn't choose any of them as the quality factors.

Group 3 (energy, entropy and homogeneity): measures homogeneity related texture. The more homogeneous the image is, the larger the value of energy/homogeneity is. Comparing to texture homogeneity, texture energy is more sensitive and directly to the focusing quality. While the entropy is in inverse to the energy, one can choose either of them as the quality factor. In this paper, the texture energy is used.

Group 4 (contrast): depicts local image variations. In general, local image with clear border/structure always shows larger texture contrast.

Then texture features used in this paper as quality factors are energy, contrast and mean as follows [11]:

Energy:

$$\sum_i P_s(i)^2 \sum_j P_d(j)^2 \tag{1}$$

Contrast:

$$\sum_j j^2 P_d(j) \tag{2}$$

Mean:

$$\frac{1}{2}\sum_i iP_s(i) \tag{3}$$

where P_s and P_d are defined respectively by

$$P_s(i;d_x,d_y) = Card\{(k,l) \mid y_{k,l} + y_{k+d_x,l+d_y} = i\} \tag{4}$$

and

$$P_d(i;d_x,d_y) = Card\{(k,l) \mid y_{k,l} - y_{k+d_x,l+d_y} = i\} \tag{5}$$

In Eqs. (4) and (5), Card refers to the number of elements of a set (k, l) and $y_{k,l} = j$ means that the intensity at the position (k,l) is equal to the grey level of j. Notation (d_x, d_y) denotes the direction vector along x and y direction, respectively. In this paper, we compute texture parameters of Eqs. (1), (2) and (3) along lateral direction (i.e., perpendicular to ultrasound beam direction) because the ultrasound speckle always shows longer size in lateral compared with that in axial direction [12].

2.2 The fuzzy system model of sound speed optimization

To improve the decision-making results of texture analysis for sound speed prediction, we applied the concept of the fuzzy sets because fuzzy logic can be used to address the issues associated with intrinsic imprecision or inexactness [13]. We define the fuzzy sets from the relationships between the selected image textures and image quality derived from a set of images with trial sound speed values. Mismatched speed of sound would produce a shift in axial direction which will involve confusion regions in ultrasound image; especially the decimation of pulse-echo signal for display would exaggerate the degradation of detail resolution. And in lateral direction, the mis-focus makes the effective pulse-echo beam width become wider, it means the edge of tissue structure in ultrasound image will be blurred and the contrast will be also reduced. It means the larger texture contrast is related to the better focused image. The simulation results in [20] shows it will be broadening in terms of the mis-focus target due to the error of assumed sound speed, which is consistent with lower texture energy. And better focused image leads to larger local image intensity that can be represented by larger texture mean. Based on the characteristics of focusing image, we can have a basic principle to judge a sample corresponding to an acceptable sound speed value:

"If contrast is LARGE, mean is LARGE, and energy is SMALL, then the focusing quality is GOOD.".

The basic linguistic rules can be used to do the approximate reasoning with suitable linguistic variables and these variables are described as follows:

Let F be the linguistic variable for focusing quality with the universe of discourse (UoD) U = {contrast, energy, mean} = {u1, u2, u3}. The terms of this linguistic variable are set to be:

T (F) = {VERY GOOD (VG), GOOD (G), FAIR (F), POOR (P), VERY POOR (VP)}.

Also, the UoD of contrast, energy and mean are defined as UC = {c1, c2… cn}, UE = {e1, e2… en} and UM = {m1, m2… mn}, where ci, ei, and mi are values of the contrast, energy and mean of the ith images with a trial sound speed, respectively. The terms of these linguistic variables are all set to be:

T(C | E | M) = {LARGE, MEDIUM, SMALL}.

In this paper we collect the fuzzy rules as the fuzzy associative memory which should preserve the basic properties of the individual's qualitative preference structure [14], as shown in Table 1. As we defined T (F), the linguistic variable in this section including P, VP as the terms, however, the fuzzy rules in Table 1 don't cover them because the decision can't be the one that produced the image with poor focusing quality.

Table 1. Fuzzy rules

Rules	Energy	Contrast	Mean	Focusing quality	Rules	Energy	Contrast	Mean	Focusing quality
1	small	large	large	very good	7	medium	large	small	fair
2	medium	large	large	good	8	large	large	medium	fair
3	small	large	medium	good	9	small	large	small	fair
4	small	medium	large	good	10	small	medium	medium	fair
5	medium	large	medium	good	11	medium	medium	large	fair
6	larger	large	large	fair					

The membership functions used in fuzzy processing, in general, should meet the concept of optimal interfaces defined in [15]. The triangular membership function overlapping at a degree of 0.5 satisfies the concept of optimal interfaces in the sense of an error-free conversion between the numeric and the linguistic terms. Each membership function is defined by its centers, hereafter denoted by a tuple X, e.g. X(X0, X1, X2) for contrast, energy or mean; and Xc(X0, X1, X2) means the centers of each terms for contrast. The choice of these threshold vectors won't change the partial order in decision group. Approximate reasoning is a type of reasoning which is neither very exact nor very inexact [16]. Our general framework for reasoning is based on a linguistic model as follows:

Using the logical formula for "and" (intersection) in [17], see Eq. (6), and then obtain the approximate reasoning and decision-making function, see Eq. (7).

$$B_1 \text{ and } B_2 = B_1 \wedge B_2 = \min\{B_1, B_2\} \tag{6}$$

$$\mu_D(s) = \mu_E(s) \wedge \mu_C(s) \wedge \mu_M(s) \tag{7}$$

where μE, μC and μM are the degrees of membership in the term sets for the decision-making terms energy, contrast and mean, respectively, and is computed from the triangular membership function. In Eq. (7), S denotes a trial sound speed set, i.e., the decision group.

After producing all the μE, μC and μM of each one in S, we apply the fuzzy rules in Table 1 with Eq. (6). Then the degrees of membership of VG, G and F sets in T (F) in S are determined by the max value of the approximate reasoning results for each term set, e.g. the one for G is the max one in the reasoning results from Rules 2 to 5. Moreover, the decision for the best sound speed s* is to find the max μD in the term sets of focusing quality. It means that we first search the super μD in VG with the rule 1; if no one in that term set, then continue the search in G, and then in F.

2.3 The flowchart of the Sound Speed Optimization (SSO) system

To simulate our sound speed optimization (SSO) techniques, we developed a computer program for the decision making with fuzzy sets, Figure1, the flowchart of the SSO system.

The beamformed I/Q data is the input for our simulation program. Firstly, each beamformed I/Q acoustic frame is square law and log detected but without scan conversion as the scan conversion will change the texture analysis consequences. Then, compute parameters of image textures in selected region of interest (ROI). The ROI should cover enough areas excluding the far and near field with better signal-to-noise signals and meaningful statistics. To get rid of signals from anechoic areas, the anechoic area is detected with pre-setting thresholds such as mean and variance. Then, the values of textures parameters must be normalized by the pre-setting extreme value in ROI. Next, the degrees of membership of each texture to the corresponding term sets in T(C | E | M) are calculated by the standard triangular membership function [18] with the pre-setting tuple X on account of the examined part. After fuzzifying the texture data with the triangular and shouldered membership functions, we compute the degrees of membership of the term sets in T(F) following the approximate reasoning rules defined in Table I. That is, the degrees of membership about the focusing quality of ultrasound images produced by varied speeds of sound are decided. Finally, we search the sound speed s* that produced the best focusing quality. It means we orderly check degrees of membership of the term sets in T (F) and finds the best one.

3 Results

This section will show the efficacy of the proposed SSO system with three typical cases from ultrasound phantom imaging. First, we take the speckle image as an example to show the relationships between nine image textures and speed of sound and verify the argument from our previous analysis. Then the performance of point targets and cyst images will be discussed.

Test data was acquired from Saset iMago C21 ultrasound scanner using SH5L38 linear array probe at 5MHz center frequency. Also, we used the CIRS Model 054 ultrasound phantom to acquire the speckle, point targets and cyst images. The speed of sound in this phantom is set to be 1544 m/s.

3.1 Speckle image

Figure 2.1 shows speckle images with a depth of 6 cm at a focal depth of 2.8cm. The ROI covers the area of the focal region with about 60 speckle resolution cells in axial direction (0.35mm per cell for 5MHz probe). The direction (dx, dy) for texture analysis is (2, 0) and the window size for texture computations is 16×16.

The normalized texture value in Figure 3 is the average of sample images with different speed of sound (15 sample images for each sound speed at the same location). From Figure 3, one can see that all textures except the group 2 show either peak or valley in the case of matched trial sound speed to the real one.

Three samples in one group of the speckle image are shown in Figure 2.1 where the speed of sound in the phantom is 1544m/s. Using the unmatched system sound speed, the speckle image of Figs. 2.1.b and 2.1.c are worse than Figure 2.1.a in terms of speckle size around the focal region. This has been verified from the spectrum analysis described in [6]: the -10dB spectrum bandwidth for the 1400m/s (Figure 2.1.b) and 1700 m/s (Figure 2.1.c) images are 84% and 92% of the correct one 1544 m/s (Figure 2.1.a), respectively.

The performance of three selective textures on speckle image is shown in Figure 4.a as a function of trial sound speed. One can see that none of the extreme points of the contrast, mean and energy is located at 1544m/s. Thus, it's not reliable to identify the best speed of sound with any one of them, though there are intrinsic relationships to the best detection in general.

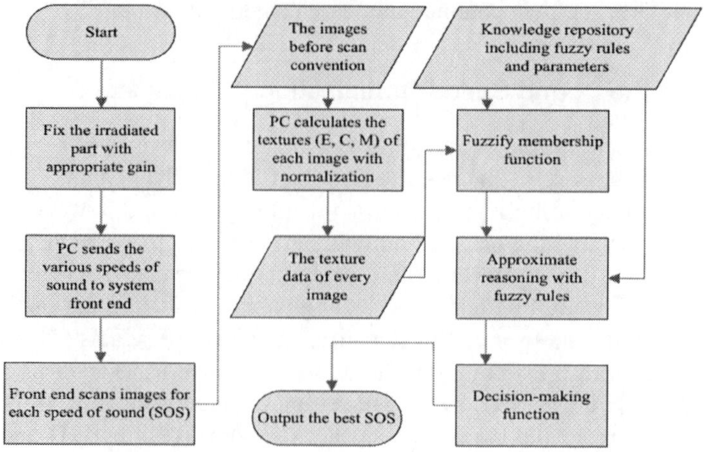

Figure 1. The flowchart of the sound speed optimization system

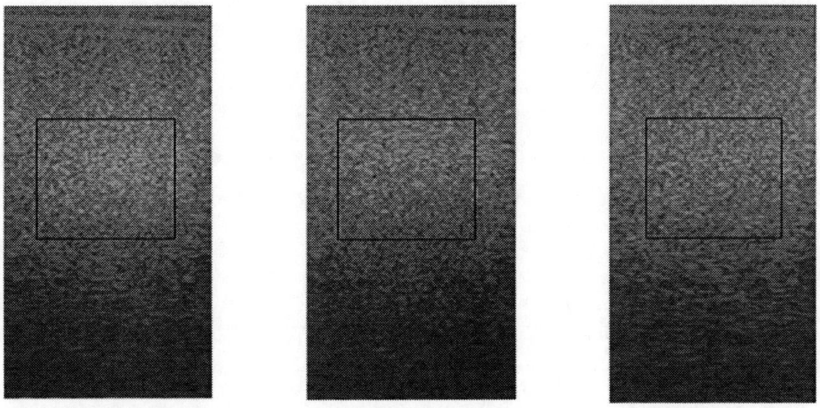

(2.1.a) 1544m/s (2.1.b) 1400m/s (84%) (2.1.c) 1700m/s (92%)

Figurea 2.1. Speckle images with different preset sound speeds

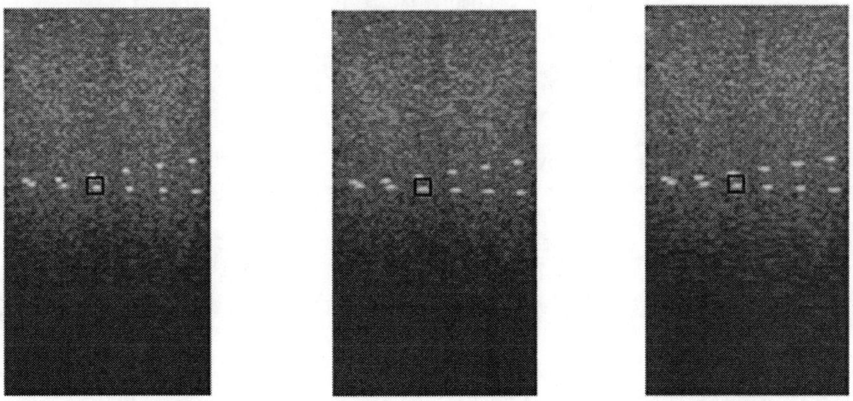

(2.2.a) 1544m/s (2.2.b) 1400m/s (150%) (2.2.c) 1700m/s (142%)

Figure 2.2. Point target images with different preset sound speeds

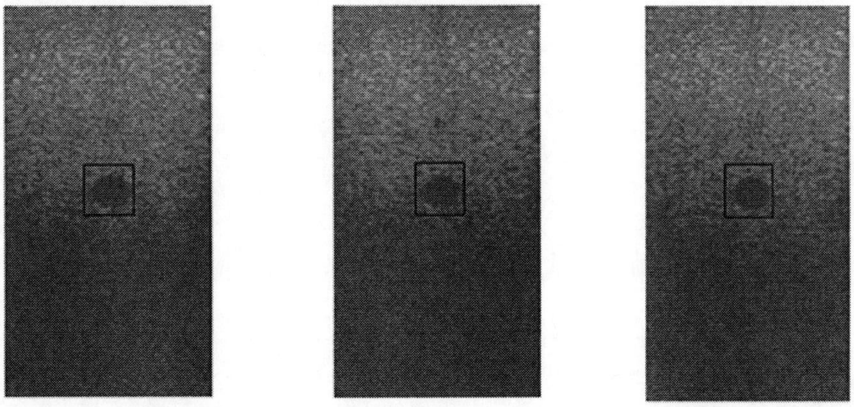

(2.3.a) 1544m/s (9.64) (2.3.b) 1400m/s (9.24) (2.3.c) 1700m/s (9.27)

Figure 2.3. Cyst images with different preset sound speeds

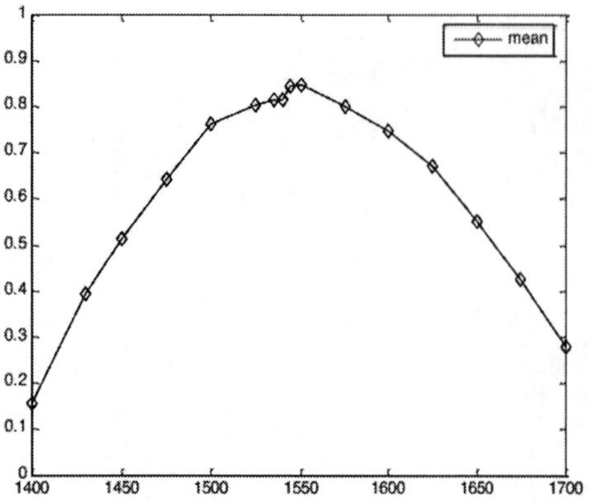

(3.a) Group 1

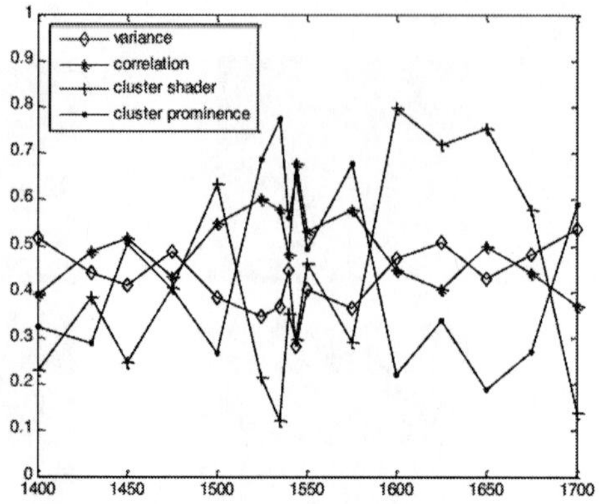

(3.b) Group 2

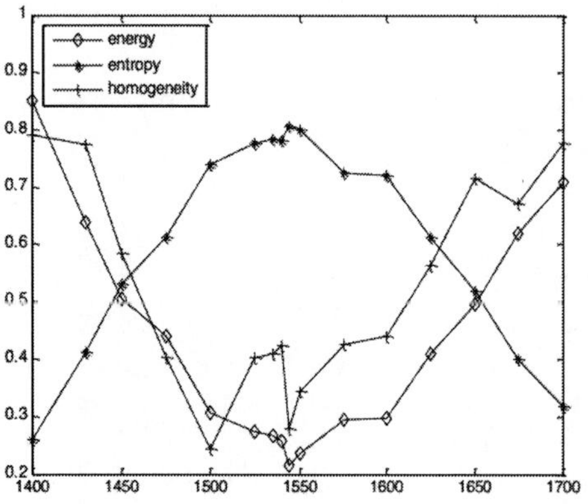

(3.c) Group 3

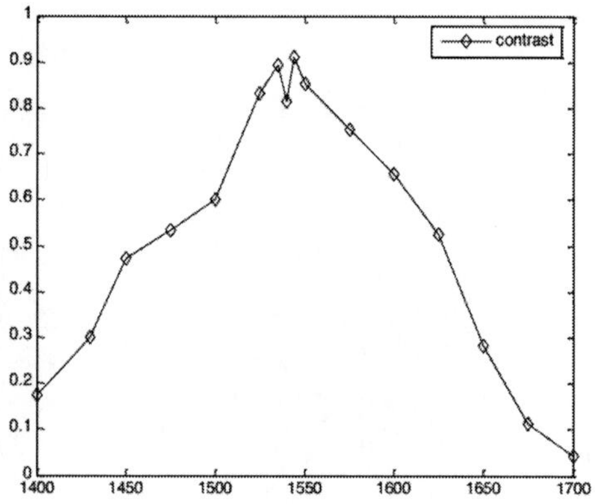

(3.d) Group 4

Figure 3. Nine texture features in four groups as mentioned in Section 2.1 on speckle image as a function of varied speed of sound.

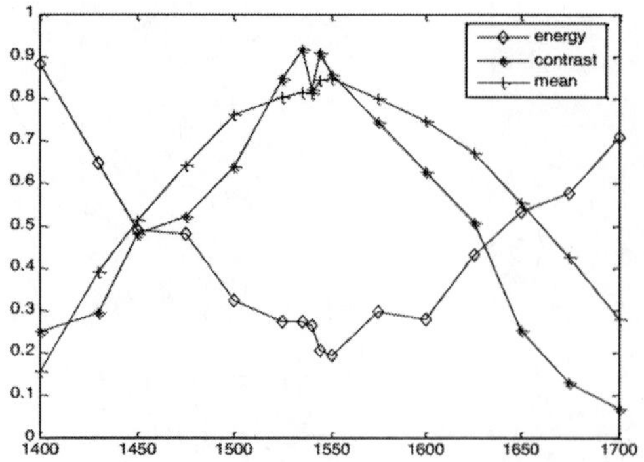

(4.a) speckle image

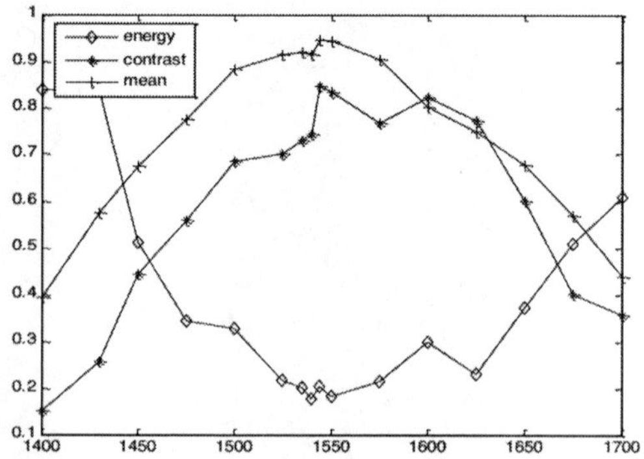

(4.b) point targets image

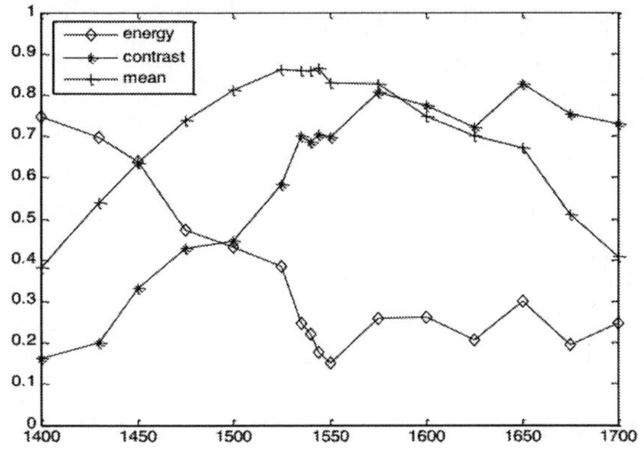

(4.c) cyst image

Figure 4. Three texture features on speckle, point targets and cyst images as a function of sound speed (○: Energy; *: Contrast; +: Mean).

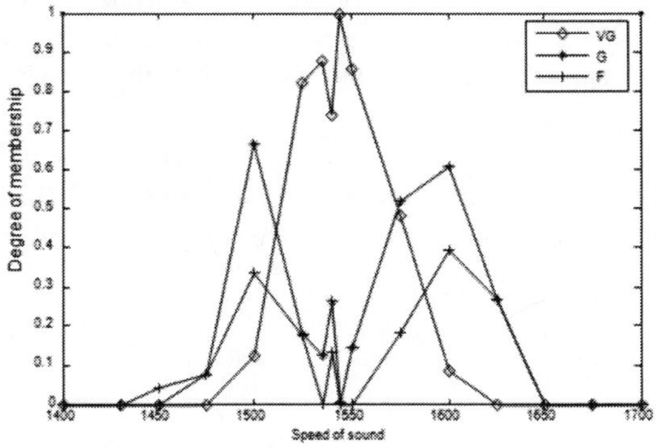

(5.a) speckle image

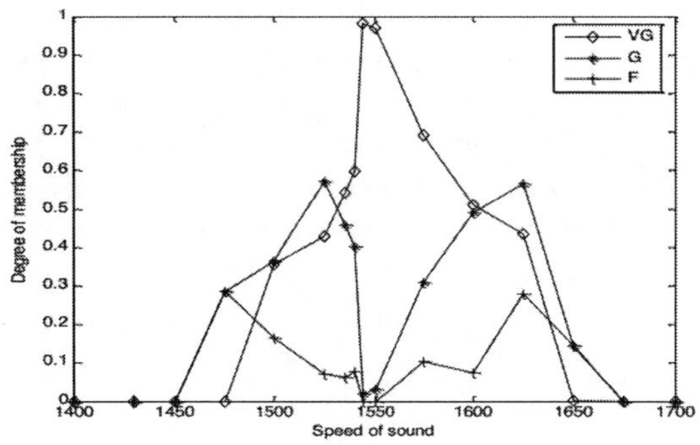

(5.b) point targets image

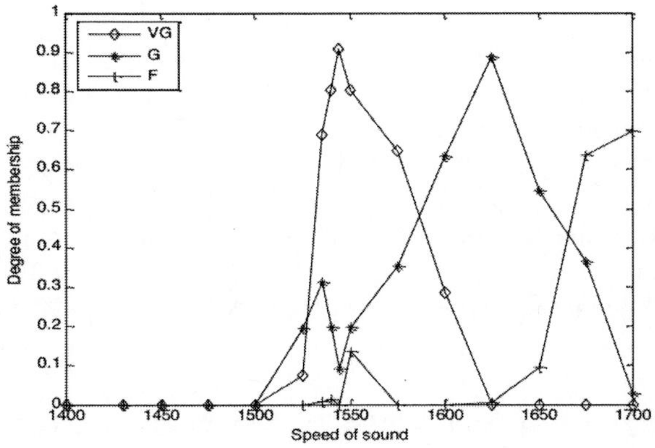

(5.c) cyst image

Figure 5. Outputs of the proposed SSO system with our fuzzy system (○: VG; *: G; +: F)

From our fuzzy processing, we set the XE(0.24,0.5,0.7) (energy), XC(0.3,0.6,0.9) (contrast), XM(0.3,0.6,0.84) (mean) for the membership function of T(C | E | M) on the basis of the texture values and the principle of triangle overlaps between 10 to 50%. Note that both the min and max of each texture value for normalization are fixed according to the location and the gain of imaging settings.

Figure 5.a shows the output of our SSO system where the y-axis represents the degrees of membership of every trial sound speed for each term set in T (F). Here we list all the degrees of membership between the trial speed of sound and every term set in T (F). Following the previous description of our algorithm, we can find the best one in VG is 1544m/s. Besides, we can decide the focusing quality of each trial sound speed by finding its maximum degree of membership.

3.2 Point targets image

The second group is obtained by the same way as the first group but there are several point targets in focal depth. Three samples of one group of the point targets images are shown in Figs. 2.2.a to 2.2.c. It's clear that the wrong system sound speeds of 1400m/s (Figure 2.2.b) and 1700m/s (Figure 2.2.c) degrade the image quality in terms of broadening lateral resolution at point targets compared with the correct one of 1544m/s (Figure 2.2.a). The size of the point target in the rectangular area from 1400m/s and 1700m/s are increased by 50% and 42% respectively, compared to the one from 1544m/s

Because of the point targets spreading, the peaks of contrast and mean of this group are both located at 1544m/s, while the valley of energy is not; see Figure 4.b. Meanwhile, from the output of our SSO system, Figure 5.b with XE(0.2,0.4,0.7), XC(0.3,0.6,0.85) and XM(0.3,0.6,0.92) for the membership functions of T(C | E | M), we find that the best one in VG is the correct one of 1544m/s. Compared with the speckle image group, the broadening of the point targets due to unmatched sound speed, leads a significant decrease in detail and contrast resolution, while the texture of the speckle does not produce such a big change which meets with the results in [8].

3.3 Cyst image

The cyst images are also obtained by the same way as the above group but the cyst exists in focal depth. The region of texture analysis for this group covers about 50 speckle resolution cells in axial direction. Unlike point targets, the cyst shows both reflection and anechoic that is more realistic for tissue analysis. Three samples of the cyst images are shown in Figs. 2.3.a to 2.3.c. From image contrast-to-noise ratio (CNR),

$$CNR = \frac{2(\mu_t - \mu_b)^2}{\sigma_t^2 + \sigma_b^2} \tag{8}$$

we can analyze the performance of our SSO system on cyst images. In Eq. (8) μ_t and μ_b denote the mean of target and background; and σ_t and σ_b are the standard deviation of target and background, respectively. A larger CNR helps better tissue differentiation from target (or tissue lesion) and image background.

In Figure 2.3, the CNR of the correct 1544m/s (Figure 2.3.a) in the rectangular box is 9.64 that is higher than the other two groups: 9.24 from 1400m/s (Figure 2.3.b) and 9.27 from 1700m/s (Figure 2.3.c). It means that the image tissue differentiation degrades due to unmatched speed of sound. As a result of the intrinsic inexactness, the peak of contrast and also the mean and the valley of energy in this group, see Figure 4.c; fairly keep consistent to the focusing quality.

For our SSO system, we can find that the best speed of sound in VG is also accurate, 1544m/s, and the distribution of VG is fairly consistent with the sound speed errors, see the output in Figure 5.c with XE(0.18,0.4,0.6), XC(0.3,0.55,0.75) and XM(0.4,0.7,0.85) for the membership functions of T(C | E | M). It shows that the capability of the fuzzy processing to make the decision is more robust, and also verifies our fuzzy rules to be feasible for different situations on phantom images.

 DOI: 10.1051/bioconf/20170803014

4 Discussions

From test results, the proposed SSO system could choose the accurate speed of sound based on a set of texture parameters. For the point target, it is clear that the sound speed error will cause the point targets broadening and lead to significant decrease in detail resolution. The off-axis response of the lateral point-spread-function (PSF) on speckle images also changes the texture mean. Unlike them, the texture energy does not depend on the brightness in ROI, but relate to the image homogeneity [19]. From point targets images, one can see the amplitude distribution of image pixels in the local window at the focal region doesn't change significantly, though the contrast and the mean brightness apparently changed with the sound speed errors. That's why texture energy shows less sensitive to the focusing quality of the point targets.

On the other hand, all of the selected textures have similar ability to depict the focusing quality in diffuse regions. Since contrast resolution and the homogeneity in local regions don't change significantly, none of them are completely sensitive to the sound speed errors. Compared with the texture contrast and energy, the texture mean is a little more sensitive which meets the theory in [4].

Unlike above twos, the sound speed errors won't make the cyst changed in terms of detail resolution. However, from CNR analysis, it seems that matched sound speed will help both background speckle and anechoic cyst inside with better mean brightness and clear cyst border in texture energy.

In this paper, we construct the fuzzy rules to do the approximate reasoning shown in Table 1. From this approach, we found that the status of each quality factor is not identical. Actually, the texture contrast has little preference for the group decision-making because of its higher sensitive to the sound speed errors which can be seen in [9]. So, not all the rules need to be used in actual processing, e.g. it is not necessary to perform the approximate reasoning when texture contrast definitely belongs to SMALL.

5 Conclusion

In this paper, a method of sound speed optimization based on fuzzy sets using image textures as focusing quality factors has been presented. The method is applicable to ultrasonic imaging system where the texture of the B-mode image is available. We investigated the properties of images textures with sound speed errors and built a decision-making system on the basic fuzzy rules. Results on phantom experiments for speckle, point targets and cyst image groups showed that the best sound speed from a set of images with trial speed of sound can be determined in our SSO system. From the sample images of each group, it's clear that the fuzzy process decision-making approach with texture based focusing quality factors behaves as a reliable and robust tool for optimal sound speed detection. Future works include in vivo tests and possible parameter optimization.

Acknowledgements

This study was supported by the project of Education Department of Sichuan Province (nos.16ZB0381).

References

1. Thomas L. Szabo, *Diagnostic ultrasound imaging: inside out. 1st ed.*, 230-260 (Elsevier Academic Press, 2004)
2. F. A. Duck, *Physical Properties of Tissues*, 73-135 (Academic Press, London, 1990)
3. N.J. Dudley, N.M. Gibson, M.J. Fleckney, P.D. Clark, The effect of speed of sound in ultrasound test objects on lateral resolution, Ultrasound in Med. & Biol, **28**, 1561–1564 (2002)

4. D. Zhao and G. E. Trahey, Comparisons of image quality factors for phase aberration correction with diffuse and point targets: theory and experiments, IEEE Transl. on Ultrasonics, Ferroelectrics, and Frequency control, **38** ,125-132 (1991)
5. D. Napolitano, C. H. Chou, et. al., Sound speed correction in ultrasound imaging, Ultrasonics, **44**, 43-46 (2006)
6. Huanhuan He, Dong C. Liu, *Sound Speed Optimization Based on Acoustic Point Spread Function, IEEE International Conference on Bioinformatics and Biomedical Engineering*, 2112-2115 (2009)
7. Xiaoying Li, Dong C. Liu,An adaptive ultrasound sound speed optimization based on image contrast analysis, J. Biomedical Engineering, **30**, 1094-1097 (2011)
8. M. E. Anderson, M.S. McKeag and G.E. Trahey, The impact of sound speed errors on medical ultrasound imaging, J. Acoust. Soc. Am, **107**, 3540-3548 (2000)
9. S. W. Flax and M. O'Donnell, Phase-Aberration Correction Using Signals from Point Reflectors and Diffuse Scatters: Basic Principles, IEEE Transl. on Ultrasonics, Ferroelectrics, and Frequency control, **35**,758-767 (1988)
10. Wen Liu, Yangjie Cheng and Dong C. Liu, Filter Based Receive-Side Spatial Compounding for Veterinary Ultrasound BMode Imaging, Bio-Medical Materials and Engineering, **24**, 1193-1199 (2014)
11. M. Unser,Sum and difference histograms for texture classification, IEEE Trans. Pattern Anal. Mach. Intell., PAMI-8, 118-125 (1986)
12. J. W. Goodman, Some fundamental properties of speckle, Journal of Optical Society of America, **66**, 1145-1150 (1976)
13. L. A. Zadeh, Fuzzy logic, Computer, **21**, 83-93 (1988)
14. E. Cox, The Fuzzy Systems Handbook: A Practitioner's Guide to Building, Using, and Maintaining Fuzzy System, AP Professional, 340-600 (1994)
15. H.Tanaka,T.Okuda and K.Asai, *Decision-making and its goal in a fuzzy environment, in:L.A.Zadeh,et al.(Eds.), Fuzzy Sets and Their Applications to Cgnitive and Decision Processes,* 257-277 (Academic Press,1975)
16. J.Valente de Oliveira, A design methodology for fuzzy system interfaces, IEEE Trans. Fuzzy Systems, **3**, 404-414 (1995)
17. R. E. Bellman and L. A. Zadeh, decision-making in a fuzzy environment, Management science, **17**, 141-164 (1970)
18. R. R. Yager and D. Basson, Decision making with fuzzy sets, Decision science, **6**, 590-600 (1975)
19. Shuang Gao, Yuhua Peng, Huizhi Guo, Weifeng Liu, Tianxin Gao, Yuanqing Xu and Xiaoying Tang, Texture analysis and classification of ultrasound liver images, Bio-Medical Materials and Engineering, **24**, 1209-1216 (2014)
20. Quan Chen and James A. Zagzebski, "Simulation study of effects of speed of sound and attenuation on ultrasound lateral resolution", Ultrasound in Med. & Biol., **30**, 1297-1306 (2004)
21. Michael Jaeger, Elise Robinson, H Gunhan Akarcay, Martin Frenz, "Full correction for spatially distributed speed-of-sound in echo ultrasound based on measuring aberration delays via transmit beam steering", Physics in Medicine and Biology, **60**, 4497-4505 (2015)
22. Liu Jin-ying, Zhou Qi, Jiang Jue, Song Wei, "Application value of sound speed correction to estimating disagree of homogeneous fatty liver", Journal of Chinese Practical Diagnosis and Therapy, **5**, 445-449 (2015)
23. Guenther Paltauf, Gerhild Wurzinger, Robert Nuster, "Speed-of-sound correction for photoacoustic and laser-ultrasound imaging with an integrating cylindrical detector", Proc. SPIE, Opto-Acoustic Methods and Applications in Biophotonics II, **16**, 1117-1127 (2015)
24. J.F. Krűcker, J.Brian Fowlkes, P.L.Carson, "Sound speed estimation using automatic ultrasound image registration", IEEE Translation on ultrasonics, ferroelectric, and frequency control, **51**, 1095-1106 (2004)

An answer summarization method based on keyword extraction

Qiaoqing Fan[1,a] and Yu Fang[1]

[1]*Department of Computer Science, Tongji University, Shanghai, China*

Abstract. In order to reduce the redundancy of answer summary generated from community q&a dataset without topic tags, we propose an answer summarization algorithm based on keyword extraction. We combine tf-idf with word vector to change the influence transferred ratio equation in TextRank. And then during summarizing, we take the ratio of the number of sentences containing any keyword to the total number of candidate sentences as an adaptive factor for AMMR. Meanwhile we reuse the scores of keywords generated by TextRank as a weight factor for sentence similarity computing. Experimental results show that the proposed answer summarization is better than the traditional MMR and AMMR.

1 Introduction

Due to the users of community question answering (QA) service have different language habits and knowledge levels, the focuses of different users' answers are various especially in consulting type and advising type questions. Different users' answers can be a complement to each other. So we can take advantage of multi-document summarization to merge these answers, but we have to pay more attention on reducing redundancy because these answers are highly similar to each other. This paper takes the keywords of a number of related answers as topic information and propose an answer summarization method based on keyword extraction from the perspective of reducing redundancy.

The key point of TextRank [1] is to construct an undirected graph where the influence of a node is transferred evenly. The work of this paper is to combine tf-idf[b] with word vector to change the influence transferred ratio equation in TextRank considering the impact of semantic and term frequency(TF) on keyword extraction. The answer summarization algorithm in this paper is based on Adaptive Maximum Marginal Relevance (AMMR) [2] and we take the keywords as the topic information for AMMR. We also reuse the influence scores of keywords derived from TextRank as a weight factor for sentence similarity computing to improve the quality of result.

The remainder of this paper is organized as follows. Section (2) provides a review of related works. Section (3) describes our keyword extraction method. Section (4) describes the answer summarization method based on keyword extraction. Section (5) shows the experimental result. The last section outlines conclusions and future directions.

[a] Qiaoqing Fan: 1433238@tongji.edu.cn

[b] https://en.wikipedia.org/wiki/Tf–idf

2 Related works

Research on keyword extraction has been ongoing for years. Mihalcea R et al. proposed TextRank based on PageRank. And then on the basis of TextRank, Li et al. [3] explored the use of tags for improving the performance of webpage keyword extraction task, Xia et al. [4] used the influence of word location to construct a word graph and calculate the probability transition matrix. In addition, there are some machine learning methods for keyword extraction, typically LDA [5]. The biggest difference between TextRank and LDA is that training corpus ahead is not required. Although LDA analyse the linkages between words form semantic level, it is required to retrain the model when new corpus comes. So TextRank is more simple in use. TextRank select only lexical units of a certain part of speech to construct an influence-evenly-transferred word graph. According to Matthew Effect[a], a word should get much more attention from neighbouring synonyms to highlight the importance of itself, so do high frequency words. Therefore, this paper will take term frequency and word sense into account for keyword extraction task.

The summarization is defined into two ways according to the number of documents to be summarized, respectively are Single Document Summarization and Multiple Document Summarization. Extractive summarization and abstractive summarization approach are used [6]. The research object of this paper is Extractive Multiple Document Summarization.

Answer summarization was firstly proposed by Liu et al. [7]. Yin et al. developed a hierarchical clustering method to group similar questions and a ranking based summarization for representing an answer [8]. Wang et at. proposed a topic-centric answer summarization method called Adaptive Maximum Marginal Relevance(AMMR) in 2013 [2]. AMMR takes the e-mail headers and webpage tags as topic information to automatically adjust the weighting of topic relativity and redundancy when picking relevant sentences from candidate answers. However, not all QA datasets are labelled with such tags or have headers. Therefore, this paper proposes a keyword-extraction-centric answer summarization method based on Wang' work.

3 Keyword extractions

TextRank is derived from PageRank [9]. It splits text into a number of units. In keyword extraction the units are a set of words. All these words that are added into a graph and an edge are added between those words that co-occur within a window of n words. After the graph is constructed, the ranking algorithm described in [1] is run on the graph for several iterations until it converges. Once a final score is obtained for each unit, select the top m units with highest scores as keywords.

However, TextRank only construct an influence-evenly-transferred word graph. Considering that Figure 1 is a word graph of a certain document. In TextRank, the influence word b contributes to word a is one third of b because three nodes are linked with b. In fact, one word should transfer much more influence to the other if the other word is closer to it in semantic space. So we need an method to evaluate the semantic relationship between these words.

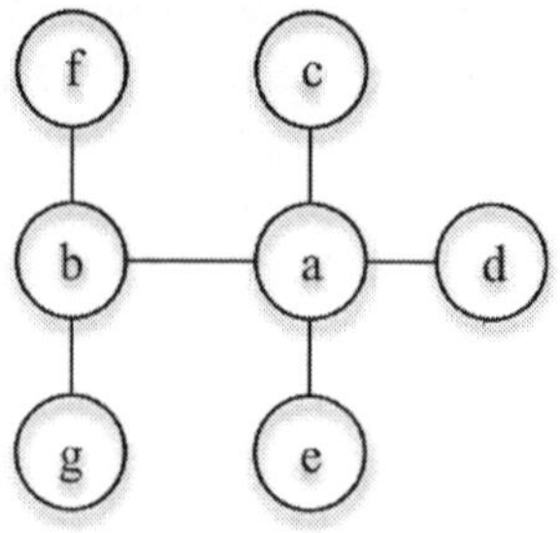

[a] https://en.wikipedia.org/wiki/Matthew_effect

Figure 1. A word graph of a document

Word2vec provides such a kind of evaluation method. It was proposed by Mikolov et al. in 2013 to compute continuous vector representations of words from very large datasets [10,11]. The vectors can be used to compute semantic similarity since it captures the context of a word during the process of training.

Formally, let $W=\{w_0,w_1,\ldots\ldots,w_{N-1}\}$ be the word set of document d. N is the number of words in W. $W' \in W$ is the word set extracted from W, including nouns, verbs and words in the user dictionary. P is the numer of words in W'. Each word in W can be represented in K-dimensional vector after training the corpus with word2vec. Let $\mathbf{vec_a}=[v_0,v_2,\ldots v_k\ldots,v_{K-1}], 0 \ll a<N, 0 \ll k<K$ denotes the vector of w_a. Then the semantic similarity of w_a and w_b can be measured by cosine similarity:

$$sem_{ab} = \frac{\sum_{k=0}^{K-1} vec_a[k] * vec_b[k]}{\sqrt{\sum_{k=0}^{K-1} vec_a[k]^2}\sqrt{\sum_{k=0}^{K-1} vec_b[k]^2}} \tag{1}$$

And the influence transferred ratio euqation in TextRank is re-weighted by the semantic similarity and re-normalized as shown below:

$$S(b,a) = \frac{sem_{ab}}{\sum_{c \in Out(b)} sem_{cb}} S(b) \tag{2}$$

Let $Out(b)$ be the index set of words that w_b points to(successors) in word graph, $S(b)$ is the influence score of w_b. In addition, considering TF is lost in TextRank and a word should transfer much more influence to its adjacent high-frequency words. So we also introduce tf-idf for the influence transferred ratio equation (IDF is used to inhibit the high-frequency words which are widely exist in corpus). The variant of IDF we adopt is the same as BM25[a]. The final influence transferred ratio equation is shown below:

$$S(b,a) = \frac{sem_{ab} * tfidf_a}{\sum_{c \in Out(b)} sem_{cb} * tfidf_c} S(b) \tag{3}$$

And the influence score of w_a is :

$$\mathrm{S(a)} = (1-\mathrm{d}) + \mathrm{d} * \sum_{b \in In(a)} \mathrm{S}(b,a) \tag{4}$$

$In(a)$ is the index set of words that point to w_a.

Let m be the number of keywords to be extracted. The whole process to extract keywords is shown below:

1. Do word segmentation, part-of-speech tagging and removing stop words on document set to generate W。
2. Filter other words out except nouns, verbs and other words (in user dictionary) from W to generate W'.
3. Construct word graph G for W'.
4. The algorithm described above is run on G for several iterations until it converges.
5. Once a final score is obtained for each word in the graph, words are sorted in reversed order of their score, and the top m words in the ranking list are returned as keywords.

4 Answer summarization

Maximum Marginal Relevance (MMR) was proposed by Carbonell et al. in 1998 [12]. Generally, the diversity distribution over text content is different from one the other. Fixed linear interpolation of MMR cannot be well adapted to this, for selecting a sentence from candidate set whether tends to enhance diversity or topic relevance to user requirements. So Wang proposed AMMR from the perspective of topic cohesion of candidate sentences, which is based on an adaptive factor. The equation is shown below:

---

[a] https://en.wikipedia.org/wiki/Okapi_BM25

$$\text{AMMR} = \text{Arg}\max_{s_i \in R\setminus S}[(1-\frac{M_T}{\text{M}}) * (Sim_1(s_i, q) - \frac{M_T}{\text{M}} * max_{s_j \epsilon S} Sim_2(s_i, s_j)] \quad (5)$$

Here q is a query or user profile, R is the set of candidate summary sentences. S is the subset of sentences in R already selected. $R\setminus S$ is the set difference. Sim_1 and Sim_2 are the similarity metric used in sentences retrieval and relevance ranking between sentences. M is the number of the sentences in R. M_T is the number of the sentences containing any topic word (tags or headers instead in [2]) in R. It means that the topic cohesion of S is more compact when the ratio $\frac{M_T}{M}$ is larger and AMMR selecting a sentence from R will focus on reducing redundancy. Similarly, if $\frac{M_T}{M}$ is smaller, it means that the topic cohesion of S is less compact, AMMR will focus on enhancing topic relevance.

Keywords can specifically reflect topic information of a text because keyword contains key information of it. Here the M_T in AMMR is set to the number of the sentences containing any keyword where M is not changed. We can see that different answers have different M_T value. So it is an adaptive procedure and $\frac{M_T}{\text{M}}$ is the adaptive factor.

The similarity metric Sim_1 and Sim_2 also have an important influence on result quality beside the adaptive factor above. The similarity metric of this paper is the same as [12] used:

$$\text{sim}_{sem}(s_1, s_2) = \frac{\sum_{w_a \epsilon s_1} sim_m(w_a, s_2) + \sum_{w_b \epsilon s_2} sim_m(w_b, s_1)}{|s_1| + |s_2|} \quad (6)$$

$sim_m(w_a, s_1)$ is the word in s_1 that has the highest semantic similarity to the word w_a. $|s_1|$ is the number of words in s_1. The equation (1) is continue being used to compute word semantic similarity.

Two sentences are similar or not always depends on the nouns and verbs in them. Therefore, in order to highlight the impact of these words in computing sentences semantic similarity, this paper further makes use of the influence score of keyword derived from TextRank and proposes the following similarity calculation equation:

$$\text{sim}_{semk}(s_1, s_2) = \frac{\sum_{w_a \epsilon s_1} sim_m(w_a, s_2) * Score(w_a) + \sum_{w_b \epsilon s_2} sim_m(w_b, s_1) * Score(w_b)}{\sum_{w_a \epsilon s_1} Score(w) + \sum_{w_b \epsilon s_2} Score(w_b)} \quad (7)$$

$Score(w_a)$ is the influence score of w_a, it is derived from TextRank, just the value $S(a)$ (see equation 4). However, a part of words doesn't have such a score due to the algorithm we used to extract keywords only focus on nouns, verbs, and user dictionary words. So we do a smooth operation for these words. The meaning of m, W', W, P and N is not changed. $W'' = W - W'$, means the set difference of W and W'. Let $score[P]$ be the array of influence score in reversed order. The influence score of all words $Score(w_i)$ can be represented as :

$$Score(w_i) = \begin{cases} S(i) & w_i \epsilon W' \\ \frac{\sum_{j=m}^{P-1} score[j]}{P-m} & w_i \epsilon W'' \end{cases} \quad (8)$$

From the equation we can see that we used the average score of non-keywords to present the score of word in W''. This smooth operation further highlights the impact of keywords on sentence similarity computing.

Let l be the number of summary sentences to be extracted. The whole process of our answer summarization is:

1. Use the method we introduced in section (3) to extract keywords from candidate answer texts.

2. Set the value $\frac{M_T}{M}$ for AMMR and compute influence score for all words in candidate answer texts through equation (8).

3. Score all the sentences in candidate answer texts for l iterations using equation (5), at each iteration, sentence with highest score is selected and added into summary list。

5 Experimental results and analysis

 DOI: 10.1051/bioconf/20170803015

The tokenizer we used is ICTCLAS, a Chinese word segmentation system. The data set used in the experiments is a collection of 44254 QA pairs from the 120ask website. In order to ensure the quality of word segmentation, we collected 132,371 medical terminologies from Sogou thesaurus, covering medicine name, disease name, anatomy, pharmaceutical company name, etc. Then after the text participle and removing stop words, we trained word2vec model with skip-gram using hierarchical softmax. The dimension of word vector was 50 and the window size was 5.

To evaluate our system, we selected 100 questions from the dataset, and got four to five most relevant questions and corresponding answers by Lucene[a] of each question. All these answers of each question were merged into one text as a candidate summary sentences set. Each candidate summary sentences set has an average of 19 sentences, and the 100 candidate summary sentences sets were distributed to five trained annotators to extract keywords and summaries. Finally, each candidate summary text has 4 keywords and 9 summary sentences on average.

5.1 Result of Keyword Extraction

All the other words were filtered out beside nouns, verbs, and user dictionary words. The *d* value of TextRank is set to 0.85 and window size is 10. ST-TextRank and S-TextRank are the keyword extraction method this paper introduced. The difference between them is that ST-TextRank adopts equation (3) as influence transferred ratio equation, but S-TextRank adopts equation (2). We compared our method with traditional TextRank and [8] (denoted as P-TextRank). P-TextRank achieved the best result after several attempts when set λ to 80, α to 0.6, β to 0.15 and γ to 0.25. The evaluation metrics we used are Precision, Recall and F-measure and the experimental results are compared in the extraction of 3, 5 and 10 key words. Results are shown in Table 1.

Table 1. Comparison of the performances of different keyword extraction algorithms

Algorithm	Numbers of keywords	Precision	Recall	F-measure
ST-TextRank	3	0.520	0.359	0.425
	5	0.406	0.463	0.433
	10	0.266	0.601	0.369
S-TextRank	3	0.433	0.304	0.357
	5	0.350	0.402	0.374
	10	0.258	0.582	0.358
P-TextRank	3	0.510	0.362	0.423
	5	0.400	0.468	0.431
	10	0.280	0.643	0.390
TextRank	3	0.407	0.290	0.339
	5	0.354	0.415	0.382
	10	0.255	0.583	0.355

From Table 1 we can observe that S-TextRank is better than TextRank, it proves the validity of word vector we use in influence transferred ratio equation. ST-TextRank outperforms all the other algorithms in all metrics when the number of keywords is 3, it proves the validity of the combination of word vector and tf-idf in influence transferred ratio equation. Although the result P-TextRank is not as good as ST-TextRank, but better than S-TextRank, evenly better than ST-TextRank when the

[a] https://lucene.apache.org/

number of keywords is 5. We think the main reason is that it is difficult for P-TextRank to coordinate all the weight factors to make the model achieve best.

5.2 Result of answer summarization

We followed the same evaluation metrics for answer summarization as Wang adopted in [2] (Accuracy, Redundancy and Summary Quality). Accuracy is a proportion of all correctly predicted sentences compared to all sentences. Redundancy reflects how much information an answer summary contains. Summary Quality is the difference between Accuracy and Redundancy. Here equation (6) was adopted as similarity computation for Redundancy. We compared our method with MMR in different λ values and Wang's AMMR . For the reason that Wang took the topic tags and e-mail headers as the topic information and our QA data set has not been labelled with such tags, or has headers. So a similar approach was adopted, we took the words from question after segmentation as headers when we implemented the algorithm (denoted as Q-AMMR). QF-AMMR is the implementation of Q-AMMR with removing stop words from headers which just contain nouns, verbs and user dictionary words. KW-AMMR and K-AMMR are the algorithms based on keyword extraction we introduced in this paper. They are different in sentence similarity computation, where KW-AMMR takes equation (7) and K-AMMR takes equation (6). In addition, according to table 1, ST-TextRank got best when the number of keywords was set to 3, therefore we chose 3 as the number of keywords for answer summarization. The experimental results are shown in table 2.

From table 2, we can see that KW-AMMR and K-AMMR outperform all the other algorithms, it proves the validity of our thought to take keywords as topic information in AMMR. KW-AMMR outperforms K-AMMR not only in precision, but also redundancy, it proves the usefulness of the equation (7) in similarity calculation for reducing redundancy. Although the redundancy of KW-AMMR is higher than MMR when the λ is set to 0.1 (or 0.2), but the accuracy is too low. This just shows the efficiency of adaptive factor used in AMMR compared with MRR in different λ values from 0.1 to 0.9.

We also notice that although the accuracy of Q-AMMR and QF-AMMR are maintained at a high level, the redundancy and summary quality do not achieve a better result. This is mainly caused by the following reasons: a) the M_t is too small (even zero) after segmenting the question into words and removing stop words; b) when the users write down their answers, a lot of synonyms and other words are expanded, which are widely exist in other sentences, and such a class of words that can reflect the topic information are not captured during keywords extraction. To summarize, the words of question do not have the ability to fully reflect the topic information, so there is a high rate of redundancy in QF-AMMR and Q-AMMR.

Table 2. Comparison of the performances of different answer summarization algorithms

Algorithm		Precision	Redundancy	Summary Quality
KW-AMMR		0.618	0.1361	0.4819
K-AMMR		0.604	0.1362	0.4678
Q-AMMR		0.582	0.1510	0.431
QF-AMMR		0.574	0.1568	0.4172
MMR	λ=0.1	0.554	0.1089	0.4451
	λ=0.2	0.564	0.1104	0.4536
	λ=0.3	0.562	0.1122	0.4498
	λ=0.4	0.580	0.1207	0.4593
	λ=0.5	0.574	0.1367	0.4373
	λ=0.6	0.570	0.1543	0.4157

 DOI: 10.1051/bioconf/20170803015

λ=0.7	0.586	0.1633	0.4227
λ=0.8	0.574	0.1688	0.4052
λ=0.9	0.570	0.1735	0.3965

6 Conclusion

An answer summary method is proposed this paper. We combine tf-idf with word vector to change the influence transferred ratio equation in TextRank and take the ratio of the number of sentences containing any keyword to the total number of candidate sentences as an adaptive factor for AMMR. Also, we reuse the scores of keywords generated from TextRank as a weight factor for sentence similarity computing. The experimental results show that the method based on keyword extraction in this paper has achieved great results in improving accuracy and reducing redundancy.

The next step we will focus on the dynamic determination of keywords' number, rather than a fixed value and optimizing the quality of the answer summary, for instance providing context information for the summary and making the sentences more fluent.

References

1. R. Mihalcea, P. Tarau, TextRank: Bringing order into texts, *Conference on Empirical Methods in Natural Language Processing*, 404(2004)
2. B. Wang, Research on the semantic mining of question-answer pairs in web communities, Harbin Institute of Technology (Doctoral dissertation, 2013)
3. P. Li, B. Wang, Z. Shi, et al, Tag-TextRank: A Webpage Keyword Extraction Method Based on Tags, Journal of Computer Research and Development, **49**, 2344 (2012)
4. T. Xia, Study on Keyword Extraction Using Word Position Weighted TextRank, New Technology of Library & Information Service, 30 (2013)
5. L. Zhang, H. Wang, Research on Sentence Extraction in Text Summarization, Journal of Chinese Information Processing, **26**, 97 (2012)
6. Y. Liu, S. Li, Y. Cao, et al, Understanding and summarizing answers in community-based question answering services, Proceedings of the 22nd International Conference on Computational Linguistics-Volume 1, Association for Computational Linguistics, **1**, 497 (2008)
7. Y. Yin, Y. Zhang, Liu X, et al, HealthQA: A Chinese QA Summary System for Smart Health, *ICSH*, 51 (2014)
8. L. Page, S. Brin, Motwani R, et al, The PageRank citation ranking: bringing order to the web, Stanford InfoLab, **9**,1 (1999)
9. T. Mikolov, K. Chen, G. Corrado, et al, Efficient estimation of word representations in vector space, arXiv preprint arXiv:1301 (2013)
10. T. Mikolov, I. Sutskever, K. Chen, et al, Distributed representations of words and phrases and their compositionality, Advances in neural information processing systems, **26**, 3111 (2013)
11. J. Carbonell, J. Goldstein, "The use of MMR, diversity-based reranking for reordering documents and producing summaries", *Proceedings of the 21st annual international ACM SIGIR conference on Research and development in information retrieval*, ACM, 335 (1998)
12. R. Malik, L. Subramaniam, S. Kaushik, Automatically Selecting Answer Templates to Respond to Customer Emails, IJCAI, **7**, 1659 (2007)

DOI: 10.1051/bioconf/20170803016

Lateral ultrasound strain imaging using subband processing

Hui Peng[a], and Juhong Tie

School of Software Engineering, Chengdu University of Information Technology, Chengdu, China

Abstract. Since most biological tissues are nearly incompressible, the axial compression leads to expansion in the lateral and elevational directions. Although axial strain is the main component of three dimensional strain field, estimation of the lateral strain may provide important additional information on the tissue mechanical properties. In this paper, we employed the idea and principle of image compounding and proposed a subband processing method to estimate lateral strain. To keep lateral radio freqency (RF) signal bandwidth and strain resolution, we split axial RF signal into several subband signals and then estimate lateral strains of these subband signals along lateral direction, finally average these strains to get a lateral compounded strain image. The simulation results demonstrate that the elastographic signal-to-noise ratio of the lateral compounded strain image is improved by 48% using this subband processing method, compared with the conventional method.

1 Introduction

Ultrasound strain imaging is a new medical imaging modality that can provide relative stiffness information of biological tissue for clinician to diagnose deceases such as tumors or cancers. However, poor quality of strain imaging may degrade its clinical value [1]. Therefore, improving the quality of strain image is an important and meaningful research direction [2-5].

For ultrasound B-mode imaging, spatial compound method is used to reduce speckle noise. Several different angular B-mode images from same scanning tissue region are averaged to produce one compounded B-mode image which SNR can be improved by $\sqrt{N}$ times [6]. Reference [4] first applied spatial compound method to strain imaging. Reference [5] proposed a transmit-side frequency compound method in which several sub transmit-side frequency strain images are averaged to produce a compounded strain image to reduce elastography nosie. We first proposed a subband processing method to estimate axial strains [3].

Since most biological tissues are nearly incompressible, the axial compression leads to expansion in the lateral and elevational directions [7]. Although axial strain is the main component of three dimensional strain field, estimation of the lateral displacement and strain may provide important additional information on the tissue mechanical properties. By taking advantage of the lateral estimation technique, shear strains and the Poisson's ratio have been shown capable of being estimated [8]. In this paper, we employed the idea and principle of image compounding and proposed a subband processing method to estimate lateral strain. We also tested this method in simulations.

[a] Corresponding author: penghui@cuit.edu.cn

2 Method

2.1 Strain image compounding theory

The correlation of two sub-elastograms P and Q, with the size of m*n pixels can be evaluated by the normalized correlation coefficient, p, 0<=p<=1 [5]:

$$\rho = \frac{\sum_{i=1}^{m}\sum_{j=1}^{n}(P_{i,j}-\bar{P})(Q_{i,j}-\bar{Q})}{\sqrt{\sum_{i=1}^{m}\sum_{j=1}^{n}(P_{i,j}-\bar{P})^2\sum_{i=1}^{m}\sum_{j=1}^{n}(Q_{i,j}-\bar{Q})^2}} \tag{1}$$

where $P_{i,j}$ and $Q_{i,j}$ are strains at location (i, j), $\bar{P}$ and $\bar{Q}$ are mean strains for P and Q respectively

The quality of strain image is usually evaluated by elastographic signal-to-noise ratio (SNRe) as:

$$\mathrm{SNRe} = \frac{\mu}{\sigma} \tag{2}$$

where μ and σ are, respectively, the mean and standard deviation of the strain estimates in region of uniform elasticity.

Define the random variables Xi(i=1,2,...,N, N>=2) of N sub-elastograms. The compounded elastogram, $\bar{X}$, is obtained by averaging N sub-elastograms. The SNRe of the compounded elastogram will be [5]:

$$\mathrm{SNRe}' = \frac{E(\bar{X})}{\sqrt{Var(\bar{X})}} = \frac{\mu}{\sigma}\sqrt{\frac{N}{1+(N-1)\rho}} \tag{3}$$

where $E(\bar{X})$ and $Var(\bar{X})$ denote the expectation and the variance of the compounded elastogram respectively. Therefore, from eqn. (3) we know the SNRe of the compounded elastogram will be increased by a factor of $a=\sqrt{N/(1+(N-1)\rho)}$ which has a maximum of $a=\sqrt{N}$ when all sub elastograms are completed decorrelated (i.e. p=0).

2.2 Subband processing

To achieve high axial strain resolution, some short pulses, such as two circle of sine wave, are sent into tissue. So, the echo radio frequency signal is usually broadband signal. The idea of subband processing is that a broadband signal is divided into many narrowband signals and these narrow band signals should cover whole frequency band to exploit all energy of the broadband signal [9-10]. To keep lateral RF signal bandwidth and strain resolution, we split axial RF signal into several subband signals and then estimate lateral strains of these subband signals along lateral direction, finally average these strains to get a compounded strain image. Bandwidth is important to strain imaging, because SNRe decreases with bandwidth according Cramer–Rao lower bound theory which states that the best achievable displacement variance is related to the center frequency, band width, correlation coefficient, window length and eSNR [11]. There are many subband methods. We adopt bandpass filter method to split axial RF signal to several subband signal. This method is shown in eqn. (4).

$$x_p(t) = BPF\{r(t),\ f_{p1}, f_{p2}\} \tag{4}$$

where r(t) is axial RF signal; f_{p1} and f_{p2} are the starting and ending frequency of pth subband respectively; $x_p(t)$ is pth subband RF signal; BPF{} denotes a function of bandpass filter. The center frequency of the pth subband is $f_p=(f_{p1}+f_{p2})/2$. For a axial RF signal of center frequency 5MHz, we split it into three subband signals which center frequencies are 3MHz, 5MHz and 7MHz respectively.

2.3 Estimation of lateral sub strain image

Each subband signal will be estimated to get its sub lateral strain image after Rf signal was devided into three subband signals. A strain image can be produced by displaying spatial derivatives from the displacement field. So, the key step of strain estimator is to estimate displacement of every correlation window according pre- and post deformation RF signals. There are many strain estimate methods such as cross-correlation, phase zero method, for producing axial strain image. Since lateral RF signal has more decorrelation noise inducing from axial displacements, some displacement estimate methods are not fit to estimate lateral displacement. We adopt phase zero method expressed in equation (5) to estimate sub lateral displacements [12].

$$\hat{d}_{k+1,n} = \hat{d}_{k,n} + \frac{\arg(\exp(-jw_0\hat{d}_{k,n})\sum_{t=n\Delta t}^{n\Delta t+T} x_{1b}^{*}(t)x_{2b}(t-\hat{d}_{k,n}))}{w_c} \quad (5)$$

where w_0 is the center frequency of sub lateral RF signal; w_c is the local frequency centroid; $\triangle t$ is the window spacing, T is the window length, $n\triangle t$ is the position of the start of the nth window. x_{1b} (t) and x_{2b} (t) are the baseband analytic signals of the sub lateral RF signals; $\hat{d}_{k,n}$ is the displacement estimate of the nth window at the kth iteration. At the first window, the search is initialized with $\hat{d}_{0,0}=0$. Subsequent searches are each initialized with the final displacement estimate from the previous window, i.e., $\hat{d}_{0,n}=\hat{d}_{K,n-1}$. We utilize the amplitude modulation correction method to correct amplitude modulation noise [13].

2.4 Lateral RF signal interpolation method

Since most clinical systems are using low-cost transducers with a small number of channels, the sampling rate of RF signal in lateral direction is greatly lower than that of axial RF signal. For a linear array transducer, the lateral sampling rate is determined by the number of scanning lines, and the axial sampling rate is determined by the axial sampling frequency. To achieve high accurate displacement estimates, the axial sampling frequency is generally at least 40MHz, so the sampling spacing in axial direction is generally at most 0.0385mm (supposed the sound speed is 1540m/s). For a 4 cm width of transducer with 192 channels, the sampling spacing in lateral direction is 0.208mm. So the sampling rate in lateral direction is lower five times than that in axial direction. Low sampling rate in lateral direction will reduce the accuracy of lateral displacement estimates [8]. If lateral sampling frequency satisfies the Nyquist sampling theorem, the lateral signal can be interpolated to improve the accuracy of lateral displacement estimation. If the pitch of a linear array transducer is smaller than half beamwidth (as expressed in formula (6)), the lateral sampling frequency satisfies the Nyquist sampling theorem. Many interpolation methods like linear interpolation and cubic spline interplolation, etc. can be used in lateral RF signal. In this paper, we use cubic spline interplolation method, because cubic spline interplolation can increases the correlation coefficient compared with linear interpolation [8]. The interpolation coefficient is 1:10.

$$PITCH \le \frac{BW}{2} = \frac{2F\lambda}{2} = F\lambda \quad (6)$$

where BW is beamwidth; F is F-number of tranducer; λ is the wavelength of the emitted pulse.

3 Simulation method and results

3.1 Simulation method

The simulated RF data is obtained by the convolution of 2D PSF and tissue scatters.

$$r(x,z) = p(x,z) * s(x,z) + n(x,z) \quad (7)$$

where r(x,z) is the 2D RF signal ; x and z are the lateral and axial coordinates respectively; p(x,z) is the psf which is a Gaussian modulated sine function; s(x,z) is the scatters function; n(x,z) is white noise function. Scatterers are deformed by moving the location of the scatterers in the lateral direction using the analytic results of a simplified model developed in reference [14]. While both lateral and axial displacement is calculated in reference [14], we are only concerned with lateral displacement. The lateral displacement u(x,z) is:

$$u(x,z)=\frac{x}{r}\left(\frac{x^2}{r^2}Q(r)+\frac{z^2}{r}\frac{\partial Q(r)}{\partial r}\right) \tag{8}$$

and Q(r) is defined as:

$$Q(r)=\begin{cases} P\dfrac{2}{1+K}r & r\le R \\ P\left[1+\dfrac{1-K}{1+K}\left(2\dfrac{R^2}{r^2}-\dfrac{R^4}{r^4}\right)\right]r & r>R \end{cases} \tag{9}$$

where r is the distance between a sample volume and the center of the coordinate system, and $r^2=x^2+z^2$;P is a deformation scalar related to the magnitude of the applied force; R is the radius of the hard inclusion centered in the medium; K is the ratio of the young's modulus of the hard inclusion to that of the background soft medium; In this paper, we choose P=0.01, K=10,R=5mm.

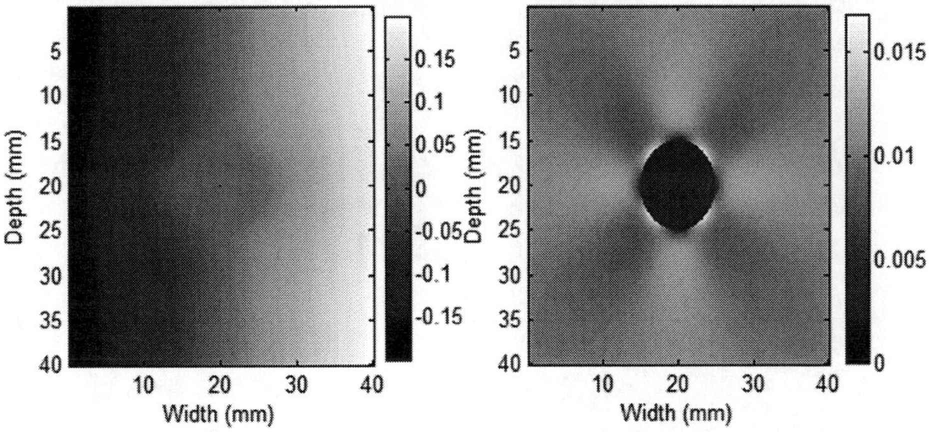

(a) Lateral displacements (b) Lateral strains

Figure.1 Lateral displacement and strain of scatter model

The center frequency of the conventional short pulse is 5MHz. The center frequencies of three subband signals are 3MHz, 5MHz and 7MHz respectively.

3.2 Simulation results

We compared the performance of subband method and conventional method in terms of SNRe and CNRe. For a pair of pre- and post-compressed RF fame, we use subband method and conventional method on them respectively to get the lateral strain image. Fig. 2 shows the results. Fig.2a shows lateral strain image using conventional method in which SNRe of background is 3.67 and SNRe of lesion is 3.77. Fig.2b , Fig.2c and Fig.2d show the lateral strain images using 3MHz,5MHz and 7MHz subband respectively. Fig.2e shows the compounded lateral strain image in which SNRe of background is 5.43 and SNRe of lesion is 5.90. Therefore, Compared with the conventional method, SNRe of background, SNRe of lesion and CNRe are increased by 48%, 56% and 120% respectively, using subband processing.

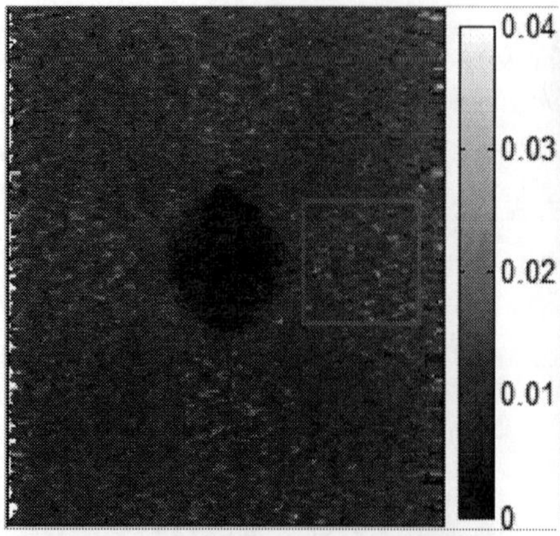

(a)Conventional method

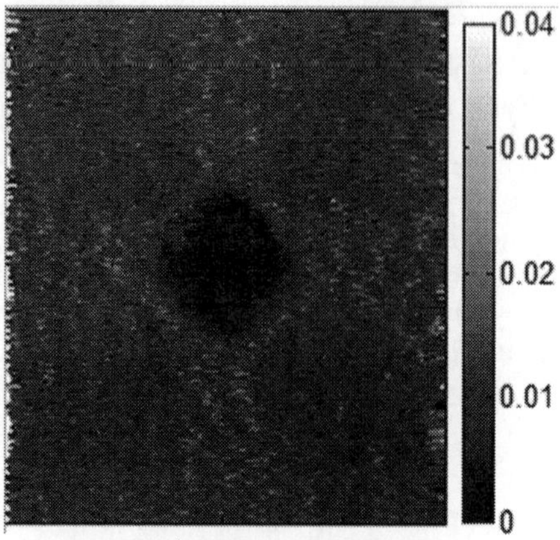

(b)3MHz subband

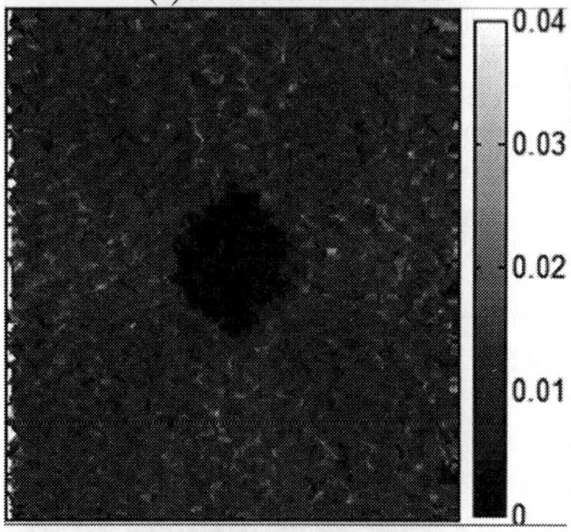

(c) 5MHz subband

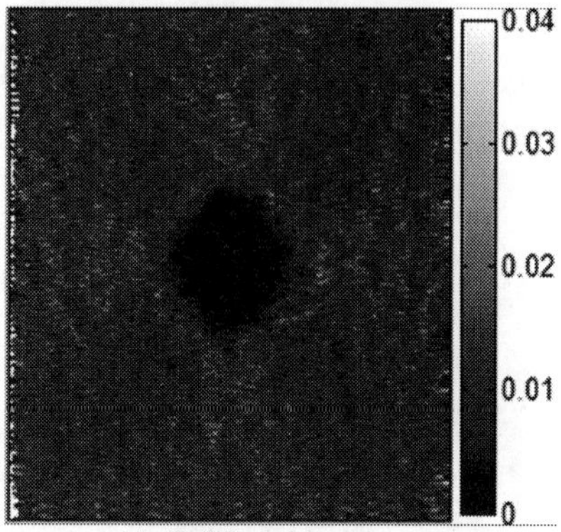

(d) 7MHz subband

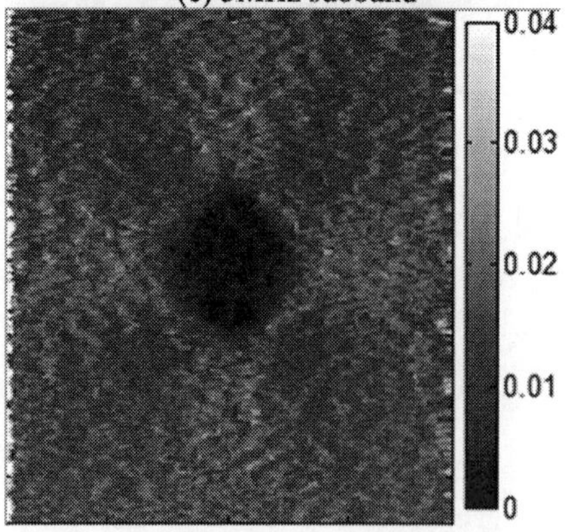

(e)Compounded strain image

Fig.2 Comparision of conventional method and subband method. (a)Strain image using conventional method, SNRe of background is 3.67, SNRe of lesion is 3.77, CNRe is 4.33;(b)Strain image for 3MHz subband, SNRe of background is 3.69, SNRe of lesion is 4.09, CNRe is 14.55; (c) Strain image for 5MHz subband, SNRe of background is 3.56, SNRe of lesion is 3.96, CNRe is 13.59; (d) Strain image for7MHz subband, SNRe of background is 3.59, SNRe of lesion is 3.59, CNRe is 13.77; (e)Compounded strain image, SNRe of background 5.43, SNRe of lesion is 5.90, CNRe is 31.57.

Table 1 correlation coefficients of three subband images

	3MHz subband	5MHz subband	7MHz subband
3MHz subband	1.0	0.0165	0.0096
5MHz subband	0.0165	1.0	0.1096
7MHz subband	0.0096	0.1096	1.0

4 Discussion and conclusion

SNRe gain and CNRe gain by using subband processing can be explained from strain image compounding theory. The correlation coefficients between three sub strain images are listed in table 1 in which the average correlation coefficient is 0.045. The SNRe of the compounded strain image which is averaged by three sub strain images can be improved by $\sqrt{3/(1+2\rho)}-1$ times, according to strain image compounding theory. Therefore, the SNRe of the compounded strain image in Fig.2e should be improved by 66% in theory. The SNRe gain of the lesion of the compounded strain image in Fig.2e is 57% which is close to the theoretical value. Although the SNRe gain of the background of the compounded strain image in Fig.2 e is 48%, it differs from the theoretical value 18%. This is because the strain field of the whole background is not uniform, and the variances of three sub strain images are not quit identical, which will result in certain deviation from the theoretical value of the SNRe of the compounded strain image. It can be seen from in Fig.2 that the SNRes of three sub strain images are nearly same with that of the stain imaing using conventional method. This result indirectly shows that the subband splitting of RF signal along the axial direction does not result in the decrease of the bandwidth of the lateral RF signal, which is the reason why we splitted the RF signals into three subbands along the axial direction. In conclusion, the quality of lateral strain image can be improved significantly using subband processing.

Acknowledgments

This work is in part supported by the Scientific Research Foundation of CUIT (Project No. KYTZ201320), and also in part supported by the Scientific Research Foundation of Education Department of Sichuan Province (Grant No. 14ZB0185).

References

1. K.M. Hiltawsky,M. Kruger, C. Starke, et al. Ultrasound Med. Biol., **27**, 1461(2001)
2. H. Peng, D. C. Liu, Biomed. Signal Process. Contr., **8**,130(2013)
3. H. Peng, D.C. Liu, J. Sichuan Univ. (Eng. Sci. Ed.), **44**, 127(2012)
4. M. Tanter,J. Bercoff, L. Sandrin,M. Fink, IEEE Trans. Ultrason., Ferroelect., Freq. Contr., **49**,1363(2002)
5. S. Cui, D.C. Liu, IEEE Trans. Ultrason., Ferroelect., Freq. Contr., **58**,509(2011)
6. H. Ping, Ultrason Imag., **19**, 251(1997)
7. E. Konofagou, and J. Ophir,Ultrasound in Med. & Biol., **24**, 1183(1998)
8. J. Luo, and E.E. Konofagou,Ultrasound Med Biol., **35**, 1352(2009)
9. F. Gran, J. Udesen, M. B. Nielsen, J.A. Jensen, IEEE Trans. Ultrason., Ferroelect., Freq. Contr., **55**, 2211(2008)
10. J. Udesen,F.Gran, M. B. Nielsen, J.A. Jensen, Procedings of the IEEE Ultrasonics Symposium, 2006
11. W.F. Walker, G.E. Trahey, IEEE Trans. Ultrason., Ferroelect., Freq. Contr., **42**, 301(1995)

DOI: 10.1051/bioconf/20170803016

12. A. Pesavento, C. Perrey, M. Krueger, H. Ermert, IEEE Trans. Ultrason., Ferroelect., Freq. Contr., **46**, 1057(1999)
13. J. E. Lindop, G. M. Treece, A. H. Gee, R. W. Prager, IEEE Trans. Ultrason., Ferroelect., Freq. Contr., **54**, 1751(2007)
14. L. Huang, Y. Petrank, S.W. Huang, C. Jia and M. O'Donnell, IEEE Trans. Ultrason., Ferroelect., Freq. Contr., **56**,1368(2009)

Ultrasound elasticity imaging using Golay code

Hui Peng[a], Juhong Tie

School of Software Engineering, Chengdu University of Information Technology, Chengdu, China

Abstract. Ultrasound elasticity imaging is becoming a new diagnostic tool for clinicians to detect lesions or cancers in tissues. In this paper, Golay code is applied to elasticity imaging to improve its quality. Phase-zero algorithm is employed as the displacement estimator, and the amplitude modulation correction location estimate method is used as the location estimator. We compared the performance of Golay code and the conventional short pulse in simulation method. The simulation results demonstrate that Golay code can achieve higher elastographic signal-to-noise ratio (SNRe) than the short pulse in low echo signal-to-noise ratio (eSNR) conditions, because the eSNR gain with Golay code increases the accuracy of the displacement estimates. However, in high eSNR conditions, Golay code performs worse than the short pulse, because the range sidelobe level of Golay code will decrease the SNRe and the performance of Golay code depends mainly on its range sidelobe level in high eSNR conditions. Therefore, the optimal conditions for Goaly code to be used in elasticity imaging are the low eSNR, great depth or high attenuation conditions.

1 Introduction

Ultrasound elasticity imaging is becoming a new diagnostic tool for clinicians to detect lesions or cancers in tissues. However, in some practical ultrasonic elasticity imaging systems, poor quality of strain images degrades its value of clinical application [1]. One of major quality metrics of strain images is elastographic signal-to-noise ratio (SNRe) affected by many factors, such as echo signal-to-noise ratio (eSNR), applied strain, attenuation, etc. Many methods [2-6] were proposed to improve the quality of ulrasound strain imaging.

O'Donnell [7] predicted that the transmitted energy can be increased up to 15-20dB by coded excitation before reaching particular intensity limits (Isppa, Ispta). The additional eSNR can be used to improve the performance of elasticity imaging. Jie Liu et al. [4] first applied coded excitation to ultrasonic strain imaging. Reference [5] studied the effects of various factors on chirp-coded strain imaging. In this paper, we applied Golay codes to elasticity imaging to improve its quality. We investigated the design considerations for Goaly-coded strain imaging, and compared the performance of Golay code with the conventional short pulse in different eSNR conditions in simulation method. The optimal condition for Golay code to be used in strain imaging is discussed.

[a] Corresponding author: penghui@cuit.edu.cn

2 Method

2.1 Golay code

Golay codes are pairs of bi-phase modulation codes which consist of two sequences of the same length N whose auto-correlation functions have sidelobes equal in magnitude but opposite in sign. Summing them up results in a composite auto-correlation function with a peak of 2N and zero sidelobes [8]. A pair of bi-phase sequences A (a (n), n = 1,. . .,N) and B (b(n), n = 1, . . .,N) of length N is a Golay code if, and only if, it satisfies [9]:

$$a(n)*a(-n)+b(n)*b(-n)=2N\delta(n) \tag{1}$$

which is known as the complementarity condition.

Fig.1 shows the decoding process of Golay codes of length 4 using matched filter. It can be seen from Fig.1 that Golay codes can cancel perfectly the range sidelobes after summing the matched filter outputs up and the amplitude of the output sequence is 2N times of that of Golay codes.

Unlike the available longest Baker code which length is 13, Golay pairs of any length can be constructed recursively from other Golay pairs. Given a Golay pair A, B, another Golay pair of twice the length can be generated by [AB, A (–B)]. The recursion can be initiated with a Golay code of length 2: A = [1, 1] and B = [1, –1].

Golay codes cannot be transmitted directly, and they need be modulated by a base pulse sequence in order to achieve better spectral control and pulse shaping as well as greater robustness to motion for multitransmission code [9]. The encoding process of the Golay-code-modulated base sequence can be described as a convolution of base pulse sequence with an oversampled Golay code sequence:

$$\begin{aligned} x_a(n) &= \hat{a}(n)*s(n) \\ x_b(n) &= \hat{b}(n)*s(n) \end{aligned} \tag{2}$$

where $\hat{a}(n)$ and $\hat{b}(n)$ are a pair of oversampled Golay sequences; $s(n)$ is the base pulse sequence; $x_a(n)$ and $x_b(n)$ are a pair of Golay-code-modulated base sequences which are transmitted; and $*$ is the convolution operator.

The received Golay-code-modulated sequences are often decoded by the matched filter to achieve maximum eSNR and same axial resolution of the base pulse sequence [10]. The decoding process of the received Golay-code-modulated sequences can be expressed as the formula (3):

$$\begin{aligned} y(n) &= x_a(n)*\hat{a}(-n)+x_b(n)*\hat{b}(-n) \\ &= \hat{a}(n)*s(n)*\hat{a}(-n)+\hat{b}(n)*s(n)*\hat{b}(-n) \\ &= [\hat{a}(n)*\hat{a}(-n)+\hat{b}(n)*\hat{b}(-n)]*s(n) \\ &= 2\text{N}s(n) \end{aligned} \tag{3}$$

where y(n) is the output pulse sequence that is the sum of a pair matched filter outputs. It can be seen from the above formula (3) that the amplitude of the output pulse sequence is 2N times of that of the base pulse sequence, and the output pulse sequence is without any range sidelobes. In practice, the displacement motion of scatters during two transmission of a pair of Golay codes will produce the range sidelobes of the output pulse sequence.

 DOI: 10.1051/bioconf/20170803017

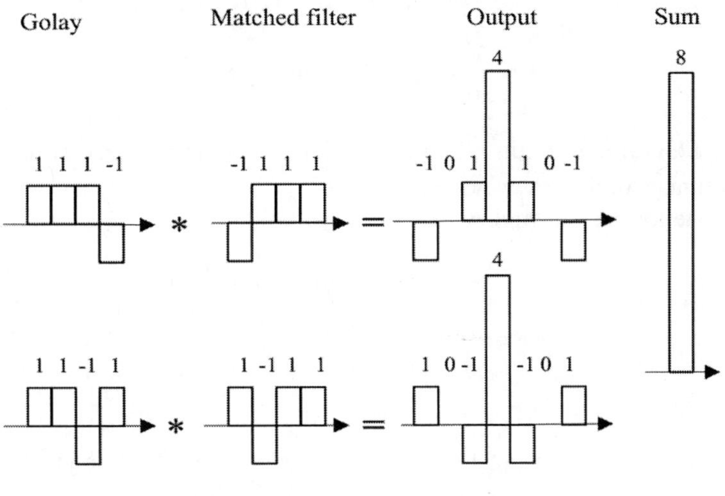

Figure 1. Illustration of decoding process of Golay codes of length 4.

2.2 Strain estimator

For a pair of decoded RF echo signals (pre- and post-deformation) of A-line, we first divided the pre-deformation signal into many partly overlapped windows spacing at a fixed interval. The displacement estimate of a window in pre- deformation signal is the difference of two locations of the window and its matched window in post-deformation signal. The basic algorithm of displacement estimate is the exhaustive search cross-correlation algorithm (ESCCA) which gets the displacement of a pre-deformation window through seeking the absolute maximum of the cross-correlation function of two pre- and post-deformation windows within the range of post-deformation echo signal. However, ESCCA is not fit to a real-time ultrasound system because of its time-consuming computing. In this paper, we employ phase-zero algorithm [11] (PZA) to estimate displacements, whose main advantage is its computational efficiency without losing accuracy compared with ESCCA. PZA is an iterative algorithm expressed by Pesavento et al. [11] in the form of (4), where w_0 is the centre frequency of echo signal, w_c is the local frequency centroid, □t is the window spacing, T is the window length, n□t is the position of the start of the nth window. x_{1b} (t) and x_{2b} (t) are the baseband analytic signals of the decoded echo signals x_1(t) and x_2(t), respectively, $\hat{d}_{k,n}$ is the displacement estimate of the nth window at the kth iteration. At the first window, the search is initialized with $\hat{d}_{0,0}$=0. Subsequent searches are each initialized with the final displacement estimate from the previous window, i.e., $\hat{d}_{0,n}=\hat{d}_{K,n-1}$. The number of iterations is usually set to 6. In practice, we choose w_c=w_0.

$$\hat{d}_{k+1,n} = \hat{d}_{k,n} + \frac{\arg(\exp(-jw_0\hat{d}_{k,n})\sum_{t=n\Delta t}^{n\Delta t+T} x_{1b}^*(t)x_{2b}(t-\hat{d}_{k,n}))}{w_c} \tag{4}$$

It is not generally true that a displacement estimate most closely tracks displacement at the window center. C´espedes and Ophir noted that the actual location is skewed towards sections where the windowed ultrasound signal has high amplitude. This causes artifacts at boundaries between regions of differing echogenicity where the strain estimates are corrupted by unwanted amplitude modulation (AM). We utilize the amplitude modulation correction (AMC) [12] technique to correct these artifacts. The location estimate algorithm using AMC is given as:

$$\hat{\tau}_n = \frac{\sum_{t=n\Delta t}^{n\Delta t+T} Env\{x_{1b}(t)\}Env\{x_{2b}(t-\hat{d}_n)\}t}{\sum_{t=n\Delta t}^{n\Delta t+T} Env\{x_{1b}(t)\}Env\{x_{2b}(t-\hat{d}_n)\}} \tag{5}$$

where $\hat{\tau}_n$ is the location estimate of the nth window, Env{a} is the envelope of the signal a, and $\hat{d}_n$ is the displacement estimate of the nth window.

The strain estimate for the nth window is:

$$\hat{s}_n = \frac{(\hat{d}_{n+1} - \hat{d}_n)}{(\hat{\tau}_{n+1} - \hat{\tau}_n)} \tag{6}$$

2.3 SNRe

One often uses SNRe to evaluate the quality of a strain image. The definition of SNRe:

$$SNR_e = \frac{\mu_{\hat{s}}}{\sigma_{\hat{s}}} \tag{7}$$

where $\mu_{\hat{s}}$ and $\sigma_{\hat{s}}$ are, respectively, the mean and standard deviation of the strain estimates in a region of uniform elasticity.

3 Simulation method and results

3.1 Simulation method

RF data of each frame were simulated using Field II [13]. The uniform phantom consists of 1.152×105 scatterers (fully developed) distributed uniformly throughout 40 × 36 × 6 mm volumes with scattering strengths gaussianly distributed. As the tissue scatterers move mainly along axial direction when the tissue is compressed for strain imaging, we only considered, for simplicity, axial strains which were simulated by rescaling the axial coordinates of the scatterer positions between consecutive frames. Gaussian white noise is added the received waveform. eSNR is measured under the condition of no attenuation by the ratio of the echo signal power of the short pulse to the power of electronic noise. The Golay codes of length 32 are used. The short pulse is a sine wave of two cycles with 5MHz center frequency. A linear array with 128 transducer elements is used to scan the software phantom. Other fixed parameters include scanning depth 6cm, attenuation coefficient 0.5 dB/cm/MHZ, applied strain 2%, window length 2.4 μs, and transmit and receive f-number 1.5.

3.2 Simulation results

We first compare the performance of the conventional short pulse and Golay code in eSNR 20dB condition. The strain image in Fig.2a is obtained using the short pulse, which SNRe is 6.02. The strain image in Fig.2b is obtained using the Golay code, which SNRe is 13.03. Therefore, the performance of Golay code is much better than matched filter in eSNR 20dB condition, and the SNRe is improved by 116%. This is because Golay code will increase eSNR by its time-bandwidth-product and the increased eSNR will improve SNRe according to the CRLB theory [14].

We second test the performances of the short pulse and Goaly code in different eSNR conditions. Fig.3 shows that the SNRe is as a function of eSNR. It is anticipated that the Golay code performs much better than the short pulse in low eSNR(<35dB) condition. However, it can be also seen that in high eSNR (>35dB) condition, the short pulse performs better than Golay code. This is because SNRe will nearly cease to increase with eSNR in high eSNR conditions according the CRLB theory [14], and the the range sidelobe level of the Golay code will cause the decorrelation noise which will degrade the accuracy of the displacement estimates.

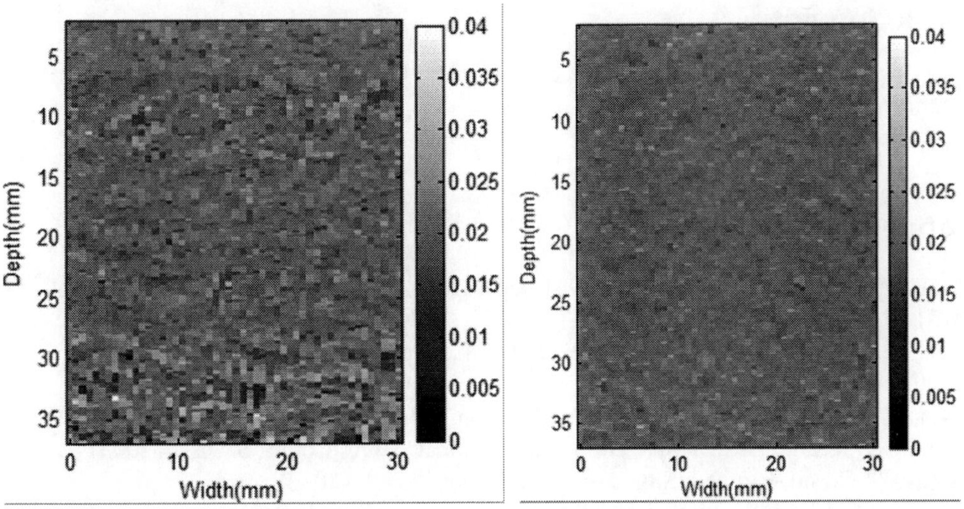

(a) short pulse (b) Golay code

Figure 2. (a)A strain image using short pulse in eSNR 20dB. (b)A strain image using Golay code in eSNR 20dB.

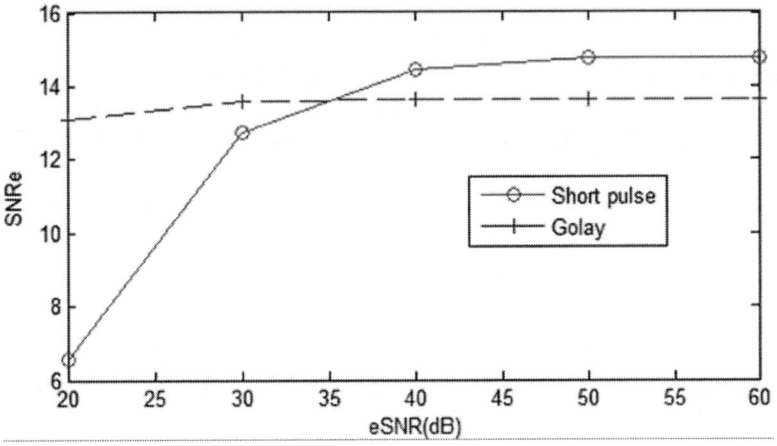

Figure 3. Comparison of short pulse and Golay code in different eSNR conditions.

4 Conclusions

Golay code as a code pulse excitation is often used to increase the eSNR in ultrasound imaging. In this paper, we applied Golay code to strain imaging to improve its quality. Instead of being transmitted directly, Golay codes are modulated by the base pulse sequence to achieve better spectral control and greater robustness to motion for multitransmission code. PZA method with AMC is used as strain estimator. We compared the performance of Golay code and the conventional short pulse in simulation method. The simulation results demonstrate that compared with the short pulse, Golay code can improve the SNRe by 116% in eSNR 20dB condition, because the eSNR gain with Golay code increases the accuracy of the displacement estimates. However, in high eSNR conditions, Golay code performs worse than the short pulse, because the performance of Golay code depends mainly on its RSLL in high eSNR conditions. Therefore, the optimal conditions for Goaly code to be used in elasticity imaging are the low eSNR, great depth or high attenuation conditions.

Acknowledgments

This work is in part supported by the Scientific Research Foundation of CUIT (Project No. KYTZ201320), and also in part supported by the Scientific Research Foundation of Education Department of Sichuan Province (Grant No. 14ZB0185).

References

1. R.G. Barr, K. Nakashima, D. Amy, D. Cosgrove, A. Farrokh, Ultrasound Med. Biol., **41**, 1148 (2015)
2. A. Kuzmin, A.M. Zakrzewski, B.W. Anthony, V. Lempitsky, IEEE Trans. Ultrason., Ferroelect., Freq. Contr., **62**, 1486(2015)
3. R. Xia, G. Tao, AK. Thitta, IEEE Trans. Ultrason., Ferroelect., Freq. Contr., **61**, 979 (2014)
4. J. Liu, M. F. Insana, IEEE Trans. Ultrason., Ferrolect., Freq. Contr., **52**, 231 (2005)
5. H. Peng, D. C. Liu, Biomed. Signal Process. Contr., **8**, 130 (2013)
6. H. Peng, D.C. Liu, J. Sichuan Univ. (Eng. Sci. Ed.), **44**, 127 (2012)
7. M. O'Donnell, IEEE Trans. Ultrason., Ferroelec., Freq. Contr., **39**,341(1992)
8. T. Misaridis, *Ultrasound Imaging Using Coded Signals* (Ph.D. thesis, Technical University of Denmark, 2001)
9. R. Y. Chiao, X. Hao, IEEE Trans. Ultrason., Ferroelect., Freq. Contr., **52**, 160 (2005)
10. C.Leavens, R. Williams, F.S. Foster, S. Member, P.N. Burns, M.D.Sherar, IEEE Trans. Ultrason. Ferroelectr. Freq. Control, **54**, 2082 (2007)
11. A. Pesavento, C. Perrey, M. Krueger, H. Ermert, IEEE Trans. Ultrason., Ferroelect., Freq. Contr., **46**, 1057 (1999)
12. J. E. Lindop, G. M. Treece, A. H. Gee, R. W. Prager, IEEE Trans. Ultrason., Ferroelect., Freq. Contr., **54**, 1751 (2007)
13. J. A. Jensen, *Proc. 10th Nordic-Baltic Conf. Biomed. Imag.*, **4**, 351 (1996)
14. W.F. Walker, G.E. Trahey, IEEE Trans. Ultrason., Ferroelect., Freq. Contr., **42**, 301 (1995)

Effects of the intraoperative application of dexmedetomidine on hemodynamics and cerebral oxygen metabolism of patients with cerebrovascular malformations

Feng Liang, Peng Chen, Chunlu Zhang and Fanglei Han [a]

China–Japan Union Hospital of Jilin University

Abstract. Circulatory stability of patients with cerebrovascular malformations during the surgery is critical to their prognosis. Anesthesia-induced intubation, tumor separation, clamping and other operations may cause severe fluctuations in blood pressure and even result in aneurysm rupture. As a highly efficient and selective adrenergic α2 receptor agonists, dexmedetomidine hydrochloride is able to regulate the release of catecholamine by means of negative feedback so as to control blood pressure. This study aims to assess the effects of dexmedetomidine hydrochloride on hemodynamics and cerebral oxygen metabolism of intraoperative patients with cerebrovascular malformations.

1 Materials and methods

1.1 Clinical information

This study has been approved by the Ethics Committee of our hospital. In addition, patients and their families all signed the informed consent. 60 patients with cerebrovascular malformations under general anesthesia were selected with age of 23-67 and weight of 52-84kg. 17 of them were female and 43 were male. ASA grading were I or II. The patients were randomly divided into 2 groups with 30 each. 30min before anesthesia, they were treated with intramuscular injection of 1mg penehyclidine hydrochloride, After entering the room, they were injected with Lactated Ringer's solution (infusion rate was 68ml / kg.hr). In addition, they're under the routine monitor of ECG, SpO2 and HR. Radial artery puncture manometry was adopted under local anesthesia. Patients in group I received a single infusion of 1/kg dexmedetomidine lasting for 10 minutes. Afterwards, they were treated with continuous infusion of 0.5/μg kg. Hr. Group II patients were infused with the same volume of normal saline. Those with less than 50 times of heart rate were given atropine treatment. After the single infusion, patients were treated with intravenous infusion of 2mg midazolam, 2mg/kg propofol, 0.5μg/kg sufentanil as well as 0.1mg/kg vecuronium induction, In addition, they were inserted with endotracheal tube to control the respiration. The PetCo2 should be maintained at 30-35mmHg.

1.2 Anesthesia maintenance

Target-controlled infusion of propofol was adopted with an initial concentration of 4.0μg/ml. It's

[a] Corresponding author: 23454675@qq.com

required to adjust the infusion concentration to maintain the bispectral index (BIS) at 40-50. Appropriate amount of sufentanil and rocuronium would be given to patients based on practical situation. Internal jugular vein puncture was used for blood collection.In addition, it's essential to record the blood pressure and heart rate at 7 different timings, namely, pre-operation (T1), immediately after intubation (T2), 3 min after intubation (T3), 5 min after intubation (T4), completion of tumor separation (T5), immediately after clamping （T6）as well as 30 min after clamping (T7). Meanwhile, it's demanded to analyze the blood gas of radial artery and internal jugular vein and to record the arterial oxygen difference (PaO2), jugular bulb oxygen difference (PjvO2), arterial oxygen saturation (SaO2), jugular bulb oxygen saturation (SjvO2), arterial hemoglobin concentration (Hb) as well as hemoglobin concentration of internal jugular vein (Hjvb).

Calculation in accordance with the formula: 1. The oxygen content difference between radial artery and internal jugular vein (DajvO2) = CaO2 Cj vO2 = (Hb × 1.34 × SaO2 + 0.031 × PaO2) - (Hjvb × 1.34 × S j vO2 + 0.31 × P j VO2), 2. Cerebral oxygen extraction rate (COER): COER = (CaO2CjvO2) / CaO2×100%

1.3 Statistical method

SPSS 16.0 statistical software was used for analysis. The measurement data were expressed as mean ± standard deviation (x ± s). One-way ANOVA was used to compare the two groups. P <0.05 indicated a statistically difference.

2 Results

The BP and HR of patients in Group II at timing of T2, T3 and T4 were significantly higher than the patients of group I (P <0.05) as shown in Table 1. The DajvO2 of patients in group Iat the timing of T5, T6 and T73 was significantly higher than those of group II while the SjvO2 was tremendously lower(P <0.05) as shown in Table 2.

Table 1. Perioperative hemodynamic changes in two groups of patients

		T1	T2	T3	T4	T5	T6	T7
SBP(MMHg)	I group	145.6±13.2	131.6±16.7	145.1±27.6	139.4±14.3	140.8±19.2	142.6±13.7	141.4±24.8
	II group	141.5±17.7	156.8±22.9	157.7±19.8	149.6±21.1	139.7±22.6	145.7±19.8	140.1±20.3
DBP(mmHg)	I group	90.5±13.5	91.6±11.7	87.6±10.1	89.9±17.3	91.6±12.8	89.7±20.4	90.7±16.4
	II group	89.5±14.3	99.5±13.9	8.6±14.2	97.6±15.9	90.2±13.8	90.5±18.9	89.3±17.1
HR(bpm)	I group	72.7±9.3	67.6±11.2	71.8±8.7	78.4±7.9	73.5±12.3	78.3±9.0	74.2±10.7
	II group	71.7±10.2	87.9±17.6	89.5±13.4	86.2±11.9	72.7±9.6	73.6±14.2	72.8±11.7

Table 2. Oxygen supply and demand of two groups of patients with cerebral aneurysms before and after clamping

		T5	T6	T7
$SjvO_2$(%)	I group	62.1±7.1	63.2±8.0	62.9±8.5
	II group	71.2±6.9	69.2±7.2	70.3±9.7
Da-jv O_2(ml/L)	I group	55.6±11.4	55.8±13.7	57.8±10.1
	II group	43.8±11.3	44.5±12.1	43.6±10.8
COER(%)	I group	37.5±7.1	36.9±6.2	39.3±8.0
	II group	31.1±5.7	31.6±5.5	31.4±6.7

DOI: 10.1051/bioconf/20170803018

3 Discussion

As far as patients with cerebrovascular malformations were concerned, before the tumor was clamped, the dramatic changes in blood pressure might lead to rupture of the tumor and bleeding. While after the clamping, the fluctuations in blood pressure might result in cerebral vasospasm as well, thus exerting a significant negative impact on the prognosis of patients. The endotracheal intubation during anesthesia, separation of the tumor, clamping aneurysm were all highly risky. The commonly used clinical method was to deepen the anesthesia or to use antihypertensive drugs so as to control blood pressure, which, however, might cause a significant reduction in blood pressure if not used properly, ultimately leading to cerebral ischemia and hypoxia.

The brain tissues were extremely sensitive to ischemia and hypoxia, and cerebral hypoxia was considered the main reason for brain damage. It's of great significance to avoid cerebral hypoxia and brain damage in the management of anesthesia. SjvO2 was able to indirectly reflect the cerebral oxygen consumption and the balance between cerebral oxygen supply and demand. Moreover, it could suggest that whether the cerebral blood flow met the oxygen demand of brain tissue and provide clinical basis for the early hypoxic state [1].

This study shows that in the timing of T2, T3, T4, the BP and HR of group I patients were not significantly different from the ones before anesthesia, while the BP and HR of group II at the same time increased significantly, which suggests a more stable hemodynamics of patients in group I during the induction of anesthesia.

The reasons may be as follows: 1. as a highly efficient and highly selective adrenergic α2 receptor agonists, dexmedetomidine hydrochloride is able to regulate the release of catecholamine by means of negative feedback, thus causing dose-dependent reduction in heart rate and blood pressure as well as effective decrease in the hemodynamic response and plasma catecholamine response during the intubation [2].

Dexmedetomidine can act on the α2 receptors between posterior horn of spinal cord of presynaptic membrane and interneuron postsynaptic membrane to hyperpolarize cells and inhibit the conduction of pain signals to the brain or the descending signals to the medulla oblongata and spinal cord. Noradrenaline is able to trigger the release of substance P in presynaptic membrane as well as other injury-like peptide, resulting in analgesic effect [3]. Dexmedetomidine itself has a mild, moderate analgesic effect [4] and can be used with sufentanil for synergistic analgesia, which can confirm that dexmedetomidine is able to reduce the amount of intraoperative opioids as stated in the previous report [4,5]. Opioids is used to ease the tracheal intubation response.

The SjvO2 values of patients in both groups were within the normal range of 60% -75%, indicating that there's no oxygen imbalance in patients of both two groups. SjvO2 <50% for more than 5min was considered abnormal [6]. The SjvO2 of group I in timing of T5, T6 and T7 was significantly lower than that of group II, which showed a higher oxygen utilization rate in group I. Moreover, compared with group II, the DajvO2 of patients in group I significantly increased, indicating that preoperative and intraoperative infusion of dexmedetomidine hydrochloride not only ensured the stability of patients' hemodynamics, but also significantly improved the brain tissue's extraction and utilization of oxygen so as to protect the brain of patients with cerebrovascular malformations to a certain extent.

In summary, dexmedetomidine hydrochloride can be used to ensure the stable perioperative hemodynamics of patients with cerebrovascular malformations, improve cerebral oxygen extraction rate and cerebral oxygenation.

References

1. Brassard P, Seifert T, Secher NH. Is cerebral oxygenation negatively affected by infusion of norepinephrine in healthy subject [J]. Br J aAnaesth , 2009,102(6):800
2. Bhanan , Coa KL,MeCLFIIAN KJ. Dexmedetomidine [J]. Drugs. 2000,59(2):263
3. Nelson LE, Lu J, Guo T , et al. The alpha2-adrenoeeptor agonist Dexmedetomidine converges

DOI: 10.1051/bioconf/20170803018

on endogrenous sleeppromotine path-way to exert its sedative effects. [J]. Anesthesiology 2003,98(2):428
4. Bhana N, Goa KL, McClellan K. Dexmedetomidine[J] .Drugs, 2000,59:263
5. Venn RM, Grounds RM. Comparison between Dexmedetomidine and Propofol for sedation in the intensive care unit: patient and clinician perceptions [J].Br J Anaesth,2001,87(5):684.
6. Oh YJ, Kim JY, ShimJK, et al. Diabets mellitus does not affect jugular bulb oxygen saturation in patients undergoing off-pump coronary artery bypass graft surgery [J].Circ J,2008,72(8):1259

The study of different attention states under different background music based on Event-Related potential analysis

Yun Tang, Xin Deng [a], Chenglin Zhang, Qiaosong Chen, Jin Wang and Qingyun Yang

Chongqing Key Laboratory of Computational Intelligence, Chongqing University of Posts and Telecommunications, Chongqing 400065, China

Abstract. This paper conducts the research on the attention sates based on ERP analysis when the subjects are under the quiet, flute and zither background music. The amplitude and latency of P300 are analyzed. The results show that there are greater P300 amplitudes and smaller P300 latencies of CZ, PZ, OZ and CP3 in music background than those in quiet background. The PCA and ICA achieve to select the effective data components and the head model is reconstructed. The active degree of brain areas are analyzed by using the source location methods. The result shows that the brain's excitement is very obvious under the Bach's flute background. The study also indicates that some background music might help to improve the attention.

1 Introduction

Event-Related potential (ERP) and Functional Magnetic Resonance Imaging (fMRI) have been combined for using in the discussion of relationship between music and human cognition. When studying musical scale structures, Neuhaus [1] discovered that music and culture were closely related. In experiment, he found that the EEG signal containing a clear P300 component would present after the seventh deviant Thai tone, and P300 component with a longer latency could be brought when the eighth deviant Thai tone was given to the subjects. The results showed that not only Thai ton, but also other tones from different countries would induce a P300 component with larger amplitude and longer latency [1]. In addition, some researchers have showed that the cognitive processes on musical synchronization [2], music phase perception [3] and music scale structure [1] differed due to different cultures. However, some researchers have an opposite opinion that culture was not linked to music. They believed that the perception on music for people had the universality and dependence on the time [4], emotion [5], and feeling [6]. Morrison [7] used fMRI to compare the neural states of subjects when listening to familiar music from their native country and unfamiliar music from other countries. The result indicated that different cultural music might activate similar neurons [8]. Morrison hold the opinion that the subjects used the same way to comprehend the music, so they obtained the same understanding of its meaning.

We have briefly mentioned the attention, memory and observation above. Among them, the research on attention started several decades. Kerns [9] discovered that the early direct intervention could improve the attention of child with Attention Deficit Hyperactivity Disorder (ADHD). Noble [10] observed that the children in higher income family had the advantages of cognitive abilities

[a] Corresponding author :tangyunluck@163.com

DOI: 10.1051/bioconf/20170803019

throughvarious neurocognitive systems. Kuniecki [11] proved that the red color was more capable to attract people's attention by the observation of EEG, especially in given context.

Graceful and soft music has the positive effect on psychology and physiology [12]. This paper discusses the relation between the music and the attention by analyzing the amplitude and latency of P300 component according to oddball paradigm. In our experiment, the brain active areas are located by using the source location methods and the active degree of each brain areas were analyzed. We obtain the result that the music might have a significant impact on the attention.

2 Materials and methods

2.1 Subjects

15 college students aged 20-30 years old are selected as the subjects randomly. They have equivalent educational background without a professional training about the music. Each subject is dextromanuality. Their intelligence mental and physical health are normal, and no one has mental disorders. All subjects volunteer to participate in this study.

2.2 Background music selection

A flute music and a zither music are chosen as the background music. The flute background music is the Bach's Sonata, and the zither background music is The Profound Mood between Clouds and Flowing. Another background is to keep quiet. In order to eliminate the physical differences, the time length of the selected music is consistent, so the different experimental data from the same subjects are produced from the different background music instead of other unrelated physical stimulation [13]. Thus, we set the power of the music around 100Hz (Fig.1) in our experiment, which is presented to the subjects through the headphones.

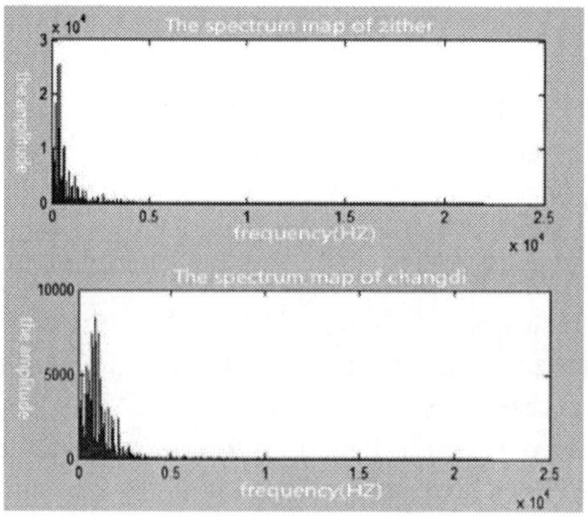

Figure 1. Spectrum map

2.3 Experimental data acquisition

The EEG recording device (Neuroscan) includes an electrode cap with 64 channels. The positions of electrodes are based on the international 10-20 system. The montage includes 8 midline sites (FPZ,FZ,FCZ,CZ,CPZ,PZ,POZ,OZ), 27 left hemisphere sites (FP1,AF3,F1,F3,F5,F7,FC1,FC3,FC5,FT7,C1,C3,C5,T7,CP1,CP3,CP5,TP7,P1,P3,P5,P7,PO3,PO5,PO7,O1,CB1), and 27 right hemisphere sites (FP2,AF4,F2,F4,F6,F8,FC2,FC4,FC6,FT8,C2,C4,C6,T8,CP2,CP4,CP6,TP8,P2,P4,P6,P8,PO4,PO6,PO8,O2,CB2) [11]. To avoid the external disturbance, the tip of the nose is the reference for all electrodes and the blink response of the subjects is monitored by the Vertical-Electrooculogram

DOI: 10.1051/bioconf/20170803019

(VEOG) signals and the Horizion-Electrooculogram (HEOG) signals. The resistance between the electrodes is less than 5KΩ. The sampling rate is 1 KHz.

2.4 Experimental process

12 pictures are displayed to the subjects on a computer screen, including 4 words and 8 numbers (Fig. 2). The probability of numbers' appearance in this experiment is 80%, and the probability of words' appearance is 20%. The subjects need to click the left mouse button when watching the words. The pictures randomly flash every 125ms. The time of dark screen is 100ms and one cycle is 800-1200ms.

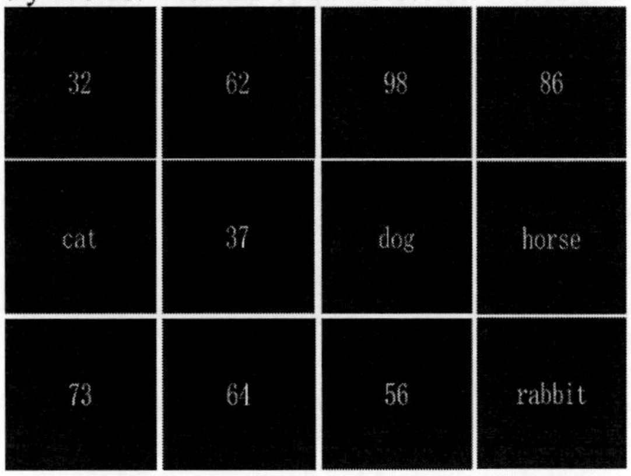

Figure 2. Pictures used for evoking the P300

2.5 Data reprocessing

In the experiment, the EEG signals are reprocessed including filtering and so on. Thus, a single experiment containing the ERP signal can be extracted. The process is as follows.

Baseline correction. Remove the Direct Current (DC) drift in the experiment.

Filtering. The signal of each channel is filtered by the Chebyshev band-pass filter with a frequency range of 0.1-30Hz. Then the power supply interference is removed by a notch filter for 50Hz.

Data adjustment. The blink or the rotation of the eyes, the slight movement of the subjects all can cause a larger amplitude of the EEG signals. However, the range of the normal amplitude of the EEG signal is between -100uV to 100uV. So the data whose amplitude are not in this range are eliminated to exclude the interference of the Electrocardiogram (ECG), the EOG, and the subjects' movement etc.

Single experimental data extraction. The single data segment of 700ms are extracted from the whole data. The data starting from 100ms before a single stimulus to 600ms after the stimulus are overlaid to extract the ERP signal.

2.6 ERP signal processed by PCA and ICA

As the advanced neutral center of human, the brain depends on the mutual relation of different regions to control the body including all high-grade activities of human. The experiment chooses 64 electrodes recording EEG signals and 1 KHz as the sampling rate. The dimensionality of the feature vector is Ne×Nt, where Ne denotes the number of the electrodes and Nt denotes the number of the temporal samples in one trail. Thus, the large dimensions of EEG data are needed to remove the correlated data. The Principal Component Analysis (PCA) is used to descend the dimensions. Many original data with a certain correlation are recombined into a new set of irrelevant integrated data. In all linear combinations, the first principal component F1 is selected, which has the largest variance. If F1 is not enough to represent the original information of t index, the second linear combination should be selected. Through the PCA algorithm, the unnecessary components are eliminated and the left can effectively reflect the information of the original signal.

After the PCA processing, the data are consisted of a large number of unrelated source signals. The Independent Component Analysis (ICA) is a multi-channel signal decomposition method and derived from blind source separation. It is used to eliminate the unnecessary components within the time ranges of the analysis. Through the ICA, the data are separated and the independent components are obtained. In all of the processing, the effective components of the SNR are higher than the noise level, and the left components have a major impact on the Mean Global Field Power (MGFP).

The PCA and ICA are used to eliminate the unnecessary components, so the number of data are decreased. Finally, the source location methods including the moving dipole localization method and the current density method are applied at the independent EEG data.

2.7 Head models reconstructing

A realistic head model based on each subjects' anatomical information is created, which can increase the accuracy of source location. Based on the measured EEG, we also obtain the sensor positions, the head model, and a fit of one dipoles. For each fitted dipole, a confidence ellipsoid is computed both in the time domain and the frequency domain, which visualizes the localization accuracy.

After the head model is reconstructed, possible sites of excitation in different brain regions are tracked by using moving dipole localization method. Meanwhile, the excitation degree of each brain region is analyzed by using the current density method.

3 Result analysis

3.1 ERP data analysis

As CZ, PZ, OZ located in the sagittal line [14] and CP3 located in the auditory central region [15], these 4 electrodes are chosen to be observed in this experiment. The ERP is the averaged evoked potentials, which contains some components closely related to the psychological research, such as P300 and MMN (Mismatch Negativity). The P300 component is analyzed in this experience, which can be clearly noticed (Fig.3, 4, 5). The positive peak is appeared around 481ms in the quiet background (Fig.3), around 469ms in the flute background (Fig.4), and around 464ms in the zither background (Fig.5). In order to analyze the influence of different background music on the ERP, the ERP data are analyzed statistically.

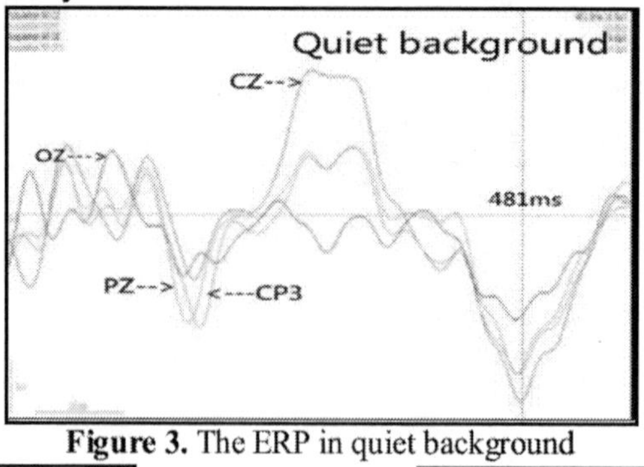

Figure 3. The ERP in quiet background

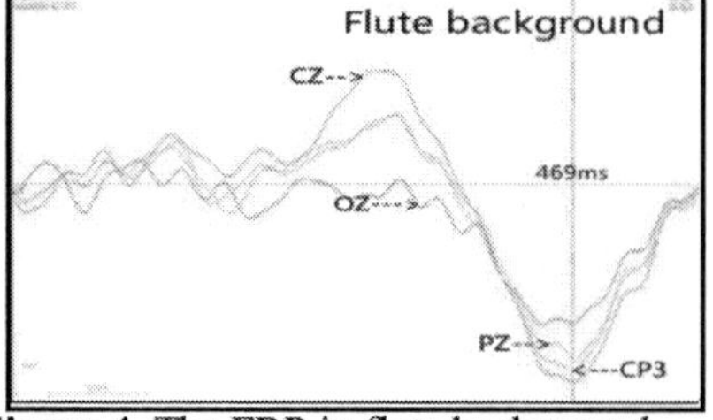

Figure 4. The ERP in flute background

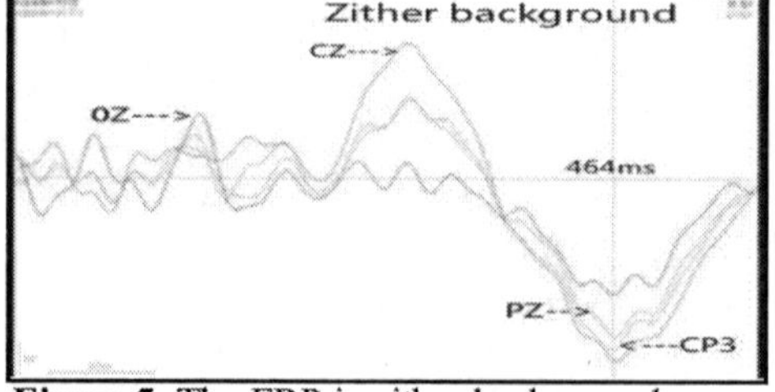

Figure 5. The ERP in zither background

3.2 Analysis of the latency and amplitude of P300

In different background music, the latency of P300 for CZ and PZ varies a little, and the difference is within 10ms (Fig. 6). The latencies of P300 for OZ and CP3 vary a lot. For OZ and CP3, the latencies are longest in a quiet background, and are shortest in the zither background music. These results indicate that music might have a certain effect on the latency of P300. The shorter latency in the flute and zither background music showed that the subjects might have a better ability to response in this environment. In conclusion, the background music might have an influence on people's attention [16], and the zither background music might have a better effect than the flute background music. Fig.7 also illustrates these conclusions.

In the flute background, the amplitudes of P300 of CZ, PZ, OZ, CP3 are bigger than those in the zither background music, and both of them are higher than those in the quiet background (Fig. 8). The amplitudes of P300 reflect the psychological involvement level of subjects [16]. Thus, the subjects' attention may differ in different background music. In summary, the attention of subjects might be improved in zither and flute background. Fig.9 also illustrates these conclusions.

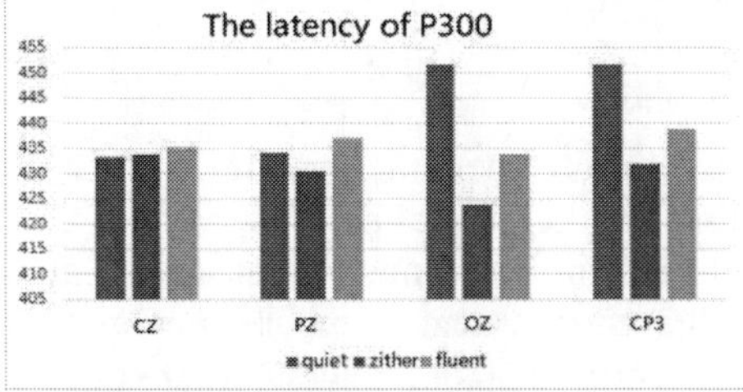

Figure 6. The latency of P300

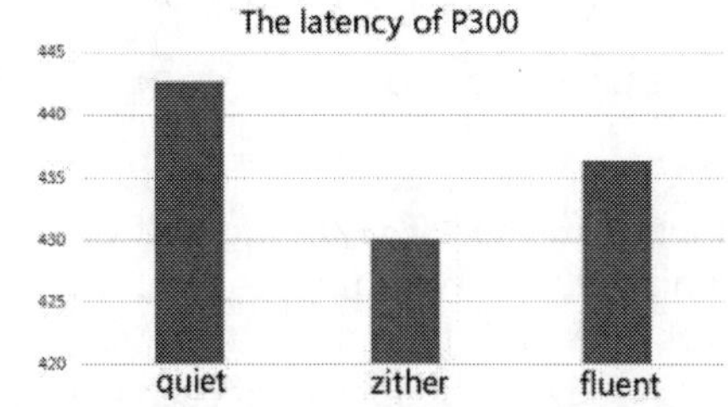

Figure 7. The latency of P300 statically

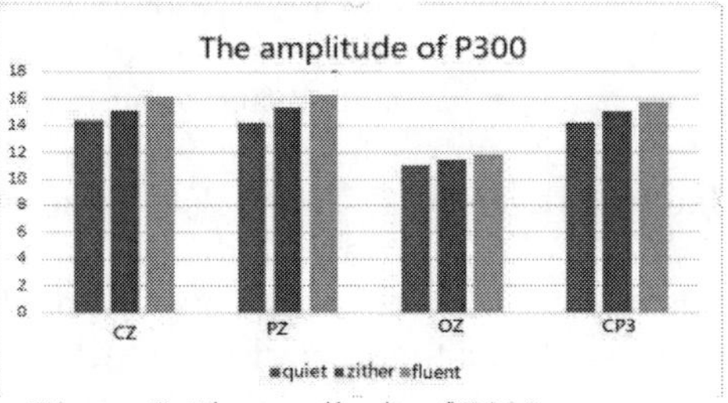

Figure 8. The amplitude of P300

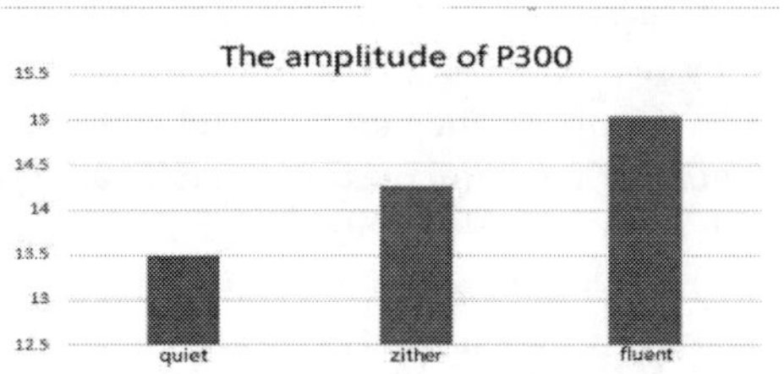

Figure 9. The amplitude of P300 statically

3.3 Spectrum analysis

The EEG signals can be divided into δ wave (Delta), θ wave (Theta), α wave (Alpha), β wave (Beta) and γ wave (Gamma). These five waves can reflect the human's consciousness and the high level of mental activities. The δ wave is relevant to the information fusion, the attention and the logic language information processing in cortical layer. The θ wave is concerned with a variety of cognitive functions. The α wave is related to the activity in cortical layer, especially the attention and the memory. The β wave is interrelated with the attention, which can improve the cognitive abilities. The γ wave emphasizes the integration of visual attention and the synchronization information of the brain rhythm [17].

The amplitudes of different waves in different backgrounds are shown in the table1. Except the four waves, other spectral amplitudes are too small to analyze. Obviously, in the three backgrounds, the changes mostly take place among the δ wave, the θ wave and the α wave. The δ wave and the θ wave, commonly known as "sleepy wave" [19], can reflect the arousal level [18] to some extent, and often disappears in the sleep and the trance state. Usually, the δ wave emerges in the deep sleep, and the θ wave appears in the shallow sleep. At this time, the subject's body is in a relaxed state and has a conscious awareness of alertness when the θ wave is produced. In our experiment, the amplitude of the δ in the quiet background is bigger than those in other two background music. The amplitude of θ wave in the quiet background is slightly smaller than those in other two background music. Compared with the θ wave, the δ wave is bigger. Therefore, the θ wave can be regarded as a quantitative, which means that the minor differences of θ might be caused by the psychological cues in different backgrounds. The result shows that subjects are more likely to be sleepy and fall into a deep sleep in a quiet background. Meanwhile, the subjects are easily fall into a light sleep and their alertness increases slightly in the zither and flute background. Besides, the greater amplitude of α wave, which reflects the relaxed state, shows that the subjects are more relaxed in the zither and flute background.

Table 1. The amplitudes of different waves in different backgrounds

wave (uV)	δ	θ	α	β
quiet	8.0	0.5	0.4	0.3
zither music	6.0	0.8	0.6	0.3
flute music	5.0	0.8	0.6	0.3

3.4 The source location

The source location methods include the dipole localization method and the current density reconstruction method. The position, direction and intensity of the dipoles are observed by the dipole localization method. With the position, direction and intensity information, we can obtain the active brain areas and track the variety of EEG signals. Then the current density reconstruction method is used to gain the current density of each brain and to analyze the excitation degree. There three factors are used to describe the result, which are background music, brain sites, and lateralization.

In the quiet background, the excitement appears on both sides of the auditory central region, and obviously the left excitation degree is greater than that on the right side (Fig.10). This result might be caused by two reasons. Firstly, the subjects who are dextromanuality have an advantage in the left of the brain. Secondly, the stereo music could make the listening asymmetry. In the zither background, the excitement only emerges on the right side of the auditory center (Fig.11). In the flute background, the excitement strongly occurs in the whole back of the head (Fig.12). Compared with these two result, the flute background brings the bigger excitation degree than those in zither background. Through the investigation of the subjects, the reason may be that the subjects are more familiar with flute. The result shows that the different backgrounds evoke the different excitement degree of brain areas.

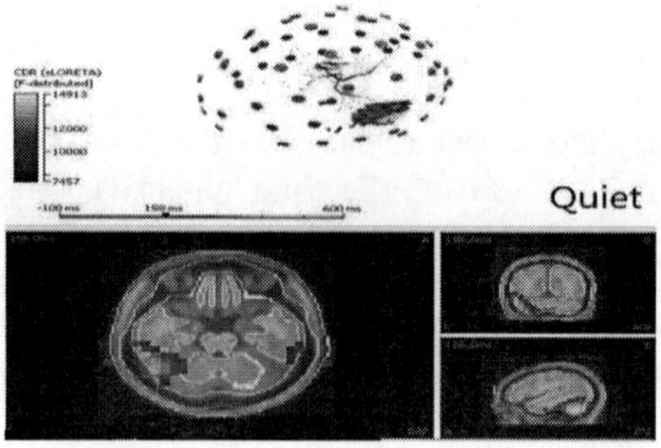

Figure 10. The excitement degree in quiet

 DOI: 10.1051/bioconf/20170803019

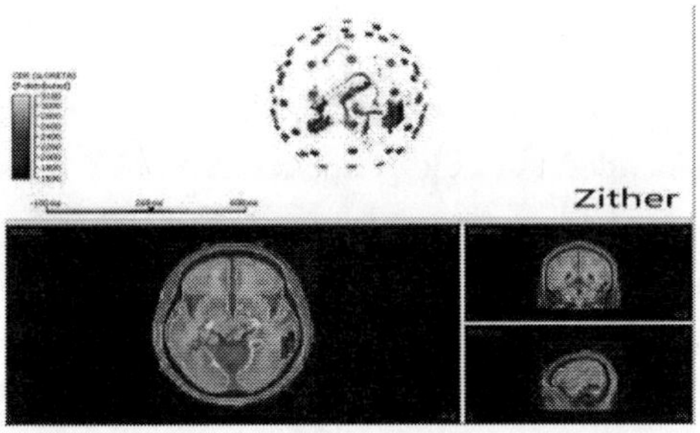

Figure 11. The excitement degree in zither

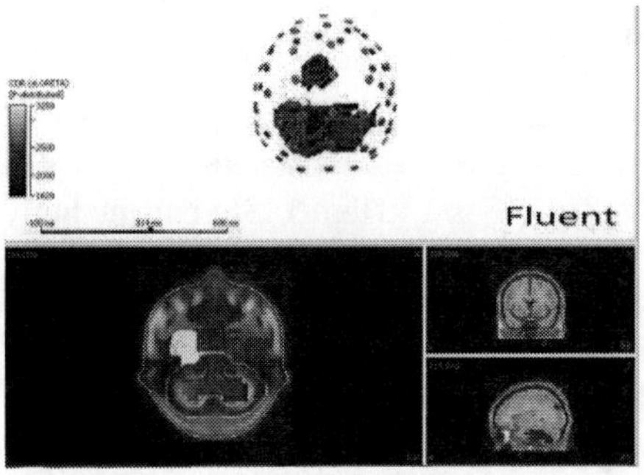

Figure 12. The excitement degree in flute

4 Discuss

The data and results of this experiment are analysed and summarized as follows.

Latency: The latencies of P300 are different in three different background music. However, the differences among them are small. Usually, the latency can reflect the reaction rate of subjects. But the small differences cannot prove that the three different background music would yield the different reaction rate. Here we conclude that the reaction rate is mainly depended on the individual difference and does not have the relationship with the environment.

Amplitude: In the flute background, the amplitude of P300 is bigger than those in the zither background music, and both of them are higher than those in the quiet background. We explain this result as the interference of the sound background. Compared with the quiet background, the subjects are more relaxed, and have a more concentrated attention in the zither and the flute background music. Meanwhile, the spectrum analysis equivalently illustrates the result. In the quiet background, there are bigger δ wave (value=8uV), smaller α wave (value=0.4uV) and β wave (value=0.3uV) than those in other backgrounds. The result shows that the subjects are more likely to be sleepy and fall into a deep sleep in the quiet background. Contrarily, in the flute and zither background, the subjects are easy to fall into a light sleep and their alertness increase slightly. At the same time, they are more relaxed or easy to relieve tension.

Source location: It appears that the largest brain areas involved and the strongest excitement degree of brain are in the flute background. They are explained as following. Every object has its own natural vibration frequency, the same as every brain area of human. The music waves utilize the resonance with some brain regions, which can yield the relevant brain areas more active. It is noteworthy that Bach's sonata is selected as the flute background. According to the relevant research, Bach's music is more likely to resonate with the human organs.

In summary, the study indicates that some background music might help to improve the attention.

In this work, the subjects' EEG signals are recorded in three backgrounds. We have shown that the certain background music may help to improve the subjects' attention. However, there are some limitations in this work. One is the number of the subjects is limited. The other is that the recording time is not long enough. Thus, the future improvements to the work presented here might be consisted of two parts. One is to prolong the time of experiment, such as 45 minutes, to verify whether the conclusion of this paper works well. The other is to increase the types of music, such as jazz, Waltz, etc.

Acknowledgment

This work was partially supported by National Natural Science Foundation of China (No.61403054), the Fundamental and Frontier Research Project of Chongqing under Grants (No.cstc2014jcyjA40001, No.cstc2014jcyjA40022) and the Science Foundation Project of CQ Education Commission under Grant (No. KJ1400436).

References

1. Neuhaus C. Perceiving musical scale structure: A cross-cultural event-related brain potentials study. Ann N Y Acad Sci, **999**, 184-188(2003).
2. Drake C, Ben EI Heni J. Synchronizing with music: intercultural differences. Ann N Y Acad Sci, **999**, 429-437(2003).
3. Nan Y, Knösche TR, Friederici AD. The perception of musical phrase structure: a cross-culturals study. Brain Res. **1094**, 179-19(2006).
4. Drake C, Bertrand D. The quest for universals in temporal processing in music. Ann N Y Acad Sci, **930**, 17-27(2001).
5. Balkwill LL, Thompson WF. A cross-cultural investigation of the perception of emotion in music: psychophysical and cultural cues. Music Percept, **17**, 43-64(1999).
6. Smith LD, Williams RN. Children's artistic responses to musical intervals. Am J Psychol, **112**, 383-410(1999).
7. Morrison SJ, Demorest SM, Aylward EH, Cramer SC, Maravilla KR. FMRI investigation of cross-cultural music comprehension. Neuroimage, **20**, 378-384(2003).
8. Demorest SM, Morrison SJ. Exploring the influence of cultural familiarity and expertise on neurological responses to music. Ann N Y Acad Sci, **999**, 112-117(2003).
9. Kimberly A. Kerns, Karen Eso, Jennifer Thomson. Investigation of a Direct Intervention for Improving Attention in Young Children with ADHD. Developmental Neuropsychology. **16**(16), 273-295(1999).
10. KG Noble, MF Norman, MJ Farah. Neurocognitive correlates of socioeconomic status in kindergarten children. Developmental Science, **8**(1), 74–87(2005).
11. M Kuniecki, J Pilarczyk, S Wichary. The color red attracts attention in an emotional context. An ERP study. Frontiers in Human Neuroscience, **9**,212(2015).
12. Hwang, R. H. & Shih, Y. N. Effects of background music on concentration of workers. WORK: A Journal of Prevention, Assessment, & Rehabilitation, **38**(4), 383-387(2011).
13. HaiWei Liu. Brain Research of Music Cognition and EEG Computational Analysis [D]. Xiamen University, 13-23 (2008).
14. Wei-Na ZHU, Jun-Jun ZHANG. Differential cognitive responses to guqin music and piano music in Chinese subjects: an event-related potential study. Neurosci Bull, **24**(1), 21-28 (2008).
15. WeiCheng Lin. Research about Pattern Classification of Recognition Tasks Based on EEG Signals [D]. Hangzhou Dianzi University, 9-20 (2013).
16. Yu Ji. Research of P300 Processing Algorithm Based on Independent Component Analysis [D]. Zhejiang University, 10-14 (2013).
17. QiuJing Xu. Research on Visual Attention Classification Based on EEG Entropy Parameters [D]. Tianjin University University, 4-6 (2009).
18. Tran Y, Craig A, Boord P, et al. Personality trails and its association with resting regional brain activity[J]. International Journal of Psychophysiology, **60**(3), 215-224 (2006).
19. MaoMao. Psychologically and Physiologically Measuring Musichinduced Emotion and its Application in Service Design [D]. Tsinghua University, 11-18 (2013).

Construction of Yeast One-hybrid Bait Reporter Vector for Screening the Binding Proteins of Cassava *MeCWINV1* Promoter

Jiao Liu[1, ¶,a], Yuqiang Guo[1, 2, ¶], Ruijun Duan[1,¶], Shuo Wang[1, 2], Shaoping Fu[1], Guanghong Kong[1, 2], Ruimei Li[1], Yuan Yao[1], Jianchun Guo[1]

[1]*Institute of Tropical Biosciences and Biotechnology, Chinese Academy of Tropical Agricultural Sciences / Key Laboratory of Biology and Genetic Resources of Tropical Crops, Ministry of Agriculture, Haikou 571101, Hainan, P.R. China*
[2]*College of Agriculture, Hainan University, Haikou 570228, Hainan, P.R. China*
[¶]*Jiao Liu, Yuqiang Guo and Ruijun Duan contributed equally to this work*

Abstract. Cellwall invertase (CWIN) hydrolyzes sucrose into glucose and fructose irreversibly, playing key roles in carbohydrate partitioning and plant defence. *MeCWINV1* is one of the CWINs in cassava, which contains several light-responsive elements and stress-responsive elements in promoter region. To analyze the regulatory function of *MeCWINV1* in the cassava starch accumulation and stress defense response, a 865 bp *MeCWINV1* promoter fragment was cloned and inserted into yeast one-hybrid bait vector to construct p*CW1*-AbAi vector, then was transformed to Y1HGold yeast strains to screen the binding proteins. It might provide a framework for further investigation on the regulation mechanism of *MeCWINV1* gene in cassava.

1 Introduction

Soluble sugars, mostly hexoses and sucrose, which not only provide energy and carbon resources, but also act as signalling molecules in plant defence. Invertase (INV) hydrolyzes sucrose into glucose and fructose, thereby playing crucial roles in primary metabolism, plant development and defence response [1]. In plant cells, invertases are found located in the cell wall, vacuole and cytosol, which are named as CWIN, VIN and CIN, respectively. CWIN and VIN are acid invertases with optimum pH 4.5-5.5, while CIN is neutral/alkaline invertases [2]. CWINs serve as key metabolic enzymes in the carbohydrate supply of sink tissues and the regulation of source-sink transitions. Meanwhile CWINs are regulated by sugar-, hormone-, stress-related signals, and generate sugar-based signals via the modulation of sucrose/hexose ratios. Hence, several signals converge at the site of CWIN regulation. CWIN acts at an integration point of hormonal, sugar, defence and diurnal/circadian regulation [3].

Cassava has six CWIN genes *MeCWINV1~6*, these genes show different spatial and temporal patterns of expression during the cassava development. *MeCWINV1* and *MeCWINV3* had higher activity than other *MeCWINVs* in cassava leaves and tubers, which suggested *MeCWINV1* and *3* were the key CWINs for catalyzing sucrose in the apoplastic space of cassava source and sink organs [4]. In previous research we found that the *MeCWINV1* gene promoter contains several light-responsive

[a] Corresponding author: liujiao@itbb.org.cn

elements and stress-responsive elements, which might be associated to the invertase gene function of cassava starch accumulation and stress defense [5].

In this study we cloned a 865 bp *MeCWINV1* promoter fragment and constructed the yeast one-hybrid bait vector p*CW1*-AbAi vector, and further transformed to Y1HGold yeast strains then determined the minimal inhibitory concentration of Aureobasidin A for Y1HGold[*CW1*/AbAi] strains. It can be used to screen the binding proteins of *MeCWINV1* gene promoter and provide a framework to further investigate the regulation mechanism and signal pathways of *MeCWINV1* gene.

2 Materials and Methods

2.1 Materials and reagents

pAbAi Y1H bait vector, p53-AbAi vector, Y1HGold yeast strains and Aureobasidin A were purchased from Clontech Inc., DH5α/E.coli strains and p*CW1*-19T plasmid were obtained from our laboratory.

2.2 Cloning and synthesizing p*CW1*-AbAi plasmid

The forward primer with *Kpn* I site (5'-GGGGTACCATGCGACTTGAAACTT-3') and the reverse primer with *Sal* I site (5'-ACGCGTCGACCTCTATTTTCCCTCT-3') was designed according to the known *MeCWINV1* gene promoter sequences in the NCBI GenBank databasc (KC465190). The *MeCWINV1* promoter region was amplificated from the p*CW1*-19T plasmid by PCR using the primers. The PCR product and pAbAi Y1H bait vector were digested with *Kpn* I and *Sal* I, then ligated at 16℃ overnight with T4 DNA ligase. The promoter was inserted into the pAbAi vector to generate a recombinant plasmid p*CW1*-AbAi, then was identified by PCR, restriction analysis, and sequencing.

2.3 Generating the bait-reporter yeast strains

Linearize 1 μg of p*CW1*-AbAi plasmid and p53-AbAi plasmid (positive control) by *BstB*I restriction enzyme. Transformed Y1HGold yeast strains with 1 μg of each of the linear plasmids using the protocol for the Yeastmaker Yeast Transformation System 2. Diluted each transformation reaction 1/10, 1/100, and 1/1000, then plated 100 μl from each dilution on SD/-Ura agar medium. After 3 days picked 5 colonies and analyzed by colony PCR using the Matchmaker Insert Check PCR Mix 1 (Clontech, Cat. No.630496). Used untransformed Y1HGold colonies as negative controls. Pick one colony for each confirmed bait, and one from the p53-AbAi control, and streak them onto SD/-Ura agar medium. After 3 days at 30℃, store at 4℃ for up to 1 month. For long term storage, growed an overnight culture in YPDA broth, collect the cells by centrifugation, and resuspend them in 1 ml of freezing medium. Quick freezed, and stored at –70℃.

2.4 Testing the bait strain for AbA^r expression

Picked a large healthy colony from the p*CW1*-AbAi bait strains and p53-AbAi control strains, resuspended each colony in 0.9% NaCl and adjusted the OD600 to ~0.002 (for approximately 2000 cells per 100 μl). Plated 100 μl on each of the following media: SD/-Ura, SD/-Ura with AbA (50 ng/ml), SD/-Ura with AbA (60 ng/ml), SD/-Ura with AbA (80 ng/ml), SD/-Ura with AbA (100 ng/ml), SD/-Ura with AbA (150 ng/ml). Allowed colonies grow at 30℃for 2–3 days, counted the number of colonies on each above media.

3 Results and Discussion

3.1 Constructon and identification of p*CW1*-AbAi bait reporter vector

A 865 bp *MeCWINV1* promoter fragment was isolated from the p*CW1*-19T plasmid (Fig.1a) and inserted into the pAbAi vector to construct the yeast one-hybrid bait vector p*CW1*-AbAi. The recombinant plasmid p*CW1*-AbAi was digested by *Kpn* Ⅰ and *Sal* Ⅰ, approximate 865 bp fragments were obtained, which was consistent with *MeCWINV1* promoters in length (Fig.1b). Sequencing results showed that we obtained the recombinant p*CW1*-AbAi vector successfully.

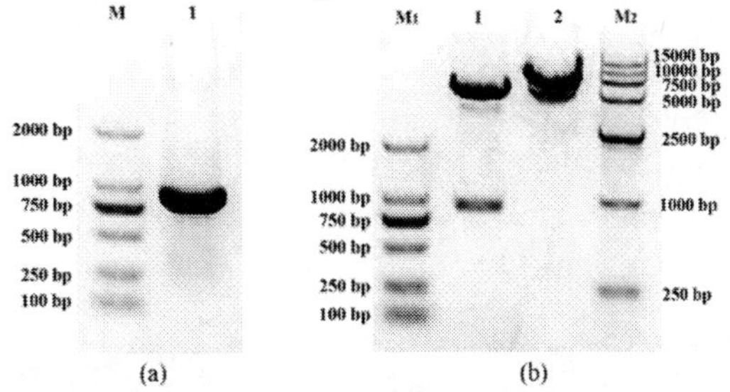

Figure 1. Clone and identification of p*CW1*-AbAi bait reporter vector
(a) The cloning of *MeCWINV1* promoter. M: 2 kb DNA ladder marker; 1: PCR amplification product of *MeCWINV1* promoter. (b) Restriction digestion of p*CW1*-AbAi vector. M1: 2 kb DNA ladder marker; M2: 15 kb DNA ladder marker; 1: p*CW1*-AbAi plasmid digested with *Kpn* Ⅰ /*Sal* Ⅰ; 2: p*CW1*-AbAi plasmid.

3.2 Generating and identifying the bait-reporter yeast strains Y1HGold[*CW1*/AbAi]

The *BstB*I linearized p*CW1*-AbAi plasmid, p53-AbAi positive control plasmid were transformed into Y1HGold yeast strains and selected for transformants on SD/-Ura media. The plasmids confirmed to integrate correctly by using a colony PCR analysis. The primers in the Matchmaker Insert Check PCR Mix 1 were located in the AbA^r gene and in the Y1Hgold genome, downstream of the *URA3* locus. The positive control Y1HGold[p53-AbAi] amplified a region of ~1.4 kb (Fig.2, line 4), each bait strain Y1HGold[*CW1*/AbAi] amplified a ~2 kb band containing 1.35 kb + insert size (865 bp) that encompassed the bait sequence and confirmed the presence of the integrated plasmid (Fig.2, line 1-3). Thus the bait-reporter strain Y1HGold[*CW1*/AbAi] is generated by homologous integration into Y1Hgold. The integrated plasmids are very stable, overnight broth media culturing without *URA3* selection will not result in loss of the integrant.

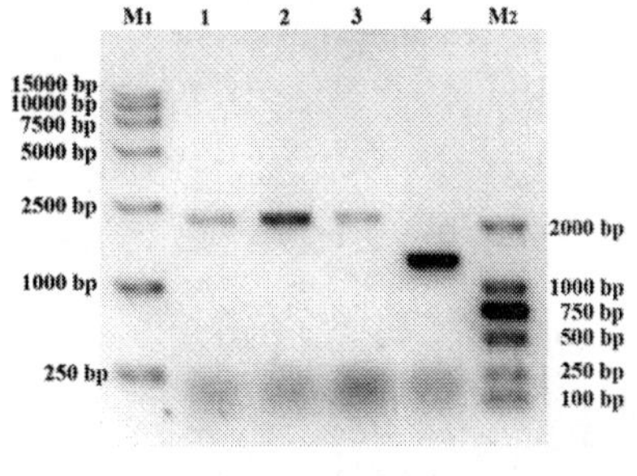

Figure 2. Confirming p*CW1*-AbAi integration by colony PCR
M1:15 kb DNA ladder marker; M2: 2 kb DNA ladder marker; 1~3: PCR product of bait plasmid p*CW1*-AbAi transformed to Y1HGold yeast genome; 4: Positive control p53-AbAi.

3.3 Determining the Minimal Inhibitory Concentration of Aureobasidin A for Y1HGold[*CW1*/AbAi] strains

Successful use of any yeast one-hybrid system depends upon no/low recognition of target sequence by endogenous yeast transcription factors. For this reason, it is critical to test the construct for AbA^r expression before screening the library. The basal expression of the bait reporter stain in the absence of prey can vary, depended on the bait sequences cloned into pAbAi vector. If 1000 ng/ml Aureobasidin A (AbA) does not suppress AbA^r in the absence of prey, the bait DNA sequence is likely being recognized by endogenous yeast transcription factors and therefore the sequence cannot be used for a yeast one hybrid screen. As recorded in the protocol, the positive control Y1HGold[p53-AbAi] has a minimal inhibitory concentration of 100 ng/ml AbA (Fig. 3). Our bait strain Y1HGold[*CW1*/AbAi] plated on SD/-Ura media with 60~80 ng/ml AbA had less clones compared to the control and no clones were appear when the concentration of AbA above100 ng/ml. Therefore, the minimal inhibitory concentration of AbA for our bait strain Y1HGold[*CW1*/AbAi] is 100 ng/ml (Fig. 3). For the follow-up library screening, we can use the minimal concentration of AbA, or a concentration that is slightly higher (by 50~100 ng/ml) to suppress the growth of the bait strain completely.

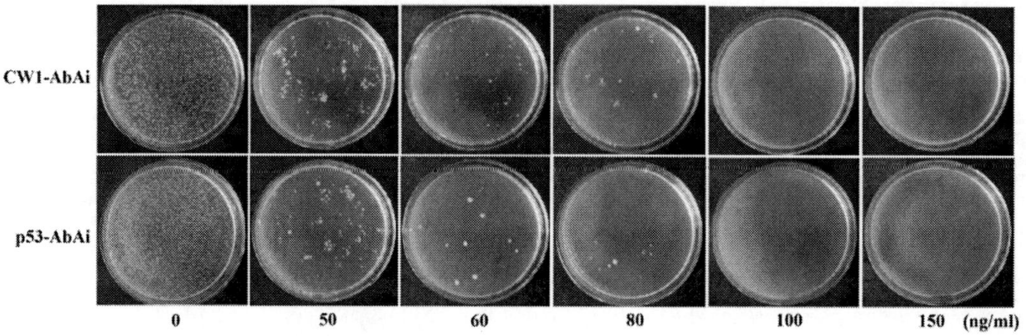

Figure 3. Testing the Y1HGold[*CW1*/AbAi] strain for AbA^r expression on SD/-Leu/AbA (0~150ng/ml) medium

Acknowledgments

This work was supported by the National Natural Science Foundation of China (31371706), Young Elite Scientist Sponsorship Program of CAST, the Fundamental Scientific Research Funds for Chinese Academy of Tropical Agricultural Sciences (CATAS-1630052014004), the Earmarked Fund for Modern Agro-industry Technology Research System (CARS-12-hnwwq), the Provincial Natural Science Foundation of Hainan (20153137), the Special Project for Application Technology Development and Demonstration of Hainan Province (ZDXM2015048).

References

1. A.S. Tauzin, T. Giardina. *Front Plant Sci*, **5**: 293 (2014).
2. A. Sturm. *Plant Physiol*, **121**(1): 1-8 (1999).
3. R.K. Proels, R. Hückelhoven. *Mol Plant Pathol*, **15**(8): 858-864 (2014).
4. Y. Yao, M.T. Geng, X.H. Wu, J. Liu, R.M. Li, X.W. Hu, J.C. Guo. *Int J Mol Sci*, **15**(5): 7313-7331 (2014).
5. J. Liu, Y.P. Hu, Y.Zhou, R.M. Li, R.J. Duan, J.C. Guo, S.P. Fu. *Molecular Plant Breeding*, **12**(6): 1169-1174 (2014).

DOI: 10.1051/bioconf/20170803021

Synthesis and Characterization of Magnetic-Graphene Molecularly Imprinted Polymers for Selective Recognition of L-tryptophan

Yi Zhang[1], Xingguo Tian[2], Xiaoyan Xu[1,a], Chuwen Liu[1], and Wanting Liu[1]

[1]*Guangdong Provincial Key Laboratory of Food Quality and Safety, College of Food Science, South China Agricultural University*
[2]*New Rural Development Institute of South China Agricultural University*

Abstract. In this paper, a novel molecular imprinted polymer using L-tryptophan(L-Trp) as the template, dopamine(DA) as both functional monomer and cross linking agent, magnetic graphene as the supporting matrix was synthesized. The prepared magnetic-graphene molecularly imprinted polymers(Fe_3O_4@GO-MIPs) were characterized by Fourier transform infrared spectrometer(FT-IR), transmission electron microscopy (TEM), vibrating sample magnetometer (VSM), respectively. The results showed that when the molar ratio of L- tryptophan and dopamine was 1:4 and the reaction temperature was 60 °C, Fe_3O_4@GO-MIPs had the best adsorption quantity of 31.9 mg/g. The rebinding experiments indicated that Fe_3O_4@GO-MIPs not only have outstanding affinity and selectivity towards L-Trp over structurally related compounds but also easily reach the magnetic separation under an external magnetic field.

1 Introduction

Molecular imprinting has been widely recognized as a potential technique for the synthesis of tailor-made recognition materials by the formation of a polymer network around a template molecule. Compared with natural receptors, the molecularly imprinted polymers have advantages including good physical and chemical stability, high selectivity of recognition and its reusability. Owing to these fantastic properties, the molecularly imprinted polymers have been widely used in the fields of biosensors[1], solid-phase extraction[2], LC[3], biosensors[4], chiral recognition and separation[5], type enzyme catalyst [6], pharmacy[7] and other fields.

Graphene has attracted considerable attention in recent years because of its large specific surface area and excellent mechanical, electrical, thermal, optical properties. This unique nanostructure with extremely (theoretical value 2630 $m^2\ g^{-1}$) and small dimension makes graphene an excellent candidate as supported material for preparing molecularly imprinted materials[8]. The magnetic molecularly imprinted polymer can be influenced under an external magnetic field and it enables separation of analyte easier and faster and also provides large surface to volume ratio. Fe_3O_4@GO-MIPs have both specific recognition and rapid separation characteristics for the template molecule based on integration of the virtues of graphene, magnetic separation technology and imprinted polymer. L-Trp is known to be metabolized to several neuroactive substances such as serotonin, kynurenic acid, and quinolinic

[a] Corresponding author: yanzixu_2011@scau.edu.cn

acid [9-11]. Analyses of L-Trp and its metabolites in biological fluids, such as serum or plasma and urine, are of clinical importance for diagnosis of several diseases in humans as several reports have indicated that concentrations of L-Trp and its metabolites are altered in various disease states [12]. Furthermore, L-Trp is used for insomnia, sleep apnea, depression, anxiety, facial pain, a severe form of premenstrual syndrome called premenstrual dysphoric disorder (PMDD), smoking cessation, grinding teeth during sleep (bruxism), attention deficit-hyperactivity disorder (ADHD), Tourette's syndrome and to improve athletic performance[13-15]. On the other hand, no medicinal properties for D-tryptophan has been reported [16]. The aim of the paper is to provide a novel synthesis method of Fe_3O_4@GO-MIPs that may be used for determination of L-Trp in presence of D-tryptophan.

2 Experimental

2.1 Reagents

DA and L-Trp purchased from Shanghai Macklin Biochemical Technology; Graphite powder, potassium permanganate purchased from Tianjin Hongda Chemical Reagent Factory; All other reagents and solvents were of analytical grade and used without further purification unless otherwise noted. All aqueous solutions were prepared using ultrapure water from US Millipore company.

2.2 Apparatus

The FT-IR spectra of the prepared nanospheres in the region 4000-400cm^{-1} were recorded using a BRUKER OPTICS FT-IR spectrometer (KBr pellets). Magnetic properties were measured on vibrating sample magnetometer (VSM, Quantum Design, Beijing). The size, morphology, and structure of the nanospheres were characterized by transmission electron microscopy (TEM, JEOL, Japan). The ground samples were dissolved in anhydrous ethanol, washed with copper mesh, dried, and then observed by transmission electron microscopy.

2.3 Preparation of Fe_3O_4@GO-MIPs

Oxidized graphene was synthesized from nature graphite via an improved Hummers' method[17]. In a typical Fe_3O_4@GO-MIPs Synthesis, GO(50 mg) was dispersed in 100 mL of ultrapure water by ultrasonication. Under the condition of stirring, 4 mmol of $FeCl_3 \cdot 6H_2O$ and 2 mmol of $FeSO_4 \cdot 7H_2O$ were added into 100 mL of oxidized graphene dispersion in sequence. After the two ferric salts were dissolved respectively, the rotation speed was increased. Then, 25 mL of ammonia was added, the pH was adjusted to 9.0. After reacting for 30 min at room temperature, 50 mg of dopamine hydrochloride was added and stirring was continued for 1 h. The cooled solution was separated with a magnet, and washed with ultrapure water until the pH was stabilized to 8.0. The black synthetic products were washed several times with ethanol and ultrapure water and dried at 70 °C for 12 h. The Fe_3O_4@GO (400 mg) were dispersed in 80 mL of Tris buffer (10 mM, pH 8.0) by ultrasonication. L-Trp (30 mg)was then added and stirred for 0.5 h in room temperature followed by 120 mg of DA the reaction was continued at 60 °C for 6 h. The obtained product was washed with water to remove the unreacted monomer, then was washed with a mixture of hydrochloric acid solution (1 mol/L) and methanol (1:4, v/v) to extract the template L-Trp, and was rewashed thoroughly with ultrapure water. For comparison, Fe_3O_4@GO-NIPs prepared for the same procedures without the addition of the template L-Trp.

 DOI: 10.1051/bioconf/20170803021

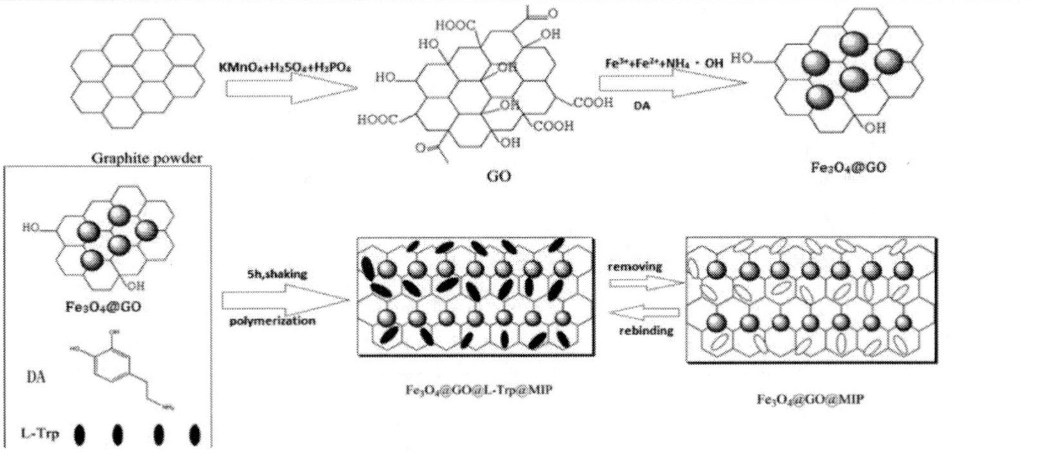

Figure 1. Preparation of Fe_3O_4@GO-MIPs

2.4 Binding Experiments

For the kinetics experiments, 10.0 mg of Fe_3O_4@GO-MIPs or Fe_3O_4@GO-NIPs was ultrasonically dispersed in 20 mL of L-Trp solution for 1 min the mixture was shaken for 2 h, respectively. Fe_3O_4@GO-MIPs or Fe_3O_4@GO-NIPs were withdrawn at appropriate time at intervals. For the adsorption isotherm experiments, an amount of 10 mg of Fe_3O_4@GO-MIPs or Fe_3O_4@GO-NIPs was added to 20 mL of L-Trp solution of different concentrations. After shaking at room temperature for 1 h. The concentration of L-Trp were measured by UV−vis spectrophotometer at 278 nm. Adsorption quantity called Q was calculated from the equation: $Q = (C_0 - C)V/W$, where C_0 and C were the template concentrations (mg•mL^{-1}) in the solutions which were measured initially and after sorption, respectively, V (mL) was the volume of the bulk solution, and W (g) was the weight of the Fe_3O_4@GO-MIPs or Fe_3O_4@GO-NIPs. The selectivity of the Fe_3O_4@GO-MIPs was investigated using 0.04 mmol/L of D-tryptophan, L-phenylalanine, D-phenylalanine, L-tyrosine, and D-tyrosine as comparatives.

3 Results and discussion

3.1 Preparation of Fe_3O_4@GO-MIPs

In order to obtain efficient L-Trp recognition, the preparation conditions of Fe_3O_4@GO-MIPs, including the molar ratio of the template L-Trp to the monomer DA, reaction temperature, were optimized as discussed below in detail. The best conditions were investigated in a set of batch experiments according to the determination of adsorption quantity (Q) of L-Trp onto Fe_3O_4@GO-MIPs. It can be seen from Fig.2 that when the mole ratio of L-Trp template molecule to dopamine was 1:4, the adsorption quantity of L-Trp reached the highest(about 15 mg/g). We think that the increasing DA amount could increase the thickness of PDA(polydopamine) film on the surface of RGO which can accommodate more L-Trp molecule and thus lead to the increase the recognition cavities. On the other hand, when the DA amount increases to a certain value, the over thickness of the PDA film could block the adsorption sites accessibility and result in the decrease of the binding amount. Thus, 1:4 was selected to be the optimum mole ratio in this work.

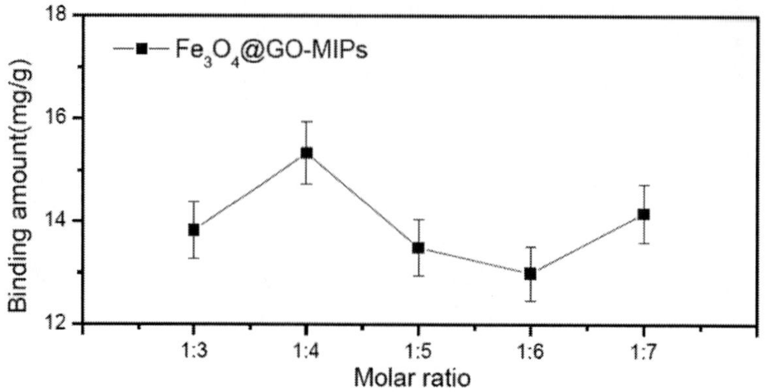

Figure 2. Effect of molar ratio on adsorption capacity

The effect of the reaction temperature on adsorption capacity was investigated from 25 °C to 80 °C . As shown in Fig.3 , with increasing temperature values between 25 °C to 60 °C, the adsorption capacity of Fe_3O_4@GO-MIPs increased gradually and then decreases slightly from 60 °C to 80 °C. It can be inferred that the synthesis temperatures have an important effect on the synthesis of Fe_3O_4@GO-MIPs from the point of increasing adsorption capacity and saving energy. Therefore, it is appropriate to use 60 °C as the optimum reaction temperature.

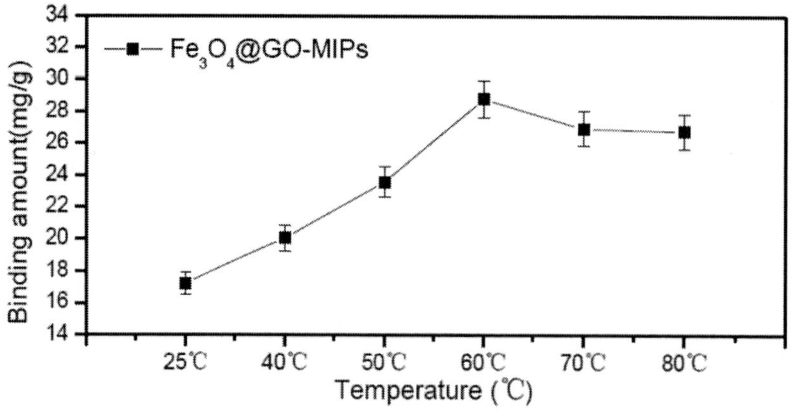

Figure 3. Effect of reaction temperature on adsorption capacity

3.2 Characterization of Fe_3O_4@GO-MIPs

The FT-IR spectra of GO(a), Fe_3O_4@GO(b), Fe_3O_4@GO-MIPs(c), Fe_3O_4@GO-NIPs(d) are shown in Fig.4. As shown in the spectrum of GO(a), the characteristic peak at 1740 cm^{-1} is attributed to Cdouble bond. The feature peaks at 588 cm^{-1} observed from spectra b-d belong to Fe_3O_4. For the FTIR spectrum of Fe_3O_4@GO-MIP(c), the appearance of the stretching vibration of $-NH_2$ at 3399 cm^{-1} provides evidence of the formation of PDA layer. In addition, compared to GO(a), the disappearance of the C=O peak at 1740 cm^{-1} provides a solid indication of GO reduction. All these bands demonstrate that Fe_3O_4@GO-MIPs were synthesized successfully.

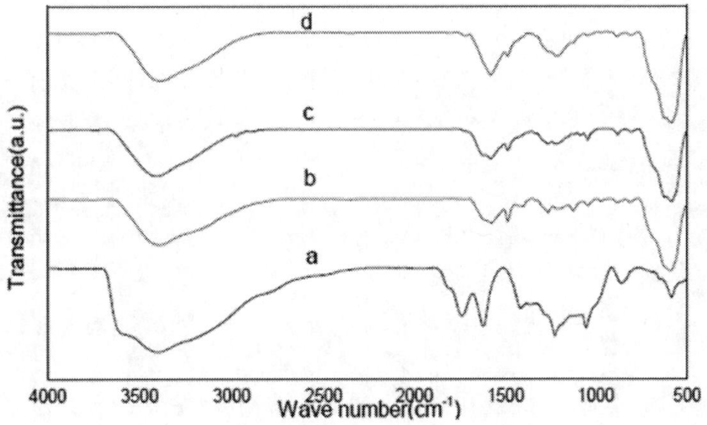

Figure 4. FT-IR spectra of GO(a), Fe_3O_4@GO(b), Fe_3O_4@GO-MIPs (c), Fe_3O_4@GO-NIPs(d)

It is vitally important for magnetic separation that the sorbent is sufficiently magnetic. VSM was employed to characterize the magnetic properties of the obtained magnetic materials, and their VSM magnetization curves are shown in Fig.5. Based on the VSM data, the Fe_3O_4@GO-MIPs exhibited a saturation magnetization value of 36.7 emu g^{-1}. In the absence of an external magnetic field, a black homogeneous dispersion exists. When an external magnetic field was applied, the black particles were attracted to the wall of vial in a short time.

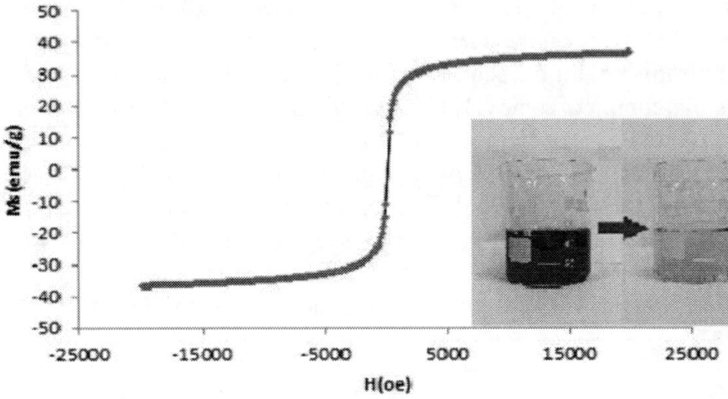

Figure 5. The hysteresis loop of Fe_3O_4@GO-MIPs beads. The insert shows the separation and redispersion process of a solution of Fe_3O_4@GO-MIPs in the absence (left) and presence (right) of an external magnetic field.

The morphological structures of Fe_3O_4@GO(a), Fe_3O_4@GO-MIPs(b), Fe_3O_4@GO-NIPs(c) were examined by TEM. It can be seen from Fig.6 that Fe_3O_4@GO(a) has a spherical structure with a diameter of about 10 nm, which is evenly distributed on the graphene oxide, with some slight agglomeration and overlap. Fe_3O_4@GO-MIPs has some larger pores on the basis of the original Fe_3O_4@GO. The pore size is about 20 nm, which is larger than that of Fe_3O_4@GO. It could be concluded that L-Trp left imprinted sites in the polymer, indicating that the imprinted polymer was modified on Fe_3O_4@GO surface. Fe_3O_4@GO-NIPs is similar to Fe_3O_4@GO in the size of about 10 nm magnetic microspheres, while it is quite different from that of Fe_3O_4@GO-MIPs due to the absence of the template in Fe_3O_4@GO-NIPs. Owing to that no L-Trp template molecules were added in the synthesis of Fe_3O_4@GO-NIPs, the agglomeration of Fe_3O_4@GO-NIPs was greater than that of Fe_3O_4@GO. Furthermore, Fe_3O_4@GO-NIPs (Figure (c)) were in a similar "tribal" form.

 DOI: 10.1051/bioconf/20170803021

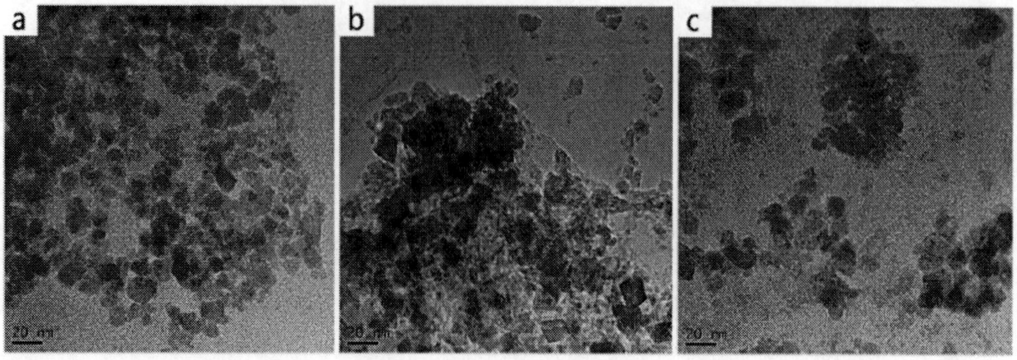

Figure 6. TEM images of Fe_3O_4@GO(a), Fe_3O_4@GO-MIPs(b), Fe_3O_4@GO-NIPs(c)

3.3 Absorption isotherms and Adsorption Kinetics

The L-Trp recognition ability of the Fe_3O_4@GO-MIPs was investigated by the static absorption experiments. The binding isotherms of L-Trp onto Fe_3O_4@GO-MIPs and Fe_3O_4@GO-NIPs are shown in Fig.7. It can be seen that the Fe_3O_4@GO-MIPs superior to those of most imprinted polymers using L-Trp as the template as reported in the literature. This remarkably high binding capacity may lie in a combination of the thin PDA layers on the surface of graphene and the large surface area of graphene. Due to the presence of Graphene, more recognition sites could be created at its surface, whereas the thin PDA layer ensures the complete removal of the template molecules. The adsorption kinetics process of L-Trp was investigated by varying the adsorption time from 1 to 700 min, and the initial concentration of L-Trp was kept at 1.0 mmol/L. As shown in Fig.8 , the Fe_3O_4@GO-MIPs took up 88% of the equilibrium amount of the template molecules from solution within a period of 40 min, and the adsorption equilibrium was reached within 100 min. The results revealed a remarkable rapid adsorption dynamics of L-Trp molecules. This fast dynamics adsorption might originate from the high ratio of surface-imprinted sites, large surface-to-volume ratio, and complete removal of the L-Trp templates.

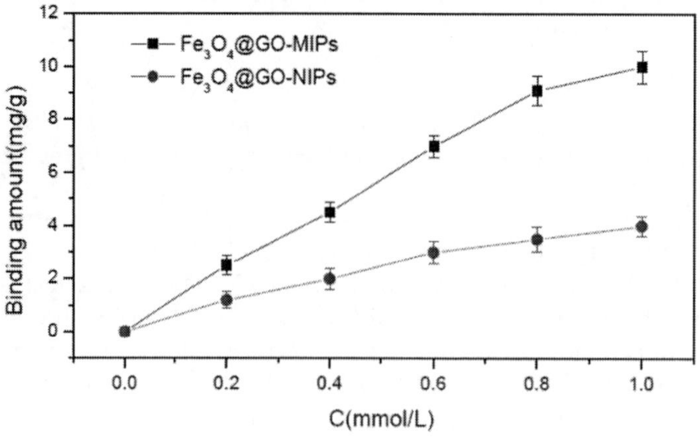

Figure 7. Absorption isotherms of L-Trp on Fe_3O_4@GO-MIPs and Fe_3O_4@GO-NIPs

DOI: 10.1051/bioconf/20170803021

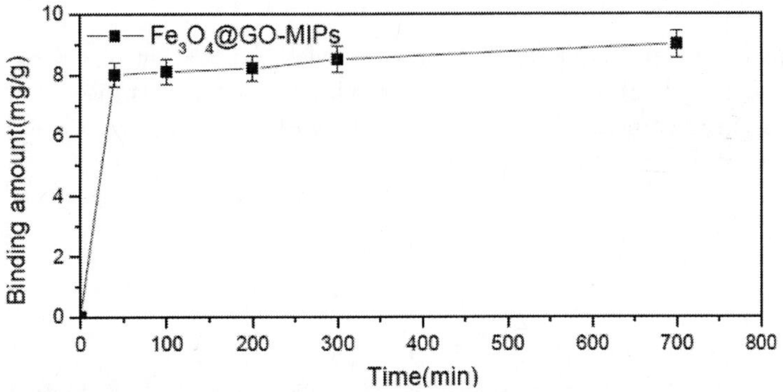

Figure 8. Adsorption kinetics of L-Trp on Fe_3O_4@GO-MIPs

3.4 Selectivity of Fe_3O_4@GO-MIP

The experiments were carried out using D-tryptophan(D-Trp), L-phenylalanine(L-Phe), D-phenylalanine(D-Phe), L-tyrosine(L-Tyr), and D-tyrosine(D-Tyr) as control molecules. Fig.9. shows the rebinding capacities of the Fe_3O_4@GO@MIPs and Fe_3O_4@GO-NIPs for these amino acid with a feed concentration of 0.4 mg/mL of. It is obvious that Fe_3O_4@GO@MIPs exhibits a much higher adsorption amount of L-Trp than other analogues suggesting a better adsorption and binding capacity toward the template molecules. In addition, the imprinting factor α, taken as the binding ratio of Fe_3O_4@GO@MIPs/ Fe_3O_4@GO@NIPs, was determined as 2.3 for binding the template L-Trp, whereas the values of α for binding D-Trp, L-Phe, D-Phe and L-Tyr, D-Tyr were 1.2, 1.1, 1.0, and 1.0, 1.1, respectively, showing that the adsorbed amounts of the non-template amino acid did not have great differences between the imprinted Fe_3O_4@GO-MIPs and Fe_3O_4@GO-NIPs.

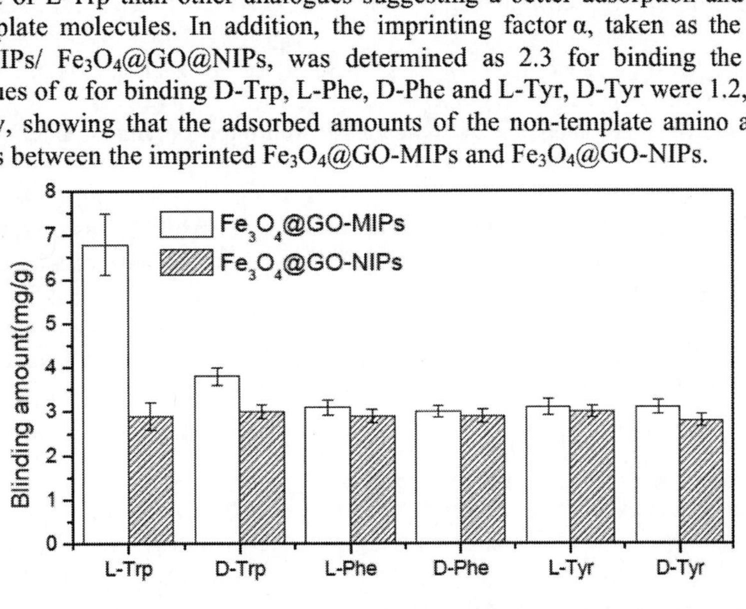

Figure 9. Binding amount of different amino acid on Fe_3O_4@GO-MIPs and Fe_3O_4@GO-NIPs

4 Conclusion

In summary, a novel strategy of synthesizing Fe_3O_4@GO-MIPs for selective recognition of L- tryptophan has been described. The prepared Fe_3O_4@GO-MIPs exhibited high adsorption capacity and good selectivity that showed the formation of the specific binding sites. The above-mentioned synthetic method of Fe_3O_4@GO-MIPs was rather simple and environment friendly, because of the little solvent consumption and the nontoxic stabilizers. Thus it may have a prosperous application because of its simpleness and low cost. We also visualize that the as-produced strategy could be applied in various fields such as bioseparation and food safety.

Acknowledgements

The projects were supported by National Natural Science Foundation of China (No. 31501554) and Guangdong Provincial Natural Science Foundation (No.2016A030310446) and Guangdong Provincial Major Scientific Research Project: Study on novel magnetic ion imprinted polymer for selective determination of trace amounts of cadmium ions.

References

1. R.J. Ansell, O. Ramström, K. Mosbach. *Towards artificial antibodies prepared by molecular imprinting.* Clin. Chem. 42(9): 1506-1512 (1996)
2. Behbahani, M., Bagheri, S., Amini, M. M., Abandansari, H. S., Moazami, H. R., Bagheri, A. Application of molecular imprinted polymer nanoparticles as a selective solid phase extraction for preconcentration and trace determination of 2,4-dichlorophenoxyacetic acid in the human urine and different water samples. Journal of Environmental Health Science and Engineering. 37: 1610-1612 (2014)
3. O. Baltrons, M. Lopez-Mesas, C. Palet, F. Le Derf, F. Portet-Koltalo. Molecularly imprinted polymer-liquid chromatography/fluorescence for the selective clean-up of hydroxylated polycyclic aromatic hydrocarbons in soils. Anal. Method, 5(22): 6297-6305 (2013)
4. A. Aghaei, M.R.M. Hosseini, M. Najafi. A novel capacitive biosensor for cholesterol assay that uses an electro polymerized molecularly imprinted polymer. Electro him. Acta, 55: 1503-1508 (2010)
5. W.J. Cheong, F. Ali, J.H. Choi, J.O. Lee, K.Y. Sung. Recent applications of molecular imprinted polymers for enantio-selective recognition. Talanta, 106: 45-59 (2013)
6. Wulff, G. Enzyme-like catalysis by molecularly imprinted polymers. Chem. Rev, 102(1):1-28 (2002)
7. P. Bures, Y. Huang, E. Oral, N.A. Peppas. Surface modifications and molecular imprinting of polymers in medical and pharmaceutical applications. J. Control. Release, 72:25-33 (2001)
8. Stoller, M. D, Park, S. J, Zhu, Y. W, An, J. H, Ruoff, R. S. Graphene-Based Ultracapacitors. Nano Lett. 8:3498–3502 (2008)
9. T. W. Stone. Endogenous neurotoxins from tryptophan. Toxicon, 39: 61-73 (2001)
10. T. W. Stone, G.M. Mackay, C.M. Forrest, J. Clark, L.G. Darlington. Tryptophan metabolites and brain disorders, Clin. Chem. Lab Med, 41: 852-859 (2003)
11. T. W. Stone, C.M. Forrest, G.M. Mackay, N. Stoy, L.G. Darlington. Tryptophan, adenosine, neurodegeneration and neuroprotection. Metab. Brain Dis, 22: 337-352 (2007)
12. C. J. McDougle, S.T. Naylor, D.J. Cohen, G.K. Aghajanian, G.R. Heninger, L.H. Price. Effects of trptophan depletion in drug-free adults with autistic disorder. Archives of General Psychiatry, 53: 993–1000 (1996)
13. N Hrboticky , LA Leiter & GH Anderson. Effects of L-tryptophan on short-term food-intake in lean men. Nutrition Research, 5: 595–607 (1985)
14. M. Garrido, A.B. Rodríguez, M.P. Terrón , Chapter 13 – tryptophan and melatonin-enriched foodstuffs to improve antioxidant status in aging. V.R. Preedy (Ed.), Aging, Academic Press, San Diego 129–136 (2014)
15. W. Kochen , H. Steinhart. L-tryptophan—Current Prospects in Medicine and Drug Safety. Walter de Gruyter, Berlin (1994)
16. H. Bagheri, M. Ahmadi, T. Madrakian, A. Afkhami. Preconcentration and spectrofluorometric determination of L-tryptophan in the presence of d-tryptophan using a chiral magnetic nanoselector. Sensors Actuators B Chem, 221: 681–687 (2015)
17. Marcano, D. C , Kosynkin, D. V , Berlin, J. M , Sinitskii, A , Sun, Z , Slesarev, A , Alemany, L. B , Lu, W , Tour, J. M , Im*proved synthesis of graphene oxide.* ACS Nano 4(8): 4806– 4814 (2010)

AUTHOR INDEX

AUTHOR INDEX

AUTHOR INDEX